AF783289

Claus Hertling
Matilde Marcolli (Eds.)

Frobenius Manifolds

Aspects of Mathematics

Edited by Klas Diederich

Vol. E 6: G. Faltings/G. Wüstholz et al.: Rational Points*

Vol. E 15: J.-P Serre: Lectures on the Mordell-Weil Theorem

Vol. E 16: K. Iwasaki/H. Kimura/S. Shimemura/M. Yoshida: From Gauss to Painlevé

Vol. E 19: R. Racke: Lectures on Nonlinear Evolution Equations

Vol. E 21: H. Fujimoto: Value Distribution Theory of the Gauss Map of Minimal Surfaces in $\mathbf{R}^m$

Vol. E 22: D. V. Anosov/A. A. Bolibruch: The Riemann-Hilbert Problem

Vol. E 27: D. N. Akhiezer: Lie Group Actions in Complex Analysis

Vol. E 28: R. Gérard/H. Tahara: Singular Nonlinear Partial Differential Equations

Vol. E 34: I. Lieb/J. Michel: The Cauchy-Riemann Complex

Vol. E 36: C. Hertling/M. Marcolli (Eds.): Frobenius Manifolds*

*A Publication of the Max-Planck-Institute for Mathematics, Bonn

Bibliografische information published by Die Deutsche Bibliothek
Die Deutsche Bibliothek lists this publication in the Deutschen Nationalbibliografie; detailed bibliographic data is available in the Internet at <http://dnb.ddb.de>.

Prof. Dr. Klaus Hertling
Universität Mannheim
Institut für Mathematik
A 5, 6
D-68131 Mannheim

hertling@hilbert.math.uni-mannheim.de

Prof. Dr. Matilde Marcolli

Max-Planck-Institut für Mathematik
Vivatsgasse 7
D-53111 Bonn

marcolli@mpim-bonn.mpg.de

Prof. Dr. Klas Diederich (Series Editor)
Fachbereich Mathematik
Bergische Universität Wuppertal
Gaußstraße 20
D-42119 Wuppertal

klas@math.uni-wuppertal.de

Mathematics Subject Classification
Primary: 53D45, 14N35
Secondary: 37K10, 37K20, 34M35, 32S40, 14B05

First edition, September 2004

All rights reserved
© Friedr. Vieweg & Sohn Verlag/GWV Fachverlage GmbH, Wiesbaden 2004

Softcover reprint of the hardcover 1st edition 2004

Vieweg is a company in the specialist publishing group Springer Science+Business Media.
www.vieweg.de

No part of this publication may be reproduced, stored in a retrieval system or transmitted, mechanical, photocopying or otherwise without prior permission of the copyright holder.

Cover design: Ulrike Weigel, www.CorporateDesignGroup.de

Printed on acid-free paper

ISBN-13: 978-3-322-80238-5 e-ISBN-13: 978-3-322-80236-1
DOI: 10.1007/978-3-322-80236-1

Claus Hertling
Matilde Marcolli
(Eds.)

Frobenius Manifolds

Quantum Cohomology and Singularities

A Publication of the Max-Planck-Institute
for Mathematics, Bonn

vieweg

Dedicated to Yuri I. Manin, on the occasion of his 65th birthday

Preface

1. Frobenius manifolds, quantum cohomology, and singularities

Frobenius manifolds are complex manifolds with a multiplication and a metric on the holomorphic tangent bundle, satisfying several natural conditions. The notion was defined in 1991 by Dubrovin, motivated by results of the physicists Witten, Dijkgraaf, E. Verlinde, H. Verlinde and others.

These initial results played an important role in the mathematical development of *quantum cohomology*, pioneered in the symplectic setting by Ruan–Tian and in the algebro-geometric setting by Kontsevich–Manin. It gave rise to the *Gromov-Witten invariants*, which morally count curves with constraints in a given (symplectic/projective) manifold. In the algebro-geometric setting, this "counting" requires the use of moduli spaces of stable maps and virtual fundamental classes. Gromov-Witten invariants exist for curves of any genus, and with *gravitational descendents*. A small part of them, those of genus zero and without gravitational descendents can be encoded in a (formal germ of a super) Frobenius manifold.

Another source of Frobenius manifolds is *singularity theory*. Already in the 70's and beginning of the 80's, K. Saito studied semiuniversal unfoldings of isolated hypersurface singularities. Together with M. Saito, he showed in 1983 that the base spaces of such unfoldings can be equipped with the structure of Frobenius manifolds. Such unfoldings of singularities and their base spaces turned up in the 90's in work of the same physicists mentioned above, as well as Cecotti, Vafa and others, in the theory of Landau-Ginzburg models. In 1998 Barannikov and Kontsevich gave a construction of *extended moduli spaces* as Frobenius manifolds, using differential Gerstenhaber-Batalin-Vilkovisky algebras. This is related to the construction in singularity theory. Duality between string theories lies behind the phenomenon of *mirror symmetry*. One mathematical formulation can be given in terms of the isomorphism of certain Frobenius manifolds from quantum cohomology (A-side) and from singularity theory or the Barannikov-Kontsevich construction (B-side). A first case of mirror symmetry for Calabi-Yau 3-folds was considered in 1991 by the physicists Candelas, de la Ossa, Green, Parkes. This and more general cases were proved by Givental in work since 1996. Despite the growing number of rigorous mathematical results, mirror symmetry remains mysterious.

A third source of Frobenius manifolds is given by *integrable systems*, more precisely by bihamiltonian hierarchies of evolutionary PDE's. This approach is pursued since 1991 by Dubrovin, and with him, Zhang. As in the case of quantum cohomology, here Frobenius manifolds are part of an a priori much richer structure. In both cases, because of strong constraints, a good part of the richer structures is determined implicitly by the underlying Frobenius manifolds. Dubrovin and

Zhang, Getzler, and Givental have done essential contributions to this aspect of the subject. A direct relation between integrable systems and quantum cohomology was conjectured by Witten and proved by Kontsevich. The quantum cohomology of a point is related to the *KdV hierarchy*. A tau-function of the KdV hierarchy is the generating function of all Gromov-Witten invariants of a point. Quantum cohomology, the theory of Frobenius manifolds, and the relations to integrable systems have been flourishing areas since the early 90's. Saito's work in singularity theory is reviewed and extended nowadays due to the growing interest and new inputs from physics, quantum cohomology and mirror symmetry.

An activity was organized at the MPIM in 2002, with the purpose of bringing together the main experts in these areas. One aim was to make the specialists in quantum cohomology aware of the related structures in singularity theory and vice versa. The activity culminated in a two weeks workshop, held at the MPIM, July 8-19, 2002, which presented the state of the art in the subject. This volume contains the proceedings of the workshop.

The activity and the workshop were financed by the Sofja Kowalevskaya Prize awarded to Marcolli in 2001 by the Humboldt Stiftung and the German Government. We thank the participants of the conference, the contributors of the articles and the referees, our co-organizers Boris Dubrovin, Yuri Manin and Kyoji Saito, and the staff of the MPIM, for making this activity and this volume possible.

2. This volume

We introduce this volume by a brief overview of the individual contributions.

Douai and Sabbah construct very interesting examples of Frobenius manifolds arising from Gauss–Manin systems of unfoldings of regular functions on affine manifolds, from certain classes of functions on complex tori. The resulting Frobenius manifolds are semisimple and determined by initial data consisting of the Gauss–Manin system, the Brieskorn lattice for the function, the higher residue pairing and the opposite filtration in the sense of M. Saito. This very concrete and beautiful class of examples illustrates all the main features of the general theory previously developed by the authors.

Fernandez and Pearlstein give a broad survey of results of Cattani and Fernandez, Deligne and Pearlstein on variations of pure or mixed Hodge structures and the deep connection to mirror symmetry and quantum cohomology. Deformations of graded Frobenius modules by a quantum potential are equivalent to special variations of pure Hodge structures. There are specific criteria that govern the extension of such deformations to (germs of) Frobenius manifolds. This provides a general frame for the small quantum cohomology of Calabi–Yau manifolds, generalizing Deligne's approach via limiting mixed Hodge structures to mirror symmetry for Calabi–Yau 3-folds.

Getzler gives a formulation of the differential equations satisfied by the generating functions of higher genus Gromov–Witten invariants in an intrinsic form, which is independent of the choice of a fundamental solution. The equations considered form an important class of differential equations in Gromov–Witten theory, such as topological recursion relations and Virasoro constraints. The intrinsic formulation is closer to the geometry of moduli spaces of stable curves, and leads to a proof of a conjecture of Eguchi and Xiong.

Elaborating on joint work with T. Coates, Givental recasts Frobenius manifolds via the symplectic loop space formalism, thus illustrating the important role of an underlying hidden twisted loop group of symmetries. The theory of genus zero Gromov–Witten invariants, organized via a potential function subject to certain partial differential equations, admits a geometric interpretation in terms of symplectic geometry of a loop space and Lagrangian submanifolds, where one can see explicitly the action of a twisted loop group. This formalism with Lagrangian cones is then applied to a variety of topics including gravitational ancestors, twisted Gromov–Witten invariants, quantum Lefschetz and Riemann–Roch theorems, Virasoro constraints, Frobenius manifolds arising from singularities, quantum K-theory and quantum cobordism.

Hertling and Manin develop a new criterion for the existence of a universal unfolding of certain germs of meromorphic connections. This provides a method for the construction of Frobenius manifolds, which unifies different existing approaches. The uniqueness part of the result is a reconstruction theorem, which identifies a more restrictive set of data that are sufficient to recover the full Frobenius structure, while the existence part is the new feature of this general construction result.

Ralph Kaufmann gives an overview of his recent work on group Frobenius algebras. These are modules over the Drinfel'd double of the group ring, with natural symmetries called discrete torsion and supertwisting. The formulation fits in with orbifold geometry and the main application is to the global orbifold cohomology of symmetric products of K3 surfaces, where the operations of discrete torsion and supertwisting are canonically defined and have a geometric interpretation in terms of the Hilbert scheme resolution of the symmetric product.

An important problem in Gromov–Witten theory is the relation between the known universal differential equations that govern Gromov–Witten invariants. Xiaobo Liu investigates relations between universal genus two relations. He shows that the three known relations of Getzler and Belorousski–Pandharipande can be reduced to two relations, up to certain genus one relations which, in turn, follow from a known genus one relation obtained by Getzler and genus zero and one topological recursion relations.

Based on their previous construction of the moduli space of *painted* stable curves, Losev and Manin develop an operadic formalism for quantum cohomology with gravitational descendants. They introduce a linear operad whose algebras are formal solutions of the commutativity equations, namely Frobenius manifolds without metric and pencils of flat connections, depending on parameters that correspond to gravitational descendants.

The operadic viewpoint is also adopted in Merkulov's paper, which develops in a systematic way techniques for the study of deformation problems for large classes of moduli spaces, providing generalizations and existence results for the F-manifolds of Hertling–Manin, and results of Merkulov on Frobenius$_\infty$ manifolds. The new operadic approach to deformation theory is developed here in four steps, involving the construction of a controlling deformation algebra, a minimal cofibrant model, a cohomological splitting, and a geometric interpretation in terms of extended moduli spaces.

An essential part of the theory of higher spin curves is Witten's virtual top Chern class, which gives the virtual fundamental class of the relevant moduli space.

Polishchuk shows that the algebraic virtual top Chern class constructed in his previous joint work with Vaintrob satisfies two constraints, the vanishing axiom and the Ramond factorization axiom. The first is a vanishing condition related to the markings, while the second is a vanishing condition on pushforwards of the class restricted to the so called Ramond sector of the moduli space. These are expected to be the only two nontrivial relations.

Saito presents a systematic account of his earlier results on Frobenius manifold structures arising from orbit spaces of the irreducible Coxeter groups. The results are described under the point of view of universal unfoldings of singularities, of Gauss–Manin connections and of simple Lie algebras. He also contributes a conjectural description of the period domain and the inverse map to a period map associated to these Frobenius manifolds.

Satake defines a holomorphic Laplacian for generically semisimple Frobenius manifolds, for which he provides explicit formulas in terms of the potential. He uses this technique to characterize the flat coordinates for the universal unfoldings of the simple elliptic singularities, relating them to theta functions on a period domain, so that the period mapping induces an isomorphism between the parameter space of the unfolding and the quotient of the period domain by the monodromy action.

The theory of virtual fundamental classes is the focus of Siebert's paper, where he develops a 'yoga of cones bundles'. This provides a very explicit and direct approach to virtual fundamental classes that bridges between the different existing approaches of Li–Tian and Behrend–Fantechi.

The Gopakumar–Vafa conjecture claims that a generating function for the Gromov–Witten invariants of Calabi–Yau 3-folds is equal to a formula containing certain integers, the BPS invariants, which have an interpretation in terms of D-brane moduli spaces in string theory. Takahashi proposes a mathematical definition of the BPS-invariants, elaborated in joint work with Hosono and M.-H. Saito, and discusses mathematical evidence for the Gopakumar-Vafa conjecture.

3. Yuri Ivanovich Manin

The activity and this resulting volume are dedicated to Yuri Ivanovich Manin, on the occasion of his 65th birthday.

Naturally, the topics of this conference only touch on some of the manifold contributions of Manin to mathematical research. It is impossible to summarize in a few sentences the entire spectrum of results, new perspectives, and the underlying philosophy that his activities encompass. Manin's amazing mathematical production is the outcome of a highly sophisticated interplay of many different fields: number theory, algebraic and arithmetic geometry, mathematical physics. Often it is precisely where the distinction between different fields fades and perspectives merge that one finds the most impressive results.

Over the past ten years, roughly covering the time since he joined the board of scientific members of the MPIM, Manin has been at the center of many of the developments mentioned above, involving the dawn and successive development of new perspectives in algebraic geometry: Gromov-Witten invariants, quantum cohomology, Frobenius manifolds, mirror symmetry. All these aspects fit into a broad perspective, through which Manin contributed fundamentally to the development of the algebraic and geometric language of modern theoretical physics, which also includes Supergeometry, Quantum Groups, and Noncommutative Geometry. We can only emphasize the global vision and philosophy on the interactions between mathematics and physics, which emerge throughout this large body of work. It is a vision over which new generations of mathematicians will be formed.

Claus Hertling and Matilde Marcolli

Contents

Gauss-Manin systems, Brieskorn lattices and Frobenius structures (II)

Antoine Douai and Claude Sabbah

1. Introduction

1.a. This article explains a detailed example of the general result developed in the first part [**3**]. We were motivated by [**1**], where S. Barannikov describes a Frobenius structure attached to the Laurent polynomial $f(u_0, \ldots, u_n) = u_0 + \cdots + u_n$ restricted to the torus $U = \{(u_0, \ldots, u_n) \in \mathbb{C}^{n+1} \mid \prod_i u_i = 1\}$, and shows that it is isomorphic to the Frobenius structure attached to the quantum cohomology of the projective space $\mathbb{P}^n(\mathbb{C})$ (as defined *e.g.*, in [**5**]).

We will freely use the notation introduced in the first part [**3**]. A reference like "§ I.3.c" will mean [**3**, § 3.c].

In the following, we fix an integer $n \geqslant 2$ and positive integers $w_0, \ldots, w_n$ such that $\gcd(w_0, \ldots, w_n) = 1$. It will be convenient to assume that $w_0 \leqslant \cdots \leqslant w_n$. We put

$$\mu := \sum_{i=0}^{n} w_i. \tag{1.1}$$

We will analyze the Gauss-Manin system attached to the Laurent polynomial

$$f(u_1, \ldots, u_n) = w_0 u_0 + w_1 u_1 + \cdots + w_n u_n$$

restricted to the subtorus $U \subset (\mathbb{C}^*)^{n+1}$ defined by the equation

$$u_0^{w_0} \cdots u_n^{w_n} = 1.$$

The case $\mu = n + 1$ (and all w_i equal to 1) was considered in [**1**]. We will not need any explicit use of Hodge Theory, as the whole computation can be made "by hand". We will use the method of § I.3.c to obtain information concerning the Frobenius structure on any germ of universal deformation space of f. As we have seen in [**3**], we have to analyze with some details the structure of the Gauss-Manin system and the Brieskorn lattice of f.

1.b. Fix a $\mathbb{Z}$-basis of $\{\sum_i w_i x_i = 0\} \subset \mathbb{Z}^{n+1}$. It defines a $(n+1)\times n$ matrix M. Denote by $m_0,\dots,m_n$ the lines of this matrix. We thus get a parametrization of U by $(\mathbb{C}^*)^n$ by putting $u_i = v^{m_i}$ for $i=0,\dots,n$ and $v=(v_1,\dots,v_n)$. The vectors $m_0,\dots,m_n$ are the vertices of a simplex $\Delta\subset\mathbb{Z}^n$, which is nothing but the Newton polyhedron of f when expressed in the coordinates v. Notice that the determinant of the $n\times n$ matrix $(m_0,\dots,\widehat{m_i},\dots,m_n)$ is $\pm w_i$.

LEMMA 1.2. *The Laurent polynomial f is convenient and nondegenerate with respect to its Newton polyhedron.*

Proof. The nondegeneracy follows from the linear independence of any n distinct vectors among $m_0,\dots,m_n$. Clearly, 0 is contained in the interior of Δ. □

We know then that f is M-tame (*cf.* § I.4) and we may therefore apply the results of § I.2 to f. An easy computation shows that f has μ simple critical points, which are the $\zeta(1,\dots,1)$ with $\zeta^\mu=1$, and thus μ distinct critical values $\mu\zeta$. We hence have $\mu(f)=\mu$.

1.c. Denote by $\mathcal{S}_w$ the *disjoint* union of the sets

$$\{\ell\mu/w_i \mid \ell=0,\dots,w_i-1\}\subset\mathbb{Q}.$$

Hence $\#\mathcal{S}_w=\mu$. Number the elements of $\mathcal{S}_w$ from 0 to $\mu-1$ in an *increasing* way, with respect to the usual order on $\mathbb{Q}$. We therefore have $\mathcal{S}_w=\{s_w(0),\dots,s_w(\mu-1)\}$ with $s_w(k)\leqslant s_w(k+1)$. In particular, we have

$$s_w(0)=\cdots=s_w(n)=0,\quad s_w(n+1)=\frac{\mu}{\max_i w_i}\leqslant n+1.$$

Moreover, using the involution $\ell\mapsto w_i-\ell$ for $\ell\geqslant 1$, one obtains, for $k\geqslant n+1$, the relation

$$s_w(k)+s_w(\mu+n-k)=\mu. \tag{1.3}$$

We consider the function $\sigma_w:\{0,\dots,\mu-1\}\to\mathbb{Q}$ defined by

$$\sigma_w(k)=k-s_w(k). \tag{1.4}$$

Hence $\sigma_w(k)=k$ for $k=0,\dots,n$. That $s_w(\bullet)$ is increasing is equivalent to

$$\forall\, k=0,\dots,\mu-1,\quad \sigma_w(k+1)\leqslant\sigma_w(k)+1, \tag{1.5}$$

where we use the convention $\sigma_w(\mu)=\sigma_w(0)=0$. We will prove:

THEOREM 1. *The polynomial $\prod_{k=0}^{\mu-1}(S+\sigma_w(k))$ is equal to the spectral polynomial $\mathrm{SP}_f(S)$ attached to f (cf. § I.2.e).*

For instance, if we take the Laurent polynomial $f(u_0,\dots,u_n)$ on the torus $\prod u_i=1$, *i.e.*, $w_0=\cdots=w_n=1$, we get $\mathrm{SP}_f(S)=\prod_{k=0}^{n}(S+k)$.

Notice that the symmetry property (1.3) is a little bit more precise than the symmetry of the spectrum (*cf.* [**9**]), which would say that, for any $j\in\{0,\dots,n\}$,

$$\#\{k\mid\sigma_w(k)=j\}=\#\{k\mid\sigma_w(k)=n-j\}.$$

Indeed, for $k\in\{n+1,\dots,\mu-1\}$, (1.3) means that $\sigma_w(k)+\sigma_w(\mu+n-k)=n$ and we clearly have $\sigma_w(k)+\sigma_w(n-k)=n$ for $k=0,\dots,n$.

1.d. Consider now the two $\mu \times \mu$ matrices

$$(1.6)\qquad A_\infty = \operatorname{diag}\big(\sigma_w(0), \dots, \sigma_w(\mu-1)\big), \quad A_0 = \mu \begin{pmatrix} 0 & & & & 1 \\ 1 & 0 & & & 0 \\ 0 & 1 & 0 & & 0 \\ \vdots & \ddots & \ddots & \ddots & \vdots \\ 0 & \cdots & 0 & 1 & 0 \end{pmatrix}.$$

Notice that A_0 is semisimple with distinct eigenvalues $\mu\zeta$, where ζ is a μ-th primitive root of 1. In the canonical basis $(e_0, \dots, e_{\mu-1})$ of the space $\mathbb{C}^\mu$ on which these matrices act, consider the nondegenerate bilinear form g defined by

$$g(e_k, e_\ell) = \begin{cases} 1 & \begin{cases} \text{if } 0 \leqslant k \leqslant n \text{ and } k+\ell = n, \\ \text{or if } n+1 \leqslant k \leqslant \mu-1 \text{ and } k+\ell = \mu+n, \end{cases} \\ 0 & \text{otherwise,} \end{cases}$$

with respect to which A_∞ satisfies $A_\infty + {}^tA_\infty = n\,\mathrm{Id}$. The data (A_0, A_∞, g, e_0) define (*cf.* [**4**, Main Theorem p. 188], see also [**5**, § II.3] or [**10**, Th. VII.4.2]) a unique germ of semisimple Frobenius manifold at the point $(\mu, \mu\zeta, \dots, \mu\zeta^{\mu-1}) \in \mathbb{C}^\mu$.

The main result of this article is then:

THEOREM 2. *The canonical Frobenius structure on any germ of a universal unfolding of the Laurent polynomial* $f(u_0, \dots, u_n) = \sum_i w_i u_i$ *on* U, *as defined in* [**3**], *is isomorphic to the germ of universal semisimple Frobenius structure with initial data* (A_0, A_∞, g, e_0) *at the point* $(\mu, \mu\zeta, \dots, \mu\zeta^{\mu-1}) \in \mathbb{C}^\mu$.

REMARK. It would be interesting to give an explicit description of the Gromov-Witten potential attached to this Frobenius structure.

2. The rational numbers $\sigma_w(k)$

Let us be now more precise on the definition of $s_w(k)$. Define inductively the sequence $(a(k), i(k)) \in \mathbb{N}^{n+1} \times \{0, \dots, n\}$ by

$$\begin{aligned} a(0) &= (0, \dots, 0), & i(0) &= 0 \\ a(k+1) &= a(k) + \mathbf{1}_{i(k)}, & i(k+1) &= \min\{i \mid a(k+1)_i/w_i = \min_j a(k+1)_j/w_j\}. \end{aligned}$$

It is immediate that $|a(k)| := \sum_{i=0}^n a(k)_i = k$ and that, for $k \leqslant n+1$, we have $a(k)_i = 1$ if $i < k$ and $a(k)_i = 0$ if $i \geqslant k$. In particular, $a(n+1) = (1, \dots, 1)$.

LEMMA 2.1. *The sequence* $(a(k), i(k))$ *satisfies the following properties:*

(1) *for all* $k \in \mathbb{N}$, $\dfrac{a(k)_{i(k)}}{w_{i(k)}} \leqslant \dfrac{a(k+1)_{i(k+1)}}{w_{i(k+1)}} \leqslant \dfrac{a(k)_{i(k)}+1}{w_{i(k)}}$,

(2) $a(\mu) = (w_0, \dots, w_n)$ *and for all* $k \in \{0, \dots, \mu-1\}$, *we have* $a(k)_{i(k)} \leqslant w_{i(k)} - 1$,

(3) *the map* $\{0, \dots, \mu-1\} \to \coprod_{i=0}^n \{0, \dots, w_i - 1\}$, *defined by* $k \mapsto [i(k), a(k)_{i(k)}]$ *is bijective.*

(4) *For* $\ell \in \mathbb{N}$, *we have* $i(k+\ell\mu) = i(k)$ *and* $a(k+\ell\mu)_{i(k)} = \ell w_{i(k)} + a(k)_{i(k)}$.

We will then put $s_w(k) := \mu a(k)_{i(k)}/w_{i(k)}$. We have $s_w(k+\ell\mu) = \ell\mu + s_w(k)$ for $\ell \in \mathbb{N}$.

Proof. (1) By induction on k. If $i(k+1) = i(k)$, the result is clear. Otherwise, we have $a(k+1)_{i(k+1)}/w_{i(k+1)} = a(k)_{i(k+1)}/w_{i(k+1)}$ and the first inequality follows from the definition of $i(k)$. Similarly, the second inequality is given by the definition of $i(k+1)$.

(2) Let us first remark the implication

$$a(k)_j \leqslant w_j \ \forall j \text{ and } \{j \mid a(k)_j < w_j\} \neq \varnothing \Longrightarrow a(k+1)_j \leqslant w_j \ \forall j.$$

[Indeed, from the assumption we have $a(k)_{i(k)} < w_{i(k)}$, hence $a(k+1)_{i(k)} = a(k)_{i(k)} + 1 \leqslant w_{i(k)}$. For $j \neq i(k)$, $a(k+1)_j = a(k)_j \leqslant w_j$.] Therefore, there exists k_0 such that $a(k_0) = (w_0, \ldots, w_n)$. Then $k_0 = |a(k_0)| = \mu$. Moreover, by what we have just seen, we have $a(k)_{i(k)} < w_{i(k)}$ for $k < \mu$.

(3) The map does exist, after (2), is clearly injective, therefore bijective as the two sets have the same number of elements.

(4) We have $a(\mu) = (w_0, \ldots, w_n)$, so that $i(\mu) = 0$, and we may apply the reasoning of (2) for $k = \mu, \ldots, 2\mu - 1$, etc.

□

REMARK 2.2. In general, the numbers $s_w(k)$ are rational. These are integers (hence the spectrum of f is integral) if and only if the following condition holds:

$$\forall i, \quad w_i \mid \mu = w_0 + \cdots + w_n. \tag{2.3}$$

Consider the simplex $\Delta(w)$ in $\mathbb{R}^n$ obtained as the intersection of the hyperplane $\mathcal{H} = \left\{\sum_{i=0}^n w_i x_i = 0\right\} \subset \mathbb{R}^{n+1}$ with the half spaces $x_i \geqslant -1$. Fix also the lattice $\mathcal{H}_{\mathbb{Z}} = H \cap \mathbb{Z}^{n+1}$. Then Condition (2.3) is equivalent to the condition that the vertices of $\Delta(w)$ are contained in the lattice $\mathcal{H}_{\mathbb{Z}}$. In other words, $\Delta(w)$ is a *reflexive simplex* in the sense of Batyrev [**2**]. For instance, if $n = 3$, one finds the following possibilities for w_i (up to a permutation):

w_0	w_1	w_2	w_3	μ
1	1	1	1	**4**
1	1	1	3	**6**
1	1	2	2	**6**
1	1	2	4	**8**
1	2	2	5	**10**
1	1	4	6	**12**

w_0	w_1	w_2	w_3	μ
1	2	3	6	**12**
1	3	4	4	**12**
1	2	6	9	**18**
1	4	5	10	**20**
1	3	8	12	**24**
2	3	10	15	**30**
1	6	14	21	**42**

For $n = 4$, here are some examples (maybe not complete):

w_0	w_1	w_2	w_3	w_4	μ
1	1	1	1	2	**6**
1	1	2	2	2	**8**
1	1	1	1	4	**8**
1	1	1	3	3	**9**
1	1	1	2	5	**10**
2	2	2	3	3	**12**
1	1	3	3	4	**12**
1	1	2	2	6	**12**
1	1	1	3	6	**12**
1	1	3	5	5	**15**

w_0	w_1	w_2	w_3	w_4	μ
1	1	2	4	8	**16**
1	1	4	4	10	**20**
1	1	4	6	12	**24**
1	1	2	8	12	**24**
1	1	3	10	15	**30**
1	1	4	12	18	**36**
1	1	8	10	20	**40**
1	1	6	16	24	**48**
1	1	8	20	30	**60**
1	2	12	15	30	**60**

3. The Gauss-Manin system

The Gauss-Manin system G of the Laurent polynomial f is a module over the ring $\mathbb{C}[\tau, \tau^{-1}]$. It is defined as in § I.2.c:

$$G = \Omega^n(U)[\tau, \tau^{-1}]/(d - \tau df \wedge)\Omega^{n-1}(U)[\tau, \tau^{-1}].$$

Put $\theta = \tau^{-1}$. The Brieskorn lattice $G_0 = \operatorname{image}(\Omega^n(U)[\theta] \to G)$ is a free $\mathbb{C}[\theta]$-module of rank μ because, by Lemma 1.2, f is convenient and nondegenerate (*loc. cit.*). We will consider the increasing filtration $G_p = \tau^p G_0$ ($p \in \mathbb{Z}$). Let ω_0 be the n-form on U defined by

$$\omega_0 = \frac{\frac{du_0}{u_0} \wedge \cdots \wedge \frac{du_n}{u_n}}{d\left(\prod_i u_i^{w_i}\right)}\Bigg|_{\prod_i u_i^{w_i}=1}.$$

Let $v \mapsto u = v^m$ be a parametrization of U as in § 1.b. The form ω_0 can be written as $\omega_0 = \pm\frac{dv_1}{v_1} \wedge \cdots \wedge \frac{dv_n}{v_n}$. The Gauss-Manin system G is then identified with the $\mathbb{C}[\tau, \tau^{-1}]$-module (putting $v = (v_1, \ldots, v_n)$)

$$\mathbb{C}[v, v^{-1}, \tau, \tau^{-1}]/\{v_j\partial_{v_j}(\varphi_j) - \tau(v_j\partial_{v_j}f)\varphi_j \mid \varphi_j \in \mathbb{C}[v, v^{-1}, \tau, \tau^{-1}],\ j = 1, \ldots, n\}.$$

It comes equipped with an action of ∂_τ: if $\psi \in \mathbb{C}[v, v^{-1}]$, let $[\psi]$ denote its class in G; then $\partial_\tau[\psi] = [-f\psi]$ (this does not depend on the representative of the class). Using the coordinate θ, we have $\theta^2\partial_\theta[\psi] = [f\psi]$; this action is extended in the usual way to Laurent polynomials in τ with coefficients like $[\psi]$.

It is convenient to use the coordinates $u = (u_0, \ldots, u_n)$. Then the previous quotient is written as

$$\mathbb{C}[u, u^{-1}, \tau, \tau^{-1}]/\big(I_w + \mathbb{C}[u, u^{-1}, \tau, \tau^{-1}](g(u) - 1)\big),$$

where we have put $g(u) = \prod_i u_i^{w_i}$ and I_w is the $\mathbb{C}[\tau, \tau^{-1}]$-submodule of $\mathbb{C}[u, u^{-1}, \tau, \tau^{-1}]$ consisting of the expressions

$$(3.1) \quad \sum_{i=0}^{n} m_{ji}\Big(u_i\frac{\partial}{\partial u_i} - \tau w_i u_i\Big)\varphi_j, \text{ with } \varphi_j \in \mathbb{C}[u, u^{-1}, \tau, \tau^{-1}],\ (j = 1, \ldots, n).$$

Consider the sequence $(a(k), i(k))$ of Lemma 2.1, and for each $k = 0, \ldots, \mu$, put

$$\omega_k = u^{a(k)}\omega_0 \in G_0,$$

Notice that $\omega_\mu = \omega_0$ and, using (3.5) below, that $f\omega_0 = \mu\omega_1$.

PROPOSITION 3.2. *The classes of $\omega_0, \omega_1, \dots, \omega_{\mu-1}$ form a $\mathbb{C}[\theta]$-basis $\boldsymbol{\omega}$ of G_0. Moreover, they satisfy the equation*

$$-\frac{1}{\mu}(\tau\partial_\tau + \sigma_w(k))\omega_k = \tau\omega_{k+1} \quad (k = 0, \dots, \mu - 1),$$

and we have Bernstein's relation in G:

$$\prod_{k=0}^{\mu-1}\big[-\tfrac{1}{\mu}(\tau\partial_\tau - s_w(k))\big]\cdot\omega_0 = \tau^\mu\omega_0.$$

The V-order of ω_k is equal to $\sigma_w(k)$ and $\boldsymbol{\omega}$ induces a $\mathbb{C}$-basis of $\oplus_\alpha \mathrm{gr}_\alpha^V(G_0/G_{-1})$.

From Theorem I.4.5, Lemma I.4.3(3), and the symmetry (1.3), we get

$$\text{(3.3)}\qquad \text{for } k = 0, \dots, \mu - 1, \quad 0 \leqslant \sigma_w(k) \leqslant n \quad \text{and} \quad \begin{cases}\sigma_w(k) = 0 \Rightarrow k = 0,\\ \sigma_w(k) = n \Rightarrow k = n.\end{cases}$$

This implies that, for any $\alpha \in]0, n[$, the length of a maximal subsequence $\alpha, \alpha+1, \dots, \alpha+\ell$ of $\sigma_w(\bullet)$ is $\leqslant n$, and even $\leqslant n-1$ if α is an integer. In other words:

COROLLARY 3.4. *The length of any maximal nonzero integral (resp. nonintegral) constant subsequence of $s_w(\bullet)$ is $\leqslant n-1$ (resp. $\leqslant n$).*

The proposition also gives a Birkhoff normal form for G_0:

$$\theta^2\partial_\theta\boldsymbol{\omega} = \boldsymbol{\omega}A_0 + \theta\boldsymbol{\omega}A_\infty$$

with A_0, A_∞ as in (1.6). The matrix A_0 is nothing but the matrix of multiplication by f on $G_0/\theta G_0$ in the basis induced by $\boldsymbol{\omega}$. Its eigenvalues are the critical values of f, as expected. In the case where $\mu = n+1$ (and all w_i equal to 1), we find that $A_\infty = \mathrm{diag}(0, 1, \dots, n)$ and A_0 is as in (1.6) with size $\mu = n+1$.

Proof of Proposition 3.2. It will be convenient to select some coordinate, say u_0. Multiplying (3.1) (applied to $\varphi_1 = \dots = \varphi_n = \varphi$) on the left by the inverse matrix of the matrix formed by the columns of $m_1, \dots, m_n$, one finds that, for any $\varphi \in \mathbb{C}[u, u^{-1}, \tau, \tau^{-1}]$, we have in G

$$\text{(3.5)}\qquad \forall i = 1, \dots, n \quad \big(\tfrac{1}{w_i}u_i\partial_{u_i} - \tfrac{1}{w_0}u_0\partial_{u_0}\big)\varphi\omega_0 = \tau(u_i - u_0)\varphi\omega_0.$$

Applying this to any monomial $\varphi = u^a$ and summing these equalities, we get the following relation for $j = 0$, hence for any $j = 0, \dots, n$ by a similar argument:

$$\text{(3.6)}\qquad -\frac{1}{\mu}\big(\tau\partial_\tau + L_j(a)\big)u^a\omega_0 = \tau u^{a+\mathbf{1}_j}\omega_0,$$

where we put $L_j(a) = \sum_{i=0}^n a_i - \mu a_j/w_j$. This is nothing but (I.4.12) in the present situation. Apply this for $a = a(k)$ and $j = i(k)$ $(k = 0, \dots, \mu-1)$ to get the first relation in the lemma (remark that $L_{i(k)}(a(k)) = \sigma_w(k)$). Bernstein's relation for ω_0 is then clear. Remark also that ω_k is given by

$$\omega_k = \tau^{-k}\prod_{j=0}^{k-1}\Big[-\frac{1}{\mu}(\tau\partial_\tau - s_w(j))\Big]\cdot\omega_0.$$

It is not difficult to derive from Bernstein's relation for ω_0 a Bernstein relation for each ω_k and conclude that ω_k has V-order $\leqslant \sigma_w(k)$. [Notice also that, as

$\sigma_w(k) = L_{i(k)}(a(k)) = \max_j L_j(a(k))$, the order of ω_k with respect to the Newton filtration is $\leqslant \sigma_w(k)$; this is compatible with Theorem I.4.5.]

Let us now show that $\omega_0, \dots, \omega_{\mu-1}$ generate G_0 as a $\mathbb{C}[\theta]$-module. Notice that Bernstein's relation for ω_0 implies that $\partial_\tau^\mu \omega_0 \in \mathbb{C}[\theta]\langle \omega_0, \dots, \partial_\tau^{\mu-1}\omega_0\rangle = \mathbb{C}[\theta]\langle \omega_0, \dots, \omega_{\mu-1}\rangle$, and this also holds for $\partial_\tau^\ell \omega_0$ for $\ell \geqslant \mu$. It is therefore enough to show that $(f^\ell \omega_0)_{\ell \geqslant 0}$ generate G_0 over $\mathbb{C}[\theta]$. Write (3.5) as

$$u^{a+\mathbf{1}_i}\omega_0 = \Big[u^a u_0 + \Big(\frac{a_i}{w_i} - \frac{a_0}{w_0}\Big)\theta u^a\Big]\omega_0. \tag{3.7}$$

The Brieskorn lattice G_0 is generated over $\mathbb{C}[\theta]$ by the $u_0^\ell \omega_0$ with $\ell \in \mathbb{N}$: indeed, it is generated by the $u^a\omega_0$; then,

- if $a_i \geqslant 1$ for some $i \geqslant 1$, one decreases a_i to 0 with (3.7);
- if $a_i \leqslant -1$ for some $i \geqslant 1$, one iterates (3.7) w_0 times and use the relation $u^w\omega_0 = \omega_0$ to express $u^b\omega_0$ (any b) as a sum (with constant coefficients) of terms $\theta^k u_0^\ell u^{b+w'}\omega_0$ and of $u^{b+w'+\mathbf{1}_i}$, with $k, \ell \geqslant 0$ and $w' = (0, w_1, \dots, w_n)$; hence if $b_i < 0$, there exists r such that $b_i + rw_i \geqslant 0$ and one iterates r times the previous process to write $u^b\omega_0$ with terms $\theta^k u^a$, with $a_i \geqslant 1$, to reduce to the previous case;
- notice that, in both previous processes, we never decrease the degree in u_0; now, we are reduced to considering $u_0^\ell\omega_0$ with $\ell < 0$; use once more the relation $u^{kw}u^b\omega_0 = u^b\omega_0$ (for any $k \geqslant 0$, any b) to replace u_0^ℓ with u^a with $a_0, \dots, a_n \geqslant 0$ and apply the first case.

A similar argument gives the result for the family $(f^\ell\omega_0)_{\ell \geqslant 0}$. As G_0 is $\mathbb{C}[\theta]$-free (*cf.* Remark I.4.8 and § I.2.c), we conclude that $\boldsymbol{\omega}$ is a $\mathbb{C}[\theta]$-basis of G_0. [Instead of using Remark I.4.8, one can directly conclude here that G_0 is $\mathbb{C}[\theta]$-free of rank μ by showing first that $\boldsymbol{\omega}$ generates G as a $\mathbb{C}[\tau, \tau^{-1}]$-module.]

Remark also that $(\omega_0, \dots, f^{\mu-1}\omega_0)$ is another basis, but the differential equation does not take Birkhoff normal form in such a basis.

We will now determine the V-filtration. Put $\omega'_k = \tau^{[\sigma_w(k)]}\omega_k$. Then $\boldsymbol{\omega}'$ is another $\mathbb{C}[\tau, \tau^{-1}]$-basis of G. The V-order of ω'_k is $\leqslant \sigma_w(k) - [\sigma_w(k)] < 1$. For $\alpha \in [0, 1[$, put

$$U_\alpha G = \mathbb{C}[\tau]\langle \omega'_k \mid \sigma_w(k) - [\sigma_w(k)] \leqslant \alpha\rangle + \tau\mathbb{C}[\tau]\langle \omega'_k \mid \sigma_w(k) - [\sigma_w(k)] > \alpha\rangle$$
$$U_{<\alpha} G = \mathbb{C}[\tau]\langle \omega'_k \mid \sigma_w(k) - [\sigma_w(k)] < \alpha\rangle + \tau\mathbb{C}[\tau]\langle \omega'_k \mid \sigma_w(k) - [\sigma_w(k)] \geqslant \alpha\rangle,$$

and $U_{\alpha+p}G = \tau^p U_\alpha G$ (*resp.* $U_{<\alpha+p}G = \tau^p U_{<\alpha}G$) for any $p \in \mathbb{Z}$. We then have

$$U_\alpha G = \mathbb{C}\langle \omega'_k \mid \sigma_w(k) - [\sigma_w(k)] = \alpha\rangle + U_{<\alpha}G.$$

Notice that, according to the formula for ω_k, the elements ω'_k satisfy

$$\begin{aligned} -\frac{1}{\mu}\big(\tau\partial_\tau + \sigma_w(k) - [\sigma_w(k)]\big)\omega'_k &= \tau^{[\sigma_w(k)]+1-[\sigma_w(k+1)]}\omega'_{k+1} \\ &= \tau^{\lceil s_w(k+1)\rceil - \lceil s_w(k)\rceil}\omega'_{k+1}, \end{aligned} \tag{3.8}$$

with $\lceil s\rceil := -[-s]$. Recall that the sequence $(s_w(k))$, hence the sequence $(\lceil s_w(k)\rceil)$, is increasing. If $\lceil s_w(k+1)\rceil > \lceil s_w(k)\rceil$, then

$$\big(\tau\partial_\tau + \sigma_w(k) - [\sigma_w(k)]\big)\omega'_k \in U_{<0}G.$$

Otherwise, we have

$$\lceil s_w(k)\rceil - s_w(k) \geqslant \lceil s_w(k+1)\rceil - s_w(k+1),$$

i.e.,

$$\sigma_w(k) - [\sigma_w(k)] \geqslant \sigma_w(k+1) - [\sigma_w(k+1)],$$

and we conclude that $U_\alpha G$ is stable under $\tau\partial_\tau$ and that $\tau\partial_\tau + \alpha$ is nilpotent on $\mathrm{gr}^U_\alpha G$. The filtration $U_\bullet G$ satisfies then the characterizing properties of $V_\bullet G$, hence is equal to it.

We may now compute $G_p \cap V_\alpha$ for $p \in \mathbb{Z}$ and $\alpha \in [0,1[$. Any element of $G_p \cap V_\alpha$ decomposes uniquely as $\sum_{k=0}^{\mu-1} p_k(\tau)\omega'_k$, with

$$p_k(\tau) \in \begin{cases} \tau^{p-[\sigma_w(k)]}\mathbb{C}[\tau^{-1}] \cap \mathbb{C}[\tau] & \text{if } \sigma_w(k) - [\sigma_w(k)] \leqslant \alpha, \\ \tau^{p-[\sigma_w(k)]}\mathbb{C}[\tau^{-1}] \cap \tau\mathbb{C}[\tau] & \text{if } \sigma_w(k) - [\sigma_w(k)] > \alpha \end{cases}$$

It follows that

$$G_p \cap V_\alpha = \sum_{k|\sigma_w(k)=\alpha+p} \mathbb{C}\cdot\omega'_k + G_p \cap V_{<\alpha} + G_{p-1} \cap V_\alpha, \tag{3.9}$$

and therefore $\mathrm{gr}^G_p \mathrm{gr}^V_\alpha G$ is generated by the classes of ω'_k with $\sigma_w(k) = \alpha + p$. These classes form a basis of $\mathrm{gr}^G_p \mathrm{gr}^V_\alpha G$, as $\dim \oplus_p \oplus_{\alpha\in[0,1[} \mathrm{gr}^G_p \mathrm{gr}^V_\alpha G = \mu$. This gives the last statement of the proposition. □

For $\alpha \in [0,1[$, let ω'_k be such that $\lceil s_w(k)\rceil - s_w(k) = \alpha$ and denote by $[\omega'_k]$ the class of ω'_k in $H_\alpha := \mathrm{gr}^V_\alpha G$. After (1.5) we have:

$$-\frac{1}{\mu}(\tau\partial_\tau + \alpha)[\omega'_k] = \begin{cases} 0 & \text{if } s_w(k+1) > s_w(k), \\ [\omega'_{k+1}] & \text{if } s_w(k+1) = s_w(k). \end{cases} \tag{3.10}$$

It follows that the *primitive elements* relative to the nilpotent operator induced by $(-1/\mu)(\tau\partial_\tau + \alpha)$ on H_α are the elements $[\omega'_k]$ such that

$$k \geqslant n+1, \quad \lceil s_w(k)\rceil - s_w(k) = \alpha \quad \text{and} \quad s_w(k-1) < s_w(k)$$

and, if moreover $\alpha = 0$, the element $[\omega'_0] = [\omega_0]$.

Therefore, the Jordan blocks of $(-1/\mu)(\tau\partial_\tau + \alpha)$ on H_α are in one-to-one correspondence with the maximal constant sequences in $\mathcal{S}_w$, and the corresponding sizes are the same. All Jordan blocks, except that of $[\omega_0]$ if $\alpha = 0$, have thus size $\leqslant n$, and even $\leqslant n-1$ if α is an integer (*cf.* Cor. 3.4). Recall also (*cf.* [**8**, **9**]) that $H := \oplus_{\alpha\in[0,1[} H_\alpha$ may be identified with the relative cohomology space $H^n(U, f^{-1}(t))$ for $|t| \gg 0$, that H_α corresponds to the generalized eigenspace of the monodromy corresponding to the eigenvalue $\exp 2i\pi\alpha$, and that the unipotent part of the monodromy operator T is equal to $\exp 2i\pi N$ with $N := -(\tau\partial_\tau + \alpha)$.

EXAMPLE 3.11. Take $n = 4$ and $w_0 = 1$, $w_1 = 2$, $w_2 = 12$, $w_3 = 15$ and $w_4 = 30$, so that $\mu = 60$. Then the only possible α is 0 and N has one Jordan block of size 5, 3 blocks of size 3, 13 blocks of size 2 and 20 blocks of size 1. On the other hand, if $\mu = n+1$ (and all w_i equal to 1), the only possible α is 0 and N has only one Jordan block (of size $n+1$).

4. Poincaré duality and higher residue pairings

Consider on $\mathbb{C}[\tau, \tau^{-1}]$ the ring involution induced by $\tau \mapsto -\tau$. We will set $\overline{p(\tau)} := p(-\tau)$ (there is no complex conjugation involved here). Given a $\mathbb{C}[\tau, \tau^{-1}]$-module G, we denote by $\overline{G}$ the $\mathbb{C}$-vector space G equipped with the new module structure $p(\tau)\cdot g = p(-\tau)g$. For convenience, we denote by $\overline{g}$ the elements of $\overline{G}$. The $\mathbb{C}[\tau,\tau^{-1}]$-structure of $\overline{G}$ is therefore given by the rule: $\overline{p(\tau)\overline{g}} = \overline{p(\tau)g}$.

If G is moreover equipped with a connection, *i.e.*, with a compatible action of ∂_τ, then so is $\overline{G}$ and we have $\partial_\tau\overline{g} := \overline{-\partial_\tau g}$. Notice that $\tau\partial_\tau\overline{g} = \overline{\tau\partial_\tau g}$.

Duality for $\mathcal{D}$-modules gives (*cf.* **[9]**) the existence of a nondegenerate $\mathbb{C}[\tau,\tau^{-1}]$-bilinear pairing

$$S : G \underset{\mathbb{C}[\tau,\tau^{-1}]}{\otimes} \overline{G} \longrightarrow \mathbb{C}[\tau,\tau^{-1}]$$

satisfying the following properties:

(1) $\dfrac{dS(g',\overline{g''})}{d\tau} = S(\partial_\tau g',\overline{g''}) + S(g',\partial_\tau\overline{g''}) = S(\partial_\tau g',\overline{g''}) - S(g',\overline{\partial_\tau g''})$,
(equivalently, $\tau\partial_\tau S(g',\overline{g''}) = S(\tau\partial_\tau g',\overline{g''}) + S(g',\overline{\tau\partial_\tau g''})$),
(2) S sends $V_0\otimes\overline{V_{<1}}$ in $\mathbb{C}[\tau]$,
(3) S sends $G_0\otimes\overline{G_0}$ in $\theta^n\mathbb{C}[\theta] = \tau^{-n}\mathbb{C}[\tau^{-1}]$,
(4) $S(g'',\overline{g'}) = (-1)^n\overline{S(g',\overline{g''})}$ (this reflects the $(-1)^n$-symmetry of the Poincaré duality on U).

Notice that (1) means that S is a horizontal section of the $\mathbb{C}[\tau,\tau^{-1}]$-module $\mathrm{Hom}_{\mathbb{C}[\tau,\tau^{-1}]}(G\otimes\overline{G},\mathbb{C}[\tau,\tau^{-1}])$ equipped with its natural connection, or also that S is a $\mathbb{C}[\tau]\langle\partial_\tau\rangle$-linear morphism $\overline{G}\to G^*$, if one endows $G^* = \mathrm{Hom}_{\mathbb{C}[\tau,\tau^{-1}]}(G,\mathbb{C}[\tau,\tau^{-1}])$ with its natural connection. Therefore, (2) follows from (1) because any $\mathbb{C}[\tau]\langle\partial_\tau\rangle$-linear morphism is strict with respect to the Malgrange-Kashiwara filtrations V and we have

$$V_\beta(G^*) = \mathrm{Hom}_{\mathbb{C}[\tau]}(V_{<-\beta+1}G,\mathbb{C}[\tau])$$

(*cf.* **[9]**).

In the case of singularities, this corresponds to the "higher residue pairings" of K. Saito **[11]**. The link with Poincaré duality is explained in **[12]**.

For our Laurent polynomial f, we will recover in an elementary way the existence of such a pairing S satisfying the previous properties. More precisely, we have:

LEMMA 4.1. *There exists a unique (up to a nonzero constant) nondegenerate pairing S satisfying Properties* (1), (2), (3). *It is given by the formula:*

$$S(\omega_k,\overline{\omega_\ell}) = \begin{cases} S(\omega_0,\overline{\omega_n}) & \begin{cases}\text{if } 0\leqslant k\leqslant n \text{ and } k+\ell=n,\\ \text{or if } n+1\leqslant k\leqslant \mu-1 \text{ and } k+\ell=\mu+n,\end{cases}\\ 0 & \text{otherwise.}\end{cases}$$

Moreover, for any k,ℓ, $S(\omega_k,\overline{\omega_\ell})$ belongs to $\mathbb{C}\tau^{-n}$ and S satisfies (4).

Proof. Assume that a pairing S satisfying (1), (2), (3) exists. For $k,\ell = 0,\dots,\mu-1$, we have $S(\omega_k,\overline{\omega_\ell})\in\tau^{-n}\mathbb{C}[\tau^{-1}]$ by (3) and $S(\omega_0,\overline{\omega_\ell})\in\tau^{-[\sigma_w(\ell)]}\mathbb{C}[\tau]$ by (2). Therefore, $S(\omega_0,\overline{\omega_\ell})\neq 0$ implies $[\sigma_w(\ell)]\geqslant n$, and if $[\sigma_w(\ell)] = n$, we have $S(\omega_0,\overline{\omega_\ell})\in\mathbb{C}\tau^{-n}$. But we know by (3.3) that

$$[\sigma_w(\ell)]\begin{cases} <n & \text{if } \ell\neq n,\\ =n & \text{if } \ell = n.\end{cases}$$

Therefore, $S(\omega_0,\overline{\omega_\ell}) = 0$ if $\ell\neq n$ and $S(\omega_0,\overline{\omega_n})\in\mathbb{C}\tau^{-n}$.

Notice also that we have by (1) and Proposition 3.2:

$$(4.2)\quad -\frac{1}{\mu}(\tau\partial_\tau+n)S(\omega_k,\overline{\omega_\ell})$$
$$=\tau\big[S(\omega_{k+1},\overline{\omega_\ell})-S(\omega_k,\overline{\omega_{\ell+1}})\big]+\frac{\sigma_w(k)+\sigma_w(\ell)-n}{\mu}S(\omega_k,\overline{\omega_\ell}),$$

if we put as above $\omega_\mu=\omega_0$.

Argue now by induction for $k<n$: as $S(\omega_k,\overline{\omega_\ell})\in\mathbb{C}\tau^{-n}$, the LHS in (4.2) vanishes. This shows that $S(\omega_{k+1},\overline{\omega_\ell})=0$ if $\ell\neq n-k,n-1-k$. Moreover, if $\ell=n-k$, we have $\sigma_w(k)+\sigma_w(\ell)-n=0$, hence $S(\omega_{k+1},\overline{\omega_{n-k}})=0$. Last, we have $S(\omega_{k+1},\overline{\omega_{n-1-k}})=S(\omega_k,\overline{\omega_{n-k}})$.

Argue similarly for $k\geqslant n+1$. □

Notice that, if A_∞^* denotes the adjoint of A_∞ with respect to S, then $A_\infty+A_\infty^*=n\,\mathrm{Id}$, *i.e.*, $A_\infty-(n/2)\,\mathrm{Id}$ is skewsymmetric with respect to S.

5. M. Saito's solution to the Birkhoff problem

One step in constructing the Frobenius structure associated to f consists in solving Birkhoff's problem for the Brieskorn lattice G_0 in the Gauss-Manin system G, that is, in finding a $\mathbb{C}[\tau]$-lattice E of G, which glues with G_0 to a trivial vector bundle on $\mathbb{P}^1$. Recall (*cf.* [**3**, App. B] for what follows) that there is a one-to-one correspondence between such lattices E which are *logarithmic*, and decreasing filtrations $\oplus_{\alpha\in[0,1[}H_\alpha^\bullet$ of $H=\oplus_{\alpha\in[0,1[}H_\alpha$ which are stable under N and which are *opposite* to the filtration

$$G_p(H):=\underset{\alpha\in[0,1[}{\oplus}(G_p\cap V_\alpha)/(G_p\cap V_{<\alpha})=\mathbb{C}\langle[\omega_k']\mid[\sigma_w(k)]\leqslant p\rangle\quad\text{after (3.9).}$$

This is analogous to [**12**, Th. 3.6].

In [**12**, Lemma 2.8], M. Saito defines a canonical decreasing filtration $H^\bullet_{\mathrm{Saito}}$ in terms of the monodromy filtration $M_\bullet$ of the nilpotent endomorphism $2i\pi N$ of H and of the filtration conjugate to $G_p(H)$, the conjugation being taken with respect to the real structure on H coming from the identification with $H^n(U,f^{-1}(t))$. This defines therefore a canonical solution to Birkhoff's problem for G_0.

Consider now the decreasing filtration $H^\bullet$ of H explicitly defined by

$$(5.1)\qquad H^p=\mathbb{C}\langle[\omega_k']\mid[\sigma_w(k)]\geqslant p\rangle,$$

where $[\,]$ denotes the class in $H=\oplus_{\alpha\in[0,1[}V_\alpha G/V_{<\alpha}G$. Then $H^\bullet$ is opposite to $G_\bullet(H)$. It satisfies

$$H^0=H,\ H^{n+1}=0,\quad NH^p\subset H^{p+1}$$

and, for $k=0,\dots,\mu-1$ and $\alpha\in[0,1[$,

$$(H^p_\alpha)^\perp=\begin{cases}H^{n-p}_{1-\alpha}&\text{if }\alpha\neq0,\\ H_0^{n+1-p}&\text{if }\alpha=0,\end{cases}$$

where $^\perp$ means taking the orthogonal with respect to the symmetric bilinear form g on H induced by S. If $\mu=n+1$ (and all w_i equal to 1), then $H^p=M_{n-2p}$ (this implies that the mixed Hodge structure on H is "Hodge-Tate").

PROPOSITION 5.2. *The filtration $H^\bullet$ is equal to the opposite filtration $H^\bullet_{\mathrm{Saito}}$. The associated logarithmic lattice is $E:=\mathbb{C}[\tau]\langle\omega_0,\dots,\omega_{\mu-1}\rangle$.*

Proof. Let us begin with the second statement. The lattice E introduced in the proposition is logarithmic, by Proposition 3.2. A computation analogous to that of $G_p \cap V_\alpha$ shows that the filtration $\tau^p E \cap V_\alpha / \tau^p E \cap V_{<\alpha}$ of H_α is equal to H_α^p. Therefore, E is the logarithmic lattice corresponding to $H^\bullet$ by the correspondence recalled above.

For the first statement, put $F^\bullet(H) = G_{n-\bullet}(H)$. This is a decreasing filtration. Consider also the increasing filtration

$$W_\bullet(H_\alpha) = \begin{cases} M_{\bullet-n-1}(H_\alpha) & \text{if } \alpha \neq 0, \\ M_{\bullet-n}(H_0) & \text{if } \alpha = 0, \end{cases}$$

where $M_\bullet(H)$ denotes the monodromy filtration of the nilpotent endomorphism $2i\pi N$ on H. Recall that $W_\bullet(H)$ is defined over $\mathbb{R}$ (even over $\mathbb{Q}$) as $2i\pi N$ is so. Then the opposite filtration given by M. Saito is

$$H^\bullet_{\text{Saito}} = \sum_q \overline{F^q} \cap W_{n+q-\bullet}(H),$$

where $\overline{E}$ denotes the conjugate of the subspace E of H with respect to the complex conjugation coming from the identification

$$H \xrightarrow{\sim} H^n(U, f^{-1}(t), \mathbb{C}) = \mathbb{C} \otimes_{\mathbb{R}} H^n(U, f^{-1}(t), \mathbb{R}).$$

We therefore need to give a description of the conjugation in term of the basis $[\omega'_k]$.

Let k_0 be such that $[\omega'_{k_0}]$ is a primitive element with respect to N, and denote by ν_{k_0} its weight. Then $N^{\nu_{k_0}+1}[\omega'_{k_0}] = 0$. For $j = 0, \dots, \nu_{k_0}$, put $k = k_0 + j$. Then $[\omega'_k] = (\frac{1}{\mu} N)^j [\omega'_{k_0}]$ has order $\nu_{k_0} + 1 - j$ with respect to N, and weight $\nu_k := \nu_{k_0} - 2j$. Moreover, we have $\sigma_w(k) = \sigma_w(k_0) + j$, as $j \mapsto s_w(k_0 + j)$ is constant. The space $B_{k_0} := \langle N^j [\omega'_{k_0}] \mid j = 0, \dots, \nu_{k_0} \rangle$ is a Jordan block of N.

Assume that $k_0 \geqslant n+1$. Then $[\omega'_{\mu+n-k_0}]$ is primitive with respect to ${}^tN$, hence $[\omega'_{\mu+n-k_0-\nu_{k_0}}]$ is primitive with respect to N. It will be convenient to put $\overline{k}_0 = \mu + n - k_0 - \nu_{k_0}$ and, for $k = k_0 + j$ with $j = 0, \dots, \nu_{k_0}$, $\overline{k} = \overline{k}_0 + j$. We therefore have $\overline{k} = \mu + n - k - \nu_k$. Notice that, for such a k, we have $s_w(k) = s_w(k_0) = \mu - s_w(\overline{k}_0) = \mu - s_w(\overline{k})$. We also have $\sigma_w(\mu + n - k) - \nu_k = \sigma_w(\overline{k})$ if $k \geqslant n+1$.

For $k \in [0, n]$, we simply put $\overline{k} = k$.

The proof of the following lemma will be given in § 6.

Lemma 5.3. *For $k_0 \geqslant n+1$, the conjugate of the Jordan block B_{k_0} is the Jordan block $B_{\overline{k}_0}$, and B_0 is self-conjugate.*

It follows from this lemma that, for k as above, we have

$$\overline{[\omega'_k]} = \sum_{\ell=k}^{k_0+\nu_0} a_\ell [\omega'_{\overline{\ell}}] \tag{5.4}$$

with $a_k \neq 0$.

Let us now end the proof of Proposition 5.2. We have

$$F^q \cap W_{n+q-p} = G_{n-q} \cap M_{q-p(-1)} = \langle [\omega'_k] \mid [\sigma_w(k)] \leqslant n - q \text{ and } \nu_k \leqslant q - p(-1) \rangle,$$

where (-1) is added if $\sigma_w(k) \notin \mathbb{Z}$ and not added otherwise. Therefore,

$$\sum_q F^q \cap W_{n+q-p} = \langle [\omega'_k] \mid [\sigma_w(k)] + \nu_k \leqslant n - p(-1) \rangle.$$

Remark now that, if $k \geqslant n+1$,

$$\begin{aligned}[]
[\sigma_w(k)] + \nu_k \leqslant n - p(-1) &\iff [n - \sigma_w(\mu + n - k)] + \nu_k \leqslant n - p(-1)\\
&\iff \lceil \sigma_w(\mu + n - k - \nu_k) \rceil \geqslant p(+1)\\
&\iff [\sigma_w(\mu + n - k - \nu_k)] \geqslant p\\
&\iff [\sigma_w(\overline{k})] \geqslant p.
\end{aligned}$$

Arguing similarly for $k \leqslant n$, we conclude from Lemma 5.3 and (5.1) that

$$\begin{aligned}
\sum_q \overline{F}^q \cap W_{n+q-p} &= \langle \overline{[\omega'_k]} \mid [\sigma_w(k)] + \nu_k \leqslant n - p(-1) \rangle \\
&= \langle [\omega'_{\overline{k}}] \mid [\sigma_w(\overline{k})] \geqslant p \rangle \\
&= H^p.
\end{aligned} \tag{5.5}$$

Notice that (5.5) follows from (5.4), as σ_w is increasing on each $B_{\overline{k}_0}$. □

6. Some topology of f and proof of Lemma 5.3

6.a. Lefschetz thimbles. Denote by Δ the subset $(\mathbb{R}_+^*)^{n+1} \cap U \subset U$, defined by $u_i > 0$ for $i = 0, \dots, n$. The restriction $f_{|\Delta}$ of f to Δ takes values in $[\mu, +\infty[$ and has only one critical point (which is a Morse critical point of index 0), namely $(1, \dots, 1)$, with critical value equal to μ. Notice also that $f_{|\Delta}$ is proper. Therefore, Δ is a Lefschetz thimble for f with respect to the critical point $(1, \dots, 1)$. Other Lefschetz thimbles at $\zeta^\ell(1, \dots, 1)$ are $\zeta^\ell \Delta$ ($\ell = 0, \dots, \mu - 1$).

Fix $\tau \neq 0$. The morphisms

$$H_n(U, \mathrm{R\acute{e}}\, \tau f > C'; \mathbb{Q}) \longrightarrow H_n(U, \mathrm{R\acute{e}}\, \tau f > C; \mathbb{Q})$$

for $C' > C$ are isomorphisms if C is big enough. We denote by $H_n(U, \mathrm{R\acute{e}}\, \tau f \gg 0)$ the limit of this inverse system. This is the germ at τ of a local system $\mathcal{H}$ of rank μ on $\mathbb{C}^* = \{\tau \neq 0\}$. Notice that Δ defines a nonzero element of the germ $\mathcal{H}_\tau$ at any τ with $\mathrm{R\acute{e}}\, \tau > 0$, *i.e.*, a section $\Delta(\tau)$ of $\mathcal{H}$ on $\{\mathrm{R\acute{e}}\, \tau > 0\}$. Therefore, it defines in a unique way a multivalued section of $\mathcal{H}$ on $\mathbb{C}^*$.

Let $\varepsilon > 0$ be small enough. As f is a C^∞ fibration over the open set $\mathbb{C} \smallsetminus \{\mu\zeta^\ell \mid \ell = 0, \dots, \mu - 1\}$, it is possible to find a basis of sections $\Delta_0(\tau), \dots, \Delta_{\mu-1}(\tau)$ of $\mathcal{H}$ on the open set

$$\mathcal{S} = \{\tau = |\tau|\, e^{2i\pi\theta} \mid \theta \in \,]\varepsilon - 1, \varepsilon[\}$$

in such a way that, for any $\ell \in \{0, \dots, \mu-1\}$ and $|\tau| > 0$, we have $\Delta_\ell(\zeta^{-\ell}\, |\tau|) = \zeta^\ell \Delta$. Of course, this basis extends as a basis of multivalued sections of $\mathcal{H}$ on $\mathbb{C}^*$.

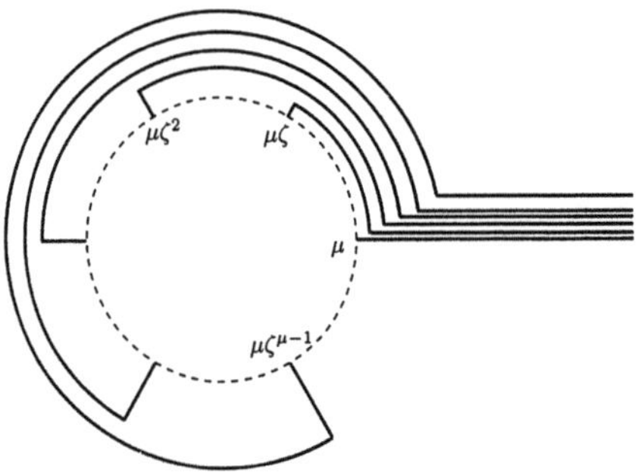

The images $f(\Delta_\ell(|\tau|))$

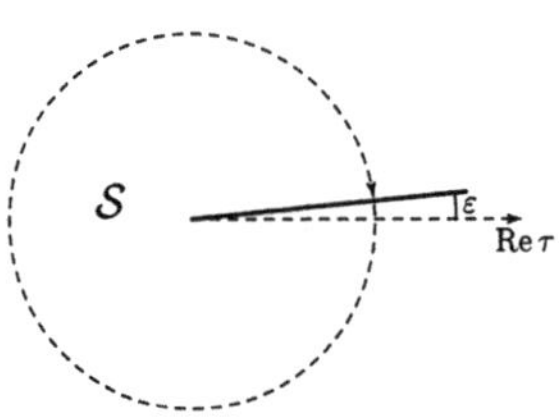

The domain $\mathcal{S}$

6.b. Integrals along Lefschetz thimbles. Let $\eta \in G$ and let $\widetilde{\eta}$ be a representative of η in $\Omega^n(U)[\tau, \tau^{-1}]$. Then the function

$$\mathcal{S} \ni \tau \longmapsto \int_{\Delta_\ell(\tau)} e^{-\tau f}\widetilde{\eta}$$

only depends on η and is holomorphic on $\mathcal{S}$. It is denoted by $\varphi_{\ell,\eta}(\tau)$. Moreover, we have

$$\frac{\partial \varphi_{\ell,\eta}(\tau)}{\partial \tau} = \varphi_{\ell,\partial_\tau \eta}(\tau).$$

It follows that, if $\eta \in V_\alpha G$, then

$$\varphi_{\ell,\eta}(\tau) = \tau^{-\alpha}\Big[\sum_{m=0}^{m_\eta} c_{\eta,\ell}^{(m)} \frac{(-\frac{1}{2i\pi}\log\tau)^m}{m!} + o(1)\Big] := \tau^{-\alpha}\big[\psi_{\eta,\ell}(\tau) + o(1)\big], \tag{6.1}$$

where $\tau^{-\alpha} = |\tau|^{-\alpha} e^{-2i\pi\alpha\theta}$, $\frac{1}{2i\pi}\log\tau = (\frac{1}{2i\pi}\log|\tau|) + \theta$, with $\theta \in\,]\varepsilon - 1, \varepsilon[$ and $c_{\eta,\ell}^{(m)} \in \mathbb{C}$. The coefficients $c_{\eta,\ell}^{(m)}$ only depend on the class $[\eta]$ of η in $\mathrm{gr}_\alpha^V G$, so we will denote them by $c_{[\eta],\ell}^{(m)}$, and we have

$$c_{[\eta],\ell}^{(m)} = c_{(2i\pi N)^m[\eta],\ell}^{(0)}. \tag{6.2}$$

We will now characterize the Jordan blocks B_{k_0} in H_α. Such a Jordan block is characterized by the constant value s of $s_w(\bullet)$, so that it will be convenient to denote such a block by $B_{\alpha,s}$.

Lemma 6.3. *For $[\eta] \in H_\alpha$, we have $[\eta] \in B_{\alpha,s}$ if and only if, for any $\ell = 0, \dots, \mu - 1$ and any $j \geqslant 0$, we have*

$$c_{(2i\pi N)^j[\eta],\ell}^{(0)} = \zeta^{\ell s} \sum_m c_{(2i\pi N)^{j+m}[\eta],0}^{(0)} \frac{(-\ell/\mu)^m}{m!}.$$

Proof. For $\eta = \omega_k$, denote $c_{[\omega_k],\ell}^{(m)} = c_{[\omega_k'],\ell}^{(m)}$ by $c_{k,\ell}^{(m)}$. Then we have

$$\int_{\Delta_\ell(\tau)} e^{-\tau f}\omega_k = \tau^{-\sigma_w(k)}\Big(\Big[c_{k,\ell}^{(m_k)} \frac{(-\frac{1}{2i\pi}\log\tau)^{m_k}}{m_k!} + \cdots + c_{k,\ell}^{(0)}\Big] + o(1)\Big), \tag{6.4}$$

where $m_k + 1$ denotes the order of $[\omega_k]$ in $\mathrm{gr}_{\sigma_w(k)}^V G$ with respect to $2i\pi N$. Remark now that, as $\omega_k = u^{a(k)}\omega_0$ and $|a(k)| = k$, we have

$$\int_{\Delta_\ell(\zeta^{-\ell}|\tau|)} e^{-\zeta^{-\ell}|\tau| f}\omega_k = \int_{\zeta^\ell\Delta} e^{-\zeta^{-\ell}|\tau| f}\omega_k = \zeta^{k\ell}\int_\Delta e^{-|\tau| f}\omega_k.$$

Hence, we get

$$c_{k,\ell}^{(m_k)} \frac{(-\frac{1}{2i\pi}\log|\tau| + \ell/\mu)^{m_k}}{m_k!} + \cdots + c_{k,\ell}^{(0)} = \zeta^{\ell s_w(k)}\Big[c_{k,0}^{(m_k)} \frac{(-\frac{1}{2i\pi}\log|\tau|)^{m_k}}{m_k!} + \cdots + c_{k,0}^{(0)}\Big],$$

and in particular

$$c_{k,\ell}^{(0)} = \zeta^{\ell s_w(k)} \cdot \sum_{m=0}^{m_k} \frac{(-\ell/\mu)^m}{m!} c_{k,0}^{(m)}. \tag{6.5}$$

Therefore, any element $[\eta]$ in $B_{\alpha,s}$ satisfies the equality of Lemma 6.3 for $j = 0$, hence for any j.

Conversely, remark first that, if $[\eta]$ is fixed, then the equality of Lemma 6.3 for any $j \geqslant 0$ is equivalent to

$$\psi_{[\eta],\ell}(\zeta^{-\ell}|\tau|) = \zeta^{\ell s}\psi_{[\eta],0}(|\tau|),$$

where $\psi_{[\eta],\ell}$ is defined by (6.1) (two polynomials are equal iff all the corresponding derivatives at 0 are equal).

Write $[\eta] = \sum \lambda_k[\omega'_k]$ in H_α, denote $m'_{[\eta]} = \max_{k|\lambda_k\neq 0} m_k$ and put $K_{[\eta]} = \{k \mid m_k = m'_{[\eta]}\}$. Notice that, for $k, k' \in K_{[\eta]}$, we have $s_w(k) \neq s_w(k')$. If $m'_{[\eta]} > m_{[\eta]}$, we have $\sum_{k\in K_{[\eta]}} \zeta^{\ell s_w(k)}\lambda_k c_{k,0}^{(m_k)} = 0$ for any $\ell = 0,\dots,\mu-1$. It follows that $\lambda_k c_{k,0}^{(m_k)} = 0$, hence $\lambda_k = 0$, for any $k \in K_{[\eta]}$, a contradiction. Therefore $m'_{[\eta]} = m_{[\eta]}$. Argue similarly to show that $K_{[\eta]}$ is reduced to one element, denoted by $k_{[\eta]}$, and that $s_w(k_{[\eta]}) = s$. Apply the lemma by induction on $m_{[\eta']}$ to $[\eta'] = [\eta] - \lambda_{k_{[\eta]}}[\omega'_{k_{[\eta]}}]$. □

6.c. Isomorphism between nearby cycles. The multivalued cycles $\Delta_\ell(\tau)$ form a basis of the space of multivalued global sections of $\mathcal{H}$, that we denote by $\psi_\tau\mathcal{H}$. This basis defines the integral (hence the real) structure on $\psi_\tau\mathcal{H}$.

Denote by $\mathcal{N}_{\alpha,p}$ the space of linear combinations with meromorphic coefficients of germs at $\tau = 0$ of the multivalued functions $e_{\alpha,q} = \tau^\alpha(-\frac{1}{2i\pi}\log\tau)^q/q!$ ($q \leqslant p$). For p large enough (here $p \geqslant n+1$ is enough), the map

$$\begin{aligned}\varphi : V_\alpha G &\longrightarrow V_0(G\otimes\mathcal{N}_{\alpha,p})\\ \eta &\longmapsto \sum_{j=0}^{p}[2i\pi(\tau\partial_\tau+\alpha)]^j\eta\otimes e_{\alpha,j}\end{aligned}$$

induces an isomorphism

$$\mathrm{gr}^V_\alpha G \xrightarrow{\ \sim\ } \mathrm{Ker}\left[\tau\partial_\tau : \mathrm{gr}^V_0(G\otimes\mathcal{N}_{\alpha,p}) \longrightarrow \mathrm{gr}^V_0(G\otimes\mathcal{N}_{\alpha,p})\right].$$

As G is regular at $\tau = 0$, there exists a perturbation $\eta \mapsto \psi(\eta) \in V_{<0}(G\otimes\mathcal{N}_{\alpha,p})$ such that $\varphi(\eta)+\psi(\eta) \in \mathrm{Ker}\left[\tau\partial_\tau : G\otimes\mathcal{N}_{\alpha,p} \to G\otimes\mathcal{N}_{\alpha,p}\right]$.

Recall (see, *e.g.*, [**7**]) that $\mathrm{Ker}\left[\tau\partial_\tau : G\otimes\mathcal{N}_{\alpha,p} \to G\otimes\mathcal{N}_{\alpha,p}\right]$ is identified with H_α. Set $\mathcal{N} = \oplus_{\alpha\in[0,1[}\mathcal{N}_{\alpha,n+1}$. Given a section λ of $H = \mathrm{Ker}\left[\tau\partial_\tau : G\otimes\mathcal{N} \to G\otimes\mathcal{N}\right]$ and a section δ of $\psi_\tau\mathcal{H}$, choose a representative $\widetilde{\lambda}$ of λ in $\Omega^n(U)\otimes_{\mathbb{C}}\mathcal{N}$. Then $\int_\delta e^{-\tau f}\widetilde{\lambda} \in \mathbb{C}$. Then (see Appendix) λ belongs to $H_{\mathbb{Q}}$ if and only if, for any $\ell = 0,\dots,\mu$ and some nonzero τ, we have

$$\int_{\Delta_\ell(\tau)} e^{-\tau f}\widetilde{\lambda} \in \mathbb{Q}. \tag{6.6}$$

For $\eta \in V_\alpha G$ and $\widetilde{\lambda} = \varphi(\eta)+\psi(\eta)$, and using (6.1), one finds

$$\int_{\Delta_\ell(\tau)} e^{-\tau f}\varphi(\eta) = c^{(0)}_{\eta,\ell} + o(1).$$

As a consequence, the conjugate $\overline{[\eta]}$ of $[\eta]$ satisfies

$$c^{(0)}_{\overline{[\eta]},\ell} = \overline{c^{(0)}_{[\eta],\ell}}.$$

It follows now from Lemma 6.3 that

$$\overline{B_{\alpha,s}} = \begin{cases} B_{1-\alpha,\mu-s} & \text{if } \alpha \in \,]0,1[, \\ B_{0,\mu-s} & \text{if } \alpha = 0. \end{cases}$$

As $s_w(\overline{k}_0) = \mu - s_w(k_0)$, this ends the proof of Lemma 5.3. □

Appendix

In this appendix, we explain with some details why the real structure on H as defined by (6.6) is indeed the real structure used in [**8**] to define the Hodge structure on H. We will need to recall some notation and results of [**8**].

We will denote by U a smooth quasi-projective variety and by $f : U \to \mathbb{A}^1$ a regular function on U. We denote by t the coordinate on the affine line $\mathbb{A}^1$. We also fix an embedding $\kappa : U \hookrightarrow \mathcal{X}$ into a smooth projective variety such that there exists an algebraic map $F : \mathcal{X} \to \mathbb{P}^1$ extending f. We have a commutative diagram, where the right part is Cartesian, thus defining X as a fibred product,

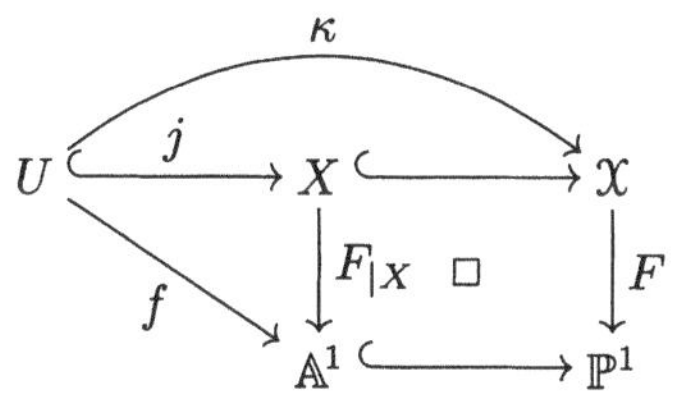

Denote by $\varepsilon : \widetilde{\mathbb{P}}^1 \to \mathbb{P}^1$ the real blow-up of $\mathbb{P}^1$ centered at ∞ ($\widetilde{\mathbb{P}}^1$ is diffeomorphic to a closed disc) and by $\widetilde{F} : \widetilde{\mathcal{X}} \to \widetilde{\mathbb{P}}^1$ the fibre-product of F with the blowing-up ε. Denote by $\widetilde{\kappa}$ the inclusion $U \hookrightarrow \widetilde{\mathcal{X}}$.

Denote by S^1 the inverse image of ∞ by the blowing-up ε. Let $\widehat{\mathbb{A}}^1$ be an affine line with coordinate τ. Denote by L'^+ the closed set of $S^1 \times \widehat{\mathbb{A}}^1 \subset \widetilde{\mathbb{P}}^1 \times \widehat{\mathbb{A}}^1$ defined by $\mathrm{Ré}(e^{i\theta}\tau) \geqslant 0$, with $\theta = \arg t$ and where t is the coordinate on $\mathbb{A}^1 = \mathbb{P}^1 \smallsetminus \{\infty\}$, and set $L^- = \widetilde{\mathbb{P}}^1 \times \widehat{\mathbb{A}}^1 \smallsetminus L'^+$. For $\tau \neq 0$, denote by $L'^+_\tau, L^-_\tau \subset \widetilde{\mathbb{P}}^1$ the fibre of L'^+, L^- over τ.

Denote similarly by $L'^+_{\widetilde{\mathcal{X}}}, L^-_{\widetilde{\mathcal{X}}} \subset \widetilde{\mathcal{X}} \times \widehat{\mathbb{A}}^1$ (*resp.* $L'^+_{\widetilde{\mathcal{X}},\tau}, L^-_{\widetilde{\mathcal{X}},\tau} \subset \widetilde{\mathcal{X}}$) the inverse image of the corresponding sets by $\widetilde{F} \times \mathrm{Id}_{\widehat{\mathbb{A}}^1}$ (*resp.* by $\widetilde{F}$).

We denote by $\alpha : \mathbb{A}^1 \times \widehat{\mathbb{A}}^1 \hookrightarrow L^-$ and $\beta : L^- \hookrightarrow \widetilde{\mathbb{P}}^1 \times \widehat{\mathbb{A}}^1$ (*resp.* $\alpha_\tau : \mathbb{A}^1 \hookrightarrow L^-_\tau$ and $\beta_\tau : L^-_\tau \hookrightarrow \widetilde{\mathbb{P}}^1$) the inclusions, and by the same letters the corresponding inclusions

$$\alpha : X \times \widehat{\mathbb{A}}^1 \hookrightarrow L^-_{\widetilde{\mathcal{X}}} \quad \text{and} \quad \beta : L^-_{\widetilde{\mathcal{X}}} \hookrightarrow \widetilde{\mathcal{X}} \times \widehat{\mathbb{A}}^1,$$

resp.

$$\alpha_\tau : X \hookrightarrow L^-_{\widetilde{\mathcal{X}},\tau} \quad \text{and} \quad \beta_\tau : L^-_{\widetilde{\mathcal{X}},\tau} \hookrightarrow \widetilde{\mathcal{X}}.$$

Therefore we have $\beta_\tau \circ \alpha_\tau \circ j = \widetilde{\kappa}$.

In [**8**, (1.8)], we have defined the Fourier transform $\mathfrak{F}_F(\boldsymbol{R}j_*\mathbb{C}_U)$ as the following complex on $\mathcal{X} \times \widehat{\mathbb{A}}^1$ (there is a shift by 1 in *loc. cit.*, that we do not introduce here):

$$\mathfrak{F}_F(\boldsymbol{R}j_*\mathbb{C}_U) := \boldsymbol{R}\varepsilon_*\, \beta_!\, \boldsymbol{R}\alpha_*\, \boldsymbol{R}j_*\mathbb{C}_{U \times \widehat{\mathbb{A}}^1},$$

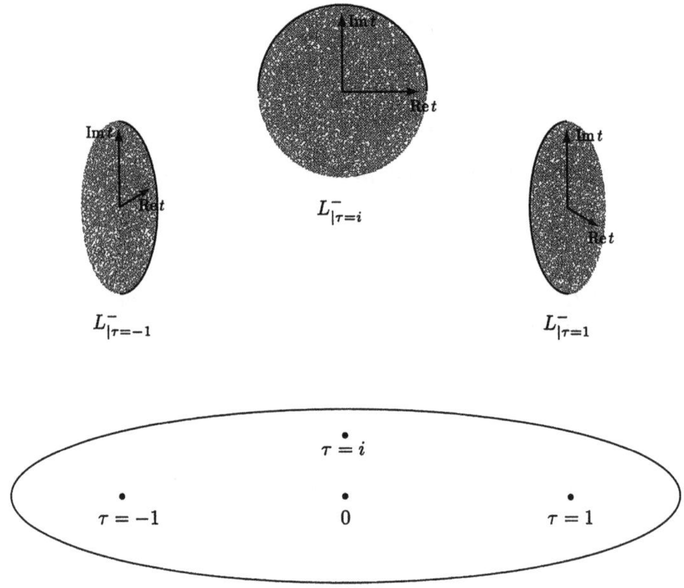

Affine line $\widehat{\mathbb{A}}^1$ with coordinate τ

where we still denote by j (*resp.* κ) the inclusion $U \times \widehat{\mathbb{A}}^1 \hookrightarrow X \times \widehat{\mathbb{A}}^1$ (*resp.* $U \times \widehat{\mathbb{A}}^1 \hookrightarrow \mathfrak{X} \times \widehat{\mathbb{A}}^1$). This complex has a natural $\mathbb{Q}$-structure (replace $\mathbb{C}_U$ with $\mathbb{Q}_U$). This induces a $\mathbb{Q}$-structure on the nearby cycle complex $\psi_\tau \mathfrak{F}_F(\boldsymbol{R}j_*\mathbb{C}_U) = \psi_\tau \mathfrak{F}_F(\boldsymbol{R}j_*\mathbb{Q}_U) \otimes_{\mathbb{Q}} \mathbb{C}$.

Denote by $\mathcal{E}^{-\tau f}$ the algebraic $\mathcal{D}_{U\times\widehat{\mathbb{A}}^1}$-module $\mathcal{O}_{U\times\widehat{\mathbb{A}}^1} e^{-\tau f}$ (*i.e.*, the $\mathcal{O}_{U\times\widehat{\mathbb{A}}^1}$-module $\mathcal{O}_{U\times\widehat{\mathbb{A}}^1}$ with connection $e^{\tau f} \circ d \circ e^{-\tau f}$). The quasi-isomorphism

$$\mathrm{DR}^{\mathrm{an}}_{\mathfrak{X}\times\widehat{\mathbb{A}}^1}(\kappa_+ \mathcal{E}^{-\tau f}) \xrightarrow{\ \sim\ } \mathfrak{F}_F(\boldsymbol{R}j_*\mathbb{C}_U) \tag{A.1}$$

constructed in [**8**, Th. 2.2] is then used to define the $\mathbb{Q}$-structure on the complex (of sheaves on $\mathfrak{X}$) $\psi_\tau \mathrm{DR}^{\mathrm{an}}_{\mathfrak{X}\times\widehat{\mathbb{A}}^1}(\kappa_+\mathcal{E}^{-\tau f})$ [on the other hand one uses the V-filtration relative to $\tau = 0$ on $\kappa_+\mathcal{E}^{-\tau f}$ to construct the Hodge filtration on this complex]. By DR we mean the usual de Rham complex, starting in degree 0.

Denote by $\widetilde{\mathcal{O}}^{\mathrm{an}}_{\widehat{\mathbb{A}}^1}$ the sheaf of multivalued holomorphic functions on $\widehat{\mathbb{A}}^1 \smallsetminus \{0\}$. Then

$$\psi_\tau \mathrm{DR}^{\mathrm{an}}_{\mathfrak{X}\times\widehat{\mathbb{A}}^1}(\kappa_+\mathcal{E}^{-\tau f}) = i^{-1}_{\tau=0} \mathrm{DR}^{\mathrm{an}}_{\mathfrak{X}\times\widehat{\mathbb{A}}^1}(\kappa_+\mathcal{E}^{-\tau f} \otimes_{\widehat{p}^{-1}\mathcal{O}^{\mathrm{an}}_{\widehat{\mathbb{A}}^1}} \widetilde{\mathcal{O}}^{\mathrm{an}}_{\widehat{\mathbb{A}}^1}),$$

where $\widehat{p} : \mathfrak{X} \times \widehat{\mathbb{A}}^1 \to \widehat{\mathbb{A}}^1$ denotes the projection and $i_{\tau=0} : \mathfrak{X} \times \{0\} \hookrightarrow \mathfrak{X} \times \widehat{\mathbb{A}}^1$ denotes the inclusion (see, *e.g.*, [**7**, (4.9.4)]). We are interested in analyzing the $\mathbb{Q}$-structure on the cohomology of $\boldsymbol{R}\Gamma\big(\mathfrak{X}, \psi_\tau \mathrm{DR}^{\mathrm{an}}_{\mathfrak{X}\times\widehat{\mathbb{A}}^1}(\kappa_+\mathcal{E}^{-\tau f})\big)$. Use C^∞ forms on $\mathfrak{X}$ to identify it with

$$\Gamma\big(\mathfrak{X}, i^{-1}_{\tau=0}\mathcal{E}^{\bullet}_{\mathfrak{X}\times\widehat{\mathbb{A}}^1}(\kappa_+\mathcal{E}^{-\tau f} \otimes_{\widehat{p}^{-1}\mathcal{O}^{\mathrm{an}}_{\widehat{\mathbb{A}}^1}} \widetilde{\mathcal{O}}^{\mathrm{an}}_{\widehat{\mathbb{A}}^1})\big),$$

with $n = \dim \mathfrak{X}$. Similarly, denote by $\widetilde{\mathbb{C}}_{\widehat{\mathbb{A}}^1}$ the sheaf of multivalued local sections of $\mathbb{C}_{\widehat{\mathbb{A}}^1}$ (*i.e.*, local sections on the universal covering of $\widehat{\mathbb{A}}^1 \smallsetminus \{0\}$). Then $\psi_\tau \mathfrak{F}_F(\boldsymbol{R}j_*\mathbb{C}_U)$ is equal to $i^{-1}_{\tau=0}\big(\mathfrak{F}_F(\boldsymbol{R}j_*\mathbb{C}_U) \otimes \widehat{p}^{-1}\widetilde{\mathbb{C}}_{\widehat{\mathbb{A}}^1}\big)$.

In order to know that the cohomology class of a closed multivalued section of $\widehat{p}_*\mathcal{E}^\bullet_{\mathscr{X}\times\widehat{\mathbb{A}}^1}(\kappa_+\mathcal{E}^{-\tau f}\otimes_{\widehat{p}^{-1}\mathcal{O}^{\mathrm{an}}_{\widehat{\mathbb{A}}^1}}\widetilde{\mathcal{O}}^{\mathrm{an}}_{\widehat{\mathbb{A}}^1})$ is rational, one has to compute its image in $\boldsymbol{R}\widehat{p}_*\mathfrak{F}_F(\boldsymbol{R}j_*\mathbb{C}_U)\otimes\widetilde{\mathbb{C}}_{\widehat{\mathbb{A}}^1}$ and decide whether its class is rational or not. As the section is closed, it is enough to verify this after restricting to some (or any) $\tau\neq 0$. Therefore, we need to compute the map (A.1) after restricting to some fixed nonzero τ. In (6.6), we apply this computation to the multivalued form $e^{-\tau f}\widetilde{\lambda}$.

Denote by $\mathcal{E}_{\widetilde{\mathscr{X}}}$ the sheaf of C^∞ functions (in the sense of Whitney) on $\widetilde{\mathscr{X}}$, by $\mathcal{E}^{\mathrm{mod}}_{\widetilde{\mathscr{X}}}$ the sheaf on $\widetilde{\mathscr{X}}$ of C^∞ functions on U which have moderate growth along $\widetilde{\mathscr{X}}\smallsetminus U$, and by $\mathcal{E}^{\mathrm{mod},-\tau}_{\widetilde{\mathscr{X}}}$ the subsheaf of functions which moreover are infinitely flat along $L'^{+}_{\widetilde{\mathscr{X}},\tau}$.

On the other hand, denote by $\mathcal{C}^\bullet_{U\cup L'^{+}_{\widetilde{\mathscr{X}},\tau},L'^{+}_{\widetilde{\mathscr{X}},\tau}}$ the complex of sheaves on $\widetilde{\mathscr{X}}$, consisting of germs on $\widetilde{\mathscr{X}}$ of relative singular cochains (*i.e.*, germs of singular cochains in $U\cup L'^{+}_{\widetilde{\mathscr{X}},\tau}$ with boundary in $L'^{+}_{\widetilde{\mathscr{X}},\tau}$).

By the de Rham theorem, the integration of forms induces a quasi-isomorphism of complexes $\int:\mathcal{E}^\bullet_U\to\mathcal{C}^\bullet_U\otimes_{\mathbb{Z}}\mathbb{C}$; moreover, the natural morphism $\mathcal{E}^{\mathrm{mod},\bullet}_{\widetilde{\mathscr{X}}}\to(\alpha_\tau\circ j)_*\mathcal{E}^\bullet_U$ is a quasi-isomorphism, so the integration morphism $\int:\mathcal{E}^{\mathrm{mod},\bullet}_{\widetilde{\mathscr{X}}}\to(\alpha_\tau\circ j)_*\mathcal{C}^\bullet_U\otimes_{\mathbb{Z}}\mathbb{C}$, which is obtained by composing both morphisms, is a quasi-isomorphism.

Similarly, we have a commutative diagram

$$\begin{array}{ccccc}
\beta_{\tau,!}\mathcal{E}^{\mathrm{mod},\bullet}_{\widetilde{\mathscr{X}}} & \xrightarrow{\ \sim\ } & \mathcal{E}^{\mathrm{mod},-\tau,\bullet}_{\widetilde{\mathscr{X}}} & & \\
\wr\downarrow & & & \searrow{\scriptstyle\int} & \\
\beta_{\tau,!}(\alpha_\tau\circ j)_*\mathcal{E}^\bullet_U & \underset{\int}{\xrightarrow{\ \sim\ }} & \beta_{\tau,!}(\alpha_\tau\circ j)_*\mathcal{C}^\bullet_U\otimes_{\mathbb{Z}}\mathbb{C} & \xrightarrow{\ \sim\ } & \mathcal{C}^\bullet_{U\cup L'^{+}_{\widetilde{\mathscr{X}},\tau},L'^{+}_{\widetilde{\mathscr{X}},\tau}}\otimes_{\mathbb{Z}}\mathbb{C}
\end{array}$$

Hence we get:

PROPOSITION A.2 (A variant of the de Rham theorem). *Both complexes* $\mathcal{C}^\bullet_{U\cup L'^{+}_{\widetilde{\mathscr{X}},\tau},L'^{+}_{\widetilde{\mathscr{X}},\tau}}\otimes_{\mathbb{Z}}\mathbb{C}$ *and* $\mathcal{E}^{\mathrm{mod},-\tau,\bullet}_{\widetilde{\mathscr{X}}}$ *are quasi-isomorphic to* $\beta_{\tau,!}\boldsymbol{R}\alpha_{\tau,*}\boldsymbol{R}j_*\mathbb{C}_U$. *Moreover, the integration of forms induces a natural quasi-isomorphism of complexes*

$$\int:\mathcal{E}^{\mathrm{mod},-\tau,\bullet}_{\widetilde{\mathscr{X}}}\xrightarrow{\ \sim\ }\mathcal{C}^\bullet_{U\cup L'^{+}_{\widetilde{\mathscr{X}},\tau},L'^{+}_{\widetilde{\mathscr{X}},\tau}}\otimes_{\mathbb{Z}}\mathbb{C}.$$

Now, given a section of $\mathcal{E}^{\mathrm{mod},\bullet}_{\widetilde{\mathscr{X}}}\otimes j_+\mathcal{E}^{-\tau f}$, *i.e.*, a section of $\mathcal{E}^{\mathrm{mod},\bullet}_{\widetilde{\mathscr{X}}}$ multiplied by $e^{-\tau f}$, it is also a section of $\mathcal{E}^{\mathrm{mod},-\tau,\bullet}_{\widetilde{\mathscr{X}}}$, and its image by (A.1) is nothing but its integral, according to the previous commutative diagram.

References

[1] S. BARANNIKOV – "Semi-infinite Hodge structures and mirror symmetry for projective spaces", preprint, math.AG/0010157, 2000.

[2] V.V. BATYREV – "Dual polyhedra and mirror symmetry for Calabi-Yau hypersurfaces in toric varieties", *J. Algebraic Geom.* **3** (1994), p. 493–535.

[3] A. DOUAI & C. SABBAH – "Gauss-Manin systems, Brieskorn lattices and Frobenius structures (I)", *Ann. Inst. Fourier (Grenoble)* **53** (2003), no. 4, p. 1055–1116.

[4] B. Dubrovin – "Geometry of 2D topological field theory", in *Integrable systems and quantum groups* (M. Francaviglia & S. Greco, éds.), Lect. Notes in Math., vol. 1260, Springer-Verlag, 1996, p. 120–348.

[5] Yu.I. Manin – *Frobenius manifolds, quantum cohomology and moduli spaces*, Colloquium Publ., vol. 47, American Mathematical Society, 1999.

[6] Z. Mebkhout – *Le formalisme des six opérations de Grothendieck pour les $\mathcal{D}$-modules cohérents*, Travaux en cours, vol. 35, Hermann, Paris, 1989.

[7] Z. Mebkhout & C. Sabbah – "§ III.4 $\mathcal{D}$-modules et cycles évanescents", in *Le formalisme des six opérations de Grothendieck pour les $\mathcal{D}$-modules cohérents* [**6**], p. 201–239.

[8] C. Sabbah – "Monodromy at infinity and Fourier transform", *Publ. RIMS, Kyoto Univ.* **33** (1998), p. 643–685.

[9] ———, "Hypergeometric period for a tame polynomial", *C. R. Acad. Sci. Paris Sér. I Math.* **328** (1999), p. 603–608, & preprint math.AG/9805077.

[10] ———, *Déformations isomonodromiques et variétés de Frobenius*, Savoirs Actuels, CNRS Éditions & EDP Sciences, Paris, 2002.

[11] K. Saito – "The higher residue pairings $K_F^{(k)}$ for a family of hypersurfaces singular points", in *Singularities*, Proc. of Symposia in Pure Math., vol. 40, American Mathematical Society, 1983, p. 441–463.

[12] M. Saito – "On the structure of Brieskorn lattices", *Ann. Inst. Fourier (Grenoble)* **39** (1989), p. 27–72.

UMR 6621 du CNRS, Laboratoire J.A. Dieudonné, Université de Nice, Parc Valrose, 06108 Nice cedex 2, France

E-mail address: douai@math.unice.fr

UMR 7640 du CNRS, Centre de Mathématiques, École polytechnique, F–91128 Palaiseau cedex, France

E-mail address: sabbah@math.polytechnique.fr

URL: http://www.math.polytechnique.fr/cmat/sabbah/sabbah.html

Opposite filtrations, variations of Hodge structure, and Frobenius modules

Javier Fernandez
Gregory Pearlstein

1. Introduction

Let X be a complex manifold. Then, an *unpolarized complex variation of Hodge structure* $(E, \nabla, \mathcal{F}, \bar{\Phi})$ *over* X consists of a flat, C^∞ complex vector bundle (E, ∇) over X equipped with a decreasing Hodge filtration $\mathcal{F}$ and an increasing filtration $\bar{\Phi}$ such that

(1.1) $\mathcal{F}$ is holomorphic with respect to ∇, and $\nabla(\mathcal{F}^p) \subseteq \mathcal{F}^{p-1} \otimes \Omega^1_X$;

(1.2) $\bar{\Phi}$ is anti-holomorphic with respect to ∇, and $\nabla(\bar{\Phi}_q) \subseteq \bar{\Phi}_{q+1} \otimes \overline{\Omega^1_X}$;

(1.3) $E = \mathcal{F}^p \oplus \bar{\Phi}_{p-1}$ for each index p (i.e. $\mathcal{F}$ is opposite to $\bar{\Phi}$).

Alternatively, conditions (1.1)–(1.3) are equivalent to the assertion that the C^∞ decomposition

$$E = \bigoplus_p \mathcal{U}^p \tag{1.4}$$

defined by the Hodge bundles $\mathcal{U}^p = \mathcal{F}^p \cap \bar{\Phi}_p$ satisfy Griffiths' transversality:

$$\nabla : \mathcal{E}^0(\mathcal{U}^p) \to \mathcal{E}^{0,1}(\mathcal{U}^{p+1}) \oplus \mathcal{E}^{0,1}(\mathcal{U}^p) \oplus \mathcal{E}^{1,0}(\mathcal{U}^p) \oplus \mathcal{E}^{1,0}(\mathcal{U}^{p-1}) \tag{1.5}$$

In the present article, we describe three constructions of complex variations of Hodge structure, and their relationship with quantum cohomology and Frobenius manifolds. More precisely, in §2 we recall that:

(a) Given a variation of graded-polarized mixed Hodge structure $\mathcal{V}$, the Hodge filtration $\mathcal{F}$ of $\mathcal{V}$ pairs with the convolution

$$\bar{\Phi} = (\bar{\mathcal{F}}^\vee * \mathcal{W})$$

of the weight filtration $\mathcal{W}$ of $\mathcal{V}$ and $\bar{\mathcal{F}}^\vee_q = \bar{\mathcal{F}}^{-q}$ to define an unpolarized $\mathbb{C}$VHS on the underlying C^∞ flat vector bundle of $\mathcal{V}$.

(b) Near a maximally unipotent boundary point, the B-model variation of Hodge structure attached to a degenerating family of Calabi–Yau threefolds gives rise to a variation of mixed Hodge structure of Hodge–Tate type, whose extension data encodes the quantum cohomology of the mirror.

(c) The Higgs field θ obtained by application of construction (a) to (b) also determines the quantum product of the mirror, i.e. in this case mirror symmetry can be understood as a duality of $\mathbb{C}$VHS.

Following [**D3**], in §3 we generalize the construction (b) by showing that given any admissible variation of graded-polarized mixed Hodge structure $\mathcal{V} \to \Delta^{*n}$, there exists a canonical filtration $\underline{\Psi}$ which is opposite to the Hodge filtration $\mathcal{F}$ near zero, and hence defines an unpolarized $\mathbb{C}$VHS. Moreover, when the limiting mixed Hodge structure is Hodge–Tate, this opposite filtration coincides with the relative weight filtration of $\mathcal{V}$.

The even cohomology of a compact smooth manifold X of dimension k is naturally a Frobenius algebra with respect to the cup product and intersection pairing. For Kähler manifolds, the quantum product provides a deformation of this algebra. These constructions restrict to $V = \oplus_p H^{p,p}(X)$ if X is Calabi-Yau. A Frobenius module of weight k is an abstract version of the module structure obtained when the product operation on V is restricted to a module over $\operatorname{Sym} H^{1,1}(X)$. In §4 we review these notions as well as the equivalence between families of Frobenius modules and variations of Hodge structure with special degenerating behavior (Theorem (4.14)).

In general, the construction of a Frobenius module arising from an algebra as described above implies a loss of information. Yet, in some cases, it is possible to recover the full product structure. In §5 we discuss how this is the case when the weight of the module is at most 5 and, also, when the algebra is "generated by $H^{1,1}$", a condition that has already been used in the context of quantum cohomology [**KM,Kr**] as well as in the recent work of C. Hertling and Yu. Manin [**HM**].

If a family of Frobenius modules generates a family of Frobenius algebras, a natural question to ask is if it is possible to unfold this last family into a Frobenius manifold. We study some cases where families of Frobenius modules can be unfolded into (germs) of Frobenius manifolds. We do this in two ways: an explicit construction is given for low weight families (5.6), and a general argument (Theorem (5.8)) is presented using the techniques of Sections 2 and 3, as well as [**HM**].

ACKNOWLEDGMENT. Part of the research presented in this paper was done at the Max Planck Institute and we are thankful for their hospitality and support. We would also like to thank Claus Hertling for very helpful discussions.

2. Variations of Mixed Hodge Structure

Let V be a finite dimensional vector space. Then, a decreasing filtration of V is an exhaustive sequence of subspaces

$$0 \subseteq \cdots \subseteq F^p \subseteq F^{p-1} \subseteq \cdots \subseteq V$$

of V such that $F^p \subseteq F^{p-1}$ for all p. Likewise, an increasing filtration Ψ of V is an exhaustive sequence of subspaces of V such that $\Psi_q \subseteq \Psi_{q+1}$ for each index q. As alluded to in the introduction, a decreasing filtration F is then said to be *opposite* to an increasing filtration Ψ if and only if

$$V = F^p \oplus \Psi_{p-1} \tag{2.1}$$

for each index p. Similarly, given an integer k, a pair of decreasing filtrations F and G are said to be k-opposed if and only if F is opposite to the increasing filtration $\Psi_q := G^{k-q}$.

DEFINITION 2.2. Let A be a subfield of $\mathbb{R}$. Then, a *pure A–Hodge structure of weight k* is pair (V_A, F) consisting of a finite dimensional A-vector space V_A and a decreasing Hodge filtration F of $V_{\mathbb{C}} = V_A \otimes \mathbb{C}$ by complex subspaces such that the conjugate filtration $\bar{F}$ defined by the real structure $V_{\mathbb{R}} = V_A \otimes \mathbb{R}$ is *k-opposed* to F.

By the Hodge decomposition theorem, the primitive[1] k'th cohomology of a smooth complex projective variety X carries a pure Hodge structure of weight k. Accordingly, one defines a polarization of a pure Hodge structure (V_A, F) of weight k to be a $(-1)^k$–symmetric bilinear form $Q : V_A \otimes V_A \to A$ such that

1. $Q(F^p, F^{k-p+1}) = 0$;
2. $i^{p-q}Q(v, \bar{v})$ is positive definite on $F^p \cap \bar{F}^{k-p}$;

for each index p.

DEFINITION 2.3. Let S be a complex manifold and A be a subfield of $\mathbb{R}$. Then, a *variation of pure, polarized A–Hodge structure of weight k over S* consists of a local system of $\mathcal{V}_A$ of finite dimensional A-vector spaces over S equipped with a decreasing Hodge filtration $\mathcal{F}$ of $\mathcal{V} = \mathcal{V}_A \otimes \mathcal{O}_S$ by holomorphic subbundles, and a flat $(-1)^k$-symmetric bilinear form $Q : \mathcal{V}_A \otimes \mathcal{V}_A \to A$ such that

1. $\mathcal{F}$ and $\bar{\mathcal{F}}$ are k-opposed;
2. $\mathcal{F}$ is horizontal, i.e. $\nabla(\mathcal{F}^p) \subseteq \mathcal{F}^{p-1} \otimes \Omega^1_S$;
3. Q polarizes each fiber of $\mathcal{V}$.

In particular, by the work of Griffiths [**G**], each smooth projective morphism $f : X \to S$ of complex algebraic varieties gives rise to a variation of pure $\mathbb{Q}$–Hodge structure of weight k on $\mathcal{V}_{\mathbb{Q}} = R^k_{f*}(\mathbb{Q})$. On the primitive part of $R^k_{f*}(\mathbb{Q})$ the variation is, also, polarized. Dropping the requirement that f be smooth and/or projective, one then obtains the notion of a variation of graded-polarized mixed Hodge structure:

DEFINITION 2.4 [**D2**]. Let A be a subfield of $\mathbb{R}$. Then, an *A–mixed Hodge structure* consists of a finite dimensional A-vector space equipped with a decreasing Hodge filtration F of $V_{\mathbb{C}}$ together with an increasing weight filtration $W(V_A)$ of V_A such that F induces a pure A–Hodge structure of weight k on each non-trivial quotient $Gr^W_k = W_k/W_{k-1}$ of the complexification W of $W(V_A)$ via the rule

$$F^p Gr^W_k = \frac{F^p \cap W_k + W_{k-1}}{W_{k-1}}$$

Likewise, a *graded-polarized mixed Hodge structure* is just a mixed Hodge structure endowed with a choice of polarization Q_k for each non-trivial quotient Gr^W_k.

DEFINITION 2.5 [**SZ**]. Let S be a complex manifold and A be a subfield of $\mathbb{R}$. Then, a *variation of graded-polarized A–mixed Hodge structure over S* consists of a local system $\mathcal{V}_A$ of finite dimensional A-vector spaces over S equipped with a decreasing Hodge filtration $\mathcal{F}$ of $\mathcal{V} = \mathcal{V}_A \otimes \mathcal{O}_S$ by holomorphic subbundles, and an increasing weight filtration $\mathcal{W}(\mathcal{V}_A)$ of $\mathcal{V}_A$ by flat subbundles, together with a collection of flat, non-degenerate bilinear forms Q_k such that:

1. $\nabla(\mathcal{F}^p) \subseteq \mathcal{F}^{p-1} \otimes \Omega^1_S$;
2. $(Gr^{\mathcal{W}}_k(\mathcal{V}_A), \mathcal{F}Gr^{\mathcal{W}}_k, Q_k)$ is a variation of pure polarized A–Hodge structure of weight k.

[1] In fact, the same is true without restricting to the primitive cohomology (see Chapter 5, Section 6 of [**We**]).

REMARK. In definitions (2.2)–(2.5) one may replace A by a finitely generated noetherian subring of $\mathbb{R}$ such that $A \otimes \mathbb{Q}$ is a field. However, in this context, one only requires W to be a filtration of $V_A \otimes \mathbb{Q}$.

Let $\mathcal{V}$ be a variation of pure, polarized Hodge structure of weight k and $\bar{\Phi}$ be the increasing filtration of $\mathcal{V}$ defined by the rule

$$\bar{\Phi}_p = \bar{\mathcal{F}}^{k-p}$$

Then, the fact that $\bar{\nabla} = \nabla$ and $\mathcal{F}$ is holomorphic, horizontal and k-opposed to $\bar{\mathcal{F}}$ implies that $\bar{\Phi}$ is an anti-holomorphic filtration of $\mathcal{V}$ which is opposite to $\mathcal{F}$ such that

$$\nabla(\bar{\Phi}_q) \subseteq \bar{\Phi}_{q+1} \otimes \bar{\Omega}^1 \tag{2.6}$$

i.e. $(\mathcal{F}, \bar{\Phi})$ defines an unpolarized $\mathbb{C}$VHS on the underlying flat C^∞ bundle of $\mathcal{V}$. Accordingly **[Si]**, one obtains an associated Higgs bundle structure $\bar{\partial} + \theta$ on the underlying C^∞ vector bundle of $\mathcal{V}$ as follows: Let

$$\mathcal{V} = \bigoplus_{p+q=k} \mathcal{H}^{p,q}$$

denote the Hodge decomposition (1.4) defined by the Hodge bundles

$$\mathcal{U}^p = \mathcal{H}^{p,q} = \mathcal{F}^p \cap \bar{\Phi}_p$$

Then, by virtue of equation (1.5), we can write

$$\nabla = \tau + \bar{\partial} + \partial + \theta \tag{2.7}$$

where

$$\begin{array}{ll} \tau(\mathcal{U}^p) \subseteq \mathcal{E}^{0,1}(\mathcal{U}^{p+1}), & \bar{\partial}(\mathcal{E}^0(\mathcal{U}^p)) \subseteq \mathcal{E}^{0,1}(\mathcal{U}^p) \\ \partial(\mathcal{E}^0(\mathcal{U}^p)) \subseteq \mathcal{E}^{1,0}(\mathcal{U}^p), & \theta(\mathcal{U}^p) \subseteq \mathcal{E}^{1,0}(\mathcal{U}^{p-1}) \end{array} \tag{2.8}$$

Moreover, upon expanding out the flatness condition $\nabla^2 = 0$ relative to (2.7) and (2.8), one obtains the Higgs field condition

$$(\bar{\partial} + \theta)^2 = \bar{\partial}^2 + \bar{\partial}\theta + \theta \wedge \theta = 0 \tag{2.9}$$

More generally, mutatis mutandis, given an unpolarized $\mathbb{C}$VHS, one obtains a corresponding Higgs bundle structure $\bar{\partial} + \theta$ on the underlying C^∞ vector bundle of the variation via the above constructions.

Let $\mathcal{V}$ be a variation of graded-polarized mixed Hodge structure. Then, as consequence of the following result, one obtains a system of Hodge bundles (1.4) on the underlying C^∞ bundle of $\mathcal{V}$ which satisfy the transversality condition (1.5), and hence a corresponding Higgs bundle structure $\bar{\partial} + \theta$.

THEOREM 2.10 **[D2,D4]**. *Let (F, W) be a mixed Hodge structure. Then, there exists a unique, functorial bigrading*

$$V_{\mathbb{C}} = \bigoplus_{p,q} I^{p,q} \tag{2.11}$$

of the underlying complex vector space $V_{\mathbb{C}}$ such that
(a) $F^p = \oplus_{a \geq p} I^{a,b}$ *for each index p;*
(b) $W_k = \bigoplus_{a+b \leq k} I^{a,b}$ *for each index k;*

(c) *For each bi-index* (p,q),

$$\overline{I^{p,q}} = I^{q,p} \mod \bigoplus_{a<q,\, b<p} I^{a,b} \tag{2.12}$$

Moreover,

$$\bigoplus_{r\leq p} I^{r,s} = \sum_k \bar{F}^{k-p} \cap W_k \tag{2.13}$$

PROOF. The existence of a direct sum decomposition (2.11) satisfying conditions (a)–(c) is shown in [**D2**]. That this decomposition also satisfies (2.13) is shown in [**D4**] (cf. [**P**]). Namely, in order to establish (2.13), observe that

$$\bigoplus_{r\leq p} I^{r,s} = \bigoplus_{r\leq p} \overline{I^{s,r}} \tag{2.14}$$

since, for any bi-index (r,s) with $r \leq p$, the left hand side of (2.14) contains every bi-index $(r',s') < (r,s)$. Thus, it is sufficient to show that

$$\bigoplus_{r\leq p} \overline{I^{s,r}} = \sum_k \bar{F}^{k-p} \cap W_k \tag{2.15}$$

By conditions (a) and (b), for each pair of indices p and k,

$$\bar{F}^{k-p} \cap W_k = \overline{F^{k-p} \cap W_k} = \bigoplus_{r+s\leq k,\, s\geq k-p} \overline{I^{s,r}} \tag{2.16}$$

and hence

$$\bar{F}^{k-p} \cap W_k \subseteq \bigoplus_{r\leq p} \overline{I^{s,r}} \tag{2.17}$$

since $r+s \leq k$ and $s \geq k-p \implies r+s \leq k \leq s+p \implies r \leq p$. Thus, fixing p and taking the sum of (2.17) over all k yields:

$$\sum_k \bar{F}^{k-p} \cap W_k \subseteq \bigoplus_{r\leq p} \overline{I^{s,r}}$$

Conversely, by equation (2.16),

$$\bigoplus_{r\leq p} \overline{I^{s,r}} \subseteq \sum_k \bar{F}^{k-p} \cap W_k \tag{2.18}$$

Indeed, given any bi-index (s,r) with $r \leq p$, there exists an index k such that $r+s \leq k \leq s+p$ (i.e. $r+s \leq k$ and $s \geq k-p$), and hence $\overline{I^{s,r}} \subseteq \bar{F}^{k-p} \cap W_k$.

COROLLARY 2.19. *Let $\mathcal{V}$ be a variation of mixed Hodge structure, and*

$$\mathcal{V} = \bigoplus_{p,q} \mathcal{I}^{p,q} \tag{2.20}$$

denotes the C^∞ decomposition of $\mathcal{V}$ into a sum of C^∞ subbundles defined by the pointwise application of Theorem (2.10). *Then, the Hodge filtration $\mathcal{F}$ of $\mathcal{V}$ pairs with the increasing filtration*

$$\bar{\Phi}_q = \sum_k \bar{\mathcal{F}}^{k-q} \cap \mathcal{W}_k \tag{2.21}$$

to define an unpolarized $\mathbb{C}$*VHS for which the resulting* C^∞ *decomposition* (1.4) *is given by the formula*

$$\mathcal{V} = \bigoplus_p \mathcal{U}^p, \qquad \mathcal{U}^p = \bigoplus_q \mathcal{I}^{p,q} \tag{2.22}$$

PROOF. As in the pure case, the fact that $\mathcal{F}$ is holomorphic and horizontal, implies that $\bar{\Phi}$ is an anti-holomorphic filtration of $\mathcal{V}$ such that $\nabla(\bar{\Phi}_p) \subseteq \bar{\Phi}_{p+1} \otimes \bar{\Omega}^1$. Moreover, by Theorem (2.10),

$$\mathcal{F}^p = \bigoplus_{a \geq p} \mathcal{I}^{a,b}, \qquad \bar{\Phi}_p = \bigoplus_{a \leq p} \mathcal{I}^{a,b} \tag{2.23}$$

Consequently, $\mathcal{V} = \mathcal{F}^p \oplus \bar{\Phi}_{p-1} = \bigoplus_{p,q} \mathcal{I}^{p,q}$ and hence $(\mathcal{F}, \bar{\Phi}, \nabla)$ satisfy the axioms of an unpolarized $\mathbb{C}$VHS. Likewise, by equation (2.23), $\mathcal{U}^p := \mathcal{F}^p \cap \bar{\Phi}_p = \bigoplus_q \mathcal{I}^{p,q}$

REMARK. Given a pair of increasing filtrations A and B of a vector space V, one defines the convolution $A * B$ to be the increasing filtration of V given by the rule

$$(A * B)_q = \sum_{r+s=q} A_r \cap B_s = \sum_k A_{q-k} \cap B_k \tag{2.24}$$

In particular, if for any decreasing filtration F of V one defines $F^\vee_r = F^{-r}$ then the increasing filtration (2.21) is given by the formula $\bar{\Phi} = (\bar{\mathcal{F}}^\vee * \mathcal{W})$

Following [**D3**], we now relate the above constructions with mirror symmetry. To this end, let $\mathcal{V} \to \Delta^{*r}$ be a variation of pure, polarized Hodge structure of weight k over a product of punctured disks which has unipotent monodromy, and $T_j = e^{-N_j}$ denote the monodromy of $\mathcal{V}$ about the j'th punctured disk. Then, as a consequence of Schmid's orbit theorems [**Sc**], one obtains an associated limiting mixed Hodge structure $(F_\infty, W[-k])$ on the central fiber [**D1**] of the canonical extension of $\mathcal{V}$, where $W[-k]_j = W_{j-k}$ denotes the shifted monodromy weight filtration of $\mathcal{V}$:

THEOREM 2.25 [**CK**]. *Let* $\mathcal{V} \to \Delta^{*r}$ *be a variation of pure, polarized A–Hodge structure with unipotent monodromy, and*

$$\mathcal{C} = \{\sum_j a_j N_j \mid a_j > 0\}$$

denote the corresponding monodromy cone. Then, there exists a unique increasing filtration $\mathcal{W}(\mathcal{V}_A)$ *of* $\mathcal{V}_A$ *by sub-local systems, called the monodromy weight filtration of* $\mathcal{V}$*, such that for every element* $N \in \mathcal{C}$*:*

1. $N(\mathcal{W}_j) \subseteq \mathcal{W}_{j-2}$;
2. $N^j : Gr^{\mathcal{W}}_j \xrightarrow{\cong} Gr^{\mathcal{W}}_{-j}$;

Suppose now that $\mathcal{V} \to \Delta^{*r}$ is a variation of pure Hodge structure for which the corresponding limiting mixed Hodge structure is Hodge–Tate, i.e.

$$I^{p,q}_{(F_\infty, W[-k])} = 0$$

unless $p = q$. Then, following [**D3**], $\mathcal{V}$ should be regarded as maximally degenerate. Moreover, as consequence of the following result, to each such maximally degenerate variation $\mathcal{V}$, one can associate a corresponding Hodge–Tate variation $\mathcal{V}^\circ$:

THEOREM 2.26 [**D3**]. *Let $\mathcal{V} \to \Delta^{*r}$ be a variation of pure polarized Hodge structure of weight k for which the associated limiting mixed Hodge structure is Hodge–Tate. Then, the Hodge filtration $\mathcal{F}$ pairs with the shifted monodromy weight filtration $\mathcal{W}[-k]$ of $\mathcal{V}$ to define a Hodge–Tate variation $\mathcal{V}^\circ$ over a neighborhood of zero in Δ^{*r}.*

REMARK. By shrinking Δ^{*r} as necessary, we shall henceforth assume that $(\mathcal{F}, \mathcal{W})$ is Hodge–Tate over all of Δ^{*r}.

To apply Theorem (2.26) to the study of mirror symmetry, we now recall the following definition:

DEFINITION 2.27. Let X be a polarized Calabi–Yau manifold of dimension d, and $\overline{\mathcal{M}}_X$ be a smooth partial compactification of the moduli space $\mathcal{M}_X$ [**B,Ti,To**] of complex structures on X, such that

$$D = \overline{\mathcal{M}}_X - \mathcal{M}_X \tag{2.28}$$

is a normal crossing divisor. Let $p \in D$ be such that, in a neighborhood of p, $D = \sum_{j=1}^r D_j$ with $r = \dim \mathcal{M}_X = \dim H^{d-1,1}(X)$ and $\{p\} = \cap_{j=1}^r D_j$. Then, p is said to be a *maximally unipotent boundary point of* $\mathcal{M}_X$ if and only if the variation of Hodge structure

$$\mathcal{V}_s = H^d(X_s) \tag{2.29}$$

over $\mathcal{M}_X$ has maximal unipotent monodromy on a neighborhood of p in $\mathcal{M}_X$, i.e. the associated monodromy weight filtration $\mathcal{W}$ satisfies the following two conditions:

1. $\dim Gr_d^{\mathcal{W}} = 1$, $\dim Gr_{d-1}^{\mathcal{W}} = 0$, $\dim Gr_{d-2}^{\mathcal{W}} = r$;
2. $Gr_{d-2}^{\mathcal{W}} = \bigoplus_{j=1}^r N_j(Gr_k^{\mathcal{W}})$.

Let X be a smooth Calabi–Yau threefold, and p be a maximally unipotent boundary point of $\mathcal{M}_X$. Then, it can be shown that the limiting mixed Hodge structure of the variation (2.29) at p is Hodge–Tate. Moreover, according to [**D3**], the resulting variation $\mathcal{V}^\circ$ produced by Theorem (2.26) should be viewed as the A-model variation of the mirror X° of X.

To extract the quantum cohomology of X° using this approach, we recall [**D3**] that given a pair of variations of $\mathbb{Z}$–Hodge structure $\mathcal{A}$, and $\mathcal{B}$ over S of pure type (p,p) and $(p-1,p-1)$ respectively,

$$\mathrm{Ext}^1(\mathcal{A}, \mathcal{B}) = \mathrm{Hom}(\mathcal{A}, \mathcal{B}) \otimes \mathcal{O}^*(S). \tag{2.30}$$

More generally, given a $\mathbb{Z}$-variation $\mathcal{V}$ of Hodge–Tate type, let E_k denote the extension class (2.30) attached to the extension

$$0 \to Gr_{2k-2}^{\mathcal{W}} \to \mathcal{W}_{2k}/\mathcal{W}_{2k-4} \to Gr_{2k}^{\mathcal{W}} \to 0$$

LEMMA 2.31 [**D3**]. *Let $\mathcal{V} \to \Delta^{*r}$ be a $\mathbb{Z}$-variation of Hodge-Tate type, and $\tilde{\mathcal{V}}$ denote the canonical extension of $\mathcal{V}$ over Δ^r defined in* [**D1**]. *Then, the Hodge filtration $\mathcal{F}$ of $\mathcal{V}$ extends holomorphically to $\tilde{\mathcal{V}}$, and remains opposite to the weight filtration $\mathcal{W}$ over $0 \in \Delta^r$ if and only if*

$$E_k \in \mathrm{Hom}(Gr_{2k}^{\mathcal{W}}, Gr_{2k-2}^{\mathcal{W}}) \otimes \mathcal{O}_{mer}^*(\Delta^{*r})$$

for each index k, where $\mathcal{O}_{mer}^*(\Delta^{*r})$ denotes the group of invertible holomorphic functions on Δ^{*r} which extend meromorphically to Δ^r.

REMARK. The variation $\mathcal{V}^\circ$ produced by Theorem (2.26) always satisfies the hypothesis of Lemma (2.31).

Thus, given a smooth Calabi–Yau threefold X and a maximal unipotent boundary point p of $\mathcal{M}_X$, one obtains via application of Lemma (2.31) to the resulting Hodge–Tate variation $\mathcal{V}^\circ$, a corresponding set of extension classes E_3, E_2 and E_1 which encode the quantum product of X° as follows: Let $\mathcal{W}$ be the weight filtration of $\mathcal{V}^\circ$. Then, $\mathcal{W}$ is defined over $\mathbb{Z}$ with weight graded quotients:

$$Gr_6^{\mathcal{W}}(\mathcal{V}_{\mathbb{Z}}^\circ) \cong \mathbb{Z}, \qquad Gr_4^{\mathcal{W}}(\mathcal{V}_{\mathbb{Z}}^\circ) \cong \mathbb{Z}^r, \qquad Gr_2^{\mathcal{W}}(\mathcal{V}_{\mathbb{Z}}^\circ) \cong \mathbb{Z}^r, \qquad Gr_0^{\mathcal{W}}(\mathcal{V}_{\mathbb{Z}}^\circ) \cong \mathbb{Z}$$

where $r = \dim H^{2,1}(X)$. Select a generator 1 of $Gr_6^{\mathcal{W}}(\mathcal{V}_{\mathbb{Z}}^\circ)$. Then, on account of maximal unipotent monodromy, there exists a unique system of "canonical coordinates" $(q_1, \dots, q_r)$ on the base of $\mathcal{V}^\circ$ relative to which the extension class E_3 assumes the form

$$E_3(1) = \sum_j q_j N_j(1) \tag{2.32}$$

Moreover, as discussed in **[D3]**, upon expanding the logarithmic derivative of E_2 relative to the system of canonical coordinates (2.32), one obtains a generating function for the number of rational curves in the mirror X°.

Alternatively, as described in **[P]**, given a maximal unipotent boundary point p as above, both the canonical coordinates (2.32) and the quantum product of the mirror X° can be described in terms of the Higgs field θ attached to $\mathcal{V}^\circ$ by (2.19) as follows: Let

$$H^{p,p} = Gr_{6-2p}^{\mathcal{W}}(\mathcal{V}_{\mathbb{Z}}^\circ),$$

1 be a generator of $H^{0,0}$, and $T_j = N_j(1)$. Then, θ induces a map

$$H^{p,p} \to H^{p+1,p+1} \otimes \Omega^1(\Delta^{*r})$$

such that

(a) $\theta(\xi)1 = \sum_j T_j \otimes \Omega_j(\xi)$;

(b) $T_a * T_b = \theta(\xi_a) \circ \theta(\xi_b)\, 1$;

where $(\Omega_1, \dots, \Omega_r)$ denotes the coframe of $\Omega^1(\Delta^{*r})$ defined by the 1-forms

$$\Omega_j = \frac{1}{2\pi i}\frac{dq_j}{q_j},$$

$T_a * T_b$ represents **[M,CoK]** the quantum product on $H^{1,1}(X^\circ)$, and $\Omega_j(\xi_k) = \delta_{jk}$.

In §3, we shall extend the correspondence $\mathcal{V} \leftrightarrow \mathcal{V}^\circ$ given by Theorem (2.26) to arbitrary variations of graded-polarized mixed Hodge structures $\mathcal{V} \to \Delta^{*n}$ which are admissible in the sense of **[SZ]** by constructing a suitable opposite filtration Ψ from the limiting mixed Hodge structure of $\mathcal{V}$.

3. Asymptotic Behavior

Let (E, ∇) be a flat $\mathbb{C}$-vector bundle over Δ^{*n} with unipotent monodromy. Then **[D1]**, up to isomorphism, there exists a unique extension $\tilde{E} \to \Delta^n$ of E such that ∇ has at worst simple poles with nilpotent residues along the normal crossing divisor

$$D = \Delta^n - \Delta^{*n} = \cup_j D_j \tag{3.1}$$

Equivalently, given a system of local coordinates $(s_1, \dots, s_n)$ on Δ^n relative to which the divisor (3.1) assumes the form

$$s_1 \cdots s_n = 0$$

with $s_j = 0$ on D_j, and flat multivalued frame $(\sigma_1, \dots, \sigma_m)$ of E, the canonical extension $\tilde{E}$ described above can be identified with the locally free sheaf generated by the sections

$$\tilde{\sigma}_j = e^{\frac{1}{2\pi i}\sum_j (\log s_j) N_j} \sigma_j \tag{3.2}$$

where $T_j = e^{-N_j}$ denotes the monodromy action of E about $s_j = 0$.

In particular, if $\mathcal{V} \to \Delta^{*n}$ is a variation of graded-polarized mixed Hodge structure for which the Hodge filtration $\mathcal{F}$ of $\mathcal{V}$ extends holomorphically to the canonical extension $\tilde{\mathcal{V}}$, one can define a corresponding nilpotent orbit $\mathcal{F}_{nilp}$ by simply extending $\mathcal{F}(0)$ to a filtration of $\mathcal{V}$ which is constant with respect to the frame (3.2).

Alternatively, $\mathcal{F}_{nilp}$ may be described in terms of the associated period map

$$\varphi : \Delta^{*n} \to \mathcal{M}/\Gamma \tag{3.3}$$

defined by parallel translation of the Hodge filtration $\mathcal{F}$ of $\mathcal{V}$ to a fixed reference fiber $V = \mathcal{V}_{s_o}$ of $\mathcal{V}$ as follows:

Let $\mathcal{V} \to \Delta^{*n}$ be a variation of graded-polarized mixed Hodge structure, and $V = \mathcal{V}_{s_o}$ denote a fixed reference fiber of $\mathcal{V}$. Let W denote the specialization of the weight filtration of $\mathcal{V}$ to V, and $Q = \{Q_k\}$ denote the corresponding specialization of the graded-polarizations of $\mathcal{V}$ to Gr^W. Define X to be the flag variety consisting of all decreasing filtrations F of V such that

$$\dim(F^p) = \operatorname{rank}(\mathcal{F}^p)$$

and let $\mathcal{M}$ be the classifying space [**P**] consisting of all filtrations $F \in X$ such that (F, W) is a mixed Hodge structure which is graded-polarized by Q. Then, the period map (3.3) takes values in the quotient of $\mathcal{M}$ by the action of the monodromy group Γ of $\mathcal{V}$.

THEOREM 3.4 [**P**]. *The classifying space $\mathcal{M}$ defined above is a complex manifold upon which the real Lie group*

$$G = \{\, g \in GL(V)^W \mid Gr(g) \in Aut_{\mathbb{R}}(Q) \,\}$$

acts transitively by automorphisms, where $Gr(g) : Gr^W \to Gr^W$ denotes the map induced by an element of the stabilizer $GL(V)^W$ of W on Gr^W.

In particular, on account of Theorem (3.4), the orbit

$$\check{\mathcal{M}} = G_{\mathbb{C}}.F_o \subseteq X \tag{3.5}$$

of a point $F_o \in \mathcal{M}$ under the complex Lie group

$$G_{\mathbb{C}} = \{\, g \in GL(V)^W \mid Gr(g) \in Aut_{\mathbb{C}}(Q) \,\} \tag{3.6}$$

is well defined, independent of $F_o \in \mathcal{M}$.

WARNING. In general, $G_{\mathbb{C}}$ is not the complexification of G.

To continue, observe that the period map (3.3) is locally liftable. Accordingly, there exists a holomorphic map $F : U^n \to \mathcal{M}$ which makes the following diagram commute

$$\begin{array}{ccc} U^n & \xrightarrow{F} & \mathcal{M} \\ {\scriptstyle p}\downarrow & & \downarrow \\ \Delta^{*n} & \xrightarrow{\varphi} & \mathcal{M}/\Gamma \end{array} \tag{3.7}$$

where $U^n \subset \mathbb{C}^n$ denotes the product of upper half-planes on which the imaginary parts of the standard coordinates $(z_1, \dots, z_n)$ are positive, and $p : U^n \to \Delta^{*n}$ denotes the standard covering map defined by the coordinates $s_j = e^{2\pi i z_j}$.

Thus, on account of the commutativity of (3.7), the map

$$\psi(z) = e^{-\sum_j z_j N_j}.F(z) \tag{3.8}$$

satisfies the periodicity condition

$$\psi(z_1, \dots, z_j + 1, \dots, z_n) = \psi(z_1, \dots, z_n)$$

and hence descends to a well defined holomorphic map

$$\psi(s) : \Delta^{*n} \to \check{\mathcal{M}} \tag{3.9}$$

LEMMA 3.10. *Let $\mathcal{V} \to \Delta^{*n}$ be a variation of graded-polarized mixed Hodge structure. Then, the Hodge filtration $\mathcal{F}$ of $\mathcal{V}$ extends holomorphically to the canonical extension $\tilde{\mathcal{V}} \to \Delta^n$ if and only if*

$$F_\infty := \lim_{s \to 0} \psi(s) \tag{3.11}$$

exists.

PROOF. Relative to the trivialization (3.2), $\mathcal{F}$ coincides with the filtration (3.9).

Likewise, after unraveling the above definitions, on finds that the pull back $p^*(\mathcal{F}_{nilp})$ of $\mathcal{F}_{nilp}$ to U^n coincides with the nilpotent orbit

$$F_{nilp} = e^{\sum_j z_j N_j}.F_\infty \tag{3.12}$$

REMARK. To be coordinate free, (3.11) and (3.12) should be viewed as follows: Let $(s_1, \dots, s_n)$ be a system of local coordinates on Δ^n which are compatible with the given divisor structure (3.1), and $(\lambda_1, \dots, \lambda_n)$ be the corresponding system of coordinates on $T_0(\Delta^n)$ defined by the basis vectors $e_j = (\frac{\partial}{\partial s_j})_0$. Then, the period map

$$\varphi_{nilp}(\lambda_1, \dots, \lambda_n) = e^{\frac{1}{2\pi i}\sum_j \log(\lambda_j) N_j}.F_\infty$$

determines a variation of mixed Hodge structure over the complement of the divisor $\lambda_1 \cdots \lambda_n = 0$, with monodromy action $T_j = e^{-N_j}$ about $\lambda_j = 0$, which is well defined, independent of the choice of local coordinates $(s_1, \dots, s_n)$ as above.

For variations of graded-polarized mixed Hodge structure, the analog of the monodromy weight filtration (2.25) is the relative weight filtration ${}^rW = {}^rW(N, W)$

discussed in [**SZ**]. Moreover, based upon the study of degenerating families of varieties, Steenbrink and Zucker proposed the following, now standard, definition of an admissible variation of graded-polarized mixed Hodge structure over Δ^*:

DEFINITION 3.13 [**SZ**]. Let $\mathcal{V} \to \Delta^*$ be a variation of graded-polarized mixed Hodge structure with unipotent monodromy. Then, $\mathcal{V}$ is *admissible* if

(a) The limiting Hodge filtration (3.11) exists;

(b) The relative weight filtration ${}^rW = {}^rW(N, W)$ exists.

The admissibility conditions (3.11) always hold in the pure case as a consequence of Schmid's nilpotent orbit theorem. For multivariable degenerations, one defines admissibility via curve test using (3.13) [**K**]. Moreover, one has the following result:

THEOREM 3.14 [**K**]. *Let $\mathcal{V} \to \Delta^{*n}$ be an admissible variation of graded-polarized mixed Hodge structure with unipotent monodromy, and*

$$\mathcal{C} = \{ \sum_j a_j N_j \mid a_j > 0 \}$$

denote the corresponding monodromy cone. Let W denote the specialization of the weight filtration of $\mathcal{V}$ to some fixed reference fiber $V = \mathcal{V}_{s_o}$. Then,

1. *${}^rW(N, W)$ exists for every element $N \in \mathcal{C}$;*
2. *${}^rW = {}^rW(N, W)$ is well defined, independent of $N \in \mathcal{C}$;*
3. *$(F_\infty, {}^rW)$ is a mixed Hodge structure;*
4. *$N_1, \ldots, N_n$ are $(-1,-1)$-morphisms of $(F_\infty, {}^rW)$.*

Let $\mathcal{V} \to \Delta^{*n}$ be an admissible variation of graded-polarized mixed Hodge structure with unipotent monodromy, and

$$V = \bigoplus_{p,q} I^{p,q} \tag{3.15}$$

denote the corresponding decomposition (2.11) defined by the limiting Hodge filtration F_∞ and the relative weight filtration ${}^rW$ of $\mathcal{V}$. Define,

$$\Psi_p = \bigoplus_{a \le p} I^{a,b} \tag{3.16}$$

and $\mathfrak{g}_-^\Psi = \{ \alpha \in \mathfrak{g}_\mathbb{C} \mid \alpha(\Psi_p) \subseteq \Psi_{p-1} \}$ denote the subalgebra of $\mathfrak{g}_\mathbb{C} = Lie(G_\mathbb{C})$ consisting of those elements which preserve the increasing filtration Ψ defined by (3.16), and act trivially on each layer $Gr_p^\Psi = \Psi_p/\Psi_{p-1}$ of Gr^Ψ.

LEMMA 3.17. *Ψ is opposite to F_∞. Moreover, relative to the decomposition*

$$\mathfrak{g}_\mathbb{C} = \bigoplus_{r,s} \mathfrak{g}^{r,s} \tag{3.18}$$

induced by the bigrading (3.15),

$$\mathfrak{g}_-^\Psi = \bigoplus_{r<0} \mathfrak{g}^{r,s} \tag{3.19}$$

PROOF. That Ψ is opposite to F_∞ is a simple consequence of definition (3.16), and the fact [cf. Theorem (2.10)] that $F^p_\infty = \oplus_{a\geq p} I^{a,b}$. Likewise, since

$$\mathfrak{g}^{r,s}(I^{a,b}) \subseteq I^{a+r,b+s}$$

the sum $\oplus_{r<0}\,\mathfrak{g}^{r,s}$ maps Ψ_p to Ψ_{p-1}, and hence is contained in $\mathfrak{g}^\Psi_-$. Conversely, by equation (3.16), if $\alpha \in \mathfrak{g}^\Psi_-$ then the components $\alpha^{r,s}$ with respect to the decomposition (3.18) can be non-zero only if $r < 0$.

COROLLARY 3.20. *Let $\mathcal{V} \to \Delta^{*n}$ be an admissible variation with unipotent monodromy. Then, on a neighborhood of the origin, the associated function* (3.9) *admits a unique representation of the form*

$$\psi(s) = e^{\Gamma(s)}.F_\infty \tag{3.21}$$

with respect to $\mathfrak{g}^\Psi_-$-valued holomorphic function $\Gamma(s)$ which vanishes at the origin.

PROOF. By (3.19), $\mathfrak{g}^\Psi_-$ is a vector space complement to $Lie(G^{F_\infty}_{\mathbb{C}})$ in $\mathfrak{g}_{\mathbb{C}}$. Consequently, the map $u \mapsto e^u.F_\infty$ is a biholomorphism from a neighborhood of zero in $\mathfrak{g}^\Psi_-$ onto a neighborhood of F_∞ in $\check{\mathcal{M}}$. Accordingly, $\psi(s)$ admits a unique representation of the type described above on a neighborhood of zero in Δ^n.

A priori, the filtration Ψ defined above depends on the choice of coordinates used in the construction of the limiting Hodge filtration (3.11). However, as the following result shows, this is in fact not the case, since the monodromy logarithms $N_1, \dots, N_n$ preserve Ψ.

THEOREM 3.22. *Ψ is independent of the choice of coordinates used in the definition of F_∞. Moreover,*

$$\Psi = (\overline{F^\vee_{nilp}}) * ({}^rW) = (\overline{F^\vee_\infty}) * ({}^rW) \tag{3.23}$$

PROOF. Without loss of generality, any two systems of local coordinates $(s_1, \dots, s_n)$ and $(\tilde{s}_1, \dots, \tilde{s}_n)$ on Δ^n which are compatible with the divisor structure (3.1) may be assumed to be of the form $\tilde{s}_j = f_j s_j$ for some collection of holomorphic functions $f_1, \dots, f_n$ which do not vanish at the origin. Moreover, direct calculation shows that, under such changes of coordinates, the corresponding filtrations (3.11) are related by the equation:

$$\tilde{F}_\infty = e^{-\frac{1}{2\pi i}\sum_j \log(f_j(0))N_j}.F_\infty \tag{3.24}$$

Let $\mathcal{N} = \mathrm{span}_{\mathbb{C}}(N_1, \dots, N_n)$. Then, $\mathcal{N} \subseteq \mathfrak{g}^{-1,-1}$ because each N_j is a $(-1,-1)$-morphism of $(F_\infty, {}^rW)$. Accordingly,

$$N \in \mathcal{N} \implies I^{a,b}_{(e^N.F_\infty, {}^rW)} = e^N.I^{a,b}_{(F_\infty, {}^rW)} \tag{3.25}$$

and hence

$$\Psi_{(\tilde{F}_\infty, {}^rW)} = e^{\tilde{N}}.\Psi_{(F_\infty, {}^rW)}$$

where

$$\tilde{N} = -\frac{1}{2\pi i}\sum_j \log(f_j(0))N_j \tag{3.26}$$

On the other hand,

$$e^{\tilde{N}}.\Psi_{(F_\infty, {}^rW)} = \Psi_{(F_\infty, {}^rW)}$$

since $\mathcal{N} \subseteq \mathfrak{g}^{-1,-1}$ and $\mathfrak{g}^{-1,-1} \subseteq \mathfrak{g}^{\Psi}_{-}$ by Lemma (3.17). Consequently,

$$\Psi = \Psi_{(F_\infty, {}^rW)} = e^{\tilde{N}}.\Psi_{(F_\infty, {}^rW)} = \Psi_{(\tilde{F}_\infty, {}^rW)} \tag{3.27}$$

is independent of the choice of coordinates used in the construction of the limiting Hodge filtration (3.11).

To verify (3.23), observe that $\Psi = (\overline{F^{\vee}_{\infty}}) * ({}^rW)$ by virtue of equations (3.16) and (2.13). To show that

$$\Psi = (\overline{F^{\vee}_{nilp}}) * ({}^rW)$$

observe that $\mathcal{N}$ is closed under complex conjugation since $\bar{N}_j = N_j$ for all j. Consequently, when evaluated at any particular point in U^n, $\overline{F_{nilp}} = e^N.\bar{F}_\infty$ for some element $N \in \mathcal{N}$. On the other hand, by definition [**SZ**], $\mathcal{N}$ preserves the relative weight filtration ${}^rW$. Thus,

$$(\overline{F^{\vee}_{nilp}}) * ({}^rW) = (e^N.\bar{F}^{\vee}_{\infty}) * ({}^rW) = e^N.(\bar{F}^{\vee}_{\infty}) * ({}^rW) = e^N.\Psi = \Psi$$

Finally, as a consequence of the above results, we obtain the following generalization of Theorem (2.26):

THEOREM 3.28. *Let $\mathcal{V} \to \Delta^{*n}$ be an admissible variation of graded-polarized mixed Hodge structure with unipotent monodromy. Then, Ψ extends to a filtration $\underline{\Psi}$ of $\mathcal{V}$ by flat subbundles which pairs with the Hodge filtration $\mathcal{F}$ of $\mathcal{V}$ to define an unpolarized $\mathbb{C}$ VHS on a neighborhood of the origin.*

PROOF. Since each N_j preserves Ψ, and $\mathcal{V}$ has only local monodromy, Ψ extends to a filtration $\underline{\Psi}$ of $\mathcal{V}$ by flat subbundles. To see that $\underline{\Psi}$ is opposite to $\mathcal{F}$, observe that after pulling everything back to the upper half-plane, we can write

$$F(z) = e^{\sum_j z_j N_j}\psi(s) = e^{\sum_j z_j N_j} e^{\Gamma(s)}.F_\infty \tag{3.29}$$

relative to a $\mathfrak{g}^{\Psi}_{-}$-valued holomorphic function $\Gamma(s)$ by virtue of equations (3.8) and (3.21). Moreover, since $\mathfrak{g}^{\Psi}_{-}$ is nilpotent, we can write

$$e^{\sum_j z_j N_j} e^{\Gamma(s)} = e^{X(z)} \tag{3.30}$$

for some function $X(z)$ with values in $\mathfrak{g}^{\Psi}_{-}$. Thus,

$$\begin{aligned} V = F^p_\infty \oplus \Psi_{p-1} &= e^{X(z)}.(F^p_\infty \oplus \Psi_{p-1}) \\ &= (e^{X(z)}.F^p_\infty) \oplus (e^{X(z)}.\Psi_{p-1}) = F^p(z) \oplus \Psi_{p-1} \end{aligned}$$

since $F(z) = e^{X(z)}.F_\infty$ and $\mathfrak{g}^{\Psi}_{-}$ preserves Ψ.

4. Frobenius Modules

Let X be a compact Kähler manifold. Then the even part of $H^*(X, \mathbb{C})$ is naturally a Frobenius algebra with product given by the cup product and the bilinear pairing arising from the intersection form[2]. The quantum cohomology defines a deformation of this ("classical") structure parameterized by the complexified Kähler

[2]The full $H^*(X, \mathbb{C})$ is a $\mathbb{Z}_2$-graded Frobenius algebra, as considered by Kontsevich and Manin in [**KM**].

cone of X. In this case, the deformed ("quantum") product is defined in terms of a function known as the Gromov-Witten potential.

In the case where X is also Calabi-Yau, the cohomology decomposes $H^*(X,\mathbb{C}) = \oplus_{p,q} H^{p,q}$ and the quantum product preserves this bigrading. In particular, $\oplus_p H^{p,p}$ inherits a Frobenius algebra structure. We should mention that these spaces also satisfy nondegeneracy and positivity conditions encoded in the Hard Lefschetz Theorem and the Hodge-Riemann relations.

In his analysis of mirror symmetry, D. Morrison constructed a polarized variation of pure Hodge structure using the data described above. This variation is known as the *A-model variation* ([**M**], [**CoK,Chapter 8**]). A close inspection of his construction shows that not all of the algebra structure is used: it is only the $\operatorname{Sym} H^{1,1}$-module structure on $\oplus_p H^{p,p}$ that is needed. This observation is the starting point for the construction of an equivalence between this type of structure, called a Frobenius module, and certain maximally degenerating variations of pure Hodge structures. In what follows we will describe this construction. See [**CF2**] for further details.

For $k \in \mathbb{N}$ let $V = \oplus_{p=0}^k V_{2p}$ be a graded finite dimensional $\mathbb{C}$-vector space and $\mathcal{B}$ a symmetric non-degenerate bilinear form on V pairing V_{2p} with $V_{2(k-p)}$. Let $\{T_a\}_{0\le a\le m}$ be a $\mathcal{B}$-self dual, graded basis of V. We will refer to $\{T_a\}$ as an *adapted basis*. For $0 \le a \le m$ define $\delta(a)$ by $\mathcal{B}(T_{\delta(a)}, T_b) = \delta_{ab}$ —the right hand side δ is Kronecker's δ— for all $b = 0, \dots, m$. We also set $\tilde{a} := p$ if and only if $T_a \in V_p$ and assume that the map $\sim : \{0, \dots, m\} \to \{0, \dots, 2k\}$ is increasing.

DEFINITION 4.1. $(V, \mathcal{B}, e, *)$ is a *graded V_2-Frobenius module* of weight k if

1. $e \neq 0$ and $V_0 = \langle e \rangle$.
2. V is a graded $\operatorname{Sym} V_2$-module under $*$.
3. For all $v_1, v_2 \in V$ and $w \in V_2$

$$\mathcal{B}(w * v_1, v_2) = \mathcal{B}(v_1, w * v_2) \tag{4.2}$$

4. $w * e = w$ for all $w \in V_2$.

Since $T_0 \in V_0$, it must be a non-zero multiple of e and we assume that an adapted basis satisfies $T_0 = e$. Clearly, the fact that V is a $\operatorname{Sym} V_2$-module is equivalent to

$$T_j * (T_l * T) = T_l * (T_j * T) \text{ for all } T_j, T_l \in V_2 \text{ and } T \in V. \tag{4.3}$$

We say that V is *real* if V has a real structure, $V_{\mathbb{R}}$, compatible with its grading, $*$ is real, $e \in V_{\mathbb{R}}$, and $\mathcal{B}$ is defined over $\mathbb{R}$.

To any real Frobenius module we can associate a Hodge–Tate mixed Hodge structure, split over $\mathbb{R}$, whose canonical bigrading is

$$I^{p,p} := V_{2(k-p)}. \tag{4.4}$$

The multiplication operator $L_w \in \operatorname{End}(V)$, $w \in V_2$, is an infinitesimal automorphism of the bilinear form

$$Q(v_a, v_b) := (-1)^{k+\tilde{a}/2} \mathcal{B}(v_a, v_b), \tag{4.5}$$

as well as a $(-1,-1)$-morphism of the associated mixed Hodge structure. We will say that $w \in V_2 \cap V_{\mathbb{R}}$ *polarizes* V if the mixed Hodge structure $(I^{*,*}, Q, L_w)$ is polarized [**CK**]. A real Frobenius module V is said to be *polarizable* if it contains a polarizing element. Given a polarizing element w, the set of polarizing elements is an open cone in $V_2 \cap V_{\mathbb{R}}$. We can then choose a basis $T_1, \ldots, T_r$ of $V_2 \cap V_{\mathbb{R}}$ spanning a simplicial cone $\mathcal{C}$ contained in the closure of the polarizing cone and with $w \in \mathcal{C}$. Such a choice of a basis of V_2 will be called a *framing* of the polarized Frobenius module.

EXAMPLE 4.6. If X is a compact Kähler manifold of dimension k, let $V_{2p} := H^{p,p}(X)$, $\mathcal{B}_{int}$ the intersection pairing on $V := \oplus_{p=0}^{k} V_{2p}$, and $\smile$ the restriction of the cup product to V. Then, $(V, \mathcal{B}_{int}, 1, \smile)$ defines a polarizable Frobenius module. The real structure is induced by $H^*(X, \mathbb{R})$.

Given an adapted basis $\{T_0, \ldots, T_m\}$ of V, let $z_0, \ldots, z_m$ be the corresponding linear coordinates on V and set $q_j := \exp(2\pi i z_j)$ for $j = 1, \ldots, r := \dim V_2$. If U is the upper-half plane, we may identify $U^r \cong (V_2 \cap V_{\mathbb{R}}) \oplus i\mathcal{C}$ and view the correspondence

$$\sum_{j=1}^{r} z_j T_j \in (V_2 \cap V_{\mathbb{R}}) \oplus i\mathcal{C} \mapsto (q_1, \ldots, q_r) \in (\Delta^*)^r.$$

Let V be a framed Frobenius module of weight k. Then the action of V_2 on V can be recovered from a homogeneous polynomial $\phi_0 \in \mathbb{C}[z_0, \ldots, z_m]$ of degree three, called the *classical potential*. Indeed, if we let

$$\phi_0(z_0, \ldots, z_m) := \sum_{\tilde{j}=2,\ 0 \le \tilde{a}, \tilde{b} \le 2k} z_j z_a z_b \frac{C(\tilde{a})}{12} \mathcal{B}(T_j * T_a, T_b)\,,$$

with

$$C(\tilde{a}) := \begin{cases} 2 \text{ if } k = 3 \text{ and } \tilde{a} = 2, \\ 3 \text{ if } k \neq 3 \text{ and } \tilde{a} = 2 \text{ or } \tilde{a} = 2k - 4, \\ 6 \text{ otherwise,} \end{cases}$$

then we recover the V_2-action by:

$$T_j * T_a := \sum_{\tilde{c}=\tilde{a}+2} \frac{\partial^3 \phi_0}{\partial z_j \partial z_a \partial z_{\delta(c)}} T_c\,; \quad j = 1, \ldots, r\,.$$

EXAMPLE 4.7. In the case of weight $k = 3$, the classical potential is

$$\phi_0(z_0, \ldots, z_m) = \sum_{\tilde{j}=2} z_0 z_j z_{\delta(j)} + \frac{1}{6} \sum_{\tilde{j}=2, \tilde{l}=2, \tilde{k}=4} \mathcal{B}(T_j * T_l, T_{\delta(k)}) z_j z_l z_{\delta(k)}.$$

In the special case where $\dim V_{2p} = 1$ for $p = 0, \ldots, 3$, we obtain $\phi_0(z_0, \ldots, z_3) = z_0 z_1 z_2 + \kappa z_1^3$ for $\kappa = \frac{1}{6}\mathcal{B}(T_1 * T_1, T_1)$. This is the case of, for instance, the central part of the cohomology of the quintic threefold in $\mathbb{P}^4$.

Next we consider deformations of a framed Frobenius module induced from a potential. Let $R := \mathbb{C}\{q_1, \ldots, q_r\}_0$ denote the ring of convergent power series vanishing for $q_1 = \cdots = q_r = 0$ and R' be its image under the map induced by $q_j \mapsto e^{2\pi i z_j}$ for $1 \le j \le r$.

DEFINITION 4.8. Let $(V, \mathcal{B}, e, *)$ be a framed Frobenius module of weight $k > 3$ with classical potential ϕ_0. A *quantum potential* on V is a function $\phi : V \to \mathbb{C}$ of the form $\phi = \phi_0 + \phi_\hbar$, where

$$\phi_\hbar(z) := \sum_{\tilde{a}=2k-4} z_a \phi_h^a(z_1, \ldots, z_r) + \sum_{2<\tilde{a}<2k-4,\, \tilde{a}+\tilde{b}=2k-2} z_a z_b \phi_h^{ab}(z_1, \ldots, z_r), \quad (4.9)$$

with $\phi_h^a, \phi_h^{ab} \in R'$ and such that the action

$$T_j \cdot_q T_a := \sum_{\tilde{c}=\tilde{a}+2} \frac{\partial^3 \phi}{\partial z_j \partial z_a \partial z_{\delta(c)}} T_c, \quad \text{with } q = (q_1, \ldots, q_r) \in \Delta^r. \quad (4.10)$$

turns $(V, \mathcal{B}, e, \cdot_q)$ into a graded V_2-Frobenius module for all q.

For weight $k = 3$ modules, the form of the potential is just $\phi = \phi_0 + \phi_\hbar$ where $\phi_\hbar \in R'$, and requiring that (4.10) defines a graded V_2-module imposes no constraint on $\phi_\hbar$, contrary to the situation of higher weights.

REMARK. For V of weight 1 or 2, the Frobenius module is determined by $\mathcal{B}$ and e; hence no deformations in the sense of Definition (4.8) are possible.

REMARK 4.11. The form (4.9) of the quantum potential is motivated by the Gromov-Witten potential, except that in the setup of Frobenius modules only a graded part of the potential is relevant.

The condition for a quantum potential to induce a Frobenius module can be interpreted as a WDVV type of equation, only that, compared to the original setup in quantum cohomology, here we only see a graded part of the full system.

EXAMPLE 4.12. In the weight $k = 3$ case we have, for $\tilde{j} = \tilde{l} = 2$,

$$T_j \cdot_q T_l = T_j * T_l + (2\pi i)^3 \sum_{\tilde{a}=4} \frac{\partial^3 \phi_\hbar(q_1, \ldots, q_r)}{\partial q_j \partial q_l \partial q_{\delta(a)}} q_j q_l q_{\delta(a)} T_a,$$

while the action of T_j on all the other graded parts of V is the one given by $*$. Notice that, since $\phi_\hbar$ is assumed to be convergent at $q_1 = \cdots = q_r = 0$, we have $\cdot_0 = *$.

As we stated at the beginning of this section, there is a correspondence between deformations of a framed Frobenius module and a certain type of maximally degenerating polarized variation of Hodge structure. To make precise the notion of maximally degenerating variation we reformulate Definition (2.27) in the context of abstract variations.

DEFINITION 4.13. Given a polarized variation of Hodge structure of weight k over $(\Delta^*)^r$ whose monodromy is unipotent, we say that $0 \in \Delta^r$ is a *maximally unipotent boundary point* if

1. $\dim I^{k,k} = 1$, $\dim I^{k-1,k-1} = r$ and $\dim I^{k,k-1} = \dim I^{k-2,k} = 0$, as well as $I^{p,q} = 0$ for all $p, q < 0$, where $I^{*,*}$ is the bigrading associated to the limiting mixed Hodge structure and,
2. $\operatorname{span}_{\mathbb{C}}\{N_1(I^{k,k}), \ldots, N_r(I^{k,k})\} = I^{k-1,k-1}$, where N_j are the monodromy logarithms of the variation.

THEOREM 4.14. *There is a 1-1 correspondence between*

(a) *Deformations of a framed Frobenius module $(V, \mathcal{B}, e, *)$ of weight k arising from a quantum potential, and*

(b) *Germs of polarized variations of pure Hodge structure of weight k on V degenerating at a maximally unipotent boundary point to a limiting mixed Hodge structure of Hodge-Tate type, split over $\mathbb{R}$, and together with a marked real point $e \in F^k$.*

The proof of this correspondence can be found in [**CF2, Theorem 4.1**]. The cases of weight 3 and 4 have been analyzed in [**P**] and [**CF1**]. Below we will describe the main constructions that set up the correspondence. The key technical step in proving the theorem is the asymptotic description of the Hodge filtration of an admissible variation presented in §3.

Let $\mathcal{V} \to (\Delta^*)^r$ be a variation of pure, polarized Hodge structure with unipotent monodromy. Then, by Corollary (3.20), the germ of $\mathcal{V}$ at zero is determined by the following data:

1. The nilpotent orbit (3.12);
2. The function $\psi(s) = e^{\Gamma(s)} \cdot F_\infty$.

Moreover, to such a nilpotent orbit (3.12), we can associate the corresponding limiting mixed Hodge structure $(F_\infty, W[-k])$, which turns out to be *polarized* by the monodromy cone (3.14) (cf. [**CK**]). Furthermore, this polarized mixed Hodge structure is equivalent to the nilpotent orbit [**Sc**], [**CKS**], [**CF1,Theorem 2.3**]. Also, [**CF1, Theorem 2.7**], on account of the horizontality of the period map of $\mathcal{V}$, the function $\Gamma(s)$ can be recovered from its projection Γ_{-1} onto the subspace $\mathfrak{p}_{-1} := \oplus_s \mathfrak{g}^{-1,s}$ of [cf. Lemma (3.17)]

$$\mathfrak{g}_- := \mathfrak{g}_-^\Psi = \bigoplus_{r<0,\, s} \mathfrak{g}^{r,s}$$

and the monodromy logarithms $N_1, \ldots, N_r$ of $\mathcal{V}$.

The period mapping of the variation $\mathcal{V}$ is written as $e^{X(s)} \cdot F_\infty$ for a (multivalued) map $X : (\Delta^*)^r \to \mathfrak{g}_-$, as in (3.29)-(3.30). The horizontality condition of the variation $\mathcal{V}$ can be written, in terms of X, as

$$e^{-X(s)}\, d(e^{X(s)}) = dX_{-1} \quad \text{for } X_{-1} = \sum_{j=1}^{r} \frac{\log(s_j)}{2\pi i} N_j + \Gamma_{-1}(s), \tag{4.15}$$

and, in turn, this is equivalent to dX_{-1} being a Higgs field.

REMARK. Theorem (4.14) can now be refined to show that under the same correspondence, the nilpotent orbit of the variation of Hodge structure corresponds to the framed Frobenius module and that the function Γ_{-1} corresponds to the quantum potential. Moreover, the transversality condition of the variation is equivalent to the graded part of the WDVV equation alluded to in Remark (4.11).

In what follows we will describe the two constructions that establish the equivalence between degenerating variations of Hodge structure and Frobenius modules.

CONSTRUCTION 4.16 (VARIATION OF HODGE STRUCTURE FROM A FROBENIUS MODULE). This is the transcription of the well known A-model variation of Hodge structure into the language of families of Frobenius modules. See [**CF2, §5**] for more details as well as [**CoK, Chapter 8**] for the standard A-model variation.

Consider a deformation of the framed Frobenius module $(V, \mathcal{B}, e, *)$ of weight k generated by the potential $\phi = \phi_0 + \phi_\hbar$, with the framing $\{T_1, \dots, T_r\}$; let $(q_1, \dots, q_r)$ be the corresponding system of local coordinates on $(\Delta^*)^r$. Define

1. A free sheaf $\mathcal{V} := V \otimes \mathcal{O}_{(\Delta^*)^r}$.
2. A Hodge filtration $\mathcal{F}^p := (\oplus_{a \geq p} V_{2(k-a)}) \otimes \mathcal{O}_{(\Delta^*)^r} \subseteq \mathcal{V}$.
3. The Dubrovin connection: for $T \in V$ and $j = 1, \dots r$,

$$\nabla_{\frac{\partial}{\partial q_j}} T := \frac{1}{2\pi i q_j} T_j \cdot_q T.$$

It follows from the symmetry condition (4.3) that ∇ is flat. Also, it has simple poles at $q_j = 0$ where the residue $\mathrm{Res}_{q_j=0}(\nabla)$ is, up to a constant, the endomorphism of $(V, \mathcal{B}, e, *)$ given by the action of T_j on V. The logarithms of the monodromy of ∇, N_j, are also given by that same action.

4. A polarization $\mathcal{Q}$ given by (4.5).
5. A real form $\mathcal{V}_{\mathbb{R}}$ defined as follows. Let

$$\tilde{\mathcal{V}} = V \otimes \mathcal{O}_{\Delta^r} \text{ and } \tilde{\nabla} = \nabla - \frac{1}{2\pi i} \sum_{j=1}^{r} N_j \otimes \frac{dq_j}{q_j}.$$

Then $\tilde{\nabla}$ is a flat connection on $\tilde{\mathcal{V}}$; for $v \in V$ we define $\tilde{\sigma}_v$ to be the $\tilde{\nabla}$-flat section such that $\tilde{\sigma}_v(0) = v$. Then $\mathcal{V}_{\mathbb{R}} \subseteq \mathcal{V}$ is the local system generated by the sections $\exp(-\frac{1}{2\pi i} \sum_j \log(q_j) N_j) \tilde{\sigma}_v$ for all $v \in V_{\mathbb{R}}$.

THEOREM 4.17. *The tuple $(\mathcal{V}, \nabla, \mathcal{F}, \mathcal{V}_{\mathbb{R}}, \mathcal{Q})$ constructed above is a polarized variation of Hodge structure of weight k having a maximally unipotent boundary point at $0 \in \Delta^r$ where the mixed Hodge structure is of Hodge–Tate type split over $\mathbb{R}$.*

A proof of Theorem (4.17) can be found in [**CF2, Theorem 5.10**]. See [**CoK, Theorem 8.5.11**] for a version of this result in the original setup of quantum cohomology of Calabi-Yau manifolds.

REMARK. The construction described above does not explicitly use the fact that the deformation of the Frobenius module comes from a potential. Nevertheless, the quoted proof does depend on that fact.

CONSTRUCTION 4.18 (FROBENIUS MODULE FROM A VARIATION OF HODGE STRUCTURE). The starting point is a polarized variation of Hodge structure of weight k degenerating to a maximally unipotent boundary point where the limiting mixed Hodge structure is of Hodge–Tate type, split over $\mathbb{R}$, together with a real element $e \in F^k$.

The construction of a Frobenius module uses only the information of the variation near the degeneration, so that we can restrict all considerations to the case of a variation over $(\Delta^*)^r$ having the origin as a maximal degeneration.

Define:

1. A graded vector space $V := \oplus_{p=0}^{k} V_{2p}$, with V_{2p} defined by (4.4).
2. A symmetric nondegenerate pairing $\mathcal{B}$ on V by (4.5).
3. A real structure $V_{\mathbb{R}}$ on V defined by the real structure carried, by hypothesis, by all $I^{p,p}$.
4. A framing $\{T_1, \dots, T_r\}$ of V, where $T_j := N_j(e) \in V_2$ for $j = 1, \dots, r$.
5. An action $*$ of V_2 on V by $T_j * T := N_j(T)$ for $j = 1, \dots, r$.

6. A "unit" $e \in F^k = I^{k,k} = V_0$.

Then, $(V, \mathcal{B}, *, e)$ is a V_2 Frobenius module framed by $\{T_1, \dots, T_r\}$. This follows from the commutativity of the monodromy logarithms, the fact that these are infinitesimal isometries of Q, and that they polarize the limiting mixed Hodge structure.

To define a deformation of $(V, \mathcal{B}, *, e)$, we need special coordinates $(q_1, \dots q_r)$ on Δ^r, known as *canonical coordinates.* Such a coordinate system is characterized by the fact that the function Γ_{-1} is normalized with respect to these coordinates to satisfy $\Gamma_{-1}(I^{1,1}) = 0$. Canonical coordinates exist because of the maximal unipotency condition on the variation (see [**CF1,** §**3**]). Finally, for $j = 1, \dots, r$ and $T \in V$, define

$$T_j \cdot_q T := \left(\frac{\partial X_{-1}}{\partial z_j}\right)_q (T) \tag{4.19}$$

where the right hand side is the evaluation of the endomorphism of V, $\frac{\partial X_{-1}}{\partial z_j}$, on the vector T. The fact that this action defines a family of Sym V_2 modules is equivalent to (4.15). The quantum potential $\phi_\hbar$ that generates the deformation is:

$$\begin{aligned}
\phi_h^{ab}(q) &:= \frac{1}{2}\mathcal{B}(\Gamma_{-1}(T_a), T_b) \text{ for } 2 < \tilde{a} < 2k-4 \text{ and } \tilde{a} + \tilde{b} = 2k-2 \\
\phi_h^{a}(q) &:= \mathcal{B}(-\Gamma_{-2}(T_a), T_0) \text{ for } \tilde{a} = 2k-4 \\
\phi_\hbar &:= \sum_{\tilde{a} = 2k-4} z_a \phi_h^a + \sum_{2 < \tilde{a} < 2k-4, \tilde{a}+\tilde{b} = 2k-2} z_a z_b \phi_h^{ab} \\
\phi &:= \phi_0 + \phi_\hbar,
\end{aligned} \tag{4.20}$$

where Γ_{-2} is determined by Γ_{-1} and the monodromy logarithms.

THEOREM 4.21. *$(V, \mathcal{B}, e, *)$ as constructed above is a graded Frobenius module framed by $\{T_1, \dots, T_r\}$. Moreover, the V_2 action defined by (4.19) is a deformation of $(V, \mathcal{B}, e, *)$ generated by the potential (4.20), whose quantum part is uniquely determined by the function Γ_{-1} of the variation.*

REMARK. In the weight $k = 1$ and $k = 2$ cases, the use of canonical coordinates implies the normalization $\Gamma = 0$. This choice parallels the fact that there are no deformations of Frobenius modules for the same weights.

5. Frobenius Algebras

Following Dubrovin, a *Frobenius algebra* over the finite dimensional $\mathbb{C}$-vector space V consists of a tuple $(V, \mathcal{B}, \hat{*}, e)$, where $\hat{*}$ defines a commutative and associative $\mathbb{C}$-algebra structure on V with unit e, and $\mathcal{B}$ is a non-degenerate symmetric bilinear form on V such that

$$\mathcal{B}(v_1 \hat{*} v_2, v_3) \quad = \quad \mathcal{B}(v_2, v_1 \hat{*} v_3) \text{ for all } v_1, v_2, v_3 \in V. \tag{5.1}$$

Additionally, $(V, \mathcal{B}, \hat{*}, e)$ is a *graded Frobenius algebra of weight k* if V decomposes as $V = \oplus_{p=0}^{k} V_{2p}$, $V_0 = \langle e \rangle$, the product $\hat{*}$ is graded and $\mathcal{B}$ pairs V_{2p} with $V_{2(k-p)}$ for all p.

If $(V, \mathcal{B}, \hat{*}, e)$ is a graded Frobenius algebra of weight k we can, by restricting the action of $\hat{*}$, define a graded V_2-module $(V, \mathcal{B}, *, e)$. In general, this process can not be reversed since part of the product structure is lost. Nevertheless, in some cases a full Frobenius algebra can be constructed from a graded V_2-module. We will illustrate this phenomenon in the following examples.

EXAMPLE 5.2. Let $(V, \mathcal{B}, *, e)$ be a graded V_2 Frobenius module of degree $k = 3$. Then, for homogeneous elements v_a, v_b in $V_{\tilde{a}}$ and $V_{\tilde{b}}$ we define a multiplication $\hat{*}$ by: $v_a \hat{*} e = e \hat{*} v_a = v_a$; for $\tilde{a} = 2$ ($\tilde{b}$ arbitrary), $v_b \hat{*} v_a = v_a \hat{*} v_b = v_a * v_b$ and, $v_a \hat{*} v_b = 0$ if $\tilde{a} > 2$ and $\tilde{b} > 2$. We notice that $\hat{*}$ is graded and commutative. Associativity can be seen as follows: if any vector in a triple product is (a multiple of) the identity, associativity is immediate. If all vectors are in V_2, associativity follows from (4.2). Finally, if none of these conditions hold, all triple products vanish by the grading properties. Finally, a short analysis shows that $\mathcal{B}(v_a \hat{*} v_b, v_c) = \mathcal{B}(v_b, v_a \hat{*} v_c)$.

Similar arguments show that the same result applies in the weight $k = 4$ and $k = 5$ cases. The cases of weight $k = 1$ and $k = 2$ are trivial since the product structure is determined by $\mathcal{B}$ and e.

EXAMPLE 5.3. Let $(V, \mathcal{B}, *, e)$ be a graded V_2 Frobenius module of degree k. We say that V *is generated by* V_2 if the linear map $\operatorname{Sym} V_2 \to V$ defined by $P \mapsto P * e$ is onto. In this case, the algebra structure of $\operatorname{Sym} V_2$ can be transferred to V. Explicitly, for homogeneous vectors $v_a = P_a * e$ and $v_b = P_b * e$ we have $v_a \hat{*} v_b = (P_a \cdot P_b) * e$. Clearly, P_a is homogeneous, so that $\hat{*}$ defines a graded multiplication that is commutative, associative and with unit e. Condition (5.1) follows immediately from the iteration of (4.2).

If $(V, \mathcal{B}, \cdot_q, e)$ is a (continuous) family of Frobenius modules such that $\cdot_0 = *$ generates V in V_2, then for all q near 0 the corresponding Frobenius module also generates and we obtain a family of graded Frobenius algebras defined over an open neighborhood of $q = 0$ of the same parameter space.

In quantum cohomology the standard constructions produce the Frobenius manifold structure on $H^*(X, \mathbb{C})$ whose product is the *big quantum product.* Then, considering $i : H^2(X, \mathbb{C}) \subset H^*(X, \mathbb{C})$, the product obtained by pulling back to $i^*(TH^*(X, \mathbb{C})) \simeq H^2 \times H^*$ the big quantum product is known as the *small quantum product.*

Starting from a family of graded V_2-Frobenius modules we have seen that in some cases it is possible to construct a family of Frobenius algebras over an open subset of the same base. In Section 4 we took the parameter space of such a family to be Δ^r, with $r = \dim V_2$. We can also consider the covering of the polydisk by the product of upper half planes, U^r, given by $z \mapsto q = \exp(2\pi i z)$; in this case we pull back our constructions and obtain a family of modules (or algebras) over U^r, that we view as contained in V_2 via the framing. In this case, the family is defined near the point at infinity.

One can pose then the following question: If V is generated by V_2 under $*$, the construction described in the previous paragraph resembles the small quantum product with a family of Frobenius algebras defined over (an open set contained in) V_2. *Is it possible to "unfold" this structure to obtain a full Frobenius manifold analogous to the big quantum product?* We will show that the answer to this question is positive in some cases.

Before going into the details, recall that a *Frobenius manifold* [**Du**] is a complex manifold M of dimension at least 1 with a commutative and associative multiplication $\hat{\cdot}$ on the holomorphic tangent bundle TM, a unit field e and a symmetric, nondegenerate bilinear pairing g on TM such that the Levi-Civita connection of g, ∇^g, is flat, the unit field e is ∇^g-flat and for all vector fields X, Y, Z on M the following conditions hold:

$$g(X \hat{\cdot} Y, Z) \;=\; g(Y, X \hat{\cdot} Z) \tag{5.4}$$

and

$$\nabla^g_X(Y \hat{\cdot} Z) - \nabla^g_Y(X \hat{\cdot} Z) + X \hat{\cdot} \nabla^g_Y Z - Y \hat{\cdot} \nabla^g_X Z - [X,Y] \hat{\cdot} Z \;=\; 0. \tag{5.5}$$

REMARK. Frobenius manifolds usually carry a vector field known as the "Euler field". We will not describe this field in our construction below, but such object can readily be constructed in terms of the grading that is part of a Frobenius module.

CONSTRUCTION 5.6. Let $(V, \mathcal{B}, *, e)$ be a framed Frobenius module of weight k that is deformed via the potential $\phi = \phi_0 + \phi_\hbar$ and assume that the module structure of $*$ can be extended to an algebra that we denote by $\hat{*}$. As we mentioned above the family can be lifted to a family of modules on $U^r \subseteq V_2$. An adapted basis $T_0, \ldots T_m$ of V provides coordinates $z_0, \ldots z_m$ and the algebra structure $\hat{*}$ corresponds to $\mathrm{Im}(z_j) = \infty$ for $j = 1, \ldots, r$.

The algebra structure $\hat{*}$ can be encoded into a classical potential $\hat{\phi}_0$, and we define a multiplication on the tangent bundle TV by

$$T_a \hat{\cdot}_z T_b \;:=\; \sum_c \frac{\partial^3(\hat{\phi}_0 + \phi_\hbar)}{\partial z_a \partial z_b \partial z_{\delta(c)}} T_c \;=\; T_a \hat{*} T_b + \sum_c \frac{\partial^3 \phi_\hbar}{\partial z_a \partial z_b \partial z_{\delta(c)}} T_c, \tag{5.7}$$

for all $\sum_a z_a T_a \in V$. We remark that this definition does not, in general, define a *graded* product.

We also consider the (constant) metric on TV induced by $\mathcal{B}$ and its Levi-Civita connection ∇ characterized by $\nabla T = 0$ for all $T \in V$ regarded as a constant vector field. In particular, the unit $e \in V_0$ extends to a flat vector field that we keep denoting by e.

To see if the previous data makes TV into a Frobenius manifold we have to check the following

1. Compatibility between $\hat{\cdot}$ and $\mathcal{B}$ (5.4):

$$\begin{aligned}
\mathcal{B}(T_a \hat{\cdot}_z T_b, T_c) &= \mathcal{B}(T_a \hat{*} T_b + \sum_d \frac{\partial^3 \phi_\hbar}{\partial z_a \partial z_b \partial z_{\delta(d)}} T_d, T_c) \\
&= \mathcal{B}(T_a \hat{*} T_b, T_c) + \sum_d \frac{\partial^3 \phi_\hbar}{\partial z_a z_b z_{\delta(d)}} \mathcal{B}(T_d, T_c) \\
&= \mathcal{B}(T_a \hat{*} T_b, T_c) + \frac{\partial^3 \phi_\hbar}{\partial z_a \partial z_b \partial z_c},
\end{aligned}$$

from where (5.4) follows by applying (5.1) to $\hat{*}$ and by the symmetry of the second summand.

2. The potentiality condition (5.5):

$$\begin{aligned}\nabla_{\frac{\partial}{\partial z_a}}(T_b \,\hat{\cdot}_z\, T_c) &= \nabla_{\frac{\partial}{\partial z_a}}(T_b \,\hat{*}\, T_c + \sum_d \frac{\partial^3 \phi_\hbar}{\partial z_b \partial z_c \partial z_{\delta(d)}} T_d) \\ &= \nabla_{\frac{\partial}{\partial z_a}}(T_b \,\hat{*}\, T_c) + \nabla_{\frac{\partial}{\partial z_a}}(\sum_d \frac{\partial^3 \phi_\hbar}{\partial z_b \partial z_c \partial z_{\delta(d)}} T_d) \\ &= \sum_d \frac{\partial^4 \phi_\hbar}{\partial z_a \partial z_b \partial z_c \partial z_{\delta(d)}} T_d.\end{aligned}$$

   By the linearity of (5.5) it suffices to check it for $X = T_a$, $Y = T_b$ and $Z = T_c$. The result then follows from the symmetry of the last expression.
3. Commutativity, associativity and unit: commutativity is clear from formula (5.7). That e is a unit follows from e being a unit for $\hat{*}$ and that there is no dependence on z_0 in $\phi_\hbar$. The associativity of $\hat{\cdot}_z$, will be discussed below.

We will only check the associativity in the cases of weight 3, 4 and 5. The fact that $\hat{\cdot}_z$ is not graded makes the computations hard, and this is only partially eased by the weaker property $V_{2p} \,\hat{\cdot}_z\, V_{2q} \subseteq \oplus_{a \geq (p+q)} V_{2a}$.

Weight $k = 3$. In this case $\hat{\cdot}_z$ is graded and can be computed from

$$T_a \,\hat{\cdot}_z\, T_b = \begin{cases} T_a \cdot_{z'} T_b \text{ if } \tilde{a} = \tilde{b} = 2, \\ T_a \,\hat{*}\, T_b \text{ otherwise,} \end{cases}$$

where $z' = \pi_2(z)$ is the projection of $\sum z_a T_a$ on V_2. Associativity follows immediately.

Weight $k = 4$. The more involved of the triple products is that of $T_a \,\hat{\cdot}_z (T_b \,\hat{\cdot}_z\, T_c)$ with $\tilde{a} = \tilde{b} = \tilde{c} = 2$. We proceed as follows. Remember that the quantum potential for weight 4 is $\phi_\hbar = \sum_{\tilde{a}=4} z_a \phi_\hbar^a(z_1, \dots, z_r)$. Then:

$$\begin{aligned}T_a \,\hat{\cdot}_z (T_b \,\hat{\cdot}_z\, T_c) &= T_a \,\hat{\cdot}_z (T_b * T_c + \sum_{\tilde{d}=4} \frac{\partial^3 \phi_\hbar}{\partial z_b \partial z_c \partial z_{\delta(d)}} T_d + \sum_{\tilde{d}=6} \frac{\partial^3 \phi_\hbar}{\partial z_b \partial z_c \partial z_{\delta(d)}} T_d) \\ &= T_a \,\hat{\cdot}_z (T_b \cdot_{z'} T_c + \sum_{\tilde{d}=6} \frac{\partial^3 \phi_\hbar}{\partial z_b \partial z_c \partial z_{\delta(d)}} T_d) \\ &= T_a \,\hat{\cdot}_z (T_b \cdot_{z'} T_c) + \sum_{\tilde{d}=6} \frac{\partial^3 \phi_\hbar}{\partial z_b \partial z_c \partial z_{\delta(d)}} T_a * T_d \\ &= T_a \cdot_{z'} (T_b \cdot_{z'} T_c) + \frac{\partial^3 \phi_\hbar}{\partial z_b \partial z_c \partial z_a} T_{\delta(0)}.\end{aligned}$$

where, as before, $z' = \pi_2(z)$ denotes the projection of z to V_2. Then, using the commutativity of $\hat{\cdot}_z$ we have

$$(T_a \,\hat{\cdot}_z\, T_b) \,\hat{\cdot}_z\, T_c = T_c \,\hat{\cdot}_z (T_a \,\hat{\cdot}_z\, T_b) = T_c \cdot_{z'} (T_a \cdot_{z'} T_b) + \frac{\partial^3 \phi_\hbar}{\partial z_a \partial z_b \partial z_c} T_{\delta(0)},$$

and, since $\cdot_{z'}$ defines a family of Frobenius modules, relation (4.3) implies the associativity of $\hat{\cdot}_z$.

The *weight* $k = 5$ case is similar to the previous one, with more painful computations.

REMARK. All together, the construction described in (5.6) provides unfoldings of the original algebra structure defined on $V_2 \times V$ to a full Frobenius manifold on TV for weights $k = 3, 4, 5$. We note that these results do not follow from the application of [**HM, Theorem 4.5**] because in our case we are working with germs at infinity or, in terms of families defined over Δ^r, our Higgs fields have logarithmic singularities. Perhaps, Construction (5.6) should be regarded as evidence for an extended version of [**HM, Theorem 4.5**].

It seems unlikely that associativity —hence this explicit construction— can be extended to weights $k \geq 6$. The reason for that is that the homogeneity properties of a potential defining the product on a Frobenius manifold imply that, for weight $k \leq 5$ all potentials have the form (4.9), whereas for $k \geq 6$ new homogeneous terms can be present.

Another connection between families of Frobenius modules generated in V_2 and Frobenius manifolds can be established as follows.

THEOREM 5.8. *Consider a deformation of the framed Frobenius module $(V, \mathcal{B}, *, e)$ induced by a quantum potential defined over a neighborhood of $0 \in \Delta^r$ $(r = \dim V_2)$ and such that V is generated by V_2 under $*$. Then, for each $\hat{s} \in (\Delta^*)^r$ near 0 the family of Frobenius modules generates a germ of a family of algebras at $\hat{s}$. Moreover, this germ can be unfolded to a germ of a Frobenius manifold.*

PROOF. By Theorem (4.14) the deformation of $(V, \mathcal{B}, *, e)$ generates a polarized variation of Hodge structure $(\mathcal{V}, \nabla, \mathcal{F}, \mathcal{V}_{\mathbb{R}}, \mathcal{Q})$ over a neighborhood of $0 \in \Delta^r$, with a marked element $e \in F^k$. The limiting mixed Hodge structure of this variation is given by (4.4). As was observed in (3.13), this variation defines an admissible variation of graded-polarized mixed Hodge structure with unipotent monodromy.

As in (3.16) let $\Psi_p = \oplus_{a \leq p,\, b} I^{a,b} = \oplus_{a \leq p} V_{2(k-a)}$. Then Ψ_p is $\mathfrak{g}_-$-invariant and opposite to F_∞. Moreover, by Theorem (3.28) Ψ_p extends to a ∇-flat increasing filtration $\underline{\Psi}_p$ of $\mathcal{V}$ that pairs with $\mathcal{F}^p$ to define an unpolarized $\mathbb{C}$VHS in a neighborhood of the origin. In particular, by (1.3), $\mathcal{F}^p$ and $\underline{\Psi}_p$ are opposite.

By parallel transporting the bundles and additional structure to a fixed fiber of $\mathcal{V}$ we write

$$F^p(s) = e^{X(s)} \cdot F^p_\infty \text{ with } F^p_\infty = \oplus_{a \geq p} V_{2(k-p)}.$$

Recalling that Ψ_p is $\mathfrak{g}_-$-invariant we can find the C^∞ decomposition (1.4) of the complex variation as follows:

$$\begin{aligned}\mathcal{U}^p(s) &= F^p(s) \cap \Psi_p(s) = (e^{X(s)} \cdot F^p_\infty) \cap \Psi_p = e^{X(s)} \cdot (F^p_\infty \cap \Psi_p) \\ &= e^{X(s)} \cdot V_{2(k-p)}.\end{aligned}$$

Then, using (1.5), the Higgs field θ of the $\mathbb{C}$VHS is

$$\theta(s) \ = \ \exp(X(s))\, dX_{-1} \exp(-X(s)). \tag{5.9}$$

We notice that since $X(s)$ is an infinitesimal automorphism of Q,

$$\mathcal{Q}(\mathcal{F}^p, \mathcal{F}^{k-p+1}) = Q(F^p_\infty, F^{k-p+1}_\infty) = 0$$

and, by the $\mathfrak{g}_-$-invariance of Ψ_p,

$$\mathcal{Q}(\underline{\Psi}_p(s), \underline{\Psi}_{k-p-1}(s)) = Q(\Psi_p, \Psi_{k-p-1}) = 0.$$

From (5.9) it follows that, for a fixed $\hat{s}$, the $\mathbb{C}$-span of the iterated application of θ_Y for $Y \in T_{\hat{s}}(\Delta^*)^r$ is conjugate to that of $(dX_{-1})_Y$. But the span for these last operators as $\hat{s} \to 0$ is the span of the iterated action of the monodromy logarithms which act as V_2 acts on V via $*$ as was remarked in (3) of Construction (4.16). Then, under the hypothesis that V is generated by V_2 under $*$, we conclude that the same condition holds for the Higgs field $\theta(\hat{s})$, for $\hat{s}$ near 0.

Next we use the machinery of [**HM, §5**] to construct a Frobenius manifold. We claim that $(((\Delta^*)^r, \hat{s}), \mathcal{V}, \nabla, \mathcal{F})$ is a germ of an H^2-generated variation of filtrations of weight k, in the language of [**HM, Definition 5.3**]. The only thing that remains to be proved is that the Higgs field $C : \mathcal{F}^p/\mathcal{F}^{p+1} \to \mathcal{F}^{p-1}/\mathcal{F}^p$ induced by ∇ satisfies the generation condition. But, under the isomorphism introduced by parallel transport to a fixed fiber of $\mathcal{V}$ and the canonical isomorphism $F^p(s)/F^{p+1}(s) \simeq U^p(s)$ we see that C is, in fact, isomorphic to θ and, since this last field satisfies the generation condition, so does C, and the claim is proved.

But then, the H^2-generated variation of filtrations of weight k defined above, together with the pairing $\mathcal{Q}$, the opposite filtration $\underline{\Psi}_p$ and the "fixed generator" e satisfy the conditions of [**HM, Theorem 5.6**]. Hence, we conclude that there is a unique unfolding of this structure to a germ of a Frobenius manifold.

REMARK. The filtration Ψ constructed in Theorem (5.8) is related to the relative weight filtration ${}^rW$ by the rule $\Psi_p = {}^rW_{2p}$ since the associated limiting mixed Hodge structure is Hodge–Tate.

In view of Theorem (4.14) there is the following immediate corollary to Theorem (5.8).

COROLLARY 5.10. *Let $\mathcal{V} \to (\Delta^*)^r$ be a variation of Hodge structure of weight k which satisfies the hypothesis of Theorem (4.14), and assume that the iterated action of the monodromy cone of $\mathcal{V}$ on $e \in F^k_\infty \subseteq F^0_\infty$ spans F^0_∞. Then, for each $\hat{s} \in \Delta^{*r}$ sufficiently close to the origin, there exists a corresponding germ of a Frobenius manifold $M_{\hat{s}}$, which is completely determined by the asymptotic behavior of $\mathcal{V}$.*

REMARK. Because of (5.9) we see that there is a simple connection between the Higgs field θ and the one that appears in the correspondence described in Section 4, namely, dX_{-1}. In fact,

$$\theta = \exp(X)\, dX_{-1} \exp(-X) \equiv dX_{-1} \mod F^{-2}\mathfrak{g}.$$

Alternatively, θ and dX_{-1} define isomorphic Higgs bundles.

References

[B] Bogomolov F., *Hamiltonian Kähler manifolds*, Dokl. Akad. Nauk. SSSR **243** (1978), 1101–1104.

[CF1] Cattani E. and Fernandez J., *Asymptotic Hodge theory and quantum products.*, Advances in algebraic geometry motivated by physics, Contemporary Mathematics (E. Previato, ed.), vol. 276, 2001, pp. 115–136, math.AG/0011137.

[CF2] ———, *Frobenius modules and Hodge asymptotics*, Comm. Math. Phys. (to appear), math.AG/0207279.

[CK] Cattani E. and Kaplan A., *Polarized mixed Hodge structures and the local monodromy of a variation of Hodge structure*, Invent. Math. **67** (1982), 101–115.

[CKS] Cattani E., Kaplan A. and Schmid W., *Degeneration of Hodge structures*, Ann. of Math. **123** (1986), 457–535.

[CoK] Cox D. and Katz S., *Mirror symmetry and algebraic geometry*, AMS, Providence, RI, 1999.

[D1] Deligne P., *Equations différentielles à singuliers réguliers*, Lecture Notes in Mathematics, vol. 163, 1970..

[D2] ———, *Théorie de Hodge I*, Actes, Congrés Intern. Math. Nice (1970), 425–430; II, Publ. Math I.H.E.S. **40** (1971), 5–58; III, Publ. Math I.H.E.S. **44** (1974), 5–77.

[D3] ———, *Local behavior of Hodge structures at infinity*, AMS/IP Stud. Adv. Math. **1** (1997), AMS, 683–699.

[D4] ———, *Private communication* (2000.).

[Du] Dubrovin B., *Geometry of 2D topological field theories*, Integrable systems and quantum groups (Montecatini Terme, 1993), Lecture Notes in Math., vol. 1620, Springer, 1996, pp. 120–348.

[G] Griffiths P., *Periods of Integrals on algebraic manifolds I*, Amer. J. Math. **90** (1968), 568–626; II, Amer. J. Math. **90** (1968), 805–865; III, Publ. Math I.H.E.S. (1970).

[HM] Hertling C. and Manin Yu., *Unfolding of meromorphic connections and a construction of Frobenius manifolds*, math.AG/0207089.

[K] Kashiwara M., *A study of variation of mixed Hodge structure.*, Publ. Res. Inst. Math. Sci. **22** (1986), 991-1024..

[KM] Kontsevich M. and Manin Yu., *Gromov-Witten classes, quantum cohomology, and enumerative geometry*, Comm. Math. Phys. **164** (1994), 525–562, hep-th/9402147.

[Kr] Kresch A., *Associativity relations in quantum cohomology*, Adv. in Math. **142** (1999), 151–169, alg-geom/9703015.

[M] Morrison, D., *Mathematical Aspects of Mirror Symmetry*, Complex algebraic geometry (Park City, UT, 1993), 1997, pp. 265–327, alg-geom/9609021.

[P] Pearlstein G., *Variations of mixed Hodge structure, Higgs fields and quantum cohomology*, Manuscripta Math. **102** (2000), 269–310.

[Sc] Schmid W., *Variation of Hodge structure: The singularities of the period mapping.*, Invent. Math. **22** (1973), 211 - 319.

[Si] Simpson C., *Higgs bundles and local systems.*, Publ. Math IHES **75** (1992), 5–95.

[SZ] Steenbrink J. and Zucker S., *Variation of Mixed Hodge Structure I.*, Invent. Math. **80** (1985), 489 - 542.

[Ti] Tian G., *Smoothness of the universal deformation space of compact Calabi-Yau manifolds and its Peterson-Weil metric*, Mathematical Aspects of String Theory (S.-T. Yau, Ed.), World Scientific, 1987, pp. 629–646.

[To] Todorov A., *The Weil-Petersson geometry of the moduli space of $SU(n \geq 3)$ (Calabi-Yau) manifolds I*, Comm. Math. Phys. **126** (1989), 325–346.

[We] Wells, R. O., Jr., *Differential analysis on complex manifolds*, Graduate Texts in Mathematics, vol. 65, Springer-Verlag, New York, 1980.

DEPARTMENT OF MATHEMATICS, UNIVERSITY OF UTAH, SALT LAKE CITY, UTAH 84112-0090, USA.

E-mail address: jfernand@math.utah.edu

DEPARTMENT OF MATHEMATICS, UNIVERSITY OF CALIFORNIA, IRVINE, CALIFORNIA 92697-3875, USA.

E-mail address: gpearlst@math.uci.edu

The jet-space of a Frobenius manifold and higher-genus Gromov-Witten invariants

Ezra Getzler

1. Introduction

The theory of genus zero Gromov-Witten invariants associates to a compact symplectic manifold X a Frobenius manifold H (also known as the small phase space of X) whose underlying flat manifold is the cohomology space $H^*(X,\mathbb{C})$. Higher genus Gromov-Witten invariants give rise to a sequence of generating functions $\mathcal{F}_g^X$, one for each genus $g>0$; these are functions on the large phase space

$$\mathsf{H}_\infty = H^*(X \times \mathbb{CP}^\infty, \mathbb{C}).$$

The manifold H_∞ has a rich geometric structure: it is the jet-space of curves in the Frobenius manifold H. (This identification is implicit in Dubrovin [**9**].)

Frobenius manifolds also arise naturally in a number of other geometric situations, such as singularity theory and mirror symmetry (see for example Dubrovin [**9**] and Manin [**27**]). The Frobenius manifolds of Gromov-Witten theory carry an additional geometric structure, a fundamental solution, given by the formula

$$\Theta_a^b(z) = \delta_a^b + \sum_{k=0}^\infty z^{k+1} \langle\langle \tau_k(\mathcal{O}_a)\mathcal{O}^b\rangle\rangle_0^X .$$

In this paper, we formulate the differential equations satisfied by the potentials $\mathcal{F}_g^X$, such as topological recursion relations and the Virasoro constraints, in an intrinsic fashion, that is, in such a way that the equations do not depend on the choice of fundamental solution. This effort is rewarded by a closer relationship between the resulting theory and the geometry of moduli spaces of stable curves.

A consequence of our analysis is the proof of a conjecture of Eguchi and Xiong [**17**]. (Here, we use the summation convention with respect to indices a, b, ...) Introduce the coordinates

$$u^a = \partial\langle\langle \mathcal{O}^a\rangle\rangle_0^X$$

on H, along with their derivatives

$$\partial^k u^a = \partial^{k+1}\langle\langle \mathcal{O}^a\rangle\rangle_0^X ,$$

which form a coordinate system on H_∞; here, ∂ is the vector field on H_∞ given by differentiation with respect to the puncture variable t_0^e.

THEOREM 1.1. *If $g > 0$, the Gromov-Witten potential $\mathcal{F}_g^X$ has the form*

$$\mathcal{F}_g^X = \sum_{n=0}^{3g-3} \frac{1}{n!} \sum_{\substack{k_i>0 \\ k_1+\cdots+k_n\le 3g-3}} f_{a_1\dots a_n}^{k_1\dots k_n}(u^a, \partial u^a)\, \partial^{k_1+1}u^{a_1} \dots \partial^{k_n+1}u^{a_n}.$$

The coefficients $f_{a_1\dots a_n}^{k_1\dots k_n}(u^a, \partial u^a)$ are symmetric in the indices (k_i, a_i), and homogeneous of degree $(2g-2)-(k_1+\cdots+k_n+n)$ in the variables ∂u^a.

This theorem generalizes a well-known formula for the case of pure gravity: by the Kontsevich-Witten theorem, the functions $f_{a_1\dots a_n}^{k_1\dots k_n}(u, \partial u)$ are coefficients of the Gelfand-Dikii polynomials (Hamiltonians of the KdV hierarchy), and the theorem may be reinterpreted as a result in the theory of integrable hierarchies. Of course, it may be easily proved in this case by direct methods from the definition of the Gelfand-Dikii polynomials. Dubrovin and Zhang [**13**] have greatly generalized this observation, showing that Theorem 1.1 holds, for quite different reasons, in the theory of integrable systems.

Acknowledgements. Many of the ideas of this paper come from collaboration with T. Eguchi and C.-S. Xiong, and from conversations with B. Dubrovin, X. Liu and Y. Zhang.

The results of the first four sections of this paper were obtained during the author's residence at RIMS, Kyoto, during the 1999–2000 year devoted to "Geometry Related to String Theory"; I thank Kyoji Saito and Masa-Hiko Saito for the invitation to participate. These results were presented at conferences at RIMS in June 2000, organized by M.-H. Saito, and at KIAS, Seoul, in August 2000, organized by Y.-G. Oh.

The results of the remaining sections were obtained were obtained during a membership of the Institute for Advanced Study, Princeton, in Winter, 2002.

This work was partially supported by the NSF under grants DMS-9794320 and DMS-0072508, and, through the IAS, under grant DMS-9729992.

2. The jet-space of a Dubrovin manifold

2.1. Dubrovin connections. A large part of the theory of Frobenius manifolds does not require the existence of a metric or Euler vector field. In this section, we introduce the geometric structure which is the essential part of the definition of a Frobenius manifold, a Dubrovin connection.

As in Dubrovin [**10**], we work with connections on the cotangent bundle; of course, there is a bijection between connections on the tangent and cotangent bundles. Let

$$\nabla : \Gamma(M, T^*M) \to \Omega^1(M, T^*M)$$

be a connection on the cotangent bundle of a manifold M. Denote by

$$\nabla_X = \iota(X)\nabla : \Gamma(M, T^*M) \to \Gamma(M, T^*M)$$

the operation of covariant differentiation along a vector field X on M.

The connection ∇ is torsion-free if the **torsion** $T \in \Omega^2(M, TM)$ vanishes, and flat if the **curvature** $R \in \Omega^2(M, \mathrm{End}(T^*M))$ vanishes. A **flat** manifold is a manifold M with torsion-free and flat connection ∇ on the cotangent bundle T^*M. A flat coordinate chart $M \supset U \to \mathbb{R}^N$ on a manifold is a coordinate chart $u^a : U \to M$ such that the one-forms du^a are parallel:

$$\nabla(du^b) = 0.$$

A manifold is flat if and only if it has an atlas of flat coordinate charts.

A **pencil** of torsion-free affine connections is a family of connections ∇^z depending on a complex parameter z such that

$$\nabla^z = \nabla + z\mathcal{A},$$

where $\nabla = \nabla^{z=0}$ and $\mathcal{A} \in \Omega^1(M, \mathrm{End}(T^*M))$. A **Dubrovin connection** is a pencil ∇^z of torsion-free affine connections which is flat for all values of z, together with a flat vector field e such that $\mathcal{A}_e$ is the identity. A Dubrovin manifold (M, ∇^z, e) is a manifold together with a Dubrovin connection.

Given a Dubrovin connection ∇^z, denote the vector field $(\mathcal{A}_X)^*Y$ by $X \circ Y$; this defines a product on the tangent bundle of M. Denote the components of the tensor $\mathcal{A}$ in flat coordinates (u^a) by $\mathcal{A}^c_{ab}$:

$$\partial_a \circ \partial_b = \mathcal{A}^c_{ab}\partial_c.$$

PROPOSITION 2.1. *The data (∇^z, e) form a Dubrovin connection if and only if the following conditions are satisfied:*

1) *the affine connection ∇ is flat;*
2) *the product $X \circ Y$ is commutative and associative:*

$$(X \circ Y) \circ Z = X \circ (Y \circ Z);$$

3) *the vector field e is flat, $\nabla e = 0$, and is an identity for the product $X \circ Y$:*

$$e \circ X = X \circ e = X;$$

4) $[\nabla_X, \mathcal{A}_Y] - [\nabla_Y, \mathcal{A}_X] = \mathcal{A}_{[X,Y]}$.

PROOF. If ∇ is torsion-free, the commutativity of the product $X \circ Y$ is equivalent to the vanishing of the torsion of ∇^z.

The curvature $R^z(X, Y)$ of the connection ∇^z is a quadratic polynomial in z:

$$\begin{aligned}[\nabla^z_X, \nabla^z_Y] - [\nabla^z_X, \nabla^z_Y] - \nabla^z_{[X,Y]} \\ = R(X, Y) + z\big([\nabla_X, \mathcal{A}_Y] - [\nabla_Y, \mathcal{A}_X] - \mathcal{A}_{[X,Y]}\big) + z^2[\mathcal{A}_X, \mathcal{A}_Y].\end{aligned}$$

The constant term vanishes if and only if ∇ is flat, the linear term if and only if 4) holds, and the quadratic term if and only if $X \circ Y$ is associative. □

The questions which we address in this paper are of local character; for this reason, we will work in the neighbourhood of a basepoint $p \in M$ of our Dubrovin manifold (M, ∇^z, e). Let (u^a) be a flat coordinate system on M centred at p: that is, $u^a(p) = 0$.

Since the identity vector field e is flat, we may assume that the flat coordinate system (u^a) is chosen in such a way that e is differentiation with respect to u^e, where e is one of the indices labelling the coordinates: in other words, $e = \partial/\partial u_e = \partial_e$.

The one-form

$$\omega = \mathrm{Tr}(\mathcal{A}_a) du^a \in \Omega^1(M) \tag{2.1}$$

plays a prominent role in the theory of Dubrovin manifolds; following Hertling [**22**], we call it the **socle** one-form.

2.2. Fundamental solutions. A fundamental solution of a Dubrovin manifold is a power series

$$\Theta = I + \sum_{n=0}^{\infty} z^{n+1} \Theta_n \in \Gamma(M, \mathrm{End}(T^*M))[\![z]\!]$$

such that

$$\nabla \cdot \Theta(z) = \Theta(z) \cdot \nabla^z.$$

This is equivalent to the sequence of equations

$$\nabla_X \Theta_n = \Theta_{n-1} \mathcal{A}_X. \tag{2.2}$$

Since the endomorphism Θ_0 plays a special role, we introduce the notation $\mathcal{M} = \Theta_0$. The case $n = 0$ of (2.2) shows that $\nabla_X \mathcal{M} = \mathcal{A}_X$. In particular,

$$\partial_a \mathcal{M}^b_e = \delta^b_a.$$

It follows that $\mathcal{M}^a_e = u^a + c^a$, where c^a is a constant.

PROPOSITION 2.2. *Let p be a point of a Dubrovin manifold (M, ∇^z, e), and let (u^a) be a system of flat coordinates which vanish at p. Then there exists a fundamental solution $\Theta(z)$ in a neighbourhood of p such that $\mathcal{M}^a_e = u^a$.*

PROOF. We construct endomorphisms Θ_n of the cotangent bundle T^*M inductively in n, starting with $\Theta_{-1} = I$. To carry out the induction, we must show that the one-form $\Theta_{n-1}\mathcal{A} \in \Omega^1(M, \mathrm{End}(T^*M))$ is exact. To do this, we use condition 4) of Proposition 2.1: we have

$$\begin{aligned} \nabla(\Theta_{n-1}\mathcal{A}) &= \big([\nabla_a, \Theta_{n-1}\mathcal{A}_b] - [\nabla_b, \Theta_{n-1}\mathcal{A}_a]\big)\, du^a \wedge du^b \\ &= \big(\Theta_{n-1}\mathcal{A}_{[\partial_a, \partial_b]} + \Theta_{n-2}[\mathcal{A}_a, \mathcal{A}_b]\big)\, du^a \wedge du^b = 0. \end{aligned}$$

In particular, the existence of Θ_0 is equivalent to condition 4) of Proposition 2.1, while granted the existence of Θ_0, the existence of Θ_1 is equivalent to the associativity of the product $X \circ Y$. □

If $\Theta(z)$ and $\tilde{\Theta}(z)$ are two fundamental solutions of the Dubrovin connection ∇^z, then $\tilde{\Theta}(z)^{-1}\Theta(z)$ is a flat section of $\Gamma(M, \mathrm{End}(T^*M))[\![z]\!]$, that is,

$$\nabla(\tilde{\Theta}(z)^{-1}\Theta(z)) = 0.$$

Conversely, if $\rho(z)$ is a flat section of $\Gamma(M, \mathrm{End}(T^*M))[\![z]\!]$ such that $\rho(0) = I$ and $\Theta(z)$ is a fundamental solution, then $\Theta(z)\rho(z)$ is again a fundamental solution.

2.3. The jet-space of a Dubrovin manifold. Let n be a natural number and let M be a manifold. An n-jet in M is a map from the variety $\operatorname{Spec}(\mathbb{C}[t]/(t^{n+1}))$ to M.

In local coordinates (u^a), we may write an n-jet as

$$u^a(t) = \sum_{k=0}^{n} \frac{t^k}{k!} u_k^a + O(t^{n+1}).$$

In particular, u_0^a are the coordinates of the origin of the jet, and $u_1^a\,\partial_a \in T_{u(0)}M$ is the velocity of the jet at $t = 0$.

The space of n-jets J_nM in M is a fibre bundle over M whose fibres are affine spaces of dimension $n \dim M$. For example, a 0-jet is the same as a point of M, and a 1-jet is the same as a tangent vector; thus $J_0M = M$ and $J_1M = TM$. If $m < n$, denote the projection from J_nM to J_mM defined by reduction mod t^{m+1} by $\rho_{n,m}$; we write $\rho_n : J_nM \to M$ instead of $\rho_{n,0}$.

Let $J_\infty M$ be the inverse limit

$$\varprojlim_n J_nM.$$

There is a fibration $\rho_{\infty,n}$ from $J_\infty M$ to J_nM.

If $\mathcal{V}$ is a coherent sheaf on M, denote the space of sections of the coherent sheaf $\rho_n^*\mathcal{V}$ on J_nM by $\mathcal{V}_n$; in particular, the space of sections

$$\mathcal{V}_\infty = \Gamma(J_\infty M, \rho_\infty^*\mathcal{V}) = \bigcup_{n=0}^{\infty} \mathcal{V}_n$$

is filtered by subspaces

$$\mathcal{V} \subset \mathcal{V}_1 \subset \cdots \subset \mathcal{V}_n \cdots \subset \mathcal{V}_\infty.$$

In particular, $\mathcal{O}_n$ is the algebra of functions on J_nM. Locally, we have

$$\mathcal{V}_n \cong \mathcal{V}[u_k^a \mid 1 \le k \le n].$$

Given a coordinate system (u^a) on M, define vector fields

$$\partial_{\ell,a} = \frac{\partial}{\partial u_\ell^a}, \quad 0 \le \ell \le n,$$

on the jet-space J_nM. The differential operator

$$\partial = \sum_{\ell=0}^{\infty} u_{\ell+1}^a\,\partial_{\ell,a}$$

is a vector field on $J_\infty M$; formally speaking, ∂ generates translation along the parameter t of the jet. Note that $u_k^a = \partial^k u^a$, where $u^a = u_0^a$.

Another important vector field on the jet-space $J_\infty M$ is the **dilaton** vector field

$$\mathcal{Q} = \sum_{\ell=0}^{\infty} \ell\, u_\ell^a\,\partial_{\ell,a},$$

which counts the coordinate $u_\ell^a = \partial^\ell u^a$ with weight ℓ.

We now specialize to the case where M is a Dubrovin manifold. The Dubrovin connection ∇^z on T^*M pulls back by ρ_∞ to a connection ∇^z on $\rho_\infty^*T^*M$. If X is a vector field on the jet-space $J_\infty M$, denote by X the covariant derivative ∇_X on $\rho_\infty^* T^*M$ and its associated tensor bundles. For example, we write ∂ instead of ∇_∂.

Given a vector field X on $J_\infty M$, define an endomorphism $\mathcal{A}_X$ of the bundle $\rho_\infty^* \operatorname{End}(T^*M)$ by the formula

$$\mathcal{A}_X = X(\mathcal{M}).$$

This definition is independent of the fundamental solution $\Theta(z)$, and consistent with the definition of $\mathcal{A}_X$ in the case that X is a vector field on M. We have the formula

$$X(\Theta(z)) = z\Theta(z)\mathcal{A}_X \in \operatorname{End}(T^*M)_\infty.$$

Let $\iota_n : M \to J_n M$ denote the section of the projection $\rho_n : J_n M \to M$ which sends a point $p \in M$ to the jet

$$\iota_n(p) = p + te + o(t^n).$$

Taking the limit over n, we obtain a section $\iota_\infty : M \to J_\infty M$ of the projection $\rho_\infty : J_\infty M \to M$ such that $\iota_n = \rho_{\infty,n} \circ \iota_\infty$.

In local coordinates, the endomorphism

$$\mathcal{X} = \partial\mathcal{M} = \mathcal{A}_\partial \in \operatorname{End}(T^*M)_1 \subset \operatorname{End}(T^*M)_\infty$$

is given by the formula

$$\mathcal{X}^a_b = u^c_1 \mathcal{A}^a_{bc}.$$

In particular, its restriction $\iota_\infty^* \mathcal{X}$ to M equals the identity.

Let $J_n^\circ M \subset J_n M$, $0 < n \le \infty$, be the subvariety where $\mathcal{X}$ is invertible. In particular, $J_n^\circ M$ contains the image of ι_n. If $\mathcal{V}$ is a coherent sheaf on M, let $\mathcal{V}_n^\circ$ be the localization

$$\mathcal{V}_n^\circ = \mathcal{V}_n[\det(\mathcal{X})^{-1}].$$

We think of $\mathcal{V}_n^\circ$ as the space of sections of $\iota_n^* \mathcal{V}$ over $J_n^\circ M$. In particular, we have

$$\mathcal{V}_\infty^\circ = \bigcup_{n=0}^\infty \mathcal{V}_n^\circ.$$

2.4. A frame of vector fields on the jet-space. Let (M, ∇^z, e) be a Dubrovin manifold with basepoint p, and let $\Theta(z)$ be a fundamental solution such that the flat coordinates $\mathcal{M}^a_e = u^a$ vanish at p. Consider the following noncommutative analogues of the Faà di Bruno polynomials:

$$\mathcal{P}_n(z) = z^{-1}\Theta^{-1}(z)\partial^n\Theta(z) \in \operatorname{End}(T^*M)_n.$$

For example, we have $\mathcal{P}_0(z) = z^{-1}I$ and $\mathcal{P}_1(z) = \mathcal{X}$. In general, we see that for $n > 0$,

$$\mathcal{P}_n(z) = (\partial + z\mathcal{X})^{n-1}\mathcal{X} = \sum_{j=0}^{n-1} z^j \mathcal{P}_{n,j},$$

where $\mathcal{P}_{n,j} \in \operatorname{End}(T^*M)_{n-j}$. The polynomials $\mathcal{P}_n(z)$ are intrinsic; they do not depend on the choice of fundamental solution.

Let $\partial_\ell = du^a\,\partial_{\ell,a}$. We now introduce a generating function for vector fields on $J_\infty M$, by the formula

$$\sigma(z) = \sum_{k=0}^{\infty} z^k \sigma_k = \sum_{\ell=0}^{\infty} \mathcal{P}_{\ell+1}(z)\partial_\ell : \mathcal{O}_\infty \to T^*M[\![z]\!]_\infty .$$

For example,

$$\sigma_0 = \sum_{\ell=0}^{\infty} \partial^\ell \mathcal{X}\,\partial_\ell, \qquad\qquad \sigma_1 = \sum_{\ell=1}^{\infty}\sum_{k=1}^{\ell} \partial^{\ell-k}(\mathcal{X}\,\partial^{k-1}\mathcal{X})\,\partial_\ell .$$

Observe that $[\partial + z\mathcal{X}, \sigma(z)] = 0$; it follows that $[\partial, \sigma_k] = -\mathcal{X}\sigma_{k-1}$. Together with the formula

$$\sigma_a(z)u^b = \mathcal{X}^b_a,$$

this characterizes $\sigma(z)$.

If X is vector field on M, define vector field $\sigma_{k,X}$ on $J_\infty M$ by the generating function

$$\sigma_X(z) = \langle X, \sigma(z)\rangle = \sum_{k=0}^{\infty} z^k\,\sigma_{k,X} .$$

We abbreviate the vector fields σ_{k,∂_a} to $\sigma_{k,a}$. In particular, $\sigma_{0,e} = \partial$.

Given of sequences $k = (k_1, \dots, k_n)$ and $\ell = (\ell_1, \dots, \ell_n)$, we say that $k \geq \ell$ if $k_i \geq \ell_i$ for all i, and that $k > \ell$ if in addition $k_i > \ell_i$ for at least one i.

PROPOSITION 2.3. *A function $f \in \mathcal{O}^\circ_\infty$ lies in $\mathcal{O}^\circ_\ell$ if and only if $\sigma(z)f$ is a polynomial in z of degree at most ℓ.*

More generally, suppose $K = (K_1, \dots, K_n)$ is a sequence of integers such that $K_i > 1$ for all i, and suppose that $\sigma_{k_1} \dots \sigma_{k_n} f = 0$ for all $k \geq K$; then $\partial_{k_1} \dots \partial_{k_n} f = 0$ for all $k \geq K$.

PROOF. Since $\sigma_j \mathcal{X} = 0$ for $j > 1$, we see that for $k \geq K$,

$$\sigma_{k_1,a_1} \cdots \sigma_{k_n,a_n} = (\mathcal{X}^{k_1+1})^{b_1}_{a_1} \cdots (\mathcal{X}^{k_n+1})^{b_n}_{a_n} \partial_{k_1,b_1} \cdots \partial_{k_n,b_n} + \sum_{\ell>k} a(k,\ell)\,\partial_{\ell_1} \dots \partial_{\ell_n} .$$

Since $\mathcal{X}$ is invertible on $J^\circ_\infty M$, it follows that

$$\partial_{k_1,a_1} \cdots \partial_{k_n,a_n} = (\mathcal{X}^{-k_1-1})^{b_1}_{a_1} \cdots (\mathcal{X}^{-k_n-1})^{b_n}_{a_n} \sigma_{k_1,b_1} \cdots \sigma_{k_n,b_n} + \sum_{\ell>k} a(k,\ell)\,\sigma_{\ell_1} \dots \sigma_{\ell_n} .$$

Thus, $\sigma_{k_1} \dots \sigma_{k_n} f = 0$ for $k \geq K$ if and only if $\partial_{k_1} \dots \partial_{k_n} f = 0$ for $k \geq K$. □

2.5. An affine structure on the jet-space. We now show how the choice of a fundamental solution $\Theta(z)$ on a Dubrovin manifold (M, ∇^z, e) gives rise to an affine structure on the jet-space $J^\circ_\infty M$, that is, a frame of the tangent bundle $TJ^\circ_\infty M$ consisting of commuting vector fields. This affine structure identifies the formal neighbourhood of ∂ with the large phase space of Gromov-Witten theory.

Introduce the generating function of vector fields on $J_\infty M$,

$$\tau(z) = \sum_{k=0}^{\infty} z^k \tau_k = \Theta(z)\sigma(z) = z^{-1} \sum_{\ell=0}^{\infty} \partial^{\ell+1}\Theta(z)\,\partial_\ell.$$

The introduction of these vector fields is motivated by the fact that they are evolutionary:

$$[\partial, \tau(z)] = 0.$$

LEMMA 2.4. *$[\mathcal{Q}, \sigma(z)] = \sigma(z)$ and $[\mathcal{Q}, \tau(z)] = \tau(z)$*

PROOF. Since $\mathcal{Q}\Theta(z) = 0$, it suffices to prove that $[\mathcal{Q}, \sigma(z)] = \sigma(z)$. Since $[\mathcal{Q}, \partial] = \partial$, it follows that $\mathcal{Q}\mathcal{P}_\ell = \ell\mathcal{P}_\ell$, and hence that

$$[\mathcal{Q}, \sigma(z)] = \sum_{\ell=0}^{\infty} [\mathcal{Q}, \mathcal{P}_{\ell+1}(z)\partial_\ell] = \sum_{\ell=0}^{\infty} \big((\ell+1)\mathcal{P}_{\ell+1}(z)\partial_\ell - \ell\mathcal{P}_{\ell+1}(z)\partial_\ell\big) = \sigma(z). \qquad \square$$

If X is a vector field on M, let $\tau_{k,X}$ be the vector field on $J_\infty M$ defined by the generating function

$$\tau_X(z) = \langle X, \tau(z) \rangle = \sum_{k=0}^{\infty} z^k \, \tau_{k,X},$$

and let $\tau_{k,a} = \tau_{k,\partial_a}$. Note that $\tau_0 = \sigma_0$; in particular, $\tau_{0,e} = \partial$. Also, observe that if $f \in \mathcal{O}$ is a function only of the coordinates u^a on the Frobenius manifold M, then

$$\tau_{0,a} f = \sigma_{0,a} f = \mathcal{X}_a^b \partial_b f.$$

In particular, we see that

$$\partial_a f = (\mathcal{X}^{-1})_a^b \tau_{0,b} f. \tag{2.3}$$

PROPOSITION 2.5. *The vector fields $\tau_{k,a}$ form a frame of the tangent bundle of $J_\infty^\circ M$.*

PROOF. We have

$$\sigma_{k,a} = \big(\mathcal{X}^{k+1}\big)_a^b \partial_{k,b} + \sum_{\ell > k} a(k,\ell)\,\partial_\ell,$$

and the result follows, since $\mathcal{X}$ is invertible on $J_\infty^\circ M$. $\square$

The vector fields $\tau_{k,a}$ were shown by Dubrovin **[9]** to be the commuting flows of an integrable hierarchy. In this paper, these vector fields play an auxilliary role: the vector fields $\sigma_{k,a}$ are a more natural frame for the tangent bundle of $J_\infty^\circ M$, because they are intrinsic, that is, independent of the fundamental solution $\Theta(z)$. For us, the fact the vector fields $\sigma_{k,a}$ do not commute is of secondary importance.

PROPOSITION 2.6. *$[\tau_a(z), \tau_b(y)] = 0$*

PROOF. We have $\mathcal{A}_{\sigma_a(z)} = \mathcal{A}_{\sigma_{0,a}} = \mathcal{W}_a = \mathcal{X}\mathcal{A}_a$; in other words, the components $\mathcal{W}^c_{ab}$ of $\mathcal{W}_a$ equal $\mathcal{A}^{c'}_{ab}\mathcal{X}^c_{c'}$, and are symmetric in a and b. Since $[\partial, \tau(z)] = 0$, we see that

$$\begin{aligned}
[\tau_a(y), \tau_b(z)] &= \sum_{k=0}^{\infty} \partial^{k+1}\big(z^{-1}\tau_a(y)\Theta^c_b(z) - y^{-1}\tau_b(z)\Theta^c_a(y)\big)\partial_{k,c} \\
&= \sum_{k=0}^{\infty} \partial^{k+1}\big(z^{-1}\Theta^{a'}_a(y)\sigma_{a'}(y)\Theta^c_b(z) - y^{-1}\Theta^{b'}_b(z)\sigma_b(z)\Theta^c_a(y)\big)\partial_{k,c} \\
&= \sum_{k=0}^{\infty} \partial^{k+1}\big(\Theta^{a'}_a(y)\Theta^{b'}_b(z)\big(\mathcal{W}^c_{a'b'} - \mathcal{W}^c_{b'a'}\big)\big)\partial_{k,c} = 0. \qquad \square
\end{aligned}$$

COROLLARY 2.7. *We have* $[\sigma_a(y), \sigma_b(z)] = \mathcal{W}^c_{ab}(y\sigma_c(y) - z\sigma_c(z))$; *in other words,*

$$[\sigma_{k,a}, \sigma_{\ell,b}] = \begin{cases} -\mathcal{W}^c_{ab}\sigma_{\ell-1,c}, & k = 0, \\ 0, & k > 0. \end{cases}$$

PROOF. Since $[\tau_a(y), \tau_b(z)] = [\Theta^{a'}_a(y)\sigma_{a'}(y), \Theta^{b'}_b(z)\sigma_{b'}(z)]$ vanishes, we see that

$$\begin{aligned}
\Theta^{a'}_a(y)\Theta^{b'}_b(z)[\sigma_{a'}(y), \sigma_{b'}(z)] \\
&= \Theta^{b'}_b(z)\big(\sigma_{b'}(z)\Theta^{a'}_a(y)\big)\sigma_{a'}(y) - \Theta^{a'}_a(y)\big(\sigma_{a'}(y)\Theta^{b'}_b(z)\big)\sigma_{b'}(z) \\
&= \Theta^{a'}_a(y)\Theta^{b'}_b(z)\big(y\mathcal{W}^c_{b'a'}\sigma_c(y) - z\mathcal{W}^c_{a'b'}\sigma_c(z)\big). \quad \square
\end{aligned}$$

2.6. Flat coordinates on the jet-space. The results of this section are taken from [**20**]. The main result of this section is Theorem 2.11, which is taken from Section 4 of [**20**]; it is interesting to compare this with Proposition 6.1 of Coates and Givental [**6**].

Let p be a basepoint of the Dubrovin manifold M. Since the vector fields $\tau_{k,a}$ are in involution, there is a coordinate system t^a_k centered at $\pi = \iota_\infty(p) \in J^\circ_\infty M$ such that $\tau_{k,a} = \partial/\partial t^a_k$ and $t^a_k(\pi) = 0$; we call these the flat coordinates on the jet-space. Of course, these coordinates do not lie in $\mathcal{O}^\circ_\infty$; rather, they lie in the completion $\widehat{\mathcal{O}}$ of $\mathcal{O}_\infty$ at π, and are only coordinates in a formal neighbourhood of π. This completion $\widehat{\mathcal{O}}$ may be identified with the algebra $\mathbb{C}[\![t^a_k \mid k \geq 0]\!]$.

PROPOSITION 2.8. *Let* $\Theta(z)$ *be a fundamental solution on the Dubrovin manifold* (M, ∇^z, e). *There is a unique generating function*

$$\mathbf{t}(z) = \sum_{k=-\infty}^{\infty} z^{-k-1}\,\mathbf{t}_k \in TM^\circ_\infty[\![z, z^{-1}]\!]$$

such that $\mathcal{Q}\mathbf{t}(z) = -\mathbf{t}(z)$ *and*

$$\tau(y)\mathbf{t}(z) = \frac{\Theta(y)\Theta^{-1}(z)}{z-y} = \sum_{k=0}^{\infty} y^k z^{-k-1}\,\Theta(y)\Theta^{-1}(z). \tag{2.4}$$

For $k \geq 0$, *we have* $\mathbf{t}^a_k = t^a_k - \delta_{k,1}\delta^a_e$.

PROOF. We start with a lemma of Dubrovin [**9**].

LEMMA 2.9. *Let*

$$\Omega(y,z)=\sum_{k,\ell=0}^{\infty} y^k z^\ell \Omega_{k,\ell}$$

be the generating function characterized by the equation

$$(y-z)\Omega(y,z)=\Theta(y)\Theta^{-1}(z)-I. \tag{2.5}$$

Then $\tau_{k,a}\Omega^c_{\ell,m|b}=\tau_{\ell,b}\Omega^c_{k,m|a}$.

PROOF. We must prove that $\tau_a(x)\Omega^c_b(y,z)=\tau_b(y)\Omega^c_a(x,z)$ or equivalently, that

$$(x-z)\tau_a(x)\big(\Theta(y)\Theta^{-1}(z)\big)^c_b=(y-z)\tau_b(y)\big(\Theta(x)\Theta^{-1}(z)\big)^c_a.$$

But

$$\tau_a(x)\big(\Theta(y)\Theta^{-1}(z)\big)=(y-z)\Theta(y)\mathcal{A}_{\tau_a(x)}\Theta^{-1}(z).$$

Thus, it suffices to show that $\big(\Theta(y)\mathcal{A}_{\tau_a(x)}\big)^c_b=\big(\Theta(x)\mathcal{A}_{\tau_b(y)}\big)^c_a$; this follows from the formula

$$\big(\Theta(y)\mathcal{A}_{\tau_a(x)}\big)^c_b=\Theta^{a'}_a(x)\Theta^{b'}_b(y)\mathcal{A}_{\sigma_{0,a'}}{}^c_{b'}=\Theta^{a'}_a(x)\Theta^{b'}_b(y)\mathcal{W}^c_{a'b'}$$

and the symmetry $\mathcal{W}^c_{ab}=\mathcal{W}^c_{ba}$. □

To show the existence of $\mathbf{t}(z)$, we must solve the system of equations

$$\tau_{j,a}\mathbf{t}^b_k=\delta_{j,k}\delta^b_a-\Omega^b_{j,-k-1|a}, \tag{2.6}$$

subject to the constraints $\mathcal{Q}\mathbf{t}_k=-\mathbf{t}_k$. For $k\ge 0$, this system has the unique solution

$$\mathbf{t}^a_k=t^a_k-\delta_{k,1}\delta^a_e.$$

For $k<0$, the system is integrable by Lemma 2.9. Since $[\mathcal{Q},\tau(z)]=\tau(z)$ and $\mathcal{Q}\Omega(y,z)=0$, we see that $\mathcal{Q}\mathbf{t}_k+\mathbf{t}_k$ is a constant; replacing $\mathbf{t}_k$ by $-\mathcal{Q}\mathbf{t}_k$, we obtain the desired solution. □

The vector fields $\mathcal{Q}$ and e have simple expressions in flat coordinates on the large phase space.

PROPOSITION 2.10. *The dilaton vector field* $\mathcal{Q}$ *and the identity vector field* e *are the residues of the generating functions* $-\langle\mathbf{t}(z),\tau(z)\rangle$ *and* $-\langle\mathbf{t}(z),z\tau(z)\rangle$ *respectively.*

PROOF. By Lemma 2.4, there are constants c^a_k such that

$$\mathcal{Q}=\sum_{k=0}^{\infty}\big(c^a_k-t^a_k\big)\,\tau_{k,a}.$$

By the definition of $\tau(z)$, we have

$$\tau_{k,a}u^b_n=\big(\Theta_{k-n-1}\mathcal{X}^{n+1}\big)^b_a.$$

Applying the equation for $\mathcal{Q}$ to the function u^b_n and evaluating at π, we see that

$$\delta_{n,1}\delta^b_e=\sum_{k=0}^{\infty}\Theta^b_{k-n-1,a}(\pi)\,c^a_k.$$

This system of linear equations is upper triangular, with diagonal entries equal to 1, hence has a unique solution: it is easily checked that this solution is $c_k^a = \delta_{k,1}\delta_e^a$, since $\Theta_e^a(\pi) = u^a(\pi) = 0$.

The proof of the formula for e in flat coordinates is similar. Since $e(\Theta(z)) = z\Theta(z)$ and $[\partial, e] = 0$, we see that

$$[e, \tau(z)] = z \sum_{\ell=0}^{\infty} \partial^{\ell+1} e(\Theta(z))\partial_\ell = z\tau(z).$$

In other words, $[e, \tau_{k,a}] = \tau_{k-1,a}$. It follows that there are constants c_k^a such that

$$e = \sum_{k=1}^{\infty} \left(c_k^a - t_k^a\right) \tau_{k-1,a}.$$

Applying this equation to the function u_n^b and evaluating at π, we see that

$$\delta_{n,0}\delta_e^b = \sum_{k=1}^{\infty} \Theta_{k-n-2,a}^b(\pi)\, c_k^a.$$

Again, this system has a unique solution $c_k^a = \delta_{k,1}\delta_e^b$. □

Just as the modification $\sigma = \Theta(z)^{-1}\tau(z)$ of $\tau(z)$ is intrinsic, so the modification $\mathbf{s} = \Theta^*(z)\mathbf{t}(z)$ of $\mathbf{t}(z)$ is intrinsic. This modification was introduced in [**20**], where it was denoted $z^{-1}G^*(-z^{-1})$.

THEOREM 2.11. *Let (M, ∇^z, e) be a Dubrovin manifold. The generating function*

$$\mathbf{s}(z) = \Theta^*(z)\mathbf{t}(z) = \sum_{k=-\infty}^{\infty} z^{-k-1}\mathbf{s}_k \in TM[\![z^{-1}]\!]_{\infty,0}$$

is intrinsic, that is, independent of the fundamental solution $\Theta(z)$, and

$$\mathbf{s}_k = \begin{cases} -\langle e, (\mathcal{X}^{-1}\partial)^{k-1}\mathcal{X}^{-1}\rangle, & k > 0, \\ 0, & k \le 0. \end{cases}$$

PROOF. We first show that $\mathbf{s}_0 = 0$. We have

$$\mathbf{s}_0^a = t_0^a + \sum_{k=1}^{\infty} t_k^b \Theta_{k-1,b}^a - \Theta_{0,e}^a.$$

Since $\Omega(y, 0) = y^{-1}(\Theta(y) - I)$, we see that $\Omega_{k,0} = \Theta_k$, and hence that $\tau_k \mathbf{t}_{-1} = -\Theta_k$. It follows that

$$\mathbf{s}_0^a = t_0^a + \left(\partial - \sum_{k=1}^{\infty} t_k^b \tau_{k-1,b}\right)\mathbf{t}_{-1}^a = \mathbf{t}_0^a + e\left(\mathbf{t}_{-1}^a\right) = 0.$$

A special case of (2.4) is $\partial\mathbf{t}(z) = z^{-1}\langle e, \Theta^{-1}(z)\rangle$; it follows that

$$\partial\mathbf{s}(z) = z^{-1}e + \mathcal{X}^*\mathbf{s}(z). \tag{2.7}$$

Taking the residue, we see that $0 = e + \mathcal{X}^*\mathbf{s}_1$, while taking the coefficient of z^{-k-1}, we see that $\partial\mathbf{s}_k = \mathcal{X}^*\mathbf{s}_{k+1}$ for $k > 1$; this establishes the formula for $\mathbf{s}_k$, $k \ge 0$.

In order to prove that $\mathbf{s}_k$ vanishes for $k < 0$, we need two lemmas.

LEMMA 2.12. *A power series f in the coordinates t^a_k such that $\partial f = ef = 0$ is constant.*

PROOF. Since
$$(\partial - e)f = \sum_{k=1}^{\infty} t^a_k \tau_{k-1,a} f$$
vanishes, it follows that f is a constant. (See [20], Section 3, for more details of the proof.) □

LEMMA 2.13. $(e+z)\mathbf{t}(z) = 0$

PROOF. The equations $[e, \tau(y)] = y\tau(y)$ and $(e+z-y)(\Theta(y)\Theta^{-1}(z)) = 0$ show that
$$\begin{aligned}(z-y)\tau(y)(e+z)\mathbf{t}(z) &= (z-y)(e+z-y)(\tau(y)\mathbf{t}(z)) \\ &= (e+z-y)(\Theta(y)\Theta^{-1}(z)) = 0.\end{aligned}$$
In other words, $\tau(y)(e+z)\mathbf{t}(z) = 0$, and hence $(e+z)\mathbf{t}(z) \in \mathbb{C}[\![z, z^{-1}]\!]$. Composing this equation with the dilaton vector field $\mathcal{Q}$, we see that $(e+z)\mathbf{t}(z) = 0$. □

It follows from the equations $\mathcal{Q}\mathbf{t}(z) = -\mathbf{t}(z)$ and $(e+z)\mathbf{t}(z) = 0$ that have $\mathcal{Q}\mathbf{s}(z) = -\mathbf{s}(z)$ and $e\mathbf{s}(z) = 0$.

Suppose that $\mathbf{s}_{-k}$ vanishes, for $k > 0$. Taking the coefficient of z^{k+1} in (2.7), we see that $\partial \mathbf{s}_{-k-1} = 0$. Since we know that $e\mathbf{s}_{-k-1} = 0$, it follows from Lemma 2.12 that $\mathbf{s}_{-k-1}$ is constant. Since $\mathcal{Q}\mathbf{s}_{-k-1} = -\mathbf{s}_{-k-1}$, we conclude that $\mathbf{s}_{-k-1}$ vanishes. Thus $\mathbf{s}_{-k} = 0$ for all $k > 0$, by induction on k. □

Using the series $\mathbf{s}(z)$, we may now rewrite Proposition 2.10 in an intrinsic fashion:
$$\mathcal{Q} = -\sum_{k=0}^{\infty} \langle \mathbf{s}_k, \sigma_k \rangle, \qquad e = -\sum_{k=0}^{\infty} \langle \mathbf{s}_{k+1}, \sigma_k \rangle.$$
Apart from being intrinsic, these formulas have the virtue that when applied to a function $f \in \mathcal{O}_n$, they truncate to a sum over $0 \le k \le n$.

Relationship to prior results. The Dubrovin connection was introduced by Dubrovin [9]. He also initiated the study of the jet-space of a Frobenius manifold (which he calls the loop-space). Sections 2.4 and 2.5 are taken from our joint work with Eguchi and Xiong [14], and Section 2.6 from [20], Section 4. (Note that the generating function $\mathbf{t}(z)$ used here is related to the generating function $\theta(\zeta)$ of [20] by the formula $\mathbf{t}(z) = z^{-1}\theta^*(-z^{-1})$.)

3. The jet-space of a conformal Dubrovin manifold

3.1. Conformal Dubrovin manifolds. A linear vector field on a flat manifold is a vector field E such that $\nabla E \in \Gamma(M, \mathrm{End}(TM))$ is flat. In flat coordinates (u^a), there are constants A^a_b and B^a such that
$$E = (A^a_b u^b + B^a)\partial_a.$$

An **Euler** vector field E for a Dubrovin connection is a linear vector field such that

$$[E, X \circ Y] = [E, X] \circ Y + X \circ [E, Y] + X \circ Y. \tag{3.1}$$

Equivalently, $[\mathcal{L}_E, \mathcal{A}_X] = \mathcal{A}_{[E,X]} + \mathcal{A}_X$, where $\mathcal{L}_E$ is the Lie derivative of the vector field E acting on one-forms.

A **conformal** Dubrovin manifold (M, ∇^z, e, E, r) is a Dubrovin manifold (M, ∇^z, e) together with an Euler vector field E and a real number r.

Let μ be the endomorphism of the cotangent bundle defined by the formula

$$\mu = 1 - \tfrac{r}{2} + \nabla_E - \mathcal{L}_E.$$

The adjoint of μ is the endomorphism of the tangent bundle given by the formula

$$\mu^*(X) = -\nabla_E X + [E, X] + (1 - \tfrac{r}{2})X = (1 - \tfrac{r}{2})X - \nabla_X E.$$

Let $\mathcal{U}$ be the endomorphism $\mathcal{U} = \mathcal{A}_E$ of the cotangent bundle.

PROPOSITION 3.1. *Let (M, ∇^z, e, E, r) be a conformal Dubrovin manifold, and let $\mathcal{H}$ be the bundle $T^*M[z, z^{-1}]$. Let δ_z be the endomorphism*

$$\delta_z = \partial_z + z^{-1}(\mu + \tfrac{1}{2}) + \mathcal{U}$$

of $\mathcal{H}$. Then $[\nabla^z, \delta_z] = 0$.

PROOF. If X is a vector field on M, we have

$$\begin{aligned}[\nabla^z_X, \delta_z] &= [\nabla_X + z\mathcal{A}_X, \partial_z + z^{-1}\mu + \mathcal{A}_E] \\ &= z^{-1}[\nabla_X, \mu] + \big([\nabla_X, \mathcal{A}_E] - [\nabla_E, \mathcal{A}_X] - \mathcal{A}_X + [\mathcal{L}_E, \mathcal{A}_X]\big) + z[\mathcal{A}_X, \mathcal{A}_E].\end{aligned}$$

Since E is linear, it follows that $[\nabla_X, \mu] = 0$; it is also clear that $[\mathcal{A}_E, \mathcal{A}_X] = 0$. By (3.1), we see that that

$$[\mathcal{L}_E, \mathcal{A}_X] = \mathcal{A}_{[E,X]} + \mathcal{A}_X = [\nabla_E, \mathcal{A}_X] - [\nabla_X, \mathcal{A}_E] + \mathcal{A}_X. \qquad \square$$

Let $\Theta(z)$ be a fundamental solution of the conformal Dubrovin manifold (M, ∇^z, e, E, r), and consider the conjugate of δ_z by $\Theta(z)$, defined by the formula

$$\tilde{\delta}_z = \Theta(z) \cdot \delta_z \cdot \Theta^{-1}(z) = \partial_z + z^{-1}R(z),$$

where

$$R(z) = \sum_{k=0}^{\infty} z^k R_k \in z\Gamma(M, \operatorname{End}(T^*M))[\![z]\!].$$

In particular, $R_0 = \mu + \frac{1}{2}$. It follows from Proposition 3.1 that $R(z)$ is flat.

By the Jordan decomposition, we may write μ as the sum of semisimple and nilpotent endomorphisms μ_s and μ_n of the cotangent bundle T^*M; furthermore, both μ_s and μ_n are flat. (In many cases, such as the theory of Gromov-Witten invariants, μ is semisimple.) Let π_λ be the projection onto the subbundle of T^*M on which μ_s has eigenvalue λ: we have

$$I = \sum_\lambda \pi_\lambda, \qquad \mu_s = \sum_\lambda \lambda \pi_\lambda.$$

PROPOSITION 3.2. *There exists a fundamental solution such that $[\mu_s, R_k] = kR_k$.*

PROOF. If $\lambda - \lambda' \neq k$, the endomorphism $k - \operatorname{ad}(\mu)$ is invertible on the space of endomorphisms of the form $\pi_\lambda A \pi_{\lambda'}$: we have

$$\begin{aligned}(k - \operatorname{ad}(\mu))^{-1}(\pi_\lambda A \pi_{\lambda'}) &= \sum_{i=0}^{\infty} (-\operatorname{ad}(\mu_n))^i (k - \operatorname{ad}(\mu_s)(\pi_\lambda A \pi_{\lambda'}))^{-i-1} \\ &= \sum_{i=0}^{\infty} \frac{(-\operatorname{ad}(\mu_n))^i (\pi_\lambda A \pi_{\lambda'})}{(k - \lambda + \lambda')^{i+1}},\end{aligned}$$

where of course, the sum is finite since μ_n is nilpotent.

Let $\Theta(z)$ and $\tilde{\Theta}(z) = \Theta(z)\rho(z)$ be a pair of fundamental solutions, where

$$\rho(z) = I + \sum_{k=1}^{\infty} z^k \rho_k$$

is a flat endomorphism of $\mathcal{H} = T^*M[z, z^{-1}]$. Let $\partial_z + z^{-1}R(z)$ and $\partial_z + z^{-1}\tilde{R}(z)$ be the associated endomorphisms of $\mathcal{H}$. Then we have

$$\tilde{R}_k = (k - \operatorname{ad}(\mu))G_k + \sum_{\ell=1}^{k-1} \big(G_\ell R_{k-\ell} - \tilde{R}_{k-\ell} G_\ell\big) + R_k.$$

Setting

$$G_k = - \sum_{\lambda - \lambda' \neq k} (k - \operatorname{ad}(\mu))^{-1} \Big(\sum_{\ell=1}^{k-1} \pi_\lambda \big(G_\ell R_{k-\ell} - \tilde{R}_{k-\ell} G_\ell\big) \pi_{\lambda'} + \pi_\lambda R_k \pi_{\lambda'} \Big),$$

we obtain a fundamental solution $\tilde{\Theta}(z)$ such that $[\mu_s, \tilde{R}_k] = k\tilde{R}_k$. □

DEFINITION 3.1. *A* **conformal** *fundamental solution is a fundamental solution satisfying the condition* $[\mu_s, R_k] = kR_k$

Extracting the constant term of the equation $\Theta(z) \cdot \delta_z - \tilde{\delta}_z \cdot \Theta(z) = 0$, we see that

$$\mathcal{U} = [\mu, \mathcal{M}] + \mathcal{M} + R_1. \tag{3.2}$$

Applying ∂ to this equation, we see that

$$\partial\mathcal{U} = [\mu, \mathcal{X}] + \mathcal{X}. \tag{3.3}$$

3.2. Vector fields on the jet-space of a conformal Dubrovin manifold. Let $\mathcal{L}_k$ be the vector field on the jet-space $J^\circ_\infty M$ of a conformal Dubrovin manifold (M, ∇^z, e, E, r) defined by the formula

$$\mathcal{L}_k = \operatorname{Res}_{z=0} \big\langle \mathbf{s}(z), z\delta_z^{k+1}\sigma(z) \big\rangle.$$

For example, $\mathcal{L}_{-1} = -e$. By the work of Eguchi, Hori and Xiong [**15**] and Dubrovin and Zhang [**12**], these vector fields play a fundamental role in the theory of Gromov-Witten invariants. In this section, we give simple intrinsic formulas for them.

THEOREM 3.3. *Restricted to* $\mathcal{O} \subset \mathcal{O}_\infty$, *the vector field* $\mathcal{L}_k$ *equals* $-E^{\circ k+1}$. *We have*

$$[\partial, \mathcal{L}_k] = \operatorname{Res}_{z=0} \big\langle e, \delta_z^{k+1} \sigma(z) \big\rangle,$$

and

$$\mathcal{L}_k = -\sum_{\ell=0}^{\infty}\Big(\partial^\ell\langle e,\mathcal{U}^{k+1}\rangle + \sum_{j=1}^{\ell}\partial^{j-1}\operatorname{Res}_{z=0}\big\langle e,\delta_z^{k+1}(\partial+z\mathcal{X})^{\ell-j}\mathcal{X}\big\rangle\Big)\partial_\ell. \tag{3.4}$$

PROOF. By Theorem 2.11, $\mathbf{s}_\ell$ vanishes for $\ell \le 0$. Since $\sigma(z)u^a = \mathcal{X}du^a$, it follows that

$$\begin{aligned}\mathcal{L}_k u^a &= \operatorname{Res}_{z=0}\big\langle \mathbf{s}(z), z\delta_z^{k+1}\mathcal{X}du^a\big\rangle = \langle \mathbf{s}_1, \mathcal{U}^{k+1}\mathcal{X}du^a\rangle \\ &= \langle \mathcal{X}^*\mathbf{s}_1, \mathcal{U}^{k+1}du^a\rangle = -\langle e, \mathcal{U}^{k+1}du^a\rangle = -E^{\circ k+1}(u^a).\end{aligned}$$

Here, we have used that $[\mathcal{U},\mathcal{X}] = 0$.

Choose a conformal fundamental solution $\Theta(z)$ on M. We may replace $\mathbf{s}(z)$ and $\sigma(z)$ in the formula for $\mathcal{L}_k$ by $\Theta^*(z)\mathbf{t}(z)$ and $\Theta^{-1}(z)\tau(z)$, and rewrite $\mathcal{L}_k$ as

$$\mathcal{L}_k = \operatorname{Res}_{z=0}\big\langle\Theta^*(z)\mathbf{t}(z), z\delta_z^{k+1}\Theta^{-1}(z)\tau(z)\big\rangle = \operatorname{Res}_{z=0}\big\langle\mathbf{t}(z), z\tilde{\delta}_z^{k+1}\tau(z)\big\rangle.$$

From this formula, and the fact that $[\partial,\tilde{\delta}_z] = 0$ and $[\partial,\tau(z)] = 0$, we see that

$$\begin{aligned}[\partial,\mathcal{L}_k] &= \operatorname{Res}_{z=0}\big\langle\partial\mathbf{t}(z), z\tilde{\delta}_z^{k+1}\tau(z)\big\rangle = \operatorname{Res}_{z=0}\big\langle\Theta^*(z)^{-1}e, \tilde{\delta}_z^{k+1}\tau(z)\big\rangle \\ &= \operatorname{Res}_{z=0}\big\langle e, \delta_z^{k+1}\sigma(z)\big\rangle.\end{aligned}$$

If X is a vector field on the jet space $J_\infty M$, we have

$$X(u_\ell^a) = \partial^\ell(Xu^a) - \sum_{j=1}^{\ell}\partial^{j-1}\cdot[\partial,X]u_{\ell-j}^a.$$

In the special case $X = \mathcal{L}_k$, this gives

$$\mathcal{L}_k u_\ell^a = -\partial^\ell(\mathcal{L}_k u^a) - \sum_{j=1}^{\ell}\partial^{j-1}\operatorname{Res}_{z=0}\langle e, \delta_z^{k+1}\sigma(z)u_{\ell-j}^a\rangle,$$

from which (3.4) follows. □

Using the noncommutative Faà di Bruno polynomials $\mathcal{P}_n(z)$, we may rewrite (3.4) in the elegant form

$$\mathcal{L}_k = -\sum_{\ell=0}^{\infty}\sum_{j=0}^{\ell}\partial^j\operatorname{Res}_{z=0}\big\langle e,\delta_z^{k+1}\mathcal{P}_{\ell-j}(z)\big\rangle\partial_\ell.$$

The following corollary is an immediate consequence.

COROLLARY 3.4. *The vector field $\mathcal{L}_k$ preserves the filtration of $\mathcal{O}_\infty$ by subalgebras $\mathcal{O}_j$.*

As an illustration of the utility of the formula (3.4), let us calculate $\mathcal{L}_k\mathcal{X}$; we will need this calculation later in the discussion of Virasoro constraints in genus 1.

PROPOSITION 3.5.

$$\mathcal{L}_k\mathcal{X} = \sum_{\ell=0}^{k}[\mathcal{X},\mathcal{U}^\ell\mu\mathcal{U}^{k-\ell}] - \sum_{\ell=0}^{k}(\mathcal{U}^\ell\mu\mathcal{U}^{k-\ell})_e^a\mathcal{A}_a\mathcal{X} - \tfrac{3}{2}(k+1)\mathcal{X}\mathcal{U}^k$$

PROOF. We have $\mathcal{X} = u_1^b \mathcal{A}_b$, hence $\partial_{0,a}\mathcal{X} = \partial\mathcal{A}_a$, $\partial_{1,a}\mathcal{X} = \mathcal{A}_a$, while $\partial_{n,a}\mathcal{X} = 0$ for $n > 1$. It follows that

$$\begin{aligned}\mathcal{L}_k\mathcal{X} &= -\langle e, \mathcal{U}^{k+1}\rangle\,\partial_0\mathcal{X} - \partial\langle e, \mathcal{U}^{k+1}\rangle\,\partial_1\mathcal{X} - \operatorname{Res}_{z=0}\langle e, \delta_z^{k+1}\mathcal{X}\rangle\,\partial_1\mathcal{X} \\ &= -\partial\big((\mathcal{U}^{k+1})_e^a\mathcal{A}_a\big) - \operatorname{Res}_{z=0}\big(\delta_z^{k+1}\mathcal{X}\big)_e^a\mathcal{A}_a \\ &= -\partial(\mathcal{U}^{k+1}) - \operatorname{Res}_{z=0}\big(\delta_z^{k+1}\big)_e^a\mathcal{A}_a\mathcal{X},\end{aligned}$$

since $[\mathcal{A}_X, \mathcal{A}_Y] = 0$. By (3.3), we have

$$\partial(\mathcal{U}^{k+1}) = \sum_{\ell=0}^{k}\mathcal{U}^\ell(\mathcal{X} + [\mu, \mathcal{X}])\mathcal{U}^{k-\ell} = -(k+1)\mathcal{X}\mathcal{U}^k + \sum_{\ell=0}^{k}[\mathcal{X}, \mathcal{U}^\ell\mu\mathcal{U}^{k-\ell}].$$

We also have

$$\begin{aligned}\operatorname{Res}_{z=0}\big(\delta_z^{k+1}\big)_e^a\mathcal{A}_a\mathcal{X} &= \sum_{\ell=0}^{k}\big(\mathcal{U}^\ell(\mu + \tfrac{1}{2})\mathcal{U}^{k-\ell}\big)_e^a\mathcal{A}_a\mathcal{X} \\ &= \tfrac{1}{2}(k+1)\big(\mathcal{U}^k\big)_e^a\mathcal{A}_a\mathcal{X} + \sum_{\ell=0}^{k}\big(\mathcal{U}^\ell\mu\mathcal{U}^{k-\ell}\big)_e^a\mathcal{A}_a\mathcal{X}.\end{aligned}$$

The proposition follows, since $\big(\mathcal{U}^k\big)_e^a\mathcal{A}_a = \mathcal{U}^k$. □

COROLLARY 3.6. *In terms of the socle one-form ω of (2.1),*

$$\mathcal{L}_k \log\det(\mathcal{X}) = -\sum_{\ell=0}^{k}\langle e, \mathcal{U}^\ell(\mu + \tfrac{3}{2})\mathcal{U}^{k-\ell}\omega\rangle.$$

PROOF. We see from Proposition 3.5 that

$$\begin{aligned}\mathcal{L}_k \log\det(\mathcal{X}) &= \operatorname{Tr}(\mathcal{X}^{-1}\mathcal{L}_k\mathcal{X}) \\ &= \sum_{\ell=0}^{k}\operatorname{Tr}[\mathcal{X}, \mathcal{X}^{-1}\mathcal{U}^\ell\mu\mathcal{U}^{k-\ell}] - \sum_{\ell=0}^{k}\big(\mathcal{U}^\ell\mu\mathcal{U}^{k-\ell}\big)_e^a\operatorname{Tr}(\mathcal{A}_a) - \tfrac{3}{2}(k+1)\operatorname{Tr}(\mathcal{U}^k).\end{aligned}$$

The corollary follows, since $\operatorname{Tr}[\mathcal{X}, \mathcal{X}^{-1}\mathcal{U}^\ell\mu\mathcal{U}^{k-\ell}] = 0$ and $\operatorname{Tr}(\mathcal{U}^k) = \langle e, \mathcal{U}^k\omega\rangle$. □

The vector fields $\mathcal{L}_k$ form a Lie algebra isomorphic to the Lie algebra of vector fields on the line, under the correspondence which associates to $\mathcal{L}_k$ the vector field $-z\partial_z^{k+1}$. These relations are sometimes called the Virasoro relations.

THEOREM 3.7. $[\mathcal{L}_j, \mathcal{L}_k] = (j-k)\mathcal{L}_{j+k}$

PROOF. We have

$$\begin{aligned}
&\Big[\langle \mathbf{t}(y), y\tilde{\delta}_y^{j+1}\tau(y)\rangle, \langle \mathbf{t}(z), z\tilde{\delta}_z^{k+1}\tau(z)\rangle\Big] \\
&\quad = \Big\langle \mathbf{t}(y), y\tilde{\delta}_y^{j+1} \cdot \frac{\Theta(y)\Theta^{-1}(z)}{z-y} \cdot z\tilde{\delta}_z^{k+1}\tau(z)\Big\rangle \\
&\quad - \Big\langle \mathbf{t}(z), z\tilde{\delta}_z^{k+1} \cdot \frac{\Theta(z)\Theta^{-1}(y)}{y-z} \cdot y\tilde{\delta}_y^{j+1}\tau(z)\Big\rangle \\
&\quad = \Big\langle \mathbf{s}(y), y\delta_y^{j+1} \cdot \frac{1}{z-y} \cdot z\delta_z^{k+1}\sigma(z)\Big\rangle - \Big\langle \mathbf{s}(z), z\delta_z^{k+1} \cdot \frac{1}{y-z} \cdot y\delta_y^{j+1}\sigma(z)\Big\rangle \\
&\quad = \frac{1}{z-y}\big\langle (\delta_y^*)^{j+1} y\mathbf{s}(y), z\delta_z^{k+1}\sigma(z)\big\rangle - \frac{1}{y-z}\big\langle (\delta_z^*)^{k+1} z\mathbf{s}(z), y\delta_y^{j+1}\sigma(z)\big\rangle.
\end{aligned}$$

The coefficients of y^ℓ and z^ℓ in $(\delta_y^*)^{j+1}y\mathbf{s}(y)$ and $(\delta_z^*)^{k+1}z\mathbf{s}(z)$ vanish for $\ell \geq 0$; it follows that

$$\operatorname{Res}_{y=0} \frac{(\delta_y^*)^{j+1}y\mathbf{s}(y)}{z-y} = (\delta_z^*)^{j+1}z\mathbf{s}(z), \qquad \operatorname{Res}_{z=0} \frac{(\delta_z^*)^{k+1}z\mathbf{s}(z)}{y-z} = (\delta_y^*)^{k+1}y\mathbf{s}(y),$$

and hence that

$$\begin{aligned}
[\mathcal{L}_j, \mathcal{L}_k] &= \operatorname{Res}_{y=0}\operatorname{Res}_{z=0}\Big[\langle \mathbf{t}(y), y\tilde{\delta}_y^{j+1}\tau(y)\rangle, \langle \mathbf{t}(z), z\tilde{\delta}_z^{k+1}\tau(z)\rangle\Big] \\
&= \operatorname{Res}_{z=0}\big\langle (\delta_z^*)^{j+1}z\mathbf{s}(z), z\delta_z^{k+1}\sigma(z)\big\rangle - \operatorname{Res}_{y=0}\big\langle (\delta_y^*)^{k+1}y\mathbf{s}(z), y\delta_y^{j+1}\sigma(y)\big\rangle \\
&= \operatorname{Res}_{z=0}\big\langle \mathbf{s}(z), \big(z\delta_z^{j+1}\cdot z\delta_z^{k+1} - z\delta_z^{k+1}\cdot z\delta_z^{j+1}\big)\sigma(y)\big\rangle.
\end{aligned}$$

The result follows, since $z\delta_z^{j+1} \cdot z\delta_z^{k+1} = z^2\delta_z^{j+k+2} + (j+1)z\delta_z^{j+k+1}$. □

As a corollary, we see that

$$[E^{\circ j}, E^{\circ k}] = (k-j)E^{\circ(j+k-1)}.$$

This relation was first conjectured for Frobenius manifolds by Dubrovin and Zhang [**12**], and proved by Hertling and Manin [**23**], using their theory of F-manifolds.

Relationship to prior results. The theory of conformal Dubrovin connections is due to Dubrovin. The study of the Virasoro vector fields on the jet-space of a Frobenius manifold was initiated by Dubrovin and Zhang [**12**]; the presentation in Section 3.2 is based on the methods of Section 5 of [**20**] (for the action of these vector fields on functions of the underlying Dubrovin manifold) and of [**14**] (for the commutator with the vector field ∂).

4. Topological recursion relations for Gromov-Witten invariants and the jet-space

The genus 0 Gromov-Witten invariants of a smooth projective variety X define a conformal Dubrovin manifold with conformal fundamental solution, called the small phase space of X. In this section, we recall the construction of the small phase space, following Dubrovin [**9**]. We then show the way in which the theory of topological recursion relations (Eguchi and Xiong [**17**], Kontsevich and Manin [**25**], and our own work [**19**]) takes a simpler form when rewritten using the intrinsic geometry of the jet-space of this Dubrovin manifold; this is the main new result of this paper.

4.1. Stable maps and Gromov-Witten invariants. Let us recall the definition of the Gromov-Witten invariants of a projective manifold over $\mathbb{C}$; see Cox and Katz [**5**] and Manin [**27**] for more detailed expositions. The definition which we outline is the one which works in the setting of algebraic geometry: Gromov-Witten invariants have also been defined for compact symplectic manifolds, using entirely different techniques. The Gromov-Witten invariants of a projective manifold reflect the intersection theory of Kontsevich's moduli spaces of stable maps $\overline{\mathcal{M}}_{g,n}(X,\beta)$, whose definition we now recall.

Let X be a projective manifold of dimension r. A **prestable map**

$$(f : C \to X, z_1, \dots, z_n)$$

of genus $g \geq 0$ and degree $\beta \in H_2(X,\mathbb{Z})$ with n marked points consists of the following data:

1) a connected projective curve C of arithmetic genus $g = h^1(C,\mathcal{O}_C)$, whose only singularities are ordinary double points,
2) n distinct smooth points $(z_1,\dots,z_n)$ of C;
3) an algebraic map $f : C \to V$, such that the degree of f, that is, the cycle $f_*[C] \in H^2(X,\mathbb{Z})$, equals β.

If $\tilde{C}$ is the normalization of C, the special points in $\tilde{C}$ are the inverse images of the singular and marked points of C. (Note that the degree of $f : C \to X$ equals 0 if and only if its image is a single point.)

A prestable map $(f : C \to X, z_1, \dots, z_n)$ is **stable** if it has no infinitesimal automorphisms fixing the marked points. The condition of stability is equivalent to the following: each irreducible component of $\tilde{C}$ of genus 0 on which f has degree 0 has at least 3 special points, while each irreducible component of $\tilde{C}$ of genus 1 on which f has degree 0 has at least 1 special point. In particular, there are no stable maps of genus g and degree 0 with n marked points unless $2(g-1)+n>0$.

The moduli stack of n-pointed stable maps $\overline{\mathcal{M}}_{g,n}(X,\beta)$ is the classifying stack for stable maps of genus g and of degree β; it is a complete Deligne-Mumford stack, though not in general smooth (Behrend and Manin [**4**]). The definition of Gromov-Witten invariants is based on the study of $\overline{\mathcal{M}}_{g,n}(X,\beta)$. Let $\mathrm{ev}_i : \overline{\mathcal{M}}_{g,n}(X,\beta) \to X$, $1 \leq i \leq n$, be evaluation at the ith marked point:

$$\mathrm{ev}_i : \overline{\mathcal{M}}_{g,n}(X,\beta) \ni (f : C \to X, z_1, \dots, z_n) \mapsto f(z_i) \in X.$$

Let $(f : C \to X, z_1, \dots, z_{n+N})$ be a stable map, representing a point of $\overline{\mathcal{M}}_{g,n+N}(X,\beta)$. The map $(f : C \to X, z_1, \dots, z_n)$ obtained by forgetting the last N points may not be stable, owing to the existence of rational components of C on which the sheaf $f^*TX \otimes \mathcal{O}_C(-(z_1+\dots+z_n))$ has non-vanishing sections. On such a component, the map f is constant; thus, there is a stable map $(f : C \to X, z_1, \dots, z_n)^{\mathrm{stab}}$ obtained by contracting each of these components of C. There is a morphism, constructed by Behrend and Manin [**4**], which sends

$(f : C \to X, z_1, \dots, z_{n+N})$ to $(f : C \to X, z_1, \dots, z_n)^{\mathrm{stab}}$, and which we denote

$$\pi_{n,N} : \overline{\mathcal{M}}_{g,n+N}(X,\beta) \to \overline{\mathcal{M}}_{g,n}(X,\beta).$$

In the case $N = 1$, this construction yields the **universal curve**

$$\pi = \pi_{n,1} : \overline{\mathcal{M}}_{g,n+1}(X,\beta) \to \overline{\mathcal{M}}_{g,n}(X,\beta). \tag{4.1}$$

The fibre of π at a stable map $(f : C \to X, z_1, \dots, z_n)$ is the curve C; $f = \mathrm{ev}_{n+1} : \overline{\mathcal{M}}_{g,n+1}(X,\beta) \to X$ is the universal stable map.

The sheaf $R^1\pi_* f^*TX$ on $\overline{\mathcal{M}}_{g,n}(X,\beta)$ is called the obstruction sheaf of $\overline{\mathcal{M}}_{g,n}(X,\beta)$. If it vanishes, the Grothendieck-Riemann-Roch theorem implies that the stack $\overline{\mathcal{M}}_{g,n}(X,\beta)$ is smooth, of dimension

$$\operatorname{vdim} \overline{\mathcal{M}}_{g,n}(X,\beta) = (3-r)(g-1) + \textstyle\int_\beta c_1(X) + n; \tag{4.2}$$

this is called the **virtual dimension** of $\overline{\mathcal{M}}_{g,n}(X,\beta)$. This hypothesis is rarely true; however, there is an algebraic cycle

$$[\overline{\mathcal{M}}_{g,n}(X,\beta)]^{\mathrm{virt}} \in H_{2\operatorname{vdim}\overline{\mathcal{M}}_{g,n}(X,\beta)}(\overline{\mathcal{M}}_{g,n}(X,\beta),\mathbb{Q}),$$

the **virtual fundamental class**, which stands in for $[\overline{\mathcal{M}}_{g,n}(X,\beta)]$ in the general case. The most important property of the virtual fundamental class is the formula

$$[\overline{\mathcal{M}}_{g,n+1}(X,\beta)]^{\mathrm{virt}} = \pi^![\overline{\mathcal{M}}_{g,n}(X,\beta)]^{\mathrm{virt}} \tag{4.3}$$

for integration over the fibres of the stabilization map.

If the obstruction sheaf $R^1\pi_* f^*TX$ is locally free of rank e, the moduli stack $\overline{\mathcal{M}}_{g,n}(X,\beta)$ is smooth of dimension $\operatorname{vdim}\overline{\mathcal{M}}_{g,n}(X,\beta) + e$, and the virtual fundamental class satisfies the equation

$$[\overline{\mathcal{M}}_{g,n}(X,\beta)]^{\mathrm{virt}} = c_e(R^1\pi_* f^*TX) \cap [\overline{\mathcal{M}}_{g,n}(X,\beta)].$$

The universal curve (4.1) over $\overline{\mathcal{M}}_{g,n}(X,\beta)$ has n canonical sections

$$\sigma_i : \overline{\mathcal{M}}_{g,n}(X,\beta) \to \overline{\mathcal{M}}_{g,n+1}(X,\beta),$$

corresponding to the n marked points of the curve C. Consider the line bundles

$$\Omega_i = \sigma_i^*\omega, \quad 1 \le i \le n,$$

where $\omega = \omega_{\overline{\mathcal{M}}_{g,n+1}(X,\beta)/\overline{\mathcal{M}}_{g,n}(X,\beta)}$ is the relative dualizing sheaf of the universal curve. Thus, the fibre of Ω_i at the stable map $f : C \to X$ equals the cotangent line $T^*_{z_i}C$ of C at the ith marked point z_i of C. Let

$$\Psi_i = c_1(\Omega_i) \in H^2(\overline{\mathcal{M}}_{g,n}(X,\beta),\mathbb{Z})$$

be the Chern class of Ω_i.

The Gromov-Witten invariants of a projective manifold are defined by integrating tautological cohomology classes against the virtual fundamental class $[\overline{\mathcal{M}}_{g,n}(X,\beta)]^{\mathrm{virt}}$. Given rational cohomology classes $x_1, \dots, x_n$ on X, we define the Gromov-Witten invariants by the formula

$$\langle \tau_{k_1}(x_1) \dots \tau_{k_n}(x_n) \rangle^X_{g,\beta} = \int_{[\overline{\mathcal{M}}_{g,n}(X,\beta)]^{\mathrm{virt}}} \Psi_1^{k_1} \dots \Psi_n^{k_n} \, \mathrm{ev}_1^* x_1 \dots \mathrm{ev}_n^* x_n \in \mathbb{Q}.$$

Perhaps the most important property of the Gromov-Witten invariants is that they are invariant under deformation of the variety X.

The **Novikov ring** Λ of X is the commutative graded algebra consisting of all formal sums

$$\sum_{\beta\in H_2(X,\mathbb{Z})} a_\beta q^\beta$$

such that for all $C>0$, the set of $\beta\in H_2(X,\mathbb{Z})$ such that $a_\beta\neq 0$ and $\int_\beta\omega<C$ is finite; the product is defined by $q^{\beta_1}q^{\beta_2}=q^{\beta_1+\beta_2}$, and the grading is defined by $|q^\beta|=-2c_1(X)\cap\beta$. For example, for $\mathbb{CP}^d$, we have $\Lambda=\mathbb{Q}[q]$, where $|q|=-2(d+1)$. By working over the Novikov ring, we may combine the Gromov-Witten invariants in different degrees into a single generating function:

$$\langle\tau_{k_1}(x_1)\dots\tau_{k_n}(x_n)\rangle_g^X=\sum_{\beta\in H_2(X,\mathbb{Z})} q^\beta\langle\tau_{k_1}(x_1)\dots\tau_{k_n}(x_n)\rangle_{g,\beta}^X.$$

4.2. The small and large phase spaces. Let

$$\{\gamma_a\in H^{p_a,q_a}(X)\mid a\in A\}$$

be a homogeneous basis of the Dolbeault cohomology of X, such that for a distinguished element $e\in A$, $\gamma_e=1$. The small phase space H of X is the formal neighbourhood of 0 in the vector space $H^\bullet(X,\mathbb{C})$. Denote by u^a the coordinates on H dual to the basis γ_a. Let η be the flat metric on H associated to the Poincaré form on $H^\bullet(X,\mathbb{C})$, with components

$$\eta_{ab}=\int_X\gamma_a\cup\gamma_b,$$

and by η^{ab} the inverse matrix of η_{ab}.

Dubrovin **[9]** shows that H is a conformal Dubrovin manifold, and construct from the genus 0 Gromov-Witten invariants of X a conformal fundamental solution $\Theta(z)$ on H. The flat connection ∇ of this Dubrovin manifold is the Levi-Cività connection associated to the flat metric η. We defer the definition of the remaining geometric structures on H, namely the tensor $\mathcal{A}$, the identity and Euler vector fields e and E and the fundamental solution Θ, to below.

The large phase space H_∞ of X is the formal neighbourhood of 0 in the vector space $H^\bullet(X\times\mathbb{CP}^\infty,\mathbb{C})$; it is a formal manifold with coordinates $\{t_k^a\mid a\in A,k\geq 0\}$, where t_k^a has degree $-p_a-2k$. (If X has cohomology of odd degree, H and H_∞ are actually supermanifolds; this detail does not materially change the theory.)

The genus g potential $\mathcal{F}_g$ of X is the function on the large phase space H_∞ given by the formula

$$\mathcal{F}_g^X=\sum_{n=0}^\infty\frac{1}{n!}\sum_{\substack{k_1\dots k_n\\ a_1\dots a_n}} t_{k_1}^{a_1}\dots t_{k_n}^{a_n}\langle\tau_{k_1,a_1}\dots\tau_{k_n,a_n}\rangle_g^X, \tag{4.4}$$

where we write $\tau_{k,a}$ in place of $\tau_k(\gamma_a)$.

Denote the constant coefficient vector field $\partial/\partial t_k^a$ on H_∞ by $\tau_{k,a}$, and $\tau_{0,e}$ by ∂. The partial derivatives of the potential $\mathcal{F}_g$ are denoted

$$\langle\langle\tau_{k_1,a_1}\dots\tau_{k_n,a_n}\rangle\rangle_g^X=\tau_{k_1,a_1}\dots\tau_{k_n,a_n}\mathcal{F}_g^X.$$

In particular, $\langle\tau_{k_1,a_1}\dots\tau_{k_n,a_n}\rangle_g^X$ is the value at $0 \in \mathsf{H}_\infty$ of the function $\langle\langle\tau_{k_1,a_1}\dots\tau_{k_n,a_n}\rangle\rangle_g^X$. Explicitly, we have

$$\langle\langle\tau_{k_1,a_1}\dots\tau_{k_n,a_n}\rangle\rangle_g^X = \sum_{\beta\in H_2(X,\mathbb{Z})} q^\beta \sum_{N=0}^{\infty}\frac{1}{N!} \sum_{\substack{b_1\dots b_N\\ k_{n+1}\dots k_{n+N}}} t_{k_{n+1}}^{b_1}\dots t_{k_{n+N}}^{b_N}$$

$$\int_{[\overline{\mathcal{M}}_{g,n+N}(X,\beta)]^{\mathrm{virt}}} \Psi_1^{k_1}\dots\Psi_{n+N}^{\ell_N}\,\mathrm{ev}_1^*\gamma_{a_1}\dots\mathrm{ev}_n^*\gamma_{a_n}\,\mathrm{ev}_{n+1}^*\gamma_{b_1}\dots\mathrm{ev}_{n+N}^*\gamma_{b_N}.$$

4.3. Dual graphs and their associated cycles. The moduli space of stable maps $\overline{\mathcal{M}}_{g,n}(X,\beta)$ is stratified by boundary strata, which are most conveniently parametrized by dual graphs.

The dual graph $C(G)$ of a pointed prestable map $(f : C \to X, z_1,\dots,z_n)$ is a graph $G = G(C)$ with one vertex v for each component $C(v)$ of the normalization $\tilde{C}$ of the curve C, labelled by the genus $g(v)$ of this component and the degree $\beta(v)$ of the restriction $f(v) : C(v) \to X$ of f to $C(v)$.

The edges of the dual graph $G(C)$ correspond to double points of the curve C; the two ends of an edge are attached to the vertices associated to the components on which the two branches of the double point lie. (If both branches lie in the same component of C, then the edge is a loop.)

Finally, to each marked point z_i of the curve corresponds a leg of the graph, labelled by i, at the vertex corresponding to the component of C on which z_i lies (which is uniquely determined, since z_i is a smooth point).

In drawing dual graphs, we denote vertices of genus 0 either by a solid circle • or leave them unmarked and vertices of genus $g > 0$ by ⓖ.

Define the genus $g(G)$ of a dual graph G to be the sum of the genera $g(v)$ over the vertices $v \in \mathrm{Vert}(G)$ of G and the first Betti number of the graph G. Then $g(G)$ equals the arithmetic genus of the curve C. Define the degree $\beta(G)$ of a dual graph G to be the sum of the degrees $\beta(v)$ over the vertices v of G. The valence $n(G)$ equals the number of external legs of G.

A dual graph is **stable** if each vertex v such that $\beta(v) = 0$ satisfies the additional condition $2(g(v) - 1) + n(v) > 0$; a pointed prestable curve is stable if and only if the associated dual graph is stable. The set of isomorphism classes of stable dual graphs G of fixed genus $g(G)$, valence $n(G)$, and degree $\beta(G)$ is finite.

If G is a stable graph of genus g, valence n and degree β, let $\mathcal{M}(G) \subset \overline{\mathcal{M}}_{g,n}(X,\beta)$ be the moduli stack of stable maps with dual graph G, and let $\overline{\mathcal{M}}(G)$ be its closure in $\overline{\mathcal{M}}_{g,n}(X,\beta)$. Let $\widetilde{\mathcal{M}}(G)$ be the product

$$\widetilde{\mathcal{M}}(G) = \prod_{v\in\mathrm{Vert}(G)} \overline{\mathcal{M}}_{g(v),n(v)}(X,\beta(v)),$$

with virtual fundamental class

$$[\widetilde{\mathcal{M}}(G)]^{\mathrm{virt}} = \prod_{v\in\mathrm{Vert}(G)} [\overline{\mathcal{M}}_{g(v),n(v)}(X,\beta(v))]^{\mathrm{virt}}.$$

There is a natural covering map

$$\pi(G) : \widetilde{\mathcal{M}}(G) \longrightarrow \overline{\mathcal{M}}(G),$$

with covering group $\mathrm{Aut}(G)$. Let $[G] = [\overline{\mathcal{M}}(G)]^{\mathrm{virt}}$ be the cycle

$$[\overline{\mathcal{M}}(G)]^{\mathrm{virt}} = \frac{\pi(G)_*[\widetilde{\mathcal{M}}(G)]^{\mathrm{virt}}}{|\operatorname{Aut}(G)|}.$$

In formulas, we will often symbolize the cycle $[G]$ by the dual graph G itself.

A dual graph G with one edge determines a Cartier divisor $D(G)$ supported by $\overline{\mathcal{M}}(G)$. Axiom V for virtual fundamental classes in Behrend [**3**] implies that

$$D(G) \cap [\overline{\mathcal{M}}_{g(G),n(G)}(X, \beta(G))]^{\mathrm{virt}} = [G].$$

The following formula is due to Witten [**29**].

PROPOSITION 4.1. *Let* $\pi : \overline{\mathcal{M}}_{g,n+1}(X,\beta) \to \overline{\mathcal{M}}_{g,n}(X,\beta)$ *be the universal curve* (4.1). *For* $1 \le i \le n$, *let* $D_i = D(G_i)$ *be the Cartier divisor on* $\overline{\mathcal{M}}_{g,n+1}(X,\beta)$ *associated to the dual graph*

$G_i =$ 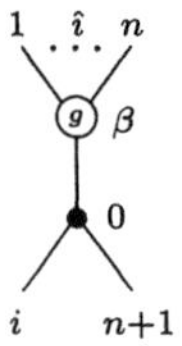

Then on $\overline{\mathcal{M}}_{g,n+1}(X,\beta)$, *we have the formula* $\pi^*\Psi_i = \Psi_i + D_i$.

4.4. The puncture, dilaton and divisor equations on the large phase space. For $\omega \in H^2(X,\mathbb{C})$, let $R_a^b(\omega)$ be the matrix of multiplication by ω on $H(X)$:

$$\omega \cup \gamma_a = R_a^b(\omega)\gamma_b.$$

Combining Proposition 4.1 with (4.3), we obtain the **puncture**, **divisor** and **dilaton** equations:

$$\langle \tau_{0,e}\tau_{k_1,a_1}\dots\tau_{k_n,a_n}\rangle_{g,\beta}^X = \sum_{i=1}^n \langle \tau_{k_1,a_1}\dots\tau_{k_i-1,a_i}\dots\tau_{k_n,a_n}\rangle_{g,\beta}^X, \tag{4.5}$$

$$\langle \tau_0(\omega)\tau_{k_1,a_1}\dots\tau_{k_n,a_n}\rangle_{g,\beta}^X = \sum_{i=1}^n R_{a_i}^b(\omega)\langle \tau_{k_1,a_1}\dots\tau_{k_i-1,b}\dots\tau_{k_n,a_n}\rangle_{g,\beta}^X + \textstyle\int_\beta \omega \cdot \langle \tau_{k_1,a_1}\dots\tau_{k_n,a_n}\rangle_{g,\beta}^X, \tag{4.6}$$

$$\langle \tau_{1,e}\tau_{k_1,a_1}\dots\tau_{k_n,a_n}\rangle_{g,\beta}^X = (2g-2+n)\langle \tau_{k_1,a_1}\dots\tau_{k_n,a_n}\rangle_{g,\beta}^X. \tag{4.7}$$

Owing to the nonexistence of the stabilization maps $\pi : \overline{\mathcal{M}}_{0,3}(X,0) \to \overline{\mathcal{M}}_{0,2}(X,0)$ and $\pi : \overline{\mathcal{M}}_{1,1}(X,0) \to \overline{\mathcal{M}}_{1,0}(X,0)$, we have the following exceptional cases:

$$\langle \tau_{0,e}\tau_{0,a}\tau_{0,b}\rangle_{0,0}^X = \eta_{ab}, \qquad \langle \tau_{1,e}\rangle_{1,0}^X = \tfrac{1}{24}\chi(X),$$

$$\langle \tau_0(\omega)\tau_{0,a}\tau_{0,b}\rangle_{0,0}^X = R_{ab}(\omega), \qquad \langle \tau_0(\omega)\rangle_{1,0}^X = \tfrac{1}{24}\textstyle\int_X \omega \cup c_{r-1}(X).$$

Introduce the vector fields

$$e = -\sum_{k=0}^{\infty} \mathbf{t}_{k+1}^{a} \partial_{k,a} = \partial - \sum_{k=0}^{\infty} t_{k+1}^{a} \partial_{k,a}, \quad \mathcal{Q} = -\sum_{k=0}^{\infty} \mathbf{t}_{k}^{a} \partial_{k,a} = \partial_{1,e} - \sum_{k=0}^{\infty} t_{k}^{a} \partial_{k,a}$$

on the large phase space H_∞. The puncture and dilaton equations are equivalent to the differential equations

$$e\mathcal{F}_g^X = \begin{cases} \frac{1}{2}\eta_{ab} t_0^a t_0^b, & g = 0, \\ 0, & g > 0, \end{cases} \qquad \mathcal{Q}\mathcal{F}_g^X = \begin{cases} \frac{1}{24}\chi(X), & g = 1, \\ (2g-2)\mathcal{F}_g^X, & g \neq 1, \end{cases}$$

for the Gromov-Witten potentials $\mathcal{F}_g^X$.

We may identify the large phase space H_∞ with the jet space $J_\infty^\circ \mathsf{H}$ of the small phase space H. Consider the map u from the large phase space H_∞ to the small phase space H given by the formula

$$u^a = \eta^{ab}\, \partial \langle\langle \tau_{0,b} \rangle\rangle_0^X .$$

Applying the vector field $\partial_{0,b}$ to the genus 0 puncture equation $e\mathcal{F}_0^X = \frac{1}{2}\eta_{ab} t_0^a t_0^b$, we see that

$$u^a = t_0^a + \sum_{k=0}^{\infty} t_{k+1}^b \tau_{k,b} u^a .$$

It follows that the map $u : \mathsf{H}_\infty \to \mathsf{H}$ is a submersion at 0; this submersion has a section, which identifies H with the submanifold of H_∞ along which the coordinates t_k^a, $k > 0$, vanish.

The following result shows that the jet coordinates $\partial^n u^a$ form a coordinate system on the large phase space, and that the origin of the large phase space corresponds to the basepoint of the jet-space $J_\infty^\circ \mathsf{H}$.

PROPOSITION 4.2. *We have* $\partial^n u^a(0) = \delta_{n,1}\delta_e^a$, *and* $\tau_{k,b}(\partial^n u^a)(0) = \delta_{n,k}\delta_b^a$ *if* $k \leq n$.

PROOF. We argue by induction on $n = 0$; we have already seen that the result is true for $n = 0$. Expanding the equation $e^n(u^a) = \delta_{n,1}\delta_e^a$, which holds for $n > 0$, we see that

$$\partial^n u^a = \delta_{n,1}\delta_e^a + \sum_{i=0}^{n-1} (-1)^i \binom{n}{i} \left(\sum_{k=0}^{\infty} t_{k+1}^b \tau_{k,b} \right)^{n-i} \partial^i u^a . \tag{4.8}$$

In particular, evaluating at $0 \in \mathsf{H}_\infty$, we see that $\partial^n u^a(0) = \delta_{n,1}\delta_e^a$. Applying the vector field $\tau_{k,b}$ to (4.8) and evaluating at $0 \in \mathsf{H}_\infty$, we see that

$$\tau_{k,b}(\partial^n u^a)(0) = \sum_{i=0}^{n-1} (-1)^i \binom{n}{i} \tau_{k-n+i,b}(\partial^i u^a)(0).$$

The result now follows by the induction hypothesis. □

The following theorem is due to Hori [**24**]; we will only need it in genus 0, where it yields a construction of an Euler vector field on the small phase space.

THEOREM 4.3. *Let $\mathcal{L}_0$ be the vector field*

$$\begin{aligned}\mathcal{L}_0 &= \sum_{k=0}^{\infty}(k+p_a+\tfrac{1-r}{2})\mathbf{t}_k^a\tau_{k,a} + \sum_{k=0}^{\infty} R_a^b \mathbf{t}_{k+1}^a \tau_{k,b} \\ &= -\tfrac{1}{2}(3-r)\partial_{1,e} + \sum_{k=0}^{\infty}(k+p_a+\tfrac{1-r}{2})t_k^a\tau_{k,a} - R_0^b\tau_{0,b} + \sum_{k=0}^{\infty} R_a^b t_{k+1}^a\tau_{k,b},\end{aligned}$$

where R_a^b is the matrix $R_a^b(c_1(X))$. We have

$$0 = \mathcal{L}_0\mathcal{F}_g^X + \begin{cases} \tfrac{1}{2}R_{ab}t_0^a t_0^b, & g=0, \\ \tfrac{1}{48}\int_X\big((3-r)c_r(X) - 2c_1(X)c_{r-1}(X)\big), & g=1, \\ 0, & g>1. \end{cases}$$

PROOF. The formula (4.2) for the dimension of the virtual fundamental class of $\overline{\mathcal{M}}_{g,n}(X,\beta)$ implies the following identity:

$$\begin{aligned}\sum_{i=1}^{n}(p_{a_i}+k_i)\langle\tau_{k_1,a_1}\dots\tau_{k_n,a_n}\rangle_{g,\beta}^X &= \operatorname{vdim}\overline{\mathcal{M}}_{g,n}(X,\beta)\cdot\langle\tau_{k_1,a_1}\dots\tau_{k_n,a_n}\rangle_{g,\beta}^X \\ &= \big((3-r)(g-1)+\textstyle\int_\beta c_1(X)+n\big)\langle\tau_{k_1,a_1}\dots\tau_{k_n,a_n}\rangle_{g,\beta}^X.\end{aligned}$$

To eliminate the dependence on the genus g, subtract $\frac{1}{2}(3-r)$ times the dilaton equation (4.6); after some rearrangement, this gives

$$\begin{aligned}&\sum_{i=1}^{n}\big(p_{a_i}+k_i+\tfrac{1-r}{2}\big)\langle\tau_{k_1,a_1}\dots\tau_{k_n,a_n}\rangle_{g,\beta}^X \\ &\qquad = \tfrac{1}{2}(3-r)\langle\tau_{1,e}\tau_{k_1,a_1}\dots\tau_{k_n,a_n}\rangle_{g,\beta}^X + \textstyle\int_\beta c_1(X)\cdot\langle\tau_{k_1,a_1}\dots\tau_{k_n,a_n}\rangle_{g,\beta}^X.\end{aligned}$$

To eliminate the dependence on β, apply the divisor equation (4.6) with $\omega = c_1(X)$: this yields

$$\begin{aligned}&\sum_{i=1}^{n}\Big(\big(p_{a_i}+k_i+\tfrac{1-r}{2}\big)\langle\tau_{k_1,a_1}\dots\tau_{k_n,a_n}\rangle_{g,\beta}^X + R_{a_i}^b\langle\tau_{k_1,a_1}\dots\tau_{k_i-1,b}\dots\tau_{k_n,a_n}\rangle_{g,\beta}^X\Big) \\ &\qquad = \tfrac{1}{2}(3-r)\langle\tau_{1,e}\tau_{k_1,a_1}\dots\tau_{k_n,a_n}\rangle_{g,\beta}^X + \langle\tau_{0,e}\tau_{k_1,a_1}\dots\tau_{k_n,a_n}\rangle_{g,\beta}^X.\end{aligned}$$

Taking into account the exceptional cases in degree 0, we obtain the theorem. □

4.5. Topological recursion relations. A stable curve is a stable map with target a point; the moduli space $\overline{\mathcal{M}}_{g,n} = \overline{\mathcal{M}}_{g,n}(\mathsf{pt},0)$ of stable curves of genus g with n marked points is a smooth Deligne-Mumford stack of dimension $3g-3+n$, the Deligne-Mumford-Knudsen moduli stack of stable curves. Denote the line bundles Ω_i on $\overline{\mathcal{M}}_{g,n} = \overline{\mathcal{M}}_{g,n}(\mathsf{pt},0)$ by ω_i, and their Chern classes $c_1(\omega_i)$ by ψ_i.

If n and N are nonnegative integers, let $\overline{\mathcal{M}}_{g,n,N}(X,\beta)$ be the moduli space of stable maps $\overline{\mathcal{M}}_{g,n+N}(X,\beta)$. If $2g-2+n>0$, the stabilization map

$$\rho_{n,N} : \overline{\mathcal{M}}_{g,n,N}(X,\beta) \to \overline{\mathcal{M}}_{g,n}$$

is the morphism which takes a stable map $(f : C \to X, z_1,\dots,z_n,x_1,\dots,x_N)$ to the stabilization $(C,z_1,\dots,z_n)^{\mathrm{stab}}$. We abbreviate the pullback $\rho_{n,N}^*\psi_i$ by ψ_i.

PROPOSITION 4.4. *For* $1 \leq i \leq n$, *let* $D_i = D(G_i)$ *be the Cartier divisor on* $\overline{\mathcal{M}}_{g,n,N}(X,\beta)$ *associated to the following dual graph:*

$G_i =$ 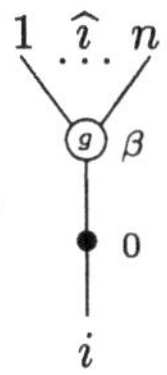

Then $\Psi_i = \psi_i + D_i$.

PROOF. See the proof of Proposition 5 in [**19**]. □

COROLLARY 4.5.

$$\frac{1}{1 - z\Psi_i} = \frac{1}{1 - z\psi_i}\left(1 + \frac{zD_i}{1 - z\Psi_i}\right)$$

PROOF. Divide the equation $1 - z\psi_i = 1 - z(\Psi_i - D_i)$ by $(1 - z\psi_i)(1 - z\Psi_i)$. □

From Proposition 4.4, we may derive many relations among Gromov-Witten invariants. For example, using the vanishing of $\psi_1^{k_1}\psi_2^{k_2}\psi_3^{k_3}$ on the zero-dimensional moduli space $\overline{\mathcal{M}}_{0,3}$ when $k_1 > 0$, we obtain the genus 0 topological recursion relation

$$\langle\langle \tau_{k_1,a_1}\tau_{k_2,a_2}\tau_{k_3,a_3}\rangle\rangle_0 = \eta^{AB}\langle\langle \tau_{k_1-1,a_1}\tau_{0,A}\rangle\rangle_0 \langle\langle \tau_{0,B}\tau_{k_2,a_2}\tau_{k_3,a_3}\rangle\rangle_0. \tag{4.9}$$

The following result of Dijkgraaf and Witten [**7**], proved using the topological recursion relation in genus 0, may be viewed as an analogue of Theorem 1.1 in genus 0.

PROPOSITION 4.6. *The function* $\langle\langle \tau_{k,a}\tau_{\ell,b}\rangle\rangle_0^X$ *on the large phase space is the pullback of a function on the small phase space; that is,*

$$\langle\langle \tau_{k,a}\tau_{\ell,b}\rangle\rangle_0^X = \sum_{n=0}^{\infty}\frac{1}{n!}u^{a_1}\dots u^{a_n}\langle \tau_{k,a}\tau_{\ell,b}\tau_{0,a_1}\dots\tau_{0,a_n}\rangle_0^X. \tag{4.10}$$

PROOF. Let $A_{k,a;\ell,b}$ and $B_{k,a;\ell,b}$ be the left and right-hand sides of (4.10). Since u^a and t_0^a are equal along $\mathsf{H} \subset \mathsf{H}_\infty$, it follows that $A_{k,a;\ell,b}$ and $B_{k,a;\ell,b}$ are too. We now calculate the derivatives of $A_{k,a;\ell,b}$ and $B_{k,a;\ell,b}$ with respect to t_m^c, $m > 0$, using (4.9). On the one hand,

$$\begin{aligned}
\tau_{m,c}B_{k,a;\ell,b} &= (\tau_{m,c}u^{a_0})\sum_{n=0}^{\infty}\frac{1}{n!}u^{a_1}\dots u^{a_n}\langle \tau_{k,a}\tau_{\ell,b}\tau_{0,a_0}\tau_{0,a_1}\dots\tau_{0,a_n}\rangle_0^X \\
&= \eta^{AB}\langle\langle \tau_{m,c}\tau_{0,e}\tau_{0,A}\rangle\rangle_0^X \sum_{n=0}^{\infty}\frac{1}{n!}u^{a_1}\dots u^{a_n}\langle \tau_{k,a}\tau_{\ell,b}\tau_{0,B}\tau_{0,a_1}\dots\tau_{0,a_n}\rangle_0^X \\
&= \eta^{AB}\eta^{CD}\langle\langle \tau_{m-1,c}\tau_{0,C}\rangle\rangle_0^X \langle\langle \tau_{0,D}\tau_{0,e}\tau_{0,A}\rangle\rangle_0^X \\
&\qquad \sum_{n=0}^{\infty}\frac{1}{n!}u^{a_1}\dots u^{a_n}\langle \tau_{k,a}\tau_{\ell,b}\tau_{0,B}\tau_{0,a_1}\dots\tau_{0,a_n}\rangle_0^X \\
&= \eta^{AB}\langle\langle \tau_{m-1,c}\tau_{0,A}\rangle\rangle_0^X \tau_{0,B}B_{k,a;\ell,b}.
\end{aligned}$$

On the other hand, we have

$$\begin{aligned}\tau_{m,c}A_{k,a;\ell,b} &= \langle\langle\tau_{k,a}\tau_{\ell,b}\tau_{m,c}\rangle\rangle_0^X = \eta^{AB}\langle\langle\tau_{m-1,c}\tau_{0,A}\rangle\rangle_0^X\langle\langle\tau_{k,a}\tau_{\ell,b}\tau_{0,B}\rangle\rangle_0^X \\ &= \eta^{AB}\langle\langle\tau_{m-1,c}\tau_{0,A}\rangle\rangle_0^X\tau_{0,B}A_{k,a;\ell,b}.\end{aligned}$$

Induction in the order of vanishing of $A_{k,a;\ell,b}-B_{k,a;\ell,b}$ in the variables $\{t_m^a \mid m>0\}$ shows that the two power series are equal. □

4.6. The Dubrovin connection on H. Let $\mathcal{M}$ be the endomorphism of the cotangent bundle of H with components $\mathcal{M}_a^b = \eta^{bc}\langle\langle\tau_{0,a}\tau_{0,c}\rangle\rangle_0^X$; by Proposition 4.6, it is a function on H. We may now define the Dubrovin connection $\nabla^z = \nabla + z\mathcal{A}$ on the small phase space H: the tensor $\mathcal{A}$ is defined by the formula $\mathcal{A}_X = X(\mathcal{M})$.

PROPOSITION 4.7. $\mathcal{A}_X = \mathcal{X}^{-1}\tau_{0,X}\mathcal{M} = (\tau_{0,e}\mathcal{X})^{-1}(\tau_{0,X}\mathcal{M})$

PROOF. By (2.3), we have $\mathcal{A}_a = (\mathcal{X}^{-1})_a^b\tau_{0,b}\mathcal{M} = \mathcal{X}^{-1}\tau_{0,a}\mathcal{M}$. □

We now check that $\nabla^z = \nabla + z\mathcal{A}$ is a Dubrovin connection. The equation $[\nabla_a,\mathcal{A}_b] = [\nabla_b,\mathcal{A}_a]$ follows from the fact that the tensor $\mathcal{A}$ is the covariant derivative of $\mathcal{M}$. The equation $[\mathcal{A}_a,\mathcal{A}_b]=0$ is equivalent to the Witten-Dijkgraaf-Verlinde-Verlinde (WDVV) equation

$$\eta^{AB}\langle\langle\tau_{0,a}\tau_{0,b}\tau_{0,A}\rangle\rangle_0^X\langle\langle\tau_{0,B}\tau_{0,c}\tau_{0,d}\rangle\rangle_0^X = \eta^{AB}\langle\langle\tau_{0,a}\tau_{0,c}\tau_{0,A}\rangle\rangle_0^X\langle\langle\tau_{0,B}\tau_{0,b}\tau_{0,d}\rangle\rangle_0^X.$$

This equation is proved by applying the vector field $\tau_{0,b}$ to the topological recursion relation (4.9):

$$\langle\langle\tau_{1,a}\tau_{0,c}\tau_{0,d}\rangle\rangle_0 = \eta^{AB}\langle\langle\tau_{0,a}\tau_{0,A}\rangle\rangle_0\langle\langle\tau_{0,B}\tau_{0,c}\tau_{0,d}\rangle\rangle_0.$$

This gives

$$\begin{aligned}&\eta^{AB}\langle\langle\tau_{0,a}\tau_{0,b}\tau_{0,A}\rangle\rangle_0\langle\langle\tau_{0,B}\tau_{0,c}\tau_{0,d}\rangle\rangle_0 \\ &\qquad = \langle\langle\tau_{1,a}\tau_{0,b}\tau_{0,c}\tau_{0,d}\rangle\rangle_0 - \eta^{AB}\langle\langle\tau_{0,a}\tau_{0,A}\rangle\rangle_0\langle\langle\tau_{0,B}\tau_{0,b}\tau_{0,c}\tau_{0,d}\rangle\rangle_0;\end{aligned}$$

the left-hand side of this equation is symmetric in the indices b and c, since the right-hand side is, and this is precisely the WDVV equation.

The identity vector field e of the Dubrovin connection is the constant vector field ∂_e in the direction of the coordinate u^e. The equation $\mathcal{A}_e = e\mathcal{M} = I$ is obtained by applying the differential operator $\tau_{0,a}\tau_{0,b}$ to the puncture equation $e\mathcal{F}_0^X = \frac{1}{2}\eta_{ab}t_0^at_0^b = 0$.

4.7. The Euler vector field on H. Using Hori's equation, we may construct an Euler vector field E on H which along with the Dubrovin connection ∇^z and the identity vector field e makes it into a conformal Dubrovin manifold. To see this, apply the differential operator $\tau_{0,b}\tau_{0,c}$ to the equation $\mathcal{L}_0\mathcal{F}_0^X + \frac{1}{2}R_{ab}t_0^at_0^b = 0$: we obtain

$$\mathcal{L}_0\langle\langle\tau_{0,b}\tau_{0,c}\rangle\rangle_0^X + (p_b+p_c+1-r)\langle\langle\tau_{0,b}\tau_{0,c}\rangle\rangle_0^X + R_{bc} = 0. \tag{4.11}$$

In particular,

$$\mathcal{L}_0u^a + (1-p_a)u^a + R_e^a = \mathcal{L}_0u^a + Eu^a = 0,$$

where E is the vector field

$$E = \sum_a \big((1 - p_a)u^a + R^a_e\big)\partial_a.$$

Let μ be the endomorphism of the cotangent bundle of H with components

$$\mu^b_a = \delta^b_a(p_a - r/2),$$

corresponding to the identification of the constant r in the definition of the conformal Dubrovin manifold M with the dimension of X. By (4.11), we see that

$$\mathcal{L}_0(\mathcal{M}) + [\mu, \mathcal{M}] + \mathcal{M} + R = 0.$$

Applying the vector field ∂ to (4.11), we see that $\mathcal{L}_0(\mathcal{X}) + [\mu, \mathcal{X}] + \frac{3-r}{2}\mathcal{X} = 0$, and hence that

$$\mathcal{L}_0(\mathcal{X}^{-1}) + [\mu, \mathcal{X}^{-1}] - \tfrac{3-r}{2}\mathcal{X}^{-1} = 0.$$

It follows that

$$\begin{aligned}
E(\mathcal{A}_a) &= -\mathcal{L}_0(\mathcal{X}^{-1}\tau_{0,a}\mathcal{M}) \\
&= -\mathcal{L}_0(\mathcal{X}^{-1})\tau_{0,a}\mathcal{M} - \mathcal{X}^{-1}\tau_{0,a}\mathcal{L}_0(\mathcal{M}) - \mathcal{X}^{-1}[\mathcal{L}_0, \tau_{0,a}]\mathcal{M} \\
&= \big([\mu, \mathcal{X}^{-1}]\tau_{0,a}\mathcal{M} - \tfrac{3-r}{2}\mathcal{A}_a\big) + \big(\mathcal{X}^{-1}\tau_{0,a}[\mu, \mathcal{M}] + \mathcal{A}_a\big) + \big(p_a + \tfrac{1-r}{2}\big)\mathcal{A}_a \\
&= [\mu, \mathcal{A}_a] + p_a\mathcal{A}_a.
\end{aligned}$$

In other words, $[E, \mathcal{A}^c_{ab}] = (p_a + p_b - p_c)\mathcal{A}^c_{ab}$; this implies (3.1), since on the one hand,

$$[E, \partial_a \circ \partial_b] = [E, \mathcal{A}^c_{ab}\partial_c] = \big(E(\mathcal{A}^c_{ab}) - (1 - p_c)\big)\partial_c,$$

while on the other hand,

$$[E, \partial_a] \circ \partial_b + \partial_a \circ [E, \partial_b] + \partial_a \circ \partial_b = \big(-(1 - p_a) - (1 - p_b) + 1\big)\mathcal{A}^c_{ab}\partial_c.$$

This establishes that E is an Euler vector field on the small phase space H.

Observe that Hori's vector field $\mathcal{L}_0$ agrees with the vector field

$$\mathcal{L}_0 = \operatorname{Res}\langle \mathbf{s}(z), z\delta_z\sigma(z)\rangle$$

associated to the conformal Dubrovin manifold $(\mathsf{H}, \nabla^z, e, E)$. Historically, Hori's equation was an important step in the discovery of the Lie algebra of vector fields $\mathcal{L}_k$.

4.8. The fundamental solution on H. It is remarkable that the small phase space H has a canonical conformal fundamental solution; this was first shown by Dubrovin (see [**9**], Chapter 6).

PROPOSITION 4.8. *The matrix*

$$\Theta^b_a(z) = \delta^b_a + \eta^{bc}\sum_{k=0}^{\infty} z^{k+1}\langle\langle \tau_{k,a}\tau_{0,c}\rangle\rangle^X_0$$

is a conformal fundamental solution on H.

PROOF. The equation $\partial_a \Theta_k = \Theta_{k-1} \mathcal{A}_a$ is an instance of the genus 0 topological recursion relation:

$$\langle\langle \tau_{k,a} \tau_{0,b} \tau_{0,c} \rangle\rangle_0^X = \eta^{AB} \langle\langle \tau_{k-1,a} \tau_{0,A} \rangle\rangle_0^X \langle\langle \tau_{0,B} \tau_{0,b} \tau_{0,c} \rangle\rangle_0^X .$$

It remains to show that the fundamental solution $\Theta(z)$ is conformal. Applying the differential operator $\tau_{k,a}\tau_{0,b}$ to Hori's equation in genus 0, we see that

$$\mathcal{L}_0 \langle\langle \tau_{k,a} \tau_{0,b} \rangle\rangle_0 + (k+1+p_a+p_b-r) \langle\langle \tau_{k,a} \tau_{0,b} \rangle\rangle_0 + \delta_{k,0} R_{ab} = 0.$$

Multiplying by z^{k+1} and summing over k, we see that

$$E\Theta(z) = z\partial_z \Theta(z) + [\mu, \Theta(z)] + zR.$$

On the other hand, $E\Theta(z) = z\Theta(z)\mathcal{U}$, where $\mathcal{U} = E\mathcal{M}$. From this equation, we see that the operator $\tilde{\delta}_z$ equals

$$\tilde{\delta}_z = \partial_z + z^{-1}(\mu + \tfrac{1}{2}) + R.$$

Since $[\mu, R] = R$, we see that $\Theta(z)$ is conformal; this equation is the statement that multiplication by $c_1(X)$ in the Dolbeault cohomology raises degree by $(1,1)$. □

4.9. Intrinsic formulation of topological recursion relations. We now come to the main result of this paper, an expression for the topological recursion relations among Gromov-Witten invariants in terms of the intrinsic vector fields $\sigma_{k,a}$ on the jet-space $J_\infty^\circ \mathsf{H}$ of the small phase space H.

Introduce functions $\langle\langle \sigma_{k_1,a_1} \dots \sigma_{k_n,a_n} \rangle\rangle_g^X$ on the large phase space by the formula

$$\langle\langle \sigma_{k_1,a_1} \dots \sigma_{k_n,a_n} \rangle\rangle_g^X = \sum_{\beta \in H_2(X,\mathbb{Z})} q^\beta \sum_{N=0}^\infty \frac{1}{N!} \sum_{\substack{b_1 \dots b_N \\ \ell_1 \dots \ell_N}} t_{\ell_1}^{b_1} \dots t_{\ell_N}^{b_N}$$
$$\int_{[\overline{\mathcal{M}}_{g,n,N}(X,\beta)]^{\mathrm{virt}}} \psi_1^{k_1} \dots \psi_n^{k_n} \Psi_{n+1}^{\ell_1} \dots \Psi_{n+N}^{\ell_N}$$
$$\operatorname{ev}_1^* \gamma_{a_1} \dots \operatorname{ev}_n^* \gamma_{a_n} \operatorname{ev}_{n+1}^* \gamma_{b_1} \dots \operatorname{ev}_{n+N}^* \gamma_{b_N} .$$

THEOREM 4.9. *If $k_i > 0$, then $\langle\langle \sigma_{k_1,a_1} \dots \sigma_{k_n,a_n} \rangle\rangle_g^X = \sigma_{k_1,a_1} \dots \sigma_{k_n,a_n} \mathcal{F}_g^X$.*

PROOF. Form the generating function

$$\mathcal{F}_{g|a_1\dots a_n}^X(z_1,\dots,z_n) = \sum_{k_1,\dots,k_n=0}^\infty z_1^{k_1} \dots z_n^{k_n} \langle\langle \sigma_{k_1,a_1} \dots \sigma_{k_n,a_n} \rangle\rangle_g^X$$
$$= \sum_{\beta \in H_2(X,\mathbb{Z})} q^\beta \sum_{N=0}^\infty \frac{1}{N!} \sum_{\substack{b_1 \dots b_N \\ \ell_1 \dots \ell_N}} t_{\ell_1}^{b_1} \dots t_{\ell_N}^{b_N}$$
$$\int_{[\overline{\mathcal{M}}_{g,n,N}(X,\beta)]^{\mathrm{virt}}} \frac{\Psi_{n+1}^{\ell_1} \dots \Psi_{n+N}^{\ell_N} \operatorname{ev}_1^* \gamma_{a_1} \dots \operatorname{ev}_n^* \gamma_{a_n} \operatorname{ev}_{n+1}^* \gamma_{b_1} \dots \operatorname{ev}_{n+N}^* \gamma_{b_N}}{(1-z_1\psi_1)\dots(1-z_n\psi_n)} .$$

Corollary 4.5 implies that

$$\Theta_{a_1}^{b_1}(z_1) \dots \Theta_{a_n}^{b_n}(z_n) \mathcal{F}_{g|b_1\dots b_n}^X(z_1,\dots,z_n) = \tau_{a_1}(z_1) \dots \tau_{a_n}(z_n) \mathcal{F}_g^X .$$

By the relation $\sigma_a(z) = \Theta_a^b(z)\tau_b(z)$, it follows that

$$\Theta_{a_2}^{b_2}(z_2)\dots\Theta_{a_n}^{b_n}(z_n)\mathcal{F}^X_{g|a_1b_2\dots b_n}(z_1,\dots,z_n) = \sigma_{a_1}(z_1)\tau_{a_2}(z_2)\dots\tau_{a_n}(z_n)\mathcal{F}^X_g.$$

Since

$$\sigma_{a_1}(z_1)\cdot\Theta_{a_2}^{b_2}(z_2) = \Theta_{a_2}^{b_2}(z_2)\cdot\big(\sigma_{a_1}(z_1)+z_2\mathcal{A}_{a_1}\big),$$

we see that

$$\Theta_{a_3}^{b_3}(z_3)\dots\Theta_{a_n}^{b_n}(z_n)\mathcal{F}^X_{g|a_1a_2b_3\dots b_n}(z_1,\dots,z_n) \\ = \big(\sigma_{a_1}(z_1)+z_2\mathcal{A}_{a_1}\big)\sigma_{a_2}(z_2)\tau_{a_3}(z_3)\dots\tau_{a_n}(z_n)\mathcal{F}^X_g.$$

Continuing in the same vein, we see that

$$\mathcal{F}^X_{g|a_1\dots a_n}(z_1,\dots,z_n) \\ = \big(\sigma_{a_1}(z_1)+(z_2+\dots+z_n)\mathcal{A}_{a_1}\big)\big(\sigma_{a_{n-1}}(z_{n-1})+z_n\mathcal{A}_{a_{n-1}}\big)\sigma_{a_n}(z_n)\mathcal{F}^X_g.$$

The proposition follows on expanding in the parameters z_i. □

This theorem immediately implies Theorem 1.1. Let $g>0$, so that the moduli space $\overline{\mathcal{M}}_{g,n}$ has dimension $3g-3+n$; it follows that when $k_1+\dots+k_n>3g-3+n$,

$$\sigma_{k_1}\dots\sigma_{k_n}\mathcal{F}^X_g = \langle\langle\sigma_{k_1,a_1}\dots\sigma_{k_n,a_n}\rangle\rangle^X_g = 0.$$

Hence, by Proposition 2.3, we see that when $k_1+\dots+k_n>3g-3+n$,

$$\tau_{k_1,a_1}\dots\tau_{k_n,a_n}\mathcal{F}^X_g = 0.$$

Theorem 4.9 gives rise to a simple procedure for translating identities among cycles on Deligne-Mumford moduli spaces into differential equations among Gromov-Witten potentials. For example, the identity on $\overline{\mathcal{M}}_{1,1}$

(4.12) $$\psi_1\cap[\overline{\mathcal{M}}_{1,1}] = \frac{1}{24}$$ 

where the vertices denoted by open circles are of genus 1, translates into the topological recursion relation

$$\langle\langle\sigma_{1,a}\rangle\rangle^X_1 = \tfrac{1}{24}\langle\langle\mathcal{O}_a\mathcal{O}_b\mathcal{O}^b\rangle\rangle^X_0.$$

Here, $\mathcal{O}_a$ is an abbreviation for $\sigma_{0,a}=\tau_{0,a}$. This is the case $k=1$ of the following equation:

(4.13) $$\sigma_{k,a}\mathcal{F}_1 = \begin{cases}\frac{1}{24}\,\sigma_{0,a}\operatorname{Tr}(\mathcal{M}) & k=1,\\ 0 & k>1.\end{cases}$$

(The cases $k>1$ reflect the fact that $\overline{\mathcal{M}}_{1,1}$ has dimension 1.)

LEMMA 4.10. *The equation* (4.13) *has a particular solution*

$$\tfrac{1}{24}\log\det(\mathcal{X}) \in \widehat{\mathcal{O}}.$$

(Note that while $\log\det(\mathcal{X})$ *is not itself an element of* $\mathcal{O}^\circ_\infty$*, its derivatives are.)*

PROOF. Clearly, $\sigma_{k,a} \log\det(\mathcal{X}) = 0$ for $k > 1$, while

$$\begin{aligned}\sigma_{1,a} \log\det(\mathcal{X}) &= \operatorname{Tr}\big(\mathcal{X}^{-1}(\mathcal{X}^2)^b_a \partial_{1,b}\mathcal{X}\big) = \operatorname{Tr}\big(\mathcal{X}^{-1}(\mathcal{X}^2)^b_a \mathcal{A}_b\big) \\ &= \operatorname{Tr}\big(\mathcal{X}\mathcal{A}_a\big) = \sigma_{0,a} \operatorname{Tr}(\mathcal{M}). \qquad \square\end{aligned}$$

Let $\mathcal{G} = \mathcal{F}_1 - \frac{1}{24}\log\det(\mathcal{X})$; we see that $\sigma_{k,a}\mathcal{G} = 0$ for all $k > 0$; hence, by Corollary 2.3, we recover a result of Dijkgraaf and Witten [**7**]: there is a function $\mathcal{G} \in \mathcal{O}$ such that

$$\mathcal{F}_1 = \mathcal{G} + \tfrac{1}{24}\log\det(\mathcal{X}).$$

The dilaton equation $\mathcal{Q}\mathcal{F}_1 = \frac{1}{24}\chi(X)$ follows automatically from the formulas $\mathcal{Q}\mathcal{G} = 0$ and

$$\mathcal{Q}\log\det(\mathcal{X}) = \operatorname{Tr}(\mathcal{X}^{-1}\mathcal{Q}\mathcal{X}) = \chi.$$

To give a second example, the identity on $\overline{\mathcal{M}}_{2,1}$

$\psi_1^2 \cap [\overline{\mathcal{M}}_{2,1}] =$ 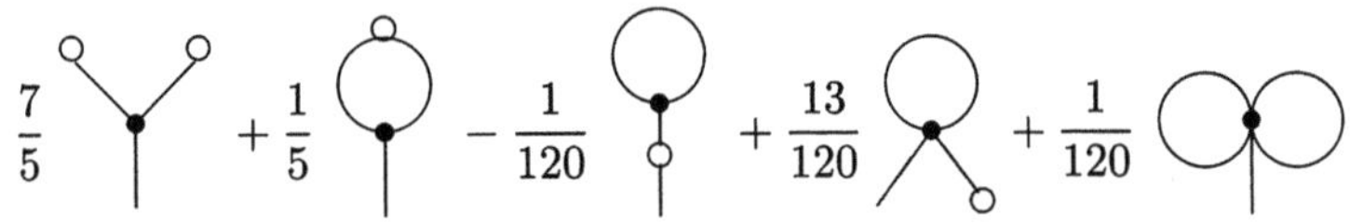

translates into the topological recursion relation

$$\begin{aligned}\langle\langle\sigma_{2,a}\rangle\rangle_2^X = \tfrac{7}{5}&\langle\langle\mathcal{O}_a\mathcal{O}^b\mathcal{O}^c\rangle\rangle_0^X\langle\langle\mathcal{O}_b\rangle\rangle_1^X\langle\langle\mathcal{O}_c\rangle\rangle_1^X + \tfrac{1}{5}\langle\langle\mathcal{O}_a\mathcal{O}^b\mathcal{O}^c\rangle\rangle_0^X\langle\langle\mathcal{O}_b\mathcal{O}_c\rangle\rangle_1^X \\ &- \tfrac{1}{120}\langle\langle\mathcal{O}_a\mathcal{O}_b\rangle\rangle_1^X\langle\langle\mathcal{O}^b\mathcal{O}^c\mathcal{O}_c\rangle\rangle_0^X + \tfrac{13}{120}\langle\langle\mathcal{O}_a\mathcal{O}_b\mathcal{O}_c\mathcal{O}^c\rangle\rangle_0^X\langle\langle\mathcal{O}^b\rangle\rangle_0^X \\ &+ \tfrac{1}{120}\langle\langle\mathcal{O}_a\mathcal{O}_b\mathcal{O}^b\mathcal{O}_c\mathcal{O}^c\rangle\rangle_0^X.\end{aligned}$$

5. Virasoro constraints on the jet-space of a Frobenius manifold

5.1. Frobenius manifolds. A Frobenius manifold is a conformal Dubrovin manifold

$$(M, \nabla^z, e, E, r)$$

together with a non-degenerate symmetric bilinear form η on the tangent bundle (also know as a pseudo-Rimannian metric) such that

1) ∇ is the Levi-Cività connection associated to η;
2) $\eta(X \circ Y, Z) = \eta(X, Y \circ Z)$, or equivalently, $\mathcal{A}_X^* = \mathcal{A}_X$;
3) $E\eta(X,Y) = \eta([E,X],Y) + \eta(X,[E,Y]) + (2-r)\eta(X,Y)$, that is, $\mathcal{L}_E\eta = (2-r)\eta$.

On a Frobenius manifold, the endomorphism μ of the cotangent bundle is skew-adjoint:

$$\begin{aligned}&\eta(\mu^*(X),Y) + \eta(X,\mu^*(Y)) = (2-r)\eta(X,Y) + \eta(\nabla_X E, Y) + \eta(X, \nabla_Y E) \\ &= (2-r)\eta(X,Y) + \eta(\nabla_E X, Y) + \eta(X, \nabla_E Y) - \eta([E,X],Y) - \eta(X,[E,Y]) = 0.\end{aligned}$$

Here, we have used that ∇ is torsion-free and compatible with the metric η.

If A is a linear operator on the bundle $\mathcal{H} = T^*M[z,z^{-1}]$, denote by A^+ its adjoint with respect to the symmetric bilinear form

$$\boldsymbol{\eta}(f,g) = \operatorname{Res}_{z=0}\eta(f(-z),g(z))\frac{dz}{z}.$$

For example, $\Theta^+(z) = \Theta^*(-z)$, $\partial_z^+ = \partial + z^{-1}$ and $z^+ = -z$. Since $\mu^* = -\mu$ and $\mathcal{U}^+ = \mathcal{U}$, it follows that

$$\delta_z^+ = (\partial_z + z^{-1}) - z^{-1}(\mu^* + \tfrac{1}{2}) + \mathcal{U}^* = (\partial_z + z^{-1}) + z^{-1}(\mu - \tfrac{1}{2}) + \mathcal{U} = \delta_z.$$

If $\Theta(z)$ is a fundamental solution, recall that the endomorphism $R(z)$ of $\mathcal{H} = T^*M[z]$ is defined by the formula

$$\partial_z\Theta(z) + z^{-1}R(z)\Theta(z) = z^{-1}(\mu + \tfrac{1}{2})\Theta(z) + \mathcal{U}\Theta(z).$$

DEFINITION 5.1. *A fundamental solution $\Theta(z)$ on a Frobenius manifold M is symmetric if it is conformal and $\Theta^+(z)\Theta(z) = I$.*

It follows from the condition $\Theta^+(z)\Theta(z) = I$ that $R(z) + R^+(z) = 0$, in other words, $R_k^* = (-1)^{k+1}R_k$ for all $k > 0$.

PROPOSITION 5.1. *There exists a symmetric fundamental solution on any Frobenius manifold,*

PROOF. Let $\Theta(z)$ be a conformal fundamental solution. Since $\nabla \cdot \Theta(z) = \Theta(z) \cdot \nabla^z$, we have

$$\begin{aligned} 0 &= \eta(\alpha(-z), \Theta(z) \cdot \nabla^z\beta(z)) - \eta(\alpha(-z), \nabla \cdot \Theta(z)\beta(z)) \\ &= \eta(\Theta^*(z)\alpha(-z), \nabla^z\beta(z)) - d\eta(\alpha(-z), \Theta(z)\beta(z)) + \eta(\nabla\alpha(-z), \Theta(z)\beta(z)) \\ &= -\eta(\nabla^{-z} \cdot \Theta^*(z)\alpha(-z), \beta(z)) + \eta(\Theta(z) \cdot \nabla\alpha(-z), \beta(z)). \end{aligned}$$

This shows that $\nabla^{-z} \cdot \Theta^*(z) = \Theta^*(z) \cdot \nabla$; conjugating by $\Theta^*(z)$, we obtain the equation

$$\nabla \cdot \Theta^+(-z) = \Theta^+(z) \cdot \nabla^{-z}.$$

Replacing z by $-z$, we see that $\Theta^+(z)^{-1}$ is also a fundamental solution.

It follows that $\rho(z) = \Theta^+(z)\Theta(z)$ is a flat section of $\Gamma(M, \mathrm{End}(TM))[\![z]\!]$. Replacing $\Theta(z)$ by $\tilde{\Theta}(z) = \Theta(z)\rho(z)^{-1/2}$, we obtain a new fundamental solution. Since $\rho^+(z) = \rho(z)$, we see that

$$\tilde{\Theta}^+(z)\tilde{\Theta}(z) = \rho^+(z)^{-1/2}\Theta^+(z)\Theta(z)\rho(z)^{-1/2} = I,$$

hence $\tilde{\Theta}(z)$ is symmetric. □

In the remainder of this section, we only consider symmetric fundamental solutions.

5.2. Gromov-Witten invariants and Frobenius manifolds. In Section 4, we showed that the small phase space H associated to the genus 0 Gromov-Witten invariants of a projective manifold X is conformal Dubrovin manifold. It also carries a flat metric η; with respect to this metric, it is a Frobenius manifold. The equation $\mathcal{L}_E\eta = (2 - r)\eta$ is equivalent to $(p_a + p_b)\eta_{ab} = r\eta_{ab}$, which is a basic property of the Poincaré form of a projective manifold of dimension r.

PROPOSITION 5.2. *The conformal fundamental solution*

$$\Theta_a^b(z) = \delta_a^b + \eta^{bc}\sum_{k=0}^{\infty} z^{k+1}\langle\langle\tau_{k,a}\tau_{0,c}\rangle\rangle_0^X$$

on the small phase space is symmetric.

PROOF. Let $\rho(z) = \Theta^+(z)\Theta(z)$. We have

$$\begin{aligned}\partial_a\rho^+(z) &= \partial_a\Theta^+(z)\Theta(z) + \Theta^+(z)\partial_a\Theta(z)\\ &= -z\Theta^+(z)\mathcal{A}_a\Theta(z) + z\Theta^+(z)\mathcal{A}_a\Theta(z) = 0.\end{aligned}$$

Thus $\rho(z)$ is constant on H. It remains to show that it equals I at the origin $0 \in \mathsf{H}$. We may write

$$\rho(z)(0) = I + \sum_{k=0}^{\infty} z^{k+1} \sum_{\beta\neq 0} q^\beta \rho_{k,\beta}.$$

For $\omega \in H^2(X,\mathbb{C})$, the divisor equation implies that

$$\textstyle\int_\beta \omega\, \rho_{k,\beta} = [R(\omega), \rho_{k-1,\beta}],$$

from which it follows that $\rho_{k,\beta} = 0$ for $\beta \neq 0$. On the other hand, in the limit $q \to 0$ in the Novikov ring, the value of Θ_k at the origin of H is an integral over $\overline{\mathcal{M}}_{0,2}(X,0)$, hence vanishes, since $\overline{\mathcal{M}}_{0,2}(X,0)$ is empty. □

5.3. The genus 0 potential on the large phase space of a Frobenius manifold. On a Frobenius manifold, we may define a power series $\mathcal{F}_0$ on the large phase space by integrating the power series $\mathbf{t}(z)$:

$$\tau_{k,a}\mathcal{F}_0 = (-1)^k \eta_{ab} \mathbf{t}^b_{-k-1}.$$

Denote $\eta_{bc}\Omega^c_{j,k|a}$ by $\Omega_{j,k|a,b}$; taking another derivative, we have

$$\tau_{j,a}\tau_{k,b}\mathcal{F}_0 = (-1)^k \Omega_{j,k|a,b}, \tag{5.1}$$

This determines $\mathcal{F}_0$ up to an affine function on the large phase space, which may be fixed by the dilaton equation

$$\mathcal{Q}\mathcal{F}_0 = -2\mathcal{F}_0.$$

The following formula $\mathcal{F}_0$ is due to Dubrovin [**9**].

PROPOSITION 5.3. $\mathcal{F}_0 = \frac{1}{2}\sum_{\ell,m=0}^{\infty} (-1)^m\, \mathbf{t}^p_\ell\, \mathbf{t}^q_m\, \Omega_{\ell,m|p,q}$

PROOF. Since the fundamental solution $\Theta(z)$ is symmetric, we see that

$$(y-z)\Omega^*(y,z) = \Theta^*(z)^{-1}\Theta^*(y) - I = \Theta(-z)\Theta^{-1}(-y) - I = (y-z)\Omega(-z,-y),$$

in other words, $\Omega_{j,k|a,b} = (-1)^{j+k}\Omega_{k,j|b,a}$. It follows by Lemma 2.9 that

$$\tau_{k,b}\Omega_{\ell,m|p,q} = (-1)^{\ell+m}\tau_{k,b}\Omega_{m,\ell|q,p} = (-1)^{\ell+m}\tau_{m,q}\Omega_{k,\ell|b,p} = (-1)^{k+m}\tau_{m,q}\Omega_{\ell,k|p,b}.$$

We have

$$\begin{aligned}
\tau_{k,b}\mathcal{F}_0 &= \tfrac{1}{2}\sum_{\ell,m}(-1)^m \mathbf{t}_\ell^p\,\mathbf{t}_m^q\,\tau_{k,b}\Omega_{\ell,m|p,q} \\
&\quad + \tfrac{1}{2}(-1)^k\sum_\ell \mathbf{t}_\ell^p\,\Omega_{\ell,k|p,b} + \tfrac{1}{2}\sum_m(-1)^m\mathbf{t}_m^q\,\Omega_{k,m|b,q} \\
&= \tfrac{1}{2}(-1)^k\sum_{\ell,m} \mathbf{t}_\ell^p\,\mathbf{t}_m^q\,\tau_{m,q}\Omega_{\ell,k|p,b} \\
&\quad + \tfrac{1}{2}(-1)^k\sum_\ell \mathbf{t}_\ell^p\,\Omega_{\ell,k|p,b} + \tfrac{1}{2}\sum_m(-1)^m\mathbf{t}_m^q\,\Omega_{k,m|b,q} \\
&= \tfrac{1}{2}(-1)^k\sum_\ell \mathbf{t}_\ell^p\,(1-\mathcal{Q})\Omega_{\ell,k|p,b} + \tfrac{1}{2}\sum_m(-1)^m\mathbf{t}_m^q\,\Omega_{k,m|b,q} \\
&= \tfrac{1}{2}(-1)^k\sum_\ell \mathbf{t}_\ell^p\,\Omega_{\ell,k|p,b} + \tfrac{1}{2}\sum_m(-1)^m\mathbf{t}_m^q\,\Omega_{k,m|b,q},
\end{aligned}$$

since $\mathcal{Q}\Omega(y,z)=0$. It follows that

$$\begin{aligned}
\tau_{j,a}\tau_{k,b}\mathcal{F}_0 &= \tfrac{1}{2}(-1)^k\Omega_{j,k|a,b} + \tfrac{1}{2}(-1)^k\sum_\ell \mathbf{t}_\ell^p\,\tau_{j,a}\Omega_{\ell,k|p,b} \\
&\quad + \tfrac{1}{2}(-1)^j\Omega_{k,j|b,a} + \tfrac{1}{2}\sum_m(-1)^m\mathbf{t}_m^q\,\tau_{j,a}\Omega_{k,m|b,q} \\
&= \tfrac{1}{2}(-1)^k\Omega_{j,k|a,b} + \tfrac{1}{2}(-1)^k\sum_\ell \mathbf{t}_\ell^p\,\tau_{\ell,p}\Omega_{j,k|a,b} \\
&\quad + \tfrac{1}{2}(-1)^j\Omega_{k,j|b,a} + \tfrac{1}{2}(-1)^j\sum_m \mathbf{t}_m^q\,\tau_{m,q}\Omega_{k,j|b,a} \\
&= \tfrac{1}{2}(-1)^k(1-\mathcal{Q})\Omega_{j,k|a,b} + \tfrac{1}{2}(-1)^j(1-\mathcal{Q})\Omega_{k,j|b,a} \\
&= \tfrac{1}{2}(-1)^k\Omega_{j,k|a,b} + \tfrac{1}{2}(-1)^j\Omega_{k,j|b,a} = (-1)^k\Omega_{j,k|a,b}.
\end{aligned}$$

This shows that $\mathcal{F}_0$ satisfies (5.1). □

5.4. The Virasoro operators on the large phase space of a Frobenius manifold. Let $\Phi(z)$ be the generating function (or *free field*) whose coefficients are linear differential operators acting on functions on the large phase space

$$\Phi_a(z) = \hbar\sum_{k=0}^\infty z^k\tau_{k,a} + \sum_{k=0}^\infty(-z)^{-k-1}\eta_{ab}\mathbf{t}_k^b.$$

The conjugate of $\Phi(z)$ by $Z_0 = e^{\mathcal{F}_0/\hbar}$ is given by the formula

$$Z_0^{-1}\cdot\Phi_a(z)\cdot Z_0 = \hbar\,\tau_a(z) + \eta_{ab}\mathbf{t}^b(-z).$$

The normal ordering of quadratic expressions in the field $\Phi(z)$ is defined by the formulas

$$:\mathbf{t}_k^a\,\mathbf{t}_\ell^b: = \mathbf{t}_k^a\,\mathbf{t}_\ell^b,\quad :\mathbf{t}_k^a\,\tau_{\ell,b}: = \mathbf{t}_k^a\,\tau_{\ell,b},\quad :\tau_{k,b}\,\mathbf{t}_\ell^a: = \mathbf{t}_\ell^a\,\tau_{k,b},\quad :\tau_{k,a}\,\tau_{\ell,b}: = \tau_{k,a}\,\tau_{\ell,b}.$$

If $k\ge -1$, let L_k be the second-order differential operator on the large phase space

$$L_k = \tfrac{1}{2\hbar}\operatorname{Res}_{z=0}:\eta(\Phi(-z), z\tilde{\delta}_z^{k+1}\Phi(z)): - \tfrac{1}{4}\delta_{k,0}\operatorname{Tr}\big(\mu^2-\tfrac{1}{4}\big).$$

These operators were introduced in the theory of Gromov-Witten invariants by Eguchi, Hori and Xiong [**15**], who proved the following result.

PROPOSITION 5.4. *The operators L_k satisfy the commutation relations*

$$[L_j, L_k] = (j-k)L_{j+k}.$$

PROOF. We give an outline of the proof; see [**20**], Section 2, for the complete details.

Using the canonical commutation relations among the coefficients of $\Phi(z)$, the formula $[L_k, \Phi(z)] + z\tilde{\delta}_z^{k+1}\Phi(z) = 0$ is easily proved. It follows that

$$\begin{aligned}[[L_j, L_k], \Phi(z)] &= [L_j, [L_k, \Phi(z)]] - [L_k, [L_j, \Phi(z)]] \\ &= -[L_j, z\tilde{\delta}_z^{k+1}\Phi(z)] + [L_k, z\tilde{\delta}_z^{j+1}\Phi(z)] \\ &= -z\tilde{\delta}_z^{k+1}[L_j, \Phi(z)] + z\tilde{\delta}_z^{j+1}[L_k, \Phi(z)] \\ &= [z\tilde{\delta}_z^{k+1}, z\tilde{\delta}_z^{j+1}]\Phi(z) = (k-j)\tilde{\delta}_z^{j+k+1}\Phi(z),\end{aligned}$$

and hence that $[L_j, L_k] - (j-k)L_{j+k}$ commutes with $\Phi(z)$.

Any differential operator commuting with $\Phi(z)$ must lie in the centre of the algebra of differential operators, and hence

$$[L_j, L_k] = (j-k)L_{j+k} + c(j,k)1$$

for some two-cocyle $c(j,k)$. Evaluating both sides of this equation at the basepoint of the large phase space, the result follows. □

In the following theorem, we show how the Virasoro operators L_k give rise to intrinsic differential operators on the jet-space of the Frobenius manifold. The explicit formula for $\mathcal{H}_k$ is equivalent to a formula of Liu ([**26**], Theorem 4.4).

THEOREM 5.5. *We have*

$$L_k \cdot Z_0 = Z_0 \cdot \big(\hbar\,\Delta_k + \mathcal{L}_k + \mathcal{H}_k\big),$$

where $\Delta_k = \frac{1}{2}\operatorname{Res}_{z=0}\eta\big(\sigma(-z), z\delta_z^{k+1}\sigma(z)\big)$, *and*

$$\mathcal{H}_k = -\tfrac{1}{4}\sum_{\ell=0}^{k}\operatorname{Tr}\big((\mu-\tfrac{1}{2})\mathcal{U}^\ell(\mu+\tfrac{1}{2})\mathcal{U}^{k-\ell}\big).$$

PROOF. We have

$$\begin{aligned}Z_0^{-1}\cdot\tfrac{1}{2\hbar}:\eta(\Phi(-z), z\tilde{\delta}_z^{k+1}\Phi(z)):\cdot Z_0 &= \tfrac{1}{2}\hbar\,\eta\big(\tau(-z), z\tilde{\delta}_z^{k+1}\tau(z)\big) \\ &+ \tfrac{1}{2}\big\langle \mathbf{t}(z), z\tilde{\delta}_z^{k+1}\tau(z)\big\rangle + \tfrac{1}{2}\big\langle z\tilde{\delta}_z^{k+1}\mathbf{t}_-(-z), \tau(-z)\big\rangle + \tfrac{1}{2}\big\langle\tau(-z), z\tilde{\delta}_z^{k+1}\mathbf{t}_+(-z)\big\rangle \\ &+ \tfrac{1}{2\hbar}\eta\big(\mathbf{t}(z), z\tilde{\delta}_z^{k+1}\mathbf{t}(-z)\big).\end{aligned}$$

On taking the residue, the first term on the right-hand side yields

$$\begin{aligned}&\tfrac{1}{2}\operatorname{Res}_{z=0}\eta\big(\tau(-z), z\tilde{\delta}_z^{k+1}\tau(z)\big) \\ &\qquad = \tfrac{1}{2}\operatorname{Res}_{z=0}\eta\big(\Theta^{-1}(-z)\sigma(-z), \Theta^{-1}(z)z\delta_z^{k+1}\sigma(z)\big) = \Delta_k.\end{aligned}$$

The residue of the last term on the right-hand side vanishes by Theorem 2.11 (this is the genus 0 Virasoro constraint), since

$$\begin{aligned}\eta\big(\mathbf{t}(z), z\tilde{\delta}_z^{k+1}\mathbf{t}(-z)\big) &= \eta\big(\Theta(z)\mathbf{t}(z), \Theta^{-1}(z)z\tilde{\delta}_z^{k+1}\mathbf{t}(-z)\big)\\ &= \eta\big(\mathbf{s}(z), z\delta_z^{k+1}\mathbf{s}(-z)\big) = O(z^{-3}).\end{aligned}$$

It follows that

$$\begin{aligned}&Z_0^{-1}\cdot L_k\cdot Z_0 - \hbar\,\Delta_k + \tfrac14\delta_{k,0}\operatorname{Tr}\big(\mu^2-\tfrac14\big)\\ &\quad= \tfrac12\operatorname{Res}_{z=0}\big\langle \mathbf{t}(z), z\tilde{\delta}_z^{k+1}\tau(z)\big\rangle\\ &\qquad+\tfrac12\operatorname{Res}_{z=0}\Big(\big\langle z\tilde{\delta}_z^{k+1}\mathbf{t}_-(-z),\tau(-z)\big\rangle + \big\langle \tau(-z), z\tilde{\delta}_z^{k+1}\mathbf{t}_+(-z)\big\rangle\Big)\\ &\quad= \tfrac12\operatorname{Res}_{z=0}\Big(\big\langle \mathbf{t}(z), z\tilde{\delta}_z^{k+1}\tau(z)\big\rangle + \big\langle \mathbf{t}_-(z), z\tilde{\delta}_z^{k+1}\tau(z)\big\rangle + \big\langle z\tilde{\delta}_z^{k+1}\tau(z), \mathbf{t}_+(z)\big\rangle\Big)\\ &\quad= \operatorname{Res}_{z=0}\big\langle \mathbf{t}(z), z\tilde{\delta}_z^{k+1}\tau(z)\big\rangle - \tfrac12\operatorname{Res}_{z=0}\operatorname{Tr}\big(y\tilde{\delta}_y^{k+1}\Omega(y,z)\big)_{y=z},\end{aligned}$$

since $\tau(y)\mathbf{t}_+(z) = -\Omega(y,z)$. It remains to show that

$$\operatorname{Res}_{z=0}\operatorname{Tr}\big(y\tilde{\delta}_y^{k+1}\Omega(y,z)\big)_{y=z} = \begin{cases} 0, & k\le 0,\\ -\tfrac12\displaystyle\sum_{\ell=0}^{k}\operatorname{Tr}\big((\mu-\tfrac12)\mathcal{U}^\ell(\mu+\tfrac12)\mathcal{U}^{k-\ell}\big), & k>0.\end{cases}$$

Applying the operator $y\tilde{\delta}_y^{k+2}$ to (2.5), we see that

$$(k+2)y\tilde{\delta}_y^{k+1}\Omega(y,z) + (y-z)y\tilde{\delta}_y^{k+2}\Omega(y,z) = y\tilde{\delta}_y^{k+2}\big(\Theta(y)\Theta^{-1}(z)-I\big).$$

Taking the trace, we see that

$$\begin{aligned}(k+2)\operatorname{Tr}\big(y\tilde{\delta}_y^{k+1}\Omega(y,z)\big) &+ (y-z)\operatorname{Tr}\big(y\tilde{\delta}_y^{k+2}\Omega(y,z)\big)\\ &= \operatorname{Tr}\big(\Theta^{-1}(z)y\tilde{\delta}_y^{k+2}\Theta(y)\big) - \operatorname{Tr}\big(y\tilde{\delta}_y^{k+2}I\big).\end{aligned}$$

Next, we restrict to the diagonal $y=z$:

$$\begin{aligned}(k+2)\operatorname{Tr}\big(y\tilde{\delta}_y^{k+1}\Omega(y,z)\big)_{y=z} &= \operatorname{Tr}\big(\Theta^{-1}(z)z\tilde{\delta}_z^{k+2}\Theta(z)\big) - \operatorname{Tr}\big(z\tilde{\delta}_z^{k+2}I\big)\\ &= \operatorname{Tr}\big(z\delta_z^{k+2}I\big) - \operatorname{Tr}\big(z\tilde{\delta}_z^{k+2}I\big).\end{aligned}$$

Taking the residue, we see that

$$\begin{aligned}(k+2)&\operatorname{Res}_{z=0}\operatorname{Tr}\big(y\tilde{\delta}_y^{k+1}\Omega(y,z)\big)_{y=z}\\ &= \sum_{0\le i+j\le k}\operatorname{Tr}\big(\mathcal{U}^i(\mu-\tfrac12)\mathcal{U}^j(\mu+\tfrac12)\mathcal{U}^{k-i-j}\big) - \operatorname{Res}_{z=0}\operatorname{Tr}\big(z\tilde{\delta}_z^{k+2}I\big).\end{aligned}$$

Since $[\mu_s, R_k] = kR_k$, we see that

$$\operatorname{Tr}\big(z\tilde{\delta}_z^{k+1}I\big) = z^{-k-1}\operatorname{Tr}\big((\mu-\tfrac12-k)\dots(\mu-\tfrac12)(\mu+\tfrac12)\big),$$

and hence $\operatorname{Res}_{z=0}\operatorname{Tr}\big(z\tilde{\delta}_z^{k+1}I\big)=\delta_{k,0}\operatorname{Tr}\big(\mu^2-\frac14\big)$. We have

$$\begin{aligned}\sum_{0\le i+j\le k}&\operatorname{Tr}\big(\mathcal{U}^i(\mu-\tfrac12)\mathcal{U}^j(\mu+\tfrac12)\mathcal{U}^{k-i-j}\big)\\&=\sum_{0\le i+j\le k}\operatorname{Tr}\big(\mathcal{U}^i\mu\mathcal{U}^j\mu\mathcal{U}^{k-i-j}\big)-\tfrac14\sum_{0\le i+j\le k}\operatorname{Tr}\big(\mathcal{U}^k\big)\\&=\big(\tfrac{k}{2}+1\big)\sum_{\ell=0}^{k}\operatorname{Tr}\big(\mu\mathcal{U}^\ell\mu\mathcal{U}^{k-\ell}\big)-\tfrac14\tbinom{k+2}{2}\operatorname{Tr}\big(\mathcal{U}^k\big)\\&=\big(\tfrac{k}{2}+1\big)\sum_{\ell=0}^{k}\operatorname{Tr}\big((\mu-\tfrac12)\mathcal{U}^\ell(\mu+\tfrac12)\mathcal{U}^{k-\ell}\big),\end{aligned}$$

and the formula for $\mathcal{H}_k$ follows. □

As a corollary of this theorem and the Virasoro relations for L_k, we obtain the following formula of Liu ([**26**], Section 6).

COROLLARY 5.6. $\mathcal{L}_j\mathcal{H}_k-\mathcal{L}_k\mathcal{H}_j=(j-k)\mathcal{H}_{j+k}$

5.5. The Virasoro constraints on a Frobenius manifold. The Virasoro constraints are differential equations among a sequence of functions $\mathcal{F}_g$, $g>0$, on the large phase space of a Frobenius manifold M such that the dilaton equation holds:

$$\mathcal{Q}\mathcal{F}_g=(2g-2)\mathcal{F}_g+\tfrac{1}{24}\delta_{g,1}\chi, \tag{5.2}$$

were $\chi=\operatorname{Tr}(I)$. As originally introduced by Eguchi, Hori and Xiong [**15**], in the case where M is the Frobenius manifold associated to the genus 0 Gromov-Witten theory of a projective manifold X and $\mathcal{F}_g$ is the genus g Gromov-Witten potentials X, these differential equations are given by the vanishing of functions $z_{k,g}$ on the large phase space defined as follows:

$$\sum_{g=0}^{\infty}\hbar^{g-1}z_{k,g}=\exp\big(-\textstyle\sum_{g=0}^{\infty}\hbar^{g-1}\mathcal{F}_g\big)\cdot L_k\cdot\exp\big(\textstyle\sum_{g=0}^{\infty}\hbar^{g-1}\mathcal{F}_g\big).$$

Using Theorem 5.5, we may write the vanishing of $z_{k,g}$ intrinsically on the jet-space of the Frobenius manifold M: in fact, we have

$$\begin{aligned}\sum_{g=1}^{\infty}\hbar^{g-1}z_{k,g}=\exp\big(-\textstyle\sum_{g=1}^{\infty}\hbar^{g-1}\mathcal{F}_g\big)\cdot\hbar\,\Delta_k\cdot\exp\big(\textstyle\sum_{g=1}^{\infty}\hbar^{g-1}\mathcal{F}_g\big)&\\+\sum_{g=1}^{\infty}\hbar^{g-1}\mathcal{L}_k\mathcal{F}_g+\mathcal{H}_k&.\end{aligned}$$

This proves the following theorem.

THEOREM 5.7. *For genus $g>1$, the Virasoro constraint $z_{k,g}=0$ is equivalent to the equation*

$$\mathcal{L}_k\mathcal{F}_g+\Delta_k\mathcal{F}_{g-1}+\tfrac12\sum_{i=1}^{g-1}\operatorname{Res}_{z=0}\eta\big(\sigma(-z)\mathcal{F}_i,z\delta_z^{k+1}\sigma(z)\mathcal{F}_{g-i}\big)=0,$$

while the Virasoro constraint $z_{k,1}=0$ is equivalent to the equation

$$\mathcal{L}_k\mathcal{F}_1 = \tfrac{1}{4}\sum_{\ell=0}^{k}\operatorname{Tr}\big((\mu-\tfrac{1}{2})\mathcal{U}^{\ell}(\mu+\tfrac{1}{2})\mathcal{U}^{k-\ell}\big).$$

Using Proposition 3.5, we see that the genus 1 Virasoro constraint $z_{k,1}=0$ is equivalent to the equation

$$\mathcal{L}_k\mathcal{G} = \sum_{\ell=0}^{k}\Big(\tfrac{1}{4}\operatorname{Tr}\big(\mu\mathcal{U}^{\ell}\mu\mathcal{U}^{k-i}\big) + \tfrac{1}{24}\langle e, \mathcal{U}^{\ell}\mu\mathcal{U}^{k-\ell}\omega\rangle\Big) \tag{5.3}$$

for $\mathcal{G} = \mathcal{F}_1^X - \frac{1}{24}\log\det(\mathcal{X})$. This equation was first proved in the special case where M is a semisimple Frobenius manifold by Dubrovin and Zhang [**12**], and extended to general Frobenius manifolds by Liu [**26**].

Let M be a Frobenius manifold. There is a second-order differential operator

$$\Psi : \Gamma(M,\mathcal{O}) \to \Gamma(M, S^4(T^*M)),$$

given by the explicit formula

$$\Psi(X_1,X_2,X_3,X_4) = \tfrac{1}{24}\sum_{\pi\in S_4}\sum_{a_1,a_2,a_3,a_4} X_{\pi_1}^{a_1}X_{\pi_2}^{a_2}X_{\pi_3}^{a_3}X_{\pi_4}^{a_4} \tag{5.4}$$

$$\begin{aligned}\big(3\,\mathcal{A}^{\mu}_{a_1a_2}\mathcal{A}^{\nu}_{a_3a_4}\partial_\mu\partial_\nu - 4\,\mathcal{A}^{\mu}_{a_1a_2}\mathcal{A}^{\nu}_{a_3\mu}\partial_{a_4}\partial_\nu + 2\,\partial_{a_4}\mathcal{A}^{\mu}_{a_2a_3}\mathcal{A}^{\nu}_{a_1\mu}\partial_\nu - \mathcal{A}^{\mu}_{a_1a_2}\partial_\mu\mathcal{A}^{\nu}_{a_3a_4}\partial_\nu \\ + \tfrac{1}{6}\partial_{a_3}\mathcal{A}^{\mu}_{a_1a_2}\partial_{a_4}\mathcal{A}^{\nu}_{\mu\nu} + \tfrac{1}{24}\partial_{a_3}\partial_{a_4}\mathcal{A}^{\mu}_{a_1a_2}\mathcal{A}^{\nu}_{\mu\nu} - \tfrac{1}{4}\partial_\nu\mathcal{A}^{\mu}_{a_1a_2}\partial_\mu\mathcal{A}^{\nu}_{a_3a_4}\big),\end{aligned}$$

such that the generating function $\mathcal{G}$ of genus 1 Gromov-Witten invariants of a compact symplectic manifold X satisfies the equation $\Psi\mathcal{G}=0$ (see [**18**]). Let

$$y_k = \mathcal{L}_k\mathcal{G} - \sum_{\ell=0}^{k}\Big(\tfrac{1}{4}\operatorname{Tr}\big(\mu\mathcal{U}^{\ell}\mu\mathcal{U}^{k-i}\big) + \tfrac{1}{24}\langle e, \mathcal{U}^{\ell}\mu\mathcal{U}^{k-\ell}\omega\rangle\Big).$$

By a lengthy calculation, Liu [**26**] proves the formula

$$y_k = \tfrac{1}{2}\sum_{i=1}^{k}\Psi(E^{\circ k-i+1}, E^{\circ i}, E, E)\mathcal{G} - \tfrac{k+1}{2}\mathcal{L}_{k-1}y_1.$$

In this way, he obtains the following theorem.

THEOREM 5.8. *If $\Psi\mathcal{G}=0$ and $\mathcal{G}$ satisfies* (5.3) *for $k=1$, then $\mathcal{G}$ satisfies* (5.3) *for all $k>1$.*

Relationship to prior results. The results of Sections 5.1 and 5.2 are due to Dubrovin. The formula for the Virasoro operators on the large phase space in terms of free fields (Section 5.3) is a modification of the formalism of Section 2 of [**20**]. (Related formulas have been used by Givental [**21**].)

6. The jet-space of a semisimple Dubrovin manifold

6.1. Semisimple Dubrovin manifolds. A Dubrovin manifold is **semisimple** if the subset $M_0\subset M$ on which the commutative algebra $(T_xM,\circ)$ is semisimple is dense in M. These manifolds were extensively studied by Dubrovin [**9**], who proved the following theorem.

THEOREM 6.1. *Around each point semisimple point in a semisimple Dubrovin manifold, there are coordinates v^i such that the vector fields $\pi_i = \partial/\partial v^i$ satisfy*

$$\pi_i \circ \pi_j = \delta_{ij}\pi_i.$$

PROOF. Locally in M_0, there is a frame $\{\pi_1, \dots, \pi_n\}$ of the tangent bundle TM such that $\pi_i \circ \pi_j = \delta_{ij}\pi_i$. We must show that the vector fields π_i satisfy $[\pi_i, \pi_j] = 0$. If i and j are distinct, the equation

$$\nabla_i(\pi_j \circ \pi_k) - \pi_j \circ (\nabla_i \pi_k) - \nabla_j(\pi_i \circ \pi_k) + \pi_i \circ (\nabla_j \pi_k) = [\pi_i, \pi_j] \circ \pi_k \tag{6.1}$$

is a consequence of condition 4) of Proposition 2.1. If k is not equal to i or j, (6.1) implies that $[\pi_i, \pi_j] \circ \pi_k$ vanishes. On the other hand, if k equals i, we see that

$$-\pi_j \circ \nabla_i \pi_i - \nabla_j \pi_i + \pi_i \circ (\nabla_j \pi_i) = [\pi_i, \pi_j] \circ \pi_i,$$

from which we conclude that $[\pi_i, \pi_j] \circ \pi_i$ vanishes. This shows that $[\pi_i, \pi_j] = 0$. It follows that there are coordinates v^i defined locally on M_0 such that $\pi_i = \partial/\partial v^i$. □

The canonical coordinates (v^i) are seen to be defined up to addition of a constant and permutation. Let $J = (J_i^a)$ be the Jacobian

$$J_i^a = \partial u^a / \partial v^i,$$

and let $J^{-1} = (J_a^i)$ be its inverse

$$J_a^i = \partial v^i / \partial u^a.$$

Denote by v_n^i the functions

$$v_n^i = \partial^n v^i = \partial^{n-1}(J_a^i u_1^a) = \sum_{k=1}^n \binom{n-1}{k-1} (\partial^{n-k} J_a^i) u_k^a.$$

The identity vector field on M_0 is given in canonical coordinates by the formula

$$e = \sum_i \pi_i.$$

6.2. Conformal semisimple Dubrovin manifolds. Let (M, ∇^z, e, E, r) be a semisimple Dubrovin manifold which is conformal. Taking $X = Y = \pi_i$ in (3.1), we see that

$$[E, \pi_i] = 2\pi_i \circ [E, \pi_i] + \pi_i,$$

hence $[E, \pi_i] = -\pi_i$. This shows that, after shifting the canonical coordinates v_i by constants c^i, the Euler vector field on M_0 is given by the formula

$$E = \sum_i v^i \pi_i.$$

In other words, the canonical coordinates are the eigenvalues of $\mathcal{U}$.

PROPOSITION 6.2. *The semisimple locus $M_0 \subset M$ contains the* **tame locus** *M_{00} of points at which the eigenvalues of $\mathcal{U}$ are distinct and nonzero.*

PROOF. Locally in M_{00}, the endormorphism $\mathcal{U}^*$ has eigenvectors π_i with eigenvalues v_i. We have

$$\mathcal{U}^*(\pi_i \circ \pi_j) = (\mathcal{U}^* \pi_i) \circ \pi_j = v^i \pi_i \circ \pi_j.$$

Antisymmetrizing in i and j gives $(v^i - v^j)\pi_i \circ \pi_j = 0$, showing that $\pi_i \circ \pi_j = 0$ if i and j are distinct. Since $E \circ \pi_i = v^i \pi_i$ is nonzero, we see that $\pi_i \circ \pi_i = f_i \pi_i$, where f_i is nowhere vanishing, hence $(T_x M, \circ)$ is semisimple in M_{00}. □

We will restrict attention to conformal semisimple Dubrovin manifolds satisfying the following condition.

CONDITION 6.1. *There is a section* $\gamma \in \Gamma(M_0, \mathrm{End}(T^*M))$ *such that*

$$\mu = [\gamma, \mathcal{U}].$$

Let μ_i^j and γ_i^j be the matrix elements of the tensors μ and γ in the frame (π_i); Condition 6.1 amounts to the relation $\mu_i^j = (v^i - v^j)\gamma_i^j$.

6.3. Semisimple Frobenius manifolds. A Frobenius manifold is semisimple if the underlying Dubrovin manifold is semisimple. We now show, borrowing from the exposition of Manin [**27**], that semisimple Frobenius manifolds satisfy Condition 6.1.

In Gromov-Witten theory, it is known that the Frobenius manifolds associated to Grassmannians [**1**], to Del Pezzo varieties [**2**], and to arbitrary products of these, are semisimple. Another class of examples are the quotients V/W of a Euclidean vector space V by a Coxeter group W of reflections, considered by Saito [**28**] (see also Dubrovin [**8**]).

Let M be a semisimple Frobenius manifold. If $\varepsilon \in \Omega^1(M)$ is the one-form on M defined by the formula $\varepsilon(X) = \eta(e, X)$, then

$$\eta(X,Y) = \eta(e \circ X, Y) = \varepsilon(X \circ Y);$$

thus $\eta(X,Y) = 0$ if $X \circ Y = 0$. It follows that the idempotent eigenvectors π_i on M_0 are orthogonal. Let $\eta_i = \eta(\pi_i, \pi_i)$, let $\eta_{ij} = \pi_j(\eta_i)$, and let $\eta_{ijk} = \pi_k(\eta_{ij})$. We now list the main properties of these functions; these are known as the Darboux-Egoroff equations.

LEMMA 6.3.

1) $\eta_{ij} = \eta_{ji}$
2) $\eta_{ijk} = \frac{1}{2}\left(\frac{\eta_{ki}\eta_{ij}}{\eta_i} + \frac{\eta_{ij}\eta_{jk}}{\eta_j} + \frac{\eta_{jk}\eta_{ki}}{\eta_k}\right)$ *when* i, j *and* k *are distinct*
3) $e(\eta_i) = 0$
4) $E(\eta_i) = -r\eta_i$

PROOF. The coefficients $\Gamma_{ijk} = \eta(\nabla_i \pi_j, \pi_k)$ of the Levi-Cività connection ∇ are given by the formula

$$\Gamma_{ijk} = \tfrac{1}{2}\big(\delta_{ik}\eta_{ij} + \delta_{jk}\eta_{ji} - \delta_{ij}\eta_{ik}\big).$$

Taking the inner product of (6.1) with π_ℓ, we obtain the equation

$$\tfrac{1}{2}(\delta_{i\ell}\delta_{jk} + \delta_{j\ell}\delta_{ik})\big(\eta_{ij} - \eta_{ji}\big) = 0,$$

which shows that $\eta_{ij} = \eta_{ji}$.

A lengthy calculation shows that

$$\begin{aligned} R_{ijk\ell} &= \tfrac{1}{4}\big(\delta_{i\ell} - \delta_{j\ell}\big)\Big(\frac{\eta_{ki}\eta_{ij}}{\eta_i} + \frac{\eta_{ij}\eta_{jk}}{\eta_j} + \frac{\eta_{jk}\eta_{ki}}{\eta_k} - 2\eta_{ijk}\Big) \\ &+ \tfrac{1}{4}\big(\delta_{jk} - \delta_{ik}\big)\Big(\frac{\eta_{\ell i}\eta_{ij}}{\eta_i} + \frac{\eta_{ij}\eta_{j\ell}}{\eta_j} + \frac{\eta_{j\ell}\eta_{\ell i}}{\eta_\ell} - 2\eta_{ij\ell}\Big) \\ &+ \tfrac{1}{4}\big(\delta_{j\ell}\delta_{ik} - \delta_{i\ell}\delta_{jk}\big)\sum_m \frac{\eta_{im}\eta_{mj}}{\eta_m}. \end{aligned}$$

Part 2) follows, by the flatness of the connection ∇.

Since the vector field e is flat, we have

$$\begin{aligned} 0 = \eta(\pi_i, \nabla_i e) &= \sum_j \eta(\pi_i, \nabla_i \pi_j) = \sum_j \Gamma_{iji} \\ &= \tfrac{1}{2}\sum_j \big(\eta_{ij} + \delta_{ij}\eta_{ji} - \delta_{ij}\eta_{ii}\big) = \tfrac{1}{2}e(\eta_i). \end{aligned}$$

Since $\mathcal{L}_E\eta = (2-r)\eta$, we see that

$$E(\eta_i) = (\mathcal{L}_E\eta)(\pi_i, \pi_i) - 2\eta([E, \pi_i], \pi_i) = (2-r)\eta(\pi_i, \pi_i) - 2\eta(\pi_i, \pi_i),$$

which shows that $E(\eta_i) = -r\eta_i$. □

PROPOSITION 6.4. *We have* $\mu = [\gamma, \mathcal{U}]$, *where* $\gamma \in \Gamma(M_0, \operatorname{End}(T^*M))$ *is the endomorphism of the tangent bundle of* M_0 *defined by*

$$\gamma dv^i = \tfrac{1}{2} d\log\eta_i.$$

PROOF. By the definition of γ,

$$\eta(\pi_i, \gamma^*\pi_j) = \tfrac{1}{2}\eta_{ij}.$$

We must show that

$$\eta(\pi_i, \mu^*\pi_j) = \eta(\mathcal{U}^*\pi_i, \gamma^*\pi_j) - \eta(\gamma^*\pi_i, \mathcal{U}^*\pi_j) = \tfrac{1}{2}(v^i - v^j)\eta_{ij}. \tag{6.2}$$

We have

$$\begin{aligned} \eta(\pi_i, \mu^*\pi_j) &= (1 - \tfrac{r}{2})\eta(\pi_i, \pi_j) - \eta(\pi_i, \nabla_j E) \\ &= (1 - \tfrac{r}{2})\delta_{ij}\eta_i - \eta(\pi_i, [\pi_j, E]) - \eta(\pi_i, \nabla_E \pi_j) \\ &= -\tfrac{r}{2}\delta_{ij}\eta_i - \sum_k v^k \Gamma_{kji} \\ &= -\tfrac{1}{2}\Big(r\delta_{ij}\eta_i + \sum_k v^k\big(\delta_{ik}\eta_{kj} + \delta_{ij}\eta_{jk} - \delta_{jk}\eta_{ik}\big)\Big) \\ &= -\tfrac{1}{2}\big(\delta_{ij}\big(r\eta_i + E(\eta_i)\big) + (v^i - v^j)\eta_{ij}\big), \end{aligned}$$

and (6.2) follows. □

6.4. The vector fields $\mathcal{L}_k$ on a semisimple Dubrovin manifold. We now give a new proof of an important recent theorem of Dubrovin and Zhang ([**13**], Theorem 3.10.20). This proof is perhaps a little simpler than theirs, though it is in the same spirit.

THEOREM 6.5. *Let M be a conformal semisimple Dubrovin manifold satisfying Condition 6.1. If $f \in \mathcal{O}_\infty$ satisfies $\mathcal{L}_k f = 0$ for all $k \geq -1$, then f is a constant.*

PROOF. The theorem is proved by induction: if $f \in \mathcal{O}_n$ satisfies $\mathcal{L}_k f = 0$ for all $k \geq -1$, then $f \in \mathcal{O}_{n-1}$. In order to carry out this induction, we must calculate the action of $\mathcal{L}_k$ on $\mathcal{O}_n$ modulo $\mathcal{O}_{n-1}$. By (3.4), we see that

$$\begin{aligned}\mathcal{L}_k u_n^a &= -\sum_{j=1}^{n} \partial^{j-1} \cdot \operatorname{Res}_{z=0}\langle e, \delta_z^{k+1}(\partial + z\mathcal{X})^{n-j}\mathcal{X} du^a\rangle - \partial^n\langle e, \mathcal{U}^{k+1} du^a\rangle \\ &= -n \operatorname{Res}_{z=0}\langle e, \delta_z^{k+1} \cdot \partial^{n-1}\mathcal{X} du^a\rangle - \partial^n\langle e, \mathcal{U}^{k+1} du^a\rangle \pmod{\mathcal{O}_{n-1}} \\ &= -n\sum_{\ell=0}^{k}\langle e, \mathcal{U}^\ell(\mu + \tfrac{1}{2})\mathcal{U}^{k-\ell}\partial^{n-1}\mathcal{X} du^a\rangle - \partial^n\langle e, \mathcal{U}^{k+1} du^a\rangle \pmod{\mathcal{O}_{n-1}}.\end{aligned}$$

Form the generating function

$$\begin{aligned}\sum_{k=-1}^{\infty} \lambda^{-k-2}\mathcal{L}_k u_n^a &= -n\,\langle e, (\lambda - \mathcal{U})^{-1}(\mu + \tfrac{1}{2})(\lambda - \mathcal{U})^{-1}\partial^{n-1}\mathcal{X} du^a\rangle \\ &\quad - \partial^n\langle e, (\lambda - \mathcal{U})^{-1} du^a\rangle \pmod{\mathcal{O}_{n-1}}.\end{aligned}$$

We now invoke the hypothesis that M is semisimple. Since $\mathcal{U} dv^i = v^i\, dv^i$, we see that

$$\langle e, (\lambda - \mathcal{U})^{-1} du^a\rangle = \sum_{i,j}\langle \pi_j, (\lambda - v^i)^{-1} J_i^a dv^i\rangle = \sum_i \frac{J_i^a}{\lambda - v^i}.$$

It follows that

$$\partial^n\langle e, (\lambda - \mathcal{U})^{-1} du^a\rangle = \sum_i \frac{v_n^i J_i^a}{(\lambda - v^i)^2} + \sum_i \frac{\partial^n J_i^a}{\lambda - v^i} \pmod{\mathcal{O}_{n-1}}.$$

We also have

$$\begin{aligned}\partial^{n-1}\mathcal{X} &= \partial^{n-1}(u_1^c \mathcal{A}_{bc}^a)\partial_a \otimes du^b \\ &= u_n^c \mathcal{A}_{bc}^a\, \partial_a \otimes du^b = \sum_i v_n^i\, \pi_i \otimes dv^i \pmod{\mathcal{O}_{n-1}},\end{aligned}$$

hence

$$\partial^{n-1}\mathcal{X} du^a = \sum_i v_n^i J_i^a dv^i \pmod{\mathcal{O}_{n-1}}.$$

It follows that

$$\begin{aligned}\langle e, (\lambda - \mathcal{U})^{-1}(\mu + \tfrac{1}{2})(\lambda - \mathcal{U})^{-1}\partial^{n-1}\mathcal{V} du^a\rangle &= \sum_{i,j}\frac{v_n^i J_i^a(\mu_j^i + \frac{1}{2}\delta_j^i)}{(\lambda - v^i)(\lambda - v^j)} \\ &= \frac{1}{2}\sum_i \frac{v_n^i J_i^a}{(\lambda - v^i)^2} + \sum_{i,j} v_n^i J_i^a \gamma_j^i\left(\frac{1}{\lambda - v^i} - \frac{1}{\lambda - v^j}\right) \pmod{\mathcal{O}_{n-1}}.\end{aligned}$$

Assembling these results, we see that

$$\sum_{k=-1}^{\infty} \lambda^{-k-2} \mathcal{L}_k u_n^a = -\left(\tfrac{n}{2}+1\right) \sum_i \frac{v_n^i J_i^a}{(\lambda - v^i)^2} - \sum_i \frac{c_i^a}{\lambda - v^i} \pmod{\mathcal{O}_{n-1}}$$

for some coefficients $c_i^a \in \mathcal{O}_n$.

Now suppose that $f \in \mathcal{O}_n$ satisfies $\mathcal{L}_k f \in \mathcal{O}_{n-1}$ for all $k \geq 0$. Extracting the coefficient of $(\lambda - v^i)^{-2}$ in

$$\sum_{k=-1}^{\infty} \lambda^{-k-2} \mathcal{L}_k f = \partial_{n,a} f \sum_{k=-1}^{\infty} \lambda^{-k-2} \mathcal{L}_k u_n^a \pmod{\mathcal{O}_{n-1}},$$

we see that $\left(\tfrac{n}{2}+1\right) v_n^i J_i^a \, \partial_{n,a} f = 0$, and hence that $\partial_n f = 0$. In other words, $f \in \mathcal{O}_{n-1}$. □

COROLLARY 6.6. *On a semisimple Frobenius manifold, any solution*

$$\{\mathcal{F}_g \in \mathcal{O}_\infty \mid g \geq 1\}$$

of the Virasoro constraints (Theorem 5.7) and the dilaton equation (5.2) is unique (except in genus 1, where it is determined up to an additive constant).

PROOF. Consider two solutions $\{\mathcal{F}_g\}$ and $\{\tilde{\mathcal{F}}_g\}$ of the Virasoro constraints; we argue by induction on g that $\mathcal{F}_g = \tilde{\mathcal{F}}_g$. Suppose that $\mathcal{F}_h = \tilde{\mathcal{F}}_h$ for $h < g$. We see that $\mathcal{L}_k(\mathcal{F}_g - \tilde{\mathcal{F}}_g) = 0$ for all $k \geq -1$, hence by Theorem 6.5, $\mathcal{F}_g - \tilde{\mathcal{F}}_g$ is a constant. If $g > 1$, the constant vanishes by the dilaton equation, which implies that

$$\mathcal{Q}(\mathcal{F}_g - \tilde{\mathcal{F}}_g) = (2g-2)(\mathcal{F}_g - \tilde{\mathcal{F}}_g).$$

□

6.5. The function $\mathcal{G}$ on a semisimple Frobenius manifold. A solution of the genus 1 Virasoro constraints (5.3) on a semisimple Frobenius manifold was found by Dubrovin and Zhang [**12**]; by Theorem 6.5, their solution is the unique one. Their formula for $\mathcal{G}$ involves a function on the semisimple locus M_0 called the isomonodromic τ-function τ_I; in defining it, we follow the exposition of Hertling [**22**].

LEMMA 6.7. *The differential form*

$$\alpha = \tfrac{1}{8} \sum_{i,j} (v^i - v^j) \frac{\eta_{ij}^2}{\eta_i \eta_j} dv^i \in \Omega^1(M_0)$$

is closed, and $\alpha(X) = -\tfrac{1}{4} \operatorname{Tr}(\mu[\gamma, \mathcal{A}_X])$

PROOF. We have

$$\begin{aligned} d\alpha &= \tfrac{1}{4} \sum_{i,j,k} \frac{\eta_{ij}}{\eta_i \eta_j} \left(\eta_{ijk} - \tfrac{1}{2} \frac{\eta_{ij}\eta_{ik}}{\eta_i} - \tfrac{1}{2} \frac{\eta_{ij}\eta_{jk}}{\eta_j} \right) (v^i - v^j)\, dv^k \wedge dv^i \\ &= \tfrac{1}{8} \sum_{i,j,k} \frac{\eta_{ij}\eta_{jk}\eta_{ki}}{\eta_i \eta_j \eta_k} (v^i - v^j)\, dv^k \wedge dv^i. \end{aligned}$$

This vanishes, since $(v^i - v^j)\, dv^k \wedge dv^i$ is antisymmetric in the indices i, j and k.

The matrix $\mathcal{A}_i$ has components $(\mathcal{A}_i)^k_j = \delta_{ij}\delta^k_i$. Since the matrices $\mathcal{U}$ and $\mathcal{A}_i$ are diagonal in the frame dv^i of T^*M_0, we have

$$\begin{aligned}\partial_i \log \tau_I &= -\tfrac{1}{4}\operatorname{Tr}([\gamma,\mathcal{U}][\gamma,\mathcal{A}_i]) \\ &= -\tfrac{1}{4}\operatorname{Tr}(\gamma\mathcal{U}\gamma\mathcal{A}_i - \gamma\mathcal{U}\mathcal{A}_i\gamma - \mathcal{U}\gamma^2\mathcal{A}_i + \mathcal{U}\gamma\mathcal{A}_i\gamma) \\ &= -\tfrac{1}{4}\sum_j(\gamma^i_j\mathcal{U}^j_j\gamma^j_i - \gamma^i_j\mathcal{U}^i_i\gamma^j_i - \mathcal{U}^i_i\gamma^j_i\gamma^i_j + \mathcal{U}^j_j\gamma^i_j\gamma^j_i) \\ &= \tfrac{1}{2}\sum_j(v^i - v^j)\gamma^i_j\gamma^j_i = \langle\pi_i,\alpha\rangle. \qquad \square\end{aligned}$$

The **isomonodromic tau-function** τ_I is defined by the equation

$$d\log\tau_I = \alpha.$$

Note that just like $\mathcal{G}$, $\log\tau_I$ is only determined up to an additive constant.

THEOREM 6.8. $\mathcal{G} = \log\tau_I - \frac{1}{48}\sum_i \log\eta_i$

PROOF. We calculate the derivative of the two terms contributing to $\mathcal{L}_k\mathcal{G}$ separately: we will see that they respectively contribute the two terms of (5.3). We have

$$\begin{aligned}\mathcal{L}_k \log\tau_I &= \tfrac{1}{4}\operatorname{Tr}(\mu[\gamma,\mathcal{A}_{E^{\circ(k+1)}}]) = \tfrac{1}{4}\operatorname{Tr}(\mu[\gamma,\mathcal{U}^{k+1}]) \\ &= \tfrac{1}{4}\sum_{\ell=0}^{k}\operatorname{Tr}(\mu\mathcal{U}^\ell[\gamma,\mathcal{U}]\mathcal{U}^{k-\ell}) = \tfrac{1}{4}\sum_{\ell=0}^{k}\operatorname{Tr}(\mu\mathcal{U}^\ell\mu\mathcal{U}^{k-\ell}).\end{aligned}$$

Further, since $e(\eta_i) = 0$, we have

$$\begin{aligned}\tfrac{1}{2}\mathcal{L}_k(\log\eta_i) &= \tfrac{1}{2}(\mathcal{L}_k + (v^i)^{k+1}e)(\log\eta_i) = \sum_j((v^i)^{k+1} - (v^j)^{k+1})\gamma^i_j \\ &= \sum_j \frac{(v^i)^{k+1} - (v^j)^{k+1}}{v^i - v^j}\mu^i_j = \sum_{\ell=0}^{k}\sum_j (v^j)^\ell\mu^i_j(v^i)^{k-\ell} \\ &= \sum_{\ell=0}^{k}\langle e, \mathcal{U}^\ell\mu\mathcal{U}^{k-\ell}dv^i\rangle.\end{aligned}$$

Summing over i, the Virasoro constraints $z_{k,1}$ follows, since $\omega = \sum_i dv^i$. $\square$

It is proved in [**11**] that $\mathcal{G}$ automatically satisfies the differential equation $\Psi\mathcal{G} = 0$ of (5.4). Ideally, this should be part of a larger phenomenon, whereby all of the known differential equations satisfied by the Gromov-Witten potentials $\{\mathcal{F}^X_g\}_{g\geq 1}$, namely the Virasoro constraints and the topological recursion relations, have a (necessarily unique) solution on every semisimple Frobenius manifold.

References

[1] L. Abrams, *The quantum Euler class and the quantum cohomology of the Grassmannians.* Israel J. Math. **117** (2000), 335–352.

[2] A. Bayer and Yu. Manin, *(Semi)simple exercises in quantum cohomology.* <math.AG/0103164>

[3] K. Behrend, *Gromov-Witten invariants in algebraic geometry,* Invent. Math. **127** (1997), 601–617. <alg-geom/9601011>

[4] K. Behrend and Yu. Manin, *Stacks of stable maps and Gromov-Witten invariants,* Duke Math. J. **85** (1996), 1–60. <alg-geom/9506023>

[5] D. Cox and S. Katz, "Mirror symmetry and algebraic geometry." Mathematical Surveys and Monographs, 68. American Mathematical Society, Providence, RI, 1999.

[6] T. Coates and A. Givental, *Quantum Riemann-Roch, Lefschetz and Serre.* <math.AG/0110142>

[7] R. Dijkgraaf and E. Witten, *Mean field theory, topological field theory, and multi-matrix models,* Nucl. Phys. **B342** (1990), 486–522.

[8] B. Dubrovin, *Differential geometry of the space of orbits of a Coxeter group.* Surveys in differential geometry: integrable systems, 181–211, Surv. Differ. Geom., IV, Int. Press, Boston, MA, 1998.

[9] B. Dubrovin, *Geometry of 2D topological field theories,* in "Integrable systems and quantum groups, Montecalini Terme, 1993," eds. M. Francaviglia and S. Greco, Lect. Notes Math., vol. 1620, Springer-Verlag, Berlin, 1996, pp. 120–348.

[10] B. Dubrovin, *Flat pencils of metrics and Frobenius manifolds.* In "Integrable systems and algebraic geometry (Kobe/Kyoto, 1997)," World Sci. Publishing, River Edge, NJ, 1998, pp. 47–72. <math.DG/9803106>

[11] B. Dubrovin, Y. Zhang, *Bihamiltonian hierarchies in 2D topological field theory at one-loop approximation,* Commun. Math. Phys., **198** (1998), 311–361. <hep-th/9712232>

[12] B. Dubrovin and Y. Zhang, *Frobenius manifolds and Virasoro constraints* Selecta Math. (N.S.) **5** (1999), 423–466. <math.AG/9808048>

[13] B. Dubrovin and Y. Zhang, *Normal forms of hierarchies of integrable PDEs, Frobenius manifolds and Gromov-Witten invariants.* <math.DG/0108160>

[14] T. Eguchi, E. Getzler and C.-S. Xiong, *Topological gravity in genus 2 with two primary fields.* Adv. Theor. Math. Phys. **4** (2000), 981-998. <hep-th/0007194>

[15] T. Eguchi, K. Hori and C.-S. Xiong, *Quantum cohomology and Virasoro algebra,* Phys. Lett. **B 402** (1997), 71–80. <hep-th/9703086>

[16] T. Eguchi, Y. Yamada and S.-K. Yang, *On the genus expansion in the topological string theory,* Rev. Math. Phys. **7** (1995), 279–309. <hep-th/9405106>

[17] T. Eguchi and C.-S. Xiong, *Quantum cohomology at higher genus: topological recursion relations and Virasoro conditions,* Adv. Theor. Math. Phys. **2** (1998), 219–228. <hep-th/9801010>

[18] E. Getzler, *Intersection theory on $\overline{\mathcal{M}}_{1,4}$ and elliptic Gromov-Witten invariants,* J. Amer. Math. Soc. **10** (1997), 973–998. alg-geom/9612004

[19] E. Getzler, *Topological recursion relations in genus 2.* In "Integrable systems and algebraic geometry (Kobe/Kyoto, 1997)," World Sci. Publishing, River Edge, NJ, 1998, pp. 73–106. <math.AG/9801003>

[20] E. Getzler, *The Virasoro conjecture for Gromov-Witten invariants,* "Algebraic geometry: Hirzebruch 70 (Warsaw, 1998)," Contemp. Math. **241**, Amer. Math. Soc., Providence, RI, 1999, pp. 147–176. <math.AG/9812026>

[21] A. Givental, *Gromov-Witten invariants and quantization of quadratic Hamiltonians,* Mosc. Math. J. **1** (2001), 551–568.

[22] C. Hertling, "Frobenius manifolds and moduli spaces for singularities." Cambridge Tracts in Mathematics, 151. Cambridge University Press, Cambridge, 2002.

[23] C. Hertling and Yu. Manin, *Weak Frobenius manifolds,* Internat. Math. Res. Notices (1999), 277–286. <math.QA/9810132>

[24] K. Hori, *Constraints for topological strings in $D \geq 1$,* Nucl. Phys. **B439** (1995) 395–420. hep-th/9411135.

[25] M. Kontsevich and Yu. I. Manin, *Relations between the correlators of the topological sigma-model coupled to gravity,* Comm. Math. Phys. **196** (1998), 385–398. <alg-geom/9708024>.

[26] X. Liu, *Elliptic Gromov-Witten invariants and Virasoro conjecture,* Comm. Math. Phys. **216**, 705–728. <math.AG/9907113>

[27] Yu. Manin, "Frobenius manifolds, quantum cohomology, and moduli spaces," American Mathematical Society Colloquium Publications, 47. American Mathematical Society, Providence, RI, 1999.
[28] K. Saito, *On a linear structure of the quotient variety by a finite reflection group*. Publ. Res. Inst. Math. Sci. **29** (1993), 535–579.
[29] E. Witten, *Two dimensional gravity and intersection theory on moduli space*, Surveys in Differential Geom. **1** (1991), 243–310.

DEPT. OF MATHEMATICS, NORTHWESTERN UNIVERSITY, EVANSTON, IL 60208, USA
E-mail address: getzler@math.northwestern.edu

Symplectic geometry of Frobenius structures

Alexander B. Givental

The concept of a Frobenius manifold was introduced by B. Dubrovin [**9**] to capture in an axiomatic form the properties of correlators found by physicists (see [**8**]) in two-dimensional topological field theories "coupled to gravity at the tree level". The purpose of these notes is to reiterate and expand the viewpoint, outlined in the paper [**7**] of T. Coates and the author, which recasts this concept in terms of linear symplectic geometry and exposes the role of the twisted loop group $\mathcal{L}^{(2)}GL_N$ of hidden symmetries.

We try to keep the text introductory and non-technical. In particular, we supply details of some simple results from the axiomatic theory, including a several-line proof of the genus 0 Virasoro constraints, but merely quote and refer to the literature for a number of less trivial applications, such as the quantum Hirzebruch–Riemann–Roch theorem in the theory of cobordism-valued Gromov–Witten invariants. The latter is our joint work in progress with Tom Coates, and we would like to thank him for numerous discussions of the subject.

The author is also thankful to Claus Hertling, Yuri Manin and Matilde Marcolli for the invitation to the workshop on Frobenius structures held at MPIM Bonn in Summer 2002.

Gravitational descendents. In Witten's formulation of topological gravity [**44**] one is concerned with certain "correlators" called *gravitational descendents.* The totality of genus 0 gravitational descendents is organized as the set of Taylor coefficients of a single formal function $\mathcal{F}$ of a sequence of vector variables $\mathbf{t} = (t_0, t_1, t_2, ...)$. The vectors t_i are elements of a finite-dimensional vector space[1] which we will denote H. One assumes that H is equipped with a symmetric non-degenerate bilinear form $(\cdot,\cdot)$ and with a distinguished non-zero element $\mathbf{1}$. The axioms are formulated as certain partial differential equations on $\mathcal{F}$: an infinite set of *Topological Recursion Relations* (TRR), the *String Equation* (SE) and the *Dilaton Equation* (DE). To state the axioms we are using the following coordinate notation: introduce a basis $\{\phi_\alpha\}$ in H such that $\phi_1 = \mathbf{1}$, put $t_k = \sum t_k^\alpha \phi_\alpha$, $(\phi_\alpha, \phi_\beta) = g_{\alpha\beta}$, $(g^{\alpha\beta}) = (g_{\alpha\beta})^{-1}$ and write $\partial_{\alpha,k}$ for the partial derivative $\partial/\partial t_k^\alpha$. Then the axioms read:

(DE) $$\partial_{1,1}\mathcal{F}(\mathbf{t}) = \sum_{n=0}^{\infty}\sum_{\nu} t_n^\nu \partial_{\nu,n}\mathcal{F}(\mathbf{t}) \; - \; 2\mathcal{F}(\mathbf{t}),$$

Research is partially supported by NSF Grant DMS-0072658.

[1]or super-space, but we will systematically ignore the signs that may come out of this possibility.

(SE) $$\partial_{1,0}\mathcal{F}\ (\mathbf{t}) = \frac{1}{2}(t_0, t_0) + \sum_{n=0}^{\infty}\sum_{\nu} t_{n+1}^{\nu}\partial_{\nu,n}\mathcal{F}\ (\mathbf{t}),$$

(TRR) $$\partial_{\alpha,k+1}\partial_{\beta,l}\partial_{\gamma,m}\mathcal{F}\ (\mathbf{t}) = \sum_{\mu\nu} \partial_{\alpha,k}\partial_{\mu,0}\mathcal{F}\ (\mathbf{t})\ g^{\mu\nu}\ \partial_{\nu,0}\partial_{\beta,l}\partial_{\gamma,m}\mathcal{F}\ (\mathbf{t})$$

for all α, β, γ and any $k, l, m \geq 0$. We will see soon that the system DE+SE+TRR admits a transparent geometrical interpretation.

Let $\mathcal{H}$ denote the *loop space* $H((z^{-1}))$ consisting of Laurent series in $1/z$ with vector coefficients. Introduce the symplectic form Ω on $\mathcal{H}$:

$$\Omega(\mathbf{f}, \mathbf{g}) = \frac{1}{2\pi i}\oint (\mathbf{f}(-z), \mathbf{g}(z))\ dz.$$

The polarization $\mathcal{H} = \mathcal{H}_+ \oplus \mathcal{H}_-$ by the Lagrangian subspaces $\mathcal{H}_+ = H[z]$ and $\mathcal{H}_- = z^{-1}H[[z^{-1}]]$ provides a symplectic identification of $(\mathcal{H}, \Omega)$ with the cotangent bundle $T^*\mathcal{H}_+$. We encode the domain variables $\mathbf{t} = (t_0, t_1, t_2, ...)$ of the function $\mathcal{F}$ in the coefficients of the vector polynomial

$$\mathbf{q}(z) = -z + t_0 + t_1 z + t_2 z^2 + ... \qquad (\text{where } -z := -\mathbf{1}z)\ .$$

Thus the formal function $\mathcal{F}(\mathbf{t})$ near $\mathbf{t} = 0$ becomes a formal function on the space $\mathcal{H}_+$ near the shifted origin $\mathbf{q} = -z$. This convention is called the *dilaton shift*. Denote by $\mathcal{L}$ the graph of the differential $d\mathcal{F}$:

$$\mathcal{L} = \{(\mathbf{p}, \mathbf{q}) \in T^*\mathcal{H}_+ :\ \mathbf{p} = d_{\mathbf{q}}\mathcal{F}\}.$$

This is a formal germ at $\mathbf{q} = -z$ of a Lagrangian section of the cotangent bundle $T^*\mathcal{H}_+$ and can therefore be considered as a formal germ of a Lagrangian submanifold in the symplectic loop space $(\mathcal{H}, \Omega)$.

Theorem 1. *The function $\mathcal{F}$ satisfies TRR, SE and DE if and only if the corresponding Lagrangian submanifold $\mathcal{L} \subset \mathcal{H}$ has the following property:*

($\star$) *$\mathcal{L}$ is a Lagrangian cone with the vertex at the origin and such that its tangent spaces L are tangent to $\mathcal{L}$ exactly along zL.*

More precisely, the last condition is interpreted as follows. First, it shows that the tangent spaces $L = T_{\mathbf{f}}\mathcal{L}$ satisfy $zL \subset L$ (and therefore $\dim L/zL = \dim \mathcal{H}_+/z\mathcal{H}_+ = \dim H$). Second, it implies that $zL \subset \mathcal{L}$. Third, the same L is the tangent space to $\mathcal{L}$ not only along the line of $\mathbf{f}$ but also at all smooth points in zL. This is meant to include $\mathbf{f}$, i.e. $L = T_{\mathbf{f}}\mathcal{L}$ implies $\mathbf{f} \in zL$.

The tangent spaces L vary therefore in a $\dim H$-parametric family. In particular they generate a *variation of semi-infinite Hodge structures* in the sense of S. Barannikov [1], i.e. a family of semi-infinite flags $\cdots zL \subset L \subset z^{-1}L \cdots$ satisfying the Griffiths integrability condition.

Thus another way to phrase the property ($\star$) is to say that
$\mathcal{L}$ is a cone ruled by the isotropic subspaces zL varying in a $\dim H$-parametric family with the tangent spaces along zL equal to the same Lagrangian space L.

Finally, the property should be understood in the sense of formal geometry near a point $\mathbf{f}_0$ different from the origin. In particular, the fact that cones develop singularities at the origin does not contradict the fact that $\mathcal{L}$ has Lagrangian tangent spaces as a section of $T^*\mathcal{H}_+$ near $\mathbf{q} = -z$.

Proof. Let us begin with the cone $\mathcal{L}$ satisfying $(\star)$. Using the polarization $\mathcal{H} = T^*\mathcal{H}_+$ we recover $\mathcal{F}$ as the restriction of the quadratic form $\mathbf{p}(\mathbf{q})/2$ (evaluation of the covector on a vector) to the cone $\mathcal{L}$ parameterized by $\mathcal{H}_+$ as a section of the cotangent bundle. The resulting function $\mathcal{F}$ is homogeneous of degree 2 (since $\mathcal{L}$ is a cone) which is exactly what DE expresses after the dilaton shift.

Multiplication by z^{-1} in $\mathcal{H}$ is anti-symmetric with respect to the symplectic form Ω and thus defines a linear hamiltonian vector field $\mathbf{f} \mapsto z^{-1}\mathbf{f}$ corresponding to the quadratic hamiltonian $\Omega(z^{-1}\mathbf{f}, \mathbf{f})/2$. The vector field is tangent to $\mathcal{L}$ since $z^{-1}\mathbf{f}$ is contained in the tangent space L at the point $\mathbf{f} \in zL$, The cone $\mathcal{L}$ is therefore situated in the zero level of the quadratic hamiltonian. It is straightforward to check that this condition written as the Hamilton–Jacobi equation for the generating function $\mathcal{F}$ coincides with the string equation.

Next, the property of $\mathcal{L}$ to have constant tangent spaces L along subspaces zL of codimension $\dim H$ shows that the quadratic differentials $d^2_{\mathbf{t}}\mathcal{F}$ depend only on $\dim H$ variables. In particular this applies to $\tau^\delta := \sum g^{\delta\mu}\partial_{\mu,0}\partial_{1,0}\mathcal{F}$. Differentiating the string equation in t_0^μ we find

$$\tau^\delta = t_0^\delta + \sum_{n,\mu,\nu} g^{\delta\mu}\, t^\nu_{n+1}\, \partial_{\nu,n}\partial_{\mu,0}\mathcal{F}. \tag{1}$$

In particular $\tau^\delta(\mathbf{t})|_{t_1=t_2=\ldots=0} = t_0^\delta$ are independent and thus τ^δ can be taken in the role of the $\dim H$ variables.

Note that the tangent spaces L are transverse to $\mathcal{H}_-$ and so the spaces zL are transverse to $z\mathcal{H}_-$. Therefore the projection of zL to the base $\mathcal{H}_+$ along $\mathcal{H}_-$ is transverse to the slice $q_1 = -\mathbf{1}, q_2 = q_3 = \ldots = 0$, and the intersection with the slice has $q_0 = \tau$. Making use of the correlator notation

$$\langle \phi_\alpha\psi^k, \phi_\beta\psi^l, \ldots, \phi_\gamma\psi^m\rangle(\tau) := (\partial_{\alpha,k}\partial_{\beta,l}\ldots\partial_{\gamma,m}\mathcal{F})|\ t_0 = \tau, t_1 = t_2 = \ldots = 0,$$

we can write

$$\partial_{\beta,l}\partial_{\gamma,m}\mathcal{F}(\mathbf{t}) = \langle \phi_\beta\psi^l, \phi_\gamma\psi^m\rangle(\tau(\mathbf{t})). \tag{2}$$

Consider (1) as an implicit equation for $\tau(\mathbf{t})$:

$$G^\delta(\tau, \mathbf{t}) := \tau^\delta - t_0^\delta - \sum_{n,\mu,\nu} g^{\delta\mu}\, t^\nu_{n+1}\, \langle \phi_\nu\psi^l, \phi_\mu\rangle(\tau) = 0. \tag{3}$$

Using the rules of implicit differentiation we conclude that the matrix $(\partial_{\lambda,0}\tau^\delta)$ is inverse to $(\partial G^\delta/\partial\tau^\lambda)$ at $\tau = \tau(\mathbf{t})$. Using this together with implicit differentiation once again, we find

$$\partial_{\alpha,k+1}\tau^\delta = \sum_{\mu,\nu} \partial_{\nu,0}\tau^\delta\ g^{\mu\nu}\ \langle \phi_\alpha\psi^k, \phi_\mu\rangle(\tau(\mathbf{t})).$$

This is a special case of TRR whose general form is obtained now by differentiation of (2):

$$\begin{aligned}
\partial_{\alpha,k+1}\partial_{\beta,l}\partial_{\gamma,m}\mathcal{F} &= \sum_\delta \partial_{\alpha,k+1}\tau^\delta\ \langle \phi_\delta, \phi_\beta\psi^l, \phi_\gamma\psi^m\rangle \\
&= \sum_{\delta,\mu,\nu} \langle \phi_\alpha\psi^k, \phi_\mu\rangle\ g^{\mu\nu}\ \partial_{\nu,0}\tau^\delta\ \langle \phi_\delta, \phi_\beta\psi^l, \phi_\gamma\psi^m\rangle \\
&= \sum_{\mu,\nu} \langle \phi_\alpha\psi^k, \phi_\mu\rangle\ g^{\mu\nu}\ \partial_{\nu,0}\partial_{\beta,l}\partial_{\gamma,m}\mathcal{F}.
\end{aligned} \tag{4}$$

In the opposite direction, suppose that $\mathcal{F}$ satisfies TRR, SE and DE. It is straightforward to see that given $\langle\phi_\beta\psi^l, \phi_\gamma\psi^l\rangle(\tau)$, the TRR uniquely determines all the other correlators with at least two seats.

We use (3) to implicitly define the (formal) map $\mathbf{t} \mapsto \tau(\mathbf{t})$ and try the right hand side of (2) in the role of $\partial_{\beta,l}\partial_{\gamma,m}\mathcal{F}$. Repeating the above derivation (4) we conclude that TRR is satisfied and hence, by the uniqueness, (2) is indeed true.

We see therefore that the tangent spaces to the graph $\mathcal{L}$ of $d\mathcal{F}$ are constant along the fibers of the map $\mathbf{t} \mapsto \tau$. [These fibers are affine subspaces (of codimension $\dim H$) since G is linear inhomogeneous in $\mathbf{t}$ and $\tau(t_0, 0, 0, ...) = t_0$. Taking $\mathbf{t} = (0, \mathbf{1}, 0, 0, ...)$ in G we find $G^\delta = \tau^\delta - \sum g^{\delta,\mu}\langle\phi_\mu, \mathbf{1}\rangle(\tau)$. This equals 0 for all τ because $\langle\phi_\mu, \mathbf{1}\rangle(\tau) = (\phi_\mu, \tau)$ as it follows from the string equation. Thus the fibers of the map $\mathbf{t} \mapsto \tau$ become linear subspaces after the dilaton shift.]

Due to the dilaton equation, $\mathcal{F}$ as a function on $\mathcal{H}_+$ is homogeneous of degree 2, and hence $\mathcal{L}$ is homogeneous of degree 1. Thus $\mathcal{L}$ is a cone, so the tangent spaces $L = T_\mathbf{f}\mathcal{L}$ contain the application points $\mathbf{f}$. Recalling that the string equation expresses the tangency of $z^{-1}\mathbf{f}$ to $\mathcal{L}$ at the points $\mathbf{f} \in \mathcal{L}$ we see that in fact $\mathbf{f} \in L \cap zL$. This is true for all those $\mathbf{f}$ where the tangent space is L. Therefore the projection of zL to $\mathcal{H}_+$ along $\mathcal{H}_-$ contains one of the fibers of the map $\mathbf{t} \mapsto \tau$. In fact the projection coincides with the fiber since both have codimension $\dim H$ in $\mathcal{H}_+$. For the space zL itself this means that it coincides with the set of points $\mathbf{f} \in \mathcal{L}$ where $T_\mathbf{f}\mathcal{L} = L$. In particular $zL \subset L$. □

The property $(\star)$ is formulated in terms of the symplectic structure Ω and the operator of multiplication by z but it does not depend on the choice of the polarization. This shows that the system DE+SE+TRR has a huge group of hidden symmetries. Let $\mathcal{L}^{(2)}GL(H)$ denote the *twisted loop group* which consists of End(H)-valued formal Laurent series M in the indeterminate z^{-1} satisfying $M^*(-z)M(z) = 1$. Here * denotes the adjoint with respect to $(\cdot,\cdot)$. The following corollary shows how the action of the twisted loop group can yield new functions $\mathcal{F}$ which satisfy the system DE+SE+TRR whenever the original one does.

Corollary. *The action of the twisted loop group preserves the class of the Lagrangian cones $\mathcal{L}$ satisfying $(\star)$.*

Frobenius structures. By a result of S. Barannikov [1], the set of tangent spaces L to the cone $\mathcal{L}$ satisfying $(\star)$ carries a canonical Frobenius structure. We outline Barannikov's construction.

Consider the intersection of the cone $\mathcal{L}$ with the affine space $-z + z\mathcal{H}_-$. The intersection is parameterized by $\tau \in H$ via the projection to $-z + H$ along $\mathcal{H}_-$ and can be considered as the graph of a function from H to $\mathcal{H}$ called the *J-function*:

$$\tau \mapsto J(-z, \tau) = -z + \tau + \sum_{k>0} J_k(\tau)(-z)^{-k}.$$

The derivatives $\partial J/\partial\tau^\delta = \phi_\delta + ...$ form a basis in $L \cap z\mathcal{H}_-$ which is a complement to zL in L. Since $z\partial J/\partial\tau^\delta \in zL \subset \mathcal{L}$, the 2nd derivatives $z\partial^2 J/\partial\tau^\delta\partial\tau^\lambda$ are in L again. On the other hand, these 2nd derivatives are in $z\mathcal{H}_-$ as well. Thus they are uniquely representable via the basis as $\sum A^\mu_{\delta\lambda}\partial J/\partial\tau^\mu$ with the coefficients $A^\mu_{\delta\lambda}(\tau)$ independent of z. Thus $\partial J/\partial\tau^\delta$ form a fundamental solution of the pencil of flat connections depending linearly on z^{-1}:

$$z\frac{\partial}{\partial \tau^\lambda}\frac{\partial J}{\partial \tau^\delta} = \sum_\mu A^\mu_{\delta\lambda}(\tau)\frac{\partial J}{\partial \tau^\mu}. \tag{5}$$

Obviously $A^\mu_{\delta\lambda} = A^\mu_{\lambda\delta}$. Moreover, the string flow $\mathbf{f} \mapsto \exp(u/z)\mathbf{f}$ preserves $-z+z\mathcal{H}_-$ and projects along $\mathcal{H}_-$ to $-z+\tau \mapsto -z+\tau-u\mathbf{1}$. This implies that $z\partial J/\partial\tau^1 = J$ and respectively $(A^\mu_{1\lambda})$ is the identity matrix. Furthermore, it is not hard to derive now that the multiplications on the tangent spaces $T_\tau H = H$ given by

$$\phi_\delta \bullet \phi_\lambda = \sum_\mu A^\mu_{\delta\lambda}(\tau)\phi_\mu$$

define associative commutative algebra structures with the unit $\mathbf{1}$ which are Frobenius with respect to the inner product on H (i.e. $(a\bullet b, c) = (a, b\bullet c)$) and satisfy the integrability condition imposed by (5). This is what a Frobenius structure is (see for instance [**39**]).[2]

Conversely, given a Frobenius structure one recovers a J-function by looking for a fundamental solution to the system (*cf.* (5))

$$z\frac{\partial}{\partial\tau^\lambda}S = \phi_\lambda \bullet S$$

in the form of an operator-valued series $S = \mathbf{1} + S_1(\tau)z^{-1} + S_2(\tau)z^{-2} + \ldots$ satisfying $S^*(-z)S(z) = \mathbf{1}$. Such a solution always exists and yields the corresponding J-function $J^\delta(z,\tau) = z[S^*(z,\tau)]^\delta_1$ and the Lagrangian cone $\mathcal{L}$ enveloping the family of Lagrangian spaces $L = S^{-1}(z,\tau)\mathcal{H}_+$ and satisfies the condition $(\star)$. A choice of the fundamental solution S is called in [**25**] a *calibration* of the Frobenius structure. The calibration is unique up to the right multiplication $S \mapsto SM$ by elements $M = \mathbf{1} + M_1z^{-1} + M_2z^{-2} + \ldots$ of the "lower-triangular" subgroup in the twisted loop group. Thus the action of this subgroup on our class of cones $\mathcal{L}$ changes calibrations but does not change Frobenius structures (while more general elements of $\mathcal{L}^{(2)}GL(H)$, generally speaking, change Frobenius structures as well).

As an example, consider the translation invariant Frobenius manifold defined by a given Frobenius algebra structure $\circ$ on H. The system (5) has constant coefficients and allows for the following obvious solution

$$J(z,\tau) = z\; e^{(\tau\circ)/z} = \sum_{k\geq 0} z^{1-k}\frac{\tau^{\circ k}}{k!}.$$

The corresponding cone is $\mathcal{L} = \{\ \mathbf{e}^{(\tau\circ)/z}\mathbf{q}(z) : \mathbf{q} \in zH[z]\ \}$. Infinitesimal automorphisms of $\mathcal{L}$ in the twisted loop Lie algebra contains the subalgebra

$$z^{-1}(a_0\circ) + z(a_1\circ) + z^3(a_2\circ) + \ldots,\ \ a_i \in H$$

and actually coincides with it when the algebra $(H,\circ)$ is semisimple.

Quantum cohomology. In Gromov – Witten theory, the genus 0 descendent potential $\mathcal{F}$ of a compact (almost) Kähler manifold X is defined in terms on intersection theory on moduli spaces $X_{0,n,d}$ of degree $d \in H_2(X,\mathbb{Z})$ stable maps $\Sigma \to X$

[2]To match this with Dubrovin's original definition [**9**] one can use the generating function $\mathcal{F}$ to show that in fact $A^\mu_{\delta\lambda} = \sum g^{\mu\nu}\ \langle\phi_\nu,\phi_\delta,\phi_\lambda\rangle(\tau)$ and therefore $\mathcal{F}(\tau,0,0,\ldots)$ satisfies the WDVV-equation.

of genus 0 complex curves with n marked points [**34**]. Namely

$$\mathcal{F} = \sum_{d,n} \frac{Q^d}{n!} \int_{[X_{0,n,d}]} \prod_{i=1}^{n} (\sum_{k=0}^{\infty} \mathrm{ev}_i^*(t_k)\psi_i^k). \tag{6}$$

Here $[X_{0,n,d}]$ is the virtual fundamental cycle [**38**], $t_k \in H^*(X, \mathbb{Q})$ are cohomology classes of the target space X to be pulled back to the moduli spaces by the evaluation maps $\mathrm{ev}_i : X_{0,n,d} \to X$ at the i-th marked points, and ψ_i is the 1-st Chern class of the *universal cotangent line bundle* L_i. The fiber of L_i is defined as the cotangent line $T^*_{\sigma_i}\Sigma$ to the curve at the i-th marked point.

Thus the cohomology of the target space equipped with the Poincaré intersection pairing

$$g_{\mu\nu} = \int_X \phi_\mu \phi_\nu$$

plays the role of the inner product super-space H. Yet the situation here does not quite fit our axiomatic set-up since an appropriate *Novikov ring* $\mathbb{Q}[[Q]]$ takes the place of the ground field. Moreover, one often has to consider the Lagrangian section $\mathcal{L} = \mathrm{graph}\, d\mathcal{F}$ in the completion of the space $\mathcal{H}$ consisting of the series $\sum f_k z^k$ infinite in both directions but such that $f_k \to 0$ as $k \to +\infty$ in the Q-adic topology of $H = H^*(X, \mathbb{Q}[[Q]])$. In what follows we will allow this and similar generalizations of the framework without special notice.

The genus 0 descendent potential is known to satisfy the axioms DE, SE and TRR and therefore defines on H a Frobenius structure with the multiplication $\bullet$ known as *quantum cup-product.* This Frobenius structure comes therefore equipped with the canonical calibration S and the J-function:

$$J(z,\tau) = z + \tau + \sum_{d,n} \frac{Q^d}{n!} \, (\mathrm{ev}_{n+1})_* [\frac{\prod_{i=1}^n \mathrm{ev}_i^*(\tau)}{z - \psi_{n+1}}], \tag{7}$$

where $(\mathrm{ev}_{n+1})_*$ is the push-forward along the evaluation map $\mathrm{ev}_{n+1} : X_{0,n+1,d} \to X$. The present description of this expression as the defining section of the whole cone $\mathcal{L}$ seems more satisfactory than the *ad hoc* construction in the literature on mirror formulas.

Below we briefly recall several results from [**7**] exploiting the language of Lagrangian cones in the context of genus 0 Gromov – Witten theory.

Gravitational ancestors. Forgetting the maps $\Sigma \to X$ (and maybe several last marked points) and stabilizing the remaining marked curves one defines the *contraction maps* $\mathrm{ct} : X_{g,m+n,d} \to \overline{\mathcal{M}}_{g,m}$ of moduli spaces of stable maps to the Deligne – Mumford spaces. The latter carry their own universal cotangent line bundles and their 1-st Chern classes whose pull-backs to $X_{g,m+n,d}$ will be denoted $\bar{L}_i$ and $\bar{\psi}_i$ respectively. We call *ancestors* the Gromov – Witten invariants which "mistakenly" use the classes $\bar{\psi}$ instead of ψ. A theorem of Kontsevich – Manin (see Theorem 2.2 in [**35**]) expresses descendents via ancestors and vice versa. [3] It was E. Getzler [**17, 19**] who emphasized a primary role of ancestors in the theory of topological recursion relations. The following theorem interprets their results geometrically in the case $g = 0$.

[3]The underlying relationship between ψ and $\bar{\psi}$ is described also in [**17**], Proposition 5.

Introduce the genus 0 *ancestor potential* $\bar{\mathcal{F}}_\tau$ depending on the parameter $\tau \in H$:

$$\bar{\mathcal{F}}_\tau = \sum_{d,m,n} \frac{Q^d}{m!n!} \int_{[X_{0,n,d}]} \prod_{i=1}^{m} (\sum_{k=0}^{\infty} \mathrm{ev}_i^*(\bar{t}_k)\bar{\psi}_i^k) \prod_{j=m+1}^{m+n} \mathrm{ev}_j^*(\tau).$$

Let $\mathcal{L}_\tau$ denotes the Lagrangian section in $\mathcal{H} = T^*\mathcal{H}_+$ representing the graph of $d\bar{\mathcal{F}}_\tau$ subject, as usual, to the dilaton shift $\mathbf{q}(z) = -z + \sum \bar{t}_k z^k$.

Theorem 2 (see Appendix 2 in [**7**]). *We have* $\mathcal{L}_\tau = S(\tau)\mathcal{L}$, *where* $\mathcal{L}$ *represents the descendent potential, and* $S(\tau) \in \mathcal{L}^{(2)}GL(H)$ *is the corresponding calibration.*

The ancestor potential $\bar{\mathcal{F}}_0$ has zero 2-jet when $\bar{t}_0 = 0$ since $\dim \overline{\mathcal{M}}_{0,2+m} = m-1 < m$. Therefore $\mathcal{L}_\tau$ is tangent to $\mathcal{H}_+$ along $z\mathcal{H}_+$. Thus Theorem 2 implies directly (i.e. by-passing Theorem 1) that the graph $\mathcal{L}$ of $d\mathcal{F}$ satisfies $(\star)$.

Corollary. *The sections* $\mathcal{L}_\tau$ *are themselves Lagrangian cones satisfying the condition* $(\star)$ *(and isomorphic to each other and to* $\mathcal{L}$*).*

The Frobenius structures defined by the cones $\mathcal{L}_\tau$ are isomorphic to that for $\mathcal{L}$ since the calibration S is "lower-triangular".

On the other hand, Theorem 2 tells us how to define the ancestor potentials $\bar{\mathcal{F}}_\tau$ in the axiomatic theory: restrict the quadratic form $\mathbf{p}(\mathbf{q})/2$ to $S(\tau)\mathcal{L}$. In particular, the genus 0 ancestor potentials $\bar{\mathcal{F}}_\tau$ satisfy SE, DE and TRR (which agrees with Theorem C in [**43**]).

Moreover, let us call a function $\mathcal{G}(\bar{t}_0, \bar{t}_1, ...)$ *tame* if

$$\frac{\partial}{\partial \bar{t}_{k_1}^{\alpha_1}} \cdots \frac{\partial}{\partial \bar{t}_{k_r}^{\alpha_r}} \mathcal{G} \mid_{\bar{\mathbf{t}}=0} = 0 \quad \text{whenever } k_1 + ... + k_r > r - 3. \tag{8}$$

The ancestor potentials in Gromov – Witten theory are tame on the dimensional grounds: $\dim \overline{\mathcal{M}}_{0,r} = r-3$. The following result shows that in the axiomatic theory there is no need to add this condition as a new axiom.

Proposition. *The ancestor potentials* $\bar{\mathcal{F}}_\tau$ *(defined in the axiomatic theory as the generating functions for the Lagrangian cones* $\mathcal{L}_\tau := S(\tau)\mathcal{L}$*) are tame.*

Proof. The ancestor potentials themselves satisfy DE+SE+TRR but in addition have zero 2-jet at $\bar{\mathbf{t}} = 0$. Let us apply the derivation $\partial^r/\partial \bar{t}_{n_1}^{\alpha_1} ... \partial \bar{t}_{n_r}^{\alpha_r}$ to the identity TRR for $\bar{F}_\tau$ and assume that (8) is satisfied for the terms on the right hand side. If at least one of these terms is non-zero at $\bar{\mathbf{t}} = 0$, then we must have $k + l + m + n_1 + ... + n_r \le r - 1$. Thus (8) is satisfied for the left hand side. This allows us to derive (8) by induction on $k_1 + ... + k_r$. □

Apparently, a more general result applicable to higher genus theory is contained in [**19**].

Twisted Gromov – Witten invariants. Let E be a holomorphic vector bundle over a Kähler target space X. One can pull it back to a moduli space $X_{g,n+1,d}$ by the evaluation map $\mathrm{ev}_{n+1} : X_{g,n+1,d} \to X$ and then push the result forward to $X_{g,n,d}$ by the map $\mathrm{ft}_{n+1} : X_{g,n+1,d} \to X_{g,n,d}$ forgetting the $n+1$-st marked point. This way one gets an element in $K^0(X_{g,n,d})$:

$$E_{g,n,d} = \text{ a virtual bundle with the fibers } H^0(\Sigma; f^*E) \ominus H^1(\Sigma; f^*E).$$

Fixing an invertible multiplicative characteristic class $S(\cdot) = e^{\sum s_k ch_k(\cdot)}$ of complex vector bundles, one defines *twisted* GW-invariants by systematically using intersection indices against the *twisted* virtual fundamental cycle $S(E_{g,n,d}) \cap [X_{g,n,d}]$.

When E is a negative line bundle, and S is the inverse total Chern class, the twisted theory represents GW-invariants of the total space of the bundle. When E is a positive line bundle, and S is the Euler class, the twisted genus 0 theory is very close to the GW-theory of the hypersurface defined by a section of the line bundle. The following "Quantum Riemann – Roch theorem" describes the twisted version $\mathcal{F}^{tw}$ of the genus 0 descendent potential (6) via the untwisted one.

Given E and S, represent the the twisted genus 0 descendent potential by the Lagrangian submanifold $\mathcal{L}^{tw} \subset \mathcal{H}$ obtained from the graph of $d\mathcal{F}^{tw}$ using the dilaton shift convention $\mathbf{q}(z) = \sqrt{S(E)}(\mathbf{t}(z) - z)$.

Theorem 3 (see [7]). *The Lagrangian submanifold $\mathcal{L}^{tw}$ is obtained from the Lagrangian cone $\mathcal{L}$ of the untwisted theory by a linear symplectic transformation:*

$$\mathcal{L}^{tw} = \triangle \mathcal{L}, \quad \triangle \sim \sqrt{S(E)} \prod_{m=1}^{\infty} S(E \otimes L^{-m})$$

where L is a line bundle with $c_1(L) = z$.

More precisely, put $s(x) = \sum s_k x^k/k!$ and let x_i be the Chern roots of E, so that $S(E) = \exp \sum_i s(x_i)$. Then the expression

$$\ln\{S(E)^{1/2} \prod_{m=1}^{\infty} S(E \otimes L^{-m})\} = \sum_i [\frac{s(x_i)}{2} + \sum_{m=1}^{\infty} s(x_i - mz)]$$

has the asymptotical expansion

$$\frac{1}{2} \sum_i \frac{1 + e^{-z\partial/\partial x_i}}{1 - e^{-z\partial/\partial x_i}} s(x_i) = \sum_{0 \le m} \sum_{0 \le l \le \dim_{\mathbb{C}} X} s_{2m-1+l} \frac{B_{2m}}{(2m)!} ch_l(E) z^{2m-1},$$

where B_{2m} are the Bernoulli numbers, and $ch_l(E)$ are understood as operators of multiplication in the cohomology algebra H of the target space X. The operator $\ln \triangle$ is the infinitesimal symplectic transformation defined by this series.

When expressed in the original terms of the generating functions, the relationship stated in the theorem would have the form

$$\mathcal{F}^{tw}(\mathbf{q}) = \text{ critical value in } \mathbf{q}' \text{ of } \mathcal{G}(\mathbf{q}, \mathbf{q}') + \mathcal{F}(\mathbf{q}'),$$

where $\mathcal{G}$ is the quadratic generating function of the above linear symplectic transformation. The operation of computing the critical value is complicated, and it seems not obvious if the new function $\mathcal{F}^{tw}$ satisfies, say, the TRR. In fact it is not hard to see directly that the twisted genus 0 GW-theory satisfies the same axioms DE, SE and TRR. Then another question arises: are these equations for $\mathcal{F}^{tw}$ formal consequences of those for $\mathcal{F}$ or they carry some additional information about the untwisted theory? The use of Lagrangian cones makes the answer obvious: since the linear transformation in the theorem belongs to the group $\mathcal{L}^{(2)}GL(H)$, the condition ($\star$) for $\mathcal{L}^{tw}$ is formally equivalent to that for $\mathcal{L}$. This elementary example illustrates possible advantages of the geometrical viewpoint.

Quantum Lefschetz Theorem. The following result is a formal (but combinatorially non-trivial) consequence of Theorem 3.

Note that the germ of a Lagrangian cone $\mathcal{L}$ satisfying ($\star$) is determined by a $\dim H$-dimensional submanifold (at least when it is sufficiently close to the J-function). We begin with the J-function $J(-z, \tau)$ representing the Lagrangian cone $\mathcal{L}$ of the untwisted theory on X and generate a new Lagrangian cone $\mathcal{L}'$ satisfying

($\star$) by using the following function $I(-z,\tau)$ in the role of "a $\dim H$-dimensional submanifold":

$$I(z,\tau) \sim \int_0^\infty e^{x/z} J(z,\tau+\rho\ln x)\, dx \Big/ \int_0^\infty e^{(x-\rho\ln x)/z} dx,$$

where $\rho = \lambda + c_1(E)$ is the total Chern class of a complex line bundle E over X, and the integrals should be understood in the sense of stationary phase asymptotics as $z \to 0$.

Theorem 4 (see [**7**]). *The Lagrangian cone $\mathcal{L}'$ determined by the function I coincides with the cone $\mathcal{L}^{tw}$ of the genus 0 Gromov – Witten theory on X twisted by the line bundle E and by the total Chern class*

$$S(\cdot) = \lambda^{\dim(\cdot)} + c_1(\cdot)\lambda^{\dim(\cdot)-1} + \dots + c_{\dim(\cdot)}(\cdot).$$

This is an abstract formulation of a very general result which in the limit $\lambda \to 0$ allows one to compute genus 0 GW-invariants of a hypersurface in X defined by a section of E in a form generalizing Quantum Lefschetz Theorems of B. Kim, A. Bertram, Y.-P. Lee and A. Gathmann [**2, 3, 36, 16**]. Namely, Theorem 4 applies beyond "small" quantum cohomology theory, to general type complete intersections as well, and also offers a new insight on the nature of *mirror maps* as a very special case of *Birkhoff factorization* in loop groups. In particular, given the J-function of a toric Fano manifold X, the theorem yields the mirror formulas of [**22**] for the J-function of a Fano or Calabi – Yau complete intersection in X (including the celebrated example of quintic three-folds in $\mathbb{C}P^4$).[4]

Singularity theory. Although not named so, Frobenius structures first emerged around 1980 from K. Saito's theory of primitive forms in singularity theory (see [**29**] for a modern exposition).

Let $f : (\mathbb{C}^m, 0) \to (\mathbb{C}, 0)$ be the germ of a holomorphic function (for simplicity — weighted homogeneous) at an isolated critical point of multiplicity N. Let $F(x,\tau)$ be a miniversal deformation of f, i.e. a family of functions in $\mathbf{x} \in \mathbb{C}^m$ depending on the N-dimensional parameter $\tau \in \mathcal{T}$ and such that $F(x,0) = f(x)$ and $\partial F/\partial\tau^\alpha|_{\tau=0}$, $\alpha = 1, \dots, N$, represent a basis in the local algebra H of the critical point:

$$H = \mathbb{C}[x]/(f_{x_1}, \dots, f_{x_m}).$$

The tangent spaces $T_\tau\mathcal{T}$ are canonically identified with the algebras $\mathbb{C}[x]/(F_x)$ of functions on the critical sets. To make the algebras $T_\tau\mathcal{T}$ Frobenius, one picks a holomorphic volume form ω on $\mathbb{C}^m$ (possibly depending on τ) and introduces the *residue pairing* on $T_\tau\mathcal{T}$ via

$$(a,b) = (\frac{1}{2\pi i})^m \oint \dots \oint a(y)\, b(y)\, \frac{dy_1 \wedge \dots \wedge dy_m}{F_{y_1} \dots F_{y_m}}.$$

(We assume here that y is a unimodular coordinate system on $\mathbb{C}^m$, i.e. $\omega = dy_1 \wedge \dots \wedge dy_m$.) To make $\mathcal{T}$ a Frobenius manifold one takes the volume form ω to be

---

[4]The initial proof [**21, 22**] of the mirror formulas for toric complete intersections was based on fixed point localization while the new approaches [**36, 16, 7**] seem to be free of this restrictive technique. It is worth mentioning however that these approaches require the J-function of the ambient toric manifold as an input. Ironically, the only currently available method of computing J-functions of general Fano toric manifolds is to specialize the result of [**22**] based on fixed point localization to toric complete intersections of codimension 0.

primitive, i.e. satisfying very special conditions which in particular guarantee that the residue metric is flat.

Many attributes of abstract Frobenius structures make more sense when interpreted in terms of singularity theory. For example, the system (5) is satisfied by the complex oscillating integrals

$$\mathcal{J}_{\mathfrak{B}}(\tau) = (-2\pi z)^{-m/2} \int_{\mathfrak{B}} e^{F(x,\tau)/z} \omega$$

written in flat coordinates τ of the residue metric on $\mathcal{T}$.

The oscillating integrals also satisfy the following weighted homogeneity condition:

$$(9) \qquad (z\frac{\partial}{\partial z} + \sum (\deg \tau^\lambda)\tau^\lambda \frac{\partial}{\partial \tau^\lambda})\, z\frac{\partial}{\partial \tau^\delta}\mathcal{J} = -\mu_\delta\, z\frac{\partial}{\partial \tau^\delta}\mathcal{J},$$

where $\mu_\delta = \deg(\partial_\delta F) + \deg(\omega) - m/2$, $\delta = 1, ..., N$, is the *spectrum* of the singularity symmetric about 0. The pencil (5) of flat connections over $\mathcal{T}$ can therefore be extended in the z-direction by the operators

$$\nabla_\tau := \frac{\partial}{\partial z} + \frac{\mu}{z} + \frac{E\bullet}{z^2}.$$

Here $E = \sum (\deg \tau^\lambda)\tau^\lambda \partial/\partial\tau^\lambda$ is the *Euler field* and $\mu = \operatorname{diag}(\mu_1, ..., \mu_N)$ is the *Hodge grading operator* (anti-symmetric with respect to the metric and diagonal in a graded basis of H).

The extended connection is flat (since the oscillating integrals $z\partial\mathcal{J}_{\mathfrak{B}}/\partial\tau^\delta$ provide a basis of flat sections) and can be considered as an *isomonodromic* family of connections ∇_τ over $z \in \mathbb{C}\backslash 0$ depending on the parameter $\tau \in \mathcal{T}$. In this situation, the calibration $S(z,\tau) = \mathbf{1} + S_1 z^{-1} + S_2 z^{-2} + ...$ can be chosen as a gauge transformation near $z = \infty$ of the operator ∇_τ to the normal form $\nabla_0 = \partial/\partial z + \mu/z$. In other words, the fundamental solution matrix $(z\partial\mathcal{J}_{\mathfrak{B}_\lambda}/\partial\tau^\delta)$ consisting of oscillating integrals can be written as $S(z,\tau)\, z^{-\mu}\, C$ (where C is a constant matrix), and the J-function can be extracted from $\mathcal{J}$ in a similar way.

In singularity theory, gravitational descendents are not defined (at least mathematically) in an intrinsic way, but rather recovered from the axioms, i.e. the total descendent potential is defined as the generating function for the Lagrangian cone $\mathcal{L} \subset \mathcal{H}$ determined by the J-function.

Semisimplicity. In the case $\dim H = 1$ (interpreted either as the Gromov – Witten theory of the point target space or in mirror terms as A_1 singularity) the J-function $ze^{\tau/z}$ generates the descendent potential (in dilaton-shifted variables)

$$(10) \qquad \mathcal{F}(\mathbf{q}) = \text{ the critical value of } \frac{1}{2}\int_0^\tau (q_0 + q_1 x + ... + q_k \frac{x^k}{k!} + ...)^2\, dx.$$

Consider now the model Cartesian product case of a semisimple translation invariant Frobenius structure defined by the *algebra* $\mathbb{C}^N$. Let $\mathcal{L}^{(N)}$ be the corresponding Lagrangian submanifold. It is isomorphic to $(\mathcal{L}^{(1)})^N$ and is generated by direct sum $\mathcal{F}(\mathbf{q}^{(1)}) + ... + \mathcal{F}(\mathbf{q}^{(N)})$ of N copies of the function (10).

Theorem 5. *The Lagrangian submanifold $\mathcal{L} \subset \mathcal{H}$ representing the germ of a semisimple N-dimensional Frobenius manifold is locally isomorphic to $\mathcal{L}^{(N)} \subset \mathbb{C}^N((z^{-1}))$.*

Proof. This is a reformulation of a result from [**24**] (*cf.* Exercise 3.7 in [**9**] though) about existence of an asymptotical fundamental solution

$$(z\frac{\partial J^{\lambda}}{\partial \tau^{\delta}}) \sim \Psi(\tau)R(z,\tau)e^{U(\tau)/z}$$

to the system (5) at semisimple points τ. Here $R = \mathbf{1}+R_1z+R_2z^2+...$ is an "upper-triangular" element of the (completed) twisted loop group $\mathcal{L}^{(2)}GL_N$, $\Psi : \mathbb{C}^N \to H$ is an isomorphism of the inner product spaces, and $U = \mathrm{diag}(u_1,...,u_N)$ is the diagonal matrix of Dubrovin's *canonical coordinates* [**9**] on the Frobenius manifold.

In the case of K. Saito's Frobenius structures of singularity theory, $u_\lambda(\tau)$ are critical values of the Morse functions $F(\cdot,\tau)$, and an asymptotical solution $\Psi R\exp(U/z)$ can be found as the stationary phase asymptotics of the complex oscillating integrals $z\partial\mathcal{J}_{\mathfrak{B}_\lambda}/\partial\tau^\delta$.

In general, consider the cones $\mathcal{L}_\tau = S(\tau)\mathcal{L}$ representing the ancestor potentials. The operators $\exp(-U(\tau)/z)R^{-1}(\tau)\Psi^{-1}(\tau)$ act on the cones $\mathcal{L}_\tau$ despite the fact that $R = \mathbf{1} + R_1z + ...$ are infinite z-series. This is because the ancestor potentials are tame and in particular can be considered as series in $\bar{t}_0, \bar{t}_1$ with coefficients *polynomial* in $\bar{t}_2, \bar{t}_3, ...$ (see Section 8 in [**26**] for more details). The resulting cone $\mathcal{L}'$ does not depend on τ since the operators S and $\Psi R\exp(U/z)$ satisfy the same differential equations (5) in τ. On the other hand, the operators R preserve the isotropic space $z\mathbb{C}^N[[z]]$ contained in (the completion of) the cones $\mathcal{L}_\tau$. This implies that the cone $\mathcal{L}'$ together with the point $-z$ contains all points $\exp(-U/z)(-z)$. Thus $z\exp(U/z)$ is the J-function of $\mathcal{L}'$ which shows that $\mathcal{L}' = \mathcal{L}^{(N)}$. □

The operator $R(z,\tau)$ is unique up to the right multiplication by the automorphisms $\exp(a_0/z + a_1z + a_2z^3 + ...)$, $a_i \in \mathbb{C}^N$, of $\mathcal{L}^{(N)}$[5]. The ambiguity in is eliminated by imposing the additional homogeneity condition (9) (or equivalently $L_ER = 0$) available in the presence of the Euler vector field E in the definition of the Frobenius structure. This is the case in singularity theory and in (non-equivariant!) Gromov – Witten theory.

Virasoro constraints. The so called "Virasoro conjecture" was invented by T. Eguchi, K. Hori, M. Jinzenji, C.-S. Xiong and S. Katz [**12, 13**] and upgraded to the rank of an axiom in abstract topological field theory by B. Dubrovin and Y. Zhang [**10**]. In the axiomatic context, the Virasoro constraints are defined in the presence of the additional grading axiom expressing the role of the Euler field E in the definition of Frobenius structures. The genus 0 Virasoro constraints are known to contain no information in addition to the grading axiom and the system DE+SE+TRR. The geometrical argument below replaces the original proof of this fact given by X. Liu – G. Tian [**38**] as well as the shorter proof given by E. Getzler [**18**].

Suppose that linear operators A and B on a vector space are anti-symmetric with respect to a bilinear form. Then the operators $l_m = ABABA...BA$ (B repeated m times) are also anti-symmetric. On the other hand, if $AB - BA = B$, then $l_m, m = 0,1,2,...$, commute as the vector fields $x^{m+1}\partial/\partial x$ on the line: $[l_m, l_n] = (n-m)l_{m+n}$.

[5]and up to reversing or renumbering coordinates in $\mathbb{C}^N$.

In the symplectic space $(\mathcal{H},\Omega)$, consider infinitesimal symplectic transformations A, B satisfying $[A,B] = B$, where $A = l_0 = zd/dz + 1/2 + a(z)$ with $a^*(-z)+a(z) = 0$, and B is multiplication by z.

The additional grading axiom for the descendent potential $\mathcal{F}$ is formulated as the invariance of the cone $\mathcal{L}$ under the flow of the linear hamiltonian vector field on $\mathcal{H}$ defined by the operator l_0 with a special choice of a (determined by the the Euler field E). For instance, in the case of weighted - homogeneous singularities $a = \mu$, and in Gromov – Witten theory take $a = \mu + \rho/z$ where μ is the Hodge grading operator on $H^*(X;\mathbb{Q})$ and ρ is the operator of multiplication by $c_1(T_X)$ in $H^*(X;\mathbb{Q})$. However the explicit form of a is irrelevant for the present discussion. Note that the vector field defined by the operator $l_{-1} = B^{-1} = z^{-1}$ (corresponding to $\partial/\partial x$) is also tangent to $\mathcal{L}$ due to the string equation. The following theorem expresses the genus 0 Virasoro constraints for $\mathcal{F}$.

Theorem 6. *Suppose the vector field on $\mathcal{H}$ defined by the operator l_0 is tangent to the Lagrangian cone $\mathcal{L}$ satisfying the condition $(\star)$. Then the same is true for the vector fields defined by the operators $l_m = l_0 z l_0 z ... z l_0$, $m = 1,2,....$*

Proof. Let L be the Lagrangian space tangent to $\mathcal{L}$ along zL. We know therefore that for any $\mathbf{f} \in zL$ the vector $l_0\mathbf{f}$ is in L. This implies $zl_0\mathbf{f} \in zL$, implies $l_0 z l_0 \mathbf{f} \in L$, implies $zl_0zl_0\mathbf{f} \in zL$, ..., and therefore for all $m = 1,2,...$ the vectors $l_m\mathbf{f}$ are tangent to the cone at the point $\mathbf{f}$. $\square$.

Higher genus and quantization. Correlators of the higher genus axiomatic theory are arranged into the Taylor coefficients of a sequence of formal functions $\mathcal{F}^{(g)}(\mathbf{t})$, $g = 0,1,2,...$, called *genus g descendent potentials*, which includes the genus 0 theory with $\mathcal{F} = \mathcal{F}^{(0)}$. In this section, we collect some results from [**24, 25, 7, 26, 27, 30, 31**] illustrating the following general observation: *the higher genus theory is quantization of the genus 0 theory and inherits the twisted loop group of symmetries.*

The polarization $\mathcal{H} = \mathcal{H}_+ \oplus \mathcal{H}_- = T^*\mathcal{H}_+$ allows one to quantize quadratic hamiltonians on the symplectic space $(\mathcal{H},\Omega)$ using the following standard rules:

$$(q_\alpha q_\beta)\hat{} = q_\alpha q_\beta/\hbar,\ (q_\alpha p_\beta)\hat{} = q_\alpha \partial/\partial q_\beta,\ (p_\alpha p_\beta)\hat{} = \hbar\partial^2/\partial q_\alpha \partial q_\beta.$$

Quantized quadratic hamiltonians act on functions of $\mathbf{q} \in \mathcal{H}_+$ depending on the parameter $\hbar$. They form a *projective* representation of the Lie algebra of infinitesimal symplectic transformations and of its Lie subalgebra $\mathcal{L}^{(2)}gl(H)$. The notation $\hat{A}$ for a simplectic transformation $A \in \mathcal{L}^{(2)}GL(H)$ will be understood as the operator $\exp(\ln A)\hat{}$.

The functions $\mathcal{F}^{(g)}$ are assembled into the *total descendent potential*

$$\mathcal{D} = e^{\mathcal{F}^{(0)}/\hbar + \mathcal{F}^{(1)} + \hbar\mathcal{F}^{(2)} + \hbar^2\mathcal{F}^{(3)} + ...}$$

Expressions of this form where $\mathcal{F}^{(g)}$ are formal functions of $\mathbf{t}$ will be called *asymptotical functions* and, after the dilaton shift $\mathbf{q}(z) = -z + t_0 + t_1 z + ...$, considered as "asymptotical elements of the Fock space". This means that we treat the 1-dimensional subspace $\langle\mathcal{D}\rangle$ spanned by $\mathcal{D}$ as a point in the projective space — the potential domain of our quantization representation.

• Theorem 3 above is derived in [**7**] as the quasi-classical limit $\hbar \to 0$ of the higher genus "Quantum Riemann – Roch Theorem" expressing twisted Gromov – Witten invariants via the untwisted ones:

$$\langle \mathcal{D}^{tw} \rangle = \hat{\Delta} \langle \mathcal{D} \rangle$$

where Δ is the symplectic transformation in Theorem 3. This result is based on Mumford's Grothendieck – Riemann – Roch formula [**41**] applied to the universal families of stable maps and generalizes the result of Faber – Pandharipande [**15**] for Hodge integrals corresponding to the case of the trivial twisting bundle E.

• The higher genus formula of Kontsevich – Manin [**35**] (see also [**19**]) relating descendents and ancestors in Gromov – Witten theory has been rewritten in [**25**] (see also [**7**]) as the quantization of the calibration operators $S(\tau)$ of Theorem 2:

$$\mathcal{A}_\tau = \hat{S}(\tau) \mathcal{D},$$

where $\mathcal{A}_\tau = \exp \sum \hbar^{g-1} \bar{\mathcal{F}}_\tau^{(g)}(\bar{\mathbf{t}})$ is the total ancestor potential.

• According to [**24, 25**], the machinery of fixed point localization and summation over graphs used by M. Kontsevich [**34**] for computation of equivariant Gromov – Witten invariants in the case of tori acting on the target spaces with isolated fixed points gives rise to the following formula for the equivariant total ancestor potentials:

$$\langle \mathcal{A}_\tau \rangle = \Psi(\tau)\ \hat{R}(\tau)\ \exp(U/z)\hat{}\ \langle \mathcal{D}_{pt}^{\otimes N} \rangle. \tag{11}$$

Here the rightmost function is the product $\mathcal{D}_{pt}(\mathbf{q}^{(1)})...\mathcal{D}_{pt}(\mathbf{q}^{(N)})$ of $N = \dim H$ copies of the total descendent potential $\mathcal{D}_{pt}$ of the one-point space $X = pt$, and $\Psi R \exp(U/z)$ is the asymptotical solution to the system (5).[6]

• The formula (11) makes sense for any *semisimple* Frobenius structure and thus can be used as a definition of the higher genus ancestor potentials in the abstract theory. According to [**25**] this definition meets a number of expectations. Namely:

∘ The total ancestor potential $\mathcal{A}_\tau$ thus defined is *tame* in the sense that the corresponding functions $\bar{\mathcal{F}}_\tau^{(g)}(\bar{\mathbf{t}})$ satisfy

$$\frac{\partial}{\partial \bar{t}_{k_1}^{\alpha_1}} \cdots \frac{\partial}{\partial \bar{t}_{k_r}^{\alpha_r}}\ \bar{\mathcal{F}}_\tau^{(g)}\ |_{\bar{\mathbf{t}}=0} = 0 \ \text{ whenever } k_1 + ... + k_r > r + 3g - 3.$$

∘ The corresponding total descendent potential *defined* by $\langle \mathcal{D} \rangle = \hat{S}^{-1}(\tau) \langle \mathcal{A}_\tau \rangle$ does not depend on τ. (Together with the previous property this means that $\mathcal{D}$ satisfies the $3g-2$-jet condition of Eguchi – Xiong [**14, 19**] playing the role of TRR in the higher genus theory.)

∘ The string equation $\hat{l}_{-1}\mathcal{D} = 0$ always holds true, and the higher Virasoro constraints $\hat{l}_m \mathcal{D} = 0, m = 1, 2, ...$ hold true if the grading condition $\hat{l}_0 \mathcal{D} \in \langle \mathcal{D} \rangle$ is satisfied.

• In non-equivariant Gromov–Witten theory, the formula

$$\langle \mathcal{D} \rangle = \hat{S}^{-1}\ \Psi\ \hat{R}\ e^{(U/z)\hat{}}\ \langle \mathcal{D}_{pt}^{\otimes N} \rangle \tag{12}$$

is known to be correct for complex projective spaces and other Fano toric manifolds [**25**] and for complete flag manifolds [**31**] (which in particular implies that the Virasoro conjecture holds true for these manifolds). Also, it is an easy exercise

---

[6] See [**24, 25**] for the explanation of how to fix the ambiguity in the construction of the asymptotical solution in the equivariant quantum cohomology theory lacking the Euler vector field E.

on application of the dilaton equation for $\mathcal{D}_{pt}$ to check that the result of Jarvis – Kimura [**30**] computing gravitational descendents in the Gromov – Witten theory on the quotient *orbifold* of the one - point space by a finite group agrees with (12). In this example, the Frobenius structure is translation invariant, $(H, \circ, (\cdot,\cdot))$ is the center of the group ring, Ψ describes its orthonormal diagonalization while $S = R = \mathbf{1}$.

• According to the part of Witten's conjectures [**44**] proved by M. Kontsevich [**33**] (see also [**42**]) the asymptotical function $\mathcal{D}_{pt}$ describing intersection theory on the Deligne – Mumford spaces $\overline{\mathcal{M}}_{g,m}$ satisfies the KdV hierarchy of evolution equations with the time variables $t_0, t_1, t_2, ...$ The remaining part of the conjectures similarly identifies the total descendent potential in the moduli theory of curves equipped with *n-spin structures* with solutions to the nKdV (or *Gelfand – Dickey*) hierarchies of integrable systems. Corresponding Frobenius structure coincides [**44**] with K. Saito's structure on the miniversal deformation of the A_{n-1}-singularity. It is a result of [**26**] that in the case of A_{n-1}-singularity the function (12) indeed satisfies the equations of the nKdV hierarchy. In [**27**] this result is generalized to the ADE-singularities.

Conventional wisdom says that the structure of the axiomatic higher genus theory should reflect the geometry of the Deligne – Mumford spaces, and vice versa. Yet both subjects are far from being clear. In view of the above examples it is tempting to impose the following two requirements on conjectural axioms of the higher genus theory.

A. The set of asymptotical elements of the Fock space satisfying the axioms has to be invariant under the quantization representation of the twisted loop Lie algebra.

In particular, the stabilizer of a Lagrangian cone $\mathcal{L}$ with the generating function $\mathcal{F}$ satisfying the axioms of the genus 0 theory has to act on the set of asymptotical elements satisfying the axioms of the higher genus theory and having the same genus 0 part $\mathcal{F}^{(0)} = \mathcal{F}$.

B. The action of the stabilizer is transitive.

The latter requirement is inspired by the uniqueness lemma of Dubrovin – Zhang [**11**] which says that in the *semisimple* case an asymptotical function $\mathcal{D}$ satisfying the higher genus TRR and the Virasoro constraints is unique (and thus coincides with (12) according to the results quoted earlier). It would be preferable however to derive the Virasoro constraints from other axioms (such as or similar to A+B) just the way it is done in the proof of Theorem 6 in the genus 0 theory.

Quantum K-theory. The material of this section represents our joint work with Tom Coates and shows that the description of the correlators and their properties in geometrical terms of Lagrangian cones and quantization remains valid in the context of quantum K-theory.

In quantum K-theory, a basis of "observables" has the form $\Phi_\alpha L_i^k$ where $\{\Phi^\alpha\}$ is a linear $\mathbb{Q}$-basis in $K^*(X)$ (here $K^0(X)$ is the Grothendieck group of complex vector bundles over the target space X) and $L_1, ..., L_n$ are the universal cotangent line bundles over the moduli spaces of stable maps $X_{g,n,d}$.

Correlators can be defined in cohomological terms using the *virtual tangent bundles* $T_{g,n,d}$ of the moduli spaces $X_{g,n,d}$ and with holomorphic Euler characteristics in mind:

$$(13)\quad \langle \Phi^{\alpha_1}L^{k_1},...,\Phi^{\alpha_n}L^{k_n}\rangle_{g,n,d} = \int_{[X_{g,n,d}]} \mathrm{td}(T_{g,n,d})\ \mathrm{ch}_*[\otimes_{i=1}^n \mathrm{ev}_i^*(\Phi^{\alpha_i}) \otimes L_i^{\otimes k_i}].$$

First properties of K-theoretic Gromov – Witten invariants are discussed in **[23]**, and the foundations in the setting of algebraic target spaces have been laid down by Yuan-Pin Lee in **[37]**. We have to stress that the correlators (13) are only approximations to the actual holomorphic Euler characteristics which in the orbifold / orbibundle context are given by Kawasaki's Hirzebruch – Riemann – Roch Theorem **[32]**. The correlators (13) differ therefore from those used in the papers **[23, 37]**. However the general *properties* of the correlators (as opposed to their *values*) remain the same as in **[23, 37]**. This is because the bundles, sheaves and their properties used in those papers are the same as in the "fake" version of K-theory considered here.

In complete analogy with the cohomology theory, one can introduce K-theoretic descendent potentials $\mathcal{F}^{(g)}(\mathbf{t})$ as formal functions of $t_0 + t_1 L + t_2 L^2 + ...$ with $t_i \in K := K^*(X)\otimes \mathbb{Q}[[Q]]$. As it was found in **[23]**, K-theoretic genus 0 Gromov – Witten invariants define a "Frobenius-like" structure on K which however exhibits the following peculiar distinction from the case of cohomology theory. The constant coefficient metric

$$g_{\alpha\beta} = \langle \Phi_\alpha, \mathbf{1}, \Phi_\beta\rangle_{0,3,0} = \chi(X;\Phi_\alpha\otimes\Phi_\beta) = \int_X \mathrm{td}(T_X)\ \mathrm{ch}_*(\Phi_\alpha)\ \mathrm{ch}_*(\Phi_\beta)$$

has to be replaced in all formulas by the non-constant one:

$$G_{\alpha\beta}(\tau) = g_{\alpha\beta} + \sum_{n\geq 0,d} \frac{Q^d}{n!}\langle \Phi_\alpha,\tau,...,\tau,\Phi_\beta\rangle_{0,n,d},\quad \text{where } \tau = \sum \tau^\mu \Phi_\mu \in K.$$

The multiplication $\Phi_\delta \bullet \Phi_\lambda = \sum_\mu A^\mu_{\delta\lambda}\Phi_\mu$, where

$$A^\mu_{\delta\lambda}(\tau) = \sum_\nu G^{\mu\nu}(\tau)\ \partial_{\delta,0}\partial_{\lambda,0}\partial_{\nu,0}\mathcal{F}^{(0)}|_{\mathbf{t}=(\tau,0,...)},\quad \text{and } (G^{\mu\nu}) = (G_{\mu\nu})^{-1},$$

makes the tangent spaces $T_\tau K$ Frobenius algebras with respect to the inner product $(G_{\mu\nu})$. The K-theory version of the J-function (7)

$$(14)\qquad J = (1-q) + \tau + \sum_{n\geq 0,d} \frac{Q^d}{n!}(\mathrm{ev}_{n+1})_*^K\left[\frac{\otimes_{i=1}^n \mathrm{ev}_i^*(\tau)}{\mathbf{1}-qL_{n+1}}\right]$$

satisfies the system

$$(15)\qquad (1-q)\frac{\partial}{\partial\tau^\delta}\frac{\partial}{\partial\tau^\lambda}J = \sum A^\mu_{\delta\lambda}\ \frac{\partial}{\partial\tau^\mu}\ J.$$

The system (15) at $q=-1$ yields the Levi – Civita connection of the metric $(G_{\mu\nu})$ which is therefore flat.

We will show that it is possible to adapt our language of symplectic loop spaces to absorb these changes at least in the version of quantum K-theory discussed here.[7]

[7] In the original version of quantum K-theory studied in **[23, 37]** the orbifold features of the moduli spaces $X_{g,n,d}$ play prominent role. For example, the classes $L_i \in K^0(X_{g,n,d})$ no longer behave unipotently. This causes some difficulties which seem formal at first (so that an analogue of Theorem 7 below still holds true), but become overwhelming in the more sophisticated Riemann – Roch problems discussed in the next section.

First, note that the J-function (14) considered as a $1/(1-q)$-series becomes finite when reduced modulo Q^d since $L_{n+1}-1$ is nilpotent in $K^0(X_{0,n,d})$. We will call such Laurent $1/(1-q)$-series *convergent* away from $q=1$ in the Q-adic topology (i.e. for those $q \in \mathbb{Q}[[Q]]$ whose Q-adic distance to 1 is ≥ 1).

Next, consider the loop space $\mathcal{K}$ of all vector Laurent series with coefficients in K convergent in this sense away from $q=1$. Equip K with the inner product $(\Phi_\alpha, \Phi_\beta) = g_{\alpha\beta}$ (the original constant one!) and define in $\mathcal{K}$ the symplectic form

$$\Omega(\mathbf{f},\mathbf{g}) = \frac{1}{2\pi i}\oint (\mathbf{f}(q^{-1}),\mathbf{g}(q))\,\frac{dq}{q} := -[\mathrm{Res}_{q=0}+\mathrm{Res}_{q=\infty}](\mathbf{f}(q^{-1}),\mathbf{g}(q))\,\frac{dq}{q}.$$

Substituting $q=\exp z$ (as motivated by $\mathrm{ch}_*(L_{n+1}) = \exp \psi_{n+1}$) we find that this Ω is in a sense the same as in cohomology theory. Consider the polarization $\mathcal{K} = \mathcal{K}_+ \oplus \mathcal{K}_-$ where $\mathcal{K}_+ = K[q]$ and $\mathcal{K}_-$ consists of the series convergent away from $q=1$ *and vanishing at* $q=\infty$. The polarization is Lagrangian (although not invariant with respect to $z \mapsto -z$) and identifies $\mathcal{K}$ with (a topologized version of) $T^*\mathcal{K}_+$. We use the dilaton shift $\mathbf{q}(q) = 1-q+t_0+t_1q+t_2q^2+\ldots$ to encode the genus 0 descendent potential by a Lagrangian submanifolds $\mathcal{L}$, and furthermore — to identify the total descendent and ancestor potentials of higher genus quantum K-theory with the appropriate asymptotical functions $\mathcal{D}$ near $\mathbf{q}=(1-q)\in\mathcal{K}_+$.

Now, repeat the construction for ancestors $\mathcal{L}_\tau$, $\mathcal{A}_\tau$ using the inner product $(\Phi_\alpha,\Phi_\beta)_\tau = G_{\alpha\beta}(\tau)$ and the corresponding symplectic structure Ω^τ in the space $\mathcal{K}^\tau = \mathcal{K}$.

It turns out that in quantum K-theory the calibration operators $S(q,\tau)$ defined via the J-function (14) as

$$S^\alpha_\beta(q,\tau) = \sum_{\mu\nu} G^{\alpha\mu}(\tau)\,\frac{\partial J^\nu}{\partial \tau^\mu}(q,\tau)\,g_{\nu\beta}$$

satisfy

$$\sum_{\alpha\beta} S^\alpha_\mu(q,\tau)G^{\alpha\beta}(\tau)S^\beta_\nu(q^{-1},\tau) = g_{\mu\nu}.$$

This shows that $S(\tau)$ is symplectic as a linear map from $(\mathcal{K},\Omega)$ to $(\mathcal{K}^\tau,\Omega^\tau)$. The following result is the K-theoretic version of the formula [**25, 7**] relating descendents and ancestors.

Theorem 7. *We have* $e^{F^{(1)}(\tau)}\mathcal{A}_\tau = \hat{S}(\tau)\mathcal{D}$ *and in particular* $\mathcal{L}_\tau = S(\tau)\mathcal{L}$.

Since $\mathrm{ch}_*(\bar{L}_i - 1) = e^{\bar{\psi}_i}-1$ are nilpotent in $H^*(\overline{\mathcal{M}}_{0,n};\mathbb{Q})$, the Lagrangian sections $\mathcal{L}_\tau$ are tangent to $\mathcal{K}^\tau_+$ along $(1-q)\mathcal{K}^\tau_+$.

Corollary. *The Lagrangian submanifold $\mathcal{L}\subset\mathcal{K}$ is a cone with the vertex at the origin and satisfies the condition that its tangent spaces L are tangent to $\mathcal{L}$ exactly along* $(1-q)L$.

The conic property of $\mathcal{L}$ is equivalent to the genus 0 case of the K-theoretic *dilaton equation*

$$\langle \mathbf{t}^{(1)}(L),\ldots,\mathbf{t}^{(n)}(L),1-L\rangle_{g,n+1,d} = (2-2g-n)\langle \mathbf{t}^{(1)}(L),\ldots,\mathbf{t}^{(n)}(L)\rangle_{g,n,d},$$

where $\mathbf{t}^{(i)}$ are arbitrary polynomials of the universal cotangent line bundles $L=L_i$.

The operator of multiplication by

$$\frac{1}{1-q}-\frac{1}{2} = \frac{1}{2}\frac{1+q}{1-q}$$

is anti-symmetric with respect to Ω and defines a linear hamiltonian vector field on $(\mathcal{K}, \Omega)$ which is tangent to the cone $\mathcal{L}$. This property of $\mathcal{L}$ expresses the following genus 0 K-theoretic *string equation*

$$\langle \mathbf{t}^{(1)}(L), ..., \mathbf{t}^{(n)}(L), 1\rangle_{0,n+1,d} =$$
$$\langle \mathbf{t}^{(1)}(L), ..., \mathbf{t}^{(n)}(L)\rangle_{0,n,d} + \sum_{i=1}^{n} \langle \mathbf{t}^{(1)}(L), ..., \frac{\mathbf{t}^{(i)}(L) - \mathbf{t}^{(i)}(1)}{L-1}, ..., \mathbf{t}^{(n)}(L)\rangle_{0,n,d}.$$

As we mentioned earlier, the same results hold true in the original version of the quantum K-theory studied in [**23, 37**].

Quantum cobordism theory. The complex cobordism theory $MU^*(\cdot)$ is defined in terms of homotopy classes of maps to the spectrum $MU(k)$ of the Thom spaces of universal $U_{k/2}$-bundles:

$$MU^n(B) = \lim_{k \mapsto \infty} \pi(\Sigma^k B, MU(n+k)).$$

When B is a stably almost complex manifold of real dimension m, the famous Pontryagin – Thom construction identifies elements of $MU^n(B)$ with appropriately framed bordism classes of maps $M \to B$ of stably almost complex manifolds M of real dimension $m-n$. This identification plays the role of the Poincaré isomorphism. Similarly to the complex K-theory, there is the *Chern – Dold character* which provides natural multiplicative isomorphisms

$$\mathrm{Ch} : MU^*(M) \otimes \mathbb{Q} \to H^*(M, \Lambda^*).$$

Here $\Lambda^* = MU^*(pt) \otimes \mathbb{Q}$ is the coefficient ring of the theory and is isomorphic to the polynomial algebra on the generators of degrees $-2k$ Poincaré – dual to the bordism classes $[\mathbb{C}P^k]$. The theory MU^* is known to be the universal extraordinary cohomology theory where complex vector bundles are oriented. Orientation of complex bundles in MU^* is uniquely determined by the cobordism-valued Euler class $u \in MU^2(\mathbb{C}P^\infty)$ of the universal complex line bundle. Explicitly, the Euler class of $\mathcal{O}(1)$ over $\mathbb{C}P^N$ is Poincaré – dual to the embedding $\mathbb{C}P^{N-1} \to \mathbb{C}P^N$ of a hyperplane section. The image of u under the Chern – Dold character has the form

$$u(z) = z + a_1 z^2 + a_2 z^3 + ... \tag{16}$$

where z is the cohomological 1-st Chern of the universal line bundle $\mathcal{O}(1)$, and $\{a_k\}$ is another set of generators in Λ^*. The operation of tensor product of line bundles with the Euler classes v and w defines a formal group law $F(v, w)$ on $MU^*(\mathbb{C}P^\infty) = \Lambda^*[[u]]$. The series $u(z)$ is interpreted as an isomorphism with the additive group $(x, y) \mapsto x + y$: $F(v, w) = u(z(v) + z(w))$. Here $z(\cdot)$ is the series inverse to $u(z)$. It is known as the *logarithm* of the formal group and explicitly takes on the form

$$z = u + [\mathbb{C}P^1]\frac{u^2}{2} + [\mathbb{C}P^2]\frac{u^3}{3} + [\mathbb{C}P^3]\frac{u^4}{4} + ...$$

Specialization of the parameters $[\mathbb{C}P^k] \mapsto 0$ yields the cohomology theory, and $[\mathbb{C}P^k] \mapsto 1$ yields the complex K-theory. In the latter example, $z = -\ln(1-u)$ and hence $u(z) = 1 - \exp(-z)$. Similarly to the K-theory, one can compute push-forwards in MU^*-theory in terms of cohomology theory. In particular, for a stably almost complex manifold B, we have the *Hirzebruch – Riemann – Roch formula*

$$\forall c \in MU^*(B), \quad \pi_*^{MU}(c) = \int_B \operatorname{Ch}(c)\ \operatorname{Td}(T_B)\ \in\ \Lambda^*, \tag{17}$$

where $\pi : B \to pt$, and $\operatorname{Td}(T_B)$ is the *Todd genus* of the tangent bundle. It is characterized as the only multiplicative characteristic class which for the universal line bundle is equal to

$$\operatorname{Td} = \frac{z}{u(z)} = \exp \sum_{k=1}^{\infty} s_k \frac{z^k}{k!}.$$

Here $s_1, s_2, \ldots$ is one more set of generators in Λ^*. To round up the introduction, let us mention the Landweber – Novikov algebra of stable cohomological operations in complex cobordism theory. The operations correspond to cobordism-valued characteristic classes σ of complex vector bundles. To apply such an operation to the cobordism element of B Poincaré-dual to a map $\pi : M \to B$ of stably almost complex manifolds, one takes the relative normal bundle $\pi^* T_B \ominus T_M$ over M and pushes-forward its characteristic class σ to B. According to Buchshtaber – Shokurov [**5**], after tensoring with $\mathbb{Q}$ the Landweber – Novikov algebra becomes isomorphic to the algebra of left-invariant differential operators on the group of diffeomorphisms (16). The Landweber – Novikov operations commute with the Chern – Dold character and can therefore be expressed in the cohomology theory $H^*(\cdot, \Lambda^*)$ as certain differential operators on the algebra Λ^* of functions on the group. The algebra is generated by the derivations L_n whose action on the generators a_k is given by

$$(L_n u)(z) = u(z)^{n+1}, \quad \text{or, equivalently,} \quad L_n = u^{n+1}\partial/\partial u, \ n = 1, 2, \ldots.$$

These generators correspond to the characteristic classes defined by the Newton polynomials

$$u_1^n + u_2^n + \ldots \in MU^*(\mathbb{C}P^\infty \times \mathbb{C}P^\infty \times \ldots).$$

In a sense, the idea of Gromov – Witten invariants with values in cobordisms is already present in Gromov's original philosophy [**28**] of symplectic invariants to be constructed as bordism invariants of spaces of pseudo-holomorphic curves. The possibility to define Gromov – Witten invariants with values in the cobordism ring Λ^* is mentioned in Kontsevich's work [**34**]. This proposal was further advanced by Morava [**40**] in a hope to explain the Virasoro constraints in terms of the Landweber – Novikov operations. Following suggestions of these authors, we define cobordism-valued Gromov – Witten invariants via (17) using the virtual tangent bundles $T_{g,n,d}$ of the moduli spaces $X_{g,n,d}$. We formulate below the genus 0 version of the "Quantum Hirzebruch – Riemann – Roch Theorem" which expresses these new Gromov – Witten invariants via the old ones. The discussion below represents joint work of Tom Coates and the author. The thesis [**6**] contains many details omitted here.

In quantum cobordism theory, a basis of observables has the form $\Phi_\alpha u_i^k$ where $\{\Phi^\alpha\}$ is a basis of $H\Lambda := MU^*(X) \otimes \mathbb{Q}[[Q]]$ over $\Lambda^* \otimes \mathbb{Q}[[Q]]$, and u_i is the Euler class of the universal cotangent line bundle L_i. The correlators are defined by the formula

(18)

$$\langle \Phi_{\alpha_1} u^{k_1}, \ldots, \Phi_{\alpha_n} u^{k_n} \rangle_{g,n,d} := \int_{[X_{g,n,d}]} e^{\sum_{k=1}^{\infty} s_k \operatorname{ch}_k(T_{g,n,d})} \prod_{i=1}^{n} \left[\operatorname{ev}_i^* \operatorname{Ch}(\Phi_{\alpha_i})\ u(\psi_i)^{k_i}\right].$$

One can introduce Λ^*-valued descendent potentials $\mathcal{F}^{(g)}(\mathbf{t})$ in complete analogy with the cohomology theory as formal functions of $t_0+t_1u+t_2u^2+...$ with $t_i \in H\Lambda$.

In order to express the cobordism-valued Gromov – Witten invariants in terms of cohomological ones one needs to control the classes $\mathrm{ch}_k(T_{g,n,d})$. Roughly speaking, the virtual tangent bundles $T_{g,n,d}$ consist of two parts. One of them, representing deformations of holomorphic *maps* of Riemann surfaces with a fixed complex structure, coincides with $E_{g,n,d}$ (as in Theorem 3 about twisted Gromov – Witten invariants) with $E = T_X$. The effect of this part can be described therefore via Theorem 3. The other part represents deformations of complex structures and contributes to the correlators in a complicated fashion. Surprisingly, the very formalism of the symplectic loop space takes effective care of these contributions.

We denote by $\Lambda = \mathbb{Q}\{s_1, s_2, ...\}$ the coefficient ring of complex cobordism theory completed with respect to the $\mathbf{s}$-adic norm defined by the grading $\deg s_k = -k$. Let $H\Lambda$ be the cobordism group of the target space X over Λ equipped with the inner product

$$(a,b)_{\mathbf{s}} := \pi_*^{MU}(ab) = \int_X \mathrm{Ch}(a)\,\mathrm{Ch}(b)\exp\left(\sum s_k\,\mathrm{ch}_k(T_X)\right).$$

Denote $\mathcal{H}\Lambda$ the loop space $H\Lambda\{\{u^{-1}\}\}$ of Laurent series $\sum f_k u^k$ with coefficients $f_k \in \mathcal{H}\Lambda$ which can be non-zero for all $k \in \mathbb{Z}$ but should satisfy the condition that $\mathbf{f}_k \to 0$ in the $\mathbf{s}$-adic topology as $k \to +\infty$. Introduce the symplectic form with values in Λ:

$$\Omega_{\mathbf{s}}(\mathbf{f},\mathbf{g}) := \frac{1}{2\pi i}\oint \pi_*^{MU}(\mathbf{f}(u^*)\,\mathbf{g}(u)) \sum_{k\geq 0}[\mathbb{C}P^k]u^k\,du = \frac{1}{2\pi i}\oint (\mathbf{f}(u(-z),\mathbf{g}(u(z))_{\mathbf{s}}\,dz.$$

The following *quantum Chern – Dold character* identifies the symplectic structure $\Omega_{\mathbf{s}}$ with its cohomological version $\Omega = \Omega_{\mathbf{0}}$:

$$\mathrm{qCh}: \mathcal{H}\Lambda \to \mathcal{H} = \Lambda \otimes H((z^{-1})): \ \ \mathrm{qCh}(\mathbf{f}) = \sqrt{\mathrm{Td}(T_X)}\sum_{k\in\mathbb{Z}}\mathrm{Ch}(f_k)u^k(z).$$

Assuming $|\psi| < |z|$ write

$$\frac{1}{u(-z-\psi)} = \sum_{k\geq 0} u(\psi)^k v_k(u(z)) = \sum_{k\geq 0}\psi^k w_k(z).$$

Put

$$\mathcal{H}\Lambda_+ = H\Lambda\{u\} = \{\text{ power series } \sum q_k u^k \text{ with } q_k \to 0\ \}$$

and

$$\mathcal{H}\Lambda_- = \{\text{ arbitrary infinite series } \sum p_k v_k(u)\ \}.$$

The following residue computation shows that $\{..., u^k, ..., v_k(u), ...\}$ (as well as $\{..., z^k, ..., w_k(z), ...\}$) is a Darboux basis in $(\mathcal{H}\Lambda, \Omega_{\mathbf{s}})$ and implies that the spaces $\mathcal{H}\Lambda_\pm$ form a Lagrangian polarization (depending on $\mathbf{s}$):

$$\frac{1}{2\pi i}\oint\frac{dz}{u(z-x)\,u(-z-y)} = \begin{cases} 0 & \text{if } \ |z|<|x|,|y| \text{ or } |x|,|y|<|z| \\ \pm\frac{1}{u(-x-y)} & \text{if } \ |x|<|z|<|y| \text{ or } |y|<|z|<|x| \end{cases}.$$

Introduce the dilaton shift convention

$$\mathbf{q}(u) = \mathbf{t}(u) + u^*(u), \ \text{ where } \ u^*(u(z)) = u(-z).$$

We consider the genus 0 descendent potential $\mathcal{F}^{(0)}$ of the quantum cobordism theory as a function depending formally on the parameters $\mathbf{s} = (s_1, s_2, ...)$ and define the corresponding family of Lagrangian submanifolds $\mathcal{L}_{\mathbf{s}}$ in the family of symplectic spaces $T^*\mathcal{H}\Lambda_+ = \mathcal{H}\Lambda$ over $Spec(\Lambda)$.

The following theorem expresses $\mathcal{L}_{\mathbf{s}}$ in terms of the cone $\mathcal{L} = \mathcal{L}_{\mathbf{0}}$ describing the quantum cohomology theory of X.

Theorem 8 (see [6]). *The image of $\mathcal{L}_{\mathbf{s}}$ under the quantum Chern – Dold character coincides with the Lagrangian cone of the Gromov – Witten theory on X twisted by the class* $\mathrm{Td}(T_X)$*:*

$$\mathrm{qCh}(\mathcal{L}_{\mathbf{s}}) = \Box\mathcal{L}, \quad \Box \sim \sqrt{\mathrm{Td}(T_X)}\ \prod_{m=1}^{\infty} \mathrm{Td}(T_X \otimes L^{-m}),$$

where L is a line bundle with $c_1(L) = z$.

In fact $\mathrm{qCh}(\mathcal{H}\Lambda_+) = \mathcal{H}_+$, but $\mathrm{qCh}(\mathcal{H}\Lambda_-)$ and the dilaton shift depend on $\mathbf{s}$. This causes a discrepancy between the descendent potentials of the quantum cobordism theory and the twisted theory which accounts for the entire contribution of variations of complex structures on Riemann surfaces into the virtual tangent bundles of the moduli spaces of stable maps.

The same happens in the higher genus theory. The quantum Chern – Dold character identifies the Heisenberg Lie algebras of the spaces $(\mathcal{H}\Lambda, \Omega_{\mathbf{s}})$. By the virtue of the Stone – von Neumann theorem and the Schur lemma, the Fock spaces corresponding to different polarizations are identified projectively. The total descendent potentials $\mathcal{D}_{\mathbf{s}}$ of the quantum cobordism theory differ from the total descendent potentials $\mathcal{D}_{\mathbf{s}}^{tw}$ of the appropriately twisted Gromov – Witten theory in such a way that the corresponding lines $\langle \mathcal{D}_{\mathbf{s}} \rangle$ and $\langle \mathcal{D}_{\mathbf{s}}^{tw} \rangle$ in the representation space of the Heisenberg algebra coincide.

Returning to the genus 0 case, notice that in the case of translation invariant Frobenius structure the cone $\mathcal{L}$ is invariant under the symplectic transformations defined by $\Box$.

Corollary. *When $X = pt$, we have* $\mathrm{qCh}(\mathcal{L}_{\mathbf{s}}) = \mathcal{L}$.

The formal group law

$$u(x+y) = u(x) + u(y) - \lambda u(x)u(y), \quad u(z) = (1 - e^{-\lambda z})/\lambda, \tag{19}$$

interpolates between cohomology and K-theory. Thus the specialization $[\mathbb{C}P^k] \mapsto \lambda^k$ allows one to adjust the quantum Hirzebruch – Riemann – Roch theorem to the case of quantum K-theory. (The correlators discussed in the previous section correspond to $\lambda = -1$.)

Corollary. *The quantum Chern character* $\mathrm{qch} : \mathcal{K} \mapsto \mathcal{H}$ *defined via* $\mathrm{qch}\, u = (1 - \exp(-\lambda z))/\lambda$ *identifies* $\mathrm{qch}(\mathcal{L}_\lambda)$ *with* $\Box\mathcal{L}$ *where* $\Box$ *is defined via the Todd class* $\mathrm{td}(L) = \lambda z/(1 - \exp(-\lambda z))$.

The quantum Chern – Dold character transforms multiplication by u to multiplication by $u(z) = z + \ldots \in \Lambda\{z\}$. Transformations defined by $\Box$ belong to the twisted loop group and thus commute with multiplication by $u(z)$.

Corollary. *The submanifolds $\mathcal{L}_{\mathbf{s}} \subset \mathcal{H}\Lambda$ are Lagrangian cones satisfying the condition $(\star)$.*

In particular, the dilaton equation holds true in quantum cobordism theory (which is easy to prove directly for any genus). Moreover, the cubical form on L/zL

defined by the correlators $\langle \Phi_\alpha, \Phi_\beta, \Phi_\gamma \rangle$ represents the *Yukawa coupling* defined on any Lagrangian submanifold in a linear symplectic space (see [**20**]). The form coincides therefore with the structure tensor of the Frobenius manifold defined by the twisted Gromov – Witten theory. However in quantum cobordism theory, there is no a simple formula for the string equation, as there is no general reasons for flatness of the metric $G_{\alpha\beta} = \langle \Phi_\alpha, 1, \Phi_\beta \rangle$ or associativity of the quantum cup-product whose constructions depend on the polarization. In fact, a key step — Barannikov's derivation of the equation (5) for the J-function — was based on the relation $u^*\mathcal{H}\Lambda_- = H\Lambda + \mathcal{H}\Lambda_-$. It is not hard to see that the group laws (19) are the only ones satisfying this condition.

Indeed, the inclusion $\forall k \geq 0,\ Hu^*w_k \subset H\Lambda + \mathcal{H}\Lambda_-$ means that the projection of all $u(-z)w_k(z)$ along the subspace $Span(w_l(z), l = 0, 1, 2, ...)$ yields constant polynomials. By definition we have $\sum_l w_l(-z)x^l = 1/u(z-x)$ when $|x| < |z|$ and $\sum_k u(-z)w_k(z)y^k = u(-z)/u(-z-y)$ when $|y| < |z|$. In terms of these generating functions the inclusion is equivalent therefore to the condition that

$$\frac{1}{2\pi i} \oint_{|x|,|y|<|z|} \frac{u(-z)\, dz}{u(z-x)\, u(-z-y)} = \frac{u(-x)}{u(-x-y)} - \frac{u(y)}{u(-x-y)}$$

is independent of x for all y. Differenting $u(-x-y)f(y) = u(-x) - u(y)$ in x we find $u'(-x-y)f(y) = u'(-x)$ which at $x = 0$ yields $f(y) = 1/u'(-y)$ (since $u'(0) = 1$) and implies $u'(-x-y) = u'(-x)u'(-y)$. Thus $u'(-x) = \exp(\lambda x)$ and respectively $u(z) = (1 - \exp(-\lambda z))/\lambda$.

It would be interesting to find out if ellipticity of $u(z)$ brings any "improvements" in properties of *elliptic quantum cohomology* in comparison with the general quantum cobordism theory.

References

[1] S. Barannikov. *Quantum periods - I. Semi-infinite variations of Hodge structures.* Preprint, alg-geom/0006193.

[2] V. Batyrev, I. Ciocan-Fontanine, B. Kim, D. van Straten. *Conifold Transitions and Mirror Symmetry for Calabi-Yau Complete Intersections in Grassmannians.* alg-geom/9710022.

[3] A. Bertram. *Another way to enumerate rational curves with torus actions.* arXiv: math.AG/9905159.

[4] R. Bott, C. Taubes. *On the rigidity theorems of Witten.* J. Amer. Math. Soc. **2** (1989), 137 – 186.

[5] V. M. Bukhshtaber, A. V. Shokurov. *The Landweber – Novikov algebra and formal vector fields on the line.* Funct. Anal. Appl. **12**, No. 3 (1978), 1 – 11.

[6] T. Coates. *Riemann–Roch theorems in Gromov–Witten theory.* PhD thesis, UC Berkeley, 2003, 145 pp.

[7] T. Coates, A. Givental. *Quantum Riemann – Roch, Lefschetz and Serre.* arXiv: math.AG/0110142.

[8] R. Dijkgraaf, E. Witten. *Mean field theory, topological field theory and multi-matrix models.* Nucl. Phys. B342 (1990), 486–522.

[9] B. Dubrovin. *Geometry of 2D topological filed theories.* In: Integrable Systems and Quantum Groups. Springer Lecture Notes in Math. 1620 (1996), 120–348.

[10] B. Dubrovin, Y. Zhang. *Frobenius manifolds and Virasoro constraints.* Selecta Math. (N.S.) 5 (1999), 423–466.

[11] B. Dubrovin, Y. Zhang. *Normal forms of hierarchies of integrable PDEs, Frobenius manifolds and Gromov – Witten invariants.* arXive: math.DG/0108160.

[12] T. Eguchi, K. Hori, C.-S. Xiong. *Quantum cohomology and Virasoro algebra.* Phys. Lett. **B 402** (1997), 71 – 80.

[13] T. Eguchi, M. Jinzenji, C.-S. Xiong. *Quantum cohomology and free field representations.* Nuclear Phys. **B 510** (1998), 608 – 622.

[14] T. Eguchi, C.-S. Xiong. *Quantum cohomology at higher genus: topological recursion relations and Virasoro conditions.* Adv. Theor. Math. Phys. **2** (1998), 219 – 229.

[15] C. Faber, R. Pandharipande. *Hodge integrals and Gromov – Witten theory.* Invent. Math. **139** (2000), 173–199.

[16] A. Gathmann. *Relative Gromov-Witten invariants and the mirror formula.* Preprint, math.AG/0009190.

[17] E. Getzler. *Topological recursion relations in genus* 2. In: Intergrable systems and algebraic geometry (Proceedings of the Taniguchi Symposium 1997), M.-H. Saito, Y. Shimizu, K. Ueno (eds.), World Scientific, Hong Kong, 1998, 73 – 106.

[18] E. Getzler. *The Virasoro conjecture for Gromov – Witten invariants.* arXiv: math.AG/9812026.

[19] E. Getzler. *The jet space of a Frobenius manifold and higher-genus Gromov – Witten invariants.* arXiv: math.AG/0211338.

[20] A. Givental. *Homological geometry I: Projective hypersurfaces.* Selecta Math. (N.S.) **1** (1995), 325 – 345.

[21] A. Givental. *Equivariant Gromov – Witten invariants.* IMRN, 1996, 613–663.

[22] A. Givental. *A mirror theorem for toric complete intersections.* In: Topological field theory, primitive forms and related topics (Kyoto, 1996), 141–175, Progr. in Math., **160**, Birkhäuser, Boston, MA, 1998.

[23] A. Givental. *On the WDVV-equation in quantum K-theory*, Mich. Math. J. **48** (2000), 295 – 304.

[24] A. Givental. *Semisimple Frobenius structures at higher genus.* Intern. Math. Res. Notices, 2001, No. 23, 1265–1286.

[25] A. Givental. *Gromov – Witten invariants and quantization of quadratic hamiltonians.* Moscow Mathematical Journal, v.1(2001), no. 4, 551–568.

[26] A. Givental. *A_{n-1}-singularities and nKdV hierarchies.* Moscow Mathematical Journal, v.3(2003), no. 2, 475 – 505.

[27] A. Givental, T. Milanov. *ADE singularities and integrable hierarchies.* arXiv: math.AG/0307176.

[28] M. Gromov. *Pseudo-holomorphic curves in symplectic manifolds.* Invent. Math. **82** (1985), 307 – 347.

[29] C. Hertling. *Frobenius manifolds and moduli spaces for singularities.* Cambridge Tracts in Mathematics. Cambridge University Press, 2002, 280 pp.

[30] T. J. Jarvis, T. Kimura. *Orbifold quantum cohomology of the classifying space of a finite group.* arXiv: math.AG/0112037.

[31] D. Joe, B. Kim. *Equivariant mirrors and the Virasoro conjecture for flag manifolds.* IMRN (2003), No. 15, 859 – 882, arXiv: math.AG/0210337.

[32] T. Kawasaki. *The signature theorem for V-manifolds.* Topology **17** (1978), 75–83.

[33] M. Kontsevich *Intersection theory on the moduli space of curves and the matrix Airy function.* Commun. Math. Phys. **147** (1992), 1 – 23.

[34] M. Kontsevich. *Enumeration of rational curves via toric actions.* In: The Moduli Space of Curves (R, Dijkgraaf, C. Faber, G. van der Geer, eds.) Progr. in Math. 129, Birkhäuser, Boston, 1995, 335–368. M. Kontsevich.

[35] M. Kontsevich, Yu. Manin. *Relations between the correlators of the topological sigma-model coupled to gravity.* Commun. Math. Phys. **196** (1998), 385 – 398. arXiv: alg-geom/9708024.

[36] Y.-P. Lee. *Quantum Lefschetz hyperplane theorem.* Invent. Math. **145**, No. 1 (2001), pp 121-149. arXiv: math.AG/0003128.

[37] Y.-P. Lee. *Quantum K-theory I: foundations.* arXiv: math.AG/0105014

[38] J. Li, G. Tian. *Virtual moduli cycles and Gromov-Witten invariants of algebraic varieties.* J. Amer. Math. Soc. **11** (1998), no. 1, 119–174.

[39] Yu. I. Manin. *Frobenius manifolds, quantum cohomology, and moduli spaces.* AMS Colloquium Publ. 47, Providence, RI, 1999, 303 pp.

[40] J. Morava. *Tate cohomology of framed circle actions as a Heisenberg group.* arXiv: math.AT/0102132.

[41] D. Mumford. *Towards enumerative geometry on the moduli space of curves.* In: Arithmetics and Geometry (M. Artin, J. Tate eds.), v.2, Birkhäuser, 1983, 271 – 328.

[42] A. Okounkov, R. Pandharipande. *Gromov-Witten theory, Hurwitz numbers, and Matrix models, I.* arXiv: math.AG/0101147.

[43] Y. Ruan, G. Tian. *Higher genus symplectic invariants and sigma model coupled with gravity.* arXiv: alg-geom/9601005.

[44] E. Witten. *Two-dimensional gravity and intersection theory on moduli space.* Surveys in Diff. Geom. 1 (1991), 243–310.

Unfoldings of meromorphic connections and a construction of Frobenius manifolds

Claus Hertling and Yuri Manin

1. Introduction

Let $\widetilde{M}$ be a complex manifold. *A structure of a Frobenius manifold on* $\widetilde{M}$ defined by B. Dubrovin consists of several pieces of data of which the most important are: (a) the choice of a flat structure on $\widetilde{M}$ represented by a subsheaf of flat vector fields $\mathcal{T}^f_{\widetilde{M}}$ of the sheaf of holomorphic vector fields $\mathcal{T}_{\widetilde{M}}$; (b) a commutative and associative $\mathcal{O}_{\widetilde{M}}$-bilinear multiplication $\circ$ on $\mathcal{T}_{\widetilde{M}}$.

Let $\nabla^{(0)}$ be the unique torsionless holomorphic connection on the holomorphic tangent bundle $T\widetilde{M}$ with kernel $\mathcal{T}^f_{\widetilde{M}}$. Then the axiom connecting (a) and (b) is the following requirement: $\nabla_{z,X} := \nabla^{(0)}_X - \frac{1}{z}X\circ$ is a pencil (with parameter z) of flat connections on $\widetilde{M}$, where $X \in \mathcal{T}_{\widetilde{M}}$. It was called the first structure connection in [**Man2**]. We use here its more sophisticated version: cf. lemma 4.4 and theorem 4.2.

Additional pieces of data include a flat holomorphic metric g, a flat identity vector field e for $\circ$, and an Euler field E: see definition 4.3 below for the complete list of their properties and interrelations.

Three general constructions of Frobenius manifolds are known: K. Saito's Frobenius structures on unfolding spaces of singularities; quantum cohomology (physicists' A-models); and Frobenius structures on the extended moduli spaces of Calabi–Yau manifolds (Barannikov–Kontsevich construction, considerably generalized in [**Ba1**][**Ba2**], physicists' B-models).

Some of the important issues of this theory require identification of Frobenius manifolds obtained by different constructions. The famous Mirror conjecture of Candelas et al. belongs to this class; it stimulated much of research.

A general strategy for establishing such an isomorphism consists in showing that two Frobenius manifolds to be compared are determined by a more restricted set of data, and then identifying these data. For example, if $\widetilde{M}$ is endowed with a point 0 at which $\circ$ is semisimple and $E\circ$ has a simple spectrum, then a finite number of numbers suffices in order to reconstruct the whole Frobenius germ $(\widetilde{M}, 0)$. This is a result of Dubrovin [**Du**, Lecture 3]. Generally, however, this is not true; in particular, A- and B-models fail to satisfy the semisimplicity restriction.

In this case it may happen that for an appropriate submanifold $M \subset \widetilde{M}$ the restriction $H := T\widetilde{M}|_M$ endowed with the restriction of the structure connection and some additional remnants of Frobenius structure ("initial conditions") carry sufficient information in order to reconstruct $\widetilde{M}$ uniquely. A statement of this kind will be especially useful if one can ensure that arbitrary initial conditions of a given type come from a Frobenius structure ("strong reconstruction theorem").

One of the main results of this paper (theorem 4.5) consists in exhibiting such a set of initial data (see definition 4.1 (b)). A special case of this result having direct applications is the theorem 5.6 which shows that one can construct a unique Frobenius manifold from any variation of Hodge filtrations over M and an opposite filtration assuming that a certain condition which we call *H^2-generation* holds (definition 5.3).

A prototype of theorem 5.6 is the first reconstruction theorem in [**KM**, Theorem 3.1]. The uniqueness statement in 5.6 is essentially that of [**KM**]. It was already used in [**Ba1**, Theorem 6.5]. The existence is the new part. The more general theorem 4.5 is applicable to the germs of semisimple Frobenius manifolds $(\widetilde{M}, 0)$ mentioned above as well; in this case M is the marked point, and one recovers Dubrovin's result.

The initial datum on M can be reformulated as a meromorphic connection on a bundle on $\mathbb{P}^1 \times M$ with a logarithmic pole along $\{\infty\} \times M$ and a pole of Poincaré rank 1 along $\{0\} \times M$. So theorem 4.5 gives a recipe for constructing Frobenius manifolds: One has to construct such connections.

The proof of theorem 4.5 relies on the study of unfoldings of germs of such connections in the chapters 2 and 3. The main result there is theorem 2.5. It shows that a germ of a meromorphic connection on a bundle on $(\mathbb{C}, 0) \times (M, 0)$ with a pole of Poincaré rank 1 along $\{0\} \times M$ has a universal unfolding if again a certain generation condition is satisfied. It generalizes the case $M = \{pt\}$ which was treated in [**Mal**, (4.1)].

With theorem 2.5 one can reduce theorem 4.5 to the case when $M = \widetilde{M}$. This case was already known. It was formulated by Sabbah [**Sab1**, Theorem (4.3.6)] [**Sab2**, Théorème VII 3.6], and independently by Barannikov [**Ba1**][**Ba2**][**Ba3**]. A major part of the initial data is equivalent to Barannikov's notion of *a semi-infinite variation of Hodge structures.* Theorem 4.5 in the case $M = \widetilde{M}$ is also implicit in the construction in singularity theory [**SK**][**SM**].

Chapter 6 is devoted to a class of variations of Hodge structures to which theorem 5.6 applies, the variations of Hodge structures on the primitive parts of the middle cohomologies of smooth hypersurfaces in $\mathbb{P}^n$ whose degrees divide $n+1$ (theorem 6.1). The corresponding Frobenius manifolds turn up as submanifolds in two different classes of Frobenius manifolds.

One class arises in singularity theory. The base space of a semiuniversal unfolding of an isolated hypersurface singularity can be equipped with the structure of a Frobenius manifold [**SK**][**SM**][**He1**]. Suppose that the singularity is a homogeneous polynomial in $\mathbb{C}[x_0, ..., x_n]$ of a degree which divides $n+1$. The submanifold of the base space which parametrizes homogeneous deformations carries a variation of Hodge structures isomorphic to the one of the projective hypersurfaces in $\mathbb{P}^n$. Then a larger submanifold which parametrizes certain semihomogeneous deformations is a Frobenius submanifold. It is the one determined by the variation of Hodge filtrations and an opposite filtration. This is discussed in chapter 7.

The other class of Frobenius manifolds arises from the Barannikov–Kontsevich construction [**BK**][**Ba1**][**Ba2**]. There for any Calabi–Yau manifold a formal germ of a space which extends the moduli space of complex structure deformations is constructed and is equipped with a semi-infinite variation of Hodge structures. Together with the choice of an opposite filtration this induces the structure of a Frobenius super manifold on the extended moduli space. In the case of a Calabi–Yau hypersurface in $\mathbb{P}^n$ it turns out [**Ba1**, Theorem 6.5] that the whole semi-infinite variation of Hodge structures is determined by the variation of Hodge structures on the moduli space of complex structure deformations. The whole chapter 8 is a discussion and reformulation of this result of Barannikov.

In the case of a homogeneous polynomial of degree $n+1$, the results in chapters 7 and 8 give for suitable opposite filtrations isomorphic Frobenius submanifolds in the two classes of Frobenius manifolds.

We thank Dennis Borisov for some discussions about these Frobenius manifolds and Claude Sabbah for remark 2.10.

Index of notations and definitions. Both basic structures, that of Frobenius manifolds and initial data, admit many useful weakenings and variations. For reader's convenience, we compiled an index of the versions used in this paper and related notions.

Frobenius manifolds: def. 4.3.
Frobenius type structure: def. 4.1 (b).
H^2-generated germ of a Frobenius manifold of weight w: def 5.4.
H^2-generated germ of a variation of filtrations of weight w: def. 5.3 (a).
Higgs field (general): lemma 2.4.
Higgs field of a Frobenius manifold: def. 4.3.
(L)-structure: def. 3.1 (c).
(LE)-structure: remark 3.3 (i).
$(LEP(w))$-structure: def. 3.1 (b).
$(LP(w))$-structure: def. 3.1 (d).
(T)-structure: def. 3.1 (c).
(TE)-structure: def. 2.1 (b).
$(TEP(w))$-structure: def. 3.1 (a).
$(TP(w))$-structure: def. 3.1 (d).
$(trTLEP(w))$-structure: def. 4.1 (a).

2. Unfoldings of meromorphic connections

Theorem 2.5 below generalizes a result of Malgrange [**Mal**, (4.1)]. We start with the same setting as in [**Mal**].

DEFINITION 2.1. (a) Consider a germ $(M,0)$ of a complex manifold, a germ $H \to (\mathbb{C},0)\times(M,0)$ of a holomorphic vector bundle, and a flat connection ∇ on the restriction of H to $(\mathbb{C}^*,0)\times(M,0)$. Let z be the coordinate on $(\mathbb{C},0)$. The connection ∇ has a pole of Poincaré rank $r\in\mathbb{Z}_{\geq 0}$ along $\{0\}\times(M,0)$ if

$$(2.1)\qquad \nabla : \mathcal{O}(H) \to \frac{1}{z^r}\left[\mathcal{O}_{\mathbb{C}\times M,0}\cdot\Omega^1_{M,0} + \mathcal{O}_{\mathbb{C}\times M,0}\cdot\frac{dz}{z}\right]\otimes\mathcal{O}(H).$$

Here $\mathcal{O}(H)$ is the $\mathcal{O}_{\mathbb{C}\times M,0}$-module of germs of holomorphic sections of H. A pole of Poincaré rank 0 is called a logarithmic pole.

(b) A (TE)-structure is a tuple $((M,0),H,\nabla)$ as in (a) with a pole of Poincaré rank 1.

REMARKS 2.2. (i) The results in [**Mal**] are formulated also for poles of higher Poincaré rank. We restrict to Poincaré rank 1 because that is the case needed in the later chapters and because a proof of a generalization of theorem 2.5, if possible, would be much more technical.

(ii) The notations (TE), $(TEP(w))$, etc. in the definitions 2.1 and 3.1 are compatible with those in [**He2**]. Here 'T' stands for Twistor, 'E' for Extension (in z-direction), 'P' for Pairing, and w is an integer.

If $((M,0),H,\nabla)$ is a (TE)-structure and $\varphi:(M',0)\to(M,0)$ a holomorphic map of germs of manifolds then one can pull back H and ∇ with $\mathrm{id}\times\varphi:(\mathbb{C},0)\times(M',0)\to(\mathbb{C},0)\times(M,0)$. One easily sees that $\varphi^*(H,\nabla)$ gives a (TE)-structure on $(M',0)$.

DEFINITION 2.3. Fix a (TE)-structure $((M,0),H,\nabla)$.

(a) An *unfolding* of it is a (TE)-structure $((M\times\mathbb{C}^l,0),\widetilde{H},\widetilde{\nabla})$ together with a fixed isomorphism

$$i:((M,0),H,\nabla)\to((M\times\mathbb{C}^l,0),\widetilde{H},\widetilde{\nabla})|_{(M\times\{0\},0)}. \tag{2.2}$$

(b) One unfolding $((M\times\mathbb{C}^l,0),\widetilde{H},\widetilde{\nabla},i)$ *induces* a second unfolding $((M\times\mathbb{C}^{l'},0),\widetilde{H}',\widetilde{\nabla}',i')$ if there is an isomorphism j from the second unfolding to the pullback of the first unfolding by a map

$$\varphi:(M\times\mathbb{C}^{l'},0)\to(M\times\mathbb{C}^l,0) \tag{2.3}$$

which is the identity on $(M\times\{0\},0)$, and if

$$i=j|_{(M\times\{0\},0)}\circ i'. \tag{2.4}$$

(Then j is uniquely determined by φ and (2.4).)

(c) An unfolding is *universal* if it induces any unfolding via a unique map φ.

By definition, a universal unfolding of a (TE)-structure is unique up to canonical isomorphism if it exists. For existence we need some conditions which are formulated in terms of the data of the following lemma.

LEMMA 2.4. [**Sab2**][**He2**, ch. 2] *Let $((M,0),H,\nabla)$ be a (TE)-structure. Define the germ of a holomorphic vector bundle $K:=H|_{\{0\}\times(M,0)}$ on $(M,0)$ with $\mathcal{O}_{M,0}$-module $\mathcal{O}(K)$ of germs of holomorphic sections. Define maps*

$$C:=[z\nabla]\ :\ \mathcal{O}(K)\to\Omega^1_{M,0}\otimes\mathcal{O}(K) \tag{2.5}$$
$$C_X[a]=[z\nabla_X a]\ \text{for}\ X\in\mathcal{T}_{M,0},\ a\in\mathcal{O}(H)$$

and

$$\mathcal{U}:=[z\nabla_{z\partial_z}]:\mathcal{O}(K)\to\mathcal{O}(K); \tag{2.6}$$

here $[\ ]$ means restriction to $\{0\}\times(M,0)$ and $X\in\mathcal{T}_{M,0}$ is lifted canonically to $(\mathbb{C},0)\times(M,0)$.

The maps C_X, $X\in\mathcal{T}_{M,0}$, and $\mathcal{U}$ are commuting $\mathcal{O}_{M,0}$-linear endomorphisms of $\mathcal{O}(K)$. Therefore the map C is a Higgs field on K.

Proof. This follows from the flatness of ∇ and the fact that ∇ has a pole of Poincaré rank 1.

For a general discussion of Higgs fields see e.g. [**Sab2**, p. 36]. A symmetric Higgs field on the tangent bundle is the same as a commutative and associative $\mathcal{O}_M$-bilinear multiplication. □

THEOREM 2.5. *Let* $((M,0),H,\nabla)$ *be a* (TE)*-structure with data* $(K,C,\mathcal{U})$ *as in lemma 2.4. Suppose that there exists a vector* $\zeta\in K_0$ *in the fiber* K_0 *of* K *at 0 such that*

(IC) (injectivity condition) the map

$$C_\bullet\zeta : T_0M\to K_0,\quad X\mapsto C_X\zeta \tag{2.7}$$

is injective and

(GC) (generation condition) ζ *and its images under iteration of the maps* $\mathcal{U}:K_0\to K_0$ *and* $C_X:K_0\to K_0$, $X\in T_0M$, *generate* K_0.

Then a universal unfolding of $((M,0),H,\nabla)$ *exists. An unfolding*

$$((M\times\mathbb{C}^l,0),\widetilde{H},\widetilde{\nabla},i)$$

with data $(\widetilde{K},\widetilde{C},\widetilde{\mathcal{U}})$ *as in lemma 2.4 is a universal unfolding if and only if the map*

$$\widetilde{C}_\bullet i(\zeta) : T_0(M\times\mathbb{C}^l)\to\widetilde{K}_0,\quad X\mapsto\widetilde{C}_X i(\zeta) \tag{2.8}$$

is an isomorphism.

The rest of this chapter is devoted to the proof of this theorem. In [**Mal**, (4.1)] the case $M=\{pt\}$ is considered. In this case the generation condition (GC) is satisfied if and only if $\mathcal{U}:K_0\to K_0$ has for each eigenvalue only one Jordan block. For example, if $\mathcal{U}:K_0\to K_0$ is semisimple, it must have simple eigenvalues. But in chapter 5 we will use theorem 2.5 in the case $\mathcal{U}=0$.

The first part of the proof follows [**Mal**], using an extension of H to a bundle on $\mathbb{P}^1\times(M,0)$ and using the rigidity properties of logarithmic poles, which are formulated in lemma 2.6.

LEMMA 2.6. *(a) (e.g.* [**Sab2**, III.1.20]*) Let* $(L'\to(\mathbb{C}^*,0)\times(M,0),\nabla)$ *be the germ of a holomorphic vector bundle with flat connection* ∇*, and let* $(L^{(0)}\to(\mathbb{C},0)\times\{0\},\nabla)$ *be an extension of* $(L',\nabla)|_{(\mathbb{C}^*,0)\times\{0\}}$ *with a logarithmic pole at 0. Then an extension* $(L\to(\mathbb{C},0)\times(M,0),\nabla)$ *of* (L',∇) *with a logarithmic pole along* $\{0\}\times(M,0)$ *exists with* $(L,\nabla)|_{(\mathbb{C},0)\times\{0\}}=(L^{(0)},\nabla)$*. It is unique up to canonical isomorphism and it is isomorphic to the pullback* $\varphi^*(L^{(0)},\nabla)$ *where* $\varphi:(M,0)\to\{0\}$.

(b) (e.g. [**He2**, 5.1]*) Let* $(L\to(\mathbb{C},0)\times(M,0),\nabla)$ *be (the germ of) a holomorphic vector bundle with a flat connection* ∇ *over* $(\mathbb{C}^*,0)\times(M,0)$ *and a logarithmic pole along* $\{0\}\times(M,0)$*. The bundle* $L|_{\{0\}\times(M,0)}$ *is equipped with a flat connection* ∇^{res}*, the residual connection, and a* ∇^{res}*-flat endomorphism* $\mathcal{V}^{res}$*, the residue endomorphism. They are defined by*

$$\nabla^{res}_X[a] := [\nabla_X a]\quad \text{for } X\in\mathcal{T}_{M,0}, a\in\mathcal{O}(L), \tag{2.9}$$

$$\mathcal{V}^{res}[a] := [\nabla_{z\partial_z}a]\quad \text{for } a\in\mathcal{O}(L). \tag{2.10}$$

LEMMA 2.7. *Let* $(H\to(\mathbb{C},0)\times(M,0),\nabla)$ *be a germ of a holomorphic vector bundle with a flat connection* ∇ *over* $(\mathbb{C}^*,0)\times(M,0)$*. There exists an extension to a trivial bundle* $H^{(gl)}\to\mathbb{P}^1\times(M,0)$ *with flat connection* ∇ *over* $(\mathbb{C}^*-\{1\})\times(M,0)$*, with logarithmic poles along* $\{1\}\times(M,0)$ *and* $\{\infty\}\times(M,0)$*, and with trivial monodromy around* $\{1\}\times(M,0)$.

Proof. We follow [**Mal**, ch. 3]. First, one extends H to a bundle $H^{(\mathbb{C})}\to\mathbb{C}\times(M,0)$ with flat connection ∇ over $\mathbb{C}^*\times(M,0)$. Then one can choose an extension of $H^{(\mathbb{C})}$ to a bundle $H^{(\mathbb{P}^1)}\to\mathbb{P}^1\times(M,0)$ with a logarithmic pole along $\{\infty\}\times(M,0)$. This bundle is not necessarily trivial.

One chooses an $\mathcal{O}_{M,0}$-basis $\sigma_1, ..., \sigma_{\mathrm{rk}\, H}$ of germs of flat sections of $\mathcal{O}(H^{(\mathbb{P}^1)})_{(1,0)}$ and defines for $r = (r_1, ..., r_{\mathrm{rk}\, H}) \in \mathbb{Z}^{\mathrm{rk}\, H}$ a bundle $H^{(r)} \to \mathbb{P}^1 \times (M,0)$ as follows: on $(\mathbb{P}^1 - \{1\}) \times (M,0)$ it coincides with $H^{(\mathbb{P}^1)}$; the germ $\mathcal{O}(H^{(r)})_{(1,0)}$ is generated by the sections $(z-1)^{r_i}\sigma_i$. The bundle $H^{(r)}$ has a logarithmic pole along $\{1\} \times (M,0)$. For some r the restriction $H^{(r)}|_{\mathbb{P}^1 \times \{0\}}$ is trivial [**Mal**, (3.2)]. Because being a trivial bundle on $\mathbb{P}^1$ is an open property, the bundle $H^{(gl)} := H^{(r)}$ for such an r is trivial. □

REMARK 2.8. If one applies lemma 2.7 to a (TE)-structure

$$(H \to (\mathbb{C},0) \times (M,0), \nabla),$$

then the extension $(H^{(gl)}, \nabla)$ has two distinguished properties:

(i) Because of lemma 2.6 (a), any unfolding $(\widetilde{H} \to (\mathbb{C},0) \times (M \times \mathbb{C}^l,0), \widetilde{\nabla})$ of $(H, \widetilde{\nabla})$ has a unique extension $(\widetilde{H}^{(gl)} \to \mathbb{P}^1 \times (M \times \mathbb{C}^l,0), \widetilde{\nabla})$ with all the properties in lemma 2.7 whose restriction to $\mathbb{P}^1 \times (M \times \{0\},0)$ coincides with $(H^{(gl)}, \nabla)$.

(ii) Denote by ∇^{res} the residual connection on $H|_{\{\infty\}\times(M,0)}$. The space

$$(2.11) \qquad V(H^{(gl)}) := \{ \text{ global hol. sections } v \text{ in } H^{(gl)} \text{ with } \nabla^{res}(v|_{\{\infty\}\times(M,0)}) = 0\}$$

is a vector space of dimension $\mathrm{rk}\, H$. A basis of it is an $\mathcal{O}_{(\mathbb{P}^1,z)\times(M,0)}$-basis of $\mathcal{O}(H^{(gl)})_{(z,0)}$ for any $z \in \mathbb{P}^1$. Below we will work with the connection matrix with respect to such a basis. For an extension $(\widetilde{H}^{(gl)}, \widetilde{\nabla})$ as in (i) the sections in $V(H^{(gl)})$ extend uniquely to the sections in $V(\widetilde{H}^{(gl)})$.

Using lemma 2.7 and these observations we can control the unfoldings of (TE)-structures as in theorem 2.5. The following lemma is the main step in its proof.

LEMMA 2.9. *Let $(H^{(gl)} \to \mathbb{P}^1 \times (M,0), \nabla)$ be a trivial holomorphic vector bundle of rank $n \geq 1$ with a flat connection over $(\mathbb{C}^* - \{1\}) \times (M,0)$, with logarithmic poles along $\{1\} \times (M,0)$ and $\{\infty\} \times (M,0)$ and with a pole of Poincaré rank 1 along $\{0\} \times (M,0)$. Define $K := H^{(gl)}|_{\{0\}\times(M,0)}$, C and $\mathcal{U}$ as in lemma 2.4. Suppose that the generation condition (GC) of theorem 2.5 is satisfied for a vector $\zeta \in K_0$. Choose a basis $v_1, ..., v_n$ of $V(H^{(gl)})$ with $v_1|_{(0,0)} = \zeta$. Choose $l \in \mathbb{N}$ and n functions $f_1, ..., f_n \in \mathcal{O}_{M\times\mathbb{C}^l,0}$ with $f_i|_{(M\times\{0\},0)} = 0$. Let $(t_1, ..., t_m, y_1, ..., y_l) = (t, y)$ be coordinates on $(M \times \mathbb{C}^l, 0)$.*

Then there exists a unique unfolding $(\widetilde{H}^{(gl)} \to \mathbb{P}^1 \times (M \times \mathbb{C}^l,0), \widetilde{\nabla})$ of $(H^{(gl)}, \nabla)$ with the following properties. $\widetilde{H}^{(gl)}$ is a trivial vector bundle with a flat connection over $(\mathbb{C}^* - \{1\}) \times (M \times \mathbb{C}^l, 0)$, with logarithmic poles along $\{1\} \times (M \times \mathbb{C}^l, 0)$ and $\{\infty\} \times (M \times \mathbb{C}^l, 0)$ and a pole of Poincaré rank 1 along $\{0\} \times (M \times \mathbb{C}^l, 0)$. Its restriction to $\mathbb{P}^1 \times (M \times \{0\}, 0)$ is $(H^{(gl)}, \nabla)$. Let $\widetilde{v}_1, ..., \widetilde{v}_n \in V(\widetilde{H}^{(gl)})$ be the canonical extensions of $v_1, ..., v_n \in V(H^{(gl)})$. Define $\widetilde{K} := \widetilde{H}^{(gl)}|_{\{0\}\times(M\times\mathbb{C}^l,0)}$, $\widetilde{C}$ and $\widetilde{U}$ as in lemma 2.4. Then

$$(2.12) \qquad \widetilde{C}_{\partial/\partial y_\alpha} \widetilde{v}_1 = \sum_{i=1}^{n} \frac{\partial f_i}{\partial y_\alpha} \widetilde{v}_i \quad \textit{for} \quad \alpha = 1, ..., l.$$

Proof. Suppose for a moment that $(\widetilde{H}^{(gl)}, \widetilde{\nabla})$ were already constructed. The connection matrix Ω with respect to the basis $\widetilde{v}_1, ..., \widetilde{v}_n$,

$$(2.13) \qquad \widetilde{\nabla}(\widetilde{v}_1, ..., \widetilde{v}_n) = (\widetilde{v}_1, ..., \widetilde{v}_n) \cdot \Omega,$$

would take the form

$$\Omega = \frac{1}{z}\sum_{i=1}^{m} C_i \mathrm{d}t_i + \frac{1}{z}\sum_{\alpha=1}^{l} F_\alpha \mathrm{d}y_\alpha + \left(\frac{1}{z^2}U + \frac{1}{z}V + \frac{1}{z-1}W\right)\mathrm{d}z \tag{2.14}$$

with matrices

$$C_i, F_\alpha, U, V, W \in M(n \times n, \mathcal{O}_{M\times\mathbb{C}^l,0}). \tag{2.15}$$

This follows from $\widetilde{\nabla}^{res}\widetilde{v}_i|_{\{\infty\}\times(M\times\mathbb{C}^l,0)} = 0$ and from the pole orders of $(\widetilde{H}^{(gl)}, \widetilde{\nabla})$ along $\{0,1,\infty\} \times (M \times \mathbb{C}^l, 0)$. The flatness condition $\mathrm{d}\Omega + \Omega \wedge \Omega = 0$ could be written as

$$[C_i, C_j] = 0 \tag{2.16}$$
$$[C_i, F_\alpha] = 0, \tag{2.17}$$
$$[F_\alpha, F_\beta] = 0, \tag{2.18}$$
$$\frac{\partial C_i}{\partial t_j} = \frac{\partial C_j}{\partial t_i}, \tag{2.19}$$
$$\frac{\partial C_i}{\partial y_\alpha} = \frac{\partial F_\alpha}{\partial t_i}, \tag{2.20}$$
$$\frac{\partial F_\alpha}{\partial y_\beta} = \frac{\partial F_\beta}{\partial y_\alpha}, \tag{2.21}$$
$$[C_i, U] = 0, \tag{2.22}$$
$$[F_\alpha, U] = 0, \tag{2.23}$$
$$\frac{\partial U}{\partial t_i} = [V, C_i] - C_i, \tag{2.24}$$
$$\frac{\partial U}{\partial y_\alpha} = [V, F_\alpha] - F_\alpha, \tag{2.25}$$
$$\frac{\partial W}{\partial t_i} = [W, C_i], \tag{2.26}$$
$$\frac{\partial W}{\partial y_\alpha} = [W, F_\alpha], \tag{2.27}$$
$$\frac{\partial V}{\partial t_i} = -[W, C_i], \tag{2.28}$$
$$\frac{\partial V}{\partial y_\alpha} = -[W, F_\alpha]. \tag{2.29}$$

The condition (2.12) would mean

$$(F_\alpha)_{i1} = \frac{\partial f_i}{\partial y_\alpha}. \tag{2.30}$$

The proof consists of three parts. In parts (I) and (II) we restrict to the case $l = \alpha = 1$. In part (I) we show inductively uniqueness and existence of matrices C_i, F_1, U, V, W with (2.16) - (2.30) and with coefficients in $\mathcal{O}_{M,0}[[y_1]]$. In part (II) their convergence will be proved with the Cauchy–Kovalevski theorem. In part (III) the general case will be proved by induction in l.

In remark 2.10 the system of equations (2.16) - (2.29) will be written in a more compact form after some integration.

Part (I). Suppose $l=\alpha=1$ and $y_1=y$. Define for $w\in\mathbb{Z}_{\geq 0}$

$$\mathcal{O}_{M,0}[y]_{\leq w} := \sum_{k=0}^{w}\mathcal{O}_{M,0}\cdot y^k, \tag{2.31}$$

$$\mathcal{O}_{M,0}[y]_{>w} := \mathcal{O}_{M,0}[y]\cdot y^{w+1}, \tag{2.32}$$

$$\mathcal{O}_{M,0}[[y]]_{>w} := \mathcal{O}_{M,0}[[y]]\cdot y^{w+1} \tag{2.33}$$

and

$$M(w) := M(n\times n,\mathcal{O}_{M,0}\cdot y^w), \tag{2.34}$$

$$M(>w) := M(n\times n,\mathcal{O}_{M,0}[y]_{>w}), \tag{2.35}$$

$$M(\leq w) := M(n\times n,\mathcal{O}_{M,0}[y]_{\leq w}). \tag{2.36}$$

Beginning of the induction for $w=0$: The connection matrix $\Omega^{(0)}$ of $(H^{(gl)},\nabla)$ with respect to the basis $v_1,...,v_n$ takes the form

$$\Omega^{(0)} = \frac{1}{z}\sum_{i=1}^{m} C_i^{(0)}\mathrm{d}t_i + \left(\frac{1}{z^2}U^{(0)}+\frac{1}{z}V^{(0)}+\frac{1}{z-1}W^{(0)}\right)\mathrm{d}z \tag{2.37}$$

with matrices $C_i^{(0)},U^{(0)},V^{(0)},W^{(0)}\in M(0)$. The flatness condition $\mathrm{d}\Omega^{(0)}+\Omega^{(0)}\wedge\Omega^{(0)}=0$ is equivalent to the equations (2.16), (2.19), (2.22), (2.24), (2.26), (2.28) for the matrices $C_i^{(0)}$, $U^{(0)}$, $V^{(0)}$, $W^{(0)}$ instead of C_i,U,V,W.

Induction hypothesis for $w\in\mathbb{Z}_{\geq 0}$: Unique matrices $C_i^{(k)},U^{(k)},V^{(k)},W^{(k)}\in M(k)$ for $0\leq k\leq w$ and $F_1^{(k)}\in M(k)$ for $0\leq k\leq w-1$ are constructed such that the matrices

$$C_i^{(\leq w)} := \sum_{k=0}^{w} C_i^{(k)}\in M(\leq w) \tag{2.38}$$

and the analogously defined matrices $U^{(\leq w)},V^{(\leq w)},W^{(\leq w)},F_1^{(\leq w-1)}$ satisfy (2.16), (2.19), (2.22), (2.24), (2.26), (2.28) modulo $M(>w)$, (2.17), (2.20), (2.23), (2.25), (2.27), (2.29) modulo $M(>w-1)$ and (2.30) modulo $\mathcal{O}_{M,0}[[y]]_{>w-1}$.

Induction step from w to $w+1$: It consists of three steps:

(i) Construction of a matrix $F_1^{(w)}\in M(w)$ such that the matrix $F_1^{(\leq w)}=F_1^{(\leq w-1)}+F_1^{(w)}$ together with the matrices $C_i^{(\leq w)},U^{(\leq w)},V^{(\leq w)},W^{(\leq w)}$ satisfies (2.17), (2.23) modulo $M(>w)$ and (2.30) modulo $\mathcal{O}_{M,0}[[y]]_{>w}$.

(ii) Construction of matrices $C_i^{(w+1)},U^{(w+1)},V^{(w+1)},W^{(w+1)}\in M(w+1)$ such that the matrices $C_i^{(\leq w+1)}=C_i^{(\leq w)}+C_i^{(w+1)}$, the analogously defined matrices $U^{(\leq w+1)},V^{(\leq w+1)},W^{(\leq w+1)}$ and the matrix $F_1^{(\leq w)}$ satisfy (2.20), (2.25), (2.27), (2.29) modulo $M(>w)$.

(iii) Proof of (2.16), (2.19), (2.22), (2.24), (2.26), (2.28) modulo $M(>w+1)$ for these matrices.

(i) The matrices $C_i^{(\leq w)}$ and $U^{(\leq w)}$ generate an algebra of commuting matrices in $M(n\times n,\mathcal{O}_{M,0}[y]/\mathcal{O}_{M,0}[y]_{>w})$. Because of the generation condition (GC), the image of the column vector $(1,0,...,0)^{tr}$ under the action of this algebra is the whole space $M(n\times 1,\mathcal{O}_{M,0}[y]/\mathcal{O}_{M,0}[y]_{>w})$. This shows two things:

(α) This algebra contains for any $i = 1, ..., n$ a matrix $E_i^{(\leq w)} \in M(n \times n, \mathcal{O}_{M,0}[y]_{\leq w})$ with first column

$$(E_i^{(\leq w)})_{j1} = \delta_{ij}. \tag{2.39}$$

(β) Any matrix in $M(n \times n, \mathcal{O}_{M,0}[y]_{\leq w})$ which commutes with the matrices $C_i^{(\leq w)}$ and $U^{(\leq w)}$ modulo $M(> w)$ is modulo $M(> w)$ a linear combination of the matrices $E_i^{(\leq w)}$ with coefficients in $\mathcal{O}_{M,0}[y]_{\leq w}$.

Therefore the matrix $F_1^{(\leq w)} \in M(\leq w)$ which is defined by

$$F_1^{(\leq w)} = \sum_{i=1}^{n} \frac{\partial f_i}{\partial y} \cdot E_i^{(\leq w)} \text{ modulo } M(n \times n, \mathcal{O}_{M,0}[[y]]_{>w}) \tag{2.40}$$

is the unique matrix which satisfies (2.17), (2.23) modulo $M(> w)$, (2.30) modulo $\mathcal{O}_{M,0}[[y]]_{>w}$ and $F_1^{(\leq w)} = F_1^{(\leq w-1)} + F_1^{(w)}$ for some $F_1^{(w)} \in M(w)$.

(ii) This step is obvious.

(iii) One checks that the derivatives by $\frac{\partial}{\partial y}$ of the equations (2.16), (2.19), (2.22), (2.24), (2.26), (2.28) modulo $M(> w + 1)$ hold. For this one uses (2.17), (2.19), (2.20), (2.23) - (2.29) modulo $M(> w)$. For example one calculates modulo $M(> w)$

$$\begin{aligned} & \frac{\partial}{\partial y}[C_i^{(\leq w+1)}, U^{(\leq w+1)}] \\ \equiv & \left[\frac{\partial F_1^{(\leq w)}}{\partial t_i}, U^{(\leq w)}\right] + [C_i^{(\leq w)}, [V^{(\leq w)}, F_1^{(\leq w)}] - F_1^{(\leq w)}] \\ \equiv & -\left[F_1^{(\leq w)}, \frac{\partial U^{(\leq w)}}{\partial t_i}\right] + [F_1^{(\leq w)}, [V^{(\leq w)}, C_i^{(\leq w)}] - C_i^{(\leq w)}] \\ \equiv & \ 0 \end{aligned} \tag{2.41}$$

The other calculations are similar or easier. This finishes the proof of the induction step from w to $w+1$. It shows uniqueness and existence of matrices $C_i, F_1, U, V, W \in M(n \times n, \mathcal{O}_{M,0}[[y]])$ with (2.16) - (2.30) and with restrictions $(C_i, U, V, W)|_{y=0} = (C_i^{(0}, U^{(0)}, V^{(0)}, W^{(0)})$.

Part (II). We have to show holomorphy of these matrices. We want to apply the Cauchy–Kovalevski theorem in the following form ([**Fo**, (1.31), (1.40), (1.41)]; there the setting is real analytic, but proofs and statements hold also in the complex analytic setting): Given $N \in \mathbb{N}$ and matrices $A_i, B \in M(N \times N, \mathbb{C}\{t_1, ..., t_m, y, x_1, ..., x_N\})$ there exists a unique vector

$$\Phi \in M(N \times 1, \mathbb{C}\{t_1, ..., t_m, y\})$$

with

$$\frac{\partial \Phi}{\partial y} = \sum_{i=1}^{m} A_i(t, y, \Phi) \frac{\partial \Phi}{\partial t_i} + B(t, y, \Phi), \tag{2.42}$$

$$\Phi(t, 0) = 0. \tag{2.43}$$

We will construct a system (2.42) - (2.43) with $N = (m+3)n^2$ such that it will be satisfied with the entries of the matrices $C_i - C_i^{(0)}, U - U^{(0)}, V - V^{(0)}, W - W^{(0)}$ as entries of Φ. The system will be built from the equations (2.20), (2.25), (2.27),

(2.29) and the following equations (2.44), (2.45) with which one can express the entries of F_1 as functions of the entries of Φ.

The commutative subalgebra of $M(n\times n,\mathcal{O}_{M,0}[[y]])$ which is generated by the matrices $C_1,...,C_m,U$ is a free $\mathcal{O}_{M,0}[[y]]$-module of rank n. Choose monomials $G^{(j)}$, $j=1,...,n$, in the matrices $C_1,...,C_m,U$ which form an $\mathcal{O}_{M,0}[[y]]$-basis of this module. Then the matrix $(G^{(j)}_{i1})_{ij}$ of the first columns of the matrices $G^{(j)}$ is invertible in $M(n\times n,\mathcal{O}_{M,0}[[y]])$. Equation (2.40) gives

$$F_1=\sum_{j=1}^{n} g_j G^{(j)} \tag{2.44}$$

with coefficients $g_j\in\mathcal{O}_{M,0}[[y]]$ such that

$$\left(\frac{\partial f_1}{\partial y},...,\frac{\partial f_n}{\partial y}\right)^{tr}=(G^{(j)}_{i1})\cdot(g_1,...,g_n)^{tr}. \tag{2.45}$$

Replacing the entries of the matrices $C_i-C_i^{(0)},U-U^{(0)},V-V^{(0)},W-W^{(0)}$ by indeterminates $x_1,...,x_N$, the coefficients of the matrices $G^{(j)}$ become elements of $\mathbb{C}\{t\}[x_1,...,x_N]$, and, because of (2.45), the coefficients g_j become elements of $\mathbb{C}\{t_1,...,t_m,y,x_1,...,x_N\}$. One obtains from (2.20), (2.25), (2.27), (2.29), (2.44), (2.45) a system (2.42) - (2.43). The theorem of Cauchy–Kovalevski shows

$$C_i,F_1,U,V,W\in M(n\times n,\mathcal{O}_{M\times\mathbb{C},0}).$$

This shows lemma 2.9 in the case $l=1$.

Part (III) By induction in l one obtains a slightly weaker version of lemma 2.9, where (2.12) is replaced by

$$\left(\widetilde{C}_{\partial/\partial y_\alpha}\widetilde{v}_1\right)|_{\{y_{\alpha+1}=...=y_l=0\}}=\left(\sum_{i=1}^{n}\frac{\partial f_i}{\partial y_\alpha}\widetilde{v}_i\right)|_{\{y_{\alpha+1}=...=y_l=0\}} \tag{2.46}$$

for $\alpha=1,...,l$. One has a connection matrix Ω as in (2.14). Condition (2.46) is equivalent to (2.30) with the same restriction to $y_{\alpha+1}=...=y_l=0$. But now (2.21) gives (2.30) and (2.12) for all y. This finishes the proof of lemma 2.9. □

Proof of theorem 2.5. Let $(H\to(\mathbb{C},0)\times(M,0),\nabla)$ be a (TE)-structure with $K,C,\mathcal{U}$ and $\zeta\in K_0$ with all the properties in theorem 2.5. We choose an extension to a trivial bundle with connection $(H^{(gl)}\to\mathbb{P}^1\times(M,0),\nabla)$ with the properties in lemma 2.7, and we choose a basis $v_1,...,v_n$ of the vector space $V(H^{(gl)})$ (cf. remark 2.8) with $v_1|_{(0,0)}=\zeta$.

By remark 2.8 (i) an unfolding $(\widetilde{H}\to(\mathbb{C},0)\times(M\times\mathbb{C}^l,0),\widetilde{\nabla},i)$ of (H,∇) extends to an unfolding $(\widetilde{H}^{(gl)}\to\mathbb{P}^1\times(M\times\mathbb{C}^l,0),\widetilde{\nabla})$ of $(H^{(gl)},\nabla)$ with the properties in lemma 2.7, and the sections $v_1,...,v_n$ extend uniquely to sections $\widetilde{v}_1,...,\widetilde{v}_n$ in the space $V(\widetilde{H}^{(gl)})$.

Consider the connection matrix Ω and the matrices C_i,F_α,U,V,W as in (2.13) - (2.15) for such an unfolding. Because of (2.19) - (2.21) there exists a unique matrix $A\in M(n\times n,\mathcal{O}_{M\times\mathbb{C}^l,0})$ with $A(0)=0$ and

$$\Omega=\frac{1}{z}\mathrm{d}A+\left(\frac{1}{z^2}U+\frac{1}{z}V+\frac{1}{z-1}W\right)\mathrm{d}z. \tag{2.47}$$

The matrices $A(t,0), U(t,0), V(t,0), W(t,0)$ are determined by $(H^{(gl)}, \nabla, v_1, ..., v_n)$. Lemma 2.9 says that for an *arbitrary* choice of functions $f_i(t,y) = A_{i1}(t,y) - A_{i1}(t,0)$ a *unique* unfolding $(\widetilde{H}^{(gl)}, \widetilde{\nabla})$ exists. The first columns of A give a map

$$\psi = (A_{11}, ..., A_{n1}) : (M \times \mathbb{C}^l, 0) \to (\mathbb{C}^n, 0). \tag{2.48}$$

The map in (2.8) is an isomorphism if and only if ψ is an isomorphism.

Fix an unfolding $(\widetilde{H}, \widetilde{\nabla})$ such that ψ is an isomorphism. This is possible thanks to the injectivity condition (IC). Consider a second unfolding $(\widetilde{H}' \to (\mathbb{C},0) \times (M \times \mathbb{C}^{l'}, 0), \widetilde{\nabla}')$ of (H, ∇) with $\widetilde{H'}^{(gl)}, \widetilde{v}'_1, ..., \widetilde{v}'_n, \Omega', A', \psi'$ defined analogously. If it is induced from $(\widetilde{H}, \widetilde{\nabla})$ via a map

$$\varphi : (M \times \mathbb{C}^{l'}, 0) \to (M \times \mathbb{C}^l, 0) \tag{2.49}$$

then $(\mathrm{id} \times \varphi)^* \Omega = \Omega'$ and therefore $A \circ \varphi = A'$ and

$$\psi \circ \varphi = \psi'. \tag{2.50}$$

This shows that the inducing map φ is unique.

If one does not yet know that $(\widetilde{H}', \widetilde{\nabla}')$ is induced from $(\widetilde{H}, \widetilde{\nabla})$ one can define φ by (2.50) and compare the unfoldings $(\widetilde{H}', \widetilde{\nabla}')$ and $\varphi^*(\widetilde{H}, \widetilde{\nabla})$. Since the first columns of the matrix A' and the corresponding matrix for $\varphi^*(\widetilde{H}, \widetilde{\nabla})$ coincide, lemma 2.9 says that the unfoldings are isomorphic. This finishes the proof of theorem 2.5.

REMARK 2.10. The system of equations (2.16) - (2.29) can be reduced: first, by (2.26) - (2.29), $V + W$ is a constant matrix which we denote in this remark by Res_∞. Second, with A as in (2.47) one finds for U the formula

$$U = U(t,0) - (W - W(t,0)) + [Res_\infty, A - A(t,0)] - (A - A(t,0)). \tag{2.51}$$

This formula and the following reduction were shown to us by C. Sabbah.

Now one can transform (2.16) - (2.29) to an equivalent system only in terms of A, W, Res_∞. We still write it with $C_i = \frac{\partial A}{\partial t_i}$, $F_\alpha = \frac{\partial A}{\partial y_\alpha}$, $V = Res_\infty - W$ and U given by (2.51). One finds with some calculations:

The equations (2.17), (2.23), (2.27), *and the restrictions to* $y = 0$ *of the equations* (2.16), (2.22), (2.24), (2.26) *are sufficient.*

The equations (2.19) - (2.21) are obvious; (2.16) and (2.26) (for all y) follow from differentiating them by $\frac{\partial}{\partial y_\alpha}$ and some transformations; the equations (2.25) and (2.24) (for all y) follow from differentiating (2.51); the equation (2.22) (for all y) follows from differentiating it by $\frac{\partial}{\partial y_\alpha}$; now (2.18) follows with the generation condition (GC) and (2.28) and (2.29) are obvious.

3. Supplements

For the applications to Frobenius manifolds in chapter 4 we need (TE)-structures (definition 2.1) with an additional ingredient, a pairing. It is also useful to consider weaker structures, (T)-structures and (L)-structures. After giving their definitions we will discuss how the concepts and results of chapter 2 extend.

DEFINITION 3.1. [**He2**, ch. 2] (a) Fix $w \in \mathbb{Z}$. A $(TEP(w))$-structure is a (TE)-structure $(H \to (\mathbb{C},0) \times (M,0), \nabla)$ together with a ∇-flat, $(-1)^w$-symmetric, nondegenerate pairing

$$P : H_{(z,t)} \times H_{(-z,t)} \to \mathbb{C} \text{ for } (z,t) \in (\mathbb{C}^* \times M, 0) \tag{3.1}$$

on a representative of H such that the pairing extends to a nondegenerate z-sesquilinear pairing

$$P : \mathcal{O}(H) \times \mathcal{O}(H) \to z^w \mathcal{O}_{\mathbb{C}\times M,0}. \tag{3.2}$$

(b) An $(LEP(w))$-structure is a germ of a bundle $H \to (\mathbb{C},0)\times(M,0)$ with a flat connection ∇ on the restriction to $(\mathbb{C}^*,0)\times(M,0)$ with a logarithmic pole along $\{0\}\times(M,0)$ and with a pairing P with the same properties as in (a).

(c) Fix $r \in \mathbb{Z}_{\geq 0}$. Consider a germ of a holomorphic vector bundle $H \to (\mathbb{C},0)\times(M,0)$ with a map

$$\nabla : \mathcal{O}(H) \to \frac{1}{z^r}\mathcal{O}_{\mathbb{C}\times M,0}\cdot \Omega^1_{M,0} \tag{3.3}$$

such that for some representative of H the restrictions $H|_{\{z\}\times(M,0)}$, $z \in (\mathbb{C}^*,0)$, are flat connections. The tuple $((M,0),H,\nabla)$ is called a (T)-structure if $r=1$, and an (L)-structure if $r=0$.

(d) A (T)-structure with a pairing P with all properties in (a) is a $(TP(w))$-structure, an (L)-structure with such a pairing is an $LP(w)$-structure.

LEMMA 3.2. *Let $(H \to (\mathbb{C},0)\times(M,0),\nabla,P)$ be a $(TEP(w))$-structure with generation condition (GC) (theorem 2.5) and let $(\widetilde{H} \to (\mathbb{C},0)\times(M\times\mathbb{C}^l,0),\widetilde{\nabla})$ be an unfolding of the underlying (TE)-structure. Then P extends to $\mathcal{O}(\widetilde{H})$ and $(\widetilde{H},\widetilde{\nabla},P)$ is a $(TEP(w))$-structure.*

Proof. It is sufficient to consider an unfolding in one parameter y. For some representative of $\widetilde{H}$ the pairing P extends to a ∇-flat pairing on the restriction to $(\mathbb{C}^*,0)\times(M\times\mathbb{C},0)$. We have to show that it takes values on $\mathcal{O}(\widetilde{H})$ in $z^w\mathcal{O}_{\mathbb{C}\times M\times\mathbb{C},0}$. A priori the values are in $\mathcal{O}_{\mathbb{C}^*\times M\times\mathbb{C},0}$. Denote $n := \operatorname{rk} H$ and let $(z,t_1,...,t_m,y)$ be coordinates on $(\mathbb{C}\times M\times\mathbb{C},0)$. Choose an $\mathcal{O}_{\mathbb{C}\times M\times\mathbb{C},0}$-basis $\widetilde{v}_1,...,\widetilde{v}_n$ of $\mathcal{O}(\widetilde{H})$ with connection matrix

$$\Omega = \frac{1}{z}\sum_{i=1}^m C_i \mathrm{d}t_i + \frac{1}{z}F\mathrm{d}y + \frac{1}{z^2}U\mathrm{d}z \tag{3.4}$$

with matrices $C_i, F, U \in M(n\times n, \mathcal{O}_{\mathbb{C}\times M\times\mathbb{C},0})$, and the matrix

$$R := (P(\widetilde{v}_i,\widetilde{v}_j)) \in M(n\times n,\mathcal{O}_{\mathbb{C}^*\times M\times\mathbb{C},0}). \tag{3.5}$$

Flatness and z-sesquilinearity of the pairing give

$$\mathrm{d}R(z,t,y) = \Omega^{tr}(z,t,y)R(z,t,y) + R(z,t,y)\Omega(-z,t,y), \tag{3.6}$$

that means,

$$z\frac{\partial}{\partial z}R(z,t,y) = \frac{1}{z}U^{tr}(z,t,y)R(z,t,y) - \frac{1}{z}R(z,t,y)U(-z,t,y), \tag{3.7}$$

$$\frac{\partial}{\partial t_i}R(z,t,y) = \frac{1}{z}C_i^{tr}(z,t,y)R(z,t,y) - \frac{1}{z}R(z,t,y)C_i(-z,t,y), \tag{3.8}$$

$$\frac{\partial}{\partial y}R(z,t,y) = \frac{1}{z}F^{tr}(z,t,y)R(z,t,y) - \frac{1}{z}R(z,t,y)F(-z,t,y). \tag{3.9}$$

Write R as a power series

$$R(z,t,y) = \sum_{k=0}^\infty R^{(k)}(z,t) \text{ with } R^{(k)} \in M(n\times n,\mathcal{O}_{\mathbb{C}^*\times M,0}\cdot y^k) \tag{3.10}$$

and define

$$R^{(\leq k)}(z,t,y) := \sum_{j=0}^{k} R^{(j)}(z,t), \tag{3.11}$$

analogously for C_i, F, U, with $C_i^{(k)}, F^{(k)}, U^{(k)} \in M(n\times n, \mathcal{O}_{\mathbb{C}\times M,0}\cdot y^k)$. Then $R^{(0)} \in M(n\times n, z^w\mathcal{O}_{\mathbb{C}\times M,0})$ because (H,∇,P) is a $(TEP(w))$-structure.

Induction hypothesis for $k\in\mathbb{Z}_{\geq 0}$*:*

$$R^{(\leq k)} \in M(n\times n, z^w\mathcal{O}_{\mathbb{C}\times M\times\mathbb{C},0}). \tag{3.12}$$

Induction step from k *to* $k+1$*:* Recall the definition of $M(>k)$ in (2.35). The equations (3.7) and (3.8) show that one has modulo $M(>k)$

$$C_i^{(\leq k)tr}(0,t,y)[z^{-w}R^{(\leq k)}(z,t,y)]|_{z=0}$$
$$\equiv [z^{-w}R^{(\leq k)}(z,t,y)]|_{z=0}C_i^{(\leq k)}(0,t,y), \tag{3.13}$$
$$U^{(\leq k)tr}(0,t,y)[z^{-w}R^{(\leq k)}(z,t,y)]|_{z=0}$$
$$\equiv [z^{-w}R^{(\leq k)}(z,t,y)]|_{z=0}U^{(\leq k)}(0,t,y). \tag{3.14}$$

Because of the generation condition (GC) the matrix $F^{(\leq k)}(0,t,y)$ is an element of the commutative subalgebra of $M(n\times n, \mathcal{O}_{M,0}[y])/M(>k)$ which is generated by $C_1^{(\leq k)},...,C_m^{(\leq k)}, U^{(\leq k)}$. Therefore modulo $M(>k)$

$$F^{(\leq k)tr}(0,t,y)[z^{-w}R^{(\leq k)}(z,t,y)]|_{z=0}$$
$$\equiv [z^{-w}R^{(\leq k)}(z,t,y)]|_{z=0}F^{(\leq k)}(0,t,y). \tag{3.15}$$

This together with (3.9) completes the induction step. □

REMARKS 3.3. (i) A bundle $(H\to(\mathbb{C},0)\times(M,0),\nabla)$ with a logarithmic pole along $\{0\}\times(M,0)$ is also called an (LE)-structure.

(ii) A $(TP(w))$-structure is essentially equivalent to Barannikov's notion of a semi-infinite variation of Hodge structures [**Ba2**][**Ba3**].

(iii) The notion of an unfolding of a (TE)-structure (definition 2.3 (a)) carries over to structures of type $(TEP(w))$, (L), $(LP(w))$, (LE), $(LEP(w))$, (T), $(TP(w))$.

(iv) Lemma 2.6 (a) says that any unfolding of an (LE)-structure is trivial. The same is true for (L)-structures. Therefore the analogue of lemma 3.2 holds for $(LEP(w))$-structures and $(LP(w))$-structures trivially. An (L)-structure comes equipped with a residual connection ∇^{res} as in lemma 2.6. In fact, an (L)-structure is just a germ of a holomorphic family of flat connections, parametrized by $(\mathbb{C},0)$ with the connection ∇^{res} for the parameter $z=0$.

(v) The analogue of lemma 2.7 for (T)-structures is easy: A (T)-structure $(H\to(\mathbb{C},0)\times(M,0),\nabla)$ can be extended to a trivial bundle $(H^{(gl)}\to\mathbb{P}^1\times(M,0),\nabla)$ with a holomorphic family of flat connections on the restrictions $H^{(gl)}|_{\{z\}\times(M,0)}$ for $z\in\mathbb{P}^1-\{0\}$. To see this, one chooses an $\mathcal{O}_{\mathbb{C},0}$-basis of sections of $\mathcal{O}(H|_{(\mathbb{C},0)\times\{0\}})$; one glues $H|_{(\mathbb{C},0)\times\{0\}}$ to a trivial bundle on $\mathbb{P}^1-\{0\}$, using this basis; one extends the trivial bundle to $(\mathbb{P}^1-\{0\})\times(M,0)$ and glues it to H with ∇.

(vi) A (T)-structure $(H\to(\mathbb{C},0)\times(M,0),\nabla)$ comes equipped with a Higgs field C on $K:=H|_{\{0\}\times(M,0)}$ as in lemma 2.4.

(vii) The analogues of theorem 2.5 and lemma 2.9 hold for (T)-structures. The proofs are the same, of course without the data encoding the part of the connection

in z-direction. The generation condition reads:
(GC)' A vector $\zeta \in K_0$ exists which together with its images under iterations of the maps $C_X : K_0 \to K_0$, $X \in T_0M$, generates K_0.

(viii) The analogue of lemma 3.2 holds for (T)-structures with the generation condition (GC)'. The proof is the same.

(ix) Lemma 3.4 below holds also for $(TP(w))$-structures and $(LP(w))$-structures, of course except (3.18) and (3.19).

LEMMA 3.4. *(a)* [**He2**, 2.5] *Let* $(H \to (\mathbb{C},0) \times (M,0), \nabla, P)$ *be a* $(TEP(w))$*-structure with* $K, C, \mathcal{U}$ *as in lemma 2.4. Define a pairing* $g : \mathcal{O}(K) \times \mathcal{O}(K) \to \mathcal{O}_{M,0}$ *by*

$$g([a],[b]) := z^{-w}P(a,b) \mod z\mathcal{O}_{\mathbb{C}\times M,0} \text{ for } a,b \in \mathcal{O}(H). \tag{3.16}$$

It is $\mathcal{O}_{M,0}$*-bilinear, symmetric, nondegenerate, and it satisfies*

$$g(C_X a, b) = g(a, C_X b) \text{ for } X \in \mathcal{T}_{M,0}, a,b \in \mathcal{O}(K). \tag{3.17}$$

$$g(\mathcal{U}a, b) = g(a, \mathcal{U}b) \text{ for } a,b \in \mathcal{O}(K). \tag{3.18}$$

(b) [**He2**, 5.1] *Let* $(H \to (\mathbb{C},0) \times (M,0), \nabla, P)$ *be an* $(LEP(w))$*-structure with* $K := H|_{\{0\}\times(M,0)}$*, residual connection* ∇^{res} *and residue endomorphism* $\mathcal{V}^{res}$ *as in lemma 2.6 (b). Define a pairing g as in (a). It is* $\mathcal{O}_{M,0}$*-bilinear, symmetric, nondegenerate,* ∇^{res}*-flat, and it satisfies*

$$g(\mathcal{V}^{res}a, b) + g(a, \mathcal{V}^{res}b) = w \cdot g(a,b) \text{ for } a,b \in \mathcal{O}(K). \tag{3.19}$$

Proof. All statements follow easily from the ∇-flatness of P and its other properties, See [**He2**, Lemma 2.14 and Lemma 5.3] for details. □

4. Construction theorem for Frobenius manifolds

Associated to a holomorphic Frobenius manifold $\widetilde{M}$ is a series of meromorphic connections, parametrized by $w \in \mathbb{Z}$, the *(first) structure connections* (lemma 4.4). Under certain assumptions theorem 2.5 allows to reconstruct anyone of them from its restriction to a submanifold $M \subset \widetilde{M}$. This restricted connection is considered as initial datum. Definition 4.1 and theorem 4.2 formalize its properties in two equivalent ways. Theorem 4.5 is a construction theorem for Frobenius manifolds, starting from such an initial datum. Definition 4.1 and theorem 4.2 are also discussed (with different notations) in [**Sab1**, I 1] [**Sab2**, VI 2].

DEFINITION 4.1. [**He2**, 5.2] (a) Fix $w \in \mathbb{Z}$. A $(trTLEP(w))$-structure is a tuple $((M,0), H, \nabla, P)$ with the following properties. $(M,0)$ is a germ of a complex manifold; $H \to \mathbb{P}^1 \times (M,0)$ is a trivial holomorphic vector bundle with a flat connection on $H|_{\mathbb{C}^*\times(M,0)}$; and P is a $(-1)^w$-symmetric, nondegenerate, ∇-flat pairing

$$P : H_{(z,t)} \times H_{(-z,t)} \to \mathbb{C} \text{ for } (z,t) \in (\mathbb{C}^*,0) \times (M,0). \tag{4.1}$$

The restriction of (H, ∇, P) to the germ $(\mathbb{C},0) \times (M,0)$ is a $(TEP(w))$-structure and the restriction to the germ $(\mathbb{P}^1,\infty) \times (M,0)$ is an $(LEP(-w))$-structure (definition 3.1).

(b) A *Frobenius type structure* is a tuple $((M,0), K, \nabla^r, C, \mathcal{U}, \mathcal{V}, g)$ with the following properties. $(M,0)$ is a germ of a complex manifold; $K \to (M,0)$ is a germ of a holomorphic vector bundle with flat connection ∇^r; the map $C : \mathcal{O}(K) \to \Omega^1_{M,0} \otimes \mathcal{O}(K)$ is a Higgs bundle, i.e., a map such that all the endomorphisms $C_X : \mathcal{O}(K) \to \mathcal{O}(K)$, $X \in \mathcal{T}_{M,0}$, commute; the endomorphism $\mathcal{U} : \mathcal{O}(K) \to \mathcal{O}(K)$

of K satisfies $[C,\mathcal{U}]=0$; the endomorphism $\mathcal{V}:\mathcal{O}(K)\to\mathcal{O}(K)$ of K is ∇^r-flat; and $g:\mathcal{O}(K)\times\mathcal{O}(K)\to\mathcal{O}_{M,0}$ is a symmetric, nondegenerate, ∇^r-flat pairing. These data satisfy

$$\nabla^r_X(C_Y)-\nabla^r_Y(C_X)-C_{[X,Y]}=0, \tag{4.2}$$

$$\nabla^r(\mathcal{U})-[C,\mathcal{V}]+C=0, \tag{4.3}$$

$$g(C_X a,b)=g(a,C_X b), \tag{4.4}$$

$$g(\mathcal{U}a,b)=g(a,\mathcal{U}b), \tag{4.5}$$

$$g(\mathcal{V}a,b)=-g(a,\mathcal{V}b) \tag{4.6}$$

for $X,Y\in\mathcal{T}_{M,0}$, $a,b\in\mathcal{O}(K)$.

THEOREM 4.2. [**Sab1**, I 1][**Sab2**, VI 2][**He2**, 5.2] *Fix $w\in\mathbb{Z}$. There is a one-to-one correspondence between (trTLEP(w))-structures and Frobenius type structures on holomorphic vector bundles. It is given by the steps in (a) and (b). They are inverse to one another.*

(a) Let $(K\to(M,0),\nabla^r,C,\mathcal{U},\mathcal{V},g)$ be a Frobenius type structure on K. Let $\pi:\mathbb{P}^1\times(M,0)\to(M,0)$ be the projection. Define $H:=\pi^*K$, and let $\psi_z:H_{(z,t)}\to K_t$ for $z\in\mathbb{P}^1$ be the canonical projection. Extend $\nabla^r,C,\mathcal{U},\mathcal{V},g$ canonically to H. Define

$$\nabla:=\nabla^r+\frac{1}{z}C+(\frac{1}{z}\mathcal{U}-\mathcal{V}+\frac{w}{2}\,\mathrm{id})\frac{\mathrm{d}z}{z}\ . \tag{4.7}$$

Define a pairing

$$\begin{aligned} P:H_{(z,t)}\times H_{(-z,t)} &\to \mathbb{C} \quad \text{for } (z,t)\in(\mathbb{C}^*,0)\times(M,0)\\ (a,b) &\mapsto z^w g(\psi_z a,\psi_{-z}b)\ . \end{aligned} \tag{4.8}$$

Then (H,∇,P) is a (trTLEP(w))-structure.

(b) Let (H,∇,P) be a (trTLEP(w))-structure. Define $K:=H|_{\{0\}\times(M,0)}$, C and $\mathcal{U}$ as in lemma 2.4 and g as in lemma 3.4 (a). Let ∇^{res} and $\mathcal{V}^{res}$ be the residual connection and the residue endomorphism on $H|_{\{\infty\}\times(M,0)}$ as in lemma 2.6 (b).

Because H is a trivial bundle, there is a canonical projection $\psi:H\to K$, and the bundles K and $H|_{\{\infty\}\times(M,0)}$ are canonically isomorphic. Structure on $H|_{\{\infty\}\times(M,0)}$ can be shifted to K. Let ∇^r on K be the shift of ∇^{res} and let $\mathcal{V}$ on K be the shift of $\mathcal{V}^{res}+\frac{w}{2}\,\mathrm{id}$. Then $(K\to M,\nabla^r,C,g,\mathcal{U},\mathcal{V})$ is a Frobenius type structure and (4.7) holds.

Proof. Part of it follows from the lemmas 2.4, 2.6 (b) and 3.4. For the rest and for details see [**He2**, Theorem 5.7] or [**Sab2**, VI 2]. □

Frobenius type structures and (*trTLEP(w)*)-structures can be restricted to any submanifold of the manifold $(M,0)$ over which they are defined.

DEFINITION 4.3. (Dubrovin) A *Frobenius manifold* $(M,\circ,e,E,g)$ is a complex manifold M of dimension ≥ 1 with a commutative and associative multiplication $\circ$ on the holomorphic tangent bundle TM , a unit field $e\in\mathcal{T}_M$, an *Euler field* $E\in\mathcal{T}_M$, and a symmetric nondegenerate $\mathcal{O}_{M,0}$-bilinear pairing g on TM with the following properties. The metric g is multiplication invariant,

$$g(X\circ Y,Z)=g(Y,X\circ Z)\quad\text{for } X,Y,Z\in\mathcal{T}_M; \tag{4.9}$$

the Levi–Civita connection ∇^g of the metric g is flat; together with the Higgs field $C : \mathcal{T}_M \to \Omega^1_{M,0} \otimes \mathcal{T}_M$ with $C_X Y := -X \circ Y$ it satisfies the potentiality condition

$$\nabla^g_X(C_Y) - \nabla^g_Y(C_X) - C_{[X,Y]} = 0 \quad \text{for } X, Y \in \mathcal{T}_M; \tag{4.10}$$

the unit field e is ∇^g-flat; the Euler field satisfies $\mathrm{Lie}_E(\circ) = \circ$ and $\mathrm{Lie}_E(g) = (2-d)\cdot g$ for some $d \in \mathbb{C}$.

LEMMA 4.4. *(Structure connections of a Frobenius manifold, e.g.* [**Du**, Lecture 3], [**Man2**, I 2.5.2], [**Sab2**, VII 1], [**He2**, Lemma 5.11]*) Let $((M,0), \circ, e, E, g)$ be the germ of a Frobenius manifold with Higgs field C, Levi-Civita connection ∇^g and $d \in \mathbb{C}$ as in definition 4.3. Define the endomorphisms $\mathcal{U} := E\circ : \mathcal{T}_{M,0} \to \mathcal{T}_{M,0}$ and*

$$\mathcal{V} := \mathcal{T}_{M,0} \to \mathcal{T}_{M,0},\ X \mapsto \nabla^g_X E - \frac{2-d}{2} X. \tag{4.11}$$

Then $(TM, \nabla^g, C, \mathcal{U}, \mathcal{V}, g)$ is a Frobenius type structure on TM. The unit field e satisfies $\nabla^g e = 0$ and $\mathcal{V}e = \frac{d}{2}e$.

For any $w \in \mathbb{Z}$ theorem 4.2 gives a $(trTLEP(w))$-structure $((M,0), H = \pi^*TM, \nabla, P)$, where $\pi : \mathbb{P}^1 \times (M,0) \to (M,0)$ is the projection. These structures are called (first) structure connections of the Frobenius manifold.

One can recover a Frobenius manifold from a structure connection and the unit field. More abstractly, one can construct from a $(trTLEP(w))$-structure and a global section with sufficiently nice properties a unique Frobenius manifold such that the $(trTLEP(w))$-structure and the global section are isomorphic to a structure connection and the unit field [**Sab1**][**Sab2**][**Ba1**][**Ba2**][**Ba3**] (cf. remark 4.6 (i)). Theorem 2.5 allows to start under certain assumptions with a $(trTLEP(w))$-structure and a global section over a *smaller* base space, to unfold them universally and then get a Frobenius manifold. This is formulated in theorem 4.5 in terms of Frobenius type structures.

THEOREM 4.5. *(Construction theorem for Frobenius manifolds) Let*

$$((M,0), K, \nabla^r, C, \mathcal{U}, \mathcal{V}, g)$$

be a Frobenius type structure and $\zeta \in K_0$ a vector with the following properties:

- *(IC) (injectivity condition) the map $C_\bullet\zeta : T_0M \to K_0$, $X \mapsto C_X\zeta$ is injective.*
- *(GC) (generation condition) ζ and its images under iteration of the maps $C_X : K_0 \to K_0$, $X \in T_0M$, and $\mathcal{U} : K_0 \to K_0$ generate K_0.*
- *(EC) (eigenvector condition) $\mathcal{V}\zeta = \frac{d}{2}\zeta$ for some $d \in \mathbb{C}$.*

Then there exist up to canonical isomorphism unique data $((\widetilde{M},0), \circ, e, E, \widetilde{g}, i, j)$ with the following properties. $((\widetilde{M},0), \circ, e, E, \widetilde{g})$ is a germ of a Frobenius manifold, $i : (M,0) \to (\widetilde{M},0)$ is an embedding, $j : K \to T\widetilde{M}|_{i(M)}$ is an isomorphism above i of germs of vector bundles which maps ζ to $e|_0$ and which identifies the Frobenius type structure on K with the natural Frobenius type structure on $T\widetilde{M}|_{i(M)}$ which is induced by that on $T\widetilde{M}$.

Proof. Choose $w \in \mathbb{Z}$. Let $((M,0), H, \nabla, P)$ be the $(trTLEP(w))$-structure which corresponds to the Frobenius type structure on $K \to (M,0)$ by theorem 4.2. Its germ over $(\mathbb{C},0) \times (M,0)$ is a $(TEP(w))$-structure with all the properties in theorem 2.5. Consider a universal unfolding of this germ with base space $(\widetilde{M},0) =$

$(M \times \mathbb{C}^l, 0)$. It extends uniquely to a $(trTLEP(w))$-structure $((\widetilde{M},0),\widetilde{H},\widetilde{\nabla},P)$ which unfolds $((M,0),H,\nabla,P)$. This follows from the rigidity of logarithmic poles (lemma 2.6 (a) and remark 3.3 (iii)) and from lemma 3.2.

Let $((\widetilde{M},0),\widetilde{K},\widetilde{\nabla}^r,\widetilde{C},\widetilde{\mathcal{U}},\widetilde{\mathcal{V}},\widetilde{g})$ be the Frobenius type structure which corresponds to this $(trTLEP(w))$-structure by theorem 4.2. There is a canonical isomorphism from the Frobenius type structure on K to the restricted one on $\widetilde{K}|_{(M\times\{0\},0)}$. Let $\widetilde{\zeta} \in \widetilde{K}_0$ be the image of $\zeta \in K_0$. It extends to a unique $\widetilde{\nabla}^r$-flat section $\widetilde{v}_1 \in \mathcal{O}(\widetilde{K})$. The map

$$v : \mathcal{T}_{\widetilde{M},0} \to \mathcal{O}(\widetilde{K}),\ X \mapsto -C_X\widetilde{v}_1 \tag{4.12}$$

is an isomorphism. It allows to shift the structure on $\widetilde{K}$ to structure on $T\widetilde{M}$. Define

$$\nabla^v := v^*\widetilde{\nabla}^r,\ g^v := v^*\widetilde{g},\ e := v^{-1}(\widetilde{v}_1),\ E := v^{-1}(\widetilde{\mathcal{U}}(\widetilde{v}_1)). \tag{4.13}$$

The connection ∇^v on $T\widetilde{M}$ is flat with $\nabla^v g^v = 0$ and $\nabla^v e = 0$. Applying (4.2) to $\widetilde{v}_1$ shows that ∇^v is torsion free; so it is the Levi–Civita connection of g^v. One defines a commutative and associative multiplication $\circ$ on $T\widetilde{M}$ by

$$v(X \circ Y) = -C_X v(Y) = C_X C_Y \widetilde{v}_1. \tag{4.14}$$

Then e is the unit field. The potentiality condition follows from (4.2). It rests to prove $\mathrm{Lie}_E(\circ) = \circ$ and $\mathrm{Lie}_E(g) = (2-d)\cdot g$. We refer to [**He2**, Theorem 5.12] for details. The calculations use (4.2), (4.3), (4.6), $\widetilde{\nabla}^r\widetilde{v}_1 = 0$ and $\widetilde{\mathcal{V}}\widetilde{v}_1 = \frac{d}{2}\widetilde{v}_1$. They show especially

$$v(\nabla^v_X E) = (\widetilde{\mathcal{V}} + \frac{2-d}{2}\mathrm{id})v(X) \ \text{ for } X \in \mathcal{T}_{M,0}. \tag{4.15}$$

One obtains a germ of a Frobenius manifold $((\widetilde{M},0),\circ,e,E,g^v)$. Each step in its construction is essentially unique. □

REMARKS 4.6. (i) Theorem 4.5 is reduced with theorem 2.5 to the case when $M = \widetilde{M}$ and when the map $C_\bullet\zeta : T_0M \to K_0$ is an isomorphism. This case was formulated by Sabbah [**Sab1**, Theorem (4.3.6)] [**Sab2**, Théorème VII.3.6], and independently by Barannikov [**Ba1**][**Ba2**][**Ba3**]. He called a major part of the initial data *semi-infinite variation of Hodge structures* (cf. remark 3.3 (ii)). Theorem 4.5 in the case $M = \widetilde{M}$ is also implicit in the construction in singularity theory [**SK**][**SM**].

(ii) A Frobenius type structure with $M = \{0\}$ is simply a vector space K with a pairing g and two endomorphisms $\mathcal{U}$ and $\mathcal{V}$ which satisfy (4.5) and (4.6). Then the condition (IC) in theorem (4.5) is empty, (GC) must be satisfied by $\mathcal{U}$ alone, and (GC) and (EC) together are still more restrictive. For example, this situation is satisfied at a point of a Frobenius manifold where the multiplication with the Euler field is semisimple with different eigenvalues. This case was first considered by Dubrovin [**Du**, Lecture 3].

(iii) The case of a Frobenius type structure with $\mathcal{U} = 0$ can be considered as opposite to the case in (ii). It will be discussed in chapter 5.

(iv) One can define $(trTLP(w))$-structures and Frobenius type structures without operators $\mathcal{U}$ and $\mathcal{V}$ [**He2**, 5.2]. Omitting the corresponding parts in theorem 4.2, one obtains a one-to-one correspondence between them. One can define Frobenius manifolds without Euler field. The analogues of lemma 4.4 and theorem 4.5

hold. This follows with the remarks 3.3. But now the generation condition (GC) requires $\dim M > 0$.

(v) All the structures in chapters 2 to 4 were convergent with respect to parameters $(t_1, ..., t_m) \in (\mathbb{C}^m, 0) \cong (M, 0)$. One can formulate everything in structures which are formal in these parameters.

(vi) A. Kresch [**Kr**, Theorem 1] proved a strong reconstruction theorem (existence and uniqueness) for formal germs of Frobenius manifolds without Euler field and with some additional properties typical for quantum cohomology. It strengthens the first reconstruction theorem in [**KM**, Theorem 3.1], which establishes only uniqueness. It looks as if his result is the special case of the analogue of theorem 4.5 without Euler field and with $(M, 0) \subset (\widetilde{M}, 0)$ being the small quantum cohomology space. The collection of $N(\beta, d)$ in [**Kr**, Theorem 1] gives the Higgs field C on $T\widetilde{M}|_M$, the conditions on A_1 and A give the generation condition (GC)' and the property $C_X C_Y = C_Y C_X$ for $X, Y \in \mathcal{T}_M$ of the Higgs bundle. The other conditions on Frobenius type structures (without $\mathcal{U}$ and $\mathcal{V}$) seem to be built-in.

5. H^2-generated variations of filtrations and Frobenius manifolds

In this chapter the special case of the construction theorem 4.5 for Frobenius manifolds is studied when the Frobenius type structure satisfies $\mathcal{U} = 0$. A Frobenius type structure with $\mathcal{U} = 0$ is equivalent to a variation of filtrations with Griffiths transversality and additional structure (lemma 5.1). Typical cases are variations of polarized Hodge structures with a condition, which is called H^2-generation condition, motivated by quantum cohomology. The definitions 5.3 and 5.4 present the relevant notions, theorem 4.5 reformulates the construction theorem in the case $\mathcal{U} = 0$.

LEMMA 5.1. *(a) The structures in* (α) *and* (β) *are equivalent.*

(α) *A Frobenius type structure* $((M, 0), K, \nabla^r, C, \mathcal{U}, \mathcal{V}, g)$ *together with an integer* w *such that* $\mathcal{U} = 0$ *and* $\mathcal{V}$ *is semisimple with eigenvalues in* $\frac{w}{2} + \mathbb{Z}$.

(β) *A tuple* $((M, 0), K, \nabla, F^\bullet, U_\bullet, w, S)$. *Here* $K \to (M, 0)$ *is a germ of a holomorphic vector bundle;* ∇ *is a flat connection on* K*;* $F^\bullet$ *is a decreasing filtration by germs of holomorphic subbundles* $F^p \subset K$, $p \in \mathbb{Z}$, *which satisfies Griffiths transversality*

$$\nabla : \mathcal{O}(F^p) \to \Omega^1_{M,0} \otimes \mathcal{O}(F^{p-1}); \tag{5.1}$$

$U_\bullet$ *is an increasing filtration by flat subbundles* $U_p \subset K$ *such that*

$$H = F^p \oplus U_{p-1} = \bigoplus_q F^q \cap U_q; \tag{5.2}$$

$w \in \mathbb{Z}$, *and* S *is a* ∇*-flat,* $(-1)^w$*-symmetric, nondegenerate pairing on* K *with*

$$S(F^p, F^{w+1-p}) = 0, \tag{5.3}$$

$$S(U_p, U_{w-1-p}) = 0. \tag{5.4}$$

(b) One passes from (α) *to* (β) *by defining*

$$\nabla := \nabla^r + C, \tag{5.5}$$

$$F^p := \bigoplus_{q\geq p} \ker(\mathcal{V} - (q - \frac{w}{2})\,\mathrm{id} : K \to K), \tag{5.6}$$

$$U_p := \bigoplus_{q\leq p} \ker(\mathcal{V} - (q - \frac{w}{2})\,\mathrm{id} : K \to K), \tag{5.7}$$

$$S(a,b) := (-1)^p g(a,b) \text{ for } a \in \mathcal{O}(F^p \cap U_p), b \in \mathcal{O}(K). \tag{5.8}$$

Proof. First we prove part (b). The connection ∇^r and the Higgs field C are maps $\mathcal{O}(K) \to \Omega^1_{M,0} \otimes \mathcal{O}(K)$. They can be extended canonically to maps $\Omega^1_{M,0} \otimes \mathcal{O}(K) \to \Omega^2_{M,0} \otimes \mathcal{O}(K)$. Then the flatness of ∇^r means $(\nabla^r)^2 = 0$, the Higgs field C satisfies $C^2 = 0$, and the potentiality condition (4.2) means $\nabla^r(C) := \nabla^r \circ C + C \circ \nabla^r = 0$. Therefore $\nabla^2 = (\nabla^r + C)^2 = 0$, the connection ∇ is flat. The filtrations $F^\bullet$ and $U_\bullet$ obviously satisfy (5.2) and

$$F^p \cap U_p = \ker(\mathcal{V} - (p - \frac{w}{2})\,\mathrm{id} : K \to K). \tag{5.9}$$

The connection ∇^r maps $\mathcal{O}(F^p \cap U_p)$ to itself because $\mathcal{V}$ is ∇^r-flat. Because of $\mathcal{U} = 0$ the condition (4.3) is $[C, \mathcal{V}] = C$. Equivalent is that C_X, $X \in \mathcal{T}_{M,0}$, maps $\mathcal{O}(F^p \cap U_P)$ to $\mathcal{O}(F^{p-1} \cap U_{p-1})$. Therefore $U_\bullet$ is ∇-flat and $F^\bullet$ satisfies Griffiths transversality.

The condition (4.6) says that for $a \in \mathcal{O}(F^p \cap U_p)$, $b \in \mathcal{O}(F^q \cap U_q)$

$$S(a,b) = (-1)^p g(a,b) = 0 \text{ if } p + q \neq w. \tag{5.10}$$

Therefore S is $(-1)^w$-symmetric and satisfies (5.3) and (5.4). It is ∇-flat because for $X \in \mathcal{T}_{M,0}$ and ∇^r-flat sections $a \in \mathcal{O}(F^p \cap U_p)$, $b \in \mathcal{O}(K)$

$$\begin{aligned}(\nabla_X S)(a,b) &= X\, S(a,b) - S(\nabla^r_X a + C_X a, b) - S(a, \nabla^r_X b + C_X b) \\ &= 0 - (-1)^{p+1} g(C_X a, b) - (-1)^p g(a, C_X b) = 0.\end{aligned} \tag{5.11}$$

This shows part (b). One passes from (β) to (α) as follows. One defines the endomorphism $\mathcal{V}$ by (5.9), the pairing g by (5.8), and one decomposes ∇ into ∇^r and C such that ∇^r maps $\mathcal{O}(F^p \cap U_p)$ to itself and C_X for $X \in \mathcal{T}_{M,0}$ maps $\mathcal{O}(F^p \cap U_p)$ to $\mathcal{O}(F^{p-1} \cap U_{p-1})$. One easily checks all conditions of a Frobenius type structure. □

REMARKS 5.2. (i) If one adds in lemma 5.1 (a) (β) real structure with suitable conditions then one obtains a germ of a variation of polarized Hodge structures of weight w.

(ii) A Frobenius type structure is equivalent to a $(trTLEP(w))$-structure (theorem 4.2), which is composed of a $(TEP(w))$-structure at $(\mathbb{C}, 0) \times (M, 0)$ and an $(LEP(w))$-structure at $(\mathbb{P}^1, \infty) \times (M, 0)$. One can refine lemma 5.1 (a). There is a correspondence between $(TEP(w))$-structures with $\mathcal{U} = 0$ and monodromy $(-1)^w\,\mathrm{id}$ on the one hand and germs of variations of filtrations with Griffiths transversality and a pairing on the other hand [**He2**, Corollary 7.14]. Under this correspondence an $(LEP(w))$-structure corresponds to a trivial, flat variation of filtrations. Putting these together, a $(trTLEP(w))$-structure corresponds to the structure in lemma 5.1 (a) (β).

But the variation of filtrations $F^\bullet$ which corresponds to a $(TEP(w))$-structure $(H \to (\mathbb{C}, 0) \times (M, 0), \nabla)$ does not live on the bundle $K = H|_{\{0\}\times(M,0)}$. It lives on a flat bundle on $(M,0)$ whose fibers are all isomorphic to the space of global flat manyvalued sections in $H|_{(\mathbb{C}^*,0)\times(M,0)}$. The bundle K is only isomorphic to $\bigoplus_p F^p/F^{p+1}$. Only in the case of a second filtration $U_\bullet$ and a splitting (5.2) one obtains an isomorphism between K and the flat bundle on which $F^\bullet$ lives.

DEFINITION 5.3. (a) A germ of an *H^2-generated variation of filtrations of weight w* is a tuple $((M,0),K,\nabla,F^\bullet,w)$ with the following properties. w is an integer, $K \to (M,0)$ is a germ of a holomorphic vector bundle with flat connection ∇ and variation of filtrations $F^\bullet$ which satisfies Griffiths transversality (5.1) and

$$0 = F^w \subset F^{w-1} \subset \ldots \subset K \tag{5.12}$$

$$\operatorname{rk} F^{w-1} = 1, \quad \operatorname{rk} F^{w-2} = 1 + \dim M \geq 2. \tag{5.13}$$

Griffiths transversality and flatness of ∇ give a Higgs field C on the graded bundle $\bigoplus_p F^p/F^{p+1}$ with commuting endomorphisms

$$C_X = [\nabla_X] : \mathcal{O}(F^p/F^{p+1}) \to \mathcal{O}(F^{p-1}/F^p) \text{ for } X \in \mathcal{T}_{M,0}. \tag{5.14}$$

H^2-generation condition: the whole module $\bigoplus_p \mathcal{O}(F^p/F^{p+1})$ is generated by $\mathcal{O}(F^{w-1})$ and its images under iterations of the maps C_X, $X \in \mathcal{T}_{M,0}$.

(b) A pairing and an opposite filtration for an H^2-generated variation of filtrations $((M,0),K,\nabla,F^\bullet)$ of weight w are a pairing S and a filtration $U_\bullet$ as in lemma 5.1 (a) (β).

DEFINITION 5.4. An H^2-generated germ of a Frobenius manifold of weight $w \in \mathbb{N}_{\geq 3}$ is a germ $((M,0),\circ,e,E,g)$ of a Frobenius manifold with the properties (I) and (II) below and with

$$E|_{t=0} = 0. \tag{5.15}$$

Let ∇^g be the Levi–Civita connection of g. The endomorphism $\nabla^g E : \mathcal{T}_{M,0} \to \mathcal{T}_{M,0}$, $X \mapsto \nabla^g_X E$, acts on the space of ∇^g-flat vector fields. In particular $e \in \ker(\nabla^g E - \mathrm{id})$.

(I) It acts semisimply with eigenvalues $\{1, 0, \ldots, -(w-3)\}$.

It turns out that then the multiplication on the algebra $T_0 M$ respects the grading

$$T_0 M = \bigoplus_{p=0}^{w-2} \ker(\nabla^g E - (1-p)\,\mathrm{id} : T_0 M \to T_0 M) \tag{5.16}$$

and that $\mathrm{Lie}_E(g) = (4-w) \cdot g$.

(II) H^2-generation condition: The algebra $T_0 M$ is generated by $\ker(\nabla^g E : T_0 M \to T_0 M)$.

REMARKS 5.5. (i) Properties (I) and (5.15) hold for even-dimensional quantum cohomology of Calabi–Yau manifolds of complex dimension $w-2$. This follows from the vanishing of the canonical class and the standard explicit formulas for the Euler field in quantum cohomology. Generally, our terminology involving "H^2-generation" is motivated by quantum cohomology, for which $\ker \nabla^g E = H^2$. For a more extended discussion of special properties of quantum cohomology Frobenius manifolds, see [**Man1**, 1.3 and 1.4, in particular Definition 1.4.1].

(ii) The uniqueness statement in theorem 5.6 that a structure as in (γ) is determined by a structure as in (β) is essentially a special case of that in [**KM**, Theorem 3.1 and 3.1.1 a) and b)]. A refined version of it (cf. lemma 8.2) was used already in the proof of [**Ba1**, Theorem 6.5]. But the existence statement that any structure as in (β) gives rise to a structure as in (γ) is new.

THEOREM 5.6. *There is a one-to-one correspondence between the structures in* (α), (β) *and* (γ).

(α) *A Frobenius type structure* $((M,0),K,\nabla^r,C,\mathcal{U},\mathcal{V},g)$ *with* $\mathcal{U}=0$ *and with a fixed vector* $\zeta\in K_0$ *which satisfies the conditions (IC), (GC) and (EC) in theorem 4.5.*

(β) *A germ of an* H^2*-generated variation of filtrations* $((M,0),K,\nabla,F^\bullet,w,S,U_\bullet)$ *of weight* $w\in\mathbb{N}_{\geq 3}$ *with pairing and opposite filtration and with a fixed generator* $\zeta\in(F^{w-1})_0\subset K_0$.

(γ) *An* H^2*-generated germ of a Frobenius manifold* $((\widetilde{M},0),\circ,e,E,\widetilde{g})$ *of weight* $w\in\mathbb{N}_{\geq 3}$.

One passes from (α) *to* (β) *by lemma 5.1 (b) and from* (α) *to* (γ) *by theorem 4.5. One passes from* (γ) *to* (α) *by defining*

$$M:=\{t\in\widetilde{M}\mid E|_t=0\},\tag{5.17}$$

$K:=T\widetilde{M}|_{(M,0)}$ *with the canonical Frobenius type structure, and* $\zeta:=e|_0$. *The eigenvector condition (EC) is* $\mathcal{V}\zeta=\frac{w-2}{2}\zeta$.

Proof. Let us start with (α). We have to show that for some $w\in\mathbb{N}_{\geq 3}$ the endomorphism $\mathcal{V}+\frac{w}{2}\,\mathrm{id}$ is semisimple with eigenvalues in $\{1,2,...,w-1\}$ and with $\mathcal{V}\zeta=\frac{w-2}{2}\zeta$.

The eigenvector condition (EC) says $\mathcal{V}\zeta=\frac{d}{2}\zeta$ for some $d\in\mathbb{C}$. Condition (4.3) reads as $[C,\mathcal{V}]=C$. This together with the generation condition (GC) shows that $\mathcal{V}$ acts semisimply on K_0 with eigenvalues in $\frac{d}{2}+\mathbb{Z}_{\leq 0}$ and that $\ker(\mathcal{V}-\frac{d}{2}\,\mathrm{id})=\mathbb{C}\cdot\zeta$. The injectivity condition (IC) implies $\dim\ker(\mathcal{V}-(\frac{d}{2}-1)\,\mathrm{id})=\dim M>0$. Condition (4.6) tells that the eigenvalues of $\mathcal{V}$ are in $(\frac{d}{2}+\mathbb{Z}_{\leq 0})\cap-(\frac{d}{2}+\mathbb{Z}_{\leq 0})$. Therefore $d\in\mathbb{N}$. Define $w:=d+2$.

The conditions (IC) and (GC) show that one passes from (α) to (β) by lemma 5.1 (b). It is also clear that one can pass back.

Theorem 4.5 applied to (α) gives a Frobenius manifold $((\widetilde{M},0),\circ,e,E,\widetilde{g})$, an embedding $i:(M,0)\to(\widetilde{M},0)$ and an isomorphism $j:K\to T\widetilde{M}|_{i(M)}$ of Frobenius type structures. It maps $\mathcal{V}$ to the restriction of $\nabla^g E+\frac{2-d}{2}\,\mathrm{id}$ on $T\widetilde{M}|_{i(M)}$ (4.15). Therefore one obtains an H^2-generated germ of a Frobenius manifold of weight w. In suitable flat coordinates the Euler field of this Frobenius manifold takes the form

$$E=\sum_{i=1}^{\dim\widetilde{M}}d_it_i\frac{\partial}{\partial t_i}\tag{5.18}$$

with $d_i\in\{1,0,...,-(w-3)\}$ and

$$\sharp(i\mid d_i=0)=\dim\ker(\nabla^g E)=\dim\ker(\mathcal{V}-(\frac{d}{2}-1)\,\mathrm{id})=\dim M.\tag{5.19}$$

Therefore $i(M)=\{t\in\widetilde{M}\mid E|_t=0\}$. One sees also easily that one passes from (γ) to (α) as described in the theorem. □

A distinguished class of H^2-generated variations of filtrations are variations of Hodge filtrations associated to certain families of homogeneous polynomials. We will discuss them and the corresponding H^2-generated Frobenius manifolds in the chapters 6, 7 and 8. The following example shows that it is easy to construct abstract H^2-generated variations of filtrations. Therefore one has a lot of freedom in constructing H^2-generated Frobenius manifolds.

EXAMPLE 5.7. Consider $(M,0) := (\mathbb{C},0)$ with coordinate t, the trivial bundle $H := \mathbb{C}^{w-1} \times (M,0) \to (M,0)$ for some $w \in \mathbb{N}_{\geq 3}$ with standard basis $v_1, ..., v_{w-1}$ of sections, a pairing S with $S(v_p, v_q) := (-1)^p \delta_{p,w-q}$, and filtrations $F^\bullet$ and $U_\bullet$ with

$$(5.20) \qquad \mathcal{O}(F^p) := \bigoplus_{q \geq p} \mathcal{O}_{\mathbb{C},0} \cdot v_{w-q}, \quad \mathcal{O}(U_p) := \bigoplus_{q \leq p} \mathcal{O}_{\mathbb{C},0} \cdot v_{w-q}.$$

Choose *any* invertible functions $b_2, ..., b_{[\frac{w-1}{2}]} \in \mathcal{O}^*_{\mathbb{C},0}$, define

$$(5.21)\ b_1 := 1,\ b_{w-1} := 0,\ b_k := b_{w-1-k} \text{ for } k = \left[\frac{w-1}{2}\right] + 1, ..., w-2$$

and define a connection ∇ on H by

$$(5.22) \qquad \nabla_{\partial/\partial t} v_i := b_i v_{i+1}.$$

Then $((M,0), K, \nabla, F^\bullet, w, S, U_\bullet)$ is a germ of an H^2-generated variation of filtrations of weight w with pairing and opposite filtration. Moreover, one can see that a second tuple $(b'_2, ..., b'_{[\frac{w-1}{2}]})$ of functions yields isomorphic data only if b_i and b'_i coincide up to multiplication by a constant. The big freedom in constructing H^2-generated Frobenius manifolds of dimension $w-1$ is in striking difference to the semisimple case, where one has only finitely many parameters [**Man2**, II 3.4.3].

6. Hypersurfaces in $\mathbb{P}^n$

In the case of certain families of smooth hypersurfaces in $\mathbb{P}^n$ one obtains variations of Hodge structures which are the prototype of H^2-generated variations of filtrations (definition 5.3). This is a simple consequence of Griffiths' description of the Hodge filtration on the primitive part of the middle cohomology of a smooth hypersurface in $\mathbb{P}^n$ in terms of rational differential forms on $\mathbb{P}^n$ [**Gr**].

Fix a degree $d \in \mathbb{N}$ and denote by $\mathbb{C}[x]^{(q)}$ for $q \in \frac{1}{d}\mathbb{Z}_{\geq 0}$ the space of homogeneous polynomials in $\mathbb{C}[x_0, ..., x_n] = \mathbb{C}[x]$ of degree $d \cdot q$. Consider a polynomial $f \in \mathbb{C}[x]^{(1)}$ with isolated singularity at 0. The grading on $\mathbb{C}[x]$ induces a grading on the Jacobi algebra

$$(6.1) \qquad R_f := \mathbb{C}[x] / \left(\frac{\partial f}{\partial x_0}, ..., \frac{\partial f}{\partial x_n}\right)$$

of f with subspaces $R_f^{(q)}$, $q \in \frac{1}{d}\mathbb{Z}_{\geq 0}$. The primitive part of the middle cohomology $H^{n-1}(X, \mathbb{C})$ of the smooth hypersurface $X := \overline{f^{-1}(0)} \subset \mathbb{P}^n$ is denoted by $H^{n-1}_{prim}(X)$, its Hodge filtration by $F^\bullet_{prim} \subset H^{n-1}_{prim}(X)$. The primitive cohomology also comes equipped with a polarizing form S.

Now consider a family of polynomials $F_t \in \mathbb{C}[x]^{(1)}$, $t \in M_0$, with isolated singularities at 0, where M_0 is a smooth parameter space with coordinates (t_α) such that for each $t \in M_0$ the map

$$(6.2) \qquad \mathbf{a} : T_t M_0 \to R^{(1)}_{F_t}, \quad \frac{\partial}{\partial t_\alpha} \mapsto \left[\frac{\partial F_t}{\partial t_\alpha}\right]$$

is an isomorphism. The bundle $H := \bigcup_{t \in M_0} H^{n-1}_{prim}(X_t)$ comes equipped with a real subbundle, a flat connection ∇, the flat pairing S and a variation of Hodge filtrations $F^\bullet_{prim}$. Together they form a variation of polarized Hodge structures of weight $n-1$.

THEOREM 6.1. *If* $\frac{n+1}{d} \in \mathbb{N}$ *then the tuple* $(M_0, H, \nabla, S, \widetilde{F}^\bullet := F_{prim}^{\bullet-2+(n+1)/d})$ *is an* H^2*-generated variation of filtrations of weight* $n+3-2(n+1)/d$ *with pairing.*

Proof. The space $\Omega^{n+1}_{alg} = \mathbb{C}[x]\mathrm{d}x_0...\mathrm{d}x_n$ of algebraic differential forms on $\mathbb{C}^{n+1}$ is graded with subspaces $(\Omega^{n+1}_{alg})^{(q)} = \mathbb{C}[x]^{(q-(n+1)/d)}\mathrm{d}x_0...\mathrm{d}x_n$ for $q \in \frac{1}{d}\mathbb{Z}_{\geq n+1}$. For $f \in \mathbb{C}[x]^{(1)}$ with isolated singularity the quotient

$$\Omega_f := \Omega^{n+1}_{alg}/\mathrm{d}f \wedge \Omega^n_{alg} \tag{6.3}$$

carries an induced grading with subspaces $\Omega_f^{(q)}$. It is a graded module of the graded algebra R_f, and it is a free module of rank 1 of R_f.

Following Griffiths [**Gr**], one obtains for $q \in \mathbb{Z}_{\geq (n+1)/d}$ a canonical isomorphism

$$\mathbf{r}_q : \Omega_f^{(q)} \to F^{n-q}_{prim}/F^{n-q+1}_{prim} \tag{6.4}$$

in the following way. Let $\mathcal{E} := \sum_{i=0}^n x_i \frac{\partial}{\partial x_i}$ be the Euler field on $\mathbb{C}^{n+1}$. For $\omega \in (\Omega^{n+1}_{alg})^{(q)}$, $q \in \mathbb{Z}_{\geq(n+1)/d}$, consider the form $i_\mathcal{E}(\frac{\omega}{f^q})$, which one obtains from $\frac{\omega}{f^q}$ by contraction with the Euler field. It extends to a rational form on $\mathbb{P}^{n+1} = \mathbb{C}^{n+1} \cup \mathbb{P}^n$ with a pole of order $\leq q$ along $f^{-1}(0) \cup X$. We denote the restriction of this form to $\mathbb{P}^n$ by $\overline{i_\mathcal{E}(\frac{\omega}{f^q})}$. It induces a class $\left[\overline{i_\mathcal{E}(\frac{\omega}{f^q})}\right] \in H^n(\mathbb{P}^n - X, \mathbb{C})$. There is a residue map

$$\mathrm{Res} : H^n(\mathbb{P}^n - X, \mathbb{C}) \to H^{n-1}(X, \mathbb{C}), \tag{6.5}$$

dual to a tube map in the homologies. One defines

$$\begin{aligned} \rho_q : (\Omega^{n+1}_{alg})^{(q)} &\to H^{n-1}(X, \mathbb{C}), \\ \omega &\mapsto \mathrm{Res}\left[\overline{i_\mathcal{E}\left(\frac{\omega}{f^q}\right)}\right] \cdot \frac{(q-1)!}{d}. \end{aligned} \tag{6.6}$$

By [**Gr**], the image of ρ_q is $F^{n-q}_{prim} \subset H^{n-1}_{prim}(X)$ and the preimage of F^{n-q+1}_{prim} is $(\mathrm{d}f \wedge \Omega^n_{alg})^{(q)}$. This gives the isomorphism $\mathbf{r}_q$.

Now for a family of polynomials $F_t \in \mathbb{C}[x]^{(1)}$, $t \in M_0$, with isolated singularities at 0 the infinitesimal variation of Hodge structures can be calculated. For each $t \in M_0$ it is a set of commuting linear maps

$$C_{\partial/\partial t_\alpha} = [\nabla_{\partial/\partial t_\alpha}] : F^p_{prim}/F^{p+1}_{prim} \to F^{p-1}_{prim}/F^p_{prim}. \tag{6.7}$$

The calculation (cf. [**Do**, 2.2])

$$\frac{\partial}{\partial t_\alpha}\frac{\omega}{F_t^q} = (-q)\frac{\frac{\partial F_t}{\partial t_\alpha} \cdot \omega}{F_t^{q+1}} \tag{6.8}$$

shows for $[\omega] \in \Omega_f^{(q)}$

$$C_{\partial/\partial t_\alpha}\mathbf{r}_q([\omega]) = \mathbf{r}_{q+1}\left(-\mathbf{a}(\frac{\partial}{\partial t_\alpha}) \cdot [\omega]\right). \tag{6.9}$$

Now one observes two facts: (i) The subring $\bigoplus_{q\in\mathbb{Z}_{\geq 0}} R_{F_t}^{(q)}$ of R_{F_t} is multiplicatively generated by $R_{F_t}^{(1)}$, because any monomial in $\mathbb{C}[x]^{(q)}$ for $q \in \mathbb{Z}_{\geq 1}$ is a product of monomials in $\mathbb{C}[x]^{(1)}$.

(ii) If $\frac{n+1}{d} \in \mathbb{N}$ then $\bigoplus_{q\in\mathbb{Z}} \Omega_{F_t}^{(q)}$ is a free module of rank 1 of the ring $\bigoplus_{q\in\mathbb{Z}_{\geq 0}} R_{F_t}^{(q)}$.

This shows that for a family of polynomials with (6.2) and $\frac{n+1}{d} \in \mathbb{N}$ the H^2-generation condition in definition 5.3 is satisfied. □

Via theorem 5.6 the variations of Hodge filtrations in theorem 6.1 together with chosen opposite filtrations give rise to Frobenius manifolds. All of them can be identified with submanifolds of Frobenius manifolds which arise in singularity theory, see chapter 7. Those with $d = n+1$ can be identified with submanifolds of Frobenius (super) manifolds in the Barannikov–Kontsevich construction, see chapter 8. The hypersurfaces $X = \overline{f^{-1}(0)} \subset \mathbb{P}^n$ are Calabi–Yau if and only if $d = n+1$.

Much of the preceding discussion generalizes to the case of quasihomogeneous singularities, but not all. Lemma 6.2 gives an example where the H^2-generation condition fails to hold.

A weight system $(w_0, ..., w_n)$ with $w_i \in \mathbb{Q} \cap (0, \frac{1}{2}]$ induces a new grading on $\mathbb{C}[x_0, ..., x_n]$ whose subspaces we also denote by $\mathbb{C}[x]^{(q)}$. A monomial $x_0^{i_0}...x_n^{i_n}$ is in $\mathbb{C}[x]^{(q)}$ for $q \in \mathbb{Q}_{\geq 0}$ if $\sum i_j w_j = q$. The Jacobi algebra R_f of $f \in \mathbb{C}[x]^{(1)}$ is defined as above and inherits a grading with subspaces $R_f^{(q)}$.

LEMMA 6.2. *Consider $n = 5$ and $(w_0, ..., w_5) = \frac{1}{9}(1,1,1,2,2,2)$ and any weighted homogeneous polynomial $f \in \mathbb{C}[x]^{(1)}$ with isolated singularity at 0. The subspace of $R_f^{(2)}$ which is generated by the set $R_f^{(1)} \cdot R_f^{(1)}$ has codimension 1 in $R_f^{(2)}$.*

Proof. A monomial in $\mathbb{C}[x]^{(2)}$ is a product of monomials in $\mathbb{C}[x]^{(1)}$ if and only if it contains x_0, x_1 or x_2. Therefore it is sufficient to show that the space

$$\mathbb{C}[x_3, x_4, x_5]^{(2)} \cap \left(\frac{\partial f}{\partial x_0}, ..., \frac{\partial f}{\partial x_n}\right) \tag{6.10}$$

has codimension 1 in $\mathbb{C}[x_3, x_4, x_5]^{(2)}$. Notice $\frac{\partial f}{\partial x_i}|_{\{x_0=x_1=x_2=0\}} = 0$ for $i = 3,4,5$ and define $\widetilde{f}_i(x_3, x_4, x_5) := \frac{\partial f}{\partial x_i}|_{\{x_0=x_1=x_2=0\}}$ for $i = 0,1,2$. The ideal $(\widetilde{f}_0, \widetilde{f}_1, \widetilde{f}_2) \subset \mathbb{C}[x_3, x_4, x_5]$ has an isolated zero at 0. The dimension of $\left(\mathbb{C}[x_3, x_4, x_5]/(\widetilde{f}_0, \widetilde{f}_1, \widetilde{f}_2)\right)^{(2)}$ is independent of the choice of $\widetilde{f}_0, \widetilde{f}_1, \widetilde{f}_2$ as long as the ideal $(\widetilde{f}_0, \widetilde{f}_1, \widetilde{f}_2)$ has an isolated zero at 0. The choice $\widetilde{f}_0 = x_3^4$, $\widetilde{f}_1 = x_4^4$, $\widetilde{f}_2 = x_5^4$ shows that the dimension is 1. □

7. Frobenius manifolds for hypersurface singularities

In theorem 7.3 the Frobenius manifolds which one obtains from theorem 6.1 combined with theorem 5.5 will be identified with submanifolds of Frobenius manifolds in singularity theory. For each holomorphic function germ $f : (\mathbb{C}^{n+1}, 0) \to (\mathbb{C}, 0)$ with an isolated singularity at 0 the base space $M_\mu \subset \mathbb{C}^\mu$ of a semiuniversal unfolding can be equipped with the structure of a Frobenius manifold [**SK**][**SM**]. A detailed account is given in [**He1**]. In [**He2**, ch. 8] the construction is recasted as the construction of a $(trTLEP(w))$-structure on M_μ and subsequent application of a special case of theorem 4.5. We restrict now to quasihomogeneous singularities and recall some facts from [**He1**].

Let $(w_0, ..., w_n)$ be a weight system with $w_i \in \mathbb{Q} \cap (0, \frac{1}{2}]$ and $f \in \mathbb{C}[x]^{(1)}$ (notation from the end of chapter 6) a weighted homogeneous polynomial with isolated singularity at 0 and Milnor number μ. ¿From the grading of the Jacobi algebra R_f one obtains the exponents $\alpha_1, ..., \alpha_\mu$, rational numbers with

$$\alpha_1 \leq ... \leq \alpha_\mu, \quad \alpha_1 = \sum w_i, \quad \sharp(i \mid \alpha_i = \alpha) = \dim R_f^{(\alpha-\alpha_1)}. \tag{7.1}$$

Choose polynomials $m_i \in \mathbb{C}[x]^{(\alpha_i-\alpha_1)}$ which represent a basis of the Jacobi algebra and such that $m_1 = 1$. The function

$$F(x_0, ..., x_n, t_1, ..., t_\mu) = f + \sum_{i=1}^{\mu} t_i m_i \tag{7.2}$$

is a semiuniversal unfolding of f. It should be seen as a family of functions F_t with parameter $t = (t_1, ..., t_\mu) \in M_\mu \subset \mathbb{C}^\mu$. Here M_μ is a suitable open neighborhood of 0. The manifold M_μ comes equipped with the unit field $e = \frac{\partial}{\partial t_1}$, the Euler field $E = \sum_{i=1}^{\mu}(1 + \alpha_1 - \alpha_i)t_i \frac{\partial}{\partial t_i}$, and a multiplication $\circ$ on the holomorphic tangent bundle. The multiplication is induced from a canonical isomorphism of T_tM_μ with the direct sum of Jacobi algebras of the singularities of the function F_t for $t \in M_\mu$. The tuple $(M_\mu, \circ, e, E)$ is an F-manifold [**Man2**, I §5][**He1**, I].

With the Gauß–Manin connection one can construct a metric g on M_μ such that $(M_\mu, \circ, e, E, g)$ is a Frobenius manifold [**SK**][**SM**]. In general the metric depends on a choice. For the construction we refer to [**He1**]. Here we merely explain the choice and state the result.

Let H^∞ be the space of global flat multivalued sections of the flat cohomology bundle $\bigcup_{z\in\mathbb{C}^*} H^n(f^{-1}(z), \mathbb{C})$. It comes equipped with a real subbundle $H^\infty_\mathbb{R}$, a semisimple monodromy operator $h : H^\infty \to H^\infty$, a monodromy invariant Hodge filtration $F^\bullet$ and a monodromy invariant pairing S^∞ [**He1**, ch. 10]. Define $H^\infty_\lambda := \ker(h - \lambda\,\mathrm{id}) \subset H^\infty$ and $H^\infty_{\neq 1} := \bigcup_{\lambda\neq 1} H^\infty_\lambda$.

Then $(H^\infty_{\neq 1}, H^\infty_\mathbb{R} \cap H^\infty_{\neq 1}, F^\bullet, S^\infty)$ and $(H^\infty_1, H^\infty_\mathbb{R} \cap H^\infty_1, F^\bullet, S^\infty)$ are polarized Hodge structures of weight n and $n+1$. An increasing monodromy invariant filtration $U_\bullet$ on H^∞ is called *opposite to* $F^\bullet$ if

$$H^\infty = \bigoplus_p F^p \cap U_p, \tag{7.3}$$

$$S^\infty(H^\infty_{\neq 1} \cap F^p \cap U_p, H^\infty_{\neq 1} \cap F^q \cap U_q) = 0 \text{ for } p + q \neq n, \tag{7.4}$$

$$S^\infty(H^\infty_1 \cap F^p \cap U_p, H^\infty_1 \cap F^q \cap U_q) = 0 \text{ for } p + q \neq n + 1. \tag{7.5}$$

THEOREM 7.1. [**SM**][**He1**, Theorem 11.1] *Any choice of an opposite filtration $U_\bullet$ induces an up to a scalar unique metric g on M_μ such that $(M_\mu, \circ, e, E, g)$ is a Frobenius manifold. The opposite filtration is uniquely determined by the metric.*

Consider the submanifold

$$M := \{t \in M_\mu \mid t_i = 0 \text{ if } \alpha_i - \alpha_1 \notin \mathbb{Z}\} \subset M_\mu. \tag{7.6}$$

It does not depend on the choice of the coordinates t_i; that means, any choice with $Et_i = (1 + \alpha_1 - \alpha_i)t_i$ gives the same submanifold M. It parametrizes the semiquasihomogeneous deformations of f by polynomials of integer degree. Unit field e and Euler field E are tangent to M. One can show that the multiplication $\circ$ on TM_μ restricts to a multiplication on TM. If $\alpha_1 = \sum w_i \in \frac{1}{2}\mathbb{N}$ then a much stronger result holds.

THEOREM 7.2. [**Man2**, III 8.7.1] *Suppose that* $\alpha_1 = \sum w_i \in \frac{1}{2}\mathbb{N}$. *For any metric* g *as in theorem 7.1, the submanifold* M *with induced multiplication, metric, Euler field* E *and unit field* e *is a Frobenius manifold.*

Going through the construction in [**He1**, 11.1] one can even see that the metric on M depends only on $U_\bullet \cap H^\infty_{(-1)^{\alpha_1}}$ and that it determines $U_\bullet \cap H^\infty_{(-1)^{\alpha_1}}$ uniquely.

Finally we restrict to the case $(w_0, ..., w_n) = \frac{1}{d}(1, ..., 1)$ for some $d \in \mathbb{N}$ with $\alpha_1 = \frac{n+1}{d} \in \mathbb{N}$. We use the notations in chapter 5. The manifold

$$M_0 := \{t \in M_\mu \mid t_i = 0 \text{ if } \alpha_i - \alpha_1 \neq 1\} \subset M \tag{7.7}$$

parametrizes the homogeneous polynomials F_t in the unfolding F. The map **a** in (6.2) is an isomorphism. Theorem 6.1 applies to the family of functions F_t, $t \in M_0$. As in chapter 6, $X_t := \overline{F_t^{-1}(0)} \subset \mathbb{P}^n$ is the hypersurface in $\mathbb{P}^n$ defined by F_t.

THEOREM 7.3. *(a) There is a canonical isomorphism*

$$\psi : \bigcup_{t\in M_0} H_1^\infty(F_t) \to \bigcup_{t\in M_0} H^{n-1}_{prim}(X_t) \tag{7.8}$$

of flat bundles with pairings S^∞ *and* S. *For each* $t \in M_0$ *it is a* $(-1,-1)$ *morphism of Hodge structures.*

(b) The Frobenius manifold structures on M *from theorem 7.2 and the Frobenius manifolds constructed in theorem 5.6 from the variation of filtrations*

$$((M_0, 0), H, \nabla, S, \widetilde{F}^\bullet, \widetilde{U}_\bullet)$$

in theorem 6.1 with opposite filtration $\widetilde{U}_\bullet$ *on* $H = \bigcup_{t\in M_0} H^{n-1}_{prim}(X_t)$ *are pairwise isomorphic. Two are isomorphic up to multiplication of the metric by a scalar if and only if the opposite filtrations* $U_\bullet \cap H_1^\infty$ *on* H_1^∞ *and* $\widetilde{U}_\bullet$ *on* H *satisfy*

$$\psi(U_{\bullet-1+(n+1)/d} \cap H_1^\infty) = \widetilde{U}_\bullet. \tag{7.9}$$

Proof. (a) This is essentially well known. The following explanations may be helpful. Consider for some $z \in \mathbb{C}^*$ a fiber $F_t^{-1}(z) \subset \mathbb{C}^{n+1}$. The hypersurface $X_t \subset \mathbb{P}^n$ is the part in $\mathbb{P}^n$ of the closure of $F_t^{-1}(z) \subset \mathbb{C}^{n+1}$ in $\mathbb{P}^{n+1} = \mathbb{C}^{n+1} \cup \mathbb{P}^n$. Consider a tubular neighborhood $T(F_t^{-1}(z))$ of X_t in $F_t^{-1}(z)$. There are canonical isomorphisms

$$H_1^\infty \leftarrow H^n(F_t^{-1}(z), \mathbb{C})_1 \to H^n(T(F_t^{-1}(z)), \mathbb{C}) \to H^{n-1}_{prim}(F_t) \tag{7.10}$$

The first is the extension to flat sections, the last is a residue map and is the dual of a tube map.

Consider a form $\omega \in (\Omega^{n+1}_{alg})^{(q)}$ for $q \in \mathbb{Z}_{\geq (n+1)/d}$ (notation from chapter 6). The restriction of the form $\frac{1}{d} \cdot i_E \frac{\omega}{f^q}$ to $F_t^{-1}(z)$ is equal to $\frac{1}{z^{q-1}} \frac{\omega}{\mathrm{d}F_t}$, because $\mathrm{d}F_t \wedge i_E\omega = d \cdot F_t \cdot \omega$. The section $z \mapsto \left[\frac{1}{z^{q-1}} \frac{\omega}{\mathrm{d}F_t}\right]$ of the bundle $\bigcup_{z\in\mathbb{C}^*} H^n(F_t^{-1}(z), \mathbb{C})_1$ is flat.

Now Varchenko's [**Va**] description of Steenbrink's Hodge filtration $F^\bullet$ on H_1^∞ reduces here to the following: The space $F^{n+1-q} \subset H_1^\infty$ is generated by such flat sections.

In $\mathbb{P}^{n+1} - (F_t^{-1}(z) \cup X_t)$ the set $T(F_t^{-1}(z))$ can be deformed to a tubular neighborhood of X_t in $\mathbb{P}^n - X_t$. The bundle $\bigcup_{z\in\mathbb{C}^*} H^n(F_t^{-1}(z), \mathbb{C})_1$ and the space $H^n(\mathbb{P}^n - X_t)$ glue to a flat bundle on $\mathbb{P}^1 - \{0\}$. The value in $H^n(\mathbb{P}^n - X_t)$ of a flat section as above is just $\left[\frac{1}{d}\cdot\overline{i_{\mathcal{E}}\frac{\omega}{f^q}}\right]$; compare the proof of theorem 6.1. This reduces Varchenko's to Griffiths' description and shows (a).

(b) This follows from part (a), theorem 5.6 and from the following nontrivial fact: the Frobenius type structure on $TM|_{M_0}$ for a Frobenius manifold M in theorem 7.2 corresponds by lemma 5.1 to a variation of filtrations and an opposite filtration which are up to the shift in (7.9) the variation of Hodge structures and an opposite filtration on $\bigcup_{t\in M_0} H_1^\infty(F_t)$.

This fact is a consequence of the construction of Frobenius manifolds in [**He1**, Theorem 11.1]. □

8. Barannikov–Kontsevich construction

The Barannikov–Kontsevich construction was initiated in [**BK**] and further developed in [**Ba1**][**Ba2**]. It yields for any Calabi–Yau manifold a family of formal germs of Frobenius submanifolds. A central part of it is the construction of a semi-infinite variation of Hodge structures (defined in [**Ba2**]). This contains the variation of Hodge structures for complex structure deformations of the given Calabi–Yau manifold. Therefore it is not surprising that in the case of a Calabi–Yau hypersurface in $\mathbb{P}^n$ certain submanifolds in the Barannikov–Kontsevich construction coincide with germs of the Frobenius manifolds which one obtains from theorem 6.1 with theorem 5.6. This will be made precise in theorem 8.1. All the results in this chapter are reformulations of results of Barannikov [**Ba1**][**Ba2**].

Let X be a compact Calabi–Yau manifold of dimension $n-1$. Let us fix a holomorphic volume form Ω. Barannikov [**Ba1**][**Ba2**] constructed a family of formal germs $(M,0)$ of Frobenius supermanifolds of dimension $m := \dim H^*(X,\mathbb{C})$. For each of them the tangent space T_0M at 0 is isomorphic to

$$H^*(X,\bigwedge^{*}\mathcal{T}_X)=\bigoplus_{p,q}H^q(X,\bigwedge^{p}\mathcal{T}_X) \tag{8.1}$$

with the canonical multiplication and with the $\mathbb{Z}_2$-grading $(p+q)$ mod 2. If $t_1,...,t_m$ are flat coordinates centered at 0 with $\frac{\partial}{\partial t_i}|_0\in H^{q_i}(X,\bigwedge^{p_i}\mathcal{T}_X)$ then

$$E=\sum_{i=1}^{m}\frac{2-p_i-q_i}{2}t_i\frac{\partial}{\partial t_i} \tag{8.2}$$

is an Euler field of the Frobenius supermanifold [**Ba1**, 5.7]. Euler field and flat metric g satisfy $\text{Lie}_E(g)=(3-n)g$ and $g(\frac{\partial}{\partial t_i},\frac{\partial}{\partial t_j})=0$ for $(p_i+p_j,q_i+q_j)\neq(n-1,n-1)$.

This family of Frobenius supermanifolds is parametrized by the set of *opposite filtrations* $W_{\leq\bullet}$ to a Hodge filtration $F^{\geq\bullet}$ on $H^*(X,\mathbb{C})$. These filtrations are defined as follows [**Ba2**, ch. 4 and 6]. Both respect the splitting

$$H^*(X,\mathbb{C})=H^{even}(X,\mathbb{C})\oplus H^{odd}(X,\mathbb{C}) \tag{8.3}$$

and are indexed by half integers. The Hodge filtration is given by

$$F^{\geq r}:=\bigoplus_{p,q:p-q\geq 2r}H^{p,q}(X)\ \text{ for } r\in\frac{1}{2}\mathbb{Z}. \tag{8.4}$$

A filtration $W_{\le\bullet}$ is *opposite* if

$$H^*(X,\mathbb{C}) = \bigoplus_{r\in\frac{1}{2}\mathbb{Z}} F^{\ge r}\cap W_{\le r+1} \quad \text{and} \tag{8.5}$$

$$(F^{\ge r}\cap W_{\le r+1}, F^{\ge \widetilde{r}}\cap W_{\le \widetilde{r}+1}) = 0 \quad \text{for } r+\widetilde{r}\neq 0. \tag{8.6}$$

Here $(,)$ is the Poincaré pairing on $H^*(X,\mathbb{C})$. Let us denote these Frobenius supermanifolds for a moment by $M_{Bar}(X, W_{\le\bullet})$.

¿From now on we suppose that $n\ge 4$ and that X is a Calabi–Yau hypersurface in $\mathbb{P}^n$, that is, $X = \overline{f^{-1}(0)} \subset \mathbb{P}^n$ where $f\in\mathbb{C}[x_0,...,x_n]$ is homogeneous of degree $n+1$ with an isolated singularity at 0. Then the cohomology $H^*(X,\mathbb{C})$ splits into two orthogonal pieces,

$$H^*(X,\mathbb{C}) = H^*_{Lef}(X)\oplus H^{n-1}_{prim}(X). \tag{8.7}$$

As in chapter 5 the second piece $H^{n-1}_{prim}(X)\subset H^{n-1}(X)$ is the primitive part of the middle cohomology; with L as standard Lefschetz operator the first piece is

$$H^*_{Lef}(X) := \bigoplus_{k=0}^{n-1} L^k H^0(X,\mathbb{C}). \tag{8.8}$$

If $n-1$ is odd then the splittings (8.3) and (8.7) coincide. Then an opposite filtration $W_{\le\bullet}$ for $F^{\ge\bullet}$ induces an opposite filtration $\widetilde{U}_\bullet$ for $\widetilde{F}^\bullet := F^{\bullet-1}H^{n-1}_{prim}(X)$ by

$$\widetilde{U}_p := W_{\le p-\frac{n-1}{2}}\cap H^{n-1}_{prim}(X), \tag{8.9}$$

and this gives a 1–1 correspondence between the opposite filtrations $\widetilde{U}_\bullet$ and the opposite filtrations $W_{\le\bullet}$. If $n-1$ is even then $H^*(X,\mathbb{C}) = H^{even}(X,\mathbb{C})$. Then formula (8.9) gives a 1–1 correspondence between the opposite filtrations $\widetilde{U}_\bullet$ and those opposite filtrations $W_{\le\bullet}$ for $F^{\ge\bullet}$ which respect the splitting (8.7).

Let us fix for a moment an opposite filtration $\widetilde{U}_\bullet$ for $\widetilde{F}^\bullet$ on $H^{n-1}_{prim}(X)$. Theorem 6.1 and theorem 5.6 yield a germ $(M',0)$ of a holomorphic Frobenius manifold of dimension $m' := \dim H^{n-1}_{prim}(X)$, which is unique up to multiplication of the metric by a scalar. Let E' be its Euler field, $\mathcal{F}'\in\mathcal{O}_{M',0}$ its potential, and $e' = \frac{\partial}{\partial t_1}$ its unit field for suitable coordinates $t_1,...,t_{m'}$.

The Frobenius manifold can be extended in the following trivial way to a Frobenius (super)manifold of dimension m. Consider $\mathbb{C}^{m-m'}$ with coordinates $(\tau_1,...,\tau_{m-m'})$ of degree $(n-1) \mod 2$. Then the germ $(M',0)\times(\mathbb{C}^{m-m'},0)$ with potential

$$\mathcal{F} := \mathcal{F}' + \frac{1}{2}\sum_{i=1}^{m-m'} t_1\tau_i\tau_{m-m'+1-i}, \tag{8.10}$$

and Euler field

$$E := E' + \sum_{i=1}^{m-m'} \frac{3-n}{2}\tau_i\frac{\partial}{\partial\tau_i} \tag{8.11}$$

is a Frobenius supermanifold for $n-1$ odd and a Frobenius manifold for $n-1$ even. We call it $M_{VHS}(X,\widetilde{U}_\bullet)$. It contains the germ $(M',0)\times\{0\}$ as a Frobenius submanifold. For $n-1$ odd this is the Frobenius submanifold in [**Man2**, Theorem 8.7.1]. The following theorem is essentially contained in [**Ba1**, Theorem 6.5].

THEOREM 8.1. *Let $X \subset \mathbb{P}^n$ be a Calabi–Yau hypersurface with $n \geq 4$.*

Consider an opposite filtration $\widetilde{U}_\bullet$ for $\widetilde{F}^\bullet := F^{\bullet-1}H^{n-1}_{prim}(X)$ on $H^{n-1}_{prim}(X)$ and an opposite filtration $W_{\leq\bullet}$ for $F^{\geq\bullet}$ on $H^*(X,\mathbb{C})$ which respects the splitting (8.7) if $n-1$ is even. Then

$$M_{Bar}(X, W_{\leq\bullet}) \cong M_{VHS}(X, \widetilde{U}_\bullet) \tag{8.12}$$

if and only if (8.9) holds.

Proof. For $n-1$ odd $M_{Bar}(X, W_{\leq\bullet})$ is a formal germ of a Frobenius supermanifold with odd part corresponding to $H^*_{Lef}(X)$ and even part corresponding to $H^{n-1}_{prim}(X)$. For $n-1$ even everything is even. This follows from the discussion at the beginning of this chapter and from the isomorphisms $\bigwedge^p \mathcal{T}_X \cong \Omega_X^{n-1-p}$ and $H^q(X, \bigwedge^p \mathcal{T}_X) \cong H^{n-1-p,q}(X)$, which one obtains by contraction of the holomorphic volume form Ω on X with holomorphic vector fields.

The condition $n \geq 4$ asserts that $H^{n-2,1}(X) \subset H^{n-1}_{prim}(X)$. Therefore a miniversal family X_t, $t \in M_0$, of complex structure deformations of $X_0 = X$ is given by a family of homogeneous polynomials f_t, $t \in M_0$, as in chapter 6 with condition (6.2) and $f_0 = f$.

By construction of $M_{Bar}(X, W_{\leq\bullet})$ there is a natural inclusion

$$(M_0, 0) \subset M_{Bar}(X, W_{\leq\bullet})$$

of formal germs (the formal germ $(M_0, 0)$ is called M^{cx} in [**Ba2**]). By lemma 4.4 and lemma 5.1 one obtains on $TM_{Bar}(X, W_{\leq\bullet})|_{(M_0,0)}$ a formal germ of a variation of filtrations with pairing and opposite filtration.

The following fact is crucial: This structure is isomorphic to the restriction to the formal germ $(M_0, 0)$ of the variation of filtrations $F^{\geq\bullet}(X_t)$ on the bundle $\bigcup_{t\in M_0} H^*(X_t, \mathbb{C})$ and the opposite filtration $W_{\leq\bullet}$.

This fact is at the heart of Barannikov's construction of semi-infinite variations of Hodge structures [**Ba2**, Theorem 4.2] and Frobenius manifolds.

Now the case $n-1$ odd is easy. The variation of filtrations $F^{\geq\bullet}(X_t)$ and the opposite filtration $W_{\leq\bullet}$ on $\bigcup_{t\in M_0} H^{n-1}_{prim}(X_t)$ correspond to the formal germ of the even Frobenius submanifold. By theorem 5.6 this germ must coincide with the germ $(M', 0)$ of the Frobenius manifold before theorem 8.1 for filtrations $\widetilde{U}_\bullet$ and $W_{\leq\bullet}$ with (8.9). Because of the degrees of the flat coordinates the extension with the odd part to a Frobenius supermanifold is rigid and is given by (8.10) and (8.11).

The case $n-1$ even is more difficult because the H^2-generation condition does not hold for the variation of filtrations on $\bigcup_{t\in M_0} H^*(X_t, \mathbb{C})$. But a slightly weaker condition holds so that lemma 8.2 below applies. It shows that the variation of filtrations $F^{\geq\bullet}(X_t)$ on $\bigcup_{t\in M_0} H^*(X_t, \mathbb{C})$ and *any* opposite filtration $W_{\leq\bullet}$ determine a unique holomorphic germ of a Frobenius manifold, whether $W_{\leq\bullet}$ respects the splitting (8.7) or not. It remains to see that precisely the filtrations $W_{\leq\bullet}$ which respect the splitting (8.7) lead to the Frobenius manifolds $M_{VHS}(X, \widetilde{U}_\bullet)$.

One can go into the proof of lemma 8.2 and start there with vector fields $\delta_1, ..., \delta_m$ and an index set $I \subset \{1, ..., m\}$ such that the isomorphism $T_0M \to H^*(X, \mathbb{C})$ maps δ_i into $H^*_{Lef}(X)$ for $i \in I$ and into $H^{n-1}_{prim}(X)$ for $i \notin I$. Then one has to check that in the case of a filtration $W_{\leq\bullet}$ which respects the splitting (8.7)

$$a_{ij}^k = 0 \text{ for } d_i = 0,\ j \in I \tag{8.13}$$

holds. Going through the induction in the proof of lemma 8.2 one finds that (8.13) holds for all $i \notin I \cup \{1\}$, $j \in I$ and that a_{ij}^k is constant for $i, j \in I$. Then the flat submanifold $\{t \in M \mid t_i = 0 \text{ for } i \in I\}$ is a Frobenius submanifold, and the whole Frobenius manifold is $M_{VHS}(\widetilde{U}_\bullet)$. □

One can generalize theorem 5.6 by relaxing the H^2-generation condition slightly. The next lemma formulates the part of this generalization which is needed in the proof of theorem 8.1. It was already used in the proof of [**Ba1**, Theorem 6.5].

LEMMA 8.2. *Consider a germ of a Frobenius manifold* $((M,0), \circ, e, E, g)$ *with a weight* $w \in \mathbb{N}_{\geq 3}$ *and all properties in definition 5.4 except the* H^2*-generation condition, which is replaced by the weaker condition (II)' on the graded algebra*

$$
\begin{aligned}
(8.14) \qquad T_0M &= \bigoplus_{p=0}^{n-2} (T_0M)_p \quad \textit{with} \\
(T_0M)_p &:= \ker(\nabla^g E - (1-p)\,\mathrm{id} : T_0M \to T_0M).
\end{aligned}
$$

(II)' $(T_0M)_1$ *generates multiplicatively a subspace of* T_0M *which contains*

$$\bigoplus_{p<(w-2)/2} (T_0M)_p.$$

Define $M_0 := \{t \in M \mid E|_t = 0\}$.

The germ of a Frobenius manifold is uniquely determined by the induced Frobenius type structure (see lemma 4.4) on $TM|_{M_0}$ *together with* $e|_0 \in T_0M$.

Proof. Choose flat coordinates $t_1, ..., t_m$ on $(M,0)$ with vector fields $\delta_i := \frac{\partial}{\partial t_i}$ such that

$$E = \sum_{i=1}^{m} (-d_i) t_i \delta_i \tag{8.15}$$

with $d_1 = -1$, $\delta_1 = e$, $d_2 = ... = d_{m_0+1} = 0$, $d_i > 0$ for $i > m_0 + 1$. Then $M_0 = \{t \in M \mid t_i = 0 \text{ for } d_i \neq 0\}$. Define matrices $A_i = (a_{ij}^k) \in M(m \times m, \mathcal{O}_{M,0})$ for $i = 1, ..., m$ by

$$\delta_i \circ \delta_j = \sum_k a_{ij}^k \cdot \delta_k. \tag{8.16}$$

Then

$$
\begin{aligned}
(8.17) \qquad A_i A_j &= A_j A_i, \\
(8.18) \qquad a_{i1}^k &= \delta_{ik}, \\
(8.19) \qquad a_{ij}^k &= a_{ji}^k,
\end{aligned}
$$

and the potentiality condition is equivalent to

$$\delta_i A_j = \delta_j A_i. \tag{8.20}$$

Especially $\delta_1 A_j = 0$ for all i. Denote for $w \in \mathbb{Z}_{\geq 0}$

$$
\begin{aligned}
(8.21) \qquad \mathcal{O}(M)_w &:= \{f \in \mathcal{O}_{M_0,0}[t_i \mid d_i > 0] \mid Ef = -wf\}, \\
(8.22) \qquad M(w) &:= M(m \times m, \mathcal{O}(M)_w), \\
(8.23) \qquad M(>w) &:= \bigoplus_{k>w} M(k).
\end{aligned}
$$

Then $t_i \in \mathcal{O}(M)_{d_i}$ for $i \geq 2$. The condition $\mathrm{Lie}_E(\circ) = \circ$ together with (8.15) shows

$$a_{ij}^k \in \mathcal{O}(M)_{d_k - d_i - d_j - 1}. \tag{8.24}$$

It is not hard to see that the Frobenius type structure on $TM|_{M_0}$ and $e|_0 \in T_0M$ provide after some choice the matrices $A_i \mod M(>0)$ with $d_i = 0$ and the coefficients $g(\delta_i, \delta_j) \in \mathbb{C}$ of the metric for all i, j. As in lemma 2.9 one has to recover all the matrices A_j in order to uniquely determine the Frobenius manifold. Again this will be done inductively.

Induction hypothesis for $w \in \mathbb{Z}$: one has determined the matrices $A_i \mod M(> w)$ for $d_i = 0$ and $A_j \mod M(> w-1)$ for $d_j > 0$.

Induction step from w to $w+1$: It consists of two steps.

(i) Determine the matrices $A_j \mod M(> w)$ for $d_j > 0$.

(ii) Determine the matrices $A_i \mod M(> w+1)$ for $d_i = 0$.

(i) The weakened generation condition (II)' together with (8.24) shows that one obtains from the matrices $A_i \mod M(> w)$ with $d_i = 0$ and from the matrices $A_j \mod M(> w-1)$ the matrices $A_k \mod M(> w)$ for k with $k < \frac{w-2}{2}$. Because of (8.19) the only unknown coefficients of the matrices $A_k \mod M(> w)$ for $k \geq \frac{w-2}{2}$ are those coefficients a_{kl}^r with $d_l \geq \frac{w-2}{2}$. Because of (8.24), in the case $d_k + d_l = w-2$ they are constant and determined by $g(\delta_k, \delta_l)$, in the case $d_k + d_l > w-2$ they vanish.

(ii) Similarly to step (iii) in the proof of lemma 2.9 one uses (8.20) in the form

$$\delta_j(A_i \mod M(> w+1)) = \delta_i(A_j \mod M(> w+1-d_j)) \tag{8.25}$$

for $d_i = 0, d_j > 0$. $\square$

REMARK 8.3. The proof of theorem 8.1 contains the three statements: for $n-1$ even any Frobenius manifold $M_{Bar}(X, W_{\leq \bullet})$ is a holomorphic germ of a Frobenius manifold; it is uniquely determined up to a scalar of the metric by the variation of filtrations $F^{\geq \bullet}$ and the opposite filtration $W_{\leq \bullet}$ on $\bigcup_{t \in M_0} H^*(X_t, \mathbb{C})$; it contains a Frobenius submanifold with tangent space at 0 isomorphic to $H^{n-1}_{prim} \subset H^*(X_t, \mathbb{C}) \cong T_0M$ precisely if the filtration $W_{\leq \bullet}$ respects the splitting (8.7)

References

[Ba1] S. Barannikov: Generalized periods and mirror symmetry in dimensions $n > 3$. preprint, math.AG/9903124.

[Ba2] S. Barannikov: Quantum periods – I. Semi-infinite variations of Hodge structures. Int. Math. Res. Notices **2001–23**: 1243–1264.

[Ba3] S. Barannikov: Semi-infinite variations of Hodge structures and integrable hierarchies of KdV-type. Int. Math. Res. Notices **2002–19**: 973–990.

[BK] S. Barannikov, M. Kontsevich: Frobenius manifolds and formality of Lie algebras of polyvector fields. Int. Math. Res. Notices **1998-14**, 201–215.

[Do] R. Donagi: Generic Torelli for projective hypersurfaces. Comp. Math **50** (1983), 325–353.

[Du] B. Dubrovin: Geometry of 2D topological field theories. In: Integrable systems and quantum groups. Montecatini, Terme 1993 (M. Francoviglia, S. Greco, eds.). Lecture Notes in Math. 1620, Springer Verlag 1996, pp. 120–348.

[Fo] G.B. Folland: Introduction to partial differential equations. Princeton University Press, 2nd ed. 1995.

[Gr] P. Griffiths: On the periods of certain rational integrals I. Annals of Math. **90** (1969), 460–495.

[He1] C. Hertling: Frobenius manifolds and moduli spaces for singularities. To appear in Cambridge Tracts in Mathematics, Cambridge University Press, August 2002.

[He2] C. Hertling: tt^* geometry, Frobenius manifolds, their connections, and the construction for singularities. J. reine angew. Math. **555** (2003), 77–161.

[KM] M. Kontsevich, Yu. Manin: Gromov–Witten classes, quantum cohomology, and enumerative geometry. Commun. Math. Phys. **164** (1994), 525–562.

[Kr] A. Kresch: Associativity relations in quantum cohomology. Advances in Mathematics **142** (1999), 151–169.

[Mal] B. Malgrange: Deformations of differential systems, II. J. Ramanujan Math. Soc. **1** (1986), 3–15.

[Man1] Yu. Manin: Three constructions of Frobenius manifolds: a comparative study. Asian J. Math. **3** (1999), 179–220.

[Man2] Yu. Manin: Frobenius manifolds, quantum cohomology, and moduli spaces. American Math. Society, Colloquium Publ. v. **47**, 1999.

[Sab1] C. Sabbah: Frobenius manifolds: isomonodromic deformations and period mappings. Expo. Math. **16** (1998), 1–58.

[Sab2] C. Sabbah: Déformations isomonodromiques et variétés de Frobenius, une introduction. Savoirs Actuels, EDP Sciences/CNRS Éditions, Paris, 2002.

[SK] K. Saito: Period mapping associated to a primitive form. Publ. RIMS, Kyoto Univ. **19** (1983), 1231–1264.

[SM] M. Saito: On the structure of Brieskorn lattices. Ann. Inst. Fourier Grenoble **39** (1989), 27–72.

[Va] A.N. Varchenko: The asymptotics of holomorphic forms determine a mixed Hodge structure. Sov. Math. Dokl. **22** (1980), 772–775.

Claus Hertling, Max-Planck-Institut für Mathematik Vivatsgasse 7, 53111 Bonn, Germany

E-mail address: `hertling@mpim-bonn.mpg.de`

Yuri Manin, Max-Planck-Institut für Mathematik Vivatsgasse 7, 53111 Bonn, Germany, and Northwestern University, Evanston , USA

E-mail address: `manin@mpim-bonn.mpg.de`

Discrete torsion, symmetric products and the Hilbert scheme

Ralph M. Kaufmann*

Introduction

Recently the understanding of the cohomology of the Hilbert scheme of points on K3 surfaces has been greatly improved by Lehn and Sorger [**18**]. Their approach uses the connection of the Hilbert scheme to the orbifolds given by the symmetric products of these surfaces. We introduced a general theory replacing cohomology algebras or more generally Frobenius algebras in a setting of global quotients by finite groups [**14**]. This is our theory of group Frobenius algebras, which are group graded non–commutative algebras whose non–commutativity is controlled by a group action. The action and the grading turn these algebras into modules over the Drinfel'd double of the group ring. The appearance of the Drinfel'd double is natural from the orbifold point of view (see also [**17**]) and can be translated into the fact that the algebra is a G–graded G–module algebra in the following sense: the G action acts by conjugation on the grading while the algebra structure is compatible with the grading with respect to left multiplication (cf. [**16, 20**]).

In the special case of the symmetric group, we recently proved existence and uniqueness for the structures of symmetric group Frobenius algebras based on a given Frobenius algebra [**15**], providing explicit formulas for the multiplication in the algebra.

This uniqueness has to be understood up to the action of two groups of symmetries on group Frobenius algebras called discrete torsion and super-twisting [**16**]. The set of G–Frobenius algebras is acted upon by both of these groups. This action only changes some defining structures of a Frobenius algebra in a projective manner while keeping others fixed.

Applying this result to the global orbifold cohomology of a symmetric product, where there is a canonical choice for the discrete torsion and super-twists, we obtain its uniqueness.

Our latest results on this topic [**16**] explain the origin of these discrete degrees of freedom. In the special case of the Hilbert scheme as a resolution of a symmetric product the choice of sign for the metric specifies a discrete torsion cocycle that in turn changes the multiplication by a much discussed sign.

Assembling our results which we review we obtain:

---

* Partially supported by NSF grant #0070681.

Theorem. *The cohomology of $Hilb^{[n]}$, the Hilbert scheme of n–points for a K3 surface, is the $\mathbb{S}_n$ invariant part of the $\mathbb{S}_n$-Frobenius Algebra associated to the symmetric product of the cohomology of the surface twisted by a discrete torsion. Or in other words the unique $\mathbb{S}_n$-Frobenius Algebra structure for the extended global orbifold cohomology twisted by the specific discrete torsion which is uniquely determined by the map of* [**18**]. *In general, the sequence of spaces $Hilb^{[n]}$ gives rise to the twisted second quantization of the underlying surface on the cohomological (motivic) level.*

Here the term associated refers to the uniqueness result of [**15**] stated above.

This result follows from a series of considerations which we will review. The logic is roughly as follows:

The theoretical background for our considerations was first presented at [**13**] and is given in [**14**] where we showed that the algebras arising from the "stringy" study of objects with a global action by a finite group are so–called G–Frobenius algebras. These algebras are non–commutative group graded extensions of their classical counterpart, Frobenius algebras, which arise for instance in the study of manifolds as cohomology algebras and in the study of singularities with an isolated critical point as a Milnor ring. A G–action by automorphisms is part of the data of a G–Frobenius algebra and taking invariants under this action yields a commutative algebra.

Given an object, such as a manifold, together with a finite group action and a functor, such as cohomology, one would like to augment this functor to take values in G–Frobenius algebras. The underlying additive structure of the G–Frobenius algebra is given by evaluating the functor on each fixed point set for each group element and forming the direct sum. This yields a collection of Frobenius algebras, one for each group element. The Frobenius algebra for the identity element is the Frobenius algebra associated to the object itself and is called the identity sector. For the other algebras, called twisted sectors, we only retain their structure as modules over the identity sector, together with their pairings – the module structure over the identity sector being induced by inclusion maps. Furthermore, there is a G action on the identity sector.

Any "stringy" extension of the original functor will respect these structures, and add a *group graded multiplication* and a group action by automorphisms on the whole algebra which is compatible with the group action on the identity sector.

It is possible to classify all such "stringy" extensions in the special case when all the twisted sectors are cyclic modules over the identity sector. Such G–Frobenius algebras are called special. The classification is in terms of group cohomological data as shown in [**14**]. The cyclicity condition is met in the situation of singularities with an isolated critical point at zero as well as for symmetric products which are the global quotient of the n-th power by the nth symmetric group $\mathbb{S}_n$. For the nth symmetric product of an object the untwisted sector is the nth tensor power of the Frobenius algebra of this object, while the twisted sector for a permutation is again a tensor power of the Frobenius algebra of this object, but to the power of the number of cycles in the permutation. It can be checked that these twisted sectors are indeed cyclic.

Imposing the cyclicity condition (i.e. restricting to special $\mathbb{S}_n$ Frobenius algebra) and an additional grading condition it is possible to make the classification concrete. The additional grading condition is satisfied in the case of symmetric products.

First, we showed in [**15**], if such a structure exists it is essentially unique.

This uniqueness is essential in the following sense: as explained in [**16**] given one "stringy" extension of the data considered above, it is possible to produce another extension with the use of a group cocycle in $Z^2(G, k^*)$. In the setting of super ($\mathbb{Z}/2\mathbb{Z}$–graded) algebras, we can also produce yet another extension for each element of $\mathrm{Hom}(G, \mathbb{Z}/2\mathbb{Z})$. These twists of the original extension can be achieved via a tensor product with a twisted group ring or a super–graded group ring [**16**]. This yields an action of both groups $Z^2(G, k^*)$ and $\mathrm{Hom}(G, \mathbb{Z}/2\mathbb{Z})$. These actions are called discrete torsion and super twist, respectively. Thus essentially unique means unique up to the action of these two groups.

Second, in the case of symmetric products, the unique structure exists, as we showed in [**15**]. There is also a canonical choice of an initial algebra structure upon which discrete torsion and super–twists act.

The proof of existence relies on a general formalism which makes use of the fact that in a setup such as a manifold with a finite group action, we can also regard the fixed point sets of all subgroups generated by several elements of the group. These fixed point sets are then the intersection of the fixed point sets for the individual generators. The general setting in which this is possible is the setting of intersection G–Frobenius algebras. In this framework, one can show that the multiplication factors through double intersections while the associativity equation is to be checked on triple intersections.

In the case where one is considering the symmetric product of a manifold, the canonical non–commutative structure coincides with the one found by [**10**] and its commutative invariants are those of the Chen–Ruan orbifold cohomology as calculated in [**21**]. Notice that our uniqueness result makes no reference to any space of maps or to any specific "stringy" extension, but only depends on the algebraic structure and is thus common to all "stringy" extensions.

Another non–commutative "stringy" multiplication based on the additive data underlying symmetric products was given in [**18**] in their consideration of Hilbert schemes of K3 surfaces. By our result this has to be related to the one stemming from symmetric products by either a twist by discrete torsion or a super–twist. Indeed there is a twist by discrete torsion, which produces the algebra of [**18**]. This discrete torsion cocycle is actually trivial on the level of cohomology, but naturally induces a sign change for the multiplication and the metric which is given in [**18**] on the level of group invariants. From our considerations of the action of discrete torsion [**16**] this cocycle is actually already fixed by the choice of sign for the metric, which by geometric reasoning (resolution of A_2 singularities) has to be negative definite.

Lastly, the family of multiplications found in [**19**] can be identified as the family of arising from twisting with discrete torsion cocycles which preserve the grading condition.

The paper is organized as follows: In §1 we present the general functorial setup for extending functors to Frobenius algebras to those with values in G–Frobenius algebras. §2 contains the basic definitions of G–Frobenius algebras and special G–Frobenius algebras [**14**], for which the possible extensions are classifiable in terms of group cohomological data. In §3 we introduce intersection Frobenius algebras which are adapted to the situation in which one can take successive intersections of fixed point sets. This structure is needed in order to show the existence of symmetric products and in general it is shown how in such a situation the multiplication can be defined via double intersections while the associativity equations are naturally given by total symmetry in the triple intersection. §4 reviews our analysis of discrete torsion [**16**]. §5 recalls our results on the structure of $\mathbb{S}_n$–Frobenius algebras [**15**] and contains the existence and uniqueness statements. §6 assembles these results in the case of any $\mathbb{S}_n$–Frobenius algebra twisted by a specific discrete torsion uniquely determined from the map [**18**]. This result applied to the situation of the Hilbert scheme yields the theorem above.

Notation

For the remainder of the paper G is a fixed finite group and k is a field of characteristic 0.

Acknowledgments

I would like to thank the Max–Planck–Institut for its kind hospitality and also gratefully acknowledge the support from the NSF. I would also like to thank the organizers of the conference. Special thanks go to Takashi Kimura, since it was a discussion with him that was the initial spark for our approach to discrete torsion. Most importantly I wish to thank Yuri I. Manin whose deep insights into the beauty of mathematics have been a continuous source of inspiration.

1. Functorial setup

1.1. General background. We will consider objects X together with the action of a finite group G. In this situation, one would classically study the invariants or the quotient of X by G. In stringy geometry for global quotients, however, it is the aim to enlarge this picture to consider the fixed point sets for all group elements, together with an induced G action on them. Here G acts on the fixed point sets by conjugation of the group elements labelling the fixed point sets. The classical part is then represented by X considered as the fixed point set of the identity in G and the G action on this fixed point set.

In particular, classical functors such as cohomology which takes values in Frobenius algebras should have an augmented counterpart including the information about all the fixed point sets. The augmented functors should take values in G–Frobenius algebras as defined in [**14**].

Physically this can be seen as the transition from topological field theory (TFT) to a finite gauge group TFT (see [**9, 8, 14**]).

The functorial setup of extending from functors with values in Frobenius algebras to those with values G Frobenius algebras in the following:

Let $\mathcal{FROB}$ be the category of Frobenius algebras, whose objects are Frobenius algebras and morphisms are maps which respect all the structures.

DEFINITION 1.1. *A G–category is a category $\mathcal{C}$ where for each object $X \in Ob(\mathcal{C})$ and each $g \in G$ there exists an object X^g and a morphism $i_g \in \mathrm{Hom}(X^g, X)$ with $X^e = X$ and $i_e = id$ and there are isomorphisms $\psi_{g,g^{-1}} \in \mathrm{Hom}(X^g, X^{g^{-1}})$.*

We call a category a G intersection category if it is a G category and for each pair $(g,h) \in G \times G$ and object $X \in Ob(\mathcal{C})$ there are isomorphisms $\psi \in \mathrm{Hom}((X^g)^h, (X^h)^g)$ and morphisms $i^{gh}_{g,h} \in \mathrm{Hom}((X^g)^h, X^{gh})$.

A G–action for a G–category is given by a collection of morphisms $\phi_g(X,h) \in \mathrm{Hom}(X^h, X^{ghg^{-1}})$ *which are compatible with the structural morphisms and satisfy* $\phi_g(X, g'hg'^{-1})\phi_{g'}(X,h) = \phi_{gg'}(X,h)$.

1.2. Examples. Examples of an intersection G–category with G–action are categories of spaces equipped with a G–action whose fixed point sets are in the same category. Actually this is the category of pairs (X, Y) with X say a smooth space with a G–action and Y a subspace of X. Then $(X,Y)^g := (X, Y \cap Fix(g, X))$ with $Fix(g), X$ denoting the fixed points of $g \in G$ in X, and $i_g = (id, \iota_g)$ with $\iota_g : Y \cap Fix(g) \to Y)$ being the inclusion. It is enough to consider pairs (X, Y) where $Y \subset X$ is the set fixed by a subgroup generated by an arbitrary number of elements of G: $H := \langle g_1, \dots, g_k \rangle$

We could also consider the action on the X^g to be trivial and set $(X^g)^h := X^g$. This will yield a G–category.

Also the category of functions $f : \mathbf{C}^n \to \mathbf{C}$ with an isolated singularity at 0 together with a group action of G on the variables induced by a linear action of G on the linear space fixing the function is an example of a G–category. This is a category of triples $(\mathbf{C}^n, f : \mathbf{C}^n \to \mathbf{C}), \rho \in \mathrm{Hom}(G, GL(n))$ such that f has an isolated singularity at zero and $f(\rho(\mathbf{z})) = f(\mathbf{z})$ for $\mathbf{z} \in \mathbf{C}^n$ with morphisms being linear between the linear spaces such that all structures are compatible. The functor under consideration is the local ring or Milnor ring. Again we set $(X^g)^h := X^g$.

Here the role of the fixed point set is played by the linear fixed point set and the restriction of the function to this fixed point set (cf.[**14**]). Again we can consider pairs of and object and a subobject as above in order to get an intersection G–category.

Our main examples are smaller categories such as a global orbifold. As a G category, the objects are the fixed point sets of the various cyclic groups generated by the element of G and the morphisms being the inclusion maps. Again we set $(X^g)^h := X^g$. For a global orbifold, we can also consider all fixed point sets of the groups generated by any number of elements of G as objects together with the inclusion maps as morphisms. This latter will render a G–intersection category.

The same is true for isolated singularities. Here the objects are the restriction of the function to the various subspaces fixed by the elements of g together with inclusion or for the G–intersection category we consider all intersections of these subspaces together with the restriction of the function to these subspaces as objects, again with the inclusion morphisms.

1.3. The classification/reconstruction program. Given a functor to Frobenius algebras (like cohomology), we would like to find its stringy counterpart for global quotients.

Now, suppose we have a G–category $\mathcal{C}$ and a contravariant functor $\mathcal{F}$ from $\mathcal{C}$ to $\mathcal{FROB}$. In this setting there might be several schemes to define a "stingy geometry" by augmenting the functor to take values in G–Frobenius algebras. But all of these schemes have to have the same additive structure provided by the "classical orbifold picture" (see 1.3.1) and satisfy the axioms of G–Frobenius algebras (see §2). Furthermore there are more structures which are already fixed in this situation, which is explained below. These data can sometimes be used to classify the possible algebra structures and reconstruct it when the classification data is known. In the case of so–called special G–Frobenius algebras a classification in terms of group cohomology classes is possible.

There are some intermediate steps which contain partial information that have been previously considered, like the additive structure, dimensions etc., as discussed in 1.3.1.

1.3.1. The "classical orbifold picture". Now, suppose we have a G–category $\mathcal{C}$ and a contravariant functor $\mathcal{F}$ from $\mathcal{C}$ to $\mathcal{FROB}$, then for each $X \in Ob(\mathcal{C})$, we naturally obtain the following collection of Frobenius algebras: $(\mathcal{F}(X^g) : g \in G)$ together with restriction maps $r_g = \mathcal{F}(i_g) : \mathcal{F}(X) \mapsto \mathcal{F}(X^g)$.

One possibility is to regard the direct sum of the Frobenius algebras $A_g := \mathcal{F}(X^g)$.

The first obstacle is presented in the presence of a grading, say by $\mathbf{N}, \mathbf{Z}$ or $\mathbf{Q}$; as it is well known that the direct sum of two graded Frobenius algebras is only well defined if their Euler dimensions (cf. e.g. [**16**]) agree. This can, however, be fixed by using the shifts s^+ discussed in 2.6. If the grading was originally in $\mathbf{N}$ these shifts are usually in $\frac{1}{2}\mathbf{N}$, but in the complex case still lie in $\mathbf{N}$.

Furthermore, if we have a G–action on the G category, it will induce the structure of a G–module on this direct sum.

Each of the Frobenius algebras A_g comes equipped with its own multiplication, so there is a "diagonal" multiplication for the direct sum which is the direct sum of these multiplications.

Using the shift s^+ it is possible to define a "classical theory" by considering the diagonal algebra structure and taking G–invariants. This is the approach used in [**2**], [**22**] and [**1**]. The paper [**2**] shows that this structure describes the G–equivariant rather than the G–invariant geometry.

One can of course forget the algebra structure altogether and retain only the additive structure. This was done e.g. in [**24**] in the setting of V–manifolds (i.e. orbifolds). Concentrating only on the dimensions one arrives for instance at the notion of "stringy numbers" [**3**].

1.3.2. The "stringy orbifold picture". The "diagonal" multiplication is however *not* the right object to study from the perspective of "stringy geometry" or a TFT with a finite gauge group [**14, 4**]. The multiplication should rather be G–graded, i.e. map $A_g \otimes A_h \to A_{gh}$. We call such a product "stringy" product.

Here the natural question is the following:

QUESTION 1.2. *Given the additive structure of a G–Frobenius algebra, what are the possible "stringy" products?*

A more precise version of this question is the setting of our reconstruction program [**15, 16**].

1.3.3. The G–action. One part of the structure of a G–Frobenius algebra is the G–action. If the G–category is already endowed with a G–action, we can use it to reconstruct the G–action on the G–Frobenius algebra, which in turn limits the choices of "stringy" products to those that are compatible with it.

1.3.4. Invariants. By definition G–Frobenius algebras come with a G action whose invariants form a commutative algebra. Due to the nature of the G action this commutative algebra is graded by conjugacy classes, and under certain conditions the metric descends and the resulting algebra is again Frobenius. The induced multiplication is multiplicative in the conjugacy classes and we call such a multiplication commutative "stringy".

1.3.5. Examples. Examples of commutative "stringy" products are orbifold (quantum) cohomology [**4**]. For cohomology of global orbifolds it was shown in [**10**] and recently in [**11**] that there is a group graded version for global orbifold cohomology which has the structure of a G Frobenius algebra, as we had previously postulated [**13**]. For new developments on quantum deformations of the G–Frobenius algebras see [**11**].

1.3.6. Special G–Frobenius algebras. The special reconstruction data reflects this situation in the case that the A_g algebras are cyclic A_e modules. This is a restriction which leads to an answer in terms of cocycles for a large class of examples. This class includes all Jacobian Frobenius algebras as well as symmetric products and special cases of geometric actions on manifolds (were the cohomology of the fixed point sets is generated by restriction from the ambient space).

The general idea can be generalized to the non–cyclic case, although computations get more involved.

DEFINITION 1.3. *Given a G-category $\mathcal{C}$, we call the tuple $(X^g) : g \in G$ a G-collection.*

The category of G-collections of a G-category is the category whose objects are G-collections and whose morphisms are collections of morphisms (f^g) s.t. the diagrams

$$\begin{array}{ccc} X^g & \xrightarrow{i_g} & X \\ \downarrow f^g & & \downarrow f \\ Y^g & \xrightarrow{i_g} & Y \end{array}$$

commute.

DEFINITION 1.4. *A G-Frobenius functor is a functor from the category of G-collections of a G-category to G-Frobenius algebras.*

1.4. Reconstruction/classification. The main question of the reconstruction/classification program is whether one can extend a functor from a G–category $\mathcal{C}$ to Frobenius algebras to a G–Frobenius functor, and if so how many ways are there to do this.

One can view this as the analogue of solving the associativity equations for general Frobenius algebras. Some of the solutions correspond to quantum cohomology, some to singularities, etc. and maybe others to other "string"–schemes. The structures of possible "stringy" products provide a common approach. The systematic consideration of all possible products confines the choices of string equivalents of classical concepts and allows to identify divers approaches.

The answer to the main question of reconstruction/classification can be answered in the special case where all of the twisted sectors are cyclic in terms of group cohomological data (see below). This is the content of the Reconstruction Theorem of [**14**].

The consequences are sometimes quite striking as in the case of symmetric products, where there is only one possible "stringy" orbifold product.

The restrictions on the possible multiplicative structures are even stricter if one is considering data stemming from a G–intersection category.

2. G–Frobenius algebras

We fix a finite group G and denote its unit element by e. We furthermore fix a ground field k of characteristic zero for simplicity. With the usual precautions the characteristic of the field does not play an important role and furthermore the group really only needs to be completely disconnected.

DEFINITION 2.1. *A G–twisted Frobenius algebra —or G–Frobenius algebra for short— over a field k of characteristic 0 is $< G, A, \circ, 1, \eta, \varphi, \chi >$, where*

G	*finite group*
A	*finite dim G-graded k–vector space* $A = \oplus_{g \in G} A_g$ *A_e is called the untwisted sector and* *the A_g for $g \neq e$ are called the twisted sectors.*
$\circ$	*a multiplication on A which respects the grading:* $\circ : A_g \otimes A_h \to A_{gh}$
1	*a fixed element in A_e–the unit*
η	*non-degenerate bilinear form* *which respects grading i.e.* $\eta\|_{A_g \otimes A_h} = 0$ *unless* $gh = e$.
φ	*an action of G on A (which will be by algebra automorphisms),* $\varphi \in \mathrm{Hom}(G, \mathrm{Aut}(A))$, *s.t.* $\varphi_g(A_h) \subset A_{ghg^{-1}}$
χ	*a character* $\chi \in \mathrm{Hom}(G, k^*)$

Satisfying the following axioms:
NOTATION: *We use a subscript on an element of A to signify that it has homogeneous group degree –e.g. a_g means $a_g \in A_g$– and we write $\varphi_g := \varphi(g)$ and $\chi_g := \chi(g)$.*

a) *Associativity*
$$(a_g \circ a_h) \circ a_k = a_g \circ (a_h \circ a_k)$$
b) *Twisted commutativity*
$$a_g \circ a_h = \varphi_g(a_h) \circ a_g$$
c) *G Invariant Unit:*
$$1 \circ a_g = a_g \circ 1 = a_g$$
and
$$\varphi_g(1) = 1$$
d) *Invariance of the metric:*
$$\eta(a_g, a_h \circ a_k) = \eta(a_g \circ a_h, a_k)$$
i) *Projective self–invariance of the twisted sectors*
$$\varphi_g | A_g = \chi_g^{-1} id$$
ii) *G–Invariance of the multiplication*
$$\varphi_k(a_g \circ a_h) = \varphi_k(a_g) \circ \varphi_k(a_h)$$
iii) *Projective G–invariance of the metric*
$$\varphi_g^*(\eta) = \chi_g^{-2} \eta$$
iv) *Projective trace axiom*
$\forall c \in A_{[g,h]}$ *and l_c left multiplication by c:*
$$\chi_h \mathrm{Tr}(l_c \varphi_h|_{A_g}) = \chi_{g^{-1}} \mathrm{Tr}(\varphi_{g^{-1}} l_c|_{A_h})$$

2.1. Remark. In the case of trivial characters the notion of G–Frobenius algebras has appeared also under the name of group–crossed algebras in [**23**], where they appeared from the point of view of homotopy field theory.

2.2. G–graded tensor product. Given two G–Frobenius algebras

$$\langle G, A, \circ, 1, \eta, \varphi, \chi\rangle \quad \text{and} \ \langle G, A', \circ', 1', \eta', \varphi', \chi'\rangle$$

we defined [**14**] their tensor product as G–Frobenius algebras to be the G–Frobenius algebra

$\langle G, \bigoplus_{g\in G}(A_g \otimes A'_g), \circ\otimes\circ', 1\otimes 1', \eta\otimes\eta', \varphi\otimes\varphi', \chi\otimes\chi'\rangle$.

We will use the short hand notation $A\hat{\otimes}A'$ for this product.

For the program outlined in §1 the following data is the starting point in order to construct a G–Frobenius algebra.

DEFINITION 2.2. *A reconstruction data is collection of Frobenius algebras*

$$(A_g, \eta_g, 1_g) : g \in G$$

together with maps of algebras $r_g : A_e \to A_g$, isomorphisms $\psi_g : A_g \xrightarrow{\sim} A_{g^{-1}}$ and a G-action φ on A_e.

In general, we would like to find the G–Frobenius algebra structures compatible with these data. For many purposes such as symmetric products it is however enough to restrict to a more specialized situation.

2.3. Special G–Frobenius algebras. We briefly review special G–Frobenius algebras. For this class of algebras which include the algebras having their origin in singularities with isolated singularities and symmetric products a classification of all possible stringy multiplications is possible in terms of group cohomological data. For details see [**14, 15**].

DEFINITION 2.3. *We call a G-Frobenius algebra special if all A_g are cyclic A_e modules via the multiplication $A_e \otimes A_g \to A_g$ and there exists a collection of cyclic generators 1_g of A_g such that $\varphi_g(1_h) = \varphi_{g,h}1_{ghg^{-1}}$ with $\varphi_{g,h} \in k^*$.*

The last condition is automatic, if the Frobenius algebra A_e only has k^* as invertibles, as is the case for cohomology algebras of connected compact manifolds and Milnor rings of quasi–homogeneous functions with an isolated critical point at zero.

Fixing the generators 1_g we obtain maps $r_g : A_e \to A_g$ by setting $r_g(a_e) = a_e 1_g$. This yields a short exact sequence

$$0 \to I_g \to A_e \stackrel{r_g}{\to} A_g \to 0 \tag{2.1}$$

It is furthermore useful to fix a section i_g of r_g.

We denote the concatenation $\pi_g := i_g \circ r_g$.

DEFINITION 2.4. *A special G reconstruction datum is a collection of Frobenius algebras $(A_g, \eta_g, 1_g) : g \in G$ together with an action of G by algebra automorphisms on A_e and the structure of a cyclic A_e module algebra on each A_g with generator 1_g such that A_g and A_g^{-1} are isomorphic as of A_e modules algebras.*

DEFINITION 2.5. *Given a Frobenius algebra A_e and a collection of cyclic A_e–modules $A_g : g \in G$ a graded cocycle is a map $\gamma : G \times G \to A_e$ which satisfies*

$$\gamma(g,h)\gamma(gh,k) \equiv \gamma(g,hk)\gamma(h,k) \bmod I_{ghk}$$

Such a cocycle is called section independent if

$$(I_g + I_h)\gamma(g,h) \subset I_{gh}$$

Two such cocycles are considered to be the same if $\gamma(g,h) \equiv \gamma'(g,h) \bmod I_{gh}$ *and isomorphic, if they are related by the usual scaling for group cocycles.*

Given non-degenerate parings η_g *on the* A_g*, a cocycle is said to be compatible with the metric, if*

$$\check{r}_g(1_g) = \gamma(g, g^{-1})$$

where $\check{r}$ *is the dual in the sense of vector spaces with non-degenerate metric.*

We will again use the notation $\gamma_{g,h} := \gamma(g,h)$.

2.4. Special G–Frobenius structure in terms of the cocycles. Fixing a cyclic generator $1_g \in A_g$, a special G–Frobenius algebra is completely characterized by two cocycles: γ a section independent graded cocycle compatible with the metric and $\varphi \in Z^1(G, k^*[G])$ where $k^*[G]$ is the group ring restricted to invertible coefficients with G–module structure induced by the adjoint action:

$$\phi(g) \cdot (\sum_h \mu_h h) = \sum_h \mu_h g h g^{-1}$$

The multiplication and G–action on the generators defines these cocycles. Set

$$1_g 1_h = \gamma_{g,h} 1_{gh} \quad \varphi_g(1_h) = \varphi_{g,h} 1_{ghg^{-1}}$$

Defining $\varphi(g) := \sum_h \varphi_{g,h} g h g^{-1}$ and $\gamma(g,h) := \gamma_{g,h}$ we obtain the desired cocycles.

The section independence follows from the fact that

$$(I_g + I_h)\gamma_{g,h} 1_{gh} = (I_g + I_h) 1_g 1_h = 0$$

In general, the multiplication is thus given by

$$a_g b_h = i_g(a_g) i_h(b_h) \gamma_{g,h} 1_{gh} \tag{2.2}$$

for any choice of sections i_g.

The cocycles furthermore satisfy the following two compatibility equations:

$$\varphi_{g,h} \gamma_{ghg^{-1},g} = \gamma_{g,h} \tag{2.3}$$

and

$$\varphi_{k,g} \varphi_{k,h} \gamma_{kgk^{-1},khk^{-1}} = \varphi_k(\gamma_{g,h}) \varphi_{k,gh} \tag{2.4}$$

which follow from the twisted commutativity and the fact that φ acts by automorphisms.

2.4.1. Remark. Notice that if $\gamma_{g,h}$ is non–zero i.e. $A_g A_h \neq 0$ then (2.3) determines $\varphi_{g,h}$. We also would like to remark that if $A_g A_h A_k \neq 0$ (2.4) follows from (2.3) (cf. [**14**]).

DEFINITION 2.6. *We call a pair of a section independent cocycle and a non–abelian cocycle compatible if they satisfy the equations (2.3) and (2.4).*

THEOREM 2.7. *(Reconstruction) Given a special G reconstruction datum the structures of special G–Frobenius algebras are in 1–1 correspondence compatible pairs of a graded, section independent G 2–cocycle with values in* A_e *that is compatible with the metric and a non–abelian G 2–cocycle with values in* K^**. Satisfying the following conditions:*

i) $\varphi_{g,g} = \chi_g^{-1}$

ii) $\eta_e(\varphi_g(a), \varphi_g(b)) = \chi_g^{-2}\eta_e(a,b)$

iii) *The projective trace axiom* $\forall c \in A_{[g,h]}$ *and* l_c *left multiplication by* c*:*

$$\chi_h \mathrm{Tr}(l_c\varphi_h|_{A_g}) = \chi_{g^{-1}}\mathrm{Tr}(\varphi_{g^{-1}}l_c|_{A_h}) \tag{2.5}$$

2.5. Remark. Changing the cyclic generators by elements of k^* leads to isomorphic G–Frobenius algebras and to cohomologous cocycles γ, φ in $Z^2(G, A_e)$ and $Z^2(G, k^*[G])$.

2.6. Grading and Shifts. Consider the graded version for Frobenius algebras, where each Frobenius A_g algebra comes naturally graded, e.g. by cohomological degree or weight in the case of a quasi–homogeneous isolated singularity. Usually this grading takes values in $\mathbf{Q}$. In this case the metric is also of a fixed degree, e.g. the dimension or the highest weight in the Milnor ring.

Then each A_g is graded as well. We denote the degree of the pairing η_g by d_g and also use the shorthand notation $d := d_e$.

2.6.1. Remark. It is well known that the direct sum of graded Frobenius algebras is a graded Frobenius algebra only if the degrees match (see e.g. [**14**]).

2.6.2. The shifts. From the previous Remark it follows that in order to form the sum $A := \bigoplus_{g\in G} A_g$, we need to shift the degrees of the elements of A_g at least uniformly, i.e. if an element a_g in A_g has degree $\deg(a_g)$ we assign to it the new shifted degree $\deg^s(a_g) = \deg(a_g) + s_g$.

This observation does not fix the shifts uniquely. Let us denote the shift in degree of A_g by s_g and set

$$s_g^+ := s_g + s_{g^{-1}}, \quad s_g^- := s_g - s_{g^{-1}}$$

Then

$$s_g^+ := d - d_g$$

for the grading reasons mentioned above.

The shift s^- is not fixed, however, there is a standard choice provided there exists a canonical choice of linear representation of G.

DEFINITION 2.8. *The standard shift for a* G*–Frobenius algebra with a choice of linear representation* $\rho : G \to GL_n(k)$ *is given by*

$$s_g^+ := d - d_g$$

and

$$s_g^- := \frac{1}{2\pi i}\mathrm{tr}(\log(g)) - \mathrm{tr}(\log(g^{-1})) := \frac{1}{2\pi i}\Big(\sum_i \lambda_i(g) - \sum_i \lambda_i(g^{-1})\Big)$$
$$= \sum_{i:\lambda_i\neq 0}\Big(\frac{1}{2\pi i}2\lambda_i(g) - 1\Big)$$

where the $\lambda_i(g)$ *are the logarithms of the eigenvalues of* $\rho(g)$ *using the branch with arguments in* $[0, 2\pi)$ *i.e. cut along the positive real axis.*

In total we obtain:

$$s_g = \frac{1}{2}(s_g^+ + s_g^-) = \frac{1}{2}(d - d_g) + \sum_{i:\lambda_i\neq 0}\Big(\frac{1}{2\pi i}\lambda_i(g) - \frac{1}{2}\Big)$$

2.6.3. Remark. This grading having its origin in physics specializes to the so–called age grading or the orbifold grading of [4] in the respective situations.

2.7. Super-grading. We can enlarge the framework by considering super–algebras rather than algebras. This will introduce the standard signs.

The action of G as well as the untwisted sector should be even. The axioms that change are

b$^\sigma$) Twisted super–commutativity

$$a_g \circ a_h = (-1)^{\tilde{a}_g \tilde{a}_h} \varphi_g(a_h) \circ a_g$$

iv$^\sigma$) Projective super–trace axiom

$\forall c \in A_{[g,h]}$ and l_c left multiplication by c:

$$\chi_h \mathrm{STr}(l_c \varphi_h|_{A_g}) = \chi_{g^{-1}} \mathrm{STr}(\varphi_{g^{-1}} l_c|_{A_h})$$

where STr is the super–trace.

Here we denoted by $\tilde{a}$ the $\mathbb{Z}/2\mathbb{Z}$ degree of a.

3. Intersection G–Frobenius algebras

We will now concentrate on the situation of functors from G–intersection categories to Frobenius algebras.

Given a G–class in such a category a functor to Frobenius algebras will provide the following structure which reflects the possibility to take fixed point sets iteratively. Say we look at the fixed points with respect to elements $g_1, \dots, g_n$. These fixed point sets will be invariant under the group spanned by the elements $g_1, \dots, g_n$ and they are just the intersection of the respective fixed point sets of the elements g_i. The underlying spaces are therefore invariant with respect to permutation of the elements g_i, and if g appears twice among the g_i then one can shorten the list by omitting one of the g_i. Also if a list g_i includes g^{-1} we may replace it by g. Finally, the fixed point set under the action of the group generated by two elements g and h is a subset of the fixed point set of the group generated by their product gh. Translating this into the categorical framework, we obtain:

DEFINITION 3.1. *A G–intersection Frobenius datum of level k is the following: For each collection $(g_1, \dots, g_n)$ with $n \leq k$ of elements of G, a Frobenius algebra $A_{g_1,\dots,g_n}$ and the following maps:*

Isomorphisms

$$\Psi_\sigma : A_{g_1,\dots,g_n} \to A_{g_{\sigma(1)},\dots,g_{\sigma(n)}}$$

for each $\sigma \in \mathbb{S}_n$ called permutations.

Isomorphisms

$$\Psi^{g_1,\dots,g_i,\dots,g_n}_{g_1,\dots,g_i^{-1},\dots,g_n} : A_{g_1,\dots,g_i,\dots,g_n} \to A_{g_1,\dots,g_i^{-1},\dots,g_n}$$

commuting with the permutations.

Morphisms

$$r^{g_1,\dots,\hat{g}_i,\dots,g_n}_{g_1,\dots,g_i,\dots g_n} : A_{g_1,\dots,\hat{g}_i,\dots,g_n} \to A_{g_1,\dots,g_n}$$

commuting with the permutations. (Here the symbol ˆ is used to denote omission.)

Isomorphisms

$$i^{g_1,\dots g,\dots,\hat{g},\dots,g_n}_{g_1,\dots,g,\dots,g,\dots,g_n} : A_{g_1,\dots,g,\dots,g,\dots,g_n} \to A_{g_1,\dots g,\dots,\hat{g},\dots,g_n}$$

commuting with the permutations.

And finally morphisms:

$$r^{g_1,\dots,g_ig_{i+1},\dots,g_n}_{g_1,\dots,g_i,g_{i+1},\dots,g_n} : A_{g_1,\dots,g_ig_{i+1},\dots,g_n} \to A_{g_1,\dots,g_i,g_{i+1},\dots,g_n}$$

commuting with the permutations.

If this data exists for all k we call the data simply G–intersection Frobenius datum.

3.1. Notation. We set $r_{g_1,\dots,g_n} := r^{g_1,\dots,g_{n-1}}_{g_1,\dots,g_n} \circ \cdots \circ r_{g_1}$ and we set $I_{g_1,\dots,g_n} :=$ Ker$(r_{g_1,\dots,g_n})$. Notice that this definition of $I_{g_1,\dots,g_n}$ is independent of the order of the g_i.

DEFINITION 3.2. *An intersection G–Frobenius algebra of level $k \geq 2$ is an intersection G–Frobenius datum of level $k \geq 2$ together with a G–Frobenius algebra structure on $A := \bigoplus A_g$.*

An intersection G–Frobenius algebra of level $k \geq 2$ is called special, if all of the $A_{g_1,\dots g_n}$ are cyclic A_e module algebras generated by the $1_{g_1,\dots g_n}$.

3.2. Remarks.

1) In order to (re)–construct a suitable multiplication on $\bigoplus A_g$ it is often convenient to use the double and triple intersections (i.e. level 3). Where the double intersection are used for the multiplication and triple intersections are used to show associativity.
2) We can use the double intersections to define G–Frobenius algebras based on each of the A_g i.e. on $\bigoplus_{h\in Z(g)} A_{g,h}$ for each fixed g–where $Z(g)$ denotes the centralizer of g.

DEFINITION 3.3. *A G–action for an intersection G–Frobenius datum of level k is given by a collection of morphisms*

$$\phi_g(A_{g_1,\dots,g_n}, h) \in \text{Hom}(A_{g_1,\dots,g_n,h}, A_{g_1,\dots,g_n,ghg^{-1}})$$

which are compatible with the structural homomorphisms and satisfy

$$\phi_g(A_{g_1,\dots,g_n}, g'hg'^{-1})\phi_{g'}(A_{g_1,\dots,g_n}, h) = \phi_{gg'}(A_{g_1,\dots,g_n}, h)$$

DEFINITION 3.4. *We call an intersection G Frobenius datum a special G intersection Frobenius datum datum, if all of the $A_{g_1,\dots,g_n}$ are cyclic A_e module algebras via the restriction maps such that the A_e module structures are compatible with the restriction morphisms r. Here the generators are given by $r_{g_1,\dots,g_n}(1)$ and the A_e module structure is given by $a \cdot b := r_{g_1,\dots,g_n}(a)b$.*

3.3. Remark. In the case of special intersection G–Frobenius algebras, there are two ways to look at the multiplication. One way is to use the restrictions r_g and sections i_g to define the multiplication as discussed in §2.4. A second possibility is to use the intersection structure. This can be done in the following way: first push forward to double intersections, second use the Frobenius algebra structure there to multiply, then pull the result back up to the invariants of the product, but allowing to multiply with an obstruction class before pulling back. This is discussed below in §3.4.

The precise relation between the two procedures is given by the following Proposition and 2.2.

PROPOSITION 3.5. [15] *Given a special G intersection algebra datum (of level 2), the following decomposition holds for section independent cocycles γ:*

$$r_{gh}(\gamma_{g,h}) = \check{r}^{gh}_{g,h}(\tilde{\gamma}_{g,h}) = i^{gh}_{g,h}(\tilde{\gamma}_{g,h})\check{r}^{gh}_{g,h}(1_{g,h}) = \bar{\gamma}_{g,h}\gamma^{\perp}_{g,h} \tag{3.1}$$

for some section $i_{g,h}$ of $r_{g,h}$, $\tilde{\gamma}_{g,h} \in (A_{g,h})^e$, $\bar{\gamma}_{g,h} \in i_g, h)(A_g, h)$ of degree e. and $\gamma^{\perp}_{g,h} := \check{r}^{gh}_{g,h}(1_{g,h})$ Here $e = s_g + s_h - s_{gh} - s^+_{g,h} + s^+_{gh}$ with $s^+_{g,h} := d - d_{g,h}$ and $d_{g,h} = \deg(\rho_{g,h})$ and we again used the unshifted degrees. (In particular if the $s^- = 0$ then $e = \frac{1}{2}(s^+_g + s^+_h + s^+_{gh}) - s^+_{g,h} = \frac{1}{2}(d - d_g - d_h - d_{gh}) + d_{g,h}$)

Here $\check{r}$ is again the dual for maps between vector spaces with non–degenerate bilinear forms.

That is using the multiplication in $A_{g,h}$

$$a_g \circ b_h = \check{r}^{gh}_{g,h}(r^g_{g,h}(a_g)r^h_{g,h}(b_h)\tilde{\gamma}_{g,h}) \tag{3.2}$$

3.3.1. Remark. The decomposition into the terms $\tilde{\gamma}$ and $\gamma^{\perp}$ can be understood as decomposing the cocycle into a part which comes from the normal bundle of $X^{g,h} \subset X^{gh}$ which is captured by $\gamma^{\perp}$ and an additional obstruction part.

Also generalizing the fact that

$$I_g\gamma_g = I_g\check{r}_g(1_g) = 0 \tag{3.3}$$

the following lemma holds:

LEMMA 3.6.

$$(I_g + I_h)\gamma^{\perp}_{g,h} \subset I_{gh} \tag{3.4}$$

3.4. Multiplication. Fix a special intersection G Frobenius algebra of level at least 2. From the section independence of γ, we see that the multiplication $A_g \otimes A_h \to A_{gh}$ can be factored through $A_{g,h}$. To be more precise, we have the following commutative diagram.

$$\begin{array}{ccc} A_g \otimes A_h & \stackrel{\mu}{\to} & A_{gh} \\ \downarrow r^g_{g,h} \otimes r^h_{g,h} & & \uparrow \check{r}^{g,h}_{gh} \circ l_{\tilde{\gamma}_{g,h}} \\ A_{g,h} \otimes A_{g,h} & \stackrel{\mu}{\to} & A_{g,h} \end{array}$$

where $l_{\tilde{\gamma}_{g,h}}$ is the left multiplication with $\tilde{\gamma}_{g,h}$.

Vice–versa we can use this diagram as an Ansatz for any G intersection Frobenius algebra of level at least 2.

3.5. Associativity equations. Fix an intersection G Frobenius algebra of level at least 3 then the associativity equations can be factored through $A_{g,h,k}$. More precisely, we have the following commutative diagram of restriction maps:

$$\begin{array}{ccccccccc} & & & & A_{ghk} & & & & \\ & & & \swarrow & & \searrow & & & \\ A_{gh} & \to & A_{gh,k} & & \downarrow & & A_{g,hk} & \leftarrow & A_{hk} \\ \downarrow & & & \searrow & & \swarrow & & & \downarrow \\ A_{g,h} & & \to & & A_{g,h,k} & & \leftarrow & & A_{h,k} \end{array} \tag{3.5}$$

More technically: Using the associativity equations for the γ, we set

$$r_{ghk}(\gamma_{g,h}\gamma_{gh,k}) := \gamma_{g,h,k} \tag{3.6}$$

Associativity dictates that also

$$r_{ghk}(\gamma_{h,k}\gamma_{g,hk}) = \gamma_{g,h,k} \tag{3.7}$$

By analogous arguments as for the decomposition of the $\gamma_{g,h}$'s one can obtain:

$$\gamma_{g,h,k} = i^{ghk}_{g,h,k}(\tilde{\gamma}_{g,h,k})\check{r}^{ghk}_{g,h,k}(1_{g,h,k}) = \check{r}^{ghk}_{g,h,k}(\tilde{\gamma}_{g,h,k}) \tag{3.8}$$

for some $\tilde{\gamma}_{g,h,k} \in A_{g,h,k}$.

Vice–versa having defined suitable $\tilde{\gamma}_{g,h}$ to show associativity one needs to show that

$$\check{r}^{ghk}_{gh,k}(r^{gh}_{gh,k}(\check{r}^{gh}_{g,h}(\tilde{\gamma}_{g,h}))\tilde{\gamma}_{gh,k}) = \check{r}^{ghk}_{g,h,k}(\tilde{\gamma}_{g,h,k}) \tag{3.9}$$

for some $\tilde{\gamma}_{g,h,k}$. This approach is actually now independent of the setup of special G intersection Frobenius algebras, where such a decomposition is guaranteed, and is suitable for all G intersection Frobenius data respectively intersection G–categories.

4. Discrete Torsion

4.1. The twisted group ring $k^\alpha[G]$. Recall that given an element $\alpha \in Z^2(G, k^*)$ one defines the twisted group ring $k^\alpha[G]$ to be given by the same linear structure with multiplication given by the linear extension of

$$g \otimes h \mapsto \alpha(g,h)gh \tag{4.1}$$

with 1 remaining the unit element. To avoid confusion we will denote elements of $k^\alpha[G]$ by $\hat{g}$ and the multiplication with $\cdot$ Thus

$$\hat{g} \cdot \hat{h} = \alpha(g,h)\widehat{gh}$$

For α the following equations hold:

$$\alpha(g,e) = \alpha(e,g) = 1, \qquad \alpha(g,g^{-1}) = \alpha(g^{-1},g) \tag{4.2}$$

Furthermore

$$\hat{g}^{-1} = \frac{1}{\alpha(g,g^{-1})}\widehat{g^{-1}}$$

and

$$\hat{g} \cdot \hat{h} \cdot \hat{g}^{-1} = \frac{\alpha(g,h)\alpha(gh,g^{-1})}{\alpha(g,g^{-1})}\widehat{ghg^{-1}} = \frac{\alpha(g,h)}{\alpha(ghg^{-1},g)}\widehat{ghg^{-1}} = \epsilon(g,h)\widehat{ghg^{-1}}$$

with

$$\epsilon(g,h) := \frac{\alpha(g,h)}{\alpha(ghg^{-1},g)} \tag{4.3}$$

4.1.1. Remark. If the field k is algebraically closed we can find a representative for each class $[\alpha] \in H^2(G,k^*)$ which also satisfies

$$\alpha(g,g^{-1}) = 1$$

4.1.2. The G–Frobenius Algebra structure of $k^\alpha[G]$. Fix $\alpha \in Z^2(G, k^*)$. Recall from [**14, 15**] the following structures which turn $k^\alpha[G]$ into a special G–Frobenius algebra:

$$
\begin{array}{ll}
\gamma_{g,h} = \alpha(g,h) & \eta(\hat{g}, \widehat{g^{-1}}) = \alpha(g, g^{-1}) \\
\chi_g = (-1)^{\tilde{g}} & \varphi_{g,h} = \dfrac{\alpha(g,h)}{\alpha(ghg^{-1}, g)} =: \epsilon(g,h)
\end{array}
\tag{4.4}
$$

4.1.3. Relations. The $\epsilon(g,h)$ which are by definition given as

$$\epsilon(g,h) := \frac{\alpha(g,h)}{\alpha(ghg^{-1},h})$$

satisfy the equations:

$$
\begin{aligned}
\epsilon(g,e) &= \epsilon(g,g) = 1 \\
\epsilon(g_1g_2, h) &= \epsilon(g_1, g_2hg_2^{-1})\epsilon(g_2, h) \\
\epsilon(k, gh) &= \epsilon(k,g)\epsilon(k,h)\frac{\alpha(kgk^{-1}, khk^{-1})}{\alpha(g,h)} \\
\epsilon(h,g) &= \epsilon(g^{-1}, ghg^{-1})\frac{\alpha([g,h],h)}{\alpha([g,h], hgh^{-1})}
\end{aligned}
\tag{4.5, 4.6}
$$

This yields for commuting elements:

$$
\begin{array}{ll}
\epsilon(g,e) = \epsilon(g,g) = 1 & \epsilon(g,h) = \epsilon(h^{-1}, g) = \epsilon(h,g)^{-1} \\
\epsilon(g_1g_2, h) = \epsilon(g_1,h)\epsilon(g_2,h) & \epsilon(h, g_1g_2) = \epsilon(h,g_1)\epsilon(h,g_2)
\end{array}
\tag{4.7}
$$

In the physics literature discrete torsion is sometimes defined to be a function ϵ defined on commuting elements of G taking values in $U(1)$ and satisfying the equations (4.7).

4.1.4. Remark. It is a nice exercise to check that the trace axiom also holds (see [**14, 16**]).

4.1.5. Remark. The function ϵ can be interpreted as a cocycle in $Z^1(G, k^*[G])$ where $k^*[G]$ are the elements of $k[G]$ with invertible coefficients regarded as a G module by conjugation (cf. [**14, 15**]). This means in particular that on *commuting elements* ϵ only depends on the class of the cocycle α.

4.2. The action of discrete Torsion.

DEFINITION 4.1. *Given a G–Frobenius algebra A and an element $\alpha \in Z^2(G,k)$, we define the α–twist of A to be the G–Frobenius algebra $A^\alpha := A\hat{\otimes}k^\alpha[G]$.*

PROPOSITION 4.2. *Notice that as vector spaces*

$$A_g^\alpha = A_g \otimes k \simeq A_g \tag{4.8}$$

Using this identification the G–Frobenius structures given by (4.8) are

$$
\begin{array}{ll}
\circ^\alpha|_{A_g^\alpha \otimes A_h^\alpha} = \alpha(g,h)\circ & \varphi_g^\alpha|_{A_h^\alpha} = \epsilon(g,h)\varphi_g \\
\eta^\alpha|_{A_g^\alpha \otimes A_{g^{-1}}^\alpha} = \alpha(g,g^{-1})\eta & \chi_g = \chi_g
\end{array}
\tag{4.9}
$$

4.2.1. Supergraded twisted group rings. *Fix*

$$\alpha \in Z^2(G,k^*), \sigma \in \mathrm{Hom}(G,\mathbb{Z}/2\mathbb{Z})$$

then there is a twisted super–version of the group ring where now the relations read

$$\hat{g}\hat{h} = \alpha(g,h)\widehat{gh} \tag{4.10}$$

and the twisted commutativity is

$$\hat{g}\hat{h} = (-1)^{\sigma(g)\sigma(h)}\varphi_g(\hat{h})\hat{g} \tag{4.11}$$

and thus

$$\varphi_g(\hat{h}) = (-1)^{\sigma(g)\sigma(h)}\alpha(g,h)\alpha(gh,g^{-1})\widehat{ghg^{-1}} =: \varphi_{g,h}\widehat{ghg^{-1}} \tag{4.12}$$

and thus

$$\epsilon(g,h) := \varphi_{g,h} = (-1)^{\sigma(g)\sigma(h)}\frac{\alpha(g,h)}{\alpha(ghg^{-1},g)} \tag{4.13}$$

We would just like to remark that the axiom iv$^\sigma$) of 2.7 shows the difference between super twists and discrete torsion.

DEFINITION 4.3. *We denote the α-twisted group ring with super–structure σ by $k^{\alpha,\sigma}[G]$. We still denote $k^{\alpha,0}[G]$ by $k^\alpha[G]$ where 0 is the zero map and we denote $k^{0,\sigma}[G]$ just by $k^\sigma[G]$ where 0 is the unit of the group $H^2(G,k^*)$.*

A straightforward calculation shows

LEMMA 4.4. $k^{\alpha,\sigma}[G] = k^\alpha[G] \otimes k^\sigma[G]$.

LEMMA 4.5. *Let $\langle G, A, \circ, 1, \eta, \varphi, \chi\rangle$ be a G–Frobenius algebra or more generally super Frobenius algebra with super grading $\tilde{}\ \in \mathrm{Hom}(A,\mathbb{Z}/2\mathbb{Z})$ then $A \otimes k^\sigma[G]$ is isomorphic to the super G–Frobenius algebra $\langle G, A, \circ^\sigma, 1, \eta^\sigma, \varphi^\sigma, \chi^\sigma\rangle$ with super grading $\tilde{}^\sigma$, where*

$$\circ^\sigma|_{A_g\otimes A_h} = (-1)^{\tilde{g}\sigma(h)}\circ \qquad \varphi^\sigma_{g,h} = (-1)^{\sigma(g)\sigma(h)}\varphi_{g,h}$$
$$\eta^\sigma_g = (-1)^{\tilde{g}\sigma(g)}\eta_g \qquad \chi^\sigma = (-1)^{\sigma(g)}\chi_g$$
$$\tilde{a}^\sigma_g = \tilde{a}_g + \sigma(g)$$

DEFINITION 4.6. *Given a G–Frobenius algebra A a twist for A is a pair of functions $(\lambda : G \times G \to k^*, \mu : G \times G \to k^*)$*

Such that A together with the new G–action

$$\varphi^\lambda(g)(a) = \oplus_h \lambda(g,h)\varphi(g)(a_h)$$

and the new multiplication

$$a_g \circ^\mu b_h = \mu(g,h)a_g \circ b_h$$

is again a G–Frobenius algebra.

A twist is called universal if it is defined for all G–Frobenius algebras.

4.2.2. Remark. We could have started from a pair of functions $(\lambda : A \times A \to k^*, \mu : G \times A \to k^*)$ in order to projectively change the multiplication and G action, but it is clear that the universal twists (i.e. defined for any G–Frobenius algebra) can only take into account the G degree of the elements.

4.2.3. Remark. These twists arise from a projectivization of the G–structures induced on a module over A as for instance the associated Ramond–space (cf. [**14**]). In physics terms this means that each twisted sector will have a projective vacuum, so that fixing their lifts in different ways induces the twist. Mathematically this means that the g twisted sector is considered to be a Verma module over A_g based on this vacuum.

THEOREM 4.7. *Given a (super) G–Frobenius algebra A the universal twists are in 1–1 correspondence with elements $\alpha \in Z^2(G, k^*)$ and the isomorphism classes of universal twists are given by $H^2(G, k^*)$. Furthermore the universal super re-gradings are in 1-1 correspondence with* $\mathrm{Hom}(G, \mathbb{Z}/2\mathbb{Z})$ *and these structures can be realized by tensoring with $k^\sigma[G]$ for $\sigma \in \mathrm{Hom}(G, \mathbb{Z}/2\mathbb{Z})$.*

Here a super re–grading is a new super grading on A with which A is a super G–Frobenius algebra and universal means that the operation of re–grading is defined for all G–Frobenius algebras.

We call the operation of forming a tensor product with $k^\alpha[G] : \alpha \in Z^2(G, k^*)$ a twist by discrete torsion. The term discrete refers to the isomorphism classes of twisted G–Frobenius algebras which correspond to classes in $H^2(G, k^*)$. Furthermore, we call the operation of forming a tensor product with $k^\sigma[G] : \sigma \in \mathrm{Hom}(G, \mathbb{Z}/2\mathbb{Z})$ super–twist.

4.3. Remark. If k is algebraically closed, then in each class of $H^2(G, k^*)$ there is a representative with $\alpha(g, g^{-1}) = 1$. Using these representatives it is possible to twist a special G–Frobenius algebra without changing its underlying special reconstruction data.

5. Symmetric group Frobenius algebra

In this section, we will consider the structure of special G–Frobenius algebras when the group is a symmetric group $\mathbb{S}_n$. The symmetric groups have two characteristics, which we will use. First they are generated by self inverse transpositions and second there is a natural grading on the elements given by the minimal number of transpositions needed to present an element.

Using the latter condition in order to fix a grading compatibility for special $\mathbb{S}_n$ Frobenius algebras, we showed [**15**] that the structure of a special $\mathbb{S}_n$–Frobenius algebra is essentially unique if it exists and can be expressed solely in terms of the dual of the restriction maps $\check{r}_\tau$ for τ a transposition. It thus depends only on the maps r_τ and the metrics η_τ. Also notice that the $\mathbb{S}_n$ action acts transitively on all A_τ since any two transpositions are conjugate.

If we specify the reconstruction data of the special $\mathbb{S}_n$ Frobenius manifold to be that stemming from a symmetric product, it can furthermore be shown that the unique structure does indeed exist [**15**].

5.1. Notation. We define the degree of $\sigma \in \mathbb{S}_n$ to be $|\sigma| :=$ the minimal length of σ as a word in transpositions. We define the length of σ as $l(\sigma) :=$ the number of cycles in a cycle decomposition. Notice $|\sigma| = n - l(\sigma)$.

We also set $l(\sigma_1, \dots, \sigma_n) = |\langle\sigma_1, \dots, \sigma_n\rangle \backslash \{1, \dots, n\}|$ where $\langle\sigma_1, \dots, \sigma_n\rangle$ is the group generated by the s_i and the quotient is by the natural permutation action.

DEFINITION 5.1. *We call two elements $\sigma, \sigma' \in \mathbb{S}_n$ transversal, if $|\sigma\sigma'| = |\sigma| + |\sigma'|$.*

5.2. Normalizability.

DEFINITION 5.2. *We call a non-abelian cocycle* φ *normalized if* $\forall \tau, \sigma \in \mathbb{S}_n, |\tau| = 1 : \varphi_{\sigma,\tau} = 1$.

We call a cocycle $\gamma : \mathbb{S}_n \times \mathbb{S}_n \to A$ *normalizable if for all transversal pairs* $\tau, \sigma \in \mathbb{S}_n, |\tau| = 1 : \gamma_{\sigma,\tau} \in A_e^*$, A^* *being the invertible elements of* A_e, *and normalized if it is normalizable and for all transversal* $\tau, \sigma \in \mathbb{S}_n, |\tau| = 1 : \gamma_{\sigma,\tau} = 1$.

In the example of symmetric products of an irreducible Frobenius algebra the invertibles are precisely k^*.

5.3. Discrete Torsion for the symmetric group. As is well known (see e.g. [**12**]) $H^2(\mathbb{S}_n, k^*) = \mathbb{Z}/2\mathbb{Z}$ and $\mathrm{Hom}(\mathbb{S}_n, k^*) = \mathbb{Z}/2\mathbb{Z}$. We denote the non–trivial element of $H^2(\mathbb{S}_n, k^*)$ by $[\Phi]$. There is a representative Φ of this class which actually satisfies $\Phi(\sigma, \sigma') = \pm 1$ for transversal σ, σ'. We denote the generator for the super–twist by Σ.

THEOREM 5.3. [**15**] *Any non–abelian cocycle* φ *after possibly twisting by the discrete-torsion* Φ *and super-twist* Σ *can be normalized.*

Any compatible pair of a normalizable $\mathbb{S}_n$ *cocycle* γ *and a normalized non–abelian cocycle* φ *can be normalized by a rescaling* $1_\sigma \mapsto \lambda_\sigma 1_\sigma$.

And vice–versa given for any normalized $\mathbb{S}_n$ *cocycle* γ *there are only two compatible non–abelian cocycle* φ *differing by the super–twist* Σ *namely:*

$$\varphi_{\sigma,\sigma'} = (-1)^{p|\sigma||\sigma'|} \tag{5.1}$$

where p *is either* 0 *or* 1.

5.4. Uniqueness.

THEOREM 5.4. [**15**] *Given a special* $\mathbb{S}_n$ *algebra datum, a choice of normalized cocycle* $\gamma : \mathbb{S}_n \times \mathbb{S}_n \to A$ *is unique up to a super–twist by* Σ.

It is determined by $\gamma_{\tau,\tau} = \check{r}_\tau(1_\tau)$ *and without the twist it is given by equation (5.2).*

5.4.1. Explicit form of the cocycles.

$$\gamma_{\sigma,\sigma'} = \pi_{\sigma\sigma'}(\gamma_{\sigma,\sigma'} \prod_{i=1}^{|\sigma'|} \gamma_{\tau'_{i+1},\prod_{j=1}^{i}\tau'_j}) = \pi_{\sigma\sigma'}(\prod_{i=1}^{|\sigma'|} \gamma_{\sigma\prod_{j=1}^{i}\tau'_{i-1},\tau'_i}) = \pi_{\sigma\sigma'}(\prod_{i\in I} \gamma_{\tau'_i,\tau'_i}) \tag{5.2}$$

where $I := \{i : |\sigma(\prod_{j=1}^{i-1}\tau'_j)\tau'_i| = |\sigma\prod_{j=1}^{i-1}\tau'_j| - 2\}$.

5.5. Existence. We would also would like to recall the following existence theorem:

THEOREM 5.5. [**15**] *The equations*

$$\begin{aligned} r_{\sigma,\sigma'}(\gamma_{\sigma,\sigma'}) &= r_{\sigma\sigma'}(\prod_{i\in I}\gamma_{\tau_i,\tau_i}) = \prod_{i\in I'}\pi_{\sigma\sigma'}(\gamma_{\tau_i,\tau_i})\prod_{j\in I''} r_{\sigma\sigma'}(\gamma_{\tau_j,\tau_j}) \\ &= \bar{\gamma}_{\sigma,\sigma'}\gamma^{\perp}_{\sigma,\sigma'} \end{aligned} \tag{5.3}$$

where

$$\begin{aligned} I' &= \{i \in I : \pi_{\sigma,\sigma'}(\gamma_{\tau_i,\tau_i}) = \pi_{\sigma\sigma'}(\gamma_{\tau_i,\tau_i})\} \\ I'' &= \{i \in I : \pi_{\sigma,\sigma'}(\gamma_{\tau_i,\tau_i}) \neq \pi_{\sigma\sigma'}(\gamma_{\tau_i,\tau_i})\} \end{aligned} \tag{5.4}$$

and $\gamma^{\perp}_{\sigma,\sigma'} = \check{r}^{gh}_{g,h}(1_{g,h})$ *are well defined and yield a group cocycle compatible with the special* $\mathbb{S}_n$ *intersection data.*

$$A_{\sigma_1,\dots,\sigma_k} = (A^{\otimes l(\sigma_1,\dots,\sigma_k)}, \eta^{\otimes l(\sigma_1,\dots,\sigma_k)}, 1^{\otimes l(\sigma_1,\dots,\sigma_k)})$$

derived from any Frobenius algebra $(A, \eta, 1)$. *The restriction maps are contractions via multiplication.*

For details on reconstruction data we refer to [**14, 15**].

5.5.1. Remark. In order to understand the form a the cocycle above let us consider the case of the second symmetric product. Thus we need to consider the $\mathbb{S}_2$-Frobenius algebra:

$$A = A_e \oplus A_\tau = A \otimes A \oplus A$$

with the metric $\eta \otimes \eta \oplus \eta$. The restriction map r_τ is just the multiplication μ : $A \otimes A \to A$. Now $\check{r}(1_\tau) = \sum a_i \otimes b_i$ and $r_\tau(\check{r}_\tau) = \sum a_i b_i = e$ where e is the Euler class.

In general, if $\tau = (ij)$ then $r_{(ij)}$ contracts the i–th and j–th component. Then $\gamma_{\tau,\tau} = \check{r}_\tau$ is a sum over elements which differ from one only in the i–th and j–th factor. Considering a product of $\gamma_{\tau,\tau}$ and restricting it to $A_{\sigma\sigma'}$ amounts to performing several contractions using the multiplication. For each individual $\gamma_{\tau,\tau}$ there are only two choices: those which get contracted and yield Euler classes – this is the set I''– and those that do not get contracted – this is the set I'.

5.5.2. Remark. For the existence proof in [**15**] we used the theory of intersection Frobenius algebras as can bee seen from the decomposition of the cocycles. This also makes it easier to compare with the results of [**18**].

5.5.3. Remarks.

1) In the case that A is the trivial one dimensional Frobenius algebra this structure coincides with the group ring $k[\mathbb{S}_n]$.
2) Applying our result to the situation where A is the Frobenius algebra associated to a variety or a compact space we recover the results of [**10**] and taking invariants those of [**21**].

DEFINITION 5.6. *Given a Frobenius algebra* A, *the series of* $\mathbb{S}_n$*-Frobenius algebras determined by Theorem 5.5 is called the second quantization of* A.

This terminology is based on [**7**].

6. The Twist for the Hilbert scheme

If one changes the sign in the metric η_τ this also changes the multiplication. These changes are such, that they are uniquely realized as a twist with a discrete torsion whose cocycle α is determined by $\alpha(\tau, \tau) = -1$ and normalization. This cocycle is actually trivial in cohomology with coefficients in $\mathbf{C}^*$, but nevertheless changes the multiplication and metric as desired.

6.1. The twisted group ring $k^\alpha[\mathbb{S}_n]$. Recall that $H^2(\mathbb{S}_n, \mathbb{C}) = \mathbb{Z}/2\mathbb{Z}$, but the twists are actually given by $\alpha \in Z^2(\mathbb{S}_n, k)$. Any normalized cocycle is fixed by $\alpha(\tau,\tau)$, τ any transposition. Thus we may regard $\alpha \in Z^2(\mathbb{S}_n, k)$ which is fixed by $\alpha(\tau,\tau) = -1$ for any transposition τ.

Notice that $[\alpha] = 0 \in H^2(\mathbb{S}_n, \mathbb{C})$ as is easily seen from the existence and uniqueness together with Remark 4.1.1. It is not trivial however in $H^2(\mathbb{S}_n, \mathbb{Q}^*)$. In any case, we can consider the twist by this class.

PROPOSITION 6.1. *Given any $\mathbb{S}_n$–Frobenius algebra A, twisting it by the normalized discrete torsion $\alpha \in Z^2(G, k^*)$ defined by $\alpha(\tau,\tau) = -1$ changes the structures via:*

$$\begin{aligned} a_\sigma \circ^\alpha b_{\sigma'} &= (-1)^{\frac{1}{2}(|\sigma|+|\sigma|-|\sigma\sigma'|)} a_\sigma \circ b_{\sigma'} \\ \varphi^\alpha_\sigma(a_{\sigma'}) &= \varphi_\sigma(a_{\sigma'}) \\ \chi^\alpha_\sigma &= \chi_\sigma \\ \eta^\alpha_\sigma &= (-1)^{|\sigma|}\eta_\sigma \end{aligned} \tag{6.1}$$

This is how the multiplication and metric get to be changed via a sign while the $\mathbb{S}_n$ action remains unchanged.

PROOF. Due to 4.2 it suffices to show that this holds for the twisted group ring i.e. the following equations hold.

$$\alpha(\sigma,\sigma') = (-1)^{\frac{1}{2}(|\sigma|+|\sigma|-|\sigma\sigma'|)} \tag{6.2}$$

$$\epsilon(\sigma,\sigma') = 1 \tag{6.3}$$

The equation (6.3) follows from $\alpha(\tau,\tau) = -1$ and the general structure of $\alpha(\sigma,\sigma')$ of 5.4 in particular from the equation (5.2) by noticing that

$$|I| = \{i : |\sigma(\prod_{j=1}^{i-1}\tau'_j)\tau'_i| = |\sigma\prod_{j=1}^{i-1}\tau'_j| - 2\} = \frac{1}{2}(|\sigma| + |\sigma'| - |\sigma\sigma'|)$$

The correction of χ is always trivial and the one for φ is given by ϵ which satisfies

$$\epsilon(\sigma,\sigma') = \frac{\alpha(\sigma,\sigma')}{\alpha(\sigma\sigma'\sigma^{-1},\sigma)} = 1$$

since

$$|\sigma\sigma'\sigma^{-1}| = |\sigma'| \text{ and so also } |\sigma'\sigma| = |\sigma\sigma'|$$

Finally for the last equation of (6.1) we read off

$$\alpha(\sigma,\sigma^{-1}) = (-1)^{\frac{1}{2}(|\sigma|+|\sigma^{-1}|-|\sigma\sigma^{-1}|)} = (-1)^{|\sigma|}$$

□

6.2. A family of multiplications. The only freedom of choice for a normalized cocycle is $\gamma_{\tau,\tau}$ which is determined uniquely from the metrics η_e and η_τ.

Given a fixed metric η_τ it is possible to change it by homothety to $\eta^\lambda_\tau = \lambda\eta_\tau$ and keeping the cocycles normalized using discrete torsion. This is achieved by twisting with the normalized discrete torsion cocycle determined by $\alpha(\tau,\tau) = \lambda$.

Vice–versa the only twists with discrete torsion cocycles that keep the cocycle γ normalized are fixed by their value $\alpha(\tau, \tau) = \lambda$. The effect on the metric $\eta_{\tau,\tau}$ is a scaling by λ.

Using the same arguments as in 6.1

PROPOSITION 6.2. *Let* $\alpha \in Z^2(\mathbb{S}_n, k^*)$ *by the normalized cocycle determined by* $\alpha(\tau, \tau) = \lambda$ *and* A *by the* $\mathbb{S}_n$*–Frobenius algebra associated to the symmetric product then the* $\mathbb{S}_n$ *Frobenius algebra* A^α *is the twisted algebra found in* [**19**].

This Proposition explains the existence and the special role of these deformed multiplications. In the complex case, the existence of this family also shows the triviality of the cocycles $[\alpha] \in H^2(\mathbb{S}_n, \mathbf{C}^*)$.

6.3. Remark. Notice that the discrete torsion ϵ is trivial. To obtain a non–trivial ϵ one could super–twist [**14, 15**], but this would also change the permutation action by tensoring on the determinant representation, which is not intended for the current application as we wish to keep symmetric invariants, not anti–symmetric ones. Another way would be to super–twist and to twist with a non–trivial cohomology class, which would restore the action to a symmetric one. These results can be compared with [**5, 6, 7, 15**].

6.4. Application to the Hilbert scheme. Comparing our analysis with that of [**18**] finally yields:

THEOREM 6.3. *The cohomology of* $Hilb^{[n]}$*, the Hilbert scheme of n–points for a K3 surface, is the* $\mathbb{S}_n$ *invariant part of the unique* $\mathbb{S}_n$*–Frobenius Algebra associated to the symmetric product of the cohomology of the surface twisted by the specific discrete torsion given above. Or in other words the* $\mathbb{S}_n$*–Frobenius Algebra structure for the extended global orbifold cohomology twisted by specific discrete torsion, which is fixed by the map of* [**18**]. *In general, the sequence of spaces* $Hilb^{[n]}$ *gives rise to the twisted second quantization of the underlying surface on the cohomological (motivic) level.*

6.4.1. Remark. The necessity of this type of twist and the particular choice are fixed by two remarks. First, the chosen isomorphism of [**18**] involves the intersection form of the resolution of the double points, which is negative definite. The intersection form of the natural symmetric group Frobenius algebra does not reflect this property. The way that discrete torsion changes the metric then fixes the cocycle as a normalized cocycle, by regarding the metric of A_τ for τ a transposition. This geometrically corresponds to a simple blow-up along the diagonal.

References

[1] A. Adem and Y. Ruan. *Twisted Orbifold K-Theory.* Comm. Math. Phys. 237 (2003), no. 3, 533–556.

[2] M. Atiyah and G. Segal. *On equivariant Euler characteristics.* J. Geom. Phys. 6, 671-677 (1989).

[3] V. Batyrev and L. Borisov *Mirror duality and string-theoretic Hodge numbers.* Invent. Math. 126, No.1, 183-203 (1996)

[4] W. Chen, Y. Ruan. *A New Cohomology Theory for Orbifold.* Commmun. Math. Phys. 248 (2004), 1-31. And *Orbifold Quantum Cohomology.* Preprint. math.AG/0005198

[5] R. Dijkgraaf *Fields, strings, matrices and symmetric products.* Moduli of curves and abelian varieties, 151–199, Aspects Math., E33, Vieweg, Braunschweig, 1999.

[6] R. Dijkgraaf *Discrete Torsion and Symmetric Products.* Preprint hep-th/9912101

[7] R. Dijkgraaf, G. Moore, E. Verlinde and H. Verlinde *Elliptic Genera of Symmetric Products and Second Quantized Strings.* Commun.Math.Phys. 185 (1997) 197-209

[8] R. Dijkgraaf, C. Vafa, E. Verlinde and H. Verlinde, *The operator algebra of orbifold models*, Comm. Math. Phys. **123** (1989), 485–526.

[9] R. Dijkgraaf and E. Witten, *Topological gauge theories and group cohomology*, Comm. Math. Phys. **129** (1990), 393–429.

[10] B. Fantechi and L. Goettsche. *Orbifold cohomology for global quotients.* Duke Math. J. 117 (2003), no. 2, 197–227.

[11] T. Jarvis, R. Kaufmann and T. Kimura. *Pointed admissible G-covers and G-cohomological Field theories.* Compositio Math., to appear.

[12] G. Karpilovsky. *Projective Representations of Finite Groups.* Dekker, 1985.

[13] R. Kaufmann. Talk "Orbifolding Frobenius algebras" at WAGP2000 conference at SISSA Trieste, Italy, October 2000.

[14] R. Kaufmann. *Orbifolding Frobenius algebras.* Int. J. of Math. Int. Journal of Math. 14 (2003), 573-619. And R. M. Kaufmann *Orbifold Frobenius Algebras, Cobordisms and Monodromies.* In: "Orbifolds in Mathematics and Physics", 135-161. Contemp. Math. (310) . Amer. Math. Soc., Providence, RI, 2002.

[15] R. M. Kaufmann. *Second quantized Frobenius algebras.* Commum. Math. Phys. 248 (2004), 33–83.

[16] R. M. Kaufmann. *The algebra of discrete torsion.* Preprint MPI 2002-112 , math.AG/0208081, 23 p.

[17] A. Kirillov Jr. *Modular categories and orbifold models.* Comm. Math. Phys. 229 (2002), 309–335.

[18] M. Lehn and C. Sorger. *The cup product of the Hilbert scheme for K3 surfaces* Invent. Math. 152 (2003), no. 2, 305–329.

[19] Z. Qin and W. Wang. *Hilbert schemes and symmetric products: a dictionary* In: "Orbifolds in Mathematics and Physics",233–257 Contemp. Math. (310). Amer. Math. Soc., Providence, RI, 2002.

[20] S. Montgomery. *Hopf algebras and their actions on rings.* CBMS Regional Conference Series in Mathematics, 82. American Mathematical Society, Providence, RI, 1993.

[21] B. Uribe. *Orbifold Cohomology of the Symmetric Product* In: "Orbifolds in Mathematics and Physics", Contemp. Math. (310) . Amer. Math. Soc., Providence, RI, 2002.

[22] B. Toen. *Théorèmes de Riemann-Roch pour les champs de Deligne-Mumford.* K-Theory 18, 33-76 (1999).

[23] V. Turaev. *Homotopy field theory in dimension 2 and group-algebras* Preprint. math.QA/9910010.

[24] I. Satake. *The Gauss-Bonnet theorem for V-manifolds.* J. Math. Soc. Japan 9, 464-492 (1957).

University of Connecticut, Department of Mathematics, Storrs, CT, USA and Max–Planck Institut für Mathematik, Bonn, Germany

E-mail address: `ralphk@mpim-bonn.mpg.de`

Relations among universal equations for Gromov-Witten invariants

Xiaobo Liu

It is well known that relations in the tautological ring of moduli spaces of pointed stable curves give partial differential equations for Gromov-Witten invariants of compact symplectic manifolds. These equations do not depend on the target symplectic manifolds and therefore are called universal equations for Gromov-Witten invariants. In the case that the quantum cohomology of the symplectic manifolds are semisimple, it is expected that higher genus Gromov-Witten invariants are completely determined by such universal equations and genus-0 Gromov-Witten invariants. This has been proved for genus-1 (cf. [**DZ**]) and genus-2 (cf. [**L2**]) cases. Universal equations also play very important role in the understanding of the Virasoro conjecture (cf. [**EHX**]). The genus-0 Virasoro conjecture for all compact symplectic manifolds follows from a universal equation called the genus-0 topological recursion relation (cf. [**LT**]). For projective varieties, we expect that such universal equations reduce higher genus Virasoro conjecture to an $SL(2)$ symmetry for the generating function of the Gromov-Witten invariants. Again this has been proved for genus-1 ([**L1**]) and genus-2 ([**L2**]) cases.

In this paper, we will discuss the relation among known universal equations for Gromov-Witten invariants. We hope that the understanding of such relations would be helpful to the study of both Gromov-Witten invariants and the topology of the moduli spaces of pointed curves. Relations among genus-2 universal equations were studied in [**L2**]. It was proved that the three universal equations in [**G2**] and [**BP**] implies certain complicated genus-1 relations (see equations (4) and (6) below). Modulo these genus-1 relations, the three known genus-2 equations can be reduced to only two equations. The main result of this paper is that the genus-1 relations derived in [**L2**] follow from a known genus-1 relation found in [**G1**] and the genus-0 and genus-1 topological recursion relations. This completes the discussion of relations among genus-2 equations in [**L2**].

Part of the work in this paper was done while the author visited Max-Planck Institute for Mathematics at Bonn during the workshop on Frobenius Manifolds, Quantum Cohomology, and Singularity Theory. The author would like to thank the organizers of the workshop for invitation. He would also like to thank E. Getzler for conversations related to this work.

Research partially supported by Alfred P. Sloan Research Fellowship and NSF research grant.

1. Generating functions for Gromov-Witten invariants

Let M be a compact symplectic manifolds. For simplicity, we assume $H^{\text{odd}}(M;\mathbb{C}) = 0$. The *big phase* is by definition the product of infinite copies of $H^*(M;\mathbb{C})$, i.e.

$$P := \prod_{n=0}^{\infty} H^*(M;\mathbb{C}).$$

Fix a basis $\{\gamma_1, \dots, \gamma_N\}$ of $H^*(M;\mathbb{C})$ with $\gamma_1 = 1$ be the identity of the ordinary cohomology ring of M. Then we denote the corresponding basis for the n-th copy of $H^*(M;\mathbb{C})$ in P by $\{\tau_n(\gamma_1), \dots, \tau_n(\gamma_N)\}$. We call $\tau_n(\gamma_\alpha)$ a *descendant* of γ_α with descendant level n. We can think of P as an infinite dimensional vector space with basis $\{\tau_n(\gamma_\alpha) \mid 1 \leq \alpha \leq N,\ n \in \mathbb{Z}_{\geq 0}\}$ where $\mathbb{Z}_{\geq 0} = \{n \in \mathbb{Z} \mid n \geq 0\}$. Let $(t_n^\alpha \mid 1 \leq \alpha \leq N,\ n \in \mathbb{Z}_{\geq 0})$ be the corresponding coordinate system on P. For convenience, we identify $\tau_n(\gamma_\alpha)$ with the coordinate vector field $\frac{\partial}{\partial t_n^\alpha}$ on P for $n \geq 0$. If $n < 0$, $\tau_n(\gamma_\alpha)$ is understood as the 0 vector field. We also abbreviate $\tau_0(\gamma_\alpha)$ as γ_α. Any vector field of the form $\sum_\alpha f_\alpha \gamma_\alpha$, where f_α are functions on the big phase space, is called a *primary vector field*. We use τ_+ and τ_- to denote the operator which shift the level of descendants, i.e.

$$\tau_\pm\left(\sum_{n,\alpha} f_{n,\alpha}\tau_n(\gamma_\alpha)\right) = \sum_{n,\alpha} f_{n,\alpha}\tau_{n\pm1}(\gamma_\alpha)$$

where $f_{n,\alpha}$ are functions on the big phase space.

We will use the following *notational conventions*: Lower case Greek letters, e.g. α, β, μ, ν, σ,..., etc., will be used to index the cohomology classes. The range of these indices is from 1 to N. Lower case English letters, e.g. i, j, k, m, n, ..., etc., will be used to index the level of descendants. Their range is the set of all non-negative integers, i.e. $\mathbb{Z}_{\geq 0}$. All summations are over the entire ranges of the indices unless otherwise indicated. Let

$$\eta_{\alpha\beta} = \int_V \gamma_\alpha \cup \gamma_\beta$$

be the intersection form on $H^*(V,\mathbb{C})$. We will use $\eta = (\eta_{\alpha\beta})$ and $\eta^{-1} = (\eta^{\alpha\beta})$ to lower and raise indices. For example,

$$\gamma^\alpha := \eta^{\alpha\beta}\gamma_\beta.$$

Here we are using the summation convention that repeated indices (in this formula, β) should be summed over their entire ranges.

Let

$$\langle \tau_{n_1}(\gamma_{\alpha_1})\tau_{n_2}(\gamma_{\alpha_2}) \cdots \tau_{n_k}(\gamma_{\alpha_k}) \rangle_g$$

be the genus-g descendant Gromov-Witten invariant associated to $\gamma_{\alpha_1}, \dots, \gamma_{\alpha_k}$ and nonnegative integers $n_1, \dots, n_k$ (cf. [**W**], [**RT**], [**LiT**], and [**M**]). Let $\overline{\mathcal{M}}_{g,k}(M;d)$ be the moduli space of stable maps from genus-g k-pointed curves to M with degree $d \in H_2(M;\mathbb{Z})$, and Ψ_i the first Chern class of the tautological line bundle over $\overline{\mathcal{M}}_{g,k}(M;d)$ whose geometric fiber over a stable map is the cotangent space of the domain curve at i-th marked point. Then the integers $n_1, \dots, n_k$ above represent powers of $\Psi_1, \ldots, \Psi_k$. The genus-g generating function is defined to be

$$F_g = \sum_{k\geq 0} \frac{1}{k!} \sum_{\substack{\alpha_1,\ldots,\alpha_k \\ n_1,\ldots,n_k}} t_{n_1}^{\alpha_1} \cdots t_{n_k}^{\alpha_k} \langle \tau_{n_1}(\gamma_{\alpha_1}) \tau_{n_2}(\gamma_{\alpha_2}) \ldots \tau_{n_k}(\gamma_{\alpha_k}) \rangle_g .$$

This function is understood as a formal power series of t_n^α.

Introduce a k-tensor $\langle\langle \underbrace{\cdots\cdots}_{k} \rangle\rangle$ defined by

$$\langle\langle \mathcal{W}_1 \mathcal{W}_2 \cdots \mathcal{W}_k \rangle\rangle_g := \sum_{m_1,\alpha_1,\ldots,m_k,\alpha_k} f_{m_1,\alpha_1}^1 \cdots f_{m_k,\alpha_k}^k \frac{\partial^k}{\partial t_{m_1}^{\alpha_1} \partial t_{m_k}^{\alpha_2} \cdots \partial t_{m_k}^{\alpha_k}} F_g,$$

for (formal) vector fields $\mathcal{W}_i = \sum_{m,\alpha} f_{m,\alpha}^i \frac{\partial}{\partial t_m^\alpha}$ where $f_{m,\alpha}^i$ are (formal) functions on the big phase space. We can also view this tensor as the k-th covariant derivative of F_g. This tensor is called the *k-point (correlation) function.*

For any vector fields $\mathcal{W}_1$ and $\mathcal{W}_2$ on the big phase space, the quantum product of $\mathcal{W}_1$ and $\mathcal{W}_2$ is defined by (cf. [**L2**])

$$\mathcal{W}_1 \bullet \mathcal{W}_2 := \langle\langle \mathcal{W}_1 \mathcal{W}_2 \gamma^\alpha \rangle\rangle_0 \gamma_\alpha .$$

Define

$$T(\mathcal{W}) := \tau_+(\mathcal{W}) - \langle\langle \mathcal{W} \gamma^\alpha \rangle\rangle_0 \gamma_\alpha$$

for any vector field $\mathcal{W}$. The operator T was introduced in [**L2**] as a convenient tool in the study of topological recursion relations. Let $\overline{\mathcal{M}}_{g,k}$ be the moduli space of genus-g stable curves with k-marked points, and ψ_i the first Chern class of the tautological line bundle over $\overline{\mathcal{M}}_{g,k}$ whose geometric fiber over a stable curve is the cotangent space of the curve at the i-th marked point. When we translate a relation in the tautological ring of $\overline{\mathcal{M}}_{g,k}$ to differential equations for generating functions of Gromov-Witten invariants, the ψ classes correspond to the operator T. The reason for such correspondence lies in the relation between the classes $\Psi_i \in H^2(\overline{\mathcal{M}}_{g,k}(M,d); \mathbb{Q})$ and $\psi_i \in H^2(\overline{\mathcal{M}}_{g,k}; \mathbb{Q})$. There is a canonical map $St : \overline{\mathcal{M}}_{g,k}(M,d) \longrightarrow \overline{\mathcal{M}}_{g,k}$ which forgets the map to the target manifold M and stabilizes the domain curve for each element of $\overline{\mathcal{M}}_{g,k}(M,d)$. The difference $\Psi_i - St^*(\psi_i)$ is represented by a cycle containing elements whose domain curves consist of one genus-g and one genus-0 components, with the i-th marked point lying on the genus-0 component (cf. [**G2**] and [**KM**]). When we interpret this relation for Gromov-Witten invariants, Ψ_i corresponds to τ_+, the i-th marked point is associated with a cohomology class, which will be extended to a vector field $\mathcal{W}$ by linearity, and the node joining the two irreducible components are associated with the diagonal class $\gamma^\alpha \otimes \gamma_\alpha$ in $H^*(M \times M; \mathbb{C})$. Consequently, $St^*(\psi_i)$ must correspond to T.

The operator T also has an algebraic interpretation. Let

$$\mathcal{S} := -\sum_{m,\alpha} \tilde{t}_m^\alpha \tau_{m-1}(\gamma_\alpha)$$

be the *string vector field*, where

$$\tilde{t}_m^\alpha := t_m^\alpha - \delta_{m,1}\delta_{\alpha,1}.$$

By the second derivative of the string equation (see Section 2.4), the operator T can be rewritten as

$$T(\mathcal{W}) = \tau_{+}(\mathcal{W}) - \mathcal{S} \bullet \tau_{+}(\mathcal{W}).$$

Therefore T measures the difference between $\mathcal{S}$ and the identity of the quantum product on the big phase space, which actually does not exist by our definition of the quantum product.

Let ∇ be the trivial flat connection on the big phase space with respect to which $\tau_n(\gamma_\alpha)$ are parallel vector fields for all α and n. Then the covariant derivative of the quantum product is given by

$$(1)\quad \nabla_{\mathcal{W}_3}(\mathcal{W}_1 \bullet \mathcal{W}_2) = (\nabla_{\mathcal{W}_3}\mathcal{W}_1) \bullet \mathcal{W}_2 + \mathcal{W}_1 \bullet (\nabla_{\mathcal{W}_3}\mathcal{W}_2) + \langle\!\langle \mathcal{W}_1\, \mathcal{W}_2\, \mathcal{W}_3\, \gamma^\alpha \rangle\!\rangle_0\, \gamma_\alpha$$

and the covariant derivative of the operator T is given by

$$(2)\qquad \nabla_{\mathcal{W}_2}\, T(\mathcal{W}_1) = T(\nabla_{\mathcal{W}_2}\mathcal{W}_1) - \mathcal{W}_2 \bullet \mathcal{W}_1$$

for any vector fields $\mathcal{W}_1, \mathcal{W}_2$ and $\mathcal{W}_3$ (cf. [**L2**, Equation (8) and Lemma 1.5]).

2. Universal equations

We will write universal equations of Gromov-Witten invariants as equations among tensors on the big phase space defined by generating functions of Gromov-Witten invariants. The set of tensors whose vanishing are equivalent to some universal equations form an ideal of contra-variant tensor algebra on the big phase space. We call this ideal the *ideal of universal relations* and denote it by U. This ideal is closed under the covariant differentiation defined by ∇, and therefore is a differential ideal.

If $T_1, \ldots, T_k$ are k contra-variant tensors, we use $I_a[T_1, \cdots, T_k]$ to denote the ideal algebraically generated by $T_1, \ldots, T_k$, i.e. $I_a[T_1, \cdots, T_k]$ is the set of tensors of the form $\sum_{i=1}^{k} S_i^{'} \otimes T_i + T_i \otimes S_i^{''}$ where $S_i^{'}$ and $S_i^{''}$ are arbitrary contra-variant tensors on the big phase space. We also use $I_d[T_1, \cdots, T_k]$ to denote the ideal differentially generated by $T_1, \ldots, T_k$, i.e. the smallest ideal which contains $T_1, \ldots, T_k$ and is closed under the covariant differentiation.

In the theory of Gromov-Witten invariants, we often need to prove two complicated expressions are equal by using certain universal equations. Since universal equations of Gromov-Witten invariants are very complicated, it is very cumbersome to write out explicitly derivations in each step. For convenience of presentation, we introduce the following phrases: If one expression can be derived from another one by using certain universal equations $T_1 = 0, \cdots, T_k = 0$, but not using their covariant derivatives, then we say that these *two expressions are equal modulo* $I_a[T_1, \cdots, T_k]$. If in the derivation, we also need covariant derivatives of the corresponding universal equations, we say that these *two expressions are equal modulo* $I_d[T_1, \cdots, T_k]$.

2.1. genus-0 universal relations.

Define

$$\rho_0(\mathcal{W}_1, \mathcal{W}_2, \mathcal{W}_3) := \langle\!\langle T(\mathcal{W}_1)\, \mathcal{W}_2\, \mathcal{W}_3 \rangle\!\rangle_0$$

where $\mathcal{W}_i$ are vector fields on the big phase space. The genus-0 topological recursion relation is equivalent to $\rho_0 = 0$. Hence $\rho_0 \in U$, and so does the tensor C_0 defined by

$$C_0(\mathcal{W}_1, \mathcal{W}_2, \mathcal{W}_3, \mathcal{W}_4) := (\nabla_{\mathcal{W}_3}\rho_0)(\mathcal{W}_1, \mathcal{W}_2, \mathcal{W}_4) - (\nabla_{\mathcal{W}_2}\rho_0)(\mathcal{W}_1, \mathcal{W}_3, \mathcal{W}_4).$$

Using equation (2), it is straightforward to show that

$$C_0(\mathcal{W}_1,\mathcal{W}_2,\mathcal{W}_3,\mathcal{W}_4) = \langle\langle\, (\mathcal{W}_1 \bullet \mathcal{W}_2)\, \mathcal{W}_3\, \mathcal{W}_4 \,\rangle\rangle_0 - \langle\langle\, \mathcal{W}_1\, (\mathcal{W}_2 \bullet \mathcal{W}_3)\, \mathcal{W}_4 \,\rangle\rangle_0$$

for any vector fields $\mathcal{W}_1,\ldots,\mathcal{W}_4$. Therefore $C_0 = 0$ implies that the quantum product "$\bullet$" on the big phase space is associative.

2.2. genus-1 universal relations. The genus-1 topological recursion relation is equivalent to $\rho_1 = 0$ where ρ_1 is defined by

$$\rho_1(\mathcal{W}) = \langle\langle\, T(\mathcal{W}) \,\rangle\rangle_1 - \frac{1}{24}\langle\langle\, \mathcal{W}\,\gamma^\alpha\,\gamma_\alpha \,\rangle\rangle_0 .$$

Define

$$\begin{aligned} G(\mathcal{W}_1,\mathcal{W}_2,\mathcal{W}_3,\mathcal{W}_4) \;:=\; & \sum_{g\in S_4} \Big\{ 3\,\langle\langle\, \{\mathcal{W}_{g(1)} \bullet \mathcal{W}_{g(2)}\}\, \{\mathcal{W}_{g(3)} \bullet \mathcal{W}_{g(4)}\} \,\rangle\rangle_1 \\ & -4\,\langle\langle\, \{\mathcal{W}_{g(1)} \bullet \mathcal{W}_{g(2)} \bullet \mathcal{W}_{g(3)}\}\, \mathcal{W}_{g(4)} \,\rangle\rangle_1 \\ & -\langle\langle\, \{\mathcal{W}_{g(1)} \bullet \mathcal{W}_{g(2)}\}\, \mathcal{W}_{g(3)}\, \mathcal{W}_{g(4)}\, \gamma^\alpha \,\rangle\rangle_0 \langle\langle\, \gamma_\alpha \,\rangle\rangle_1 \\ & +2\,\langle\langle\, \mathcal{W}_{g(1)}\, \mathcal{W}_{g(2)}\, \mathcal{W}_{g(3)}\, \gamma^\alpha \,\rangle\rangle_0 \langle\langle\, \{\gamma_\alpha \bullet \mathcal{W}_{g(4)}\} \,\rangle\rangle_1 \\ & +\frac{1}{6}\,\langle\langle\, \mathcal{W}_{g(1)}\, \mathcal{W}_{g(2)}\, \mathcal{W}_{g(3)}\, \gamma^\alpha \,\rangle\rangle_0 \langle\langle\, \gamma_\alpha\, \mathcal{W}_{g(4)}\, \gamma_\beta\, \gamma^\beta \,\rangle\rangle_0 \\ & +\frac{1}{24}\,\langle\langle\, \mathcal{W}_{g(1)}\, \mathcal{W}_{g(2)}\, \mathcal{W}_{g(3)}\, v_{g(4)}\, \gamma^\alpha \,\rangle\rangle_0 \langle\langle\, \gamma_\alpha\, \gamma_\beta\, \gamma^\beta \,\rangle\rangle_0 \\ & -\frac{1}{4}\,\langle\langle\, \mathcal{W}_{g(1)}\, \mathcal{W}_{g(2)}\, \gamma^\alpha\, \gamma^\beta \,\rangle\rangle_0 \langle\langle\, \gamma_\alpha\, \gamma_\beta\, \mathcal{W}_{g(3)}\, \mathcal{W}_{g(4)} \,\rangle\rangle_0 \Big\} \end{aligned}$$

for any vector fields $\mathcal{W}_1,\ldots,\mathcal{W}_4$. Then the genus-1 equation discovered by Getzler [**G1**] is equivalent to $G = 0$. So $\rho_1 \in U$ and $G \in U$.

2.3. genus-2 universal relations. Define

$$\begin{aligned} &\rho_{2;1}(\mathcal{W}) := \langle\langle\, T^2(\mathcal{W}) \,\rangle\rangle_2 - A_1(\mathcal{W}), \\ &\rho_{2;2}(\mathcal{W},\mathcal{V}) := \langle\langle\, T(\mathcal{W})\, T(\mathcal{V}) \,\rangle\rangle_2 - 3\,\langle\langle\, T(\mathcal{W} \bullet \mathcal{V}) \,\rangle\rangle_2 - A_2(\mathcal{W},\mathcal{V}), \\ &\rho_{2;3}(\mathcal{W}_1,\mathcal{W}_2,\mathcal{W}_3) := 2\,\langle\langle\, \{\mathcal{W}_1 \bullet \mathcal{W}_2 \bullet \mathcal{W}_3\} \,\rangle\rangle_2 - 2\,\langle\langle\, \mathcal{W}_1\, \mathcal{W}_2\, \mathcal{W}_3\, \gamma^\alpha \,\rangle\rangle_0 \langle\langle\, T(\gamma_\alpha) \,\rangle\rangle_2 \\ &\qquad + \sum_{g\in S_3} \langle\langle\, \mathcal{W}_{g(1)}\, T(\mathcal{W}_{g(2)} \bullet \mathcal{W}_{g(3)}) \,\rangle\rangle_2 - \langle\langle\, T(\mathcal{W}_{g(1)})\, \{\mathcal{W}_{g(2)} \bullet \mathcal{W}_{g(3)}\} \,\rangle\rangle_2 \\ &\qquad - B(\mathcal{W}_1,\mathcal{W}_2,\mathcal{W}_3), \end{aligned}$$

where

$$\begin{aligned} A_1(\mathcal{W}) \;=\; & \frac{7}{10}\,\langle\langle\, \gamma_\alpha \,\rangle\rangle_1 \langle\langle\, \{\gamma^\alpha \bullet \mathcal{W}\} \,\rangle\rangle_1 + \frac{1}{10}\,\langle\langle\, \gamma_\alpha\, \{\gamma^\alpha \bullet \mathcal{W}\} \,\rangle\rangle_1 \\ & -\frac{1}{240}\,\langle\langle\, \mathcal{W}\, \{\gamma_\alpha \bullet \gamma^\alpha\} \,\rangle\rangle_1 + \frac{13}{240}\,\langle\langle\, \mathcal{W}\, \gamma_\alpha\, \gamma^\alpha\, \gamma^\beta \,\rangle\rangle_0 \langle\langle\, \gamma_\beta \,\rangle\rangle_1 \\ & +\frac{1}{960}\,\langle\langle\, \mathcal{W}\, \gamma^\alpha\, \gamma_\alpha\, \gamma^\beta\, \gamma_\beta \,\rangle\rangle_0 , \end{aligned}$$

$$
\begin{aligned}
A_2(\mathcal{W},\mathcal{V}) &= \frac{13}{10}\langle\langle \mathcal{W}\mathcal{V}\gamma^{\alpha}\gamma^{\beta}\rangle\rangle_0 \langle\langle \gamma_\alpha\rangle\rangle_1 \langle\langle \gamma_\beta\rangle\rangle_1 + \frac{4}{5}\langle\langle \mathcal{W}\gamma^{\alpha}\rangle\rangle_1 \langle\langle \{\gamma_\alpha \bullet \mathcal{V}\}\rangle\rangle_1 \\
&\quad + \frac{4}{5}\langle\langle \mathcal{V}\gamma^{\alpha}\rangle\rangle_1 \langle\langle \{\gamma_\alpha \bullet \mathcal{W}\}\rangle\rangle_1 - \frac{4}{5}\langle\langle \{\mathcal{W}\bullet\mathcal{V}\}\gamma^{\alpha}\rangle\rangle_1 \langle\langle \gamma_\alpha\rangle\rangle_1 \\
&\quad + \frac{23}{240}\langle\langle \mathcal{W}\mathcal{V}\gamma^{\alpha}\gamma_\alpha\gamma^{\beta}\rangle\rangle_0 \langle\langle \gamma_\beta\rangle\rangle_1 + \frac{1}{48}\langle\langle \mathcal{W}\gamma^{\alpha}\gamma_\alpha\gamma^{\beta}\rangle\rangle_0 \langle\langle \gamma_\beta\mathcal{V}\rangle\rangle_1 \\
&\quad + \frac{1}{48}\langle\langle \mathcal{V}\gamma^{\alpha}\gamma_\alpha\gamma^{\beta}\rangle\rangle_0 \langle\langle \gamma_\beta\mathcal{W}\rangle\rangle_1 - \frac{1}{80}\langle\langle \mathcal{W}\mathcal{V}\{\gamma^{\alpha}\bullet\gamma_\alpha\}\rangle\rangle_1 \\
&\quad + \frac{7}{30}\langle\langle \mathcal{W}\mathcal{V}\gamma^{\alpha}\gamma^{\beta}\rangle\rangle_0 \langle\langle \gamma_\alpha\gamma_\beta\rangle\rangle_1 + \frac{1}{30}\langle\langle \gamma_\alpha\{\gamma^{\alpha}\bullet\mathcal{W}\}\mathcal{V}\rangle\rangle_1 \\
&\quad + \frac{1}{30}\langle\langle \gamma_\alpha\{\gamma^{\alpha}\bullet\mathcal{V}\}\mathcal{W}\rangle\rangle_1 - \frac{1}{30}\langle\langle \{\mathcal{W}\bullet\mathcal{V}\}\gamma_\alpha\gamma^{\alpha}\rangle\rangle_1 \\
&\quad + \frac{1}{576}\langle\langle \mathcal{W}\mathcal{V}\gamma^{\alpha}\gamma_\alpha\gamma^{\beta}\gamma_\beta\rangle\rangle_0,
\end{aligned}
$$

and

$$
\begin{aligned}
& B(\mathcal{W}_1,\mathcal{W}_2,\mathcal{W}_3) \\
= {} & \frac{1}{5}\langle\langle \mathcal{W}_1\mathcal{W}_2\mathcal{W}_3\gamma^{\alpha}\gamma^{\beta}\rangle\rangle_0 \langle\langle \gamma_\alpha\rangle\rangle_1 \langle\langle \gamma_\beta\rangle\rangle_1 - \frac{6}{5}\langle\langle \mathcal{W}_1\mathcal{W}_2\mathcal{W}_3\gamma^{\alpha}\rangle\rangle_0 \langle\langle \gamma_\alpha\gamma^{\beta}\rangle\rangle_1 \langle\langle \gamma_\beta\rangle\rangle_1 \\
& + \frac{1}{120}\langle\langle \mathcal{W}_1\mathcal{W}_2\mathcal{W}_3\gamma^{\alpha}\gamma_\alpha\gamma^{\beta}\rangle\rangle_0 \langle\langle \gamma_\beta\rangle\rangle_1 - \frac{1}{120}\langle\langle \mathcal{W}_1\mathcal{W}_2\mathcal{W}_3\{\gamma^{\alpha}\bullet\gamma_\alpha\}\rangle\rangle_1 \\
& + \frac{1}{10}\langle\langle \mathcal{W}_1\mathcal{W}_2\mathcal{W}_3\gamma^{\alpha}\gamma^{\beta}\rangle\rangle_0 \langle\langle \gamma_\alpha\gamma_\beta\rangle\rangle_1 - \frac{1}{20}\langle\langle \mathcal{W}_1\mathcal{W}_2\mathcal{W}_3\gamma^{\alpha}\rangle\rangle_0 \langle\langle \gamma_\alpha\gamma^{\beta}\gamma_\beta\rangle\rangle_1 \\
& - \frac{1}{5}\sum_{\sigma\in S_3}\langle\langle \mathcal{W}_{\sigma(1)}\mathcal{W}_{\sigma(2)}\gamma^{\alpha}\gamma^{\beta}\rangle\rangle_0 \langle\langle \gamma_\alpha\rangle\rangle_1 \langle\langle \gamma_\beta\mathcal{W}_{\sigma(3)}\rangle\rangle_1 \\
& + \frac{2}{5}\sum_{\sigma\in S_3}\langle\langle \{\mathcal{W}_{\sigma(1)}\bullet\gamma_\alpha\}\rangle\rangle_1 \langle\langle \gamma^{\alpha}\mathcal{W}_{\sigma(2)}\mathcal{W}_{\sigma(3)}\rangle\rangle_1 \\
& - \frac{3}{5}\sum_{\sigma\in S_3}\langle\langle \mathcal{W}_{\sigma(1)}\{\mathcal{W}_{\sigma(2)}\bullet\gamma^{\alpha}\}\rangle\rangle_1 \langle\langle \gamma_\alpha\mathcal{W}_{\sigma(3)}\rangle\rangle_1 \\
& + \frac{3}{10}\sum_{\sigma\in S_3}\langle\langle \mathcal{W}_{\sigma(1)}\gamma_\alpha\rangle\rangle_1 \langle\langle \gamma^{\alpha}\{\mathcal{W}_{\sigma(2)}\bullet\mathcal{W}_{\sigma(3)}\}\rangle\rangle_1 \\
& - \frac{1}{5}\sum_{\sigma\in S_3}\langle\langle \gamma_\alpha\rangle\rangle_1 \langle\langle \gamma^{\alpha}\mathcal{W}_{\sigma(1)}\{\mathcal{W}_{\sigma(2)}\bullet\mathcal{W}_{\sigma(3)}\}\rangle\rangle_1 \\
& - \frac{1}{80}\sum_{\sigma\in S_3}\langle\langle \mathcal{W}_{\sigma(1)}\mathcal{W}_{\sigma(2)}\gamma^{\alpha}\gamma_\alpha\gamma^{\beta}\rangle\rangle_0 \langle\langle \gamma_\beta\mathcal{W}_{\sigma(3)}\rangle\rangle_1 \\
& + \frac{1}{80}\sum_{\sigma\in S_3}\langle\langle \mathcal{W}_{\sigma(1)}\gamma^{\alpha}\gamma_\alpha\gamma^{\beta}\rangle\rangle_0 \langle\langle \gamma_\beta\mathcal{W}_{\sigma(2)}\mathcal{W}_{\sigma(3)}\rangle\rangle_1 \\
& - \frac{1}{20}\sum_{\sigma\in S_3}\langle\langle \mathcal{W}_{\sigma(1)}\gamma_\alpha\gamma_\beta\rangle\rangle_1 \langle\langle \gamma^{\alpha}\gamma^{\beta}\mathcal{W}_{\sigma(2)}\mathcal{W}_{\sigma(3)}\rangle\rangle_0 \\
& + \frac{1}{60}\sum_{\sigma\in S_3}\langle\langle \mathcal{W}_{\sigma(1)}\mathcal{W}_{\sigma(2)}\gamma_\alpha\{\gamma^{\alpha}\bullet\mathcal{W}_{\sigma(3)}\}\rangle\rangle_1 \\
& - \frac{1}{120}\sum_{\sigma\in S_3}\langle\langle \mathcal{W}_{\sigma(1)}\gamma^{\alpha}\gamma_\alpha\{\mathcal{W}_{\sigma(2)}\bullet\mathcal{W}_{\sigma(3)}\}\rangle\rangle_1 .
\end{aligned}
$$

Note that the 3 complicated tensors A_1, A_2, B are symmetric tensors which depend on genus-0 and genus-1 data.

The universal equation corresponding to a formula due to Mumford is equivalent to $\rho_{2;1} = 0$ (cf. [**G2**]). Another equation due to Getzler [**G2**] is equivalent to $\rho_{2;2} = 0$. The equation $\rho_{2;3} = 0$ is due to Belorousski and Pandharipande [**BP**]. Therefore $\rho_{2;i} \in U$ for $i = 1, 2, 3$.

2.4. String and dilaton equations. There are also universal equations which do not come from relations in the tautological ring of the moduli space of stable curves. For example, the string equation, dilaton equation and divisor equation. In this paper, we will only need the string and dilaton equations.

The *string equation* has the form

$$\langle\!\langle \mathcal{S} \rangle\!\rangle_g = \frac{1}{2}\delta_{g,0}\eta_{\alpha\beta}t_0^\alpha t_0^\beta$$

for $g \geq 0$. Since $\nabla_{\mathcal{W}}\mathcal{S} = -\tau_-(\mathcal{W})$, taking covariant derivative on both sides of the string equation, we obtain

$$\langle\!\langle \mathcal{S}\,\mathcal{W}_1 \cdots \mathcal{W}_k \rangle\!\rangle_g = \sum_{i=1}^{k} \langle\!\langle \mathcal{W}_1 \cdots \{\tau_-(\mathcal{W}_i)\} \cdots \mathcal{W}_k \rangle\!\rangle_g + \delta_{g,0}\nabla^k_{\mathcal{W}_1,\cdots,\mathcal{W}_k}\left(\frac{1}{2}\eta_{\alpha\beta}t_0^\alpha t_0^\beta\right).$$

for any vector fields $\mathcal{W}_1, \ldots, \mathcal{W}_k$. The *dilaton vector field* is defined by

$$\mathcal{D} = T(\mathcal{S}).$$

The *dilaton equation* has the form

$$\langle\langle \mathcal{D} \rangle\rangle_g = (2g-2)F_g + \frac{1}{24}\chi(V)\delta_{g,1}$$

for $g \geq 0$. Since $\nabla_{\mathcal{W}}\mathcal{D} = -\mathcal{W}$, taking covariant derivative on both sides of the dilaton equation, we obtain

$$\langle\!\langle \mathcal{D}\,\mathcal{W}_1 \cdots \mathcal{W}_k \rangle\!\rangle_g = (k + 2g - 2)\,\langle\!\langle \mathcal{W}_1 \cdots \mathcal{W}_k \rangle\!\rangle_g.$$

3. Relations among lower genus universal equations

3.1. Relation among genus-2 universal equations. Relations among genus-2 universal equations were studied in [**L2**]. Using the string and dilaton equations, it was proved that

$$\rho_{2;2}(\mathcal{S}, \mathcal{W}) = \rho_{2;1}(\tau_-(\mathcal{W}))$$

for any vector field $\mathcal{W}$ (cf. [**L2**, Theorem 2.6]). Since τ_- is surjective, this shows that the universal equation $\rho_{2;1} = 0$ follows from $\rho_{2;2} = 0$. On the other hand, it is straightforward to show that

$$\begin{aligned} & (\nabla_{T(\mathcal{W}_1)}\rho_{2;1})(\mathcal{W}_2) - \rho_{2;2}(\mathcal{W}_1, T(\mathcal{W}_2)) \\ = \; & \{3\rho_0(\mathcal{W}_2, \mathcal{W}_1, \gamma^\alpha) - \rho_0(\mathcal{W}_1, \mathcal{W}_2, \gamma^\alpha)\}\,\langle\!\langle T(\gamma_\alpha) \rangle\!\rangle_2 - \rho_0(\mathcal{W}_2, T(\mathcal{W}_1), \gamma^\alpha)\,\langle\!\langle \gamma_\alpha \rangle\!\rangle_2 \\ (3) \quad & + A_2(\mathcal{W}_1, T(\mathcal{W}_2)) - (\nabla_{T(\mathcal{W}_1)}A_1)(\mathcal{W}_2) \end{aligned}$$

for any vector fields $\mathcal{W}_1$ and $\mathcal{W}_2$. Two conclusions can be drawn from this equation: First, we obtain a genus-1 universal relation

$$A_2(\mathcal{W}_1, T(\mathcal{W}_2)) - \left(\nabla_{T(\mathcal{W}_1)} A_1\right)(\mathcal{W}_2) = 0 \qquad (4)$$

for any vector fields $\mathcal{W}_1$ and $\mathcal{W}_2$. Secondly, the universal equation $\rho_{2;2}(\mathcal{W}_1, T(\mathcal{W}_2)) = 0$ follows from $\rho_{2;1} = 0$, the genus-0 topological recursion relation and the genus-1 equation (4).

Note that a special case of the genus-0 topological recursion relation is the following

$$T(\mathcal{W}_1) \bullet \mathcal{W}_2 = \rho_0(\mathcal{W}_1, \mathcal{W}_2, \gamma^\alpha)\gamma_\alpha = 0$$

for all vector fields $\mathcal{W}_1$ and $\mathcal{W}_2$. Taking covariant derivative of this equation with respect to $\mathcal{W}_3$, we have

$$\langle\langle T(\mathcal{W}_1)\, \mathcal{W}_2\, \mathcal{W}_3\, \gamma^\alpha \rangle\rangle_0\, \gamma_\alpha = \mathcal{W}_1 \bullet \mathcal{W}_2 \bullet \mathcal{W}_3$$

for all vector fields $\mathcal{W}_1$, $\mathcal{W}_2$, and $\mathcal{W}_3$. Using these two equations, we obtain

$$\begin{aligned} & \rho_{2;3}(\mathcal{W}_1, \mathcal{W}_2, T(\mathcal{W}_3)) + (\nabla_{\mathcal{W}_1 \bullet \mathcal{W}_2} \rho_{2;1})(\mathcal{W}_3) - \rho_{2;2}(\mathcal{W}_1 \bullet \mathcal{W}_2, \mathcal{W}_3) \\ = & -B(\mathcal{W}_1, \mathcal{W}_2, T(\mathcal{W}_3)) - (\nabla_{\mathcal{W}_1 \bullet \mathcal{W}_2} A_1)(\mathcal{W}_3) + A_2(\mathcal{W}_1 \bullet \mathcal{W}_2, \mathcal{W}_3). \end{aligned} \tag{5}$$

Three conclusions can be drawn from this equation: First, we obtain a genus-1 universal equation

$$B(\mathcal{W}_1, \mathcal{W}_2, T(\mathcal{W}_3)) - A_2(\mathcal{W}_1 \bullet \mathcal{W}_2, \mathcal{W}_3) + (\nabla_{\mathcal{W}_1 \bullet \mathcal{W}_2} A_1)(\mathcal{W}_3) = 0 \tag{6}$$

for all vector fields $\mathcal{W}_1$, $\mathcal{W}_2$, and $\mathcal{W}_3$. Secondly, the universal equation

$$\rho_{2;3}(\mathcal{W}_1, \mathcal{W}_2, T(\mathcal{W}_3)) = 0$$

follows from $\rho_{2;1} = 0$, $\rho_{2;2} = 0$, the genus-0 topological recursion relation, and the genus-1 equation (6). Thirdly, since any vector field $\mathcal{W}$ can be written as

$$\mathcal{W} = \mathcal{S} \bullet \mathcal{W} + T(\tau_-(\mathcal{W})),$$

combining equation (5) and equation (3), we obtain the result that the universal equation $\rho_{2;2} = 0$ follows from $\rho_{2;3} = 0$ and $\rho_{2:1} = 0$ together with genus-0 and genus-1 universal equations. This is precisely [**L2**, Theorem 2.9].

3.2. Relation among genus-1 universal equations. In this section, we will study the relations between genus-1 universal equations in section 2.2 and equations (4) and (6). The main result is the following

THEOREM 3.1. *Equations (4) and (6) can be derived from known genus-0 and genus-1 relations described in section 2.1 and 2.2.*

To prove this theorem, we need first use genus-0 and genus-1 topological recursion relations and their covariant derivatives to get rid of the T operator in equations (4) and (6). The first three covariant derivatives of the genus-0 topological recursion $\rho_0 = 0$ gives

$$\langle\langle T(\mathcal{W}_1)\, \mathcal{W}_2\, \mathcal{W}_3\, \mathcal{W}_4 \rangle\rangle_0 = \langle\langle (\mathcal{W}_1 \bullet \mathcal{W}_2)\, \mathcal{W}_3\, \mathcal{W}_4 \rangle\rangle_0, \tag{7}$$

$$\begin{aligned} \langle\langle T(\mathcal{W}_1)\, \mathcal{W}_2 \ldots \mathcal{W}_5 \rangle\rangle_0 = & \langle\langle \{\mathcal{W}_1 \bullet \mathcal{W}_2\}\, \mathcal{W}_3\, \mathcal{W}_4\, \mathcal{W}_5 \rangle\rangle_0 \\ & + \langle\langle \{\mathcal{W}_1 \bullet \mathcal{W}_3\}\, \mathcal{W}_2\, \mathcal{W}_4\, \mathcal{W}_5 \rangle\rangle_0 \\ & + \langle\langle \mathcal{W}_1\, \mathcal{W}_2\, \mathcal{W}_3\, \{\mathcal{W}_4 \bullet \mathcal{W}_5\} \rangle\rangle_0, \end{aligned} \tag{8}$$

and

$$\begin{aligned}
\langle\langle T(\mathcal{W}_1)\,\mathcal{W}_2 \ldots \mathcal{W}_6 \rangle\rangle_0 \;=\; & \sum_{i=2}^{4} \left\langle\!\left\langle \{\mathcal{W}_1 \bullet \mathcal{W}_i\}\,\mathcal{W}_2 \ldots \widehat{\mathcal{W}_i} \ldots \mathcal{W}_6 \right\rangle\!\right\rangle_0 \\
& + \langle\langle \mathcal{W}_1 \ldots \mathcal{W}_4 \{\mathcal{W}_5 \bullet \mathcal{W}_6\} \rangle\rangle_0 \\
& + \langle\langle \mathcal{W}_1\,\mathcal{W}_2\,\mathcal{W}_3\,\gamma^\beta \rangle\rangle_0 \langle\langle \gamma_\beta\,\mathcal{W}_4\,\mathcal{W}_5\,\mathcal{W}_6 \rangle\rangle_0 \\
& + \langle\langle \mathcal{W}_1\,\mathcal{W}_2\,\mathcal{W}_4\,\gamma^\beta \rangle\rangle_0 \langle\langle \gamma_\beta\,\mathcal{W}_3\,\mathcal{W}_5\,\mathcal{W}_6 \rangle\rangle_0 \\
& + \langle\langle \mathcal{W}_1\,\mathcal{W}_3\,\mathcal{W}_4\,\gamma^\beta \rangle\rangle_0 \langle\langle \gamma_\beta\,\mathcal{W}_2\,\mathcal{W}_5\,\mathcal{W}_6 \rangle\rangle_0 . \qquad (9)
\end{aligned}$$

It's obvious that the left hand sides of these equations are symmetric with respect to all arguments except $\mathcal{W}_1$. This symmetry property is not apparent for the right hand sides of these equations. Therefore when apply these equations to get rid of operator T, we may get two seemingly different but actually equal expressions. For the convenience of presentation, in this section we will follow the rule that when applying these equations, we always take the form which gives the smallest number of terms. A typical example is the following: When applying equation (9) to $\langle\langle T(\mathcal{W})\,\mathcal{V}\,\gamma_\alpha\,\gamma^\alpha\,\gamma_\beta\,\gamma^\beta \rangle\rangle_0$, we choose $\mathcal{W}_1 = \mathcal{W}$, $\mathcal{W}_2 = \gamma_\alpha$, $\mathcal{W}_3 = \gamma^\alpha$, $\mathcal{W}_4 = \gamma_\beta$, $\mathcal{W}_5 = \gamma^\beta$, $\mathcal{W}_6 = \mathcal{V}$. Keeping in mind that we are summing over repeated indices, using apparent symmetries, we obtain

$$\begin{aligned}
& \langle\langle T(\mathcal{W})\,\mathcal{V}\,\gamma_\alpha\,\gamma^\alpha\,\gamma_\beta\,\gamma^\beta \rangle\rangle_0 \\
= \; & 3\,\langle\langle (\mathcal{W} \bullet \gamma_\alpha)\,\gamma^\alpha\,\gamma_\beta\,\gamma^\beta\,\mathcal{V} \rangle\rangle_0 + \langle\langle \mathcal{W}\,\gamma_\alpha\,\gamma^\alpha\,\gamma_\beta\,(\gamma^\beta \bullet \mathcal{V}) \rangle\rangle_0 \\
& + \langle\langle \mathcal{W}\,\gamma_\alpha\,\gamma^\alpha\,\gamma^\mu \rangle\rangle_0 \langle\langle \gamma_\mu\,\gamma_\beta\,\gamma^\beta\,\mathcal{V} \rangle\rangle_0 + 2\,\langle\langle \mathcal{W}\,\gamma_\alpha\,\gamma_\beta\,\gamma^\mu \rangle\rangle_0 \langle\langle \gamma_\mu\,\gamma^\alpha\,\gamma^\beta\,\mathcal{V} \rangle\rangle_0 . \qquad (10)
\end{aligned}$$

If we instead choose $\mathcal{W}_1 = \mathcal{W}$, $\mathcal{W}_2 = \mathcal{V}$, $\mathcal{W}_3 = \gamma_\alpha$, $\mathcal{W}_4 = \gamma^\alpha$, $\mathcal{W}_5 = \gamma_\beta$, $\mathcal{W}_6 = \gamma^\beta$, we obtain

$$\begin{aligned}
& \langle\langle T(\mathcal{W})\,\mathcal{V}\,\gamma_\alpha\,\gamma^\alpha\,\gamma_\beta\,\gamma^\beta \rangle\rangle_0 \\
= \; & \langle\langle (\mathcal{W} \bullet \mathcal{V})\,\gamma_\alpha\,\gamma^\alpha\,\gamma_\beta\,\gamma^\beta \rangle\rangle_0 + 2\,\langle\langle (\mathcal{W} \bullet \gamma_\alpha)\,\gamma^\alpha\,\gamma_\beta\,\gamma^\beta\,\mathcal{V} \rangle\rangle_0 \\
& + \langle\langle \mathcal{W}\,\mathcal{V}\,\gamma_\alpha\,\gamma^\alpha\,(\gamma_\beta \bullet \gamma^\beta) \rangle\rangle_0 + 2\,\langle\langle \mathcal{W}\,\mathcal{V}\,\gamma^\alpha\,\gamma^\mu \rangle\rangle_0 \langle\langle \gamma_\mu\,\gamma_\alpha\,\gamma_\beta\,\gamma^\beta \rangle\rangle_0 \\
& + \langle\langle \mathcal{W}\,\gamma_\alpha\,\gamma^\alpha\,\gamma^\mu \rangle\rangle_0 \langle\langle \gamma_\mu\,\mathcal{V}\,\gamma_\beta\,\gamma^\beta \rangle\rangle_0 . \qquad (11)
\end{aligned}$$

Although the right hand sides of equation (11) and equation (10) look very different from each other, they are actually equal since the left hand sides of these equations are the same. The right hand side of equation (11) has 5 terms. It is slightly more complicated than the right hand side of equation (10) which has only 4 terms. So following the rule stated earlier, we should use equation (10) rather than equation (11).

To eliminate operator T in genus-1 correlation functions, we need the first three covariant derivatives of the genus-1 topological recursion relation $\rho_1 = 0$, which have the following forms:

$$\langle\langle T(\mathcal{W})\,\mathcal{V} \rangle\rangle_1 = \langle\langle \{\mathcal{W} \bullet \mathcal{V}\} \rangle\rangle_1 + \frac{1}{24} \langle\langle \mathcal{W}\,\mathcal{V}\,\gamma^\mu\,\gamma_\mu \rangle\rangle_0 , \qquad (12)$$

$$\begin{aligned}
\langle\langle T(\mathcal{W})\,\mathcal{V}_1\,\mathcal{V}_2 \rangle\rangle_1 \;=\; & \langle\langle \{\mathcal{W} \bullet \mathcal{V}_1\}\,\mathcal{V}_2 \rangle\rangle_1 + \langle\langle \{\mathcal{W} \bullet \mathcal{V}_2\}\,\mathcal{V}_1 \rangle\rangle_1 \\
& + \langle\langle \mathcal{W}\,\mathcal{V}_1\,\mathcal{V}_2\,\gamma^\alpha \rangle\rangle_0 \langle\langle \gamma_\alpha \rangle\rangle_1 + \frac{1}{24} \langle\langle \mathcal{W}\,\mathcal{V}_1\,\mathcal{V}_2\,\gamma^\mu\,\gamma_\mu \rangle\rangle_0 , \qquad (13)
\end{aligned}$$

and

$$\begin{aligned} & \langle\!\langle T(\mathcal{W})\, \mathcal{V}_1\, \mathcal{V}_2\, \mathcal{V}_3 \rangle\!\rangle_1 \\ = & \sum_{i=1}^{3} \left\{ \left\langle\!\!\left\langle (\mathcal{W} \bullet \mathcal{V}_i)\, \mathcal{V}_1 \ldots \widehat{\mathcal{V}_i} \ldots \mathcal{V}_3 \right\rangle\!\!\right\rangle_1 \right. \\ & \left. + \left\langle\!\!\left\langle \mathcal{W}\, \mathcal{V}_1 \ldots \widehat{\mathcal{V}_i} \ldots \mathcal{V}_3\, \gamma^\alpha \right\rangle\!\!\right\rangle_0 \langle\!\langle \gamma_\alpha\, \mathcal{V}_i \rangle\!\rangle_1 \right\} \\ & + \langle\!\langle \mathcal{W}\, \mathcal{V}_1\, \mathcal{V}_2\, \mathcal{V}_3\, \gamma^\alpha \rangle\!\rangle_0 \langle\!\langle \gamma_\alpha \rangle\!\rangle_1 + \frac{1}{24} \langle\!\langle \mathcal{W}\, \mathcal{V}_1\, \mathcal{V}_2\, \mathcal{V}_3\, \gamma^\alpha\, \gamma_\alpha \rangle\!\rangle_0 \end{aligned} \tag{14}$$

for all vector fields. Note that both sides of these equations are symmetric with respect to vector fields $\mathcal{V}_i$'s. There is no ambiguity on how to apply these formulas.

Besides the genus-0 and genus-1 topological recursion relations and their covariant derivatives, we also need universal equations $C_0 = 0$ and $G = 0$ (see Section 2.1 and 2.2 for the definitions of tensors C_0 and G). The application of $C_0 = 0$ is straightforward. Most times we only need a slightly weaker version of this equation, i.e. the associativity of the quantum product. However it is less obvious how to apply the covariant derivatives of $C_0 = 0$. This turns out to be one of the most subtle points in the proof of Theorem 3.1. In the proof, we will include tensors $\nabla^i C_0$ in the relations among genus-1 universal tensors. It is then apparent where and how covariant derivatives of $C_0 = 0$ are used.

The tensors for the covariant derivatives of C_0 are defined by

$$C_i(\mathcal{W}_1, \cdots, \mathcal{W}_{i+4}) := (\nabla_{\mathcal{W}_{i+4}} C_{i-1})(\mathcal{W}_1, \cdots, \mathcal{W}_{i+3})$$

for $i \geq 1$. These tensors lie in U, the ideal of tensors given universal equations. We need the first three covariant derivatives of C_0 in the proof of Theorem 3.1. Using equation (1), we can compute these tensors more explicitly. For example,

$$\begin{aligned} C_1(\mathcal{W}_1, \cdots, \mathcal{W}_5) = & \langle\!\langle (\mathcal{W}_1 \bullet \mathcal{W}_2)\, \mathcal{W}_3\, \mathcal{W}_4\, \mathcal{W}_5 \rangle\!\rangle_0 + \langle\!\langle \mathcal{W}_1\, \mathcal{W}_2\, (\mathcal{W}_3 \bullet \mathcal{W}_4)\, \mathcal{W}_5 \rangle\!\rangle_0 \\ & - \langle\!\langle (\mathcal{W}_1 \bullet \mathcal{W}_3)\, \mathcal{W}_2\, \mathcal{W}_4\, \mathcal{W}_5 \rangle\!\rangle_0 - \langle\!\langle \mathcal{W}_1\, \mathcal{W}_3\, (\mathcal{W}_2 \bullet \mathcal{W}_4)\, \mathcal{W}_5 \rangle\!\rangle_0 \end{aligned}$$

and

$$\begin{aligned} & C_2(\mathcal{W}_1, \cdots, \mathcal{W}_6) \\ = & \langle\!\langle (\mathcal{W}_1 \bullet \mathcal{W}_2)\, \mathcal{W}_3 \cdots \mathcal{W}_6 \rangle\!\rangle_0 + \langle\!\langle \mathcal{W}_1\, \mathcal{W}_2\, (\mathcal{W}_3 \bullet \mathcal{W}_4)\, \mathcal{W}_5\, \mathcal{W}_6 \rangle\!\rangle_0 \\ & - \langle\!\langle (\mathcal{W}_1 \bullet \mathcal{W}_3)\, \mathcal{W}_2\, \mathcal{W}_4\, \mathcal{W}_5\, \mathcal{W}_6 \rangle\!\rangle_0 - \langle\!\langle \mathcal{W}_1\, \mathcal{W}_3\, (\mathcal{W}_2 \bullet \mathcal{W}_4)\, \mathcal{W}_5\, \mathcal{W}_6 \rangle\!\rangle_0 \\ & + \langle\!\langle \mathcal{W}_1\, \mathcal{W}_2\, \mathcal{W}_5\, \gamma^\alpha \rangle\!\rangle_0 \langle\!\langle \gamma_\alpha\, \mathcal{W}_3\, \mathcal{W}_4\, \mathcal{W}_6 \rangle\!\rangle_0 + \langle\!\langle \mathcal{W}_1\, \mathcal{W}_2\, \mathcal{W}_6\, \gamma^\alpha \rangle\!\rangle_0 \langle\!\langle \gamma_\alpha\, \mathcal{W}_3\, \mathcal{W}_4\, \mathcal{W}_5 \rangle\!\rangle_0 \\ & - \langle\!\langle \mathcal{W}_1\, \mathcal{W}_3\, \mathcal{W}_5\, \gamma^\alpha \rangle\!\rangle_0 \langle\!\langle \gamma_\alpha\, \mathcal{W}_2\, \mathcal{W}_4\, \mathcal{W}_6 \rangle\!\rangle_0 - \langle\!\langle \mathcal{W}_1\, \mathcal{W}_3\, \mathcal{W}_6\, \gamma^\alpha \rangle\!\rangle_0 \langle\!\langle \gamma_\alpha\, \mathcal{W}_2\, \mathcal{W}_4\, \mathcal{W}_5 \rangle\!\rangle_0 \end{aligned}$$

for any vector fields $\mathcal{W}_1, \ldots, \mathcal{W}_6$. The third derivative C_3 also has a similar expression, but more complicated. So we omit the explicit expression for C_3.

We are now ready to prove Theorem 3.1.

Proof of Theorem 3.1: First using genus-0 and genus-1 topological recursion relations and their derivatives, we can get rid of operator T in the expression

$$A_2(\mathcal{W}_1, T(\mathcal{W}_2)) - \left(\nabla_{T(\mathcal{W}_1)} A_1\right)(\mathcal{W}_2).$$

We then check that the resulting expression is equal to certain combinations of tensors G and C_1. More precisely, modulo

$$I_a[\rho_0, \nabla\rho_0, \nabla^2\rho_0, \nabla^3\rho_0, \ \rho_1, \nabla\rho_1, \nabla^2\rho_1, C_0],$$

we have

$$\begin{aligned}
&120\left\{A_2(\mathcal{W}_1, T(\mathcal{W}_2)) - \left(\nabla_{T(\mathcal{W}_1)} A_1\right)(\mathcal{W}_2)\right\} \\
= \; & -\frac{1}{24} G(\mathcal{W}_1, \mathcal{W}_2, \gamma^\alpha, \gamma_\alpha) + \left\{\frac{1}{6} C_1(\mathcal{W}_2, \gamma_\alpha, \mathcal{W}_1, \gamma^\alpha, \gamma_\beta)\right. \\
& \left. + \frac{19}{2} C_1(\mathcal{W}_1, \gamma_\beta, \gamma_\alpha, \gamma^\alpha, \mathcal{W}_2) - \frac{21}{2} C_1(\mathcal{W}_2, \gamma_\beta, \gamma^\alpha, \gamma_\alpha, \mathcal{W}_1)\right\} \langle\langle \gamma^\beta \rangle\rangle_1
\end{aligned} \tag{15}$$

for any vector fields $\mathcal{W}_1$ and $\mathcal{W}_2$. This shows that equation (4) follows from genus-0 and genus-1 universal equations in Section 2. The proof of equation (15) is straightforward, but somehow tedious.

Equation (6) can be treated in a similar fashion. We first use genus-0 and genus-1 topological recursion relations and their derivatives to get rid of operator T in the expression

$$B(\mathcal{W}_1, \mathcal{W}_2, T(\mathcal{V})) - A_2(\mathcal{W}_1 \bullet \mathcal{W}_2, \mathcal{V}) + (\nabla_{\mathcal{W}_1 \bullet \mathcal{W}_2} A_1)(\mathcal{V}),$$

then check that the resulting expression is equal to certain combinations of tensors G, ∇G, C_1, C_2, and C_3. More precisely, modulo

$$I_a[\rho_0, \nabla\rho_0, \nabla^2\rho_0, \nabla^3\rho_0, \rho_1, \nabla\rho_1, \nabla^2\rho_1, \nabla^3\rho_1, C_0],$$

we have

$$\begin{aligned}
&720\left\{B(\mathcal{W}_1, \mathcal{W}_2, T(\mathcal{V})) - A_2(\mathcal{W}_1 \bullet \mathcal{W}_2, \mathcal{V}) + (\nabla_{\mathcal{W}_1 \bullet \mathcal{W}_2} A_1)(\mathcal{V})\right\} \\
= \; & 24\, G(\mathcal{V}, \mathcal{W}_1, \mathcal{W}_2, \gamma^\alpha) \langle\langle \gamma_\alpha \rangle\rangle_1 + (\nabla_{\gamma_\alpha} G)(\mathcal{W}_1, \mathcal{W}_2, \mathcal{V}, \gamma^\alpha) \\
& + \frac{1}{4}\left\{(\nabla_{\mathcal{V}} G)(\mathcal{W}_1, \mathcal{W}_2, \gamma_\alpha, \gamma^\alpha) - (\nabla_{\mathcal{W}_1} G)(\mathcal{V}, \mathcal{W}_2, \gamma_\alpha, \gamma^\alpha)\right. \\
& \qquad \left. - (\nabla_{\mathcal{W}_2} G)(\mathcal{V}, \mathcal{W}_1, \gamma_\alpha, \gamma^\alpha)\right\} \\
& -3 \langle\langle \mathcal{V}\, \gamma_\beta \rangle\rangle_1 \left\{C_1(\mathcal{W}_1, \mathcal{W}_2, \gamma_\alpha, \gamma^\alpha, \gamma^\beta) - C_1(\mathcal{W}_1, \gamma_\alpha, \mathcal{W}_2, \gamma^\beta, \gamma^\alpha)\right. \\
& \qquad \left. - C_1(\mathcal{W}_2, \gamma_\alpha, \mathcal{W}_1, \gamma^\beta, \gamma^\alpha)\right\} \\
& + \langle\langle \mathcal{W}_1\, \gamma_\beta \rangle\rangle_1 \left\{27\, C_1(\mathcal{W}_2, \gamma_\alpha, \mathcal{V}, \gamma^\beta, \gamma^\alpha) - 34\, C_1(\mathcal{V}, \gamma_\alpha, \mathcal{W}_2, \gamma^\beta, \gamma^\alpha)\right. \\
& \qquad \left. - 7\, C_1(\mathcal{W}_2, \gamma^\beta, \gamma_\alpha, \gamma^\alpha, \mathcal{V})\right\} \\
& + \langle\langle \mathcal{W}_2\, \gamma_\beta \rangle\rangle_1 \left\{27\, C_1(\mathcal{W}_1, \gamma_\alpha, \mathcal{V}, \gamma^\beta, \gamma^\alpha) - 34\, C_1(\mathcal{V}, \gamma_\alpha, \mathcal{W}_1, \gamma^\beta, \gamma^\alpha)\right. \\
& \qquad \left. - 7\, C_1(\mathcal{W}_1, \gamma^\beta, \gamma_\alpha, \gamma^\alpha, \mathcal{V})\right\} \\
& -8 \langle\langle \gamma_\alpha\, \gamma_\beta \rangle\rangle_1 \left\{2\, C_1(\mathcal{W}_1, \mathcal{W}_2, \gamma^\alpha, \gamma^\beta, \mathcal{V}) - C_1(\mathcal{W}_1, \mathcal{V}, \mathcal{W}_2, \gamma^\alpha, \gamma^\beta)\right. \\
& \qquad \left. - C_1(\mathcal{W}_2, \mathcal{V}, \mathcal{W}_1, \gamma^\alpha, \gamma^\beta)\right\} \\
& -120 \langle\langle \gamma_\alpha \rangle\rangle_1 \langle\langle \gamma_\beta \rangle\rangle_1 \left\{3\, C_1(\mathcal{W}_1, \mathcal{W}_2, \gamma^\alpha, \gamma^\beta, \mathcal{V}) - C_1(\mathcal{W}_1, \mathcal{V}, \mathcal{W}_2, \gamma^\alpha, \gamma^\beta)\right. \\
& \qquad \left. - C_1(\mathcal{W}_2, \mathcal{V}, \mathcal{W}_1, \gamma^\alpha, \gamma^\beta)\right\} \\
& - \langle\langle \gamma_\beta \rangle\rangle_1 \left\{19\, C_2(\mathcal{W}_2, \gamma^\beta, \mathcal{W}_1, \gamma_\alpha, \mathcal{V}, \gamma^\alpha) + 19\, C_2(\mathcal{W}_1, \gamma^\beta, \mathcal{W}_2, \gamma_\alpha, \mathcal{V}, \gamma^\alpha)\right. \\
& \qquad + 19\, C_2(\mathcal{W}_1, \mathcal{W}_2, \gamma_\alpha, \gamma^\alpha, \mathcal{V}, \gamma^\beta) - 12\, C_2(\mathcal{V}, \gamma_\alpha, \gamma^\beta, \gamma^\alpha, \mathcal{W}_1, \mathcal{W}_2) \\
& \qquad - 3\, C_2(\mathcal{W}_1, \mathcal{V}, \mathcal{W}_2, \gamma^\beta, \gamma_\alpha, \gamma^\alpha) - 3\, C_2(\mathcal{W}_2, \mathcal{V}, \mathcal{W}_1, \gamma^\beta, \gamma_\alpha, \gamma^\alpha) \\
& \qquad \left. - 28\, C_2(\mathcal{W}_1, \gamma_\alpha, \mathcal{W}_2, \gamma^\beta, \mathcal{V}, \gamma^\alpha) - 28\, C_2(\mathcal{W}_2, \gamma_\alpha, \mathcal{W}_1, \gamma^\beta, \mathcal{V}, \gamma^\alpha)\right\} \\
& - C_3(\mathcal{W}_1, \mathcal{W}_2, \gamma_\alpha, \gamma^\alpha, \mathcal{V}, \gamma_\beta, \gamma^\beta)
\end{aligned} \tag{16}$$

for any vector fields $\mathcal{W}_1$, $\mathcal{W}_2$, and $\mathcal{V}$. This implies that equation (6) follows from genus-0 and genus-1 universal equations in Section 2. The proof of equation (16) is very tedious but nevertheless straightforward. All what is needed is to plug the expressions for the relevant tensors into both sides of equation (16) and check that they are equal after obvious cancellations. □

References

[BP] Belorousski, P. and Pandharipande, R., *A descendent relation in genus 2*, Ann. Scuola Norm. Sup. Pisa Cl. Sci. (4) 29 (2000) 171-191.

[DZ] Dubrovin, B., Zhang, Y., *Bihamiltonian hierarchies in 2D topological field theory at one-loop approximation*, Comm. Math. Phys. 198 (1998) 311 - 361.

[EHX] Eguchi, T., Hori, K., and Xiong, C., *Quantum Cohomology and Virasoro Algebra*, Phys. Lett. B402 (1997) 71-80.

[G1] Getzler, E., *Intersection theory on* $\bar{M}_{1,4}$ *and elliptic Gromov-Witten Invariants*, J. Amer. Math. Soc. 10 (1997) 973-998

[G2] Getzler, E., *Topological recursion relations in genus 2*, Integrable systems and algebraic geometry (Kobe/kyoto, 1997) 73-106.

[KM] Kontsevich, M. and Manin, Y., *Relations between the correlators of the topological sigma-model coupled to gravity*, Comm. Math. Phys., 196 (1998) 385-398.

[LiT] Li, J. and Tian, G., *Virtual moduli cycles and Gromov-Witten invariants of general symplectic manifolds*, Topics in symplectic 4-manifolds (Irvine, CA, 1996), 47-83.

[L1] Liu, X., *Elliptic Gromov-Witten invariants and Virasoro conjecture*, Comm. Math. Phys. 216 (2001), 705-728.

[L2] Liu, X., *Quantum product on the big phase space and Virasoro conjecture*, to appear in Advances in Mathematics (math.AG/0104030).

[LT] Liu, X. and Tian, G., *Virasoro constraints for quantum cohomology*, J. Diff. Geom. 50 (1998), 537 - 591.

[M] Manin, Y., *Frobenius manifolds, quantum cohomology, and moduli spaces*, Amer. Math. Soc. Colloq. Publ. 47, Providence, RI, 1999.

[RT] Ruan, Y. and Tian, G., *Higher genus symplectic invariants and sigma models coupled with gravity*, Invent. Math. 130 (1997), 455-516.

[W] Witten, E., *Two dimensional gravity and intersection theory on Moduli space*, Surveys in Diff. Geom., 1 (1991), 243-310.

Department of Mathematics
University of Notre Dame
Notre Dame, IN 46556, USA

E-mail address: *xliu3@nd.edu*

Extended Modular Operad

A. Losev and Yu. Manin

§0. Introduction and summary

This paper, together with [Ma2], constitutes a sequel to [LoMa] where some new moduli spaces of pointed curves were introduced and studied. We start with a review of the main results of [LoMa] and then give a summary of this paper.

0.1. Painted stable curves. Let S be a finite set. *A painting* of S is a partition of S into two disjoint subsets: *white* W and *black* B.

Let T be a scheme, S a painted set, $g \geq 0$. *An S-pointed (or labeled) curve of genus g over T* consists of the data

$$(\pi : C \to T;\, x_i : T \to C,\ i \in S) \tag{0.1}$$

where

(i) π is a flat proper morphism whose geometric fibres C_t are reduced and connected curves, with at most ordinary double points as singularities, and $g = H^1(C_t, \mathcal{O}_{C_t})$.

(ii) $x_i, i \in S$, are sections of π not containing singular points of geometric fibres.

(iii) $x_i \cap x_j = \emptyset$ for all $i \in W,\ j \in S,\ i \neq j$.

Such a curve (0.1) is called *painted stable*, if the normalization of any irreducible component C' of a geometric fibre carries ≥ 3 pairwise different special points when C' is of genus 0 and ≥ 1 special points when C' is of genus 1. Special points are inverse images of singular points and of the structure sections x_i. Equivalently, such a normalization has only a finite automorphism group fixing the special points.

We get the usual notion of a (family of) stable painted curves, if S consists of only white points.

The (dual) modular graph of a geometric fibre is defined in the same way as in the usual case (for the conventions we use see [Ma1], III.2). Tails now can be of two types, we may refer to them and their marks as "black" and "white" ones as well, and call the graph painted one. More on the geometry of such graphs and morphisms between them see in the subsections 1.1–1.2 below. The isomorphism class of a painted modular graph is called *the combinatorial type* of the respective curve.

0.2. Moduli stacks $\overline{L}_{g,S}$. In sec. 4 of [LoMa], we gave two equivalent descriptions of the stacks $\overline{L}_{g,S}$ where S is a painted set. In [Ma2], Lemma 1.2.2, a third description was given based upon a more general definition due to B. Hassett [H]. For reader's convenience, we will reproduce them here.

(i) $\overline{L}_{g,S}$ consists of families of curves of *chain stable combinatorial types.*

To define the latter, consider first combinatorial types of classical (semi)stable curves with only white points labeled by a finite set W. They are isomorphism classes of graphs, whose vertices are labeled by "genera" $g \geq 0$ and tails are bijectively labeled by elements of W. Stability means that vertices of genus 0 bound ≥ 3 flags, and vertices of genus 1 bound ≥ 1 flags. Graphs can have edges with only one vertex, that is, simple loops.

Starting with the geometric realization of such a graph Γ, we can obtain an infinite series of painted graphs, which will be called chain stable. Namely, subdivide edges and tails of Γ by a finite set of new vertices of genus zero (on each edge or tail, this set may be empty). If a tail was subdivided, move the respective label (from W) to the newly emerged tail. Distribute the black tails labeled by elements of B arbitrarily among the old and the new vertices. The resulting graph is called chain stable if it becomes stable after repainting black tails into white ones. The new vertices depict chains of $\mathbf{P}^1$'s stabilized by black points. Each end of such a chain carries either a white tail or a singular special point (or both).

(ii) The second description of $\overline{L}_{g,S}$ is based upon the inductive construction of "adjoining a new black point".

Namely, starting with an S–pointed family $(C/T, (x_i \,|\, i \in S))$ as in (0.1) we will produce another S'–pointed family $(C'/T', (x'_i \,|\, i \in S'))$ where $T' = C$, $S' = S \cup \{*\}$, $*$ being a new black label. The new curve and sections will be produced in two steps. At the first step we make in (0.1) the base change $C \to T$. We get an S–pointed curve $X := C \times_S C$, with sections $x_{i,C}$. We then add an extra section $\Delta : C \to C \times_S C$ which is the relative diagonal, and mark it by $*$. We have not yet produced an S'–pointed curve over $T' = C$, because the extra black section will intersect singular points of the fibres and white sections as well.

Now comes the second step of the construction, where we birationally modify $C \times_T C \to C$ as in [Kn], Definition 2.3. Define $C' := \mathrm{Proj}\,\mathrm{Sym}\,\mathcal{K}$ as the relative projective spectrum of the symmetric algebra of the sheaf $\mathcal{K}$ on $X = C \times_T C$ which is defined as the cokernel of the map

$$\delta : \mathcal{O}_X \to \mathcal{J}_\Delta^{\check{}} \oplus \mathcal{O}_X(\sum_{i \in A} x_{i,C}),\ \delta(f) = (f, f). \tag{0.3}$$

Here $\mathcal{J}_\Delta$ is the $\mathcal{O}_X$–ideal of Δ, and $\mathcal{J}_\Delta^{\check{}}$ is its dual sheaf considered as a subsheaf of meromorphic functions, as in [Kn], Lemma 2.2 and Appendix.

As a result, we get an S'–pointed curve, because Knudsen's treatment of his modification is local and can be directly extended to our case. If the initial curve was painted stable, the new one will be painted stable as well.

Moreover, this construction is obviously functorial, in the sense that it produces a stack if we started with a stack.

We are now ready to construct $\overline{L}_{g,S}$. For concreteness, we assume that $W = \{1, \ldots, m\}$ and $B = \{1, \ldots, n\}$ are initial segments of (two copies of) natural numbers.

If $g \geq 2$, $m \geq 0$, we start with $\overline{M}_{g;m} = \overline{M}_{g;m,\emptyset}$ and adjoin n new black points, one in turn. Denote the resulting stack by $\overline{L}_{g;m,n}$.

For $g = 1$, we repeat this construction for $m \geq 1$. We need one more sequence of stacks, corresponding to $m = 0$ (elliptic curves stabilized only with black points). Since we want to restrict ourselves to Deligne–Mumford stacks, we start with $\overline{M}_{1;0,1}$ which is defined by repainting the white point in $\overline{M}_{1;1}$, and adjoin black points to get the sequence $\overline{L}_{1;0,n}$, $n \geq 1$.

Finally, for $g = 0$ we obtain the series of spaces $\overline{L}_n = \overline{L}_{0;2,n}$, thoroughly studied in [LoMa] and moreover $\overline{L}_{0;m,n}$, for all $m \geq 3, n \geq 0$.

From our construction it follows that the genus zero stacks are actually smooth manifolds.

(iii) The third, and technically most useful, description identifies $\overline{L}_{g,S}$ with certain stacks of *weighted pointed stable curves* studied in [H], see also [Ma2], Definition 1.2.1. Roughly speaking, in Hassett's definition we endow each section x_s with a weight $a_s \in \mathbf{Q} \cap (0,1]$. *Weighted stability* means that (an appropriate power of) $\omega_{C/T}(\sum_s a_s x_s)$ is relatively ample, and moreover, the sum of weights of any subfamily of x_s coinciding on a geometric fiber does not exceed 1. Lemma 1.2.1 in [Ma2] shows that if S is painted, and if we choose weights so that $a_s = 1$ for $s \in W$ and $\sum_{t \in B} a_t \leq 1$, the two definitions of stability coincide.

0.3. Extended modular operad. The classical modular spaces $\overline{M}_{g,m+1}$ come together with the action of $\mathbf{S}_{m+1}$. Moreover, there is a family of clutching morphisms

$$\overline{M}_{g_1,m_1+1} \times \overline{M}_{g_2,m_2+1} \to \overline{M}_{g_1+g_2,m_1+m_2+1} \tag{0.4}$$

which on the level of geometric points can be described as gluing one (say, the last) labeled point of the first pointed curve to, say, the first labeled point of the second curve. Finally, there is a similar family of morphisms

$$\overline{M}_{g-1,m+1} \to \overline{M}_{g,m-1} \tag{0.5}$$

gluing together two different labeled points.

We can axiomatize the evident relations between these structures and get the abstract notion of the modular operad in the category of DM–stacks. One can also add a linear structure, by passing to the (homology) motives or any of their realizations and get the notion of the modular operad in a tensor category, as well as the basic example of this notion. This was done by E. Getzler and M. Kapranov in [GeK2]. Restricting ourselves to the $g = 0$ case, we get almost a classical operad, with two modifications: first, its m–th component $\overline{M}_{g,m+1}$ is acted upon by $\mathbf{S}_{m+1}$ rather than by $\mathbf{S}_m$; second, it lacks the $m = 1$–component since $\mathbf{P}^1$ with only two labeled points is unstable.

The first modification means that we actually deal with *a cyclic operad*, the notion which was as well introduced and studied by E. Getzler and M. Kapranov ([GeK1]). Recall that generally classical linear operads classify k–algebras of a given type, like associative, commutative, Lie, Poisson etc. The component $P(m)$ of an operad P (in our case $H_*(\overline{M}_{0,m+1})$) is the linear space of polylinear operations of m arguments that can be constructed from the basic operations (multiplications, commutators etc), modulo identical relations supplied by the axioms (associativity, Jacobi etc). The action of $\mathbf{S}_m$ upon $P(m)$ corresponds to the permutations of the arguments. If A is an algebra over P, each $p \in P(m)$ defines thus a tensor $p: A^{\otimes m} \to A$, and we know how these tensors transform under all permutations of m lower indices. If A is endowed with a scalar product, we can lower the $m+1$–th index as well identifying p with a map $A^{m+1} \to k$. It makes sense to require the scalar product and the action of $\mathbf{S}_{m+1}$ on such tensors to be a part of the structure of algebras of a given type. Cyclic operads then are well–suited for axiomatizing such a situation.

We now pass to the problem of the 1–component. It is missing because a projective line with two labeled (white) points has an infinite automorphism group G_m and so is unstable. The trick suggested in [LoMa] consists in stabilizing $\mathbf{P}^1$ with an arbitrary number of additional black points. This forces us to add degenerations which are arbitrary chains of such $\mathbf{P}^1$'s. Thus our candidate for the missing $\overline{M}_{0,2}$ is $\coprod_{n\geq 1} \overline{L}_{0,2,n}$. The remaining spaces $\overline{L}_{g,m,n}$ constitute a minimal extension of the family $\overline{M}_{g,m}$ containing all $\overline{L}_{0,2,n}$ and stable with respect to the operadic (clutching) morphisms.

Specifically, we put

$$\widetilde{L}_{g,m} := \coprod_n \overline{L}_{g,m+1,n}\,. \tag{0.6}$$

Morphisms (0.4) and (0.5) in which we allow to glue together *only white points* induce the clutching maps

$$\widetilde{L}_{g_1,m_1} \times \widetilde{L}_{g_2,m_2} \to \widetilde{L}_{g_1+g_2,m_1+m_2}\,, \tag{0.7}$$

$$\widetilde{L}_{g-1,m+1} \to \widetilde{L}_{g,m-1}\,. \tag{0.8}$$

The symmetry group $\mathbf{S}_{m+1}$ acts upon $\widetilde{L}_{g,m}$ by renumbering white points.

Moreover, each stack $\widetilde{L}_{g,m}$ is a union of infinitely many irreducible DM–stacks which are indexed by the number of black points; the group $\mathbf{S}_n$ acts on the n–th component by renumbering black points. These structures induce the respective symmetries of the linearized versions. This is the price we paid for acquiring the $m=1$ component. It has a nice meaning in the language of the respective algebras: an algebra now consists of two spaces (A,T); polylinear operations are of the form $A^{\otimes m} \otimes T^{\otimes n} \to A$. They depend on $t_i \in T$ as parameters which are associated with black points. Thus we acquired degrees of freedom allowing us to encompass certain deformations.

The key result of the theory of the (cyclic) homology operad $\{H_*(\overline{M}_{0,m+1})\}$ consists in establishing a bijection between finite–dimensional operadic algebras and formal Frobenius manifolds. A more detailed analysis which we relegate to a future paper shows that in our present context we get exactly the deformations associated with gravitational descendants.

0.4. Genus zero extended operad. It is convenient to imagine its bigraded components $\overline{L}_{0,m+1,n}$ sitting at the lattice points (m,n) of the first quadrant. Then the lowest horizontal line is the classical genus zero modular operad $\overline{M}_{0,m+1}$ whereas the leftmost vertical line represents the 1–component of the extended operad. Operadic multiplication must induce a monoid structure on the 1–component, and in our case it is supplied by the appropriate clutching morphisms described above and more formally in [Ma2], [LoMa]:

$$\widetilde{L}_{0,m_1+1} \times \widetilde{L}_{0,m_2+1} \to \widetilde{L}_{0,m_1+m_2+1}\,, \tag{0.9}$$

In [LoMa], it was proved that the appropriately defined algebras over the 1–component $\widetilde{L}_{0,1}$ bijectively correspond to the pencils of formal flat connections. The treatment in [LoMa] stressed also the toric picture of $\widetilde{L}_{0,1}$; for certain generalizations of this picture, cf. [R].

Thus at the boundary of the (m, n)–quadrant we have a pretty detailed understanding of the geometry of the respective moduli spaces, algebraic structure of their homology and cohomology including the operadic formalism, and finally, a reduction of the theory of operadic algebras to the study of specific differential equations.

The main goal of [Ma2] and this paper taken together is to extend these results from the boundary of the (m, n)–quadrant to the whole quadrant.

Here we focus mainly on the combinatorial aspects of the picture; those arguments for which algebraic geometry is indispensable, are treated in [Ma2].

0.5. Contents of the paper. In §1 and §2 we define for any painted set S purely combinatorially a graded ring H^*_S and a graded module H_{*S} over this ring and study their structure. Using the results of this study and additional geometric arguments, we establish in [Ma2] canonical isomorphisms $H^*_S = H^*(\overline{L}_{0S})$ and $H_{*S} = H_*(\overline{L}_{0S})$, together with identification of the action of H^*_S upon H_{*S} as the cap product.

In §3 we extend the construction of H^* and H_* to the labeled trees replacing S. Algebraic geometric meaning of this extension consists in studying (co)homology of moduli spaces of painted stable curves of a given combinatorial type and their degenerations. Combinatorially, these rings/modules together with various morphisms connecting them constitute a version of graded (co)operad.

In §4 we introduce several versions of algebras over this (co)operad and describe them in terms of correlators and top correlators.

Finally, in §5 we demonstrate the geometric meaning of the differential equations satisfied by the generating series for top correlators. There are two basic geometric languages, corresponding to what we call in §5 *the commutativity* and *the oriented associativity equations.* One language leads to the study of pencils of flat connections as in [LoMa1], another to a version of "Frobenius manifolds with affine flat structure but without invariant metric". Both versions are related by the induction of flat affine structure from a fiber. This procedure first appeared in [Lo1],[Lo2].

Acknoledgements. The research of A. L. was partially supported by grants RFBR 01-01-00548, by INTAS-99-590 and by support of scientific schools grant 00-15-96557.

§1. Combinatorial cohomology rings

1.1. Graphs. We define *a graph* τ as a quadruple $(V_\tau, F_\tau, \partial_\tau, j_\tau)$ where V_τ, resp. F_τ, are finite sets of vertices, resp. flags; $\partial_\tau : F_\tau \to V_\tau$ is the boundary, or incidence, map; $j_\tau : F_\tau \to F_\tau$ is an involution of the set of flags. This combinatorial definition is related to the more common one via the notion of *the geometric realization of* τ. The latter is a topological space which is obtained from F_τ copies of $[0, 1]$ by gluing together points 0 in the copies corresponding to each vertex $v \in V_\tau$, and by gluing together points 1 in each orbit of j_τ. This motivates the introduction of the following auxiliary sets and their geometric realizations: the set E_τ of *edges* of τ, formally consisting of cardinality two orbits of j_τ, and the set T_τ of *tails*, consisting of those flags, which are j_τ–invariant.

We will very often think and speak about graphs directly in terms of their geometric realizations. In particular, τ will be called a *connected* (resp. *tree*, resp. *forest*) if its geometric realization is connected (resp. connected and has no loops, resp. is a disjoint union of trees).

Situations in which one must be more careful and invoke the initial combinatorial definition usually occur when one has to consider morphisms of graphs. We will define here three types of generating morphisms which we will need. For more sophisticated definitions, cf. [BeMa].

In fact, [BeMa] considers graphs endowed with additional labeling of their vertices with integers ≥ 0. Such graphs arise as dual graphs of (semi)stable pointed curves: vertices correspond to the irreducible components, labels to the genera of their normalizations. This part of the structure can be neglected in most of this paper which treats only the combinatorics of the genus zero curves whose dual graphs are trees with all vertices labeled by zero.

Moreover, flags are in a bijection with the special points of the normalized curve. Hence when S is weighted and/or painted, we must consider graphs whose flags are weighted, resp. painted.

(i) *An isomorphism* $\sigma \to \tau$ consists of two bijections $\varphi^F : F_\tau \to F_\sigma$, $\varphi_V : V_\sigma \to V_\tau$. They must be compatible with ∂_σ, ∂_τ, and j_σ, j_τ. As a corollary, they induce bijections of edges and tails as well. (Notice that maps between flags go in reverse direction with respect to the morphisms of graphs. This convention will persist below.)

(ii) Let σ be a graph, and e its edge consisting of two flags e_1, e_2. A morphism $c : \sigma \to \tau$ *contracting* e geometrically simply deletes the interior of e if it is a loop with vertex v, or else contracts e so that its ends v_1, v_2 become a new vertex v otherwise. Combinatorially, it is given by a map $c^F : F_\tau \to F_\sigma$ which identifies F_τ with $F_\sigma \setminus \{e_1, e_2\}$ and is compatible with involutions, and a map $c_V : V_\sigma \to V_\tau$ which glues together vertices of e_1, e_2 and is a bijection elsewhere. Distribution of flags between all vertices except for those of e_1, e_2 must be compatible with these maps. Flags of τ at v (if e is not a loop) are thus $F_\sigma(v_1) \cup F_\sigma(v_2) \setminus \{e_1, e_2\}$.

(iii) Let σ be a graph and $\{f_1, f_2\}$ its two different tails. A morphism $\sigma \to \tau$ which *glues* these two tails together consists of two bijections $F_\tau \to F_\sigma$, $V_\sigma \to V_\tau$ compatible with ∂_σ, ∂_τ, and making of $\{f_1, f_2\}$ one new edge.

Composition of morphisms is the composition of the relevant maps c^F, c_V (in reverse order for flags). A composition of contractions of edges and isomorphisms will be called simply a contraction. The category of trees and contractions will reappear in §3.

1.1.1. Painted graphs and labeled graphs. *A painting* of τ is a partition of its tails into two subsets: white and black. If a painting is not given explicitly, we assume that all tails are white.

Morphisms between painted graphs are subject to the following restrictions: isomorphisms and contractions do not change color of tails; it is allowed to glue together only white tails.

Let S be a set, and τ a graph. *An S–marking* (or labeling) of τ is a bijection $S \to T_\tau$. Equivalently, we call a graph with an S–marking an S–graph, and a morphism between two S–graphs identical on S, an S–morphism. We identify a painting of τ with a painting of S.

1.1.2. Stable and stably painted trees. Let now τ be a tree. We recall that it is called *stable*, iff its every vertex is incident to at least three flags.

A painted tree is called *painted stable*, iff

(i) It is stable.

(ii) Every end vertex of τ carries at least one white tail. (An end vertex is a vertex incident to 1 or 0 edges).

In particular, an one edge S–tree is the same as an unordered 2–partition $S = \sigma_1 \cup \sigma_2$ showing how the tails are distributed between the two vertices. Painted stability means that σ_1 and σ_2 each contain at least one white point, and $|\sigma_i| \geq 2$.

We will often use the following notation. If S_1, S_2 are two subsets of S and σ is a 2–partition of S, then $S_1\sigma S_2$ is a shorthand for the following statement: all elements of S_1 lie in one part of σ whereas all elements of S_2 lie in another one. If one of the sets is empty, we may omit it.

1.2. 2–partitions and trees. Let τ be a tree, e its edge. If we delete the interior of (the geometric realization of) e, the tree will split into two connected components, and the set of its tails will break into two subsets. If τ is S–marked, we get thus an unordered 2–partition σ_e of S. If τ is painted stable, this partition will also be painted stable.

Let $\sigma = (\sigma_1, \sigma_2)$ and $\tau = (\tau_1, \tau_2)$ be two non–trivial unordered partitions of S. Denote by $\delta(\sigma, \tau)$ the number of non–empty pairwise distinct sets among $\sigma_k \cap \tau_l$ diminished by 2.

Clearly, $\delta(\sigma, \tau)$ can take only values $0, 1, 2$. Moreover,

(i) $\delta(\sigma, \tau) = 0$ *iff* $\sigma = \tau$.

(ii) $\delta(\sigma, \tau) = 1$ *iff the following condition is satisfied. There exists a partition of S into three (non-empty pairwise disjoint) subsets ρ_i, $i = 1, 2, 3$, such that one of the partitions σ, τ is $(\rho_1 \cup \rho_2, \rho_3)$, and another is $(\rho_1, \rho_2 \cup \rho_3)$. The partition (ρ_i) is determined uniquely up to a permutation of ρ_1 and ρ_3.*

To see this, notice that there must be exactly one empty intersection $\sigma_a \cap \tau_b$. Renumbering the sets if necessary we may assume that $\sigma_2 \cap \tau_1 = \emptyset$. Then $\rho_1 := \tau_1, \rho_2 := \sigma_1 \cap \tau_2$ and $\rho_3 := \sigma_2$ will do.

Moreover, assume in this situation that σ and τ are painted stable as one–edge trees. Denote by $\sigma * \tau$ the two–edge S–tree whose tails are distributed between its three vertices according to the partition (ρ_i), with ρ_2 in the middle. Then $\sigma * \tau$ is stable painted as well. In fact, its end vertices carry at least two tails of which at least one is white because this holds for σ and τ; and its middle vertex carries at least three flags because ρ_2 is non–empty.

(iii) $\delta(\sigma, \tau) = 2$ *iff there exists an ordered quadruple of pairwise distinct elements $i, j, k, l \in S$ such that $ji\sigma kl$ and $jk\tau il$. Moreover, if S is painted and σ, τ are painted stable, i, j, k, l can be chosen in such a way that the partitions induced by σ and τ on this quadruple are also painted stable.*

In fact, if one disregards stability, it suffices to choose $j \in \sigma_1 \cap \tau_1$, $i \in \sigma_1 \cap \tau_2$, $k \in \sigma_2 \cap \tau_1$, $l \in \sigma_2 \cap \tau_2$. This is possible exactly when $\delta(\sigma, \tau) = 2$.

If σ and τ are painted stable, consider the $(2,2)$ matrix of sets $(\sigma_a \cap \tau_b)$. A contemplation will convince the reader that at least one of the two diagonals of this matrix has the following property: each of its elements contains a white label. Hence either i and k, or else j and l can be chosen white. In both cases, the induced partitions of $\{i, j, k, l\}$ are painted stable.

We will call a set E of pairwise distinct (painted) stable 2–partitions $\{\sigma\}$ of a (painted) stable set S *good*, if $\delta(\sigma, \sigma') = 1$ for any $\sigma \neq \sigma'$ in E. In particular, empty set is good, and any one–element set is good.

1.2.1. LEMMA. *(a) For any (painted) stable S–tree τ, the set $\{\sigma_e \,|\, e \in E_\tau\}$ is good.*

(b) The map $\tau \mapsto \{\sigma_e \,|\, e \in E_\tau\}$ establishes a bijection between (painted) stable S–trees up to an S–isomorphism, and good families.

(c) Let E' be a non-empty good family, $\sigma \in E'$, $E = E' \setminus \{\sigma\}$. Let τ', resp. τ, be the respective S–trees, and $e \in E_{\tau'}$ the edge corresponding to σ. Then there is a unique S–morphism $\tau' \to \tau$ contracting e and inducing the tautological embedding $E \subset E'$.

Proof. If one disregards painting, statements (a) and (b) are proved in the Proposition 3.5.2 of Chapter III of [Ma1]. To deal with painted trees, it remains only to check that if one starts with a non–empty good family and produces the respective tree, it will be painted stable, that is, its every end vertex carries a white tail. In fact, such a vertex belongs to a unique edge e, and the tails carried by it constitute a part of the partition σ_e. Since σ_e is painted stable, the conclusion follows.

We leave (c) as an exercise to the reader.

1.2.2. PROPOSITION. *Let τ be a painted stable S–tree, and σ a painted stable 2–partition of S. Then exactly one of the following alternatives takes place.*

(i) There exists an edge $e \in E_\tau$ such that $\sigma = \sigma_e$. In this case we will say that σ breaks τ at e.

(ii) There exists a vertex $v \in V_\tau$, an S–tree τ' and an S–morphism $\tau' \to \tau$ contracting an edge e of τ' to the vertex v of τ such that $\sigma = \sigma_e$. In this case we will say that σ breaks τ at v.

(iii) None of the above. In this case there is an edge $e \in E_\tau$ such that $\delta(\sigma, \sigma_e) = 2$. We will say that σ does not break τ.

Proof. Consider the family of numbers $\delta(\sigma, \sigma_e)$, $e \in E_\tau$. If it contains zero, we are in the case (i), and then all other numbers are 1, because of Lemma 1.2.1 (a). If it consists only of 1's, we can add σ to $\{\sigma_e\}$ and get a good family, which produces a new tree τ' and a contracting morphism $\tau' \to \tau$ with required properties. If it contains 2, we are in the case (iii).

1.2.3. Remark. In the following it will be useful to keep in mind a more detailed picture of various possibilities that can arise in the case (ii).

If we delete a vertex v from the geometric realization of τ, it will break into a set of ≥ 3 connected components which we will call *branches* of τ at v. Their set is canonically bijective to the set of flags $F_\tau(v)$ incident to v: we can say that the branch starts with the respective flag. In the extreme case, a branch can be a single tail.

Each branch b carries its full subset of tails, or their labels, S_b, so that all S_b together form a partition of S. A 2–partition σ of S breaks τ at v precisely when each of its parts is a union of several S_b's. In this way we establish a bijection between such 2–partitions σ of S and 2–partitions α of $F_\tau(v)$. We will say that α is (painted) stable if the respective σ is. Most of α's will be automatically painted stable. Unstable α's are of two types:

(i) one part of α consists of one white tail; or

(ii) one part of α consists of only black tails.

In fact, if a branch starts with an edge, then at an end vertex of this edge there are at least two tails, of which at least one is white.

1.3. Combinatorial cohomology. In the following we fix a painted set S with $|S| \geq 3$ containing at least two white elements, and a commutative coefficient ring k. Consider the family of independent commuting variables $\{l_\sigma\}$ indexed by painted stable unordered 2–partitions σ of S and put $\mathcal{R}_S := k[l_\sigma]$.

Call an ordered quadruple of pairwise distinct elements $i, j, k, l \in S$ *allowed,* if both partitions $ij|kl$ and $kj|il$ are painted stable. For a painted stable σ put $\epsilon(\sigma; i, j, k, l) = 1$ if $\{i, j, k, l\}$ is allowed and $ij\sigma kl$; -1, if $\{i, j, k, l\}$ is allowed and $kj\sigma il$; and 0 otherwise.

Define
$$R_{ijkl} := \sum_\sigma \epsilon(\sigma; i, j, k, l)\, l_\sigma \in \mathcal{R}_S, \tag{1.1}$$
$$R_{\sigma\sigma'} := l_\sigma l_{\sigma'} \textit{ if there exists an allowed quadruple } i, j, k, l \textit{ with } ij\sigma kl,\ kj\sigma' il. \tag{1.2}$$
Here σ, σ' run over 2–partitions. Denote by $I_S \subset \mathcal{R}_S$ the ideal generated by all elements (1.1) and (1.2) and define *the combinatorial cohomology ring* by
$$H_S^* := \mathcal{R}_S / I_S. \tag{1.3}$$

The main result of [Ma2], Theorem 3.1.1, establishes a canonical isomorphism of H_S^* with the cohomology and the Chow ring of $\overline{L}_{0,S}$. The proof heavily uses calculations of this paper and Keel's theorem who proved this result in [Ke] for the basic case when there are no black points.

Returning to our combinatorial setup, let now τ be a painted stable S–tree. Put
$$m(\tau) := \prod_{e \in E_\tau} l_{\sigma_e} \in \mathcal{R}_S \tag{1.4}$$
and call such a monomial *good.* It depends only on the S–isomorphism class of τ. If τ is one–vertex tree, we put $m(\tau) = 1$.

1.3.1. Multiplication table. Below we will calculate $l_\sigma m(\tau) \bmod I_S$. The answer will depend on the mutual position of σ and τ as described in Prop. 1.2.2 which we will consider in reverse order.

Case (iii): σ does not break τ. Then
$$l_\sigma m(\tau) \equiv 0 \bmod I_S \tag{1.5}$$

because in this case $l_\sigma m(\tau)$ is divisible by one of the elements (1.2).

Case (ii): σ breaks τ at a vertex v. Then

$$l_\sigma m(\tau) = m(\sigma * \tau) \tag{1.6}$$

where $\sigma * \tau$ is the painted stable S–tree corresponding to the set of 2–partitions $\{\sigma_e \,|\, e \in E_\tau\} \cup \{\sigma\}$.

Case (i): σ breaks τ at an edge e. In this case we will write several formulas depending on additional choices. Namely, denote by v_1, v_2 two vertices of e and by e_1, e_2 the respective flags ("halves of e"). Choose two distinct flags $I, J \in F_\tau(v_1) \setminus \{e_1\}$ and similarly $K, L \in F_\tau(v_2) \setminus \{e_2\}$ with the following property:

(*) *there exist labels i, j, k, l on the branches starting with I, J, K, L respectively, forming an allowed quadruple.*

Again, we will call such I, J, K, L an allowed quadruple. It is important to notice that allowed I, J, K, L always exist: it suffices to take for J any branch at v_1 carrying a white tail, for L any branch at v_2 carrying a white tail, and complement them by arbitrary I, K.

Now we state that for each allowed quadruple of flags as above we have:

$$l_\sigma m(\tau) \equiv - \sum_{\alpha:\, IJ\alpha e_1} m(\tau(\alpha)) - \sum_{\beta:\, e_2\beta KL} m(\tau(\beta)) \bmod I_S. \tag{1.7}$$

Here α (resp. β) runs over painted stable 2–partitions of $F_\tau(v_1)$ (resp. $F_\tau(v_2)$) in the sense of 1.2.3. If α determines a 2–partition σ of S, we denote by $\tau(\alpha)$ the former $\sigma * \tau$.

To prove (1.7), we choose i, j, k, l as in (*) and write first of all

$$R_{ijkl} m(\tau) = \sum_\rho \epsilon(\rho; i, j, k, l)\, l_\rho m(\tau) \in I_S. \tag{1.8}$$

where we sum over 2–partitions ρ of S. Now, if $\epsilon(\rho; i, j, k, l) = -1$, then $kj\rho il$, and since $ij\sigma kl$, we have $l_\rho l_\sigma \in I_S$; but l_σ divides $m(\tau)$, and so $l_\sigma m(\tau) \in I_S$. Hence in the r.h.s. of (1.8) we may sum only over ρ's with $ij\rho kl$, and σ is among this set. Therefore,

$$l_\sigma m(\tau) \equiv - \sum_{\substack{\rho:\, ij\rho kl \\ \rho \neq \sigma}} l_\rho m(\tau) \bmod I_S. \tag{1.9}$$

In the sum (1.9) we may and will omit ρ's that do not break τ. Moreover, no ρ can break τ at an edge, or at a vertex distinct from v_1 or v_2: otherwise the condition $ij\rho kl$ cannot hold. Finally, ρ's which break τ at v_1 or v_2 will produce precisely the right hand side of (1.7).

1.3.2. Corollary. *The ring $\mathcal{R}_S$ is linearly spanned by I_S and good monomials $m(\tau)$. The cohomology ring H_S^* is spanned by the classes of $m(\tau)$.*

In fact, the span of I_S and $m(\tau)$ contains 1 and is stable with respect to multiplications by all l_σ.

However, the space of linear relations $I_S \cap \langle m(\tau) \rangle$ is generally non–trivial. Below we will exhibit some elements of this space which we will call *the standard ones* and which will later be shown to span all the relations.

1.4. Standard relations between good monomials. As above, we will calculate some expressions $R_{ijkl}m(\tau)$. However, formerly we started with a tree τ, an edge $e = e_\sigma$ and flags I, J and K, L at two ends of this edge.

Now we will start with a tree τ, a vertex v, an allowed quadruple of flags I, J, K, L incident to this vertex, and an allowed quadruple of labels i, j, k, l carried by the respective branches. Looking at the summands at the right hand side of (1.8), we see that if ρ does not break τ, we can omit it, and if ρ breaks τ at an edge or at a vertex different from v, then $\epsilon(\rho; i, j, k, l) = 0$.

Partitions ρ breaking τ at v correspond to 2–partitions α of $F_\tau(v)$. So finally we get the standard relation

$$\sum_\alpha \epsilon(\alpha; I, J, K, L)\, m(\tau(\alpha)) \equiv 0 \bmod I_S \tag{1.10}$$

where the sign ϵ is ± 1 depending on whether $IJ\alpha KL$ or $KJ\alpha IL$.

§2. Combinatorial homology modules

2.1. Homology space. We keep notations of §1, in particular, fix a painted set S and a coefficient ring k. Define a free k–module $\mathcal{M}_S := \oplus_\tau k\,\mu(\tau)$ freely generated by all S–isomorphism classes of painted stable S–trees τ. Denote by $J_S \subset \mathcal{M}_S$ the submodule spanned by the standard relations (1.10) in which $m(\tau)$ are replaced by $\mu(\tau)$:

$$r(\tau, v; I, J, K, L) := \sum_\alpha \epsilon(\alpha; I, J, K, L)\, \mu(\tau(\alpha)) \in J_S. \tag{2.1}$$

Finally, define the homology k–space as

$$H_{*S} := \mathcal{M}_S / J_S \tag{2.2}$$

and put $[\mu(\tau)] := \mu(\tau) \bmod J_S$.

The main content of this section is the definition of an action of H_S^* on H_{*S} which turns the homology module into a free module of rank one over the cohomology ring. To be more precise, we start with defining an action of the ring generators l_σ upon $\mathcal{M}_S$ which imitates the multiplication table (1.5)–(1.7), and proceed to show that it is compatible with the defining relations.

2.2. Multiplication table revisited. We will first formally define the action of l_σ upon generators $\mu(\tau)$ sending them to some elements of $\mathcal{M}_S$. If σ does not break τ, we put

$$l_\sigma \mu(\tau) = 0. \tag{2.3}$$

If σ breaks τ at a vertex v, we put

$$l_\sigma \mu(\tau) = \mu(\sigma * \tau). \tag{2.4}$$

Finally, if σ breaks τ at an edge e, we choose an arbitrary allowed quadruple of flags I, J, K, L as in 1.3.1, Case (i), and put in the notations explained above

$$l_\sigma \mu(\tau) = -\sum_{\alpha:\, IJ\alpha e_1} \mu(\tau(\alpha)) - \sum_{\beta:\, e_2\beta KL} \mu(\tau(\beta))\,. \tag{2.5}$$

2.3. THEOREM. *Formulas (2.3)–(2.5) induce a well defined action of H_S^* upon H_{*S}.*

2.3.1. COROLLARY. *(a) This action makes of H_{*S} a free module of rank one over H_S^*.*

(b) The standard relations (1.10) span all linear relations between the classes of good monomials $m(\tau)$ in H_S^*.

Deduction of the Corollary. Since the relations between the classes of $\mu(\tau)$ correspond to a part of linear relations between good monomials, we have a surjective k–linear map

$$s: H_{*S} \to H_S^*: \ [\mu(\tau)] \mapsto [m(\tau)].$$

Here we set $[m(\tau)] := m(\tau) \bmod I_S$.

On the other hand, denoting by $\mathbf{1} \in H_{*S}$ the class of one–vertex S–tree, thanks to Theorem 2.3 we have another linear map

$$t: H_S^* \to H{*_S}: \ [m(\tau)] \mapsto [m(\tau)]\,\mathbf{1}.$$

Induction on the number of edges of τ using (2.4) shows that $[m(\tau)]\,\mathbf{1} = [\mu(\tau)]$. Therefore s and t are mutually inverse, so that $\mathbf{1}$ is a free generator of $H{*_S}$ as an H_S^*–module. This completes the deduction.

We now turn to the proof of the theorem which is given in the subsections 2.3.2 – 2.3.6. It consists of a chain of sometimes tedious checks, and the reader may prefer to skip it.

2.3.2. Prescription (2.5) is well defined. The right hand side of (2.5) formally depends on the choice of an allowed quadruple I, J, K, L. One can pass from one allowed quadruple to another one replacing each time only one flag so that all intermediate quadruples will also be allowed (at each vertex, start with replacing a flag carrying a white label by another flag also carrying a white label).

We will now take the difference of the right hand sides of (2.5) written for quadruples I, J, K, L and I', J, K, L. The second sums will cancel. In the first sums, the terms corresponding to α's with $II'J\alpha e_1$ will also cancel. The remaining terms are

$$-\sum_{\alpha:\, IJ\alpha I'e_1} \mu(\tau(\alpha)) + \sum_{\alpha:\, I'J\alpha Ie_1} \mu(\tau(\alpha)) \bmod J_S. \tag{2.6}$$

Clearly, (2.6) is an element of the form (2.1), namely $r(\tau, v_1; I', J, I, e_1)$.

One can similarly treat a replacement of J by J'.

2.3.3. Multiplication by l_σ maps J_S into itself. As in (2.1), choose a relation $r(\tau, v; I, J, K, L)$ and a 2–partition σ. We want to show that $l_\sigma r(\tau, v; I, J, K, L)$ belongs to J_S. Consider in turn the following subcases.

(i) σ does not break τ

If σ does not break τ, it does not break any of $\tau(\alpha)$, hence $l_\sigma r(\tau, v; I, J, K, L) = 0$ in view of (2.3).

(ii) σ breaks τ at a vertex different from v

In this case we obviously have

$$l_\sigma r(\tau, v; I, J, K, L) = r(\sigma * \tau, v; I, J, K, L).$$

(iii) σ breaks τ at an edge not incident to v

In this case we calculate $l_\sigma \tau(\alpha)$ using one of the formulas (2.5) which involve surgery only at the ends of e so that it does not interact with the partitions α in (2.1). Hence we get a sum of the elements of the type (2.1) written for various $\tau(\alpha)$.

The remaining two cases are more difficult.

(iv) σ breaks τ at an edge e incident to v

We will start with some notation. Let v_0 be the vertex of e distinct from v. Denote by e_0, resp. e_1, the flags of e at v_0, resp. v. For each α appearing in (2.1), the tree $\tau(\alpha)$ has an extra edge which we denote $f = f_\alpha$; other edges, in particular, e, "are" the edges of τ. Clearly, e and f_α have in $\tau(\alpha)$ a common vertex v.

In order to calculate $l_\sigma \mu(\tau(\alpha))$ we have to use a formula of the type (2.5) written for the edge e of $\tau(\alpha)$. Notation in (2.5) conflicts with that in (2.1), so we will rewrite (2.5) as:

$$l_\sigma \mu(\tau(\alpha)) = -\sum_{\beta:\, I_\alpha J_\alpha \beta e_0} \mu(\tau(\alpha)(\beta)) - \sum_{\gamma:\, e_1 \gamma K_\alpha L_\alpha} \mu(\tau(\alpha)(\gamma)). \tag{2.7}$$

Here $I_\alpha, J_\alpha, K_\alpha, L_\alpha$ can be chosen depending on α, thanks to 2.3.2. Recall from 1.3.1 that these flags must form an allowed quadruple, and

$$I_\alpha, J_\alpha \in F_{\tau(\alpha)}(v_0) = F_\tau(v_0), \quad K_\alpha, L_\alpha \in F_{\tau(\alpha)}(v). \tag{2.8}$$

Moreover, β, resp. γ, run over 2–partitions of $F_{\tau(\alpha)}(v_0) = F_\tau(v_0)$, resp. of $F_{\tau(\alpha)}(v)$, whereas α runs over partitions of $F_\tau(v)$.

With this notation, we have

$$l_\sigma r(\tau, v; I, J, K, L) = \sum_\alpha \epsilon(\alpha; I, J, K, L)\, l_\sigma \mu(\tau(\alpha)) =$$

$$-\sum_\alpha \epsilon(\alpha; I, J, K, L) \left(\sum_{\beta:\, I_\alpha J_\alpha \beta e_0} \mu(\tau(\alpha)(\beta)) + \sum_{\gamma:\, e_1 \gamma K_\alpha L_\alpha} \mu(\tau(\alpha)(\gamma)) \right). \tag{2.9}$$

Now, in (2.9) α and β occur at different vertices of τ. Hence $\mu(\tau(\alpha)(\beta)) = \mu(\tau(\beta)(\alpha))$, and we can interchange summation on α and β if we choose I_α, J_α independent of α. This is possible: namely, choose as $J_\alpha = J_0$ an arbitrary flag in $F_\tau(v_0) \setminus \{e_1\}$ whose branch carries a white label, and for $I_\alpha = I_0$ choose any other flag in this set. The condition on J_0 will make a quadruple $I_0, J_0, K_\alpha, L_\alpha$ allowed if L_α carries a white label as well. We will care about it later.

Meanwhile we can rewrite the first half of (2.9):

$$-\sum_{\beta:\, I_0 J_0 \beta e_0} \sum_\alpha \epsilon(\alpha; I, J, K, L)\, \mu(\tau(\beta)(\alpha)) = -\sum_{\beta:\, I_0 J_0 \beta e_0} r(\tau(\beta), v; I, J, K, L) \in J_S. \tag{2.10}$$

Hence it remains to show that, after an appropriate choice of K_α, L_α,

$$(?)\quad \sum_{\alpha} \epsilon(\alpha; I, J, K, L) \sum_{\gamma:\, e_1\gamma K_\alpha L_\alpha} \mu(\tau(\alpha)(\gamma)) \in J_S. \tag{2.11}$$

Choice of K_α, L_α. For L_α we will always choose the flag of the new edge f_α incident to v. This choice assures that $I_0, J_0, K_\alpha, L_\alpha$ will form an allowed quadruple.

The choice of K_α will additionally depend on the mutual position of e and of flags I, J, K, L. Notice that in our calculation this set of data is fixed: it determines the left hand side of (2.9).

There are five logical possibilities: one or none of the flags I, J, K, L coincides with e_v, the flag of e incident to v. By symmetry, it suffices to consider three options: none is e_v, $I = e_v$, or $J = e_v$. We need only look at those α for which $IJ\alpha KL$ or $KJ\alpha IL$.

Here is the list of choices.

If none of I, J, K, L is e_v, we put $K_\alpha = I$ for $\alpha I e_v$ and $K_\alpha = K$ for $\alpha K e_v$.

If $I = e_v$, we put $K_\alpha = J$ for $\alpha J e_v$ and $K_\alpha = L$ for $\alpha L e_v$.

Finally, if $J = e_v$, we again put $K_\alpha = I$ for $\alpha I e_v$ and $K_\alpha = K$ for $\alpha K e_v$.

Now the left hand side of (2.11) is well defined, and we will start rewriting it as a sum of elements of J_S.

First of all, each term $\mu(\tau(\alpha)(\gamma))$ can be uniquely rewritten as $\mu(\tau(\gamma')(\alpha'))$ creating two new edges of $\tau(\alpha)(\gamma)$ in reverse order.

More formally, let $\tau(\gamma')$ be the result of contracting the edge $f = f_\alpha$ in $\tau(\alpha)(\gamma)$. It can be obtained from τ by inserting a new edge g breaking the set of branches at the vertex v into two parts. This edge has a common vertex with e. Denote by $w = w_{\gamma'}$ another vertex of g. The tree $\tau(\gamma')(\alpha')$ is obtained from $\tau(\gamma')$ by inserting an edge breaking the branches at w into two parts.

Second, we will show that the sign $\epsilon(\alpha; I, J, K, L)$ in (2.11) can be rewritten as $\epsilon(\alpha'; I', J', K', L')$ where the flags I', J', K', L' at w will depend on γ' but not on α' and form an allowed quadruple.

Then (2.11) can be rewritten as

$$\sum_{\gamma'} \sum_{\alpha'} \epsilon(\alpha'; I', J', K', L')\, \mu(\tau(\gamma')(\alpha')). \tag{2.12}$$

It will remain only to convince ourselves that α' runs over all 2–partitions of $F_{\tau(\gamma')}(w_{\gamma'})$.

We will dogmatically describe the choices involved. We checked that everything fits together by drawing twelve diagrams exhausting all possible mutual positions of I, J, K, L, e, f, g in the trees $\tau(\alpha)(\gamma) = \tau(\gamma')(\alpha')$.

The diagrams show that γ' breaks I, J, K, L in such a way that either all four flags get in the same part, or one of them is taken apart. In the first case we simply choose $I' = I, J' = j, K' = K, L' = L$. In the second case we replace the flag that is taken apart by the appropriate flag belonging to the edge g. One easily sees that the resulting quadruples are allowed.

(v) σ breaks τ at v

Again, we will consider several cases depending on how σ breaks I, J, K, L. There are four basic options: $\sigma IJKL$, $I\sigma JKL$, $J\sigma IKL$, $IJ\sigma KL$. All other possibilities either can be obtained from one of these by exchanging J and L, I and K, or else refer to those σ for which $l_\sigma r(\tau, v; I, J, K, L) = 0$. The latter happens when $IK\sigma JL$, because then σ cannot break any $\tau(\alpha)$ with either $IJ\alpha KL$ or $IJ\alpha KL$.

Assume first that one of the options $\sigma IJKL$, $I\sigma JKL$, $J\sigma IKL$ holds. Denote by w the vertex of $\sigma * \tau$ incident to three or more of the flags I, J, K, L.

As in (2.12), we can rewrite the nonvanishing terms of $l_\sigma r(\tau, v; I, J, K, L)$ as follows:

$$\epsilon(\alpha; I, J, K, L)\, l_\sigma \mu(\tau(\alpha)) = \epsilon(\alpha'; I', J', K', L')\, \mu((\sigma * \tau)(\alpha')). \tag{2.13}$$

Here α' is a 2–partition of $F_{\sigma*\tau}(w)$, and I', J', K', L' are defined by the following prescription. If I, J, K, L are all contained in the same part of α, $I' = I, J' = J, K' = K, L' = L$. If one of the flags is taken apart, it must be replaced by the flag of e incident to w.

A straightforward check then shows that the summation over all α produces the same result as the summation over all α', so that finally

$$l_\sigma r(\tau, v; I, J, K, L) = r(\sigma * \tau, w; I', J', K', L'). \tag{2.14}$$

It remains to consider the case $IJ\sigma KL$. In this case, among the summands of $r(\tau, v; I, J, K, L)$ exactly one has the property $\tau(\alpha_0) = \sigma * \tau$. To multiply it by l_σ, we use the formula (2.5), in which we have to replace τ by $\tau(\alpha_0)$ and rename the summation variables, say, to β, γ. However, we will use our current I, J, K, L in the same sense as in (2.5), so that finally we get

$$l_\sigma \mu(\tau(\alpha_0)) = -\sum_{\beta:\, IJ\beta e_1} \mu((\sigma * \tau)(\beta)) - \sum_{\gamma:\, e_2\gamma KL} \mu((\sigma * \tau)(\gamma)). \tag{2.15}$$

On the other hand, the nonvanishing terms $l_\sigma \mu(\tau(\alpha))$ with $\alpha \neq \alpha_0$ can be represented as $\mu((\sigma * \tau)(\delta))$ where δ breaks $\sigma * \tau$ at one of the vertices of e. A contemplation shows that these terms exactly cancel (2.15).

2.3.4. Multiplications by l_σ pairwise commute mod J_S. We want to prove that

$$l_{\sigma_1}(l_{\sigma_2}\mu(\tau)) \equiv l_{\sigma_2}(l_{\sigma_1}\mu(\tau)) \bmod J_S. \tag{2.16}$$

There is a long list of subcases that have to be treated separately: σ_1, resp. σ_2, can break τ at a vertex v_1, resp. v_2, an edge e_1, resp. e_2, or not to break τ. For each of the possible breaking combinations, if the surgery loci are not incident, say, v_1 is not a vertex of e_2, the reasoning is straightforward, but it becomes more tedious otherwise.

In the section A.3 of the Appendix to [LoMa], we displayed the relevant calculations for the case when the painted set S has exactly two white labels. In this case edges of τ have to form a linear chain, with two white labels attached at the respective end vertices.

The total number of white labels influences everything in our definitions: the total supply of painted stable trees, the formulas for multiplication, and the list of the generators of I_S and J_S. Therefore generally we cannot simply refer to the Appendix in [LoMa]. However, a check of pairwise commutativity of l_σ's does allow such a reduction.

Let us illustrate this in the case numbered $(ii)(ii')$ in A.3, [LoMa]: σ_1, resp. σ_2, breaks τ at the edges e_1, resp. e_2, having a common vertex. Let us replace τ by the linear tree τ_0 with edges e_1, e_2. Besides halves of these edges, we endow τ_0 with tails which are branches of τ incident to the respective three vertices. At the two end vertices of τ_0, choose two branches of τ which carry white labels and declare the respective tails of τ_0 white. Declare other tails black. Replace the 2–partitions σ_1, σ_2, by the induced partitions of tails of τ_0.

Now, the calculations in A.3 referring to τ_0 are compatible with the respective calculations for τ, because the formula (2.5) is "local". They allow less freedom with respect to the choice of I, J, K, L, but this is inessential, because from 2.3.2 and 2.3.3 it follows that any choice will do.

Finally, with the choices made in [LoMa], we get (2.16) as an exact equality.

We have now proved that $\mathcal{R}_S$ acts upon H_{*S}. It remains to check that I_S annihilates H_{*S}.

2.3.5. Elements R_{ijkl} annihilate H_{*S}. The case when all labels are white was treated in [Ma1], Chapter III, 4.8.4. With very minor additional precautions, the treatment can be repeated for general painted S.

The reason is that, although we have a short supply of painted stable 2–partitions σ's if not all labels are white, nevertheless for an allowed quadruple $i, j, k, l \subset S$, the relation R_{ijkl} involves with non–zero coefficients exactly those σ's which are simply stable. Painted stable trees also are simply stable. Finally, the multiplication formulas (2.5) written for allowed quadruples I, J, K, L constitute a part of the multiplication formulas in the unpainted case.

In [Ma1], III.4.8.4, we show that $R_{ijkl}[\mu(\tau)] = 0$, again for different reasons in different combinatorial situations. When no choices are involved in a calculation, it is valid in the painted case as well. If a choice of I, J, K, L is involved, one can check that for an allowed i, j, k, l, the choices made in [Ma1] are allowed as well.

2.3.6. Elements $R_{\sigma_1\sigma_2}$ annihilate H_{*S}. These elements are $l_{\sigma_1} l_{\sigma_2}$ such that σ_1 and σ_2 do not break each other. If one of them, say, σ_1, does not break τ either, then because of commutativity $l_{\sigma_1} l_{\sigma_2} \mu(\tau) = 0$.

If both of them break τ, they have to break τ at one and the same vertex v. It is then easy to check that $l_{\sigma_1} l_{\sigma_2} \mu(\tau) = 0$.

§3. Cohomology and homology as functors on trees

3.1. Groups $H^*(\tau)$ and $H_*(\tau)$. Let τ be a stably painted tree. We extend the painting of tails of τ to the painting of all flags declaring all halves of the edges white. Finally, we put

$$H^*(\tau) := \otimes_{v \in V_\tau} H^*_{F_\tau(v)}, \quad H_*(\tau) := \otimes_{v \in V_\tau} H_{*F_\tau(v)} \,. \tag{3.1}$$

For one–vertex trees, we recover the rings/modules (1.3), (2.2) with which we worked in the earlier sections. The tensor products (over k) of families of objects labeled by finite sets, as in (3.1), are defined generally in symmetric monoidal categories and allow one to make explicit the functorial properties with respect to the maps of the index sets. In particular, $H^*(\tau)$, resp. $H_*(\tau)$, has a natural structure of k–algebra, resp. $H^*(\tau)$–module, with compositions defined componentwise.

The main result of this section is:

3.2. THEOREM. *For any contraction morphism $f : \tau' \to \tau$ of stably painted trees, one can define a ring homomorphism*

$$f^* : H^*(\tau) \to H^*(\tau') \tag{3.2}$$

and a compatible homomorphism of modules

$$f_* : H_*(\tau') \to H_*(\tau) \tag{3.3}$$

which makes H^* and H_* functors on the category of trees and contractions.

Functors H^*, H_* are well determined by their actions on isomorphisms (obvious) and on the morphisms contracting a unique edge of an one–edge tree. These restrictions are explicitly described below.

Proof. Let S be a painted stable set identified with an one–vertex tree with flags S, σ an one–edge painted stable S–tree identified with a 2–partition $S = S_1 \cup S_2$. We denote by e the edge of σ, and by v_1, v_2 its vertices carrying flags S_1, S_2, and by e_1, e_2 the respective halves of e. Denote by f_e the contraction morphism. We want first of all to define a ring homomorphism $f_e^* : H_S^* \to H_{S_1\cup\{e_1\}}^* \otimes H_{S_2\cup\{e_2\}}^*$.

We start with describing its (depending on some auxiliary choices) lift $\phi_e^* : \mathcal{R}_S \to \mathcal{R}_{S_1\cup\{e_1\}} \otimes \mathcal{R}_{S_2\cup\{e_2\}}$. Put $\phi_e^*(l_\rho) = 0$ if ρ does not break σ. If ρ breaks σ at v_1, it defines a painted stable 2–partition ρ_1 of S_1 and the respective new edge f of $\rho * \sigma$. Let S_1' denote the set of flags of this tree at the common vertex of e and f, and S_1'' the flags at the other vertex of f. Denote by ρ_1 the one edge $(S_1 \cup \{e_1\})$–tree with tails $S_1' \cup \{e_1\}$, S_1''. Similarly, if ρ breaks σ at v_2 define an one–edge $(S_2 \cup \{e_2\})$–tree ρ_2. Put

$$\phi_e^*(l_\rho) := l_{\rho_1} \otimes 1, \text{ resp. } 1 \otimes l_{\rho_2} \in \mathcal{R}_{S_{1\cup\{e_1\}}} \otimes \mathcal{R}_{S_{2\cup\{e_2\}}}. \tag{3.4}$$

It remains to define ϕ_e^* in the case when $\rho = \sigma$, or equivalently, when ρ breaks σ at e. In this case we choose an allowed quadruple of flags i, j, k, l such that $i, j \in S_1$, $k, l \in S_2$, and define, similarly to (1.7), (2.5),

$$\phi_e^*(l_\sigma) = -\sum_{\alpha:\, ij\alpha e_1} l_\alpha \otimes 1 - \sum_{\beta:\, ij\beta e_1} 1 \otimes l_\beta \in \mathcal{R}_{S_{1\cup\{e_1\}}} \otimes \mathcal{R}_{S_{2\cup\{e_2\}}}. \tag{3.5}$$

Notice right away that another choice of i, j, k, l will not change the result modulo $I_{S_{1\cup\{e_1\}}} \otimes \mathcal{R}_{S_{2\cup\{e_2\}}} + \mathcal{R}_{S_{1\cup\{e_1\}}} \otimes I_{S_{2\cup\{e_2\}}}$. The same reasoning as in 2.3.2 shows this: we consider a replacement of one flag, say, i', and then the respective difference as in (2.6) can be identified with $r(\alpha, v_1; i', j, i, e_1) \otimes 1$.

3.2.1. LEMMA. *(a) There is a well defined ring homomorphism*

$$f_e^* : H_S^* \to H_{S_1\cup\{e_1\}}^* \otimes H_{S_2\cup\{e_2\}}^*, \quad [l_\rho] \mapsto [\phi_e^*(l_\rho)]. \tag{3.6}$$

(ii) f_e^* is surjective. More precisely, let τ_1, resp. τ_2, be a painted stable $(S_1 \cup \{e_1\})$–tree, resp. $(S_2 \cup \{e_2\})$–tree. Denote by $\tau_1 \bullet \tau_2$ the S–tree obtained by gluing e_1 to e_2. Then

$$f_e^*([m(\tau_1 \bullet \tau_2)]) = [m(\tau_1)] \otimes [m(\tau_2)]. \tag{3.7}$$

Proof of Lemma 3.2.1. We must show that ϕ_e^* maps the generators (1.1) and (1.2) of I_S into $I_{S_1\cup\{e_1\}}^* \otimes \mathcal{R}_{S_2\cup\{e_2\}}^* + \mathcal{R}_{S_1\cup\{e_1\}}^* \otimes I_{S_2\cup\{e_2\}}^*$.

Start with a generator R_{ijkl} which we will now endow with a superscript S. Our reasoning depends on the mutual position of i, j, k, l and σ.

Case ijklσ. If, say, $i, j, k, l \in S_2$, we have

$$\phi_e^*(R_{ijkl}^S) = 1 \otimes R_{ijkl}^{S_2 \cup \{e_2\}}.$$

Case jσikl. If $i, k, l \in S_2$, we have

$$\phi_e^*(R_{ijkl}^S) = 1 \otimes R_{ie_2kl}^{S_2 \cup \{e_2\}}.$$

Case iσjkl. If $j, k, l \in S_2$, we have

$$\phi_e^*(R_{ijkl}^S) = 1 \otimes R_{e_2jkl}^{S_2 \cup \{e_2\}}.$$

Case ikσjl. In this case

$$\phi_e^*(R_{ijkl}^S) = 0.$$

Case ijσkl. In this case one summand of R_{ijkl}^S is l_σ. Applying ϕ_e^* to it with the same choice of i, j, k, l we get the expression (3.5). Applying ϕ_e^* to all other terms of R_{ijkl}^S and using (3.4), we get the same terms as in (3.5), with reverse signs. Hence finally

$$\phi_e^*(R_{ijkl}^S) = 0.$$

Up to obvious symmetries, we have exhausted all possible alternatives. It remains to show that

$$\phi_e^*(l_{\sigma_1})\phi_e^*(l_{\sigma_2}) \in I_{S_1 \cup \{e_1\}} \otimes \mathcal{R}_{S_2 \cup \{e_2\}} + \mathcal{R}_{S_1 \cup \{e_1\}} \otimes I_{S_2 \cup \{e_2\}}$$

if $\delta(\sigma_1, \sigma_2) = 2$. We leave this as an exercise to the reader.

We have thus established that ϕ_e^* induces a ring homomorphism f_e^*. Representing $m(\tau_1 \bullet \tau_2)$ as the product of generators corresponding to the edges, and applying (3.4), we get (3.7). This shows the surjectivity of f_e^* and completes the proof of the lemma.

We now continue the proof of the Theorem 3.2 and turn to H_*. Motivated by (3.7) (and, of course, by algebraic geometry), we define (in the notations of Lemma 3.2.1 (ii)):

$$f_{e*}: H_{*S_1 \cup \{e_1\}} \otimes H_{*S_2 \cup \{e_2\}} \to H_{*S}, \quad [\mu(\tau_1)] \otimes [\mu(\tau_2)] \mapsto [\mu(\tau_1 \bullet \tau_2)]. \tag{3.8}$$

Compatibility of this prescription with the defining relations of J_S (cf. (2.1)) is straightforward.

Since $[\mu(\tau)] = [m(\tau)]\,\mathbf{1}$ (cf. the proof of 2.3.1), combining (3.7) and (3.8) we obtain

$$f_{e*}(f_e^*([m(\tau_1 \bullet \tau_2)])\,(\mathbf{1} \otimes \mathbf{1})) = [m(\tau_1 \bullet \tau_2)]\,\mathbf{1}.$$

This is the key special case of the compatibility of f_e^* and f_{e*} ("projection formula" of algebraic geometry). The general case (obtained by replacing $\mathbf{1} \otimes \mathbf{1}$ with any element of $H_* \otimes H_*$) follows from it formally, because $\mathbf{1}$ is a free generator of H_* over the respective H^*.

We can now define (3.2) and (3.3) by decomposing f into a product of edge contractions and a final isomorphism and using (3.7), (3.8). Edge contractions commute in an intuitively evident sense, so that checking that the result is independent of the decomposition involves only a careful bookkeeping. We leave this as an exercise to the reader.

§4. $\mathcal{L}$–algebras

In this section we will define several versions of the notion of "algebra over extended modular operad of genus zero". The basic definition we start with is modeled upon the functorial treatment introduced in [KoMa], sec. 6. It involves unoriented painted stable trees and (combinatorial) cohomology algebras, and it produces a version of *cyclic algebras* in the sense of Getzler and Kapranov. Tensor product of algebras appears in the most straightforward way in this construction.

Other versions involve oriented trees and/or homology algebras. The generic name we use for all these versions is *$\mathcal{L}$–algebras.*

Finally, we will explain how the classical operads appear in this context.

4.1. From trees to tensors: unoriented case. Let $\mathcal{A} = \{T; F, (\,,)\}$ be a triple consisting of two projective finite rank $\mathbf{Z}_2$–graded k–modules T, F and a (super)symmetric scalar product $(\,,)$ on F. We assume that $(\,,)$ induces an isomorphism $F \to F^t$ where F^t is the graded dual to F. We denote by $\Delta \in F \otimes F$ the respective Casimir element.

Let τ be a painted stable tree. Denote by W_τ, resp. B_τ, the set of its white, resp. black flags (recall that halves of edges are all white.) If v is a vertex of τ, $W_\tau(v)$, resp. $B_\tau(v)$, denotes the set of white, resp. black, flags incident to this vertex. Put

$$\mathcal{A}(\tau) := T^{\otimes B_\tau} \otimes F^{\otimes W_\tau} = \otimes_{v \in V_\tau} T^{\otimes B_\tau(v)} \otimes F^{\otimes W_\tau(v)}\,. \tag{4.1}$$

Thus, we attach the "white space" F at each white flag, the "black space" T at each black flag, and take the tensor product of all spaces.

Let now $f:\tau \to \sigma$ be a contraction. It identifies F_σ with a subset F^s_τ of non–contracted flags. The complement consist of the halves of contracted edges E^c_τ. We can define a natural map

$$f^*_{\mathcal{A}} : \mathcal{A}(\sigma) \to \mathcal{A}(\tau) \tag{4.2}$$

which tautologically identifies spaces attached to the flags which are paired by f and then tensor multiplies the result by $\Delta^{\otimes E^c_\tau} \in (F \otimes F)^{\otimes E^c_\tau}$.

4.1.1. PROPOSITION. *In this way $\tau \mapsto \mathcal{A}(\tau)$ becomes a contravariant functor on the category of stably painted trees and contractions.*

This is straightforward. From now on we will use $\mathcal{A}$ both as a notation for the triple $\{T; F, (\,,)\}$ and for the functor described above.

4.2. DEFINITION. *The structure of a cyclic $\mathcal{L}$–algebra upon $\mathcal{A} = \{T; F, (\,,)\}$ is a morphism of functors compatible with gluing*

$$I : \mathcal{A} \to H^*. \tag{4.3}$$

In other words, it consists of a family of maps indexed by stably painted trees

$$I(\tau) : T^{\otimes B_\tau} \otimes F^{\otimes W_\tau} \to H^*(\tau) \tag{4.4}$$

such that for any contraction morphism $f : \tau \to \sigma$ we have

$$I(\tau) \circ f_{\mathcal{A}}^* = f^* \circ I(\sigma) : \ T^{\otimes B_\sigma} \otimes F^{\otimes W_\sigma} \to H^*(\tau) \tag{4.5}$$

and moreover

$$I(\tau_1 \bullet \tau_2) = I(\tau_1) \otimes I(\tau_2) : \mathcal{A}(\tau_1) \otimes \mathcal{A}(\tau_2) \to H^*(\tau_1) \otimes H^*(\tau_2) = H^*(\tau_1 \bullet \tau_2). \tag{4.6}$$

Restricting this definition to the case of pure white painted trees, we get the structure on F which was called an operadic tree level Cohomological Field Theory in [KoMa], Def. 6.10. This is the same as an algebra over the cyclic operad $\{H_*(\overline{M}_{0,n+1}, k)\}$. The case of trees with exactly two white flags (and orientation, cf. below) produces a structure which is essentially a representation of a certain algebra H_*T on F studied in [LoMa], sec. 3.3–3.6. Thus, the notion of an $\mathcal{L}$–algebra combines both these structures.

4.2.1. Tensor product of cyclic $\mathcal{L}$–algebras. Let $\mathcal{A} = \{T_1; F_1, (\,,)_1\}$, I_1, resp. $\mathcal{B} = \{T_2; F_2, (\,,)_2\}$, I_2, be two cyclic $\mathcal{L}$–algebras. Put

$$\mathcal{A} \otimes \mathcal{B} := \{T_1 \otimes T_2;\ F_1 \otimes F_2, (\,,)_1 \otimes (\,,)_2\},$$

$$I_{\mathcal{A}\otimes\mathcal{B}}(\tau) := M \circ (I_1 \otimes I_2)(\tau) :$$

where $M : H^*(\tau) \otimes H^*(\tau) \to H^*(\tau)$ is the ring multiplication. Compatibility with (4.5), (4.6) is straightforward.

4.2.2. Economy class descriptions of cyclic algebras. Let us consider only maps (4.4) indexed by one–vertex trees, with a painted set of flags $S = W \cup B$:

$$I_S : T^{\otimes B} \otimes F^{\otimes W} \to H_S^*. \tag{4.7}$$

The axiom (4.7) puts certain restrictions upon this family of maps. Namely, for any painted stable partition $S = S_1 \cup S_2$, $S_i = W_i \cup B_i$, the following diagram must be commutative:

$$\begin{array}{ccc} T^{\otimes B} \otimes F^{\otimes W} & \xrightarrow{I_S} & H_S^* \\ {\scriptstyle S\circ(\otimes\Delta)}\downarrow & & \downarrow{\scriptstyle f^*} \\ T^{\otimes B_1} \otimes F^{\otimes W_1 \cup \{e_1\}} \otimes T^{\otimes B_2} \otimes F^{\otimes W_2 \cup \{e_2\}} & \xrightarrow{I_{S_1\cup\{e_1\}} \otimes I_{S_2\cup\{e_2\}}} & H^*_{S_1\cup\{e_1\}} \otimes H^*_{S_2\cup\{e_2\}} \end{array} \tag{4.8}$$

Here f^* is the map defined by (3.6), (3.7), and S is an obvious reshuffling of factors.

Moreover, any painted isomorphism $S \to S'$ produces isomorphisms of the respective corners of the diagram (4.8), and we require the resulting cubic diagram to be commutative.

One easily sees that, conversely, given I_S satisfying (4.8) and compatible with isomorphisms, we can uniquely reconstruct the structure of cyclic $\mathcal{L}$–algebra on $\mathcal{A} = \{T; F, (,)\}$.

We can further restrict ourselves to the subcategory of painted sets $S_{n,m}$ in which W (resp. B) consists of an initial segment of the natural numbers $\{1, \dots, n\}$ (resp. an initial segment of another copy of the natural numbers $\{\bar{1}, \dots, \bar{m}\}$. Isomorphisms will reduce to the actions of two copies of symmetric groups $\mathbf{S}_n \times \mathbf{S}_m$. The axioms we get in this way are extensions of ones that are stated in [KoMa], Definition 6.1.

4.3. From trees to tensors: oriented case. An *orientation* of a painted stable root tree τ is uniquely determined by a choice of its *root*: an arbitrary white flag r. In the geometric realization, we then orient r away from its vertex, and each other white flag in the direction of r. Put $W_\tau = W_\tau^+ \cup W_\tau^-$ where W_τ^+ consists of flags oriented towards their vertices ("in"), and W_τ^- of flags oriented in reverse direction ("out".) Notice that all black flags are oriented "in".

An oriented contraction of oriented trees is the same as a usual contraction which sends root to root.

To produce tensors associated with oriented trees we start with a pair of modules $\mathcal{A}_0 = \{T, F\}$ as in 4.1, but do not assume a scalar product given. We then denote by $\Delta \in F \otimes F^t$ the image of the identity morphism.

For an oriented tree τ we put

$$\mathcal{A}_0(\tau) := T^{\otimes B_\tau} \otimes (F^t)^{\otimes W_\tau^-} \otimes F^{\otimes W_\tau^+}.$$

Let now $f : \tau \to \sigma$ be an oriented contraction. As in 4.1, we can define a natural map

$$f^*_{\mathcal{A}_0} : \mathcal{A}_0(\sigma) \to \mathcal{A}_0(\tau) \tag{4.9}$$

which tautologically identifies spaces attached to the flags which are paired by f and then tensor multiplies the result by $\Delta^{\otimes E_\tau^c} \in (F \otimes F^t)^{\otimes E_\tau^c}$.

4.3.1. PROPOSITION. *In this way $\tau \mapsto \mathcal{A}_0(\tau)$ becomes a contravariant functor on the category of oriented stably painted trees and contractions.*

4.4. DEFINITION. *The structure of an $\mathcal{L}$–algebra upon $\mathcal{A}_0 = \{T, F\}$ is a morphism of functors compatible with gluing*

$$I : \mathcal{A}_0 \to H^*. \tag{4.10}$$

4.5. $\mathcal{L}$–algebras in terms of combinatorial homology and correlators. We will now introduce one more version of $\mathcal{L}$–algebras involving homology and unoriented trees. It is this version which is most convenient for the passage to the differential equations.

For a change, we will start with an economy class description. We fix $\mathcal{A}_0 = \{T, F\}$ as in 4.3.

4.5.1. Data. Consider one-vertex oriented stable painted trees. Such a tree is the same as a finite set of its flags represented in the form $S = (W \cup \{0\}) \cup B$, $|W \cup B| \geq 2$, where 0 is the root, $W \cup \{0\}$ (resp. B) is the subset of white (resp. black) flags. For each such S we assume given an even homomorphism of k–modules (correlator)

$$C_S : H_{*S} \otimes T^{\otimes B} \otimes F^{\otimes W} \to F. \tag{4.11}$$

This should be compared to (4.7): to pass from (4.7) to (4.11), we make a partial dualization: cohomology becomes homology and moves to the left hand side, whereas the factor F^t corresponding to the root becomes F and moves to the right hand side.

Once again, we can restrict ourselves by the sets $A = \{0, 1, \dots, n\}, B = \{\overline{1}, \dots, \overline{m}\}$ taken from two disjoint copies of the set of nonnegative integers.

Since H_{*S} is spanned by $[\mu(\tau)]$ where τ runs over unoriented stable painted S–trees, to give (4.11) is the same as to give a family of maps

$$C_S([\mu(\tau)]) : T^{\otimes B} \otimes F^{\otimes W} \to F. \tag{4.12}$$

Notice again that the orientation is encoded by S which is given together with the choice of (a label of the) root.

The intuitive meaning of (4.12) is this: given a rooted tree τ, we consider it as the flowchart of a computation. Concretely, an input assigns an element f_a of F (resp. t_b of T) to each white non–root tail a (resp. to each black tail b). Then the output is an element

$$C_S([\mu(\tau)])(\otimes_b t_b \otimes_a f_a) \in F. \tag{4.13}$$

We associate this output with the root of τ.

4.5.2. Axioms. (A) *Operadic relations.* If we graft two (or more) oriented trees by gluing root(s) to white tail(s), the resulting tree must be the flowchart for the composition of the respective functions.

Notice that since black tails are never glued, elements of T may be considered as "parameters", whereas the actual arguments (and values) of the correlators belong to F.

(B) *Symmetry relations.* They say that the product $\mathbf{S}_W \times \mathbf{S}_B$ of the permutation groups of W and B acting upon (4.11) in an evident way leaves these maps invariant. Notice that the action on H_{*S} is also involved in this requirement.

(C) *Linear relations.* They assure us that all linear relations between $[\mu(\tau)]$ in H_{*S} imply the respective relations between correlators. We can restrict ourselves by the basic relations (2.1), or their cohomological version (1.10). Namely, starting with an S–tree τ, its vertex v, and an allowed quadruple of flags I, J, K, L at this vertex, as in 1.4, we must have

$$\sum_{\alpha} \epsilon(\alpha; I, J, K, L)\, C_S([\mu(\tau(\alpha))]) = 0 \tag{4.14}$$

where we sum over all 2–partitions of the set of flags at v, and $\varepsilon = 1, -1$, or 0, depending on whether $IJ\alpha KL$, $KJ\alpha IL$ or none of it.

4.6. Top correlators. Let τ be an one vertex tree whose tails are labeled by $(W \cup \{0\}) \cup B$. For $W = \{1, \dots, n+1\}$, $B = \{\overline{1}, \dots, \overline{m}\}$, we will write $c(t_1, \dots, t_m, f_1, \dots, f_{n+1})$ in place of (4.13) and call these elements *the top correlators.* They are symmetric in T– and F–arguments separately. We will formally extend them to polylinear functions symmetric in all variables. All the operadic relations follow by iteration from this symmetry and from the following ones:

$$\begin{aligned} c(t_1, \dots, t_m, f_1, f_2, \dots, f_n, c(t'_1, \dots, t'_q; f'_1, \dots, f'_{p+1})) = \\ c(t_1, \dots, t_m, f_1, f_2, \dots, f_n, t'_1, \dots, t'_q, f'_1, \dots, f'_{p+1}) \end{aligned} \tag{4.15}$$

Now fixing all the arguments except for f_{n+1}, consider $c(t_1, \dots, t_m, f_1, \dots, f_{n+1})$ as a function of f_{n+1}, that is, an element of $\operatorname{End} F$ (even or odd, depending on the sum of parities of the fixed arguments), and denote it by

$$\langle t_1, \dots, t_m, f_1, \dots, f_n \rangle \in \operatorname{End} F.$$

Again, we will allow arbitrary permutations of arguments in this expression, followed by the standard sign change, and call the resulting expressions *top matrix correlators.* Then the grafting relations (4.15) turn into the simple *factorization relations* in End F:

$$\langle t_1, \ldots, t_m, f_1, f_2, \ldots, f_n, {t'}_1, \ldots, {t'}_q, {f'}_1, \ldots, {f'}_p\rangle =$$
$$\langle t_1, \ldots, t_m, f_1, f_2, \ldots, f_n\rangle\langle {t'}_1, \ldots, {t'}_q, {f'}_1, \ldots, {f'}_p\rangle. \tag{4.16}$$

We can now formulate the first result describing $\mathcal{L}$–algebras in terms of their top matrix correlators. In order to make connection with results of [LoMa], we introduce the following notation. Choose bases of F, T and let their union, which is a basis of $T \oplus F$, be $\{\Delta_i\}$, for some index set I. Fix a structure of the (oriented, homological) $\mathcal{L}$–algebra upon (T, F) and consider its top matrix correlators for the elements of the chosen basis.

4.6.1. THEOREM. *(i) Top matrix correlators*

$$\langle \Delta_{a_1} \ldots \Delta_{a_n}\rangle \in \operatorname{End} F, \quad n \geq 2,$$

are (super)symmetric in their arguments. Furthermore, for any allowed quadruple $i, j, k, l \in I$

$$\sum_{\sigma:\, ij\sigma kl} \varepsilon(\sigma, (a_m))\langle \prod_{m\in\sigma_1} \Delta_{a_m}\rangle \cdot \langle \prod_{m\in\sigma_2} \Delta_{a_m}\rangle$$
$$- \sum_{\sigma:\, kj\sigma il} \varepsilon(\sigma, (a_m))\langle \prod_{m\in\sigma_1} \Delta_{a_m}\rangle \cdot \langle \prod_{m\in\sigma_2} \Delta_{a_m}\rangle = 0\,. \tag{4.17}$$

Here σ runs over ordered 2–partitions of $\{1, \ldots, n\}$. We choose additionally an arbitrary ordering of both parts σ_1, σ_2 determining the ordering of Δ's in the angular brackets, and compensate this choice by the ± 1–factor $\varepsilon(\sigma, (a_k))$.

(ii) Conversely, any family of elements $\langle \Delta_{a_1} \ldots \Delta_{a_n}\rangle \in \operatorname{End} F$ defined for all $n \geq 2$ and $(a_1, \ldots, a_n) \in I^n$ and satisfying the symmetry conditions and (4.17), is the family of top matrix correlators of a unique structure of $\mathcal{L}$–algebra on (T, F).

Proof. (i) Both symmetry properties follow from the definitions, whereas (4.17) follows from (4.14), written for one–vertex trees τ and transformed with the help of (4.16).

(ii) If we know all top matrix correlators, we can recursively determine maps (4.13) for all stable trees τ with $e \geq 1$ edges. In fact, choose an end edge of τ and cut it, thus representing τ as the result of grafting an one–vertex tree ρ to a tree σ with $e - 1$ edges. Choose the flag a of σ that gets grafted as special input and, using partial dualization, rewrite (4.13) as a map $T^{\otimes B} \otimes F^{\otimes W} \to \operatorname{End} F$, that is, a matrix correlator related to the tree σ. From the grafting axiom (A) above it follows that (4.13) is uniquely calculated from the latter matrix correlator and the top matrix correlator for ρ, via a generalization of the factorization relations (4.16).

This shows that the $\mathcal{L}$–structure on (T, F) with given top matrix correlators is unique, if it exists at all. In order to prove existence, we must check that all linear relations (4.14) follow from their special cases (4.17). The argument is essentially the same as in the proof on p. 462 of [LoMa]; the difference is that in [LoMa] the relevant products of top correlators were linear ordered, whereas here they are controlled by a tree, so a more careful bookkeeping is required. We leave this as an exercise.

§5. Differential geometry and $\mathcal{L}$–algebras

5.1. Notation. Consider a structure of $\mathcal{L}$–algebra upon (T, F) determined by its top matrix correlators $\langle \Delta_{a_1} \dots \Delta_{a_n} \rangle$ as in Theorem 4.6.1 (i). Denote by (x^a) the linear coordinate system on $T \oplus F$ dual to (Δ_a). Put

$$\mathcal{B} = \sum_{n=1}^{\infty} \sum_{(a_1,\dots,a_n)} \frac{x^{a_n} \dots x^{a_1}}{n!} \langle \Delta_{a_1} \dots \Delta_{a_n} \rangle \in k[[x]] \otimes \operatorname{End} F. \qquad (5.1)$$

5.1.1. THEOREM. *a) We have*

$$d\mathcal{B} \wedge d\mathcal{B} = 0. \qquad (5.2)$$

b) Conversely, let $\mathcal{B} \in k[[x]] \otimes \operatorname{End} F$ be any even formal series without constant term satisfying the equation (5.2). Define its symmetrized coefficients $\Delta(a_1, \dots, a_n)$, $a_i \in I$, by the following conditions: first,

$$\mathcal{B} = \sum_{n=1}^{\infty} \sum_{(a_1,\dots,a_n)} \frac{x^{a_n} \dots x^{a_1}}{n!} \Delta(a_1, \dots, a_n),$$

second, the parity of $\Delta(a_1, \dots, a_n)$ coincides with that of $x^{a_n} \dots x^{a_1}$, and third,

$$\Delta(a_{s(1)}, \dots, a_{s(n)}) = \varepsilon(s, (a_i))\, \Delta(a_1, \dots, a_n)$$

for any permutation s of $\{1, \dots, n\}$.

Then there exists a unique structure of $\mathcal{L}$–algebra upon (T, F) with top correlators

$$\langle \Delta_{a_1} \dots \Delta_{a_n} \rangle = \Delta(a_1, \dots, a_n).$$

Theorem 4.6.1 shows that this is a particular case of the Proposition 3.6.1 of [LoMa]. Moreover, the discussion in 3.2 of [LoMa] establishes a bijection between the even formal solutions to the equation (5.2) and pencils of formal flat connections $\nabla_0 + \lambda \mathcal{A}$ on the trivial vector bundle with fiber F on the formal completion of $T \oplus F$ at zero.

In the remaining part of this section we will establish certain properties of solutions to (5.2) and related structures. As in 3.1–3.2 of [LoMa], we will work in a more general setting, allowing not necessarily formal manifolds as our base spaces.

5.2. Commutativity equations. Generally, let M be a (super)manifold in one of the standard categories (C^∞, analytic, formal ...). Denote by F a local system of finite–dimensional vector (super)spaces on M. This is essentially the same as a locally free sheaf $\mathcal{S}$ endowed with a flat connection $\nabla_0 : \mathcal{S} \to \Omega^1_M \otimes_{\mathcal{O}_M} \mathcal{S}$: from $(\mathcal{S}, \nabla_0)$ one gets $F := \operatorname{Ker} \nabla_0$, and from F one gets $\mathcal{S} := \mathcal{O}_M \otimes F$ (tensor product over constants), $\nabla_0(t \otimes s) = dt \otimes_{\mathcal{O}_M} s$ for local sections $t \in \mathcal{O}_M$, $s \in F$.

Here d is the de Rham differential, which extends in the standard way to Ω^*_M, whereas ∇_0 extends to the whole tensor algebra of S.

In particular, we have the induced flat connection denoted by the same letter ∇_0 upon $End\, S = \mathcal{O}_M \otimes End\, F$.

An even section B of $\mathcal{O}_M \otimes End\, F$ is called *a solution to the commutativity equations for (M, F)* if

$$\nabla_0 B \wedge \nabla_0 B = 0. \tag{5.3}$$

Our sign conventions are determined by postulating that d and ∇_0 are odd. We traditionally denote by $\wedge$ the multiplication in Ω^*_M and in the tensor products of Ω^*_M with other sheaves of algebras, although in the supergeometry this is slightly misleading: multiplication in Ω^*_M is supercommutative, not superalternate.

If we choose local coordinates (t^i) in M and a basis of flat (belonging to F) local sections of $\mathcal{S}$, B becomes a matrix function $t \mapsto B(t)$, and $\nabla_0 B$ becomes a matrix of 1–forms dB on M: $(\nabla_0 B)^j_i = dB^j_i$ so that (5.2) is a particular case of (5.3).

Putting $dB = \sum_i dt^i b_i(t)$ where $t = (t^i)$, so that

$$b_i(t) := \frac{\partial B}{\partial t^i} \in \mathcal{O}_M \otimes \operatorname{End} F$$

we can easily check that (5.3) can be written in the form of commutativity equations

$$\forall\, i, j,\ [b_i(t), b_j(t)] = 0 \tag{5.4}$$

where the brackets denote the supercommutator. Thus b_i span over $\mathcal{O}_M$ a sheaf of (super)abelian Lie subalgebras of $\mathcal{O}_M \otimes \operatorname{End} F$. This sheaf is intrinsically associated with B; we will denote it DB.

5.2.1. DEFINITION. *(i) A solution B to the commutativity equations as above is called maximal one, if DB is a maximal subsheaf of (super)abelian Lie subalgebras of $\mathcal{O}_M \otimes \operatorname{End} F$ in the following sense: any local section $c(t) \in \mathcal{O}_M \otimes \operatorname{End} F$ (super)commuting with DB belongs to DB.*

(ii) A maximal solution is called strictly maximal if DB is locally freely generated by $b_i(t)$.

Notice that if T is the formal completions of a linear space at zero, maximality of DB is equivalent to the maximality at zero.

5.2.2. Pullback. Any morphism $\varphi : M' \to M$ and any solution B to the commutativity equations for (M, F) produces the pullback solution $\varphi^*(B)$ for $(M', \varphi^*(F))$.. Of course, if M and M' are formal completions of linear spaces at zero, φ is generally given by formal series with vanishing constant terms. If φ is a closed embedding, we call $\varphi^*(B)$ the restriction of B, and conversely, we call B a continuation of $\varphi^*(B)$ (both with respect to φ).

In particular, the automorphism group of (M, F) acts upon the space of solutions. In the formal case, this is the group of formal invertible coordinate changes. Hence the linear structure of M plays no role, and we can simply speak about formal manifolds.

5.2.3. PROPOSITION. *Let B be a solution to the commutativity equations over a manifold M, and B' its continuation to M' with respect to a closed embedding $\iota : M \to M'$. Denote by $\widehat{M'}$ the formal completion of M' along M, and by $\widehat{B}'$ the solution induced by B' on it.*

(i) If B is maximal, everywhere locally over M there exists a formal projection $\varphi : \widehat{M'} \to M$, $\varphi \circ \iota = id_M$, such that $\widehat{B}' = \varphi^*(B)$.

(ii) If B is strictly maximal, these local formal projections are unique and hence glue together to a global formal projection $\hat{M}' \to M$ with the same property.

Proof. Working in a local chart, choose local coordinates

$$(t, \theta) = (t^1, \dots, t^m; \theta^1, \dots, \theta^n)$$

on M' such that M in M' is given by the equations $\theta = 0$. Choosing also a basis of sections of F, consider B and B' as matrix functions. Denote by $\theta^{(k)}$ (super)symmetric monomials in θ where (k) are polydegrees. We have

$$dB \wedge dB = 0, \; dB' \wedge dB' = 0, \; B'(t, 0) = B(t). \tag{5.5}$$

Let $|k|$ be the total degree (sum of coordinates) of the polydegree (k). To prove the statement (i), we have to find a family of local functions $(\lambda^i_{(k)}(t))$, $i = 1, \dots, m$, $|k| \geq 1$, on M, such that putting

$$B'_N(t, \theta) := B(t + \sum_{|k| \leq N} \lambda_{(k)}(t) \theta^{(k)}), \tag{5.6}$$

we have

$$B'_N(t, \theta) \equiv B'(t, \theta) \bmod \theta^{\geq N+1}, \tag{5.7}$$

where $\theta^{\geq N+1}$ is the ideal generated by $\theta^{(k)}$ with $|k| \geq N+1$. In fact, if we find such a family, the local formal projection φ we look for is given by $\varphi^*(t^i) = t^i + \sum_{(k)} \lambda^i_{(k)}(t) \theta^{(k)}$. Uniqueness of this family means uniqueness of the formal projection.

Obviously, $B'_0(t, \theta) = B(t)$ is the only possible choice. In order to find $B'_1(t, \theta) = B(t + \sum_{j=1}^n \lambda_j(t) \theta^j)$ satisfying (5.7) notice that

$$B(t + \sum_{j=1}^n \lambda_j(t) \theta^j) \equiv B(t) + \sum_{i=1}^m \sum_{j=1}^n \lambda^i_j(t) \theta^j b_i(t) \bmod \theta^{\geq 2}$$

whereas

$$B'(t, \theta) \equiv B(t) + \sum_{j=1}^n \frac{\partial B'}{\partial \theta^j} \Big|_{\theta=0} \theta^j \bmod \theta^{\geq 2}.$$

Hence $\lambda^i_j(t)$ exist iff all $\dfrac{\partial B'}{\partial \theta^j} \Big|_{\theta=0}$ belong to DB that is, (super)commute with $b_i(t)$. But this follows from $dB' \wedge dB' = 0$ evaluated at $\dfrac{\partial}{\partial t^i} \wedge \dfrac{\partial}{\partial \theta^j}$. Moreover, $\lambda^i_j(t)$ are unique, if DB is freely generated by $(b_i(t))$.

This reasoning can be generalized to provide an inductive step from N to $N+1$.

Namely, assume that we already found $B'_N(t, \theta)$ of the form (5.6) satisfying (5.7), $N \geq 1$. We need a vector $X_{N+1} = (X^j_{N+1})$, $j = 1, \dots, n$, whose coordinates are forms of degree $N+1$ in θ with coefficients depending on t such that

$$B'_{N+1}(t, \theta) := B(t + \sum_{|k| \leq N} \lambda_{(k)}(t) \theta^{(k)} + X_{N+1}) \equiv$$

$$\equiv B'_N(t, \theta) + \sum_j X^j_{N+1} b_j(t + \sum_{|k| \leq N} \lambda_{(k)}(t) \theta^{(k)}) \bmod \theta^{\geq N+2}$$

$$\equiv B'_N(t,\theta) + \sum_j X^j_{N+1} b_j(t) \bmod \theta^{\geq N+2}$$

satisfies (5.7) with N replaced by $N+1$. Let

$$B'(t,\theta) \equiv B'_N(t,\theta) + Y_{N+1} \bmod \theta^{\geq N+2}$$

where Y_{N+1} is a form of degree $N+1$ in θ with coefficients depending on t. In order to establish the existence of X_{N+1} it suffices to check as above that the coefficients of Y_{N+1} commute with $b_j(t)$. In fact, since $dB' \wedge dB' = 0$, we have

$$d(B'_N + Y_{N+1}) \wedge d(B'_N + Y_{N+1}) \equiv 0 \bmod \theta^{\geq N+1}$$

and since moreover $dB'_N \wedge dB'_N = 0$ and $N \geq 1$, it follows that

$$dY_{N+1} \wedge dB'_N + dB'_N \wedge dY_{N+1} \equiv 0 \bmod \theta^{\geq N+1}.$$

Hence the coefficients of all derivatives $\dfrac{\partial}{\partial \theta^j} Y_{N+1}$ belong to DB. From Euler's formula it follows that the same holds for the coefficients of Y_{N+1}. Uniqueness in the case of strict maximality follows for this inductive step as well.

5.2.4. Primitive vectors. We keep the notation described above. Consider fibers of F as (super)manifolds endowed with a linear structure. We can cover M by open submanifolds such that over any chart U we have a canonical trivialization $F|_U = F_0 \times U$, where F_0 is a fiber, or else the space of sections of F over U.

An even section $h \in \Gamma(M,F)$ is called *primitive (for the solution B)* if the map of (super)manifolds $U \to F_0 : t \mapsto B(t)h$ is a local isomorphism everywhere on M. An evident necessary condition for the existence of a primitive vector is the coincidence of (super)dimensions of M and fibers of F.

5.3. Oriented associativity equations. Consider now a (super)manifold M endowed with an affine flat structure.

By definition, such a structure is given by the subsheaf $\mathcal{T}^f_M \subset \mathcal{T}_M$ of flat vector fields which form a local system of linear spaces *and abelian Lie superalgebras* (with respect to the supercommutator of vector fields) such that $\mathcal{T}_M = \mathcal{O}_M \otimes \mathcal{T}^f_M$. Let $\nabla_0 : \mathcal{T}_M \to \Omega^1_M \otimes_{\mathcal{O}_M} \mathcal{T}_M$ be the associated connection.

Consider a vector field A on M. Its covariant differential $\nabla_0 A$ belongs to $\Omega^1_M \otimes_{\mathcal{O}_M} \mathcal{T}_M$. We have the standard isomorphism

$$j : \Omega^1_M \otimes_{\mathcal{O}_M} \mathcal{T}_M \to End_{\mathcal{O}_M}(\mathcal{T}_M).$$

Put $B := j(\nabla_0 A)$. We are now in a position to write the commutativity equations for $(M, \mathcal{T}^f_M)$ and B.

A concrete way of fixing a flat structure consists in giving an atlas with affine linear transition functions between local coordinates of its charts. Such local coordinates (t^i) are then called flat, and $\mathcal{T}^f_M$ is locally generated by the dual vectors $\partial_i = \partial/\partial t^i$. When we write the field $B = B(t)$ as a matrix, we always use (∂_i) as a basis of local sections.

5.3.1. DEFINITION. *(i) A solution to the oriented associativity equations is an even vector field A on M with the following property. Write $A = \sum_c A^c\partial_c$ in a local flat coordinate system and put $A_{ab}{}^c := \partial_a\partial_b A^c$. Then the composition law*

$$\partial_a \circ \partial_b = \sum_c A_{ab}{}^c \partial_c \tag{5.8}$$

extends to the associative (super)commutative $\mathcal{O}_M$-bilinear multiplication on $\mathcal{T}_M$.

(ii) An even flat vector field ∂_0 (contained among ∂_i) is called a flat identity (for A), if $A_{a0}{}^c = \delta_a^c$, or in other words, if $e \circ X = X$ for each vector field X.

5.3.2. PROPOSITION. *Let A be a solution to the oriented associativity equations with flat identity e. Define $B := j(\nabla_0 A)$ as above and put $h = e$.*

Then B is a solution to the commutativity equations with primitive vector h.

The map $t \mapsto B(t)e$ establishes a local embedding of M into a fiber of F such that the tautological flat affine structure on F induces the initial affine structure on M.

Proof. In the notations of 5.3.1, we have

$$\nabla_0 A = \sum_{b,c} dt^b \partial_b A^c \otimes \partial_c =$$

$$= \sum_{b,c} (-1)^{(|t_b|+1)(|t_b|+|t_c|)} \partial_b A^c dt^b \otimes \partial_c.$$

where $|t^a|$ denotes the parity of t^a. The last line allows us to compute $j(\nabla_0 A)$ as a matrix B: $j(dt^b \otimes \partial_c)$ considered as endomorphism of $\mathcal{T}_M^f$ maps ∂_a to $\delta_a^b \partial_c$ so that

$$B_b{}^c = (-1)^{(|t_b|+1)(|t_b|+|t_c|)} \partial_b A^c. \tag{5.9}$$

Then a direct calculation shows that the equation $dB \wedge dB = 0$ becomes

$$\forall\, a, b, c, f, \qquad \sum_e A_{ab}{}^e A_{ec}^f = (-1)^{|t^a|(|t^b|+|t^c|)} \sum_e A_{bc}{}^e A_{ea}{}^f.$$

These are precisely the associativity equations: cf. e. g. [Ma1], pp. 19–20. Since this formulation gives a coordinate free description of the operation $\circ$, it does not depend on the choice of local flat coordinates and agrees on intersections. Supercommutativity of $\circ$ follows from the fact that $\mathcal{T}_M^f$ is a superabelian Lie algebra so that $A_{ab}{}^c$ is symmetric with respect to its subscripts.

Let now $t = (t^0, t^1, \dots, t^m)$ be a flat local coordinate system such that $e = \partial_0$. We have $\partial_a(A_0^c) = \delta_a^c$, hence $A_0^c = t^c + a^c$ where a^c are constants. Therefore

$$B(t)e = \sum_c (-1)^{|t^c|}(t^c + a^c)\partial_c. \tag{5.10}$$

This proves the last statement of the Proposition.

The converse statement also holds, at least locally. The point is that for a given B, if the equation $B = j(\nabla_0 A))$ can be solved for A at all, then we can get other solutions by adding to A any flat vector field, so there may be an obstruction for finding a global solution.

5.3.3. PROPOSITION. *Let B be a solution to the commutativity equations for (M, F), and let h be a primitive vector for B. Working locally, induce a flat affine structure on M from a fibre F_0 with the help of the map $t \mapsto B(t)h$.*

Then in any local flat coordinate system (t^c) we can define a vector field $\sum_c A^c \partial_c$ such that

$$\partial_b A^c = (-1)^{(|t^b|+1)(|t^b|+|t^c|)} B_b{}^c.$$

(cf. (5.9)).

This field is a solution to the oriented associativity equations with flat identity e which is the pullback of h (considered as a tangent vector to a fiber of F) with respect to the map $t \mapsto B(t)h$.

Proof. If we prove the existence of functions A^c, the rest will follow from the proof of the previous Proposition. The equations for A^c are equivalent to

$$dA^c = \sum_b (-1)^{(|t^b|+1)(|t^b|+|t^c|)} dt^b B_b{}^c$$

and their integrability is equivalent to the closedness of all forms

$$\omega^c := (-1)^{(|t^b|+1)(|t^b|+|t^c|)} \sum_b dt^b B_b{}^c.$$

We will show that the latter in appropriate coordinates is expressed by the equations $dB \wedge d(Bh) = 0$ which follow from $dB \wedge dB = 0$ because h is flat.

In fact, choose a basis of flat vector vector fields of F_0 containing h and the dual coordinate system (t^c) such that $h = \partial_0$. Identifying M (or its local chart) with a subdomain of F via $t \mapsto B(t)h$ we can consider (t^c) as flat coordinates on M, and in these coordinates the map $t \mapsto B(t)h$ is given by the formula (5.10). Hence

$$d(B(t)h) = \sum dt^b \otimes \partial_b$$

and

$$dB \wedge d(B(h)) = \sum_c \left[\sum_b (dB)_b{}^c \wedge dt^b \right] \otimes \partial_c = \sum_c (-1)^{|t^c|} d\omega^c \otimes \partial_c$$

This completes the proof.

5.4. Formal solutions to the commutativity equations corresponding to $\mathcal{L}$–algebras. Let us now return to the situation described in the Theorem 5.1.1. The formal series $\mathcal{B}$ is a solution to the commutativity equations for the formal manifold M which is the completion of $T \oplus F$ at zero, and trivial local system with the fibre F. Denote by $\mathcal{T}$ (resp. $\mathcal{F}$) formal completions of T (resp. F) at zero. They are embedded as closed formal submanifolds in M.

We can restrict $\mathcal{B}$ to $\mathcal{T}$. Assume that this restriction $\mathcal{B}_T$ is maximal. Then in view of the Proposition 5.2.3, $\mathcal{B}$ is a pullback of $\mathcal{B}_T$ with respect to a formal projection $\varphi : M \to \mathcal{T}$.

We can also restrict $\mathcal{B}$ to $\mathcal{F}$. Clearly, $\mathcal{F}$ is endowed with a formal flat structure coming from F. The restriction $\mathcal{B}_F$ of $\mathcal{B}$ to $\mathcal{F}$ produces a solution to the oriented associativity equations with base $\mathcal{F}$. If the latter admits a primitive vector h, we will say that h is weakly primitive for $\mathcal{B}$.

5.4.1. PROPOSITION. *Fix two linear superspaces T and F and the following additional data:*

(i) A maximal solution $\mathcal{B}_1$ to the commutativity equations with the base $\mathcal{T}$ and fiber F.

(ii) A solution $\mathcal{B}_2$ to the commutativity equations with the base $\mathcal{F}$ and fiber F which comes from a solution to the oriented associativity equations on $\mathcal{F}$ with flat structure induced by F, admitting a flat identity e.

In this case there exists a pair $(\mathcal{B}, h)$, where $\mathcal{B}$ is a solution to the commutativity equations on the formal completion of $T \oplus F$ at zero, with the fiber F, and h is a weakly primitive vector corresponding to e, such that $\mathcal{B}_1 = \mathcal{B}_T$, $\mathcal{B}_2 = \mathcal{B}_F$. This pair is unique.

Proof. Let (t) denote some coordinates on $\mathcal{T}$ and (θ) flat coordinates on $\mathcal{F}$. The explicit pullback formula

$$\mathcal{B}(t, \theta) = \mathcal{B}_2(\theta + \mathcal{B}_1(t)h)$$

produces a solution with necessary properties. It is unique because of maximality of $\mathcal{B}_1$.

5.5. Compatibility of two tensor products. Given two structures I_i of (oriented) $\mathcal{L}$–algebras upon (T_i, F_i), $i = 1, 2$, we can form the tensor product structure $I = I_1 \otimes I_2$ upon $(T, F) = (T_1 \otimes T_2, F_1 \otimes F_2)$ imitating the definition 4.2.1 in the nonoriented (cyclic) case. This operation induces the tensor product $*_{COMM}$ on the formal solutions to the commutativity equations (COMM).

There is also an oriented version $*_{ASS}$ of the tensor product of the formal solutions to the associativity equations ASS (for the cyclic case, see [Ma1], p. 100).

Considering only solutions to ASS with flat identities, and the associated solutions to COMM, we may conjecture that the two operations are compatible. This question was raised in [LoPo] where it was checked that the answer is positive in several first orders.

Proposition 5.4.1 suggests a strategy for proving this conjecture at least for the case when the solutions to COMM are maximal. However, the most natural and general approach is furnished by the equivalence theorem 5.1.1 and the language of $\mathcal{L}$–algebras where the tensor product is simply induced by the product in the cohomology of moduli spaces, resp. coproduct in their homology. We hope to spell out the details elsewhere.

References

[BeMa] K. Behrend, Yu. Manin. *Stacks of stable maps and Gromov–Witten invariants.* Duke Math. Journ., 85:1 (1996), 1–60.

[GeK1] E. Getzler, M. Kapranov. *Cyclic operads and cyclic homology.* In: Geometry, Topology, and Physics for Raoul, ed. by B. Mazur, Internat. Press, Cambridge, MA, 1995, 167–201.

[GeK2] E. Getzler, M. Kapranov. *Modular operads.* Compositio Math., 110 (1998), 65–126.

[H] B. Hassett. *Moduli spaces of weighted pointed stable curves.* Preprint math.AG/0205009

[Ke] S. Keel. *Intersection theory of moduli space of stable N-pointed curves of genus zero.* Trans. AMS, 330:2 (1992), 545–574.

[Kn] Knudsen F.F. *The projectivity of the moduli space of stable curves II. The stacks $\overline{M}_{0,n}$.* Math. Scand. 52 (1983), 163–199.

[KoMa] M. Kontsevich, Yu. Manin. *Gromov–Witten classes, quantum cohomology, and enumerative geometry.* Comm. Math. Phys., 164:3 (1994), 525–562.

[KoMaK] M. Kontsevich, Yu. Manin (with Appendix by R. Kaufmann). *Quantum cohomology of a product.* Inv. Math., 124, f. 1–3 (Remmert's Festschrift) (1996), 313–339.

[Lo1] A.Losev, *On "Hodge" Topological Strings at genus zero.* Pis'ma v ZhETF 65, 374-379 (1997).

[Lo2] A.Losev, *"Hodge strings" and elements of K. Saito's theory of the primitive form.* In: Proceedings of Taniguchi Symposium on Topological Field Theory, Primitive Forms and Related Topics, Kyoto, Japan, 9-13 Dec 1996, Springer,1998. Preprint hep-th/9801179

[LoMa] A. Losev, Yu. Manin. *New moduli spaces of pointed curves and pencils of flat connections.* Michigan Journ. of Math., vol. 48 (Fulton's Festschrift), 2000, 443–472. Preprint math.AG/0001003

[LoPo] A. Losev, I. Polyubin. *On compatibility of tensor products on solutions to commutativity and WDVV equations.* JETP Letters, 73:2 (2001), 53–58.

[Ma1] Yu. Manin. *Frobenius manifolds, quantum cohomology, and moduli spaces.* AMS Colloquium Publications, vol. 47, Providence, RI, 1999, xiii+303 pp.

[Ma2] Yu. Manin. *Moduli stacks $\overline{L}_{g,S}$.* Preprint math.AG/0206123

[R] M. A. Readdy. *The Yuri Manin ring and its $\mathcal{B}_n$-analogue.* Adv. in Appl. Math., 2001.

INSTITUTE OF THEORETICAL AND EXPERIMENTAL PHYSICS, MOSCOW, RUSSIA

MAX-PLANCK-INSTITUT FÜR MATHEMATIK, BONN, GERMANY AND NORTHWESTERN UNIVERSITY, EVANSTON, USA

Operads, deformation theory and F-manifolds

S.A. Merkulov

§0. Introduction

0.1. Little disks operad and Hertling-Manin's F-manifolds. Frobenius manifolds created by Dubrovin in 1991 from rich theoretical physics material have been found since in many different fragments of mathematics — quantum cohomology and mirror symmetry, complex geometry, symplectic geometry, singularity theory, integrable systems — raising hopes for unifying them into one picture. It also became clear that the notion of Frobenius manifold is not broad enough to cover *all* objects of the associated working categories; say, on the B-side of the mirror symmetry it applies only to extended moduli spaces of *Calabi-Yau* manifolds, the latter forming a rather small subcategory of the category of complex manifolds. In 1998 Hertling and Manin [**HeMa**] introduced a weaker notion of *F-manifold* which is, by definition, a pair (M, μ_2) consisting of a smooth supermanifold M and a smooth $\mathcal{O}_M$-linear associative graded commutative multiplication on the tangent sheaf, $\mu_2 : \otimes^2_{\mathcal{O}_M} \mathcal{T}_M \to \mathcal{T}_M$, satisfying the integrability condition,

$$[\mu_2, \mu_2] = 0,$$

where $[\mu_2, \mu_2] : \otimes^4_{\mathcal{O}_M} \mathcal{T}_M \to \mathcal{T}_M$ is given explicitly by

$$\begin{aligned}
[\mu_2, \mu_2](X, Y, Z, W) \;:=\;& [\mu_2(X, Y), \mu_2(Z, W)] - \mu_2([\mu_2(X, Y), Z], W) \\
& -(-1)^{(|X|+|Y|)|Z|}\mu_2(Z, [\mu_2(X, Y), W]) \\
& -\mu_2(X, [Y, \mu_2(Z, W)]) - (-1)^{|Y|(|Z|+|W|)}\mu_2[X, \mu_2(Z, W)], Y) \\
& +(-1)^{|Y||Z|}\mu_2(X, \mu_2(Z, [Y, W])) + \mu_2(X, \mu_2([Y, Z], W)) \\
& +(-1)^{|Y||Z|}\mu_2([X, Z], \mu_2(Y, W)) \\
& +(-1)^{|W|(|Y|+|Z|)}\mu_2([X, W], \mu_2(Y, Z)).
\end{aligned}$$

A non-trivial part of the above definition is an implicit assertion that $[\mu_2, \mu_2]$ is a tensor, i.e. $\mathcal{O}_M$-polylinear in all four inputs. It is here where the assumption that μ_2 is both graded commutative an associative plays a key role.

Any Frobenius manifold is an F-manifold. Any F-manifold with semi-simple product μ_2 can be made into a Frobenius manifold [**HeMa**]. Hertling in his book [**He**] explained in detail how F-manifolds turn up in the singularity theory.

In this paper we show that (cohomology/strong homotopy, see below) F-manifolds arise naturally in every mathematical structure which admits an action of the chain operad of the little disks operad (or its more compact version, $\mathcal{G}_\infty$-operad [**GetJo**]). In particular, we prove

Theorem A. *Let* A *be either a complex or symplectic structure on a compact manifold. Then the smooth part,* M_{reg}*, of the extended moduli space* M *of deformations of* A *is canonically an* F*-manifold.*

0.2. Cohomology F-manifolds. It is often not a pleasure to work with objects like $M_{\text{reg}} \subset M$ in Theorem A; moreover, their existence is not guaranteed for many reasonable deformation problems.

The germ, $(M, *)$, of, in general, singular moduli space M (if it exists at all) at the distinguished point $*$ always admits a smooth dg resolution, $(\mathcal{M}, *, \eth)$ [**Me2**]. The latter consists of a germ of a *smooth* graded pointed manifold, $(\mathcal{M}, *)$, and a germ of a *smooth* degree 1 vector field, $\eth$, satisfying two conditions,

$$[\eth, \eth] = 0, \quad \text{and } \eth I \subset I^2,$$

where I is the ideal of the distinguished point. The relation between $(M, *)$ (which may not exist as an analytic space) and $(\mathcal{M}, *, \eth)$ (which always exists for any deformation problem!) is given by the well known formula of "nonlinear cohomology",

$$M \simeq \frac{\mathsf{Zeros}(\eth)}{\mathsf{Im}\, \eth},$$

representing M as the quotient of the zero set of the vector field $\eth$ by the integrable distribution

$$\mathsf{Im}\, \eth := \{X \in \mathcal{T}_\mathcal{M} : X = [\eth, Y] \text{ for some } Y \in \mathcal{T}_\mathcal{M}\},$$

which, as it is easy to check, is tangent to $\mathsf{Zeros}(\eth)$.

With any dg manifold $(\mathcal{M}, \eth)$ one can associate two cohomology sheaves: the cohomology structure sheaf,

$$H(\mathcal{O}_\mathcal{M}) := \frac{\mathsf{Ker}\, \eth : \mathcal{O}_\mathcal{M} \to \mathcal{O}_\mathcal{M}}{\mathsf{Im}\, \eth : \mathcal{O}_\mathcal{M} \to \mathcal{O}_\mathcal{M}},$$

and the cohomology tangent sheaf,

$$\mathcal{H}\mathcal{T}_\mathcal{M} := \frac{\mathsf{Ker}\, Lie_\eth : \mathcal{T}_\mathcal{M} \to \mathcal{T}_\mathcal{M}}{\mathsf{Im}\, Lie_\eth : \mathcal{T}_\mathcal{M} \to \mathcal{T}_\mathcal{M}},$$

which is a sheaf of $H(\mathcal{O}_\mathcal{M})$-modules (in fact, a sheaf of Lie $H(\mathcal{O}_\mathcal{M})$-algebras).

We define a *cohomology F-manifold* to be a dg manifold $(\mathcal{M}, \eth)$ together with a graded commutative associative $H(\mathcal{O}_\mathcal{M})$-polylinear product $\mu_2 : \mathcal{H}\mathcal{T}_\mathcal{M} \times \mathcal{H}\mathcal{T}_\mathcal{M} \to \mathcal{H}\mathcal{T}_\mathcal{M}$, such that the integrability condition, $[\mu_2, \mu_2] = 0$, holds. This notion also makes sense in the category of formal dg manifolds.

Theorem B. *If the operad* $\mathcal{G}_\infty$ *acts on a dg vector space* (V, d)*, then the formal graded manifold associated with the cohomology vector space* $H(V, d)$ *is canonically a cohomology* F*-manifold.*

Let us emphasize again that the notion of (cohomology) F-manifold is diffeomorphism invariant. Though the input in Theorem B belongs to the category of vector spaces which one can geometrically interpret as pointed affine (=flat) manifolds, the output lies in the category of general smooth graded manifolds with morphisms being arbitrary (not necessary, linear) smooth maps[1]. Thus the output of Theorem B belongs to the realm of differential geometry.

Recent proofs of Deligne's conjecture [**Ko2, KoSo1, McSm, Ta, Vo**] together with Theorem B imply[2]

Corollary C. (i) *Let A be an associative k-algebra. The formal manifold associated with the Hochschild cohomology $H^\bullet(A, A)$ is naturally a cohomology F-manifold.*

(ii) *Let X be a compact topological space. The formal manifold associated with its singular cohomology $H^\bullet(X, k)$ is naturally a cohomology F-manifold.*

0.3. F_∞-manifolds. Instead of passing to cohomology sheaves as above, one can adopt the notion of F-manifold to the category of dg manifolds by constructing its strong homotopy version. We do it in this paper with the help of the $\mathcal{G}_\infty$-operad (cf. Theorem B).

Let $(\mathcal{M}, \eth, *)$ be a formal dg manifold and let

$$\mu_\bullet = \{\mu_n\}_{n\geq 1} : \otimes^\bullet_{\mathcal{O}_\mathcal{M}} \mathcal{T}_\mathcal{M} \to \mathcal{T}_\mathcal{M}$$

be a structure of $\mathcal{C}_\infty$-algebra on the tangent sheaf. We call it *geometric* if $\mu_1 = Lie_\eth$ and $\mu_{\bullet\geq 2}$ are morphisms of $\mathcal{O}_\mathcal{M}$-modules, i.e. are tensors. If all μ_n except μ_2 vanish, this structure reduces to the structure of graded commutative associative product as in Sect. 0.1. Note that $(\otimes^\bullet_{\mathcal{O}_\mathcal{M}} \mathcal{T}^*_\mathcal{M} \otimes_{\mathcal{O}_\mathcal{M}} \mathcal{T}_\mathcal{M}, Lie_\eth)$ is a complex of (sheaves of) $\mathcal{O}_\mathcal{M}$-modules. Its cohomology is denoted by $H(\otimes^\bullet_{\mathcal{O}_\mathcal{M}} \mathcal{T}^*_\mathcal{M} \otimes_{\mathcal{O}_\mathcal{M}} \mathcal{T}_\mathcal{M})$.

Choosing a torsion-free affine connection ∇ on $\mathcal{M}$, one can construct an extension of the Hertling-Manin's "bracket" $[\mu_2, \mu_2]$ to geometric $\mathcal{C}_\infty$-structures,

$$[\mu_\bullet, \mu_\bullet]^\nabla(X_1, \ldots, X_\bullet, Y_1, \ldots, Y_\bullet) := [\mu_\bullet(X_1, \ldots, X_\bullet), \mu_\bullet(Y_1, \ldots, Y_\bullet)] + \text{correction terms},$$

producing thereby a collection of *tensors*[3], $[\mu_\bullet, \mu_\bullet]^\nabla : \otimes^{\bullet+\bullet}_{\mathcal{O}_\mathcal{M}} \mathcal{T}_\mathcal{M} \to \mathcal{T}_\mathcal{M}$, satisfying the following two conditions,

- $Lie_\eth [\mu_\bullet, \mu_\bullet]^\nabla = 0$,
- the cohomology class,

$$[[\mu_\bullet, \mu_\bullet]] \in H(\otimes^\bullet_{\mathcal{O}_\mathcal{M}} \mathcal{T}^*_\mathcal{M} \otimes_{\mathcal{O}_\mathcal{M}} \mathcal{T}_\mathcal{M}),$$

  produced by $[\mu_\bullet, \mu_\bullet]^\nabla$ does not depend on the choice of the connection ∇ and hence gives a well-defined invariant of the geometric $\mathcal{C}_\infty$-structure. Moreover, this invariant depends only on the homotopy class of that structure.

The correction terms to $[[\mu_\bullet, \mu_\bullet]]$ can, in principle, be read off from the structural equations of the $\mathcal{G}_\infty$-operad, as explained in Sect. 4. However, all the basic properties of the bracket $[\mu_\bullet, \mu_\bullet]^\nabla$, such as its existence, $\mathcal{O}_\mathcal{M}$-linearity, $Lie_\eth$-closedness etc., can be proved without doing this sort of explicit calculations.

---

[1] There is no way to remember the original flat structure of the input unless the combination (G_∞ action, V, d) is formal as a L_∞-algebra, and one makes a particular choice of a homotopy class of formality maps.

[2] Here and everywhere in this paper k stands for a field of characteristic 0. Every vector space is implicitly assumed to be over k.

[3] The assumption that $\mu_\bullet$ is a $\mathcal{C}_\infty$-structure is important. The construction does not work for geometric $\mathcal{A}_\infty$-structures.

Definition D. *An F_∞-manifold is a dg manifold, $(\mathcal{M}, \eth, *)$, together with a homotopy class of geometric $\mathcal{C}_\infty$-structures, $\{\mu_\bullet : \otimes^\bullet_{\mathcal{O}_\mathcal{M}} \mathcal{T}_\mathcal{M} \to \mathcal{T}_\mathcal{M}\}$, satisfying the integrability condition,*

$$[[\mu_\bullet, \mu_\bullet]] = 0.$$

Clearly, any F_∞-manifold gives naturally rise to a cohomology F-manifold. In fact, the cohomology F-manifolds discussed above in Sect. 0.2 are precisely of this type:

Theorem E. *All statements of Theorem B and Corollary C remain true if one replaces*

$$\textit{cohomology } F\textit{-manifold} \quad \longrightarrow \quad F_\infty\textit{-manifold.}$$

0.4. Content. In §1 and §2 we remind basic notions and notations of the (homotopy) theory of operads and discuss in detail some particular examples. The main result in §3 is an explicit graphical description, Proposition 3.6.1, of the cobar construction for the operad of non-commutative Gerstenhaber algebras and a surprisingly nice geometric interpretation, Theorem 3.9.2, of the derived category of algebras over that operad. In §4 we outline an operadic guide to the extended deformation theory (as a more informative alternative to the classical idea of deformation functor) and, in that context, prove all the claims made in the Introduction.

§1. Operads and their algebras

1.1. Operads. By an operad in this paper we always understand what is usually called a *nonunital or pseudo-operad* [**Mar1**], that is, a pair of collections,

$$\mathcal{O} = \left(\{\mathcal{O}(n)\}_{n\geq 1} \ , \ \{\circ_i^{n,n'}\}_{\substack{n,n'\geq 1 \\ 1\leq i\leq n}} \right),$$

where each $\mathcal{O}(n)$ is a $\mathbb{Z}$-graded vector space equipped with a linear action of the permutation group $\mathbb{S}_n$ (the collection $\{\mathcal{O}(n)\}_{n\geq 1}$ will sometimes be called an $\mathbb{S}$-*module*), and each $\circ_i^{n,n'}$ is a linear equivariant map,

$$\circ_i^{n,n'} : \mathcal{O}(n) \otimes \mathcal{O}(n') \longrightarrow \mathcal{O}(n+n'-1),$$

such that, for any $f \in \mathcal{O}(n)$, $f' \in \mathcal{O}(n')$ and $f'' \in \mathcal{O}(n'')$, one has

$$\left(f \circ_i^{n,n'} f'\right) \circ_{j+n'-1}^{n+n'-1,n''} f'' = (-1)^{|f'||f''|} \left(f \circ_j^{n,n''} f''\right) \circ_i^{n+n''-1,n'} f', \ \forall\, 1 \leq i < j \leq n,$$

and

$$f \circ_i^{n,n'+n''-1} \left(f' \circ_j^{n',n''} f''\right) = \left(f \circ_i^{n,n'} f'\right) \circ_{i+j-1}^{n+n'-1,n''} f'', \ \forall\, 1 \leq i \leq n, 1 \leq j \leq n'.$$

Equivariance of $\circ_i^{n,n'}$ above means that for any $\sigma \in \mathbb{S}_n$ and $\sigma' \in \mathbb{S}_{n'}$ one has

$$(\sigma f) \circ_i^{n,n'} (\sigma' f') = (\sigma_i \sigma')(f \circ_i^{n,n'} f')$$

where $(\sigma_i\sigma') \in \mathbb{S}_{n+n'-1}$ is given by inserting the permutation σ' into the ith place of σ.

An *ideal* in an operad $\mathcal{O}$ is a collection I of $\mathbb{S}_n$-invariant subspaces $\{\mathcal{I}(n) \subset \mathcal{O}(n)\}_{n\geq 1}$ such that $f \circ_i^{n,n'} f' \in \mathcal{I}(n+n'-1)$ whenever $f \in \mathcal{I}(n)$ or $f' \in \mathcal{I}(n')$; in particular, $\mathcal{I}$ is a suboperad of $\mathcal{O}$. It is clear that the quotient $\mathbb{S}$-module

$\{\mathcal{O}(n)/\mathcal{I}(n)\}_{n\geq 1}$ has a naturally induced structure of an operad called a *quotient operad.*

An operad $\mathcal{O}$ with $\mathcal{O}(1)=0$ is called *simply connected.*

1.2. Free operads and trees. A morphism of operads, $f:\mathcal{O}\to\mathcal{O}'$, is, by definition, a morphism of the associated $\mathbb{S}$-modules, $\{f(n):\mathcal{O}(n)\longrightarrow\mathcal{O}'(n)\}_{n\geq 2}$, which commutes in the obvious way with all the operations $\circ_i^{n,n'}$. Operads form a category.

The forgetful functor

$$\begin{array}{ccc}\text{Category of operads} & \longrightarrow & \text{Category of } \mathbb{S}\text{–modules}\\ \left(\{\mathcal{O}(n)\}_{n\geq 1}\ ,\ \{\circ_i^{n,n'}\}_{\substack{n,n'\geq 1\\ 1\leq i\leq n}}\right) & \longrightarrow & \left(\{\mathcal{O}(n)\}_{n\geq 1}\right),\end{array}$$

has a left adjoint functor, $Free$, which associates to an arbitrary collection, $\mathcal{E}=\{\mathcal{E}(n)\}_{n\geq 1}$, of graded vector $\mathbb{S}_n$-spaces the *free operad*, $Free(\mathcal{E})$. It is best described in terms of trees as follows (see **[GiKa, GetJo, KoSo1]** for more details).

An $[n]$*-tree* T is, by definition, the data (V_T, N_T, ϕ_T) consisting of

- a stratified finite set $V_T=V_T^i\sqcup V_T^t$ whose elements are called *vertices*; elements of the subset V_T^i (resp. V_T^t) are called *internal* (resp. *tail*) vertices;
- a bijection $\phi: V_T^t\to\{1,2,\dots,n\}=:[n]$;
- a map $N_T: V_T\to V_T$ satisfying the conditions: (i) N_T has only one fixed point $root_T$ which lies in V_T^i and is called the *root vertex*, (ii) $N_T^k(v)=root_T$, $\forall\ v\in V_T$ and $k\gg 1$, (iii) for all $v\in V_T^i$ the cardinality, $\#v$, of the set $N_T^{-1}(v)$ is greater than or equal to 1, while for all $v\in V_T^t$ one has $\#v=0$.

The number $\#v$ is often called the *valency* of the vertex v; the pairs $(v, N_T(v))$ are called edges.

Given an $\mathbb{S}$-module $\mathcal{E}=\{\mathcal{E}(n)\}_{n\geq 1}$, we can associate to an $[n]$-tree T the vector space

$$\mathcal{E}(T):=\bigotimes_{v\in V_T^i}\mathcal{E}(\#v).$$

Its elements are interpreted as $[n]$-trees whose internal vertices are decorated with elements of $\mathcal{E}$. The permutation group $\mathbb{S}_n$ then acts on this space via relabelling the tail vertices (i.e changing ϕ_T to $\sigma\circ\phi_T$, $\sigma\in\mathbb{S}_n$).

Now, as an $\mathbb{S}$-module the free operad $Free(\mathcal{E})$ is defined as

$$Free(\mathcal{E})(n)=\bigoplus_{[[n]-\text{trees } T]}\mathcal{E}(T),$$

where the summation goes over all isomorphism classes of $[n]$-trees. The composition, say $f\circ_i^{n,n'}f'$, is given by gluing the root vertex of the decorated $[n']$-tree $f'\in Free(\mathcal{E})(n')$ with the i-labelled tail vertex of the decorated $[n]$-tree f. The new numeration, $\phi: V_T^t\to[n+n'-1]$, of tails is clear.

Any free operad is naturally graded, $Free(\mathcal{E})=\bigoplus_{p=1}^{\infty}Free^p(\mathcal{E})$, where $Free^p(\mathcal{E})$ is the $\mathbb{S}$-submodule of $Free(\mathcal{E})$ spanned by all possible isomorphism classes of $\mathcal{E}$-decorated trees with precisely p internal vertices.

1.3. Example. Let V be a $\mathbb{Z}$-graded vector space. The associated $\mathbb{S}$-module,

$$\mathcal{E}_V = \{\mathcal{E}_V(n) := Hom(V^{\otimes n}, V)\},$$

has a natural structure of operad with compositions, $f \circ_i^{n,n'} f'$, given by inserting the output of $f' \in Hom(V^{\otimes n'}, V)$ into the i-th input of $f \in Hom(V^{\otimes n}, V)$.

An *algebra over an operad* $\mathcal{O}$ is, by definition, a $\mathbb{Z}$-graded vector space V together with a morphism of operads $\mathcal{O} \to \mathcal{E}_V$.

1.4. Example. Let $\mathcal{A}$ be an $\mathbb{S}$-module given by

$$\mathcal{A}(n) := \begin{cases} k[\mathbb{S}_2][0] & \text{if } n = 2 \\ 0 & \text{otherwise,} \end{cases}$$

where here and below the symbol $k[\mathbb{S}_n][p]$ stands for the graded vector space whose only non-vanishing homogeneous component lies in degree $-p$ and equals the regular representation $k[\mathbb{S}_n]$ of the permutation group $\mathbb{S}_n$[4]. If we identify the natural basis, *id* and (12), of $k[\mathbb{S}_2]$ with *planar* [2]-corollas,

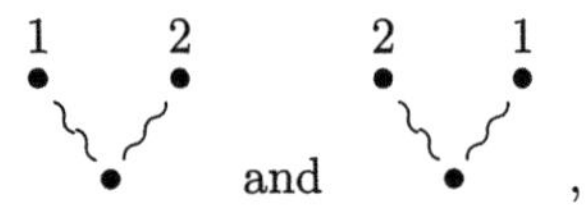

then the associated free operad $\{Free(\mathcal{A})(n), \circ_i^{n,n'}\}$ can be represented as a linear span of all possible (isomorphism classes of) binary *planar* $[n]$-trees, e.g.

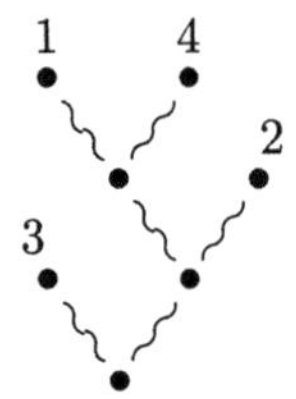

$\in Free(\mathcal{A})(4)$,

with the compositions $\circ_i^{n,n'}$ given simply by gluing the root vertex of a planar $[n']$-tree to the ith tail vertex of an $[n]$-tree (the new numeration of tails is clear). Indeed an isomorphism class of an $\{id, (12)\}$-decorated abstract($\equiv$space) binary tree of Subsect. 1.2 has a natural representative which lies in a fixed plane in the space and which is consistent with the interpretation of the labelling set $\{id, (12)\}$ as the set of planar [2]-corollas; more importantly, the resulting correspondence

$$\left\{ \begin{array}{c} \text{isomorphism classes of} \\ \{id, (12)\}\text{–decorated abstract binary trees} \end{array} \right\} \longrightarrow \left\{ \begin{array}{c} \text{isomorphism classes of} \\ \text{planar binary numbered trees} \end{array} \right\}$$

is one-to-one.

Algebras over $Free(\mathcal{A})$ are not that interesting objects — they are just graded vector spaces V together with a fixed element of $\mathrm{Hom}(V^{\otimes 2}, V)$ which can be arbitrary.

---

[4]More generally, for a $\mathbb{Z}$-graded vector space $V = \oplus_{i\in\mathbb{Z}} V^i$, the symbol $V[p]$ stands for the $\mathbb{Z}$-graded vector space with $V[p]^i := v^{i+p}$.

Let $\mathcal{I}_{\mathcal{A}}$ be the ideal in $Free(\mathcal{A})$ generated by 3! vectors of the form

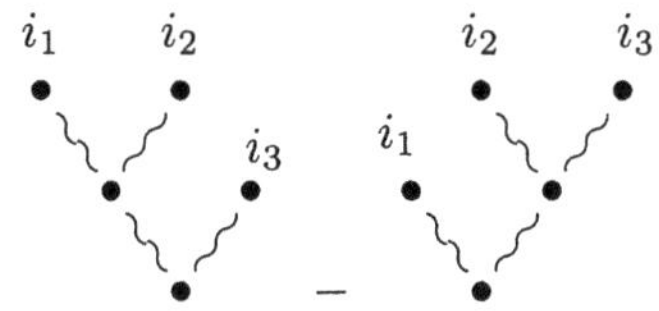

The quotient operad $Free(\mathcal{A})/\mathcal{I}_{\mathcal{A}}$ is denoted by $\mathcal{A}ss$ for the obvious reason — its algebras are nothing but the usual graded associative algebras. As an $\mathbb{S}$-module, $\mathcal{A}ss(n) \simeq k[\mathbb{S}_n]$.

1.5. Example. Consider an $\mathbb{S}$-module,

$$\mathcal{C}omm(n) := \mathbf{1}_n[0]$$

where $\mathbf{1}_n$ stands for the trivial representation of the permutation group $\mathbb{S}_n$. This $\mathbb{S}$-module can be made into an operad $\mathcal{C}omm$ by defining the compositions $\circ_i^{n,n'}$ to be the identity maps. It is not hard to check that $\mathcal{C}omm$-algebras are graded commutative associative algebras in the usual sense.

For later reference it will be convenient to represent the operad $\mathcal{C}omm$ as a quotient of a free operad. For this purpose we first consider an $\mathbb{S}$-module $\mathcal{C}$,

$$\mathcal{C}(n) := \begin{cases} \mathbf{1}_2[0] & \text{if } n = 2 \\ 0 & \text{otherwise.} \end{cases}$$

If we identify a basis vector of $\mathbf{1}_n[0]$ with the unique (up to an isomorphism) *space* corolla (i.e. the one embedded in $\mathbb{R}^3$)

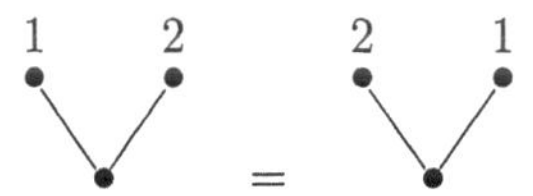

then the associated free operad $\{Free(\mathcal{C})(n)\}$ can be represented as a linear span of all possible isomorphism classes of binary *space* $[n]$-trees; for example, $Free(\mathcal{C})(3)$ is a 3-dimensional vector space spanned by the following space $[3]$-trees

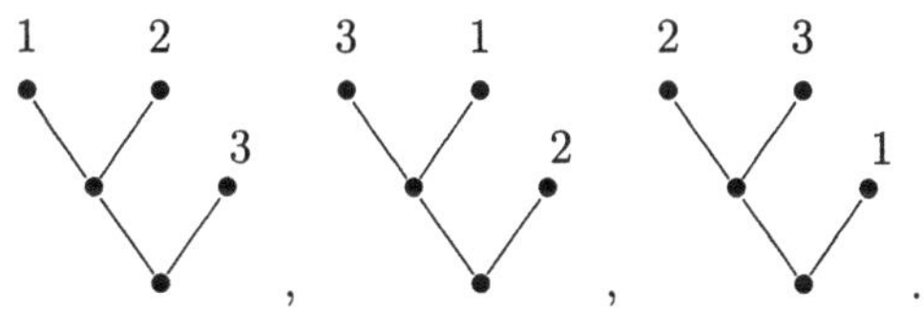

The composition in $Free(\mathcal{C})$ is given by gluing the root vertex of one space tree to a tail vertex of another one. The new numeration of tail vertices is clear.

Let $\mathcal{I}_{\mathcal{C}}$ be the ideal in $Free(\mathcal{C})$ generated by 2 vectors of the form

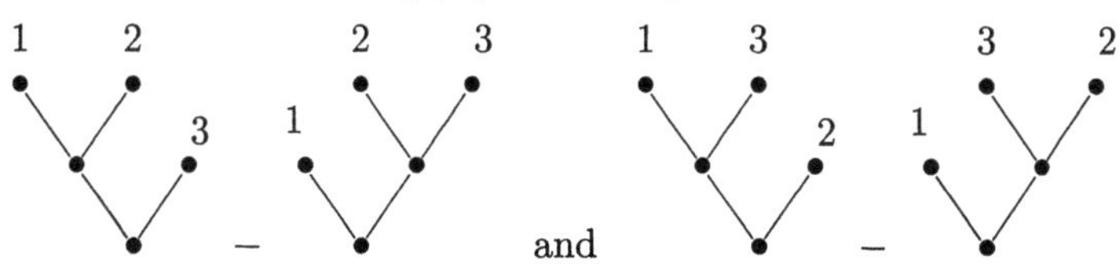

The quotient operad $Free(\mathcal{C})/\mathcal{I}_{\mathcal{C}}$ is clearly isomorphic to $\mathcal{C}omm$.

1.6. Example. Let $\mathcal{L}$ be an $\mathbb{S}$-module given by

$$\mathcal{L}(n) := \begin{cases} \mathbf{1}_2[-1] & \text{if } n = 2, \\ 0 & \text{otherwise.} \end{cases}$$

If we identify, as in Example 1.4, a basis vector of the one dimensional vector space $\mathbf{1}_2[1]$ with the unique (up to isomorphism) *space* [2]-corolla

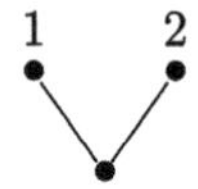

then the associated free operad $\{Free(\mathcal{L})(n)\}$ can be represented as a linear span of all possible (isomorphism classes of) binary *space* $[n]$-trees with the composition given by gluing the root vertex of one space tree to a tail vertex of another one.

Let $\mathcal{I}_\mathcal{L}$ be the ideal in $Free(\mathcal{L})$ generated by the following $\mathbb{S}_3$-invariant vector,

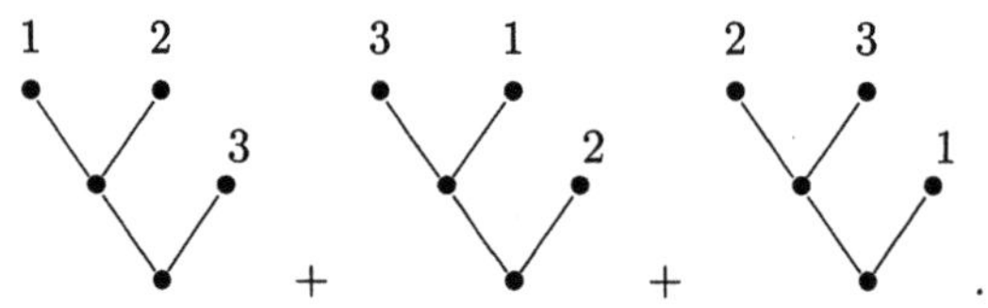

Algebras over the associated quotient operad, $\mathcal{L}ie := Free(\mathcal{L})/\mathcal{I}_\mathcal{L}$, are graded vector spaces V equipped with a degree -1 element $\nu \in Hom(\odot^2 V, V)$ satisfying the Jacobi condition,

$$\nu(\nu(v_1, v_2), v_3) + (-1)^{|v_3|(|v_1|+|v_2|)}\nu(\nu(v_3, v_1), v_2) + (-1)^{|v_1|(|v_2|+|v_3|)}\nu(\nu(v_2, v_3), v_1) = 0.$$

Setting

$$[v_1 \bullet v_2] := (-1)^{|v_1|}\nu(v_1, v_2)$$

we recover the notion of (odd) Lie algebra [**Ma**]. It is, of course, the same thing as the usual graded Lie algebra structure on the shifted graded vector space $V[1]$ but for our purposes it is more suitable *not* to make this shift; thus in the present paper by a graded Lie algebra we always understand an algebra over the operad $\mathcal{L}ie$, i.e. a pair $(V, [\ \bullet\ ])$ with $[\ \bullet\ ] : \odot^2 V \to V$ having degree -1 and satisfying (odd) Jacobi identity.

1.7. Example. Let $\mathcal{AL}$ be an $\mathbb{S}$-module given by $\mathcal{AL}(n) := \mathcal{A}(n) \oplus \mathcal{L}(n)$. Its only non-vanishing component $\mathcal{AL}(2)$ is a 3-dimensional vector space spanned by two planar corollas in degree 0 and one space corolla in degree -1,

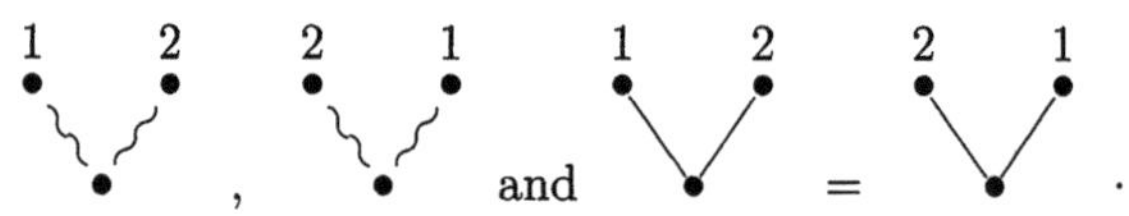

The associated free operad $Free(\mathcal{AL})$ can be represented as a linear span of all possible isomorphism classes of binary $[n]$-trees in the 3-space $\mathbb{R}^3$ with the condition that all "planar" corollas are perpendicular to a fixed line in $\mathbb{R}^3$. The composition in $Free(\mathcal{AL})$ is given again by gluing the root vertex of one such partially planar/partially space tree to a tail vertex of another one.

Let $\mathcal{I}_{\mathcal{AL}}$ be the ideal in $Free(\mathcal{AL})$ generated by the following 3! vectors,

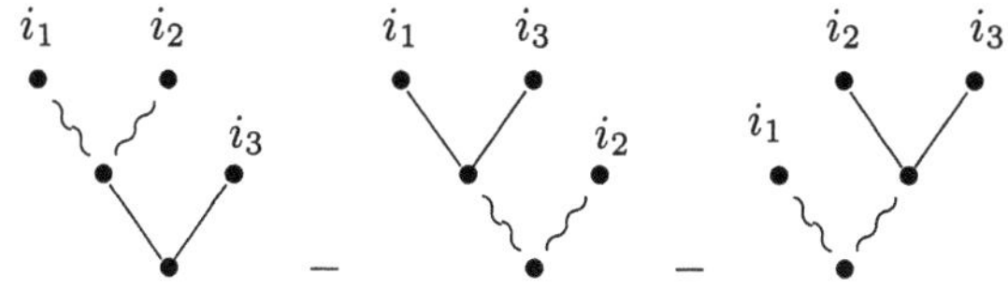

Algebras over the quotient operad,

$$\mathcal{G}erst := Free(\mathcal{L})/ < \mathcal{I}_{\mathcal{A}}, I_{\mathcal{L}}, I_{\mathcal{AL}} >,$$

are called (non-commutative) *Gerstenhaber* algebras. These are triples, $(V, \circ, [\bullet])$, consisting of a graded vector space V, a degree 0 associative product, $\circ : V \otimes V \rightarrow V$ and a degree -1 Lie bracket, $[\ \bullet\ ] : \odot^2 V \rightarrow V$ which satisfy the following compatibility condition,

$$[a \bullet (b \circ c)] = [a \bullet b] \circ c + (-1)^{(\tilde{a}+1)\tilde{b}} b \circ [a \bullet c],$$

for all homogeneous $a, b, c \in V$.

1.8. Example. A Gerstenhaber algebra $(V, \circ, [\bullet])$ is called *graded commutative* if such is the product $\circ$. Let us denote by $\mathcal{G}$ the operad which governs graded commutative Gerstenhaber algebras.

1.9. Remark. There is a canonical map of operads,

$$\mathcal{G} \longrightarrow \mathcal{G}erst,$$

corresponding to the obvious functor

$$\left\{ \begin{array}{c} \text{A category of} \\ \mathcal{G}\text{–algebras} \end{array} \right\} \longrightarrow \left\{ \begin{array}{c} \text{A category of} \\ \mathcal{G}erst\text{–algebras} \end{array} \right\}$$

which simply forgets graded commutativity of the associated product.

Both operads $\mathcal{G}$ and $\mathcal{G}erst$ are composed from a pair of simpler operads, $(\mathcal{C}omm, \mathcal{L}ie)$ and, respectively, $(\mathcal{A}ss, \mathcal{L}ie)$. The difference, however, is that the composition of $(\mathcal{C}omm, \mathcal{L}ie)$ into $\mathcal{G}$ satisfies the distributive law **[Mar2]**, while the composition of $(\mathcal{A}ss, \mathcal{L}ie)$ into $\mathcal{G}erst$ *does not.* Indeed, "opening" the expression

$$[(a_1 \circ a_2) \bullet (a_3 \circ a_4)] \quad \simeq$$

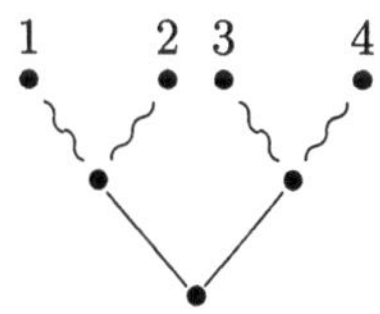

in two possible ways,

$$\begin{aligned} & [(a_1 \circ a_2) \bullet (a_3 \circ a_4)] \\ = \ & a_1 \circ [a_2 \bullet (a_3 \circ a_4)] + (-1)^{|a_2|(|a_3|+|a_4|+1)} [a_1 \bullet (a_3 \circ a_4)] \circ a_2 \\ = \ & a_1 \circ [a_2 \bullet a_3] \circ a_4 + (-1)^{|a_3|(|a_2|+1)} a_1 \circ a_3 \circ [a_2 \bullet a_4)] \\ & +(-1)^{|a_2|(|a_3|+|a_4|+1)} \Big([a_1 \bullet a_3] \circ a_4 \circ a_2 + (-1)^{|a_3|(|a_1|+1)} a_3 \circ [a_1 \bullet a_4] \circ a_2 \Big), \end{aligned}$$

$$\begin{aligned}
& [(a_1 \circ a_2) \bullet (a_3 \circ a_4)] \\
= \ & [(a_1 \circ a_2) \bullet a_3] \circ a_4 + (-1)^{|a_3|(|a_1|+|a_2|+1)} a_3 \circ [(a_1 \circ a_2) \bullet a_4] \\
= \ & a_1 \circ [a_2 \bullet a_3] \circ a_4 + (-1)^{|a_2|(|a_3|+1)} [a_1 \bullet a_3] \circ a_2 \circ a_4 \\
& +(-1)^{|a_3|(|a_1|+|a_2|+1)} \left(a_3 \circ a_1 \circ [a_2 \bullet a_4] + (-1)^{|a_2|(|a_4|+1)} a_3 \circ [a_1 \bullet a_4] \circ a_2 \right)
\end{aligned}$$

and then decomposing the associated relation in $\mathcal{G}erst(4)$ into irreducibles, one gets

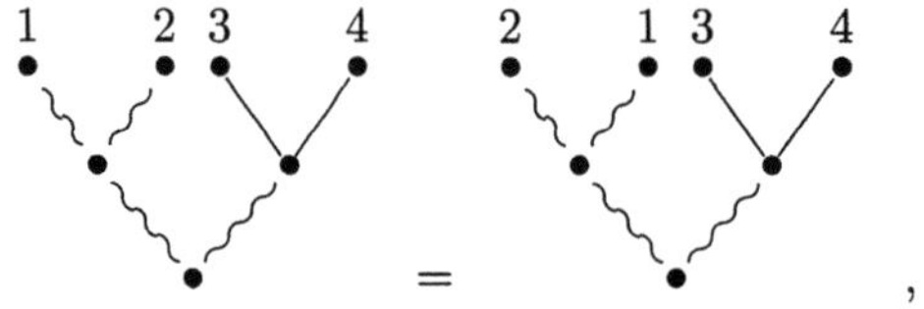

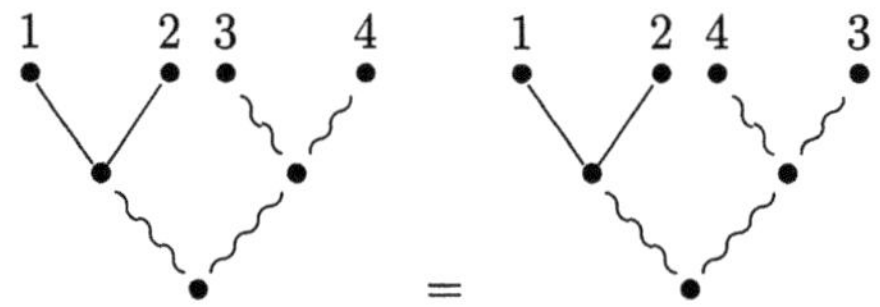

The resulting relations in $\mathcal{G}erst(4)$ are non-trivial unless the product $\circ$ is graded commutative.

1.10. Example. Let $\mathcal{AC}$ be an $\mathbb{S}$-module given by $\mathcal{AC}(n) := \mathcal{A}(n)[-1] \oplus \mathcal{C}omm(n)$. Its only non-vanishing component $\mathcal{AC}(2)$ is a 3-dimensional vector space spanned by two planar corollas of degree 1 and one space corolla of degree 0,

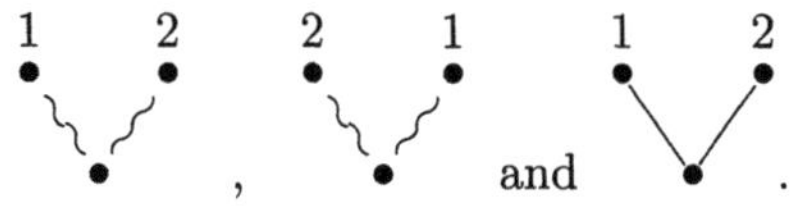

As in example 1.7, the associated free operad $Free(\mathcal{AC})$ can be represented as a linear span of all possible isomorphism classes of binary $[n]$-trees in the 3-space $\mathbb{R}^3$ with the condition that all planar corollas are perpendicular to a fixed line in $\mathbb{R}^3$. The composition in $Free(\mathcal{AC})$ is given by gluing the root vertex of one such partially planar/ partially space tree to a tail vertex of another one.

Let $\mathcal{I}_{\mathcal{AC}}$ be the ideal in $Free(\mathcal{AC})$ generated by the following 3+3=3! vectors,

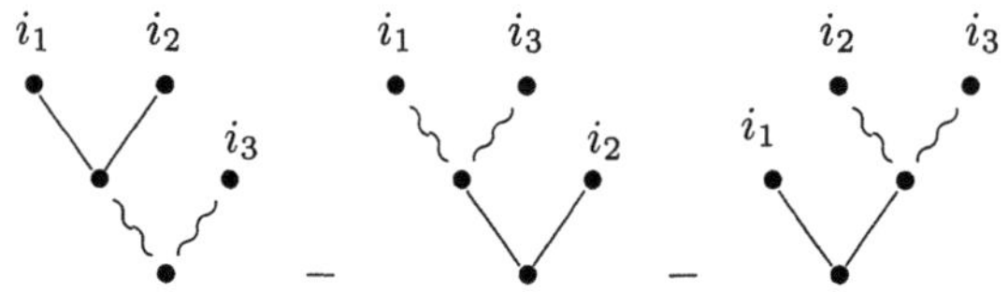

and

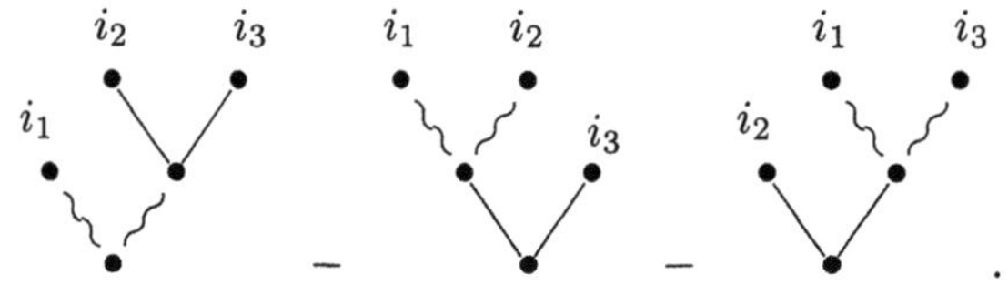

Algebras over the quotient operad,

$$\mathcal{H} := Free(\mathcal{AC}) / < \mathcal{I}_{\mathcal{A}[-1]}, I_{\mathcal{C}}, I_{\mathcal{AC}} >,$$

are triples, $(V, \circ, \bullet)$, consisting of a graded vector space V, a degree 0 associative graded commutative product, $\circ : V \odot V \to V$ and a degree 1 associative product $\bullet : V \otimes V \to V$ which satisfy the following compatibility conditions,

$$a \bullet (b \circ c) = (a \bullet b) \circ c + (-1)^{(\tilde{a}+1)\tilde{b}} b \circ (a \bullet c),$$

$$(a \circ b) \bullet c = a \circ (b \bullet c) + (-1)^{(\tilde{c}+1)\tilde{b}} (a \bullet c) \circ b,$$

for all homogeneous $a, b, c \in V$.

§2. Strongly homotopy algebras

2.1. Dg operads. A differential graded (shortly, dg) operad is an operad,

$$\mathcal{O} = \left(\{\mathcal{O}(n)\}_{n\geq 1} \;,\; \{\circ_i^{n,n'}\}_{\substack{n,n'\geq 1\\ 1\leq i\leq n}} \right),$$

in the sense of §1 together with a degree 1 equivariant linear map $d : \mathcal{O}(n) \to \mathcal{O}(n)$, $\forall n$, satisfying the conditions,

$$\begin{aligned} d^2 &= 0, \\ d\left(f \circ_i^{n,n'} f'\right) &= (df) \circ_i^{n,n'} f' + (-1)^{|f|} f \circ_i^{n,n'} df', \quad \forall f \in \mathcal{O}(n), f' \in \mathcal{O}(n'). \end{aligned}$$

The associated cohomology $\mathbb{S}$-module $H(\mathcal{O}) := \{H^\bullet(\mathcal{O}(n))\}_{n\geq 1}$ has an induced operad structure.

A morphism, $f : (\mathcal{O}, d) \to (\mathcal{O}', d')$, of dg operads is, by definition, a morphism of operads $f : \mathcal{O} \to \mathcal{O}'$ which commutes in the obvious sense with the differentials. A morphism, $f : (\mathcal{O}, d) \to (\mathcal{O}', d')$ is called a *quasi-isomorphism* if the induced morphism of the cohomology operads, $[f] : H(\mathcal{O}) \to H(\mathcal{O}')$, is an isomorphism.

If (V, d) is a dg vector space, then $(\mathcal{E}_V, d_{\text{ind}})$ is naturally a dg operad where $d_{\text{ind}} : Hom(\otimes^\bullet V, V) \to Hom(\otimes^\bullet V, V)$ is the differential which is naturally induced by d and which we denote from now on by the same symbol d.

An *algebra* over a dg operad $(\mathcal{O}, d)$ is a dg vector space (V, d) together with a morphism of dg operads $(\mathcal{O}, d) \to (\mathcal{E}_V, d)$.

It was shown in [**Mar1**] that for any simply connected dg operad $(\mathcal{O}, d)$ there exists a unique (up to an isomorphism) triple $\mathcal{O}_\infty := (Free(\mathcal{E}), d, f)$ where

(i) $Free(\mathcal{E})$ is the free operad generated by an $\mathbb{S}$-module $\mathcal{E} = \{\mathcal{E}(n)\}_{n\geq 2}$;
(ii) d is the differential in $Free(\mathcal{E})$ which is decomposable in the sense that $df \in Free^{\geq 2}(\mathcal{E})(n)$ for any $f \in \mathcal{E}(n)$, $n \geq 2$.
(iii) $f : (Free(\mathcal{E}), d) \to (\mathcal{O}, d)$ is a quasi-isomorphism of dg operads.

This operad $\mathcal{O}_\infty$ is called the *minimal resolution*[5] of the operad $\mathcal{O}$. Such minimal resolutions play a very important role in the homotopy theory of operadic algebras which we discuss below after considering a few examples.

2.2. Remark. It is clear that to define a particular dg operad, $(\mathcal{O}, d)$, is the same thing as to define its algebras, i.e. the image of the map $(\mathcal{O}, d) \to \mathcal{E}_V$ for some "variable" graded vector space V. Moreover, for this purely descriptive purpose it is enough to assume that $\dim V < \infty$. We *always* make such an assumption when

[5]An operad $\mathcal{O}_\infty := (Free(\mathcal{E}), d)$ satisfying relations (i) and (ii) is often called *minimal.*

applying this method to concrete examples; in particular, there is never a problem with replacing V by its dual, $V^* = \mathrm{Hom}(V, k)$.

2.3. Example: operad $\mathcal{A}_\infty$. Let $\hat{A}$ be an $\mathbb{S}$-module given by,

$$\mathcal{A}(n) := \begin{cases} 0 & \text{if } n = 1 \\ k[\mathbb{S}_n][n-2] & \text{if } n \geq 2. \end{cases}$$

If we identify the natural basis of $k[\mathbb{S}_n][n-2]$ with *planar* $[n]$-corollas,

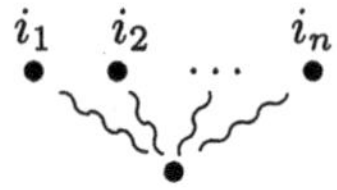

then the associated free operad $\mathcal{A}_\infty := (\{Free(\hat{\mathcal{A}})(n), \circ_i^{n,n'}\})$ can be represented as a linear span of all possible (isomorphism classes of) *planar* $[n]$-trees with the compositions $\circ_i^{n,n'}$ given simply by gluing the root vertex of a planar $[n']$-tree to the ith tail vertex of an $[n]$-tree.

One can make $\mathcal{A}_\infty$ into a dg operad with the differential d given on generators by

d 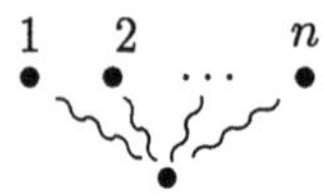

$= \sum_{\substack{l+p=n+1 \\ l,p\geq 2}} \sum_{s=0}^{p-1} (-1)^{l+s(l+1)}$ 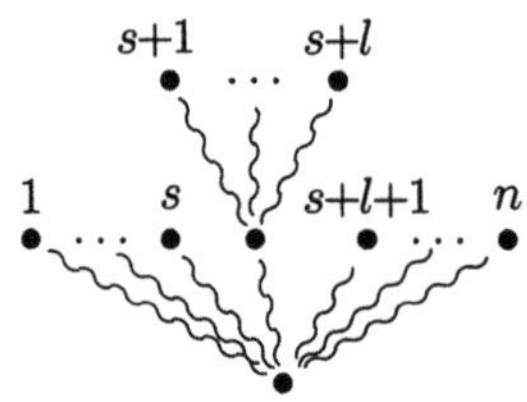

The associated cohomology operad, $H(\mathcal{A}_\infty, d)$, is in fact isomorphic to $\mathcal{A}ss$ [**GiKa**]. Hence, the natural morphism of dg operads,

$$f : (\mathcal{A}_\infty, d) \longrightarrow (\mathcal{A}ss, 0),$$

defined to be identity on [2]-corollas and zero on $[n \geq 3]$-corollas, is a quasi-isomorphism. Thus $(\mathcal{A}_\infty, d)$ is the minimal resolution of $\mathcal{A}ss$.

An algebra over the dg operad $\mathcal{A}_\infty$ is called a *strongly homotopy associative algebra* or, shortly, an *A_∞-algebra.* This is a dg vector space (V, d) equipped with degree $2-n$ multilinear operations $\mu_n : \otimes^n V \to V$, $n \geq 2$, such that for any $N \geq 1$ and any $v_1, \ldots, v_N \in V$,

$$\sum_{\substack{l+p=N+1 \\ l,p\geq 1}} (-1)^r \mu_p(v_1, \ldots, v_s, \mu_l(v_{s+1}, \ldots, v_{s+l}), v_{s+l+1}, \ldots, v_N) = 0,$$

where $\mu_1 = d$ and $r = l + s(l+1) + p(|v_1| + \ldots + |v_s|$. If all μ_n except $n = 2$ vanish, the above equation translates into the associative condition for the product $v_1 \circ v_2 := \mu_2(v_1, v_2)$. Strongly homotopy associative algebras have been invented by Stasheff [**St**] in his study of spaces homotopy equivalent to loop spaces.

2.4. Example: operad $\mathcal{C}_\infty$. The dg operad $\mathcal{A}_\infty$ has a commutative analog, $\mathcal{C}_\infty$, which provides us with the minimal resolution of the operad $\mathcal{C}omm$ from Example 1.5. Using remark 2.2 one can describe $\mathcal{C}_\infty$ as follows: a *$\mathcal{C}_\infty$-algebra* is, by definition, an $\mathcal{A}_\infty$-algebra $(V, \{\mu_n\}_{n\geq 1})$ such that every multilinear operation

$\mu_n : \otimes^n V \to V$ is a Harrison cochain, that is, vanishes on every *shuffle product* which is given on generators by the formula

$$(v_1 \otimes \ldots \otimes v_i)\mathbf{x}(v_{i+1} \otimes \ldots \otimes v_n) = \sum_{\sigma \in Sh(i,n)} (-1)^\sigma v_{\sigma(1)} \otimes \ldots \otimes v_{\sigma(n)}.$$

Here $(-1)^\sigma$ is the standard Koszul sign and $Sh(i,n)$ stands for a subset of $\mathbb{S}_n$ consisting of all permutations satisfying $\sigma^{-1}(1) < \ldots < \sigma^{-1}(i)$, $\sigma^{-1}(i+1) < \ldots < \sigma^{-1}(n)$. In particular, μ_2 must be graded commutative, $\mu_2(v_1, v_2) = (-1)^{|v_1||v_2|}\mu_2(v_2, v_1)$.

2.5. Example: operad $\mathcal{L}_\infty$. Let $\hat{\mathcal{L}}$ be an $\mathbb{S}$-module given by,

$$\hat{\mathcal{L}}(n) := \begin{cases} 0 & \text{if } n = 1 \\ \mathbf{1}_n[2n-3] & \text{if } n \geq 2. \end{cases}$$

If we identify a basis vector of the one dimensional vector space $\mathbf{1}_n[2n-3]$ with the (unique, up to an isomorphism) *space* $[n]$-corolla,

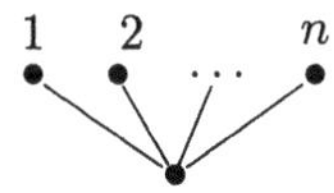

then the associated free operad $\mathcal{L}_\infty := (\{Free(\hat{\mathcal{L}})(n), \circ_i^{n,n'}\})$ can be represented as a linear span of all possible (isomorphism classes of) *space* $[n]$-trees with the compositions $\circ_i^{n,n'}$ given by gluing the root vertex of a planar $[n']$-tree to the ith tail vertex of an $[n]$-tree. The point is that $\mathcal{L}_\infty$ can be naturally made into a dg operad with the differential d given on the generators by

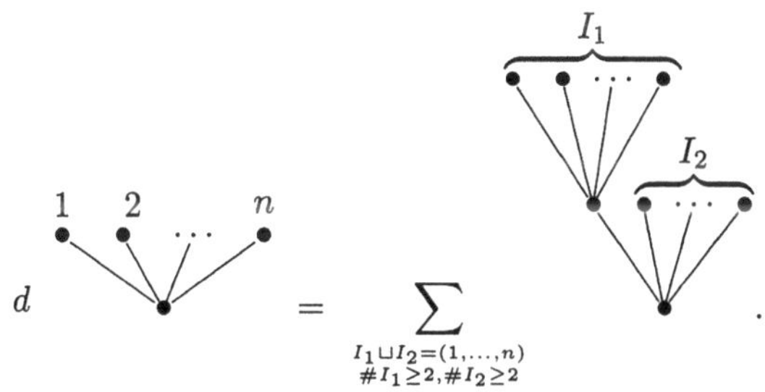

The associated cohomology operad, $H(\mathcal{L}_\infty, d)$, is isomorphic to $\mathcal{L}ie$ [**GiKa**]. Hence, the natural morphism of dg operads,

$$f : (\mathcal{L}_\infty, d) \longrightarrow (\mathcal{L}ie, 0),$$

defined to be identity on [2]-corollas and zero on $[n \geq 3]$-corollas, is a quasi-isomorphism. Thus $(\mathcal{L}_\infty, d)$ is the minimal resolution of the operad of Lie algebras.

An algebra over the dg operad $\mathcal{L}_\infty$ is called a *strongly homotopy Lie algebra* or, shortly, a $\mathcal{L}_\infty$-*algebra*. This is a dg vector space (V, d) equipped with degree $3 - 2n$ multilinear operations $\nu_n : \odot^n V \to V$, $n \geq 2$, such that for any $N \geq 1$ and any $v_1, \ldots, v_N \in V$,

$$\sum_{\substack{I_1 \sqcup I_2 = (1,\ldots,n) \\ \#I_1 \geq 1, \#I_2 \geq 0}} (-1)^\sigma \mu_{\#I_2+1}\left(\mu_{\#I_1}(v_{I_1}), v_{I_2}\right) = 0,$$

where $\mu_1 := d$, $(-1)^\sigma$ is the standard Koszul sign associated with a permutation of the elements $v_1, \ldots, v_N$, and

$$v_I := v_{i_1} \otimes \ldots \otimes v_{i_l}, \quad \text{for } I = (i_1, \ldots, i_l) \subset (1, \ldots, n).$$

2.6. Example: operad $\mathcal{G}_\infty$. Using remark 2.2 one can describe the minimal resolution, $\mathcal{G}_\infty$, of the operad, $\mathcal{G}$, of graded commutative Gerstenhaber algebras as follows [**GetJo**][6].

Let V be a finite dimensional graded vector space and

$$\mathsf{Lie}(V^*[-2]) = \bigoplus_{k=1}^{\infty} \mathsf{Lie}^k(V^*[-2])$$

the free graded Lie algebra generated by the shifted dual vector space, i.e.

$$\mathsf{Lie}^1(V^*[-2]) := V^*[-2], \quad \mathsf{Lie}^k(V^*[-2]) := \left[V^*[-2] \bullet \mathsf{Lie}^{k-1}(V^*[-2])\right].$$

The Lie bracket on $\mathsf{Lie}(V^*[-2])$ extends in the usual way to the completed (with respect to the natural filtration, $\mathsf{Lie}^{\geq k}(V^*[-2])$) graded commutative associative algebra

$$\hat{\odot}^\bullet \mathsf{Lie}(V^*[-2]) = \prod_{k=0}^{\infty} \odot^k \mathsf{Lie}(V^*[-2]),$$

making the latter into a graded commutative Gerstenhaber algebra.

2.6.1. Proposition [**GetJo**] (see also [**Ta, TaTs**]). *A G_∞-algebra structure on a finite-dimensional vector space V is, by definition, a differential, of the free $\mathcal{G}$-algebra $\hat{\odot}^\bullet \mathsf{Lie}(V^*[-2])$.*

There is a one-to-one correspondence between arbitrary derivations, D, of $\hat{\odot}^\bullet \mathsf{Lie}(V^*[-2])$ and arbitrary collections of linear maps,

$$m^*_{k_1,\ldots,k_n} : V^*[-2] \longrightarrow \mathsf{Lie}^{k_1}(V^*[-2]) \odot \ldots \odot \mathsf{Lie}^{k_n}(V^*[-2]).$$

Upon dualization the latter go into linear homogeneous maps,

$$m_{k_1,\ldots,k_n} : \frac{V^{\otimes k_1}}{\text{shuffle products}} \otimes \ldots \otimes \frac{V^{\otimes k_n}}{\text{shuffle products}} \longrightarrow V,$$

of degree $3 - n - k_1 - \ldots - k_n$. The condition $D^2 = 0$ translates into a well-defined set of quadratic equations for $m_{k_1,\ldots,k_n}$ which say, in particular, that m_1 is a differential on V and that the product, $v_1 \cdot v_2 := (-1)^{\tilde{v}_1} m_2(v_1, v_2)$, together with the Lie bracket, $[v_1 \bullet v_2] := -(-1)^{\tilde{v}_1} m_{1,1}(v_1, v_2)$, satisfy the Poisson identity up to a homotopy given by $m_{2,1}$. Hence the associated cohomology space $H(V, \mu_1)$ is a graded commutative Gerstenhaber algebra with respect to the binary operations induced by m_2 and $m_{1,1}$.

2.7. A tower of approximations to the $\mathcal{G}_\infty$ operad. Let V be a (finite-dimensional, see Remark 2.2) vector space and $\hat{\odot}^\bullet (\mathsf{Lie}V^*[-2])$ the associated free $\mathcal{G}$-algebra. It is easy to see that the multiplicative ideal, $I := < \mathsf{Lie}^{\geq 2} V^*[-2] >$, generated by the commutant of $\mathsf{Lie}V^*[-2]$, as well as its multiplicative power I^n, $n \geq 2$, are also *Lie* ideals of $\hat{\odot}^\bullet (\mathsf{Lie}V^*[-2])$. Hence the quotient, $\hat{\odot}^\bullet (\mathsf{Lie}V^*[-2]) / I^n$,

[6] Actually, in [**GetJo**] this operad was called *Fulton-MacPherson* operad. We, however, follow in this paper the terminology and notations used in [**TaTs**]

$n \geq 1$, has a canonical $\mathcal{G}$-algebra structure (note, however, that in the case $n = 1$ the induced Lie brackets vanish).

2.7.1. Definition of operad $\mathcal{G}_\infty^{(n)}$: A $\mathcal{G}_\infty^{(n)}$-*algebra* structure on a finite-dimensional vector space V is a differential of the quotient $\mathcal{G}$-algebra $\hat{\odot}^\bullet(\mathsf{Lie}V^*[-2])/I^n$, $n \geq 1$.

The following statements are obvious.

2.7.2. Lemma. $\mathcal{G}_\infty^{(1)} = \mathcal{L}_\infty$.

2.7.3. Lemma. *For any $n \geq 2$, there is a commutative diagram,*

$$\begin{array}{ccc} \mathcal{G}_\infty^{(n)} & \longrightarrow & \mathcal{G}_\infty \\ \uparrow & & \uparrow \\ \mathcal{L}_\infty & \xrightarrow{=} & \mathcal{L}_\infty \end{array}$$

with all arrows being canonical cofibrations[7] *of operads.*

We shall give below in this paper a fairly explicit description of the next two floors, $\mathcal{G}_\infty^{(2)}$ and $\mathcal{G}_\infty^{(3)}$, of the tower,

$$\mathcal{L}_\infty = \mathcal{G}_\infty^{(1)} \longrightarrow \mathcal{G}_\infty^{(2)} \longrightarrow \mathcal{G}_\infty^{(3)} \longrightarrow \ldots \longrightarrow \mathcal{G}_\infty^{(n)} \longrightarrow \ldots,$$

of cofibrant approximations to the operad $\mathcal{G}_\infty = \mathrm{colim}\mathcal{G}_\infty^{(n)}$. Interestingly enough, the operad $\mathcal{G}_\infty^{(2)}$ is closely related (through the cobar construction) to the operad $\mathcal{G}erst$ and governs Frobenius$_\infty$ manifolds introduced in [**Me2**], while the operad $\mathcal{G}_\infty^{(3)}$ governs strong homotopy generalizations of Hertling-Manin's F-manifolds.

2.8. Homotopy theory. The categories of operads and of their algebras belong to a class of so called closed model categories [**Qu**] which have a particularly nice homotopy theory. Here is a brief outline of all the relevant notions and facts we use in the paper (see, e.g., [**DS, GeMa**] for more details and proofs).

2.8.1. Definition. A *closed model category* is, by definition, a category $\mathcal{C}at$ with three distinguished classes of morphisms — (i) *weak equivalences*, $\mathfrak{E}$, (ii) *fibrations*, $\mathfrak{F}$, and (iii) *cofibrations*, $\mathfrak{F}^\circ$, — which are closed under composition and contain all identity maps. The following axioms must be satisfied:
CMC1: Finite limits and colimits exist in $\mathcal{C}at$.
CMC2: If f and g are morphisms in $\mathcal{C}at$ such that their composition fg is defined, then if any two of the three maps f,g, fg are weak equivalences then so is the third morphism.
CMC3: Given any commutative diagram of the form,

$$\begin{array}{ccccc} A & \xrightarrow{i} & B & \xrightarrow{p} & A \\ {\scriptstyle f}\downarrow & & \downarrow{\scriptstyle g} & & {\scriptstyle f}\downarrow \\ A' & \xrightarrow{i'} & B' & \xrightarrow{p'} & A' \end{array}$$

with pi and $p'i'$ being the identity maps. If g is in $\mathfrak{E}$, $\mathfrak{F}$ or $\mathfrak{F}^\circ$, then so is f.

[7]The notion of cofibration is explained in Sect. 2.8.

CMC4: Given any commutative solid arrow diagram of the form,

$$\begin{array}{ccc} A & \longrightarrow & B \\ {\scriptstyle f}\downarrow & \overset{h}{\nearrow} & \downarrow{\scriptstyle g} \\ A' & \longrightarrow & B' \end{array}$$

A dotted arrow h commuting with all other maps exists in either of the following two situations: (i) $f \in \mathfrak{F}^\circ \cap \mathfrak{E}$ and $g \in \mathfrak{F}$, or (ii) $f \in \mathfrak{F}^\circ$ and $g \in \mathfrak{F} \cap \mathfrak{E}$.
CMC5 Any morphism can be factored in two ways: (i) $F = pi$ with $i \in \mathfrak{F}^\circ$ and $p \in \mathfrak{F} \cap \mathfrak{E}$, and (ii) $f = pi$ $i \in \mathfrak{F}^\circ \cap \mathfrak{E}$ and $p \in \mathfrak{F}$.

2.8.2. Definitions. (i) If a pair of morphisms, $f : A \to A'$ and $g : B \to B'$, satisfies the condition **CMC4**, then we say that f has the *left lifting property* (LLP) with respect to g or that g has the *right lifting property* (RLP) with respect to f.

(ii) The morphisms in $\mathfrak{F} \cap \mathfrak{E}$ are called *acyclic fibrations*. The morphisms in $\mathfrak{F}^\circ \cap \mathfrak{E}$ are called *acyclic cofibrations*.

(iii) The axiom **CMC1** implies, in particular, that every closed model category $\mathcal{C}at$ has both an initial object $\emptyset$ and a terminal object $*$. An object A of $\mathcal{C}at$ is called *fibrant* (resp., *cofibrant*) if $A \to *$ is a fibration (resp., $\emptyset \to A$ is a cofibration).

2.8.3. Facts [GetJo, Hi]. (i) The category of dg operads is a closed model category with

- weak equivalences $\mathfrak{E}$ ={ the morphisms, $f : (\mathcal{O}, d) \to (\mathcal{O}', d')$, of dg operads which induce isomorphism, $[f] : H(\mathcal{O}) \to H(\mathcal{O}')$, in cohomology};
- fibrations $\mathfrak{F}$={surjective morphisms of dg operads};
- cofibrations $\mathfrak{F}^\circ$={the morphisms which have LLP with respect to all acyclic fibrations }.

(ii) Given a dg operad $\mathcal{O}$, the associated category of $\mathcal{O}$-algebras is a closed model category with the classes of morphism $\mathfrak{E}$, $\mathfrak{F}$ and $\mathfrak{F}^\circ$ defined in close analogy to (i).

(iii) Every object in the closed model categories (i) and (ii) is obviously fibrant.

2.8.4. Homotopy and derived categories of a closed model category. From now on $\mathcal{C}at$ stands for a closed model category. Moreover, we assume for simplicity that every object in $\mathcal{C}at$ is fibrant (as in the two examples above).

Two morphisms, $f, g : A \to B$, are called (right) *homotopic* if there exists a path object, for B (that is an object B^I together with a diagram

$$B \xrightarrow{i} B^I \xrightarrow{p} B \times B, \quad i \in \mathfrak{E},$$

which factors the diagonal map $B \overset{(id,id)}{\longrightarrow} B \times B$) such that the product map $(f, g) : A \to B \times B$ lifts to a map $H : A \longrightarrow B^I$. Such a map H is called a (right) homotopy from f to g. This, in fact, defines an equivalence relation $\sim$ in $\mathrm{Hom}_{\mathcal{C}at}(A, B)$ for any objects A, B. However, the associated homotopy classes of maps,

$$\pi(A, B) := \frac{\mathrm{Hom}_{\mathcal{C}at}(A, B)}{\sim},$$

do not necessarily compose, $\pi(A, B) \times \pi(B, C) \to \pi(A, C)$, unless the objects involved are cofibrant.

By **CMC5**(i), the map $\emptyset \to A$ can be factored, $\emptyset \to QA \stackrel{p_A}{\to} A$, with QA being cofibrant and p_A a weak equivalence. Such an object QA is called a *cofibrant resolution* of A. Usually, cofibrant resolutions are constructed by the method of "adding a new variable and killing a cycle". A nice illustration of the method at work is, for example, Markl's [**Mar1**] original construction of the minimal resolution $\mathcal{O}_\infty$ of a simply connected dg operad $\mathcal{O}$; Markl's minimal resolutions give an important class of cofibrant objects in the category of dg operads.

Another remarkable fact is that not only every object, but also every morphism, $f : A \to B$, has a "cofibrant resolution" Qf making the following diagram commutative,

$$\begin{array}{ccc} QA & \xrightarrow{Qf} & QB \\ {\scriptstyle p_A}\downarrow & & \downarrow{\scriptstyle p_B} \\ A & \xrightarrow{f} & B \end{array}$$

Moreover, the homotopy class of such maps $[Qf]$ is defined uniquely by the homotopy class $[f]$.

The *homotopy category*, $\mathbf{H}o(\mathcal{C}at)$, is the category with the same objects as $\mathcal{C}at$ and with morphisms given by

$$\mathrm{Hom}_{\mathbf{H}o(\mathcal{C}at)}(A, B) := \pi(QA, QB).$$

Clearly, there is a canonical functor $\alpha : \mathcal{C}at \to \mathbf{H}o(\mathcal{C}at)$, which is the identity on objects and sends morphisms f to $[Qf]$.

The derived category, $\mathbf{D}(\mathcal{C}at)$, is the category obtained from $\mathcal{C}at$ by localization with respect to weak equivalences; put another way, this is a category together with a functor $F : \mathcal{C}at \to \mathbf{D}(\mathcal{C}at)$ satisfying two conditions,

- $F(f)$ is an isomorphism for each weak equivalence f;
- every functor $G : \mathcal{C}at \to \mathcal{C}at'$ sending weak equivalences into isomorphisms factors uniquely through $(\mathbf{D}(\mathcal{C}at), F)$,

$$G : \mathcal{C}at \xrightarrow{F} \mathbf{D}(\mathcal{C}at) \xrightarrow{G'} \mathcal{C}at',$$

  for some functor G'.

Note that the definition of $\mathbf{D}(\mathcal{C}at)$ involves only one class, $\mathfrak{E}$, of the three classes which define the closed model structure in $\mathcal{C}at$. Nevertheless, one of the central results in Quillen's [**Qu**] theory of closed model categories asserts the equivalence,

$$\mathbf{H}o(\mathcal{C}at) \simeq \mathbf{D}(\mathcal{C}at),$$

of the two categories associated to $\mathcal{C}at$.

2.8.5. Transfer Theorem. *Let $\mathcal{P}$ be a cofibrant dg operad and $f : V \to V'$ a quasi-isomorphism of dg vector spaces. For any $\mathcal{P}$-algebra structure on V (resp. V'), there exists a $\mathcal{P}$-algebra structure on V' (resp. V) so that V and V' are equivalent as $\mathcal{P}$-algebras (i.e. there exists a $\mathcal{P}$-algebra V'' and a pair of quasi-isomorphisms of $\mathcal{P}$-algebras, $V \leftarrow V'' \to V'$).*

This is a well known fact. We show the proof only for completeness (cf. [**BeMo**]).

Proof. The dg operads $\mathcal{E}_V$ and $\mathcal{E}_{V'}$, if viewed only as dg $\mathbb{S}$-modules, have two natural maps,

$$\mathcal{E}_V \xrightarrow{f^*} \mathcal{E}_{V,V'}, \quad \mathcal{E}_{V'} \xrightarrow{f_*} \mathcal{E}_{V,V'},$$

to the dg $\mathbb{S}$-module $\mathcal{E}_{V,V'} := \{Hom(V^{\otimes n}, V'), d\}_{n\geq 1}$. Define a dg $\mathbb{S}$-module, $\mathcal{E}_f = \{\mathcal{E}_f(n)\}_{n\geq 1}$, by setting

$$\mathcal{E}_f(n) := \left\{(v, v') \in \mathcal{E}_V(n) \times \mathcal{E}_{V'}(n) \mid fv = v' f^{\otimes n}\right\},$$

or, alternatively, by the pullback diagram of dg $\mathbb{S}$-modules,

$$\begin{array}{ccc} \mathcal{E}_f & \xrightarrow{i_1} & \mathcal{E}_V \\ {\scriptstyle i_2}\downarrow & & \downarrow{\scriptstyle f_*} \\ \mathcal{E}_{V'} & \xrightarrow{f^*} & \mathcal{E}_{V,V'}. \end{array}$$

It is easy to check that $\mathcal{E}_f$ inherits from the $\mathcal{E}_V$ and $\mathcal{E}_{V'}$ not only the $\mathbb{S}$-module structure, but also the compositions $\circ_i^{n,n'}$. Thus $\mathcal{E}_f$ is a dg operad with maps i_1 and i_2 above being morphisms of dg operads. Moreover, i_1 and i_2 are weak equivalences (recall that we are working over a field of characteristic zero). Hence, as objects in $\mathbf{D}(\mathrm{Oper})$, the dg operads $\mathcal{E}_V$, $\mathcal{E}_f$ and $\mathcal{E}_{V'}$ are isomorphic. Since the derived category of dg operads is equivalent to the homotopy category, we get isomorphisms of sets,

$$\pi(\mathcal{P}, \mathcal{E}_V) \simeq \pi(\mathcal{P}, \mathcal{E}_f) \simeq \pi(\mathcal{P}, \mathcal{E}_{V'}).$$

If, say, V is a $\mathcal{P}$-algebra, the homotopy equivalence class of the structure map $\phi : \mathcal{P} \to \mathcal{E}_V$ gives rise to an element $[\phi_f]$ in $\pi(\mathcal{P}, \mathcal{E}_f)$. As $\mathcal{P}$ is cofibrant, the latter has a representative, $\phi_f \in \mathrm{Hom}_{\mathrm{Oper}}(\mathcal{P}, Q\mathcal{E}_f)$, where $p_f : Q\mathcal{E}_f \to \mathcal{E}_f$ is some cofibrant resolution of $\mathcal{E}_f$. By construction, the composition $i_1 \circ p_f \circ \phi_f : \mathcal{P} \to \mathcal{E}_V$ is homotopy equivalent to the original structure map $\phi : \mathcal{P} \to \mathcal{E}_V$. Finally, another composition, $i_1 \circ \partial_f \circ \phi_f : \mathcal{P} \to \mathcal{E}_{V'}$ makes V' into a $\mathcal{P}$ algebra which, in the derived category of $\mathcal{P}$-algebras, is obviously isomorphic to ϕ. Analogously one proves the dual statement.

□

2.8.6. Sh algebras. An algebra over a cofibrant operad is called a *strongly homotopy* (or, shortly, sh) algebra. By the Theorem above, sh algebraic structures can be transferred by quasi-isomorphisms of complexes.

2.9. Markl's theory of sh maps. The beauty of sh algebras, the transfer property 2.8.5, is spoiled by the fact that to compare such structures on quasi-isomorphic dg spaces V and V' (we refer to 2.8.5 again) one has to resort to a *chain* of strict $\mathcal{P}$-algebra morphisms, $V \leftarrow V'' \to V'$, involving a third party which is often hard to construct explicitly. One may try to overcome this deficiency by appropriately extending the notion of *map* between sh algebras.

Markl made in [**Mar3, Mar4**] an interesting suggestion which, in the setting of the proof of the Transfer Theorem, can be illustrated as follows. First one observes that the operad $\mathcal{E}_f$ is in fact a two coloured operad with one colour associated to V and another one to V'. Next one constructs a two coloured cofibrant resolution, $Q\mathcal{E}_f$, and then defines the set of *sh maps* between the $\mathcal{P}$-algebras V and V' as the set of all algebras over the dg operad $Q\mathcal{E}_f$. In this way Markl was able to prove stronger versions of the Transfer Theorem [**Mar3**]. The problem, however, with this approach is that it is not yet clear whether or not such sh maps can be composed making the pair (sh algebras, sh maps) into a genuine category. At present, this is known to be true only for a class of sh algebras associated with Koszul operads. In particular, it is true for $\mathcal{A}_\infty$-, $\mathcal{C}_\infty$- and $\mathcal{L}_\infty$-algebras reproducing thereby the well established theory of sh maps of these three classes of sh algebras. For later reference we review below a few basic facts (see, e.g., [**Ko1, Pr**]).

2.10. Sh maps of $\mathcal{A}_\infty$- and $\mathcal{C}_\infty$-algebras. An A_∞-structure on a vector space V can be suitably represented as a codifferential, $\mu : (T^\bullet V[1], \Delta) \to (T^\bullet V[1], \Delta)$, of the free tensor coalgebra cogenerated by $V[1]$. A *sh map*, $f : (V, \mu_\bullet) \to (\hat{V}, \hat{\mu}_\bullet)$, of $\mathcal{A}_\infty$-algebras is, by definition, a morphism of the associated differential coalgebras, $f : (T^\bullet V[1], \Delta, \mu) \to (T^\bullet \hat{V}[1], \Delta, \hat{\mu})$. Such a map is equivalent to a set of linear maps $\{f_n : V^{\otimes n} \longrightarrow \tilde{V},\ n \geq 1\}$ of degree $1 - n$ which satisfy the equations,

$$\sum_{1\leq k_1<k_2<\ldots<k_i=n} (-1)^{i+r} \hat{\mu}_i(f_{k_1}(v_1, \ldots, v_{k_1}), f_{k_2-k_1}(v_{k_1+1}, \ldots, v_{k_2}), \ldots, f_{n-k_{i-1}}(v_{k_{i-1}+1}, \ldots, v_n))$$

$$= \sum_{k+l=n+1} \sum_{j=0}^{k-1} (-1)^{l(\tilde{v}_1+\ldots+\tilde{v}_j+n)+j(l-1)} f_k(v_1, \ldots, v_j, \mu_l(v_{j+1}, \ldots, v_{j+l}), v_{j+l+1}, \ldots, v_n),$$

The pair (Ob = $\mathcal{A}_\infty$-algebras, Mor = sh maps) forms a category called the *category of $\mathcal{A}_\infty$-algebras.*

A sh map $f = \{f_n\} : (V, \mu_\bullet) \to (\hat{V}, \hat{\mu}_\bullet)$ is called a *quasi-isomorphism* if the associated map of dg vector spaces, $f_1 : (V, \mu_1) \to (\hat{V}, \hat{\mu}_1)$, induces an isomorphism in cohomology.

Two sh maps, $f, g : (T^\bullet V[1], \Delta, \mu) \to T^\bullet \hat{V}[1], \Delta, \hat{\mu})$, are said to be *homotopic* if there is a homogeneous map, $h : T^\bullet V[1] \to T^\bullet \hat{V}[1]$, of degree -1 such that

$$\Delta h = (f \otimes h + h \otimes g)\Delta, \qquad f - g = \hat{\mu} \circ h + \mu \circ h.$$

Remarkably enough, homotopy induces an equivalence relation in the set of sh maps $(V, \mu_\bullet) \to (\hat{V}, \hat{\mu}_\bullet)$ [**Pr**]. Moreover, a sh map $f = \{f_n\} : (V, \mu_\bullet) \to (\hat{V}, \hat{\mu}_\bullet)$ is a quasi-isomorphism if and only if it is a homotopy equivalence. Thus the derived category of $\mathcal{A}_\infty$-algebras is simply the quotient of the category $\mathcal{A}_\infty$-algebras by the above homotopy relation!

For $\mathcal{C}_\infty$-algebras one has a similar list of definitions and results.

2.11. Sh maps of $\mathcal{L}_\infty$-algebras. A $\mathcal{L}_\infty$-algebra structure, $\nu = \{\nu_n : \odot^n V \to V,\ |\nu_n| = 3 - 2n\}$, on a vector space V can be compactly described as a codifferential, $\nu : (\odot^\bullet V[2], \Delta) \to (\odot^\bullet V[2], \Delta)$, of the free cocommutative tensor coalgebra cogenerated by $V[2]$. A *sh map*, $f : (V, \nu_\bullet) \to (\hat{V}, \hat{\nu}_\bullet)$, of $\mathcal{L}_\infty$-algebras is, by definition, a morphism of the associated differential cocommutative coalgebras, $f : (\odot^\bullet V[2], \Delta, \nu) \to (\odot^\bullet \hat{V}[2], \Delta, \hat{\nu})$. Such a map is equivalent to a set of linear maps $\{f_n : \odot^n V \longrightarrow \tilde{V},\ n \geq 1\}$ of degree $2 - 2n$ which satisfy the equations similar to the ones in Subsect. 2.10. The notions of quasi-isomorphism and homotopy are similar as well.

The pair (Ob = $\mathcal{L}_\infty$-algebras, Mor = sh maps) forms a category called the *category of $\mathcal{L}_\infty$-algebras.*

Dualizing the above formulae for a finite-dimensional vector space V one arrives at a beautiful geometric formulation of $\mathcal{L}_\infty$-algebras and their sh maps [**Ko1**]:

- a $\mathcal{L}_\infty$-algebra structure on V can be identified with a smooth degree 1 vector field $\vec{\nu}$ on the pointed flat graded manifold $(V[2], 0)$ which satisfies the conditions $[\vec{\nu}, \vec{\nu}] = 0$ and $\vec{\nu}|_0 = 0$. Explicitly, the identification,

$$\vec{\nu} \longleftrightarrow \{\nu_n : \odot^n V \to V\},$$

is given by the formula,

$$\vec{\nu} = \sum_{n=1}^{\infty} \sum_{\alpha, \beta_1, \dots \beta_n} \frac{(-1)^r}{n!} t^{\beta_1} \dots t^{\beta_n} \mu^{\alpha}_{\beta_1, \dots, \beta_n} \frac{\partial}{\partial t^\alpha},$$

where $\{t^\alpha, \alpha = 1, \dots, \dim V\}$ is the basis of $V^*[-2]$ associated to a basis, $\{e_\alpha\}$, of V (so that $|t^\alpha| = -|e_\alpha| + 2$),

$$r = (2n-3)(|e_{\alpha_1}| + \dots + |e_{\alpha_n}|) + \sum_{k=2}^{n} |e_{\alpha_k}|(|e_{\alpha_1}| + \dots + |e_{\alpha_{k-1}}|),$$

and $\mu^{\alpha}_{\beta_1, \dots, \beta_n} \in k$ are given by

$$\mu_n(e_{\beta_1}, \dots, e_{\beta_n}) = \sum_\alpha \mu^{\alpha}_{\beta_1, \dots, \beta_n} e_\alpha.$$

- A sh map of $f : (V, \nu_\bullet) \to (\hat{V}, \hat{\nu}_\bullet)$, of $\mathcal{L}_\infty$-algebras is a smooth map of pointed graded manifolds, $f : (V^*[-2], 0) \to (\hat{V}^*[-2], 0)$ such that $f_*(\vec{\nu})$ is well defined and coincides with $\vec{\hat{\nu}}$. Put another way, a sh map of $\mathcal{L}_\infty$-algebras is just a morphism of the associated pointed dg manifolds.

A $\mathcal{L}_\infty$-algebra $(V, \{\nu_\bullet\}_{n\geq 1})$ with $\nu_1 = 0$ is called *minimal* (equivalently, the homological vector field $\vec{\nu}$ has zero at the distinguished point of order ≥ 2).

2.11.1. Facts [Ko1]. (i) *Every $\mathcal{L}_\infty$-algebra is quasi-isomorphic to a minimal one.*

(ii) *There is a one-to-one correspondence between quasi-isomorphisms of $\mathcal{L}_\infty$-algebras and diffeomorphisms of the associated dg manifolds.*

2.11.2. Fact [Me2]. *The canonical functor*

$$\left\{ \begin{array}{c} \text{The category} \\ \text{of } \mathcal{L}_\infty\text{-algebras} \end{array} \right\} \longrightarrow \left\{ \begin{array}{c} \text{The derived category} \\ \text{of } \mathcal{L}_\infty\text{-algebras} \end{array} \right\},$$

when restricted to minimal $\mathcal{L}_\infty$-algebras, becomes simply a forgetful functor,

$$(M, *, \text{flat structure}, \vec{\nu}) \longrightarrow (M, *, \vec{\nu}),$$

which forgets the flat (=affine) structure on $(M, *) = (V^*[2], 0)$.

Thus a homotopy class of minimal $\mathcal{L}_\infty$-algebras is nothing but a pointed formal dg manifold, $(M, *, \vec{\nu})$, with no preferred choice of local coordinates. Moreover, the derived (=homotopy) category of L_∞-algebras is equivalent to the purely geometric category of formal dg manifolds.

§3. Cobar construction for $\mathcal{G}erst$

3.1. Cobar construction. For an $\mathbb{S}$-module $\mathcal{O} = \{\mathcal{O}(n)\}_{n\geq 1}$ we set $\mathcal{O}\{m\}$ to be an $\mathbb{S}$-module given by the tensor product,

$$\mathcal{O}\{m\}(n) := \mathcal{O}(n) \otimes_k \Lambda_n^{\otimes m}[m(n-1)],$$

where Λ_n is the sign representation of the permutation group $\mathbb{S}_n$. If $\mathcal{O}$ is a dg operad, then $\mathcal{O}\{m\}$ is naturally a dg operad as well: a structure of $\mathcal{O}\{m\}$-algebra on a dg vector space V is the same as a structure of $\mathcal{O}$-algebra on the shifted dg vector space $V[m]$.

Let $\mathcal{O} = \{\mathcal{O}(n), \circ_i^{n,n'}, d\}$ be a simply connected dg operad and let $\mathcal{O}^*[-1]$ stand for the $\mathbb{S}$-module $\{\mathcal{O}(n)^*[-1]\}$. It was shown in [**GiKa**] that the free operad associated to the $\mathbb{S}$-module $Free(\mathcal{O}^*[-1]\{-1\})$ can be naturally made into a *differential* operad, $\mathcal{D}(\mathcal{O}) = (Free(\mathcal{O}^*[-1]\{-1\}), \delta)$, with the differential δ defined by that in $\mathcal{O}$ and the compositions $\circ_i^{n,n'}$. This construction gives rise to a functor, $\mathcal{D} : Oper_1 \to Oper_1$, on the category of simply connected dg operads with the property that $\mathcal{D}(\mathcal{D}(\mathcal{O}))$ is canonically quasi-isomorphic to the original operad $\mathcal{O}$. This functor is called a *cobar construction*.

3.2. Koszul duals. An operad $\mathcal{O}$ is called *quadratic* if it can be represented as a quotient,

$$\mathcal{O} = \frac{Free(\mathcal{E})}{< R >},$$

of the free operad generated by an $\mathbb{S}$-module $\mathcal{E}$ with $\mathcal{E}(n) = 0$ for $n \neq 2$ by an ideal generated by an $\mathbb{S}_3$-invariant subspace R in $Free(\mathcal{E})(3)$. For example, operads $\mathcal{C}omm$, $\mathcal{A}ss$ and $\mathcal{L}ie$ are quadratic.

The *Koszul dual* of a quadratic operad $\mathcal{O} = Free(\mathcal{E})/ < R >$ is, by definition, the quadratic operad $\mathcal{O}^! = Free(\hat{\mathcal{E}})/ < R^\perp >$ where $\hat{\mathcal{E}}$ is the $\mathbb{S}$-module whose only non-vanishing component is $\hat{\mathcal{E}}(2) = \mathcal{E}(2)^* \otimes \Lambda_2$ and $R^\perp$ is the annihilator of R, i.e. the kernel of the natural map $Free(\hat{\mathcal{E}})(3) \to R^*$.

Applying cobar construction to the Koszul dual of a quadratic operad $\mathcal{O}$ one gets a cofibrant dg operad $\mathcal{D}(\mathcal{O}^!)$ together with a canonical map of dg operads [**GiKa**],

$$(\mathcal{D}(\mathcal{O}^!), \delta) \longrightarrow (\mathcal{O}, 0).$$

Whatever the operad $\mathcal{O}$ is, the associated $\mathcal{D}(\mathcal{O}^!)$-algebras are strong homotopy ones.

3.3. Koszul operads. A quadratic operad $\mathcal{O}$ is called *Koszul* if the canonical map $(\mathcal{D}(\mathcal{O}^!), \delta) \longrightarrow (\mathcal{O}, 0)$ is a quasi-isomorphism. In such a case the cobar construction applied to the Koszul dual operad $\mathcal{O}^!$ provides us with the minimal resolution, $\mathcal{O}_\infty = \mathcal{D}(\mathcal{O}^!)$, of the operad $\mathcal{O}$ (see Sect. 2.1).

3.4. Examples [GetJo, GiKa]. The operads $\mathcal{A}ss$, $\mathcal{C}omm$, $\mathcal{L}ie$ and $\mathcal{G}$ are Koszul with

$$\mathcal{A}ss^! = \mathcal{A}ss, \quad \mathcal{C}omm^! = \mathcal{L}ie, \quad \mathcal{L}ie^! = \mathcal{C}omm, \quad \mathcal{G}^! = \mathcal{G}\{1\}.$$

Thus the operads $\mathcal{A}_\infty := \mathcal{D}(\mathcal{A}ss)$, $\mathcal{C}_\infty := \mathcal{D}(\mathcal{L}ie)$, $\mathcal{L}_\infty := \mathcal{D}(\mathcal{C}omm)$ and $\mathcal{G}_\infty := \mathcal{D}(G\{1\})$ are minimal resolutions of the operads, $\mathcal{A}ss$, $\mathcal{C}omm$, $\mathcal{L}ie$ and $\mathcal{G}$ respectively. This explains all the claims made in Examples 2.3—2.6.

3.5. Proposition. *$\mathcal{G}erst^! = \mathcal{H}\{1\}$ and $\mathcal{H}^! = \mathcal{G}erst\{1\}$, where $\mathcal{H}$ is the operad defined in Example 1.10.*

Proof is straightforward.

3.6. Cobar construction for $\mathcal{G}erst^!$. Surprisingly enough, the minimal operad[8] $Gerst_\infty := \mathcal{D}(\mathcal{G}erst^!)$ turns out to be a much more elementary object than its "graded commutative" analogue $\mathcal{G}_\infty = \mathcal{D}(\mathcal{G}^!)$. In this section we present an explicit and simple description of the cobar construction $Gerst_\infty$ in terms of partially planar/partially space trees (reflecting its nature as a composition of the operads $\mathcal{A}_\infty$ and $\mathcal{L}_\infty$, see below), and in the next section we show that strong homotopy $Gerst_\infty$-algebras admit a nice geometric interpretation.

The main reason behind that acclaimed simplicity of $\mathcal{D}(\mathcal{G}erst^!) = \mathcal{D}(\mathcal{H}\{1\})$ is the non-distributive nature of the operad $\mathcal{H}$ (cf. Remark 1.9): "opening" the left- and right hand sides of the following two equalities in the operad $\mathcal{H}$,

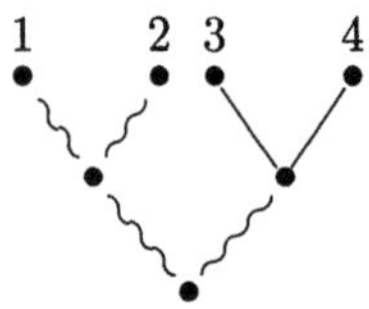

$$\simeq (a_1 \bullet a_2) \bullet (a_3 \circ a_4) = a_1 \bullet (a_2 \bullet (a_3 \circ a_4)) \simeq$$

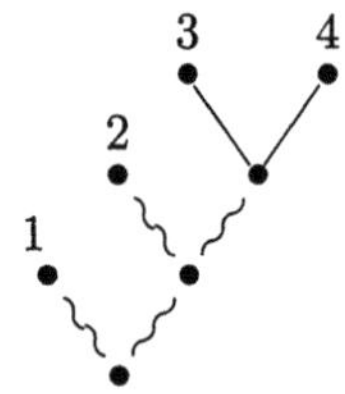

in two possible ways, one gets after the decomposition into irreducibles the following relation in $\mathcal{H}(4)$,

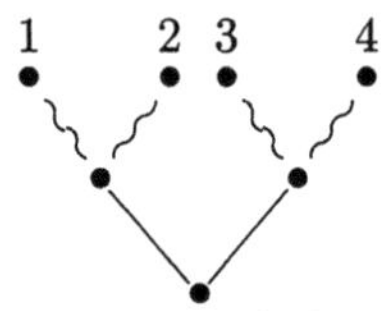

$$= 0.$$

This relation makes the computation of the $\mathbb{S}$-module structure of the operad $Gerst_\infty \simeq \mathcal{D}(\mathcal{H}\{1\}) \simeq Free(\mathcal{H}^*[-1]\{-2\})$ an easy exercise:

$$\mathcal{G}erst_\infty(n) = \mathbf{1}_n[2-3n] \quad \oplus \quad \bigoplus_{p=2}^{n} \mathrm{Ind}_{\mathbb{S}_p \times \mathbb{S}_{n-p}}^{\mathbb{S}_n} \left(k[\mathbb{S}_p] \otimes \mathbf{1}_{n-p}\right)[n+2p-2].$$

If we identify a basis vector of the one dimensional summand $\mathbf{1}_n[2n-3]$ with the (unique, up to an isomorphism) *space* $[n]$-corolla,

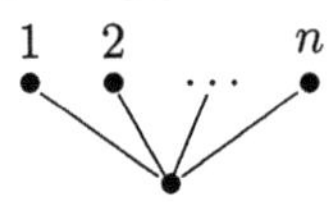

and the natural basis of all other summands with partially planar partially space $[n]$-corollas (cf. Sect. 1.7),

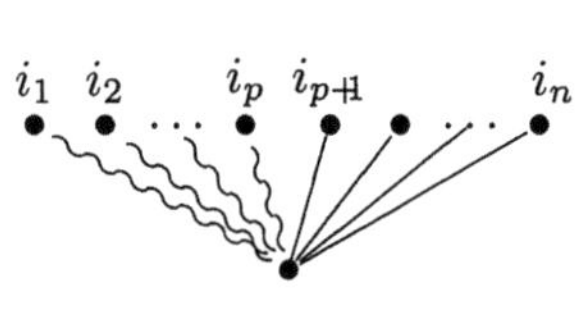

,

[8]Conjecture: $Gerst_\infty$ is a minimal resolution of $Gerst$, i.e. the operad $Gerst$ is Koszul.
In this paper we need only the fact that the operad $Gerst_\infty$ is minimal so that $Gerst_\infty$-algebras are strong homotopy ones.

then $Gerst_\infty$, as an $\mathbb{S}$-module. equals the linear span of all possible (isomorphism classes of) partially plane partially space trees formed by these corollas, while the compositions $\circ_i^{n,n'}$ are given simply by gluing the root vertex of an $[n']$-tree to the ith tail vertex of an $[n]$-tree. To complete the description of the operad $Gerst_\infty$ we need only to compute Ginzburg-Kapranov's cobar differential d.

3.6.1. Proposition. *The cobar differential in* $\mathcal{G}erst_\infty$ *is given on generators by*

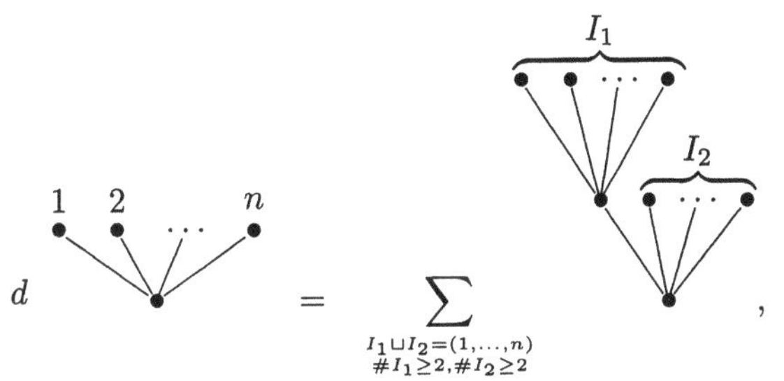

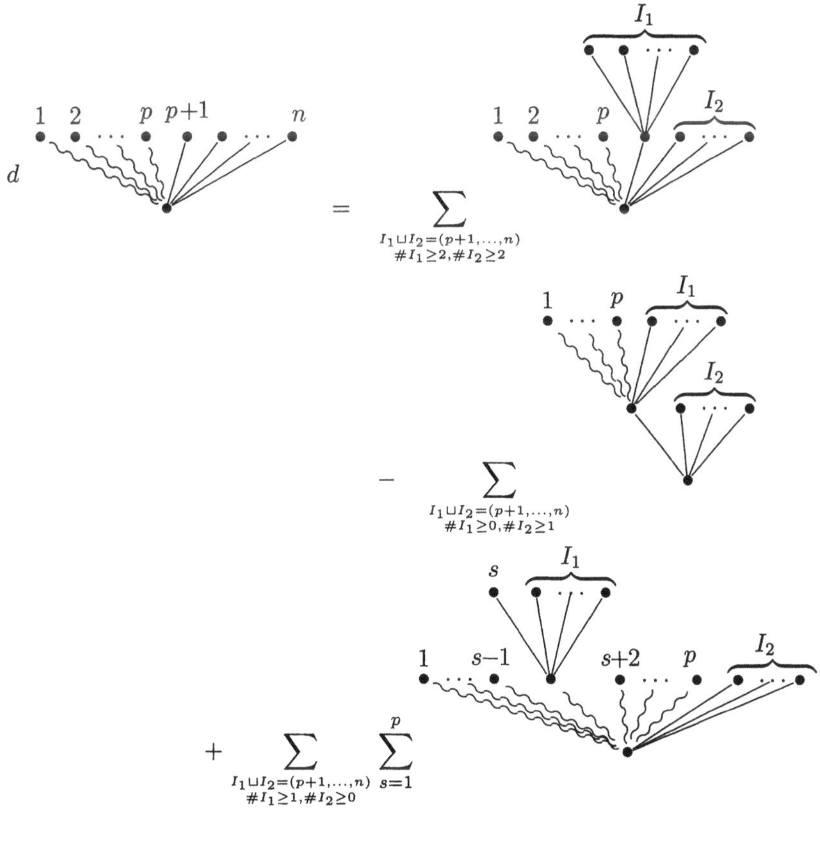

$$+ \sum_{\substack{I_1 \sqcup I_2 = (p+1,\ldots,n) \\ \#I_1 \geq 2, \#I_2 \geq 2}} \sum_{\substack{l+m=p+1 \\ l,m \geq 2}} \sum_{s=0}^{m-1} (-1)^{l+s(l+1)}$$

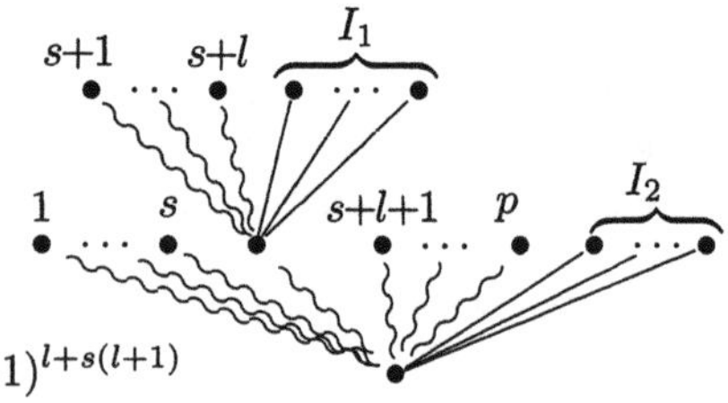

Proof is straightforward though very tedious.

Corollary 3.6.2. *There is a canonical cofibration of operads,* $\mathcal{L}_\infty \to \mathcal{G}erst_\infty$.

Corollary 3.6.3. *A* $\mathcal{G}erst_\infty$*-algebra is the data,*

$$\left(V, \ \{\nu_n\}_{n\geq 1}, \ \{\mu_{p,n}\}_{\substack{p\geq 2 \\ n \geq 1}}\right),$$

consisting of a graded vector space V *and two collections of homogeneous linear maps,*

- $\nu_n : \odot^n V \to V$ *of degree* $3-2n$, $n = 1,2,3,\ldots$, *and*
- $\mu_{n;p} : (\otimes^n V) \otimes (\odot^p V) \to V$ *of degree* $2-n-2p$, $k = 2,3,4,\ldots$, $n = 0,1,2,\ldots$,

which satisfy the equations, for any $a_1, \ldots, a_n, b_1, \ldots, b_N \in V$,

$$\sum_{S_1 \sqcup S_2 = (1,\ldots N)} (-1)^{\sqcup} \nu_{|S_2|+1}(\nu_{|S_1|}(b_{S_1}), b_{S_2}) = 0, \qquad N \geq 1,$$

and, for $n \geq 2, N \geq 0$,

$$\sum_{S_1 \sqcup S_2 = (1,\ldots N)} (-1)^{\sigma} \left\{ \nu_{|S_2|+1}(\mu_{n;|S_1|}(a_1, \ldots, a_n; b_{S_1}), b_{S_2}) \right.$$

$$-(-1)^{\tilde{a}_1 + \ldots \tilde{a}_n - n} \mu_{n;|S_2|+1}(a_1, \ldots .a_n; \nu_{S_1}(b_{S_1}), b_{S_2})$$

$$-\sum_{j=1}^{n} (-1)^{\tilde{a}_1 + \ldots \tilde{a}_{j-1} + (\tilde{a}_{j+1} + \ldots + \tilde{a}_n)\tilde{b}_{S_1} - n}$$

$$\left. \mu_{n;|S_2|}(a_1, \ldots, a_{j-1}, \nu_{|S_1|+1}(a_j, b_{S_1}), a_{j+1}, \ldots, a_n; b_{S_2}) \right\}$$

$$= \sum_{\substack{k+l=n+1 \\ k,l \geq 2}} \sum_{j=0}^{k} \sum_{S_1 \sqcup S_2 = (1,\ldots N)} (-1)^r$$

$$\mu_{k;|S_2|}(a_1, \ldots, a_j, \mu_{l;|S_1|}(a_{j+1}, \ldots, a_{j+l}; b_{S_1}), a_{j+l+1}, \ldots, a_n; b_{S_2}),$$

where $(-1)^\sigma$ *is the standard Koszul sign of the shuffle permutation* $b_1 \otimes \ldots \otimes b_N \to b_{S_1} \otimes b_{S_2}$, *and* $r = j + l(n-j-l) + l(\tilde{a}_1 + \ldots + \tilde{a}_j) + (\tilde{a}_{j+l+1} + \ldots + \tilde{a}_n)\tilde{b}_{S_1}$.

Thus the operations $\nu_\bullet$ define on V the structure of L_∞ algebra, while the operations $\mu_{\bullet;0}$ define on V the structure of A_∞ algebra; the remaining operations $\mu_{\bullet;\bullet\geq 1}$ are homotopies, and homotopies of homotopies, and ..., which make these two basic structures Poisson-type consistent. In the special case when all operations but ν_2 and $\mu_{2;0}$ vanish the $\mathcal{G}erst_\infty$-algebra V becomes nothing but a $\mathcal{G}erst$-algebra, i.e. a (non-commutative, in general) Gerstenhaber algebra.

3.7. Coalgebra interpretation. The notion of $\mathcal{H}$-algebra (which is a graded vector space V equipped with two consistent associative multiplications, $\circ : V \odot V \to V$ and $\bullet : V \otimes V \to V$, of degrees 0 and 1 respectively, see Sect. 1.10) can be naturally dualized leading to the notion of $\mathcal{H}$-coalgebra.

Let V be a graded vector space and $(\bar{B}V := \otimes^{\bullet\geq 1}V[1], \Delta_\bullet)$ the (reduced, as $\otimes^0 V[1]$ is omitted) free tensor coalgebra generated by $V[1]$. Let $(\hat{\odot}^\bullet(\bar{B}V[1]), \Delta_\circ)$ the completed (with respect to the natural filtration $F^{\geq r} := \otimes^{\geq r}V[1]$) free graded cocommutative coalgebra generated by $\bar{B}V[1]$ (the latter is $\bar{B}V$ with shifted grading). Let $I \subset \hat{\odot}^\bullet(\otimes^\bullet V[1])$ be the coideal generated by $\otimes^{\bullet\geq 2}V[1]$. The standard coassociative coproduct $\Delta_\bullet$ in $\bar{B}V$ extends naturally to a degree -1 coproduct in the quotient $\hat{\odot}^\bullet(\bar{B}V[1])/I^2$ making the latter into an $\mathcal{H}$-coalgebra. (Here I^2 is the square of the coideal, i.e. the set of elements $x \in I$ such that $\Delta_\circ x \in I \otimes I$.)

3.7.1. Proposition. *There is a one-to-one correspondence,*

$$\left\{ \begin{array}{c} \mathcal{G}erst_\infty\text{-algebra} \\ \text{structures in } V \end{array} \right\} \leftrightarrow \left\{ \begin{array}{c} \text{codifferentials in the cofree} \\ \mathcal{H}\text{-coalgebra } \left(\hat{\odot}^\bullet(\bar{B}V[1])/I^2, \Delta_\circ, \Delta_\bullet\right) \end{array} \right\}.$$

Proof. A coderivation D of the coalgebra $\hat{\odot}^\bullet(\bar{B}V[1])/I^2$ is equivalent to a degree 1 linear map,

$$\hat{\odot}^\bullet(\bar{B}V[1])/I^2 \longrightarrow V[2],$$

vanishing on k, the direct summand of constants. As

$$\hat{\odot}^\bullet(BV[1])/I^2 = \odot^\bullet(V[2]) \oplus \left(\otimes^{\bullet\geq 2}V[1]\right)[1] \otimes \odot^\bullet(V[2]),$$

this is the same as two collections of homogeneous linear maps, $\{\nu_n : \odot^n V \to V$ of degree $3-2n\}_{n\geq 1}$ and $\{\mu_{n;p} : (\otimes^n V) \otimes (\odot^p V) \to V$ of degree $2-n-2p\}_{n\geq 2, p\geq 0}$.

It is straightforward to check that the condition $D^2 = 0$ translates precisely into the equations of Corollary 3.6.3. □

3.8. Geometric interpretation. $\mathcal{L}_\infty$-structure on a finite-dimensional vector space V has a beautiful geometric interpretation **[Ko1]** as a smooth degree 1 vector field $\vec{\nu}$ on the pointed affine graded formal manifold $(V[2], 0)$ which satisfies the conditions $[\vec{\nu}, \vec{\nu}] = 0$ and $\vec{\nu}|_0 = 0$.

Surprisingly enough, $\mathcal{G}erst_\infty$-structure is also of purely geometric nature.

3.8.1. Definitions. (i) A *geometric* $\mathcal{A}_\infty$*-structure* (respectively, *geometric* $\mathcal{C}_\infty$*-structure*) on a graded manifold $\mathcal{M}$ is the data $(\vec{\nu}, \mu_\bullet)$ consisting of

(a) a smooth homological vector field $\vec{\nu}$ making $\mathcal{M}$ into a dg manifold;
(b) a collection of maps,

$$\mu_\bullet = \{\mu_n\}_{n\geq 1} : \otimes^\bullet_{\mathcal{O}_\mathcal{M}} \mathcal{T}_\mathcal{M} \to \mathcal{T}_\mathcal{M}, \quad \text{with } \mu_1 = Lie_{\vec{\nu}},$$

making the tangent sheaf $\mathcal{T}_\mathcal{M}$ into a sheaf of $\mathcal{A}_\infty$-algebras (respectively, $\mathcal{C}_\infty$-algebras).

(ii) A geometric $\mathcal{A}_\infty$-structure/$\mathcal{C}_\infty$-structure on a pointed graded manifold $(\mathcal{M}, *)$ is called *minimal* if $\vec{\nu}I \subset I^2$ where I is the ideal of the distinguished point $* \in \mathcal{M}$.

3.8.2. Theorem. *There is a one-to-one correspondence*

$$\left\{ \begin{array}{c} \mathcal{G}erst_\infty\text{-algebra} \\ \text{structures in } V \end{array} \right\} \leftrightarrow \left\{ \begin{array}{c} \text{geometric } \mathcal{A}_\infty\text{-structures on the pointed} \\ \text{affine formal manifold } \mathcal{M} = (V[2], 0) \end{array} \right\}.$$

Proof. Let $\{e_\alpha, \alpha = 1, \ldots, \dim V\}$ be basis of V and $\{t^\alpha\}$ the associated dual basis of $V^*[-2]$ which we identify with coordinate functions on $V[2]$. Set $t := \sum_\alpha t^\alpha e_\alpha$, and let τ be the isomorphism of $\mathcal{O}_\mathcal{M}$-modules

$$\begin{array}{cccc} \tau : & \mathcal{T}_\mathcal{M} & \longrightarrow & \mathcal{O}_\mathcal{M} \otimes V \\ & X = \sum_\alpha X^\alpha(t)\partial/\partial t^\alpha & \longrightarrow & \tau(X) := \sum_\alpha X^\alpha(t) e_\alpha. \end{array}$$

Let $\{\nu_n : \odot^n V \to V\}_{n\geq 1}$ and $\{\mu_{n;p} : (\otimes^n V) \otimes (\odot^p V) \to V\}_{n\geq 2, p\geq 0}$ be a $\mathcal{G}erst_\infty$-algebra structure on V. Define the vector field,

$$\vec{\nu} := \sum_{n=1}^{\infty} \frac{1}{n!} \tau^{-1} \circ \nu_n(t, \ldots, t),$$

and the tensors, for $n \geq 2$,

$$\begin{array}{cccc} \mu_n : & \otimes^n_{\mathcal{O}_\mathcal{M}} \mathcal{T}_\mathcal{M} & \longrightarrow & \mathcal{T}_\mathcal{M} \\ & X_1 \otimes \ldots \otimes X_n & \longrightarrow & \sum_{p=1}^{\infty} \frac{1}{p!} \tau^{-1} \circ \mu_{n;p}\left(\tau(X_1), \ldots, \tau(X_n); t, \ldots, t\right) \end{array}$$

on $\mathcal{M}$. Then the equations of Corollary 3.6.3 translate precisely into the statement that $(\vec{\nu}, \mu_\bullet)$ is a geometric $\mathcal{A}_\infty$-structure on $\mathcal{M}$.

This argument also works in the opposite direction through the Taylor decomposition of all the tensors at the distinguished point. □

3.9. $\mathcal{G}erst_\infty$-manifolds. Since the operad $\mathcal{G}erst_\infty$ is minimal, its algebras are strong homotopy ones, i.e. can be transferred via quasi-isomorphisms. In this subsection we essentially give a purely geometric description of the derived(=homotopy) category of $\mathcal{G}erst_\infty$-algebras.

3.9.1. Definition. A *$\mathcal{G}erst_\infty$-manifold* is a smooth manifold $\mathcal{M}$ together with a homotopy class (in the sense of 2.10), $(\eth, [\mu_\bullet])$, of minimal geometric $\mathcal{A}_\infty$-algebra structures on the tangent sheaf.

Note that we do *not* assume in the above definition that $\mathcal{M}$ is affine: the notion of $\mathcal{G}erst_\infty$- manifold is built on a collection of *tensors* satisfying a system of *diffeomorphism covariant* differential equations. The following tautologically formulated theorem is one of the main results of this section; it essentially describes a functor from the category of $\mathcal{G}erst_\infty$-algebras to a subcategory of the category of formal dg manifolds.

3.9.2. Theorem. *If the operad $\mathcal{G}erst_\infty$ acts on a dg vector space (V, d), then the formal graded manifold associated with the cohomology vector space $H(V, d)$ is canonically a $\mathcal{G}erst_\infty$-manifold.*

Proof. Since we work over a field in this paper, there always exists a quasi-isomorphism of complexes $(V, d) \to (H(V, d), 0)$ and hence an induced structure of $\mathcal{G}erst_\infty$-algebra on $H(V,d)$. This structure, however, is not canonical. What we have to show is that, first reinterpreting this induced structure as a geometric $\mathcal{A}_\infty$-structure $(\mathfrak{d}, \mu_\bullet)$ on the pointed flat formal manifold $(\mathcal{M}, *) = (H(V,d)[2], 0)$, then passing to the associated homotopy class, $(\mathfrak{d}, [\mu_\bullet])$, just in the sense of $\mathcal{A}_\infty$-algebras, and finally forgetting the flat structure one gets at the end the structure on $\mathcal{M}$ which does not depend on any choices made. It is precisely this structure which was termed in 3.9.1 an $\mathcal{G}erst_\infty$-manifold.

By Theorem 3.8.1, our input is a formal affine dg manifold,

$$(M \simeq V[2],\, \vec{\nu} = \{\nu_\bullet\} \text{ with } \nu_1 = d,\, * = 0)\,,$$

together with a geometric $\mathcal{A}_\infty$-structure $\mu_\bullet$. Let us choose a cohomological splitting of the complex (V,d), i.e. a decomposition of the $\mathbb{Z}$-graded vector space V into a direct sum,

$$V = H(V,d) \oplus B \oplus B[-1],$$

in such a way that the differential vanishes when restricted to the summands $H(V,D) \oplus B[-1]$ while on the remaining summand it equals the shifted by [1] identity map $B \to B[-1]$. According to Kontsevich [**Ko1**], such a splitting can be lifted to an isomorphism of formal affine dg manifolds,

$$(M, \vec{\nu}, *) \simeq (\mathcal{M}, \mathfrak{d}, *) \times (\mathcal{B}, \vec{d}_{DR}, *),$$

where

- $(\mathcal{M}, \mathfrak{d}, *)$ is the formal affine *minimal* dg manifold whose tangent space at $*$ is $H(V,d)[2]$,
- $(\mathcal{B}, \vec{d}_{DR}, *)$ is the formal affine dg manifold whose tangent space at $*$ is $B[2] \oplus B[1]$, homological vector field $\vec{d}_{DR}$ is linear and coincides precisely with the usual De Rham differential when one identifies the structure sheaf, $\mathcal{O}_{\mathcal{B},*}$, with the De Rham algebra of smooth formal differential forms on the vector space B. In particular, both the cohomology groups, $H(\mathcal{O}_{\mathcal{B},*}, \vec{d}_{DR})$ and $H(\mathcal{T}_{\mathcal{B},*}, Lie_{\vec{d}_{DR}})$ are trivial.

Let $\pi_1 : M \to \mathcal{M}$ and $\pi_2 : M \to \mathcal{B}$ be the projections associated with the chosen above cohomological splitting. There is an associated decomposition of complexes of vector spaces (note that differentials are *not* $\mathcal{O}_M$-linear),

$$(\mathcal{T}_{M,*}, Lie_{\vec{\nu}}) = (\pi_1^*\mathcal{T}_{\mathcal{M},*}, Lie_{\vec{\nu}}) \;\oplus\; (\pi_2^*\mathcal{T}_{\mathcal{B},*}, Lie_{\vec{\nu}}).$$

The tangent vector space at $* \in \mathcal{B}$ can be identified with $B[2] \oplus B[1]$. Let $H : B[2] \oplus B[1] \to B[2] \oplus B[1]$ be a degree -1 linear map which is equal to zero on the summand $B[2]$ and is equal to the shifted by $[-1]$ identity map $B[1] \to B[2]$ on the remaining summand. Denote by the same letter H its natural $\mathcal{O}_{\mathcal{B},*}$-linear extension to $\mathcal{T}_{\mathcal{B},*}$. It is an easy calculation to check that the identity automorphism of the tangent sheaf to M decomposes as follows,

$$Id = pr \;\oplus\; Lie_{\vec{\nu}} \circ \pi_2^*(H) \;\oplus\; \pi_2^*(H) \circ Lie_{\vec{\nu}},$$

where $\pi_2^*(H)$ is assumed to act as zero on the summand $\pi_1^*\mathcal{T}_{\mathcal{M},*}$, and pr stands for the canonical projection $\mathcal{T}_{M,*} \to \pi_1^*\mathcal{T}_{\mathcal{M},*}$. Thus we have constructed an $\mathcal{O}_M$-*linear* homotopy, $\pi_2^*(H) : \mathcal{T}_{M,*} \to \mathcal{T}_{M,*}$, associated with quasi-isomorphic complexes of k-*linear* vector spaces $(\mathcal{T}_{M,*}, Lie_{\vec{\nu}})$ and $(\pi_1^*\mathcal{T}_{\mathcal{M},*}, Lie_{\vec{\nu}})$.

Next step is to employ, say, the explicit formulae of [**Me1, KoSo2**], to construct an $\mathcal{A}_\infty$-algebra structure,

$$\left(\hat{\mu}_{n\geq 2} : \otimes^n_{\mathcal{O}_M} \pi_1^* \mathcal{T}_{\mathcal{M}} \to \pi_1^* \mathcal{T}_{\mathcal{M}},\ \mu_1 = Lie_{\vec{\nu}}\right),$$

on the lifted tangent sheaf $\pi_1^*\mathcal{T}_{\mathcal{M}}$. The key fact that the resulting $\hat{\mu}_{n\geq 2}$ are tensors is ensured by $\mathcal{O}_M$-linearity of the constructed homotopy.

Finally one repeats the above procedure using the standard contraction homotopy (which is $\pi^{-1}\mathcal{O}_{\mathcal{M}}$-linear) of the cohomologically trivial De Rham complex $(\mathcal{O}_{\mathcal{B},*}, \vec{d}_{DR})$ to induce a *geometric* $\mathcal{A}_\infty$-algebra structure, $(\eth, \tilde{\mu}_\bullet)$, on $\mathcal{M}$. It is easy to check that the associated homotopy class (in the sense of 2.11 for $\eth$, and 2.10 for $\tilde{\mu}_\bullet$) does not depend on the choice of a particular factorization of the dg manifold $(M, \vec{\nu}, 0)$ into a direct product of a minimal dg manifold and a linearly contractible one. □

3.9.3. Remark. The special case of the above Theorem when $(V, d, \mathcal{G}erst_\infty$-action) is just a (non-commutative) Gerstenhaber algebra was proved in [**Me2**] by explicit perturbative calculations.

3.10. Homotopy commutative sibling of $\mathcal{G}erst_\infty$. Theorem 3.9.2 motivates the following definition.

3.10.1. Definition. $\mathcal{G}erst^c_\infty$ is the operad whose algebras are given by

$$\left\{ \begin{array}{c} \text{A } \mathcal{G}erst^c_\infty\text{-algebra} \\ \text{structure in } V \end{array} \right\} := \left\{ \begin{array}{c} \text{A geometric } \mathcal{C}_\infty\text{-structure on the pointed} \\ \text{affine formal manifold } (V[2], 0) \end{array} \right\}.$$

Almost repeating 3.9.1 and 3.9.2 one obtains the notion of $\mathcal{G}erst^c_\infty$-manifold (which was in fact introduced earlier [**Me2**] under the name *Frobenius*$_\infty$ manifold) and the statement:

3.10.2. Theorem. *If the operad $\mathcal{G}erst^c_\infty$ acts on a dg vector space (V, d), then the formal graded manifold associated with the cohomology vector space $H(V, d)$ is canonically a $\mathcal{G}erst^c_\infty$-manifold.*

§4 Deformation theory

4.1. Deformation functor. The traditional approach to the deformation theory of a mathematical structure A is based on the idea of deformation functor, $\mathsf{Def}_{\mathfrak{g}}$, on Artinian rings.

Initially that idea was applied to Artinian rings concentrated in degree 0 so that the tangent space, $\mathsf{Def}_{\mathfrak{g}}(k[\varepsilon]/\varepsilon^2)$, to the deformation functor (which is the same as the Zariski tangent space, $\mathcal{T}_{\mathsf{A}}\mathcal{M}$, to the moduli space at the distinguished point) equals some particular homogeneous bit, $H^i(\mathfrak{g}, d)$, of the $\mathbb{Z}$-graded cohomology group, $H^\bullet(\mathfrak{g}, d) = \oplus_{i\in\mathbb{Z}} H^i(\mathfrak{g}, d)$, of the dg Lie algebra, $(\mathfrak{g}, d)$, controlling the deformations of A. The next homogeneous bit, $H^{i+1}(\mathfrak{g}, d)$, absorbs the obstructions to exponentiating infinitesimal deformations from $H^i(\mathfrak{g}, d)$ to genuine ones.

Recent studies in mirror symmetry led to an extension [**BaKo, Ma, Me2**] of the deformation functor first to $\mathbb{Z}$-graded and then differential $\mathbb{Z}$-graded Artinian rings. The dg extension of $\mathsf{Def}_{\mathfrak{g}}$ always produces *smooth* formal dg moduli spaces

$(\mathcal{M}, \eth, *)$ with Zariski tangent space, $\mathcal{T}_{\mathsf{A}}\mathcal{M}$, isomorphic to the full cohomology group $H^\bullet(\mathfrak{g}, d)$ and with obstructions encoded into a homological vector field on $\mathcal{M}$ [**Ko1, Me2**]. The table below compares the two deformation functors in three important examples,

A	$\mathcal{T}_{\mathsf{A}}\mathcal{M}$ in classical Def	$\mathcal{T}_{\mathsf{A}}\mathcal{M}$ in extended Def
Complex manifold (M, J)	$H^1(M, T_M)$	$H^\bullet(M, \wedge^\bullet T_M)$
Symplectic manifold (M, ω)	$H^2(M, \mathbb{R})$	$H^\bullet(M, \mathbb{R})$
Associative algebra $(A, \circ)$	$\mathrm{Hoch}^2(A, A)$	$\mathrm{Hoch}^\bullet(A, A)$

These developments lead naturally to a question: What happens to A when it is deformed in the generic direction in $H^\bullet(\mathfrak{g}, d)$ (rather than in $H^i(\mathfrak{g}, d)$)? Or, equivalently, what is the universal structure $\mathcal{A}$ over the extended moduli space $\mathcal{M}$? Thanks to Stasheff [**St**], we know the answer to this question in the case A = Associative algebra: deforming any given associative algebra A along a generic tangent vector in $\mathcal{T}_A\mathcal{M} = \mathrm{Hoch}^\bullet(A, A)$ one obtains, if all obstructions vanish, an $\mathcal{A}_\infty$-algebra. To author's knowledge, infinity versions of such notions as complex and symplectic structure are still a mystery, and the deformation theory in the form of deformation functor gives no clue to its solution. Moreover, the dg extension of Def makes it evident that one does not really need Artinian rings to do the deformation theory — the dg versal moduli space, $(\mathcal{M}, \eth, *)$, representing the functor $\mathsf{Def}_\mathfrak{g}$ on dg Artinian rings is nothing but the image of $(\mathfrak{g}, d)$ under the canonical functor

$$\left\{ \begin{array}{c} \text{the category of} \\ \text{dg Lie algebras} \end{array} \right\} \longrightarrow \left\{ \begin{array}{c} \text{the derived category of} \\ \text{dg Lie algebras} \end{array} \right\}.$$

Thus the Artinian functor approach to deformation theory is tantamount to a perturbative computation of the minimal $\mathcal{L}_\infty$-model for the controlling dg Lie algebra.

4.2. An operadic guide to deformation theory. We loose too much information about the deformed mathematical structure A if we naively understand the extended deformation theory as outlined above in Sect. 4.1, i.e., as the deformation functor $\mathsf{Def}_\mathfrak{g}$ extended to (differential) $\mathbb{Z}$-graded Artinian rings . Here is a suggestion on what one should do instead.

Step 1: Associate to the mathematical structure A we wish to deform a "controlling" Deformation Algebra, $(\mathfrak{g}, [\,,\,], d, \mathsf{ADD})$, consisting of a dg Lie algebra[9] $(\mathfrak{g} = \bigoplus_{i\in\mathbb{Z}} \mathfrak{g}^i, [\,,\,], d)$, and a collection of some additional algebraic operations, ADD, on $\mathfrak{g}$. For example,

- if A is a symplectic or complex structure (see Examples 4.4 and and 4.5 below) , then Deformation Algebra is a graded commutative Gerstenhaber algebra, and ADD is just a graded commutative product consistent with the dg Lie algebra structure via Poisson type identities.
- if A is an associative algebra structure (see Example 4.6 below), then Deformation Algebra is what is called in [**GeVo**] a homotopy Gerstenhaber algebra, and ADD is an infinite series of operations called braces.

[9]More generally, one can replace the dg Lie algebra structure on $\mathfrak{g}$ with a $\mathcal{L}_\infty$-algebra structure.

Step 2: Find a cofibrant resolution, $\mathcal{DA}_\infty$, of the operad $\mathcal{DA}$ describing species Deformation Algebra obtained in the previous step. In many important cases there exists *the* minimal cofibrant model $\mathcal{DA}_\infty$ of $\mathcal{DA}$ whose differential is decomposable.

Step 3: As the operad $\mathcal{DA}_\infty$ is cofibrant, its algebras are strong homotopy ones, that is, $\mathcal{P}$-algebra structures can be transferred by quasi-isomorphisms of complexes (see Theorem 2.8.5). Then choosing a cohomological splitting of $(\mathfrak{g}, d)$, one induces on the cohomology space $H(\mathfrak{g}, d)$ a canonical homotopy class of $\mathcal{DA}_\infty$-algebras (which is independent of the splitting used.)

Step 4: Try to find a geometric interpretation (called *extended moduli space*) of the homotopy class of minimal $\mathcal{DA}_\infty$-algebra structures canonically induced on $H(\mathfrak{g}, d)$ in the previous step, that is, try to interpret the latter in terms of sections of some natural vector bundles (equipped with $\eth$-connections, tensorial algebraic structures, ...) over a formal pointed dg manifold $(\mathcal{M}, \eth, *)$ whose tangent space at the unique geometric point is precisely $H(\mathfrak{g}, d)$.

4.3. Motivation and evidence. All the steps above are functorial with respect to the choice of input, Deformation Algebra. Making a particular choice means essentially a choice of precision with which we want to do the deformation theory. The most crude one is to apply first the obvious forgetful functor,

$$F : \{\text{Deformation Algebra}\} \longrightarrow \{\text{dg Lie Algebra}\}$$

and then apply the operadic algorithm. All the four steps can be easily fulfilled (see Sect. 2.11) with the final outcome at Step 4 being a smooth formal minimal dg manifold $(\mathcal{M}, \eth, *)$. Note that this outcome is well defined only up to an action of the formal diffeomorphism group $Diff(\mathcal{M}, *)$,

$$(\mathcal{M}, \eth, *) \sim (\mathcal{M}, f_*\eth, *), \quad \forall f \in Diff(\mathcal{M}, *).$$

Thus, whatever the choice of Deformation Algebra in Step 1, the outcome of Step 4 will always be at least a smooth dg manifold with its automorphism group $Diff(\mathcal{M}, *)$. As we expect ADD in the input of the deformation theory to be consistent, in some or other sense, with the underlying dg Lie algebra structure, it is natural to expect that what we get from ADD in Step 4 will also be consistent with the homological vector field $\eth$ and behave reasonably well under the action of $Diff(\mathcal{M}, *)$. The main results of this paper (see the Introduction for the list) show that these expectations are met in such important cases as, e.g., extended deformations of complex and symplectic structures. This is why we believe that Step 4, the most questionable one in the above programme, makes sense.

4.4. Example (deformations of complex structures). The dg commutative Gerstenhaber algebra controlling extended deformations of a given complex structure on a real $2n$-dimensional manifold X is given by

$$\mathfrak{g} = \left(\bigoplus_{i=0}^{2n} \mathfrak{g}^i, \ \mathfrak{g}^i = \bigoplus_{p+q=i} \Gamma(X, \wedge^p \mathcal{T}_X \otimes \Omega_X^{0,q}), \ [\bullet], \bar{\partial}, \ \circ \right)$$

where $\mathcal{T}_X$ stands for the sheaf of holomorphic vector fields, $\Omega_X^{s,q}$ for the sheaf of smooth differential forms of type (s,q), and $[\bullet] = \text{extended Schouten brackets}$, and $\circ = \text{wedge product of polyvector fields} \otimes \text{wedge product of forms}$.

4.5. Example (deformations of Poisson and symplectic structures). The dg commutative Gerstenhaber algebra controlling deformations of a given Poisson structure, $\nu \in \Gamma(X, \wedge^2 T_{\mathbb{R}})$, on a real smooth manifold X is given by

$$(\oplus_{i=0}^{\dim X}\Gamma(X, \wedge^i T_X), [\bullet] = \text{Schouten brackets}, d = [\nu \bullet \ldots], \text{wedge product of polyvector fields}),$$

where T_X stands for the sheaf of real tangent vectors.

If ν is non-degenerate, that is, $\nu = \omega^{-1}$ for some symplectic form ω on X, then the natural "lowering of indices map" $\omega^{\wedge i} : \wedge^i T_X \to \Omega_X^i$ sends $[\nu\bullet\ldots]$ into the usual de Rham differential. The image of the Schouten brackets under this isomorphism we denote by $[\bullet]_\omega$. In this way we make the de Rham complex of X into a dg commutative Gerstenhaber algebra,

$$\mathfrak{g} = \left(\bigoplus_{i=0}^{\dim X} \Gamma(X, \Omega_X^i),\ [\bullet]_\omega,\ d = \text{de Rham differential, wedge product of forms}\right),$$

which controls the extended deformations of the symplectic structure ω. More explicitly,

$$[\kappa_1 \bullet \kappa_2]_\omega := (-1)^{\tilde{\kappa}_1}[i_\nu, d](\kappa_1 \wedge \kappa_2) - (-1)^{\tilde{\kappa}_1}([i_\nu, d]\kappa_1) \wedge \kappa_2 - \kappa_1 \wedge [i_\nu, d]\kappa_2, \quad \forall \kappa_1, \kappa_2 \in \wedge^* T_{\mathbb{R}},$$

with $i_\nu : \Omega_X^\bullet \to \Omega_X^{\bullet-2}$ being the natural contraction with the 2-vector ν.

4.6. Example (deformations of holomorphic vector bundles). Let $E \to X$ be a holomorphic vector bundle on a complex manifold X. The standard Lie algebra structure in the endomorphism sheaf $End(E)$ extends naturally to $\wedge^\bullet_{\mathcal{O}_X} End(E) \simeq \odot^\bullet_{\mathcal{O}_X}(End(E)[1])$ in such a way that $(\wedge^\bullet_{\mathcal{O}_X} End(E), \wedge, [\bullet]_E)$ becomes a sheaf of graded commutative Gerstenhaber algebras.

The dg commutative Gerstenhaber algebra controlling extended deformations of a given holomorphic structure in $E \to X$ is given by

$$\mathfrak{g} = \left(\Gamma(X, \wedge^\bullet_{\mathcal{O}_X} End(E) \otimes \Omega_X^{0,\bullet}),\ [\bullet], \bar{\partial},\ \circ\right)$$

where $[\bullet] = [\bullet]_E \otimes \text{wedge product of forms}$, and $\circ = \text{wedge product of endomorphisms} \otimes \text{wedge product of forms}$.

4.7. Homotopy Gerstenhaber algebras. Let V be a graded vector space and $(BV := \otimes^{\bullet \geq 0} V[1], \Delta)$ the free tensor coalgebra cogenerated by $V[1]$.

4.7.1. Definitions. (i) A *B_∞-algebra* structure on a graded vector space V is the structure of dg bialgebra,

$$(BV, \Delta, \circ : BV \otimes BV \to BV, d : BV \to BV),$$

on the tensor coalgebra (BV, Δ) such that the element $1 \in k = \otimes^0 V[1]$ is the identity element.

(ii) A *homotopy Gerstenhaber algebra* structure on a graded vector space V is a structure of B_∞-algebra such that multiplication $\circ$ preserves the filtration $F_r :=$

$\otimes^{\bullet \leq r} V[1]$. We denote by $h\mathcal{G}$ the operad whose algebras are homotopy Gerstenhaber algebras.

The $h\mathcal{G}$-algebra structure on V can be described by a collection of homogeneous maps,

$$\begin{aligned} M_k: &\quad \otimes^k V \longrightarrow V, \\ M_{1;k}: &\quad \otimes^{k+1} V \longrightarrow V, \end{aligned}$$

satisfying a system of quadratic equations written explicitly in [**Vo**]. In particular, the operations M_2 and $M_{1;1}$ induce on $H(V, M_1)$ the structure of graded commutative Gerstenhaber algebra. Thus there is a canonical map of operads, $p: h\mathcal{G} \to \mathcal{G}$.

4.7.2. Example (higher order Steenrod operations). Let $S_\bullet X$ be the singular chain complex, of a topological space X. Elements of $S_n X$ are formal linear combinations of continuous maps, $\sigma: \Delta[n] \to X$, from the standard n-simplex $\Delta[n]$ to X. For such a map $\sigma \in S^n X$ and a k-face, $f: \Delta[k] \to \Delta[n]$, of the standard simplex spanned by vertices $n_0 = f(0), \ldots, n_k = f(k)$ (f being injective and monotone), denote by $\sigma[n_0, \ldots, n_k] \in S_k X$ the associated composition,

$$\sigma[n_0, \ldots, n_k]: \Delta[k] \xrightarrow{f} \Delta[n] \xrightarrow{\sigma} X.$$

Let $S^\bullet X := \mathrm{Hom}_k(S_\bullet, k)$ be the associated cochain complex of X. It was noted by Gerstenhaber and Voronov in [**GeVo**] using earlier results of Baues that the data,

$$\begin{aligned} M_1(\phi)(\sigma) &:= \sum_{k=0}^{n+1} (-1)^k \phi(\sigma[0, 1, \ldots, \hat{k}, \ldots, n]), \\ M_2(\phi, \psi)(\sigma) &:= \sum_{k=0}^{n} \phi(\sigma[0, \ldots, k]) \psi(\sigma[k, \ldots, n]), \quad \forall\ \phi, \psi \in S^\bullet X, \\ M_k &:= 0 \text{ for } k \geq 3, \\ M_{1;k}(\phi_0; \phi_1, \ldots, \phi_n)(\sigma) &:= \phi_0(\sigma[0, n_1, n_1 + n_2, \ldots, n_1 + \ldots + n_k]) \\ &\qquad \phi_1(\sigma[0, 1, \ldots, n_1) \phi_2(\sigma[n_1, \ldots, n_1 + n_2) \ldots \\ &\qquad \phi_k(\sigma[n_1 \ldots + n_{k-1}, \ldots, n_1 + \ldots + n_k]) \\ &\qquad \forall\ \phi_0 \in S^k X, \phi_1 \in S^{n_1} X, \ldots, \phi_k \in S^{n_k} X, \end{aligned}$$

make $S^\bullet X$ into an $h\mathcal{G}$-algebra. The operation $M_{1;1}$ is nothing but the Steenrod operation $\cup_1$.

4.7.3. Example (deformations of associative algebras). Let A be an associative algebra, and $C^\bullet(A, A) := \mathrm{Hom}_k(A^\bullet, A)$ its *Hochschild complex* with the differential,

$$\begin{aligned} (d\phi)(a_1, \ldots, a_{n+1}) &:= a_1 \phi(a_2, \ldots, a_{n+1}) \\ &\qquad \sum_{k=1}^{n} (-1)^k \phi(a_1, \ldots, a_{k-1}, a_k, a_{k+1}, \ldots, a_{n+1}) \\ &\qquad +(-1)^{k+1} \phi(a_1, a_2, \ldots, a_n) a_{n+1}. \end{aligned}$$

It was shown in [**GeVo**] that the data,

$$\begin{aligned} M_1 &:= d, \\ M_2(\phi,\psi)(a_1,\dots,a_{k+l}) &:= \phi(a_1,\dots,a_k)\psi(a_{k+1},\dots,a_{k+l}) \\ &\qquad \forall\, \phi\in C^k(A,A), \psi\in C^l(A,A) \\ M_k &:= 0 \text{ for } k\geq 3, \end{aligned}$$

and, for any $\phi_0,\phi_1,\dots,\phi_k\in C^\bullet(A,A)$,

$$M_{1;k}(\phi_0;\phi_1,\dots,\phi_n)(a_1,\dots,a_m) :=$$

$$\sum(-1)^{\sum_{p=1}^n(|\phi_p|-1)i_p}\phi_0(a_1,\dots,a_{i_1},\phi_1(a_{i_1},\dots),\dots,a_{i_n},\phi_n(a_{i_n},\dots),\dots,a_m)$$

where the summation runs over all possible ordered substitutions of $\phi_1,\dots,\phi_n$ into ϕ_0, makes the Hochschild complex into an $h\mathcal{G}$-algebra.

This $h\mathcal{G}$-algebra controls deformations of the associative algebra structure in A.

4.8. Deligne's conjecture. The operations M_2 and $M_{1;1}$ induce on the Hochschild cohomology, $H^\bullet(A,A)$, the structure of $\mathcal{G}$-algebra. Deligne conjectured that this action of the operad $\mathcal{G}$ on $H^\bullet(A,A)$ can be lifted to the action of $\mathcal{G}_\infty$ on $C^\bullet(A,A)$. This conjecture, which was recently proved in [**Ko2, KoSo1, McSm, Ta, Vo**], is essentially the same as the following statement.

4.8.1. Theorem [**Ta, TaTs, Vo**]. *There is a natural morphism of operads, $f:\mathcal{G}_\infty\to h\mathcal{G}$, such that the diagram,*

$$\begin{array}{ccc} \mathcal{G}_\infty & \xrightarrow{f} & h\mathcal{G} \\ & \searrow^{q.-iso.} & \downarrow p \\ & & \mathcal{G} \end{array}$$

commutes.

4.9. Approximations to deformation theory. The cofibrant resolution, $\mathcal{DA}_\infty$, of the operad $\mathcal{DA}$ describing species Deformation Algebra could be so complicated that it would be unrealistic to ask for an immediate geometric interpretation of the homotopy class of minimal $\mathcal{DA}_\infty$-algebra structures as in Step 4.

In Steps 3 and 4 one can therefore replace $\mathcal{DA}_\infty$ by its "approximation", a cofibrant operad $\widetilde{\mathcal{DA}}_\infty$ fitting the commutative diagram

$$\begin{array}{ccc} \widetilde{\mathcal{DA}}_\infty & \xrightarrow{i} & \mathcal{DA}_\infty \\ & \searrow_{j} & \downarrow q.-iso. \\ & & \mathcal{G} \end{array}$$

for some natural morphisms of operads i and j, i being preferably a cofibration.

4.10. Approximations to $\mathcal{G}_\infty$ deformation theory. As discussed in Sections 4.4-4.6, extended deformations of basic geometric structures are described by the operad $\mathcal{G}_\infty$. In view of theorem 4.8.1, the same operad can be applied to the deformation theory of associative algebras.

At present we have no complete picture of the geometric object behind a homotopy class of minimal $\mathcal{G}_\infty$-algebras. We appeal instead to the infinite tower of cofibrant approximations to $\mathcal{G}_\infty$ introduced in Sect. 2.7,

$$\mathcal{L}_\infty = \mathcal{G}_\infty^{(1)} \longrightarrow \mathcal{G}_\infty^{(2)} \longrightarrow \mathcal{G}_\infty^{(3)} \longrightarrow \ldots \longrightarrow \mathcal{G}_\infty^{(n)} \longrightarrow \ldots \longrightarrow \mathcal{G}_\infty,$$

and, in the rest of this paper, attempt to give such a picture for the first three floors of this tower using all the previous results. The ground floor, $\mathcal{G}_\infty^{(1)}$, corresponds, in view of 2.7.2, to the forgetful functor 4.3 so that $\mathcal{G}_\infty^{(1)}$-approximation to the deformation theory of examples 4.4-4.6 and 4.7.3 simply says that the extended moduli space is a formal minimal dg manifold $(\mathcal{M}, \eth, *)$. If its non-linear cohomology (see Introduction),

$$M \simeq \frac{\mathsf{Zeros}(\eth)}{\mathsf{Im}\,\eth},$$

makes sense in some geometric category, it is precisely the versal moduli space associated with the $\mathbb{Z}$-graded extension of the classical deformation functor $\mathsf{Def}_{\mathfrak{g}}$.

4.11. Theorem. *There is a canonical isomorphism of operads,*

$$\mathcal{G}_\infty^{(2)} = \mathcal{G}erst_\infty^{c},$$

where $\mathcal{G}erst_\infty^{c}$ is defined in Sect. 3.10.1.

Proof. The statement follows immediately from the definition 3.10.1 and the proof of Theorem 3.4.2 in [**Me2**] (see also Sect. 4.12 for a reconstruction of that argument). □

4.11.1. Corollary. *If the operad $\mathcal{G}_\infty$ acts on a dg vector space (V, d), then the formal graded manifold associated with the cohomology vector space $H(V, d)$ is canonically a $\mathcal{G}erst_\infty^{c}$-manifold.*

Proof. The statement follows from 4.11 and 3.9.2. □

4.11.2. Corollary [**Me2**]. *Let A be one of the structures 4.4-4.6 or 4.7.3. The extended moduli space of deformations of A is naturally a $\mathcal{G}erst_\infty^{c}$-manifold, i.e. a dg manifold equipped with a homotopy class of geometric C_∞-structures on the tangent bundle.*

4.11.3. Notation. For a graded module V over a graded ring $\mathcal{O}$ we set

$$\circledast_{\mathcal{O}}^{\bullet} V := (\otimes_{\mathcal{O}}^{\bullet} V[1])\,[-1]$$

and

$$\frac{\circledast_{\mathcal{O}}^{\bullet} V}{\text{shuffle products}} := \left(\frac{\otimes_{\mathcal{O}}^{\bullet} V[1]}{\text{shuffle products}}\right)[-1].$$

4.12. Proof of Theorems A, B and E. The $\mathcal{G}_\infty$-algebra structure on V induces canonically a homotopy class of minimal $\mathcal{G}_\infty$-algebra structures on its cohomology, $H(V, d)$. Let $\mathbf{\Theta}$ be any representative of this homotopy class.

Let $\{\partial_\alpha, \alpha = 1, \ldots, \dim H(V, d)\}$ be a homogeneous basis of $H(V, d)[2]$ and $\{t^\alpha\}$ the associated dual basis of $H(V, d)^*[-2]$. We identify the latter with coordinate functions on the formal manifold $(\mathcal{M}, *)$ associated with $H(V, d)[2]$, and the former with basis vector fields $\partial/\partial t^\alpha$. We can assume without loss of generality[10] that degrees of all ∂_α vanish $\bmod 2\mathbb{Z}$.

Let t^{B_i}, $i \geq 2$, stand for the Lie polynomial,

$$t^{B_i} := \left[\ldots\left[[t^{\beta_1} \bullet t^{\beta_2}] \bullet t^{\beta_3}\right] \ldots \bullet t^{\beta_i}\right],$$

and

$$\partial_{B_i} \equiv (\partial_{\beta_1}|\partial_{\beta_2}) \ldots |\partial_{\beta_i})$$

for the image of the tensor product,

$$\partial_{\beta_1} \otimes \partial_{\beta_2} \otimes \ldots \otimes \partial_{\beta_i},$$

under the degree $n-1$ composition

$$\otimes^n H(V,d)[2] \longrightarrow \circledast^n H(V,d)[2].$$

Note that under our assumption the parities of t^{B_i} and ∂_{B_i} are both equal to $(i-1) \bmod 2\mathbb{Z}$. Here and below we use B_i to denote the multi-index $\beta_1\beta_2\ldots\beta_i$.

By Proposition 2.6.1, the $\mathcal{G}_\infty$-algebra structure Θ is the same as the differential, δ, of the Gerstenhaber algebra $\hat{\odot}^\bullet \mathsf{Lie}(H(V,d)^*[-2])$. The latter is uniquely determined on the generators,

$$\delta t^\alpha = \sum_{\substack{k\geq 0, n\geq 0\\ B_{i_1},\ldots,B_{i_k}\\ \beta_1\ldots\beta_n}} \frac{1}{n!} \Theta^\alpha_{B_{i_1}\ldots B_{i_k};\beta_1\ldots\beta_n} t^{B_{i_1}} \ldots t^{B_{i_k}} t^{\beta_1} \ldots t^{\beta_n},$$

for some homogeneous constants $\Theta^\alpha_{B_{i_1}\ldots B_{i_k};\beta_1\ldots\beta_n} \in k$ (see footnote 9). Here and below juxtaposition of $t^\bullet$s means their symmetric product $\odot$.

We re-arrange the above data into a smooth "vector field",

$$\eth := \sum_{\substack{n\geq 0\\ \beta_1\ldots\beta_n}} \frac{1}{n!} \Theta^\alpha_{0;\beta_1\ldots\beta_n} t^{\beta_1} \ldots t^{\beta_n} \frac{\partial}{\partial t^\alpha}$$

and a collection of maps, for $k \geq 2$,

$$\begin{array}{llll} \Theta^{[k]}: & \odot^k_{\mathbb{C}}(\circledast^\bullet_{\mathbb{C}} \mathcal{T}_{\mathcal{M}}) & \longrightarrow & \mathcal{T}_{\mathcal{M}} \\ & \partial_{B_1} \odot \ldots \odot \partial_{B_k} & \longrightarrow & \sum_{\substack{n\geq 0\\ \beta_1\ldots\beta_n}} \frac{1}{n!} \Theta^\alpha_{B_{i_1}\ldots B_{i_k};\beta_1\ldots\beta_n} t^{\beta_1} \ldots t^{\beta_n} \frac{\partial}{\partial t^\alpha}. \end{array}$$

Thus the defining equation for the differential δ takes the form,

$$\sum_\alpha (\delta t^\alpha)\partial_\alpha = \eth + \sum_{\substack{k\geq 1\\ B_{i_1},\ldots,B_{i_k}}} \Theta^{[k]}(\partial_{B_1}, \ldots, \partial_{B_k}) t^{B_{i_1}} \ldots t^{B_{i_k}}.$$

The ideal $I := < \mathsf{Lie}^{\geq 2} V^*[-2] >$ is generated by t^{B_i}.

Using the identities,

$$\delta f(t) = \sum_\alpha (\delta t^\alpha) \frac{\partial f}{\partial t^\alpha},$$

[10]For example, we could opt to work in the category of graded manifolds over graded base spaces, "sources of $\mathbb{Z}$-graded constants"; then all the signs lost under our assumption in "natural over the base" calculations can be easily restored through the condition that the expression under study is functorial with respect to the base space change.

and

$$[t^\alpha \bullet f(t)] = \sum_\beta [t^\alpha \bullet t^\beta] \frac{\partial f}{\partial t^\beta},$$

where $f(t)$ is an arbitrary smooth function on $(\mathcal{M}, *)$, it is not hard to study the transformation properties of the defined above fields $\eth$ and $\Theta^{[k]}$ under an arbirary formal (non-linear) change of coordinates,

$$t^\alpha \longrightarrow \tilde{t}^\alpha = f^\alpha(t),$$

and conclude that

(i) $\eth$ is indeed a vector field on $\mathcal{M}$;

(ii) the map $\Theta^{[1]}$ factors through the composition

$$\Theta^{[1]} : \circledast^{\bullet}_{\mathbb{C}} \mathcal{T}_\mathcal{M} \longrightarrow \circledast^{\bullet}_{\mathcal{O}_\mathcal{M}} \mathcal{T}_\mathcal{M} \longrightarrow \mathcal{T}_\mathcal{M},$$

i.e. represents a family of *tensors*, $\mu_\bullet : \otimes^{\bullet}_{\mathcal{O}_\mathcal{M}} \mathcal{T}_\mathcal{M} \to \mathcal{T}_\mathcal{M}$, vanishing on shuffle products;

(iii) the maps $\Theta^{[k]}$ are sections of certain jet bundles (of order $k-1$) on $\mathcal{M}$.

A similar calculation shows that

(iv) the equation $\sum_\alpha (\delta^2 t^\alpha)\partial_\alpha = 0 \bmod I$ implies $[\eth, \eth] = 0$; this is essentially Lemma 2.7.2;

(v) the equation $\sum_\alpha (\delta^2 t^\alpha)\partial_\alpha = 0 \bmod I^2$ implies that the data $\Theta^{[1]} \simeq \mu_\bullet$ is nothing but a geometric $\mathcal{C}_\infty$-structure on the dg manifold $(\mathcal{M}, *, \eth)$; this is essentially Theorem 4.11.

To prove the statements A,B and E we have to move one level up and study the equation

$$\sum_\alpha (\delta^2 t^\alpha)\partial_\alpha = 0 \bmod I^3,$$

which now involves the non-tensorial object $\Theta^{[2]}$, a section of the bundle $J^1(\otimes^{\bullet}_{\mathcal{O}_\mathcal{M}} \mathcal{T}^*_\mathcal{M}) \otimes_{\mathcal{O}_\mathcal{M}} (\otimes^{\bullet}_{\mathcal{O}_\mathcal{M}} \mathcal{T}^*_\mathcal{M}) \otimes_{\mathcal{O}_\mathcal{M}} \mathcal{T}_\mathcal{M}$. It is convenient for our purposes to choose a splitting of this bundle, say, an affine torsion-free affine connection ∇ on $\mathcal{M}$. Then we can replace non-tensorial objects

$$t^{B_n} = \left[\ldots \left[[t^{\beta_1} \bullet t^{\beta_2}] \bullet t^{\beta_3} \right] \ldots \bullet t^{\beta_n} \right],$$

by tensorial ones,

$$\check{t}^{B_n} = \left[\ldots \left[[t^{\beta_1} \bullet t^{\beta_2}]' \bullet t^{\beta_3} \right]' \ldots \bullet t^{\beta_n} \right]',$$

which are given recursively by

$$\begin{aligned}
[t^{\beta_1} \bullet t^{\beta_2}]' &= [t^{\beta_1} \bullet t^{\beta_2}] \\
\left[[t^{\beta_1} \bullet t^{\beta_2}]' \bullet t^{\beta_3} \right]' &= \left[[t^{\beta_1} \bullet t^{\beta_2}]' \bullet t^{\beta_3} \right] + \Gamma^{\beta_1}_{\mu\nu} \odot [t^\mu \bullet t^{\beta_2}]' \odot [t^\nu \bullet t^{\beta_3}]' \\
&\quad + \Gamma^{\beta_2}_{\mu\nu} \odot [t^\mu \bullet t^{\beta_1}]' \odot [t^\nu \bullet t^{\beta_3}]' \\
\ldots &= \ldots \\
\check{t}^{B_n} &= \left[\check{t}^{B_{n-1}} \bullet t^{\beta_n} \right] + \sum_{i=1}^{n-1} \Gamma^{\beta_i}_{\mu\nu} \odot \check{t}^{B_{n-1}(\beta_i, \mu)} \odot [t^\nu \bullet t^{\beta_n}]',
\end{aligned}$$

where $B_{n-1}(\beta_i,\mu)$ stands for the multi-index $\beta_1\dots\beta_{i-1}\mu\beta_{i+1}\dots\beta_{n-1}$ and $\Gamma^{\beta}_{\mu\nu}$ are the Christoffel symbols of the connection ∇ in the coordinate system $\{t^\alpha\}$. Thus the defining equation for the differential δ can be written in the form,

$$\sum_\alpha(\delta t^\alpha)\partial_\alpha=\eth+\sum_{\substack{k\geq 1\\ B_{i_1},\dots,B_{i_k}}}\check{\Theta}^{[k]}(\partial_{B_1},\dots,\partial_{B_k})\odot\check{t}^{B_{i_1}}\odot\dots\odot\check{t}^{B_{i_k}}.$$

where all the coefficients, $\check{\Theta}^{[k]}$, are tensors rather than sections of the jet bundles.

Now, to prove Theorems A and B it is enough to understand the first non-trivial component,

$$[t^{\beta_1}\bullet t^{\beta_2}]\odot[t^{\gamma_1}\bullet t^{\gamma_2}],$$

of the formal power series $\sum_\alpha(\delta^2t^\alpha)\partial_\alpha$ mod I^3. The only terms of $\sum_\alpha(\delta t^\alpha)\partial_\alpha$ mod I^3 which contribute to that component are written explicitly below,

$$\begin{aligned}\sum_\alpha(\delta t^\alpha)\partial_\alpha &= \eth+\sum\check{\Theta}^{[1]}(\partial_{\beta_1}|\partial_{\beta_2})\odot[t^{\beta_1}\bullet t^{\beta_2}]\\ &\quad+\sum\check{\Theta}^{[1]}(\partial_{\beta_1}|\partial_{\beta_2}|\partial_{\beta_3})\odot\left[[t^{\beta_1}\bullet t^{\beta_2}]\bullet t^{\beta_3}\right]'\\ &\quad+\sum\check{\Theta}^{[2]}((\partial_{\beta_1}|\partial_{\beta_2}),(\partial_{\gamma_1}|\partial_{\gamma_2}))\odot[t^{\beta_1}\bullet t^{\beta_2}]\odot[t^{\gamma_1}\bullet t^{\gamma_2}]+\dots\end{aligned}$$

Applying to the shown terms the differential δ and ignoring all components of the equation $\sum_\alpha(\delta^2t^\alpha)\partial_\alpha=0$ mod I^3 except the chosen one, one gets an equation,

$$[\mu_2,\mu_2]^\nabla=Lie_\eth\mu_{2,2},$$

where

$$\begin{aligned}&[\mu_2,\mu_2]^\nabla(X,Y,Z,W)\\ =\;&[\mu_2,\mu_2]_{HM}(X,Y,Z,W)\\ &-(Lie_\eth\mu_3)(W,\nabla_ZX+\nabla_XZ,Y)-(Lie_\eth\mu_3)(Z,\nabla_WX+\nabla_XW,Y)\\ &-(Lie_\eth\mu_3)(W,\nabla_ZY+\nabla_YZ,X)-(Lie_\eth\mu_3)(Z,\nabla_WY+\nabla_YW,X)\\ &+\mu_3(W,(Lie_\eth\nabla)_ZX+(Lie_\eth\nabla)_XZ,Y)\\ &+\mu_3(W,(Lie_\eth\nabla)_ZX+(Lie_\eth\nabla)_XZ,Y)\\ &+\mu_3(W,(Lie_\eth\nabla)_ZY+(Lie_\eth\nabla)_YZ,X)\\ &+\mu_3(W,(Lie_\eth\nabla)_ZY+(Lie_\eth\nabla)_YZ,X),\end{aligned}$$

μ_2, μ_3 and $\mu_{2,2}$ are, respectively, the compositions,

$$\mu_2:\odot^2_{\mathcal{O}_\mathcal{M}}\mathcal{T}_\mathcal{M}\xrightarrow{\simeq}\frac{\circledast^2_{\mathcal{O}_\mathcal{M}}\mathcal{T}_\mathcal{M}}{\text{shuffle products}}\hookrightarrow\frac{\circledast^\bullet_{\mathcal{O}_\mathcal{M}}\mathcal{T}_\mathcal{M}}{\text{shuffle products}}\xrightarrow{\Theta^{[1]}}\mathcal{T}_\mathcal{M},$$

$$\mu_3:\frac{\otimes^3_{\mathcal{O}_\mathcal{M}}\mathcal{T}_\mathcal{M}}{\text{shuffle products}}\xrightarrow{\simeq}\frac{\circledast^3_{\mathcal{O}_\mathcal{M}}\mathcal{T}_\mathcal{M}}{\text{shuffle products}}\hookrightarrow\frac{\circledast^\bullet_{\mathcal{O}_\mathcal{M}}\mathcal{T}_\mathcal{M}}{\text{shuffle products}}\xrightarrow{\Theta^{[1]}}\mathcal{T}_\mathcal{M}$$

$$\mu_{2,2}:\wedge^2_{\mathcal{O}_\mathcal{M}}(\odot^2_{\mathcal{O}_\mathcal{M}}\mathcal{T}_\mathcal{M})\xrightarrow{\simeq}\odot^2_{\mathcal{O}_\mathcal{M}}\frac{\circledast^2_{\mathcal{O}_\mathcal{M}}\mathcal{T}_\mathcal{M}}{\text{shuffle products}}\hookrightarrow\frac{\circledast^\bullet_{\mathcal{O}_\mathcal{M}}\mathcal{T}_\mathcal{M}}{\text{shuffle products}}\xrightarrow{\check{\Theta}^{[2]}}\mathcal{T}_\mathcal{M},$$

$[\mu_2,\mu_2]_{HM}$ is the Hertling-Manin bracket (see Sect. 0.1), and X,Y,Z,W are arbitrary smooth even (for simplicity of presentation) vector fields on $\mathcal{M}$. As the r.h.s. of the equation $(*)$ is obviously a tensor, the l.h.s. must represent a tensor as well. (The original Hertling-Manin bracket can not possibly be a tensor without correction terms as the product μ_2 is associative only up to homotopy.) Changing the affine connection,

$$\nabla \to \nabla' = \nabla + S,$$

for some symmetric tensor $S : \odot^2_{\mathcal{O}_\mathcal{M}} \mathcal{T}_\mathcal{M} \to \mathcal{T}_\mathcal{M}$, we get

$$[\mu_2, \mu_2]^{\nabla'} = [\mu_2, \mu_2]^{\nabla} - 2Lie_{\eth}\nu,$$

where

$$\begin{aligned}\nu(X,Y,Z,W) &= \mu_3(W, S(Z,X), Y) + \mu_3(Z, S(W,X), Y) \\ &\quad + \mu_3(W, S(Z,Y), X) + \mu_3(Z, S(W,Y), X).\end{aligned}$$

Thus the cohomology class, $[[\mu_2, \mu_2]] \in H(\wedge^2_{\mathcal{O}_\mathcal{M}}(\odot^2_{\mathcal{O}_\mathcal{M}} \mathcal{T}_\mathcal{M}))$, associated with $[\mu_2, \mu_2]^\nabla$ does not depend on the choice of ∇. The equation $(*)$ implies Theorems A and B almost immediately.

To prove Theorem E one needs to understand a general structure of the components,

$$\check{t}^{B_k} \odot \check{t}^{B_l},$$

of the formal power series $\sum_\alpha (\delta^2 t^\alpha)\partial_\alpha \bmod I^3$. Here are the terms of $\sum_\alpha (\delta t^\alpha)\partial_\alpha \bmod I^3$ contributing to that component,

$$\sum_\alpha (\delta t^\alpha)\partial_\alpha = \eth + \sum_{k\geq 2, B_k} \breve{\Theta}^{[1]}(\partial_{B_k}) \odot \check{t}^{B_k} + \sum_{\substack{k,l\geq 2\\ B_k, B_l}} \breve{\Theta}^{[2]}(\partial_{B_k}, \partial_{B_l}) \odot \check{t}^{B_k} \odot \check{t}^{B_l} + \dots .$$

It is not hard to see that the projection of the equation $\sum_\alpha (\delta^2 t^\alpha)\partial_\alpha = 0 \bmod I^3$ to that component gives the equation of the form,

$$\begin{aligned}&[\mu_k(X_1, \dots, X_k), \mu_l(Y_1, \dots, Y_k)] + \text{correction terms} \\ &\qquad = (Lie_\eth \mu_{k,l})(X_1, \dots, X_k, Y_1, \dots, Y_k),\end{aligned}$$

where the brackets stand for the standard commutator of vector fields, and μ_k, μ_l ($\mu_{k,l}$) are homogeneous components of $\Theta^{[1]}$ (respectively, $\breve{\Theta}^{[2]}$). Again, as the r.h.s. of the above equation is obviously a tensor, the l.h.s. must be a tensor as well. This tensor, $[\mu_k, \mu_l]^\nabla$, is $Lie_\eth$-closed and, moreover, defines a vanishing cohomology class $[[\mu_k, \mu_l]]$ of the complex of $\mathcal{O}_M$-modules, $(\otimes^\bullet_{\mathcal{O}_\mathcal{M}} \mathcal{T}^*_\mathcal{M} \otimes \mathcal{T}_\mathcal{M}, Lie_\eth)$. □

Acknowledgement. A part of this work was done during author's visit to the Max Planck Institute for Mathematics in Bonn which was supported by the "Sofia Kovalevskaya" award of Matilde Marcolli. Excellent working conditions in the MPIM are gratefully acknowledged. I would like to thank E. Getzler, Yu.I. Manin, M. Markl and A.A. Voronov for valuable correspondence and discussions, and M. Marcolli for supporting the project and the visit. Thanks are also due to the referee for valuable remarks.

References

[BaKo] S. Barannikov and M. Kontsevich, *Frobenius manifolds and formality of Lie algebras of polyvector fields*, Internat. Math. Res. Notices **14** (1998), 201-215.

[BeMo] C. Berger and I. Moerdijk, *Axiomatic homotopy theory for operads*, preprint math.AT/0206094.

[DS] W.G. Dwyer and J. Sparlinski, *Homotopy theories and model categories*, In: Handbook of Algebraic Topology, edited by I.M. James, Elsevier, Amsterdam, 1995, pages 73-126.

[GeVo] M. Gerstenhaber and A. Voronov, *Homotopy G-algebras and moduli space operad*, Internat. Math. Res. Notices **3** (1995), 141-153.

[GeMa] S. Gelfand and Yu.I. Manin, Homological algebra, 2nd edition, Springer, 2002.

[GetJo] E. Getzler and J.D.S. Jones, *Operads, homotopy algebra, and iterated integrals for double loop spaces*, preprint hep-th/9403055.

[GiKa] V. Ginzburg and M. Kapranov, *Koszul duality for operads*, Duke Math. J. **76** (1994), 203-272.

[GoMi] W.M. Goldman and J.J. Millson, *The deformation theory of representations of fundamental groups of compact Kähler manifolds*, Inst. Hautes Etudes Sci. Publ. Math. **67** (1988), 43-96.

[He] C. Hertling, Frobenius manifolds and moduli spaces for singularities, Cambridge University Press 2002.

[HeMa] C. Hertling and Yu.I. Manin, *Weak Frobenius manifolds*, Intern. Math. Res. Notices **6** (1999) 277-286.

[Hi] V. Hinich, *Homological algebra of homotopy algebras*, Comm. Algebra **25** (1997) 3291 -3323.

[Ko1] M. Kontsevich, *Deformation quantization of Poisson manifolds I*, math/9709040.

[Ko2] M. Kontsevich, *Operads and motives in deformation quantization*, QA/9904055.

[KoSo1] M. Kontsevich and Y. Soibelman, *Deformations of algebras over operads and Deligne's conjecture*, math.QA/0001151.

[KoSo2] M. Kontsevich and Y. Soibelman, *Homological mirror symmetry and torus fibrations*, math.SG/0011041.

[Ma] Yu.I. Manin, Frobenius manifolds, quantum cohomology, and moduli spaces. AMS, Providence, 1999.

[Mar1] M. Markl, *Models for operads*, Commun. in Algebra, **20** (1996) 1471-1500.

[Mar2] M. Markl, *Distributive laws and Koszulness*, Ann. Inst. Fourier, Grenoble **46** (1996), 307-323.

[Mar3] M. Markl, *Homotopy algebras are homotopy algebras*, math.AT/9907138.

[Mar4] M. Markl, *Homotopy diagrams of algebras*, math.AT/0103052.

[McSm] J.E. McClure and J.H. Smith, *A solution of Deligne's conjecture*, preprint math/9910126.

[Me1] S.A. Merkulov, *Strong homotopy algebras of a Kähler manifold*, math.AG/9809172.

[Me2] S.A. Merkulov, *Frobenius*$_\infty$ *invariants of homotopy Gerstenhaber algebras I*, Duke Math. J. **105** (2000), 411–461.

[Me3] S.A. Merkulov, *Semi-infinite A-variations of Hodge structure over extended Kähler cone*, Internat. Math. Res. Notices, 21 (2001), 1111-1139.

[Pr] A. Prouté, *Algèbres différentielles fortiment homotopiquement associatives*, PhD thesis, 1984.

[Qu] D. Quillen, *Rational homotopy theory*, Ann. Math. **90** (1969), 205-295.

[St] J.D. Stasheff, *On the homotopy associativity of H-spaces, I II*, Trans. Amer. Math. Soc. **108** (1963), 272-292 & 293-312.

[Ta] D. Tamarkin, *Another proof of Kontsevich formality conjecture*, math.QA/9803025.

[TaTs] D. Tamarkin and B.L. Tsygan, *Noncommutative differential calculus, homotopy BV algebras and formality conjectures*, math.KT/0002116.

[Vo] A.A. Voronov, *Homotopy Gerstenhaber algebras*, preprint math.QA/9908040.

Matematiska institutionen, Stockholms universitet
Department of Mathematics, University of Glasgow

Witten's top Chern class on the moduli space of higher spin curves

Alexander Polishchuk

This paper is a sequel to [9]. Its goal is to verify that the *virtual top Chern class* $c^{1/r}$ in the Chow group of the moduli space of higher spin curves $\overline{\mathcal{M}}_{g,n}^{1/r}$, constructed in [9], satisfies all the axioms of *spin virtual class* formulated in [5]. Hence, according to [5], it gives rise to a cohomological field theory in the sense of Kontsevich-Manin [7]. As was observed in [9], the only non-trivial axioms that have to be checked for the class $c^{1/r}$ are two axioms that we call *Vanishing axiom* and *Ramond factorization axiom.* The first of them requires $c^{1/r}$ to vanish on all the components of the moduli space $\overline{\mathcal{M}}_{g,n}^{1/r}$, where one of the markings is equal to $r-1$. The second demands vanishing of the push-forward of $c^{1/r}$ restricted to the components of the moduli space corresponding to the so called Ramond sector, under some natural finite maps.

Recall that the virtual top Chern class is a crucial ingredient in the generalized Witten's conjecture formulated in [10], [11]. The original index-theoretic construction of this class sketched by Witten was recently extended to the compactified moduli space by T. Mochizuki [8] who also showed that the obtained class satisfies the axioms of [5]. The algebraic construction of [9] gives a class in the Chow group with rational coefficients (and axioms are satisfied on the level of Chow groups). Presumably, the algebraic construction induces the same class in cohomology of $\overline{\mathcal{M}}_{g,n}^{1/r}$ as the analytic construction.

It is interesting to note that the class $c^{1/r}$ is constructed as a characteristic class of certain supercommutative $\mathbb{Z}/2\mathbb{Z}$-graded dg-algebra over $\overline{\mathcal{M}}_{g,n}^{1/r}$ equipped with an odd closed section (where the entire data is defined up to quasi-isomorphism). This resembles Kontsevich's approach to the construction of the virtual fundamental class (see [6]). One may hope that both constructions can be embedded into a more general framework involving dg-spaces. This would be in agreement with the philosophy of derived moduli spaces promoted in [6], [2], [3].

The paper is organized as follows. In section 1 we prove two identities for localized Chern characters of specific $\mathbb{Z}/2\mathbb{Z}$-graded complexes. In section 2 we deduce Vanishing axiom from the first identity and in section 3 we deduce Ramond factorization axiom from the second identity.

This work was partially supported by NSF grant DMS-0070967.

1. Some $\mathbb{Z}/2\mathbb{Z}$-graded homological algebra

Recall (see [**9**]) that for every $\mathbb{Z}/2\mathbb{Z}$-graded complex $(V^\bullet = V^+ \oplus V^-, d)$ of vector bundles on a scheme X, which is strictly exact off a closed subset $Z \subset X$, the graph-construction associates the *localized Chern character* $\mathrm{ch}_Z^X(V^\bullet) \in A^*(Z \to X)_\mathbb{Q}$, where $A^*(Z \to X)_\mathbb{Q}$ is the bivariant Chow group with rational coefficients (the original construction given in [**1**] or [**4**, Ch. 18] deals with $\mathbb{Z}$-graded complexes). Here we use the following terminology from [**9**]: $(V^\bullet, d)$ is *strictly exact* if it is exact and $\mathrm{im}(d)$ is a subbundle of $V^\bullet$. Note that if a $\mathbb{Z}/2\mathbb{Z}$-graded complex is homotopic to zero then it is strictly exact (since in this case $\mathrm{im}(d)$ is a direct summand of $V^\bullet$). Witten's top Chern class is constructed in [**9**] by slightly modifying the localized Chern class of the complex corresponding to the action of an isotropic section of an orthogonal bundle on a spinor bundle, where the relevant orthogonal data is constructed using the higher spin structure on the universal curve over $\overline{\mathcal{M}}_{g,n}^{1/r}$ (we will recall this construction in section 2).

We are going to prove two identities for the localized Chern character that will be main ingredients for the proof of Vanishing axiom and Ramond factorization axiom respectively. In both cases the identities hold on the level of K-theory (of complexes that are strictly exact off a closed subset).

LEMMA 1.1. *Let $d(\lambda) : V^\bullet \to V^\bullet[\lambda]$ be an odd endomorphism of a $\mathbb{Z}/2\mathbb{Z}$-graded bundle $V^\bullet$ on X of the form $d(\lambda) = d_0 + d_1\lambda + \ldots + d_{r-1}\lambda^{r-1}$ for some $r \geq 2$, depending on a formal parameter λ (commuting with everything). Assume that $d(\lambda)^2 = \lambda^r$. Then $\mathrm{ch}_Z^X(V^\bullet, d_0) = 0$ for every closed subset $Z \subset X$ such that $(V^\bullet, d_0)$ is strictly exact off Z.*

Proof. We can extend $d(\lambda)$ to a $\mathcal{O}_X[\lambda]$-linear endomorphism of $V^\bullet[\lambda]$. Let $W^\bullet = V^\bullet[\lambda]/(\lambda^r)$ with the odd endomorphism $d_W : W^\bullet \to W^\bullet$ induced by $d(\lambda)$. Then $d_W^2 = 0$ and there is a natural r-step filtration on the complex $(W^\bullet, d_W)$ with all consecutive quotient-complexes isomorphic to $(V^\bullet, d_0)$. Thus, it is enough to prove that $(W^\bullet, d_W)$ is strictly exact (everywhere). We claim that in fact this complex is homotopic to zero. Indeed, for every interval of integers $[a, b]$ let us denote $V^\bullet[\lambda]_{[a,b]} = \oplus_{i=a}^b V^\bullet\lambda^i$. Let us denote by d' and d'' the following components of the restriction of $d(\lambda)$ to $V^\bullet[\lambda]_{[0,r-1]}$:

$$d(\lambda) : V^\bullet[\lambda]_{[0,r-1]} \overset{(d',d'')}{\to} V^\bullet[\lambda]_{[0,r-1]} \oplus V^\bullet[\lambda]_{[r,2r-1]}.$$

Note that $d' = d_W$ upon the natural identification of $V^\bullet[\lambda]_{[0,r-1]}$ with $W^\bullet$. Also, the image of d'' is contained in $V^\bullet[\lambda]_{[r,2r-2]}$. Extending d' and d'' to $\mathcal{O}_X[\lambda^r]$-linear endomorphisms of $V^\bullet[\lambda]$ we can write $d = d' + d''$. Then the condition $d(\lambda)^2 = \lambda^r$ implies that $d'd'' + d''d' = \lambda^r$ on $V^\bullet[\lambda]_{[0,r-1]}$. Hence

$$h = d''/\lambda^r : V^\bullet[\lambda]_{[0,r-1]} \to V^\bullet[\lambda]_{[0,r-1]}$$

gives a homotopy between the identity and zero endomorphisms of the complex

$$(V^\bullet[\lambda]_{[0,r-1]}, d') \simeq (W^\bullet, d_W).$$

□

Remark. The above lemma admits the following generalization: if the differential $d(\lambda) = d_0 + d_1\lambda + \ldots + d_{r-1}\lambda^{r-1}$ as above satisfies $d(\lambda)^2 = f(\lambda)$ for some polynomial

f of degree r then

$$\sum_{z:f(z)=0} m_z \cdot \mathrm{ch}_Z^X(V^\bullet, d(z)) = 0$$

where m_z is the multiplicity of a root z, $Z \subset X$ is a closed subset such that all the complexes $(V^\bullet, d(z))$ (for $f(z) = 0$) are strictly exact off Z.

LEMMA 1.2. *Let $d : V^\bullet \to V^\bullet$ be an odd endomorphism of a $\mathbb{Z}/2\mathbb{Z}$-graded bundle $V^\bullet$ on X such that $d^2 = -(f_1 \dots f_r) \cdot \mathrm{id}_{V^\bullet}$, where $f_1, \dots, f_r$ are functions on X. For every $i = 1, \dots, r$ let us introduce the differential d_i on $V^\bullet \oplus V^\bullet[1]$ by the formula*

$$d_i(x, x') = (d(x) + (\prod_{j \neq i} f_j) \cdot x', -d(x') + f_i \cdot x)$$

where $x \in V^\bullet$, $x' \in V^\bullet[1]$. Then

$$\sum_{i=1}^{r} \mathrm{ch}_Z^X(V^\bullet \oplus V^\bullet[1], d_i) = 0$$

for every closed subset $Z \subset X$ such that all $(V^\bullet \oplus V^\bullet[1], d_i)$ are strictly exact off Z.

Proof. Let us introduce the differential D on $W^\bullet := (V^\bullet \oplus V^\bullet[1])^{\oplus r}$ by the formula

$$D(x_i, x_i')_{i=1,\dots r} = (y_i, y_i')_{i=1,\dots,r}$$

where $x_i, y_i \in V^\bullet$, $x_i', y_i' \in V^\bullet[1]$,

$$y_i = dx_i + f_{i+1} f_{i+2} \dots f_r \cdot [x_1' + f_1 x_2' + f_1 f_2 x_3' + \dots + f_1 \dots f_{i-1} x_i'] \text{ for } i < r,$$
$$y_r = dx_r + x_1' + f_1 x_2' + f_1 f_2 x_3' + \dots + f_1 \dots f_{r-1} x_r',$$
$$y_i' = -dx_i' + f_i x_i - x_{i-1} \text{ for } i \geq 2,$$
$$y_1' = -dx_1' + f_1 x_1.$$

One can easily check that $D^2 = 0$. There is a natural decreasing filtration of $W^\bullet$ by subcomplexes $W^\bullet = F^1 W^\bullet \supset \dots \supset F^r W^\bullet \supset F^{r+1} W^\bullet = 0$, where

$$F^j W^\bullet = \{(x_i, x_i')_{i=1,\dots,r} : x_1 = \dots = x_{j-1} = 0, x_1' = \dots = x_{j-1}' = 0\}.$$

The associated graded quotients are

$$F^j W^\bullet / F^{j+1} W^\bullet \simeq (V^\bullet \oplus V^\bullet[1], d_j),$$

$j = 1, \dots, r$. Therefore,

$$\sum_{i=1}^{r} \mathrm{ch}_Z^X(V^\bullet \oplus V^\bullet[1], d_i) = \mathrm{ch}_Z^X(W^\bullet, D).$$

It remains to prove that the complex $(W^\bullet, D)$ is strictly exact on X. For this we construct a homotopy h between the identity and zero endomorphisms of $W^\bullet$. Namely, we set

$$h(x_i, x_i')_{i=1,\dots r} = (y_i, y_i')_{i=1,\dots,r},$$

where

$$y_i = -[x_{i+1}' + f_{i+1} x_{i+2}' + f_{i+1} f_{i+2} x_{i+2}' + \dots + f_{i+1} \dots f_{r-1} x_r'] \text{ for } i < r-1,$$
$$y_{r-1} = -x_r', \ y_r = 0,$$
$$y_1' = x_r, \ y_i' = 0 \text{ for } i \geq 2.$$

It is easy to check that $Dh + hD = \mathrm{id}_{W^\bullet}$. □

2. Vanishing axiom

Henceforward all our schemes are assumed to be quasiprojective over a field k. We assume that $\operatorname{char} k > r$ and that k contains all r-th roots of unity.

Let $\pi : \mathcal{C} \to X$ be a family of prestable curves over a scheme X, and let $\mathcal{T}$ be a family of rank-one torsion-free sheaves on $\mathcal{C}$ equipped with a non-zero homomorphism $b : \mathcal{T}^r \to \omega_{\mathcal{C}/X}$, where $\omega_{\mathcal{C}/X}$ is the dualizing sheaf of π. In this situation we defined in section 5.1 of [**9**] the class $c(\mathcal{T}, b) \in A^{-\chi}(X)_{\mathbb{Q}}$, where χ is the Euler-Poincaré characteristic of members of the family $\mathcal{T}$. To construct this class we consider the map $\tau : S^r R\pi_*\mathcal{T} \to \mathcal{O}_X[-1]$ induced by b and by the trace map $\operatorname{Tr} : R\pi_*\omega_{\mathcal{C}/X} \to \mathcal{O}_X[-1]$. As was proved in Proposition 4.7 of [**9**] there exists a complex $C_0 \to C_1$ of vector bundles on X representing $R\pi_*\mathcal{T}$ such that the map τ is represented by the chain map of complexes $S^r[C_0 \to C_1] \to \mathcal{O}_X[-1]$. This chain map corresponds to a morphism of vector bundles $\nu : S^{r-1}C_0 \to C_1^\vee$. We can consider the differential $d : C_0 \to C_1$ and the map ν as sections of the pull-backs of C_1 and $C_1^\vee$ to the total space of C_0. Then $s = (d, \nu)$ will be an isotropic section of the orthogonal vector bundle $p^*C_1 \oplus p^*C_1^\vee$ on C_0, where $p : C_0 \to X$ is the projection. Moreover, s vanishes exactly on X embedded into C_0 by the zero section. Then we consider the action of s on the spinor bundle $\Lambda^*p^*C_1^\vee$. The obtained $\mathbb{Z}/2\mathbb{Z}$-graded complex $(\Lambda^*p^*C_1^\vee, s)$ is exact outside $X \subset C_0$. Therefore, the localized Chern character of this complex is an element of the bivariant Chow group $A^*(X \to C_0)_{\mathbb{Q}} \simeq A^*(X)_{\mathbb{Q}}$. The class $c(\mathcal{T}, b)$ is obtained by multiplying this localized Chern character with the Todd class of C_1. Theorem 4.3 of [**9**] assures that $c(\mathcal{T}, b)$ does not depend on the choices made.

Remark. We can consider $(\mathcal{A} := p_*(\Lambda^*p^*C_1^\vee), \delta)$ as a sheaf of $\mathbb{Z}/2\mathbb{Z}$-graded dg-algebras over X, where the differential δ is induced by $d : C_0 \to C_1$. The action of the isotropic section s on $\mathcal{A}$ has form $\delta + \epsilon(e)$, where $\epsilon(e)$ is the operator of multiplication with the δ-closed odd section $e \in \mathcal{A}$ corresponding to $\nu : S^{r-1}C_0 \to C_1^\vee$. The proof of Theorem 4.3 in [**9**] can be converted into the proof of the fact that the quasi-isomorphism class of the data $(\mathcal{A}, e)$ is uniquely determined by $(\mathcal{T}, b)$.

Let $\sigma : X \to \mathcal{C}$ be a section of π such that π is smooth near $\sigma(X)$ (a marked point). By abuse of notation we will denote by $F \mapsto F(\sigma)$ the operation of tensoring with the line bundle $\mathcal{O}_{\mathcal{C}}(\sigma(X))$. The following theorem immediately implies Vanishing axiom for $c^{1/r}$ (Axiom 4 of [**5**]).

THEOREM 2.1. *Assume that b factors as a composition*

$$\mathcal{T}^r \xrightarrow{b_0} \omega_{\mathcal{C}/X}(-(r-1)\sigma) \to \omega_{\mathcal{C}/X},$$

where σ^*b_0 is an isomorphism. Then $c(\mathcal{T}, b) = 0$.

Let us set $L = \mathcal{T}(\sigma)|_\sigma$. The map σ^*b_0 gives an isomorphism

$$L^r \widetilde{\to} \omega_{\mathcal{C}/X}(\sigma)|_\sigma \simeq \mathcal{O}_X.$$

Since we are working with rational coefficients, we can replace X by its finite étale covering over which L is trivial (right now we do not need to choose a specific trivialization of L).

Let $\pi^{(r)} : \mathcal{C}^{(r)} \to X$ denote the relative r-th symmetric power of $\mathcal{C}$ over X. We denote by $\sigma^r \in \mathcal{C}^{(r)}$ the X-point corresponding to the relative divisor $r\sigma(X)$ and by $\mathcal{I}_{\sigma^r} \subset \mathcal{O}_{\mathcal{C}^{(r)}}$ the ideal sheaf of $\sigma^r(X) \subset \mathcal{C}^{(r)}$. For every coherent sheaf $\mathcal{F}$ on $\mathcal{C}$, let $\mathcal{F}^{(r)}$ denote the r-th symmetric power of $\mathcal{F}$, which is a sheaf on $\mathcal{C}^{(r)}$. We claim that b_0 induces a morphism

$$R\pi_*^{(r)}(\mathcal{I}_{\sigma^r} \otimes (\mathcal{T}(\sigma))^{(r)}) \to R\pi_*(\omega_{\mathcal{C}/X}). \tag{2.1}$$

Indeed, let $\Delta : \mathcal{C} \to \mathcal{C}^{(r)}$ be the diagonal map. Then we have a natural morphism

$$\mathcal{I}_{\sigma^r} \otimes (\mathcal{T}(\sigma))^{(r)} \to \Delta_*\Delta^*(\mathcal{I}_{\sigma^r} \otimes (\mathcal{T}(\sigma))^{(r)}) \to \Delta_*\mathcal{T}^r((r-1)\sigma) \to \Delta_*\omega_{\mathcal{C}/X},$$

where the last arrow is induced by b_0. Now the morphism (2.1) is obtained by applying the functor $R\pi_*^{(r)}$. Composing (2.1) with the trace map $R\pi_*(\omega_{\mathcal{C}/X}) \to \mathcal{O}_X[-1]$, we get a morphism

$$\widetilde{\tau} : R\pi_*^{(r)}(\mathcal{I}_{\sigma^r} \otimes (\mathcal{T}(\sigma))^{(r)}) \to \mathcal{O}_X[-1]$$

that will play a major role in the proof of Theorem 2.1. Note that we also have a natural map

$$\iota : R\pi_*^{(r)}(\mathcal{T}^{(r)}) \to R\pi_*^{(r)}(\mathcal{I}_{\sigma^r} \otimes (\mathcal{T}(\sigma))^{(r)})$$

and an isomorphism $S^r R\pi_*\mathcal{T} \simeq R\pi_*^{(r)}(\mathcal{T}^{(r)})$, such that $\tau = \widetilde{\tau} \circ \iota$ can be identified with the map $S^r R\pi_*\mathcal{T} \to \mathcal{O}_X[-1]$ used in the definition of $c(\mathcal{T}, b)$.

Let us consider the natural exact triangle

$$\mathcal{O}_X[-1] \xrightarrow{\delta} R\pi_*^{(r)}(\mathcal{I}_{\sigma^r} \otimes (\mathcal{T}(\sigma))^{(r)}) \to R\pi_*^{(r)}(\mathcal{T}(\sigma))^{(r)} \to \mathcal{O}_X, \tag{2.2}$$

where we use the canonical trivialization of L^r.

LEMMA 2.2. *The composition $\widetilde{\tau} \circ \delta$ is the identity map.*

Proof. This follows immediately from the existence of a natural morphism of exact triangles

$$\begin{array}{ccccccc} \Delta^*\mathcal{I}_{\sigma^r} \otimes \mathcal{T}^r(r\sigma) & \longrightarrow & \mathcal{T}^r(r\sigma) & \longrightarrow & \Delta^*(\sigma^r)_*\mathcal{O}_X & \longrightarrow & \cdots \\ \downarrow & & \downarrow & & \downarrow & & \\ \omega_{\mathcal{C}/X} & \longrightarrow & \omega_{\mathcal{C}/X}(\sigma) & \longrightarrow & \sigma_*\mathcal{O}_X & \longrightarrow & \cdots \end{array} \tag{2.3}$$

and from the fact that the composition

$$\mathcal{O}_X[-1] \simeq R\pi_*\sigma_*\mathcal{O}_X[-1] \to R\pi_*\omega_{\mathcal{C}/X} \to \mathcal{O}_X[-1]$$

is the identity map. □

We want to realize the maps $\widetilde{\tau}$ and ι on the level of complexes in a compatible way. We start by realizing the canonical distinguished triangle in $D^b(X)$:

$$R\pi_*\mathcal{T} \xrightarrow{\alpha} R\pi_*(\mathcal{T}(\sigma)) \xrightarrow{\beta} L \xrightarrow{\gamma} R\pi_*\mathcal{T}[1]$$

by an exact triple of complexes of vector bundles on X.

LEMMA 2.3. *Let $[C_0 \xrightarrow{d} C_1]$ be a complex of vector bundles (concentrated in degrees $[0,1]$) representing $R\pi_*\mathcal{T}$. Then there exists an extension of vector bundles*

$$0 \to C_0 \to \widetilde{C}_0 \to L \to 0, \tag{2.4}$$

and a morphism $\widetilde{d} : \widetilde{C}_0 \to C_1$ extending d, such that the morphism $\gamma : L \to R\pi_*\mathcal{T}[1]$ is represented by the chain map of complexes

$$[C_0 \to \widetilde{C}_0] \xrightarrow{(\mathrm{id},\widetilde{d})} [C_0 \to C_1], \tag{2.5}$$

hence, the complex $[\widetilde{C}_0 \xrightarrow{\widetilde{d}} C_1]$ *represents* $R\pi_*(\mathcal{T}(\sigma))$ *and the morphisms* α *and* β *are represented by the natural chain maps*

$$[C_0 \to C_1] \to [\widetilde{C}_0 \to C_1] \to L;$$

Proof. Applying the second arrow in the exact sequence

$$\mathrm{Hom}(L, C_1) \to \mathrm{Hom}(L, R\pi_*\mathcal{T}[1]) \to \mathrm{Ext}^1(L, C_0) \to \mathrm{Ext}^1(L, C_1) \tag{2.6}$$

to the element γ we get an extension class $e \in \mathrm{Ext}^1(L, C_0)$ which becomes trivial in $\mathrm{Ext}^1(L, C_1)$. Let

$$0 \to C_0 \to \widetilde{C}_0 \to L \to 0$$

be an extension with the class e, $\widetilde{d} : \widetilde{C}_0 \to C_1$ be a splitting of its push-out by $d : C_0 \to C_1$. The element in $\mathrm{Hom}(L, R\pi_*\mathcal{T}[1])$ represented by the chain map (2.5) induces the same class e in $\mathrm{Ext}^1(L, C_0)$. Now the sequence (2.6) shows that after changing a splitting $\widetilde{d}$ by an appropriate element of $\mathrm{Hom}(L, C_1)$ the chain map (2.5) will represent γ. □

Let $C = [C_0 \to C_1]$ be a complex representing $R\pi_*\mathcal{T}$ and let $\widetilde{C} = [\widetilde{C}_0 \to C_1]$ be the complex representing $R\pi_*(\mathcal{T}(\sigma))$ obtained by applying the above lemma. Then the complex S^rC (resp., $S^r\widetilde{C}$) represents

$$S^rR\pi_*\mathcal{T} \simeq R\pi_*^{(r)}\mathcal{T}^{(\sigma)}$$

(resp., $R\pi_*^{(r)}(\mathcal{T}(\sigma))^{(r)}$) and we have a natural surjective map of complexes $S^r\widetilde{C} \to L^r \simeq \mathcal{O}_X$ induced by the map $\widetilde{C}_0 \to L$. Then the kernel complex $\ker(S^r\widetilde{C} \to \mathcal{O}_X)$ represents $R\pi_*^{(r)}(\mathcal{I}_{\sigma^r} \otimes (\mathcal{T}(\sigma))^{(r)})$ in a way compatible with the exact triangle (2.2). Moreover, the map ι is represented by the natural chain map $S^rC \to \ker(S^r\widetilde{C} \to \mathcal{O}_X)$. It remains to choose our data in such a way that $\widetilde{\tau}$ would be represented by a chain map $\widetilde{\tau} : \ker(S^r\widetilde{C} \to \mathcal{O}_X) \to \mathcal{O}_X[-1]$. For this we use the following lemma analogous to Proposition 4.7 from **[9]**.

LEMMA 2.4. *There exists a complex of vector bundles* $C_0 \to C_1$ *representing* $R\pi_*\mathcal{T}$*, such that one has*

$$\mathrm{Hom}_{K^b(X)}(E, \mathcal{O}_X[n]) \simeq \mathrm{Hom}_{D^b(X)}(E, \mathcal{O}_X[n])$$

for $n \le 0$ *and* $E = \ker(S^r[\widetilde{C}_0 \to C_1] \to L)$.

Proof. We start with an arbitrary complex of vector bundles $C_0' \to C_1'$ representing $R\pi_*\mathcal{T}$ and then replace it by the quasiisomorphic complex $C_0 \to C_1$, where $C_1 = \mathcal{O}_X(-m)^{\oplus N} \to C_1'$ is a surjection (see **[9]**, Lemma 4.6), $\mathcal{O}_X(1)$ is an ample line bundle on X m is an integer (later we will need to choose m sufficiently large). The spectral sequence computing $\mathrm{Hom}_{D^b(X)}(E, \mathcal{O}_X[n])$ shows that to prove (ii) it suffices to check the vanishing

$$H^i(S^j\widetilde{C}_0^\vee(m')) = 0$$

for $i > 0$, $j < r$, $m' \ge m$. Since $\widetilde{C}_0$ is an extension of the trivial bundle by C_0, this would follow from the vanishing of $H^i(S^jC_0^\vee(m'))$ under the same conditions on i, j, m'. We know that for sufficiently large m one has

$$H^{>0}(S^j(C_0')^\vee \otimes S^{j_1}(C_1')^\vee \otimes \ldots \otimes S^{j_k}(C_1')^\vee(m')) = 0$$

for $j + j_1 + \ldots + j_k < r$ and $m' \geq m$. As was shown in Proposition 4.7 of [**9**], this implies that $H^{>0}(S^j C_0^\vee(m')) = 0$ for $j < r$, $m' \geq m$. □

Proof of Theorem 2.1. Let us choose the data $(C_0, C_1, \widetilde{C}_0, d, \widetilde{d})$ as in Lemmas 2.4 and 2.3. Let us set $K := \ker(S^r[\widetilde{C}_0 \to C_1] \to \mathcal{O}_X)$. Then the morphism $\widetilde{\tau}$ is represented by the chain map $K \to \mathcal{O}_X[-1]$ that corresponds to a morphism

$$\widetilde{\tau} : S^{r-1}\widetilde{C}_0 \otimes C_1 \to \mathcal{O}_X$$

such that the composition

$$\ker(S^r\widetilde{C}_0 \to \mathcal{O}_X) \to S^{r-1}\widetilde{C}_0 \otimes C_1 \to \mathcal{O}_X \tag{2.7}$$

is zero.

Let $X' \to X$ be the affine bundle classifying splittings of the exact sequence (2.4). Since the pull-back induces an isomorphism of Chow groups of X and X' we can make a base change of our data by the morphism $X' \to X$. Thus, we can assume that the extension (2.4) splits. Let $1 \in \widetilde{C}_0$ be a section projecting to a trivialization of L. It is easy to see that the morphism $\mathcal{O}_X[-1] \to K$ corresponding to the section $1^{r-1} \otimes \widetilde{d}(1)$ of $S^{r-1}\widetilde{C}_0 \otimes C_1$ represents the map δ from (2.2). So from Lemma 2.2 we derive that $\widetilde{\tau}(1^{r-1} \otimes \widetilde{d}(1)) = 1$. Together with the condition that the composition (2.7) vanishes this is equivalent to the equation

$$\langle \nu((x + \lambda \cdot 1)^{r-1}), \widetilde{d}(x + \lambda \cdot 1) \rangle = \lambda^r, \tag{2.8}$$

where $x \in C_0 \subset \widetilde{C}_0$, the morphism $\nu : S^{r-1}\widetilde{C}_0 \to C_1^\vee$ is induced by $\widetilde{\tau}$. It follows that the section

$$s_\lambda(x) = (\widetilde{d}(x + \lambda), \nu(x + \lambda \cdot 1))$$

of the orthogonal bundle $p^*C_1 \oplus p^*C_1^\vee$ on C_0 satisfies $s_\lambda(x) \cdot s_\lambda(x) = \lambda^r$. Applying Lemma 1.1 to the action of s_λ on the spinor bundle $\Lambda^* p^* C_1^\vee$ we derive the vanishing of localized Chern class corresponding to the isotropic section s_0 obtained from s_λ by setting $\lambda = 0$. But the latter class is precisely $c(\mathcal{T}, b)$. □

3. Ramond factorization axiom

Let the data $(\pi : C \to X, \mathcal{T}, b : \mathcal{T}^r \to \omega_{C/X})$ be as in section 2. Assume in addition that we have an X-point $\sigma : X \to C$ which is a nodal point of every fiber and that $\widetilde{\pi} : \widetilde{C} \to X$ is a fiberwise normalization of this point. We denote by $n : \widetilde{C} \to C$ the corresponding morphism and by $\sigma_1, \sigma_2 : X \to \widetilde{C}$ two disjoint points that project to $p \in C$. Finally, let us assume that $\mathcal{T}$ is locally free at σ and that that map b is an isomorphism at σ (in [**5**] this situation is referred to as "Ramond case").

For every $\lambda \in k^*$ there is a natural line bundle $\mathcal{L}_\lambda$ on C such that $n^*\mathcal{L}_\lambda \simeq \mathcal{O}_{\widetilde{C}}$ and the isomorphism $n^*\mathcal{L}_\lambda|_{\sigma_1} \to n^*\mathcal{L}_\lambda|_{\sigma_2}$ corresponds to the multiplication by λ. It is clear that $\mathcal{L}_{\lambda\lambda'} \simeq \mathcal{L}_\lambda \otimes \mathcal{L}_{\lambda'}$. In particular, if ξ is an r-th root of unity then $\mathcal{L}_\xi^r \simeq \mathcal{O}_C$. Therefore, we can twist the data $(\mathcal{T}, b)$ by considering $\mathcal{T} \otimes \mathcal{L}_\xi$ and the map $b_\xi : (\mathcal{T} \otimes \mathcal{L}_\xi)^r \to \omega_{C/X}$ induced by b and the trivialization of $\mathcal{L}_\xi^r$.

Now the Ramond case of Axiom 3 in [**5**] is implied easily by Theorem 2.1 together with the following result.

THEOREM 3.1. *One has*

$$\sum_{\xi:\xi^r=1} c(\mathcal{T}\otimes\mathcal{L}_\xi, b_\xi)=0.$$

Recall that the relative dualizing sheaves on $\mathcal{C}/X$ and $\widetilde{\mathcal{C}}/X$ are related by the isomorphism $n^*\omega_{\mathcal{C}/X}\simeq\omega_{\widetilde{\mathcal{C}}/X}(\sigma_1+\sigma_2)$ such that the following diagram is commutative

$$(3.1)\qquad \begin{array}{ccccc} n^*\omega_{\mathcal{C}/X}|_{\sigma_1} & & \longrightarrow & & n^*\omega_{\mathcal{C}/X}|_{\sigma_2} \\ \downarrow & & & & \downarrow \\ \omega_{\widetilde{\mathcal{C}}/X}(\sigma_1+\sigma_2)|_{\sigma_1} & \xrightarrow{\text{Res}} & \mathcal{O}_X & \xleftarrow{-\text{Res}} & \omega_{\widetilde{\mathcal{C}}/X}(\sigma_1+\sigma_2)|_{\sigma_2} \end{array}$$

where the top arrow is the canonical isomorphism (the sign comes from the relation $dx/x=-dy/y$ near the node $xy=0$). In particular, there is a canonical trivialization of $\omega_{\mathcal{C}/X}|_\sigma$ such that the boundary map $\delta:\mathcal{O}_X\simeq\omega_{\mathcal{C}/X}|_\sigma\to R\pi_*\omega_{\mathcal{C}/X}[1]$ from the exact triangle

$$R\pi_*\omega_{\mathcal{C}/X}\to R\pi_*n_*n^*\omega_{\mathcal{C}/X}\to\omega_{\mathcal{C}/X}|_\sigma\xrightarrow{\delta}R\pi_*\omega_{\mathcal{C}/X}[1]$$

satisfies $\text{Tr}\circ\delta=\text{id}$, where $\text{Tr}:R\pi_*\omega_{\mathcal{C}/X}[1]\to\mathcal{O}_X$ is the trace map.

We can recover $R\pi_*\mathcal{T}$ from $R\widetilde{\pi}_*\widetilde{\mathcal{T}}$, where $\widetilde{\mathcal{T}}=n^*\mathcal{T}$, together with the evaluation maps at σ_1 and σ_2. Namely, if we denote $L=\widetilde{\mathcal{T}}|_{\sigma_1}\simeq\widetilde{\mathcal{T}}|_{\sigma_2}$ then there is an exact triangle

$$(3.2)\qquad R\pi_*\mathcal{T}\to R\widetilde{\pi}_*\widetilde{\mathcal{T}}\xrightarrow{\text{ev}_1-\text{ev}_2}L\to R\pi_*\mathcal{T}[1],$$

where $\text{ev}_i: R\widetilde{\pi}_*\widetilde{\mathcal{T}}\to L$ is the evaluation map at σ_i $(i=1,2)$. Note that the morphism $b:\mathcal{T}^r\to\omega_{\mathcal{C}/X}$ induces a morphism

$$\widetilde{b}:\widetilde{\mathcal{T}}^r\to\omega_{\widetilde{\mathcal{C}}/X}(\sigma_1+\sigma_2).$$

Moreover, $\widetilde{b}$ is an isomorphism at σ_1 and σ_2, so restricting to either of these points we get a trivialization of L^r (the two trivializations are the same). Passing to an étale cover of X we can assume that L itself is trivial.

Let $[C_0\xrightarrow{d}C_1]$ be a complex of vector bundles representing $R\widetilde{\pi}_*\widetilde{\mathcal{T}}$ with C_1 a direct sum of sufficiently negative powers of an ample line bundle on X. Then the evaluation maps $\text{ev}_1,\text{ev}_2: R\widetilde{\pi}_*\widetilde{\mathcal{T}}\to L$ can be realized by morphisms $[C_0\to C_1]\to L$ in the homotopy category of complexes. Let $e_1,e_2:C_0\to L$ be the corresponding morphisms (unique up to adding morphisms that factor through C_1). Then we can choose a quasiisomorphism of $R\pi_*\mathcal{T}$ with the complex

$$\text{Cone}([C_0\to C_1]\xrightarrow{e_1-e_2}L)[-1]=[C_0\xrightarrow{(d,e_1-e_2)}C_1\oplus L]$$

compatible with the triangle (3.2), where $\text{Cone}(C\to C')$ denotes the cone of a morphism of complexes $C\to C'$.

The triangle (3.2) is obtained by applying the functor $R\pi_*$ to to the triangle

$$\mathcal{T}\to n_*\widetilde{\mathcal{T}}\xrightarrow{\text{ev}_1-\text{ev}_2}\sigma_*L\to\mathcal{T}[1]$$

on $\mathcal{C}$. To understand the map $S^rR\pi_*\mathcal{T}\to R\pi_*\omega_{\mathcal{C}/X}$ we can use the symmetric Künneth isomorphism $S^rR\pi_*\mathcal{T}\simeq R\pi_*^{(r)}(\mathcal{T}^{(r)})$, where $\mathcal{T}^{(r)}$ is the r-th symmetric power of $\mathcal{T}$ on $\mathcal{C}^{(r)}$. The maps $\mathrm{ev}_1, \mathrm{ev}_2 : n_*\widetilde{\mathcal{T}} \to \sigma_* L$ induce naturally the maps

$$\mathrm{ev}_1^r, \mathrm{ev}_2^r : (n_*\widetilde{\mathcal{T}})^{(r)} \to \sigma_*^r L,$$

where $\sigma^r : X \to \mathcal{C}^{(r)}$ is the r-tuple point of $\mathcal{C}^{(r)}$ corresponding to σ. Let us define a coherent sheaf on $\mathcal{C}^{(r)}$ as follows:

$$K := \ker((n_*\widetilde{\mathcal{T}})^{(r)} \stackrel{\mathrm{ev}_1^r - \mathrm{ev}_2^r}{\to} \sigma_*^r L^r).$$

Then we have a natural embedding $\mathcal{T}^{(r)} \to K$ which induces a morphism

$$\iota : S^r R\pi_*\mathcal{T} \simeq R\pi_*^{(r)}\mathcal{T}^{(r)} \to R\pi_*^{(r)} K.$$

Let $\Delta : \mathcal{C} \to \mathcal{C}^{(r)}$ be the diagonal embedding. We claim that there is a natural morphism $K \to \Delta_*\omega_{\mathcal{C}/X}$, such that the composition of the induced map $\widetilde{\eta} : R\pi_*^{(r)} K \to R\pi_*\omega_{\mathcal{C}/X}$ with ι coincides with the map $\eta : S^r R\pi_*\mathcal{T} \to R\pi_*\omega_{\mathcal{C}/X}$ induced by b. Indeed, $\Delta^* K$ maps to the kernel of the upper horizontal arrow in the commutative diagram

$$\begin{array}{ccc} (n_*\widetilde{\mathcal{T}})^{\otimes r} & \xrightarrow{\mathrm{ev}_1^r - \mathrm{ev}_2^r} & \sigma_* L^r \\ \downarrow & & \downarrow \\ n_* n^*\omega_{\mathcal{C}/X} & \longrightarrow & \sigma_*(\omega_{\mathcal{C}/X}|_\sigma) \end{array} \tag{3.3}$$

Therefore, we obtain the natural map from $\Delta^* K$ to the kernel of the lower horizontal arrow in this diagram, i.e., a map $\Delta^* K \to \omega_{\mathcal{C}/X}$. By adjunction we get a morphism $K \to \Delta_*\omega_{\mathcal{C}/X}$. The restriction of this map to the subsheaf $\mathcal{T}^{(r)} \subset K$ is the map induced by b which implies our claim. We also have a morphism of exact sequences

$$\begin{array}{ccccccccc} 0 & \longrightarrow & K & \longrightarrow & (n_*\widetilde{\mathcal{T}})^{(r)} & \xrightarrow{\mathrm{ev}_1^r - \mathrm{ev}_2^r} & \sigma_*^r L^r & \longrightarrow & 0 \\ & & \downarrow & & \downarrow & & \downarrow & & \\ 0 & \longrightarrow & \Delta_*\omega_{\mathcal{C}/X} & \longrightarrow & \Delta_* n_* n^*\omega_{\mathcal{C}/X} & \longrightarrow & \sigma_*^r(\omega_{\mathcal{C}/X}|_\sigma) & \longrightarrow & 0 \end{array} \tag{3.4}$$

This implies the commutativity of the following diagram:

$$\begin{array}{ccc} \mathcal{O}_X \simeq L^r & \longrightarrow & R\pi_*^{(r)} K[1] \\ \downarrow & & \downarrow{\scriptstyle \widetilde{\eta}[1]} \\ \omega_{\mathcal{C}/X}|_\sigma & \longrightarrow & R\pi_*\omega_{\mathcal{C}/X}[1] \end{array} \tag{3.5}$$

Therefore, the composition of the map $\widetilde{\tau} := \mathrm{Tr} \circ \widetilde{\eta}[1] : R\pi_*^{(r)} K[1] \to \mathcal{O}_X$ with the natural map $\mathcal{O}_X \simeq L^r \to R\pi_*^{(r)} K[1]$ is equal to the identity. On the other hand, since $\widetilde{\eta} \circ \iota = \eta$, it follows that the composition $\widetilde{\tau} \circ \iota = \tau : S^r R\pi_*\mathcal{T}[1] \to \mathcal{O}_X$ is exactly the map induced by b (which is used in the definition of the class $c(\mathcal{T}, b)$).

Note that $R\pi_*^{(r)} n_*\widetilde{\mathcal{T}} \simeq S^r R\widetilde{\pi}_*\widetilde{\mathcal{T}}$, so the object $R\pi_*^{(r)} K$ fits into the distinguished triangle

$$R\pi_*^{(r)} K \to S^r R\widetilde{\pi}_*\widetilde{\mathcal{T}} \stackrel{\mathrm{ev}_1^r - \mathrm{ev}_2^r}{\to} L^r \to R\pi_*^{(r)} K[1].$$

Therefore, it can be represented by the complex $\mathrm{Cone}(S^r[C_0 \to C_1] \stackrel{e_1^r - e_2^r}{\to} L^r)[-1]$ in a way compatible with this triangle. Furthermore, the natural morphism $S^r R\pi_* \mathcal{T} \to R\pi_*^{(r)} K$ is realized by the natural map of complexes

$$(3.6) \qquad S^r[C_0 \stackrel{(d, e_1 - e_2)}{\to} C_1 \oplus L] \to \mathrm{Cone}(S^r[C_0 \to C_1] \stackrel{e_1^r - e_2^r}{\to} L^r)[-1]$$

with the components $\mathrm{id} : S^r C_0 \to S^r C_0$,

$$S^{r-1}C_0 \otimes (C_1 \oplus L) \to (S^{r-1}C_0 \otimes C_1) \oplus L^r : x^{r-1} \otimes (y, z) \mapsto (x^{r-1} \otimes y, \sum_{i=0}^{r-1} e_1^i(x) e_2^{r-1-i}(x) z),$$

etc. Finally, we claim that for a suitable choice of the complex $C_0 \to C_1$ (as in Proposition 4.7 of [**9**]) the map $\widetilde{\tau} : R\pi_*^{(r)} K[1] \to \mathcal{O}_X$ is represented by the chain map of complexes $\mathrm{Cone}(S^r[C_0 \to C_1] \stackrel{e_1^r - e_2^r}{\to} L^r) \to \mathcal{O}_X$. This is a consequence of the following general result.

LEMMA 3.2. *Let $g : A \to B$, $f : B \to C$ be a pair of maps in the homotopy category $\mathcal{K}$ of some abelian category and let $\mathcal{D}$ be the corresponding derived category. Consider the subsets $H_{\mathcal{K}}(f) \subset \mathrm{Hom}_{\mathcal{K}}(\mathrm{Cone}(g), C)$ and $H_{\mathcal{D}}(f) \subset \mathrm{Hom}_{\mathcal{D}}(\mathrm{Cone}(g), C)$ consisting of morphisms $\mathrm{Cone}(g) \to C$ such that their composition with the canonical morphism $i : B \to \mathrm{Cone}(g)$ is equal to f (in $\mathcal{K}$ and $\mathcal{D}$ respectively). Assume that the map*

$$\mathrm{Hom}_{\mathcal{K}}(A, C) \to \mathrm{Hom}_{\mathcal{D}}(A, C)$$

is injective and the map

$$\mathrm{Hom}_{\mathcal{K}}(A[1], C) \to \mathrm{Hom}_{\mathcal{D}}(A[1], C)$$

is surjective. Then the natural map

$$\kappa : H_{\mathcal{K}}(f) \to H_{\mathcal{D}}(f)$$

is surjective.

Proof. Let us denote by $\pi : \mathrm{Cone}(g) \to A[1]$ the canonical chain map. If the set $H_{\mathcal{D}}(g)$ is empty then the assertion is clear, so we can assume that $H_{\mathcal{D}}(g) \neq \emptyset$. Then the composition $f \circ g : A \to C$ becomes zero in the derived category. By our assumption the natural map $\mathrm{Hom}_{\mathcal{K}}(A, C) \to \mathrm{Hom}_{\mathcal{D}}(A, C)$ is injective, hence $f \circ g$ is homotopic to zero. Every homotopy h from $g \circ f$ to 0 induces naturally a chain map $\mathrm{Cone}(h) : \mathrm{Cone}(g) \to C$ which coincides with f on the subcomplex $i(B) \subset \mathrm{Cone}(g)$. In fact, it is easy to see that the map $h \mapsto \mathrm{Cone}(h)$ is a bijection between homotopies from $g \circ f$ to 0 and chain maps $\mathrm{Cone}(g) \to C$ extending f on B. If we have two homotopies h_1, h_2 from $g \circ f$ to 0 then the difference $h_1 - h_2$ gives a chain map from $A[1]$ to C. It is easy to see that

$$\mathrm{Cone}(h_1) - \mathrm{Cone}(h_2) = (h_1 - h_2) \circ \pi.$$

Now let $\gamma \in H_{\mathcal{D}}(g)$ be any element. Let us pick a homotopy h_0 from $g \circ f$ to 0. Then the homotopy class $[\mathrm{Cone}(h_0)]$ is an element of $H_{\mathcal{K}}(f)$. The composition of $\kappa([\mathrm{Cone}(h_0)]) - \gamma$ with i vanishes in the derived category, hence we have $\kappa([\mathrm{Cone}(h_0)]) - \gamma = \beta \circ \pi$ for some $\beta \in \mathrm{Hom}_{\mathcal{D}}(A[1], C)$. By our assumption there exists a chain map $\widetilde{\beta} : A[1] \to C$ representing β. Then $h = h_0 - \widetilde{\beta}$ is another homotopy from $g \circ f$ to 0. We have

$$\kappa([\mathrm{Cone}(h)]) = \kappa([\mathrm{Cone}(h_0) - \widetilde{\beta} \circ \pi]) = \kappa([\mathrm{Cone}(h_0)]) - \beta \circ \pi = \gamma.$$

□

We apply the above lemma to $A = S^r[C_0 \to C_1]$, $B = L^r$ and $C = \mathcal{O}_X$, where $f : L^r \to \mathcal{O}_X$ is the canonical isomorphism. To satisfy the assumptions of the lemma we choose the complex $C_0 \to C_1$ representing $R\widetilde{\pi}_*\widetilde{\mathcal{T}}$ with C_1 a direct sum of sufficiently negative powers of an ample line bundle (one has to argue as in Proposition 4.7 of [**9**]). Hence, the map $\widetilde{\tau}$ is represented by a morphism in the homotopic category $\mathrm{Cone}(S^r[C_0 \to C_1] \stackrel{e_1^r - e_2^r}{\to} L^r) \to \mathcal{O}_X$ that we still denote by $\widetilde{\tau}$. The restriction of $\widetilde{\tau}$ to the subcomplex L^r is equal to the canonical isomorphism $L^r \to \mathcal{O}_X$, while its composition with the map (3.6) is the morphism

$$\tau : S^r[C_0 \stackrel{(d, e_1 - e_2)}{\to} C_1 \oplus L] \to \mathcal{O}_X[-1]$$

that should be used for the computation of $c(\mathcal{T}, b)$. It follows that the restriction of the corresponding morphism

$$\tau : S^{r-1}C_0 \otimes (C_1 \oplus L) \to \mathcal{O}_X$$

to $S^{r-1}C_0 \otimes L$ has form

$$\tau(x^{r-1} \otimes y) = \sum_{i=0}^{r-1} e_1(x)^i e_2(x)^{r-1-i} y \in L^r \simeq \mathcal{O}_X,$$

where $x \in C_0$, $y \in L$. Hence, the corresponding isotropic section of $p^*(C_1 \oplus L \oplus C_1^\vee \oplus L^{-1})$ (where $p : C_0 \to X$ is the projection) has form

$$s(x) = (d(x), (e_1 - e_2)(x), \nu(x), \sum_{i=0}^{r-1} e_1(x)^i e_2(x)^{r-1-i}),$$

where the last component belongs to $L^{r-1} \simeq L^{-1}$, ν is given by some morphism $S^{r-1}C_0 \to C_1^\vee$.

To compute the class corresponding to the twisted data $(\mathcal{T} \otimes \mathcal{L}_\xi, b_\xi)$ for some r-th root of unity ξ we simply have to replace the pair (e_1, e_2) by $(e_1, \xi e_2)$. Note that this will not affect the definition of K and of the morphism $\widetilde{\tau}$. Hence the corresponding isotropic section of $p^*(C_1 \oplus L \oplus C_1^\vee \oplus L^{-1})$ will take form

$$s_\xi(x) = (d(x), (e_1 - \xi e_2)(x), \nu(x), \sum_{i=0}^{r-1} \xi^{r-1-i} e_1(x)^i e_2(x)^{r-1-i}), \tag{3.7}$$

for some $\nu : S^{r-1}C_0 \to C_1^\vee$.

Now we can finish the proof of Theorem 3.1. For every ξ the class $c(\mathcal{T} \otimes \mathcal{L}_\xi, b_\xi)$ is equal to

$$\mathrm{td}(C_1 \oplus L) \cdot \mathrm{ch}_X^{C_0}(\Lambda^* p^*(C_1^\vee \oplus L^{-1}), s_\xi),$$

where $s_\xi \in p^*(C_1 \oplus L \oplus C_1^\vee \oplus L^{-1})$ is the isotropic section constructed above. Let us set $f_\xi = e_1 - \xi e_2$. We consider (f_ξ) as a collection of sections of p^*L on C_0. We have an orthogonal decomposition

$$p^*(C_1 \oplus L \oplus C_1^\vee \oplus L^{-1}) \simeq p^*(C_1 \oplus C_1^\vee) \oplus p^*(L \oplus L^{-1}),$$

so that the section s_ξ has components $s_0 = (d, \nu) \in p^*(C_1 \oplus C_1^\vee)$ and $(f_\xi, \prod_{\xi' \neq \xi} f_{\xi'})$. Recall that we can trivialize L, so the spinor bundle $\Lambda^* p^*(C_1^\vee \oplus L^{-1})$ can be identified with $\Lambda^* p^* C_1^\vee \oplus \Lambda^* p^* C_1^\vee[1]$. Under this identification the action of sections s_ξ will take form of differentials (d_i) in Lemma 1.2, where the odd endomorphism d

of $\Lambda^* p^* C_1^\vee$ is given by the action of s_0. Now the assertion of the theorem follows from Lemma 1.2.

References

[1] P. Baum, W. Fulton, R. MacPherson, *Riemann-Roch for singular varieties*, Inst. Hautes Itudes Sci. Publ. Math. **45** (1975), 101–145.

[2] I. Ciocan-Fontanine, M. Kapranov, *Derived Quot schemes*, Ann. Sci. École Norm. Sup. (4) 34 (2001), 403–440.

[3] I. Ciocan-Fontanine, M. Kapranov, *Derived Hilbert schemes*, J. Amer. Math. Soc. 15 (2002), 787–815.

[4] W. Fulton, *Intersection theory*, Springer, 1998.

[5] T. J. Jarvis, T. Kimura, A. Vaintrob, *Moduli spaces of higher spin curves and integrable hierarchies*, Compositio Math. 126 (2001), 157–212.

[6] M. Kontsevich, *Enumeration of rational curves via torus action*, in *Moduli Space of Curves (R. Dijkgraaf, C. Faber, G. van der Geer, Eds.)*, 335–368, Birkhauser, Boston, 1995.

[7] M. Kontsevich, Yu. I. Manin, *Gromov-Witten classes, quantum cohomology, and enumerative geometry*, Commun. Math. Phys. **164** (1994), 525–562.

[8] T. Mochizuki, *The virtual class of the moduli stack of r-spin curves*, preprint, see http://www.math.ias.edu/~takuro/list.html

[9] A. Polishchuk, A. Vaintrob, *Algebraic construction of Witten's top Chern class*, in *Advances in Algebraic Geometry motivated by Physics*, E. Previato, ed., 229–250. AMS, 2001.

[10] E. Witten, *The N-matrix model and gauged WZW models*, Nucl. Phys. B **371** (1992), no. 1-2, 191-245.

[11] E. Witten, *Algebraic geometry associated with matrix models of two dimensional gravity*, Topological methods in modern mathematics (Stony Brook, NY, 1991), Publish or Perish, Houston, 1993, 235-269.

Department of Mathematics, Boston University, Boston, MA 02215
E-mail address: apolish@math.bu.edu

Uniformization of the orbifold of a finite reflection group

Kyoji Saito

1. Introduction

Let W be a finite reflection group of a real vector space V. If W is crystallographic, then the quotient space $V^*//W$ appears in several contexts in geometry: i) in Lie theory as the quotient space of a simple Lie algebra by the adjoint Lie group action [Ch1,2] and ii) in complex geometry as the base space of the universal unfolding of a simple singularity [Br1]. Having these backgrounds, $V^*//W$ carries some distinguished geometric properties and structures, which, fortunately and also amusingly, can be described only in terms of the reflection group regardless whether W is crystallographic or not. We recall two of them:

1. The complexified regular orbit space $(V^*//W)_{\mathbf{C}}^{reg}$ is a $K(\pi, 1)$-space (Brieskorn [Br3], Deligne [De]). In other words, $\pi_1((V^*//W)_{\mathbf{C}}^{reg})$ is an Artin group (i.e. a generalized braid group [B-S][De]) and the universal covering space of $(V^*//W)_{\mathbf{C}}^{reg}$ is contractible (c.f. also [Sa]).

2. The quotient space $V^*//W$ carries a *flat structure* (Saito [S3][S6])[1]. This means roughly that the tangent bundle of $V^*//W$ carries a flat metric J together with some additional structures. Nowadays, a flat structure without a primitive form is also called a *Frobenius manifold structure with gravitational descendent* (Dubrovin [Du], Manin [Ma1,2]).

Apparently, these two geometries on $V^*//W$ are of a quite different nature, one topological and the other differential geometric. Nevertheless, there is already a remarkable relationship between them on a combinatorial level: the *polyhedron dual to the system of real reflection hyperplanes of* W (which is the key in [Br2][De] to determine the topology of the complex regular orbit space) is reconstructed by a use of the *formal group action* $\exp(tD)$:=*the integral of the primitive vector field* D on $(V^*//W)_{\mathbf{R}}$ (which is a basic ingredient of the flat structure) [S8].

The present article is a revised version of §5, §6, §7 of the lecture note "Geometry of finite reflection groups" delivered by the author at RIMS (1999). The author is grateful to Claus Hertling for a discussion which helped to clarify the formulation. He also thanks Susumu Tanabe for his kind help finding the references [A][S2] .

[1]The original construction of the flat structure on $V^*//W$ was given in [S3]. The description of the gravitational descendent was modified in [S6] to obtain the system of uniformization equations. The present article follows the latter style.

Inspired by the observation, the present article aims to construct a more direct relationship between the two geometries. The working hypothesis is that *a bridge between them is given by the topological behavior of the map (which, for brevity, we call the period map) obtained from solutions of the uniformization equation* $\mathcal{M}_{W,s}$ [S6] *on* $V^*//W$ *constructed from the flat structure* for a special s (see 6.1 Remark).[2]

Here we begin a program to examine this hypothesis. In the first half §2-4, we describe the uniformization equation. After fixing notation for finite reflection groups in §2, we give a detailed exposition of the flat structure in §3 and the uniformization equation $\mathcal{M}_{W,s}$ in §4. Although they are already known [S6], we renew and clarify several arguments and make them accessible for our purpose. (c.f. also [He][Sab][Ta]).

In the latter half (§5, 6), we begin to analyze the period map. In §5, solutions of the uniformization equation for the parameter $s = 1/2$ are partly given by *primitive Abelian integrals on a certain family of plane curves parameterized by* $V^*//W$. Although this fact is easy [S6], the attached period map is not studied from the view point of the primitive form, although some information is available in classical works [Th][Mu][Ko]. We examine examples of type A_1, A_2, A_3 and B_2.

In §6, first, we describe the monodromy group $\Gamma(W)$ in term of Coxeter diagram. Then we give a possible formulation of the *period domain* and the *inverse map to the period map*, and pose some conjectures. §6 is quite incomplete. It requires more work to verify or to modify the conjectures, which is beyond the scope of the present article.

2. Finite reflection group

This section gives a short summary of basic results on finite reflection groups used in the present article (see also [B]). Experienced readers are recommended to look only at the notation in 2.9 and skip to §3.

2.1. Reflection. Let V and V^* be a real vector space and its dual. An element $\alpha \in \mathrm{GL}(V) \simeq \mathrm{GL}(V^*)$ is a *reflection* if there exist a hyperplane H_α in V^* and a non-zero vector $f_\alpha \in V^*$ such that $\alpha|_{H_\alpha} = id_{H_\alpha}$ and $\alpha(f_\alpha) = -f_\alpha$. The H_α is called the *reflection hyperplane* of α. One has $\alpha(x) = x - f_\alpha(x)e_\alpha$ for $x \in V$ and $\alpha(x^*) = x^* - e_\alpha(x^*)f_\alpha$ for $x^* \in V^*$, where $e_\alpha \in V$ is a defining form of H_α with $\langle e_\alpha, f_\alpha \rangle = 2$.

---

[2] *A historical note*: Before the theory of the primitive form and the flat structure reached its present form, the author suggested in [S2] to study the uniformization of the regular orbits $(V^*//W)^{reg}_{\mathbb{C}}$ by the horizontal sections of *logarithmic flat torsion-free connections on the logarithmic tangent bundle* on $V^*//W$. The torsion-free condition implies the existence of a *primitive function*, whose derivatives give a system of fundamental solutions. The primitive function for type A_1 is the logarithm. For type A_2, it is given by the elliptic integral of the first kind which gives the universal covering (up to center) of the regular orbit space. For type A_3, the space of all logarithmic flat torsion-free connections decomposes into two one-parameter families [S2,§3]. The first family gives the uniformization equation $\mathcal{M}_{W,s}$ of the present article. The meaning of the second family is unknown: for example, what is the Fourier-Laplace transform of the second family? (c.f. [A])

We note also that there is related work on certain integrable systems defined on the quotient space $V^*//W$ ([Gi] [Tak]). However the relationship with the flat structure still needs to be worked out.

2.2. Finite reflection group W. We shall mean by a *finite reflection group* W a finite group generated by reflections acting irreducibly on a real vector space V. Put $R(W) := \{\alpha \in W \mid \text{a reflection}\}$. There exist, unique up to a constant factor, W-invariant symmetric bilinear forms I and I^* on V and V^*, respectively.[3] One has $f_\alpha = 2I(e_\alpha, \cdot)/I(e_\alpha, e_\alpha)$ and $e_\alpha = 2I^*(f_\alpha, \cdot)/I^*(f_\alpha, f_\alpha)$. A connected component C of $V^* \setminus \cup_{\alpha \in R(W)} H_\alpha$ is called a *chamber*. A hyperplane H_α ($\alpha \in R(W)$) is called a *wall* of a chamber C, if $H_\alpha \cap \bar{C}$ contains an open subset of H_α.

2.3. Coxeter group representation of W.

We may present a finite reflection group as a Coxeter group [Co1].

A *Coxeter matrix* $M := (m(\alpha, \beta))_{\alpha,\beta \in \Pi}$ is a symmetric matrix with index set Π s.t. $m(\alpha, \alpha) = 1$ ($\alpha \in \Pi$) and $m(\alpha, \beta) \in \mathbf{Z}_{\geq 2} \cup \{\infty\}$ ($\alpha \neq \beta \in \Pi$). The group $W(M)$ generated by letters a_α ($\alpha \in \Pi$) and defined by fundamental relations: $(a_\alpha a_\beta)^{m(\alpha,\beta)} = 1$ ($\alpha, \beta \in \Pi$) is called a *Coxeter group*. The pair $(W(M), \{a_\alpha \mid \alpha \in \Pi\})$ is called a *Coxeter system*.

THEOREM. *Let W be a finite reflection group acting on V and let C be a chamber of the W-action. Then the following 1.- 5. hold.*

1. *The pair $(W, \Pi(C))$ is a Coxeter system, where we put*

$$\Pi(C) := \{\alpha \in R(W) \mid H_\alpha \textit{ is a wall of the chamber } C\} \tag{2.3.1}$$

and the Coxeter matrix is given by $m(\alpha, \beta) :=$ the order of $\alpha\beta$ in W.

2. *W acts on the set of chambers simply and transitively. Hence, the Coxeter matrix does not depend on the choice of a chamber.*

3. *The closure $\bar{C}$ of a chamber is a fundamental domain for the action of W on V. That is: there is a homeomorphism: $\bar{C} \simeq V^*/W$.*

4. *Fix the sign of the vector e_α for $\alpha \in \Pi(C)$ in the manner:*

$$C = \{x \in V^* \mid \langle e_\alpha, x\rangle > 0 \textit{ for } \alpha \in \Pi(C)\}. \tag{2.3.2}$$

Then the off-diagonals of the matrix $(I(e_\alpha, e_\beta))_{\alpha,\beta \in \Pi(C)}$ are non-positive.

5. *$\Pi_W := \{e_\alpha | \alpha \in \Pi(C)\}$ forms a basis of V. The coefficients of $e_\beta = \sum_{\alpha \in \Pi} c_\alpha e_\alpha$ for $\beta \in R(W)$ are either all non-negative or non-positive.*

2.4. Classification.

We recall the classification of finite Coxeter groups ([B, ch.VI,§4]).

To a Coxeter matrix M, one attaches a Coxeter graph Γ, whose vertices are indexed by the set Π and two vertices α and β are connected by an edge iff $m(\alpha, \beta) \geq 3$. The edge is labeled by $m(\alpha, \beta)$ (omitted if $m(\alpha, \beta) = 3$). The graph is called *simply-laced* if all labels are 3.

The following is the list of graphs associated to finite Coxeter groups.

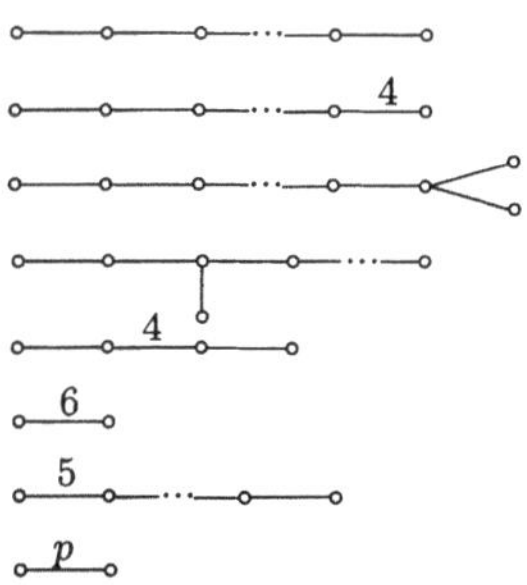

A_l ($l \geq 1$)
B_l ($l \geq 2$)
D_l ($l \geq 4$)
E_l ($l = 6, 7$ or 8)
F_4
G_2
H_l ($l = 2, 3$ or 4)
$I_2(p)$ ($p \geq 7$).

NOTE. 1. Different Coxeter diagrams define non-isomorphic groups, i.e. the same group is not attached to different Coxeter matrices.

2. The group W is called crystallographic if it preserves a full-lattice in V. This condition rules out the groups of type H_l and $I_2(p)$.

3. The irreducibility of W implies the indecomposability of the Coxeter matrix M and, hence, the connectedness of the graph Γ.

2.5. Polynomial invariants.

Let $S(V) = \mathbf{R} \oplus V \oplus S(V)_2 \oplus S(V)_3 \oplus \cdots$ be the symmetric tensor algebra of V. The action of $g \in W$ on V induces the action on $S(V)$. Define the set of invariants:

$$(2.5.1) \qquad S(V)^W := \{P \in S(V) \mid g(P) = P \text{ for } \forall g \in W\}.$$

Obviously, $S(V)^W$ is a graded subalgebra of $S(V)$.

THEOREM. *(Chevalley [Ch 2]). Let W be a finite reflection group acting irreducibly on a real vector space V of rank l. Then $S(V)^W$, as an $\mathbf{R}$-algebra, is generated by l algebraically independent homogeneous elements, say $P_1, \cdots, P_l$. The set of degrees $d_1 = \deg(P_1), \cdots, d_l = \deg(P_l)$ (with multiplicity) is independent of a choice of the generators.*

NOTE. The ring $S(V)$, viewed as a $S(V)^W$-module, is free of rank $\#W$, and $\dim(S(V)/S(V)S(V)^W_+) = \#W$, where $S(V)^W_+$ is the maximal ideal of $S(V)^W$ of all positively graded elements (c.f. (2.6.1) i)).

2.6. Poincare series.

The $S(V)^W$ is a graded subring of $S(V)$, i.e. $S(V)^W = \oplus_{d\in\mathbf{Z}_{\geq 0}} S(V)^W_d$ for $S(V)^W_d = S(V)_d \cap S(V)^W$. The Poincare series: $P_{S(V)^W}(t) := \sum_{d=0}^{\infty} \dim_{\mathbf{R}}(S(V)^W_d)\, t^d$ is calculated in two different ways : i) Using $S(V)^W \simeq \mathbf{R}[P_1] \otimes \cdots \otimes \mathbf{R}[P_l]$, one has $P_{S(V)^W}(t) = \prod_{i=1}^{l} P_{\mathbf{R}[P_i]}(t) = \prod_{i=1}^{l} \frac{1}{(1-t^{d_i})}$ (this expression reproves the uniqueness of the $d_1, \cdots, d_l$), and ii) since $\dim_{\mathbf{R}}(S(V)^W_d) = \mathrm{tr}(\frac{1}{\#W}\sum_{w\in W} w|S(V)_d)$ for $d \in \mathbf{Z}_{\geq 0}$ and $\sum_{d=0}^{\infty} \mathrm{tr}(w|S(V)_d)t^d = \frac{1}{\det(1-tw)}$ (use the extension of V to $V_{\mathbf{C}}$), one has $P_{S(V)^W}(t) = \frac{1}{\#W}\sum_{w\in W} \frac{1}{\det(1-tw)}$. Comparing the values and derivatives at t=1 of the two expressions of $P_{S(V)^W}(t)$, one obtains:

$$(2.6.1) \qquad i)\ \#W = d_1 \cdots d_l \quad \text{and} \quad ii)\ \#R(W) = \sum_{i=1}^{l}(d_i - 1).$$

2.7. Anti-invariants.

An element $P \in S(V)$ is called an *anti-invariant* if $g \cdot P = \det(g)^{-1}P$ for all $g \in W$. The set of all anti-invariant shall be denoted by $S(V)^{-W}$.

Put $\delta_W := \prod_{\alpha\in R(W)} e_\alpha$. Let $\frac{\partial(P_1,\cdots,P_l)}{\partial(X_1,\cdots,X_l)}$ be the Jacobian for generator system $P_1, \cdots, P_l$ and $X_1, \cdots, X_l$ of the algebras $S(V)^W$ and $S(V)$, respectively. It is easy to see that δ_W and the Jacobian are anti-invariants. Using (2.6.1) ii), we further show an important lemma.

LEMMA. 1. *Any anti-invariant is divisible by δ_W: $S(V)^{-W} = S(V)^W \delta_W$.*

2. *One has $\frac{\partial(P_1,\cdots,P_l)}{\partial(X_1,\cdots,X_l)} = c\, \delta_W$ for a nonzero constant $c \in \mathbf{R}$.*

2.8. Coxeter elements and exponents.

A *Coxeter element* is a product $c := \Pi_{\alpha\in\Pi(C)}\alpha$ for a linear ordering of elements of $\Pi(C)$. Its conjugacy class depends neither on C nor on the ordering (for Γ is a tree [B,Ch.V,n°6.2.]). The order h of c is called the *Coxeter number*. Put $\det(\lambda 1 - c) = \prod_{i=1}^{l}(\lambda - \exp(2\pi\sqrt{-1}m_i/h))$ for some integers $m_1, \cdots, m_l$, called the *exponents* of W, such that

$$(2.8.1) \qquad 0 < m_1 \leq m_2 \leq \cdots \leq m_l < h.$$

Here, 0 is not an exponent (i.e. 1 cannot be an eigenvalue of c), since I is non-degenerate. So $\exp(2\pi\sqrt{-1}m_i/h)$ and $\exp(2\pi\sqrt{-1}(m_{l-i+1})/h))$ should be complex conjugate to each other. Thus, we have

$$(2.8.2) \qquad m_i + m_{l-i+1} = h \quad \text{and} \quad \textstyle\sum_{i=1}^{l} m_i = \frac{1}{2}lh.$$

In the rest of this §, we assume $l \geq 2$ (i.e. W is not of type A_1) although the resulting formula (2.8.3) is valid even that case.

Since the Coxeter graph Γ is a tree, one can find a unique decomposition $\Pi(C) = \Pi_1 \cup \Pi_2$ such that any two elements in Π_i mutually commute for $i = 1, 2$ (ie. $\langle e_\alpha, f_\beta\rangle = 0$ for $\alpha \neq \beta \in \Pi_i$). Put $c_i := \prod_{\alpha\in\Pi_i}\alpha$ and $c = c_1c_2$. The mutual commutativity of elements in Π_i implies that $c_i(x) = x - \sum_{\alpha\in\Pi_i} e_\alpha(x)f_\alpha$ and that c_i is an involution, i.e. $c_i^2 = 1$.

We state a key lemma on the eigenvectors of the Coxeter elements (Kostant [K1] and Coleman [C]), which we shall use in §3-6 crucially.

LEMMA. *There exists a real 2 dimensional subspace U of V^* satisfying:*

i) *U is invariant under the actions of c_1 and c_2, and $\{c_1|U, c_2|U\}$ forms a Coxeter system for the dihedral group $W(I_2(h))$ acting on U.*

ii) *$U \cap C = \mathbf{R}_{>0}\cdot\zeta_1 + \mathbf{R}_{>0}\cdot\zeta_2$ is a chamber of the group $\langle c_1|U, c_2|U\rangle = W(I_2(h))$, where ζ_j is a c_j-fixed vector in $\bigcap_{\alpha\in\Pi_j} H_\alpha \cap \bigcap_{\alpha\in\Pi\setminus\Pi_j}\{e_\alpha > 0\}$.*

The lemma implies in particular that no reflection hyperplane of W intersects the open cone $\mathbf{R}_{>0}\cdot\zeta_1 + \mathbf{R}_{>0}\cdot\zeta_2$.

COROLLARY. **1.** *Any reflection hyperplane of W intersects U only along one of the h lines which are $W(I_2(h))$ orbits of $\mathbf{R}\zeta_1$ or $\mathbf{R}\zeta_2$. If a reflection hyperplane H_α contains the line $\mathbf{R}\zeta_i$, then $\alpha \in \Pi_i$.*

2. *Let W be a finite reflection group of rank l and Coxeternumber h.*

$$(2.8.3) \qquad i)\ m_1 = 1,\ m_l = h - 1 \quad \text{and} \quad ii)\ \#R(W) = \tfrac{1}{2}lh$$

3. *The eigenvectors of the action $c|_U$ belonging to the eigenvalues $\exp(2\pi\sqrt{-1}/h)$ and $\exp(-2\pi\sqrt{-1}/h)$, respectively, do not belong to any complexified reflection hyperplane $H_{\alpha,\mathbf{C}} := H_\alpha \otimes \mathbf{C}$ for $\alpha \in R(W)$.*

Recall that $d_1, \cdots, d_l$ are the degrees of a generator system of $S(V)^W$. A study of the Jacobian J shows $d_j - 1 \equiv m_j \bmod h$ $(1 \leq j \leq l)$ for renewed index. This together with (2.6.1) ii), (2.8.2) and (2.8.3) implies

$$(2.8.4) \qquad d_i = m_i + 1 \quad \text{for} \quad i = 1, \cdots, l.$$

Recall the W-invariant bilinear forms I and I^* on V and V^* such that $I(x, y) = I^*(I(x), I(y))$. The associated quadratic form

$$(2.8.5) \qquad P_1 := I^*(x,x)/2h = \textstyle\sum_{ij=1}^{l} X_iX_jI^*(x_i, x_j)/2h$$

(here $x = \sum_{i=1}^{l} X_ix_i$, and x_i and X_i are dual basis of V^* and V) gives an invariant in $S(V)^W$ of lowest degree $d = 2$ (unique up to constant since W-action is irreducible). This fact together with (2.8.4) implies

COROLLARY. **4.** *The multiplicity of the smallest exponent* ($= 1$) *is equal to 1. Hence, that of the largest exponent* ($= h - 1$) *is also equal to 1.*

REMARK. l and h cannot be simultaneously odd due to the second formula (2.8.2). More precisely (see 6.2 Assertion 1. for a proof): l is odd $\Rightarrow$ $\#\Pi_1 \neq \#\Pi_2$ $\Leftrightarrow$ h is even and $\frac{1}{2}h$ is an exponent $\Rightarrow$ h is even. Here, the two arrows are trivial. The converse of the first arrow does not hold for type D_l (even l). The converse of the second arrow does not hold for types B_l and C_l (even l), E_6, E_8, F_4, H_4 and $I_2(p)$ (even p).

2.9. The quotient variety $V^*//W$ and the discriminant D_W.
The categorical quotient variety of V^* by the action of W is given by

$$S_W := V^*//W := \mathrm{Spec}(S(V)^W). \tag{2.9.1}$$

It has origin 0 defined by the maximal ideal $S(V)^W_+$ (recall 2.5).
Let K be either $\mathbf{R}$ or $\mathbf{C}$.The set of K-rational points of S_W is given by

$$S_{W,K} := \mathrm{Hom}^{alg}_{\mathbf{R}}(S(V)^W, K) \tag{2.9.2}$$

where $\mathrm{Hom}^{alg}_{\mathbf{R}}(*,*)$ means the set of all $\mathbf{R}$-algebra homomorphisms.

Put $V^*_{\mathbf{C}} := V^* \otimes_{\mathbf{R}} \mathbf{C}$. The action of W on $V^* = V^*_R$ extends complex linearly to $V^*_{\mathbf{C}}$. For any point $x \in V^*_K$, the evaluation homomorphism: $P \in S(V)^W \mapsto P(x) \in K$ induces the W-invariant morphism:

$$\pi_K : V^*_K \to S_{W,K}. \tag{2.9.3}$$

Put $V^*_K/W :=\{W\text{-orbits on } V^*_K\}$, where a W-orbit on V^*_K means a subset of V^*_K of the form Wx for some $x \in V^*_K$. An element $P \in S(V)^W$ is naturally considered as a function on V^*_K/W since it is constant on each orbit. Since *for $x, y \in V_K$ one has $Wx = Wy$ if and only if $P(x) = P(y)$ for all $P \in S(V)^W$*, the morphism π_K (2.9.3) induces an injection $V^*_K/W \to S_{W,K}$ for $K = \mathbf{R}$ or $\mathbf{C}$. In fact, the $\pi_{\mathbf{C}}$ induces a homeomorphism: $V^*_{\mathbf{C}}/W \simeq S_{W,\mathbf{C}}$, but $\pi_{\mathbf{R}}$ induces an embedding $V^*_{\mathbf{R}}/W \subset S_{W,\mathbf{R}}$ onto a closed semi-algebraic set. Choosing a generator system $P_1, \cdots, P_l$ of $S(V)^W$ (with $\deg(P_1) \leq \cdots \leq \deg(P_l)$), one has a bijection $S_{W,K} \simeq K^l$ and the π_K is given by $(P_1, \cdots, P_l) : V^*_K \to K^l$.

The square δ^2_W of the anti-invariant δ_W (2.7) is an invariant. We call it the *discriminant* of W and denote by Δ_W. The *discriminant divisor* is defined by $\Delta_W = 0$. The *discriminant locus* in $S_{W,\mathbf{C}}$ is given by

$$D_{W,\mathbf{C}} := \{t \in S_{W,\mathbf{C}} \mid \Delta_W(t) = 0\}. \tag{2.9.4}$$

2.7 Lemma 2. implies i) the critical values of the morphism π lie in the discriminant D_W and ii) $(\pi_{\mathbf{C}})^{-1} D_{W,\mathbf{C}} = \bigcup_{\alpha \in R(W)} H_{\alpha,\mathbf{C}}$. Therefore,
Fact. 1. *Any W-fixed point in $V_{\mathbf{C}}$ lies in a reflection hyperplane.*

2. *The complement of the discriminant locus $S_{W,\mathbf{C}} \setminus D_{W,\mathbf{C}}$ is the space of regular (i.e. isotropy free) orbits of the W-action on $V_{\mathbf{C}}$.*

Let us express the discriminant Δ_W as a polynomial in $(P_1, \cdots, P_l)$. Since $\deg(\Delta_W) = hl$ (definition of δ_W and (2.8.3) ii)) and $\deg(P_l) = h$ ((2.8.3) i) and (2.8.4)), Δ_W is a polynomial in P_l of degree at most l:

$$\Delta_W = A_0 P_l^l + A_1 P_l^{l-1} + \cdots + A_l \tag{2.9.5}$$

where A_i is a polynomial in $P_1, \cdots, P_{l-1}$ of degree hi. Since $\Delta_W(\xi) \neq 0$ and $P_1(\xi) = \cdots = P_{l-1}(\xi) = 0$ for an eigenvector ξ of a Coxeter element belonging to $\exp(2\pi\sqrt{-1}/h)$ (use 2.8 Cor.3, 4 and (2.8.4)), one obtains the next goal of this section and the starting point of the present article:

LEMMA. 1. *A_0 is non-zero. Hence, Δ_W is normalized to a monic polynomial of degree l in P_l and D_W has multiplicity l at the origin.*

2. *The eigenspace of a Coxeter element belonging to the eigenvalue $\exp(2\pi\sqrt{-1}/h)$ is mapped by $\pi_{\mathbf{C}}$ to a line $P_1 = \cdots = P_{l-1} = 0$ in $S_{W,\mathbf{C}}$.*

3. Flat structure

We describe the flat structure, the Frobenius manifold structure and the associated flat coordinates on the variety S_W in detail. The setting and the notation are the same as in §2, that is: W is a finite reflection group of a real vector space V and S_W is the quotient variety $V^*//W$ (recall 2.9). The flat structure is obtained by Fourier transform of the Levi-Civita connection for the W-invariant form I ([S3], [S6]).

3.1. Logarithmic forms and logarithmic vector fields.

We recall ([S4]) the definition and the basic properties of the modules of logarithmic forms and vector fields for the variety S_W with the divisor $D_W = \{\Delta_W = 0\}$ (see 2.9). In the sequel, we shall use coordinates $P_1, \cdots, P_l$ of S_W satisfying the degree conditions (2.8.1)-(2.8.5) by choosing a generator system of the invariants ring $S(V)^W$.

Let Der_{S_W} and $\Omega^1_{S_W}$ be the modules of $\mathbf{R}$-derivations of $S(V)^W$ and of 1-forms on S_W over $\mathbf{R}$, respectively. They are $S(V)^W$-free modules of rank l generated by the derivations $\partial/\partial P_i$ and by the differentials dP_i ($i = 1, \cdots, l$), respectively. The logarithmic modules are defined by

$$(3.1.1)\qquad \begin{array}{rcl} Der_{S_W}(-\log\Delta) & := & \{X \in Der_{S_W} \mid X\Delta_W \in \Delta_W S(V)^W\} \\ \Omega^1_{S_W}(\log\Delta) & := & \{\omega \in \frac{1}{\Delta_W}\Omega^1_{S_W} \mid d\omega \in \frac{1}{\Delta_W}\Omega^2_{S_W}\} \end{array}$$

where d is the exterior differentiation and $\Omega^2_{S_W} = \Omega^1_{S_W} \wedge \Omega^1_{S_W}$. It is easy to see that $Der_{S_W}(-\log\Delta)$ is closed under the bracket product and that $d\Omega^1_{S_W}(\log\Delta) \subset \Omega^1_{S_W}(\log\Delta) \wedge \Omega^1_{S_W}(\log\Delta)$.

The natural pairing $\langle\ \cdot\ ,\ \cdot\ \rangle$ between Der_{S_W} and $\Omega^1_{S_W}$ induces the $S(V)^W$-perfect-pairing: $Der_{S_W}(-\log\Delta) \times \Omega^1_{S_W}(\log\Delta) \to S(V)^W$ (i.e. they are $S(V)^W$-dual to each other) ([S4,(1.6) Lemma ii)]).

By identifying the (co-)tangent spaces T_xV^* or $T^*_xV^*$ at each point $x \in V^*$ with V^* or with the dual space V, respectively, the W-invariant forms I^* and I on V^* and V (recall 2.2) induce the $S(V)^W$-bilinear forms: $I^* : Der_{S_W} \times Der_{S_W} \to \frac{1}{\Delta}S(V)^W$ and $I : \Omega^1_{S_W} \times \Omega^1_{S_W} \to S(V)^W$, $I^*(\frac{\partial}{\partial P_i}, \frac{\partial}{\partial P_j}) = \Sigma_{pq}\frac{\partial X_p}{\partial P_i}\frac{\partial X_q}{\partial P_j}I^*(\frac{\partial}{\partial X_p}, \frac{\partial}{\partial X_q})$ and $I(dP_i, dP_j) = \Sigma_{pq}\frac{\partial P_i}{\partial X_p}\frac{\partial P_j}{\partial X_q}\, I(dX_p, dX_q)$, where $X_1, \cdots, X_l$ is a linear coordinate system of V.

We now have the following important lemma.

LEMMA. *The pairings I^* and I induce $S(V)^W$-perfect pairings:*

$$(3.1.2)\qquad \begin{array}{rl} I^* : & Der_{S_W} \times Der_{S_W}(-\log\Delta) \to S(V)^W \\ I : & \Omega^1_{S_W} \times \Omega^1_{S_W}(\log\Delta) \to S(V)^W \end{array}$$

This is equivalent to say that one has $S(V)^W$-isomorphisms

$$(3.1.3)\qquad \begin{array}{rl} I^*: & Der_{S_W} \simeq \Omega^1_{S_W}(\log \Delta) \\ I: & \Omega^1_{S_W} \simeq Der_{S_W}(-\log \Delta), \end{array}$$

which make the following diagram commutative:

$$(3.1.4)\qquad \begin{array}{ccc} Der_{S_W}(-\log \Delta) & \subset & Der_{S_W} \\ \uparrow I & & \downarrow I^* \\ \Omega^1_{S_W} & \subset & \Omega^1_{S_W}(\log \Delta) \end{array}$$

Proof. We prove only the isomorphism $I : \Omega^1_{S_W} \simeq Der_{S_W}(-\log \Delta)$ since the other isomorphism I^* is obtained by taking its $S(V)^W$-dual. Recall δ_W such that $\delta_W^2 = \Delta_W$ 2.7. For any $\omega \in \Omega^1_{S_W}$, $I(\omega, d\delta) \in S(V)$ is W-anti-invariant, it is divisible by δ (2.7 Lemma 1). Thus, $I(\omega, d\Delta_W) = 2\delta_W I(\omega, d\delta_W)$ is divisible by δ_W^2, implying $I(\omega)$ belongs to $Der_{S_W}(-\log \Delta)$. To prove that the images $I(dP_i)$ $i = 1, \cdots, l$ form an $S(V)^W$-free basis of $Der_{S_W}(-\log \Delta)$, it is sufficient to show that the determinant of their coefficients matrix w.r.t. the basis $\frac{\partial}{\partial P_i}$ $i = 1, \cdots, l$ is a unit multiple of Δ_W (due to a theorem [S4,(1.7)Theorem ii)]). This is true due to 2.7 Lemma2:

$$(3.1.5)\qquad \begin{aligned} &\det((I(dP_i)(P_j))_{ij}) = \det((I(dP_i, dP_j))_{ij}) \\ &= \det\left((\tfrac{\partial P_i}{\partial X_p})_{ip} \cdot (I(X_p, X_q))_{pq} \cdot (\tfrac{\partial P_j}{\partial X_q})_{jq} \right) = c\delta^2 = c\Delta_W. \quad \square \end{aligned}$$

Recalling (2.8.5), we have the following definition of *Euler operator*:

$$(3.1.6)\qquad E := I(dP_1) = \textstyle\sum_{i=1}^{l} \frac{m_i+1}{h} P_i \frac{\partial}{\partial P_i},$$

3.2. The primitive vector field D and the invariants $S(V)^{W,\tau}$.

We fix a particular vector field: the primitive vector field D ([S3,(2.2)]). The D is transversal to the discriminant locus (see Note below). This fact gives the quite important and key role to D in the sequel.

The Der_{S_W} is naturally a graded module since $S(V)^W$ is a graded algebra such that $\deg(\delta P) = \deg(\delta) + \deg(P)$ for any homogeneous $\delta \in Der_{S_W}$ and $P \in S(V)^W$. Due to the maximality $\deg(P_l) > \deg(P_i)$ for $i = 1, \cdots, l-1$ (c.f. 2.8 Corollary **4.**), the lowest graded piece of Der_{S_W} is a vector space of dimension 1 spanned by $\frac{\partial}{\partial P_l}$. We fix a base

$$(3.2.1)\qquad D := \frac{\partial}{\partial P_l} \quad \text{with the normalization } \; DP_l = 1$$

and call it the *primitive vector field* or the *primitive derivation* (see 3.10 for the name). The D is unique, up to a scaling factor, independent of coordinates. We introduce the subring of $S(V)^W$ of D-invariants:

$$(3.2.2)\qquad S(V)^{W,\tau} := \{P \in S(V)^W \mid DP = 0\}.$$

One has $S(V)^{W,\tau} = \mathbf{R}[P_1, \cdots, P_{l-1}]$ and $S(V)^W = S(V)^{W,\tau}[P_l]$.

NOTE. The 1-parameter group action $\exp(tD)$ on S_W is denoted by τ_t [S8, (3.1)]. This justifies the notation (3.2.2) since $S(V)^{W,\tau} = \{P \in S(V)^W \mid P \circ \tau_t = P \ \forall t\}$. The τ-action is transversal to the discriminant locus D_W ([S8, (3.4) Lemma 6], recall also 2.9 Lemma 2.).

We, further, introduce the "descent" modules of Der_{S_W} and $\Omega^1_{S_W}$:

$$(3.2.3)\qquad \begin{array}{rcl} \mathcal{G} & := & \{\delta \in Der_{S_W} \mid [D,\delta]=0\} \\ \mathcal{F} & := & \{\omega \in \Omega^1_{S_W} \mid L_D\omega = 0\} \end{array}$$

where L_D is the Lie derivative given by $\langle L_D\omega, \delta\rangle = D\langle \omega, \delta\rangle - \langle \omega, [D,\delta]\rangle$. These $\mathcal{G}$ and $\mathcal{F}$ are $S(V)^{W,\tau}$-free modules of rank l with free dual basis $\frac{\partial}{\partial P_1}, \cdots, \frac{\partial}{\partial P_l}$ and $dP_1, \cdots, dP_l$, respectively. One has the expressions:

$$(3.2.4)\qquad \begin{array}{rcl} Der_{S_W} & = & \mathcal{G} \otimes_{S(V)^{W,\tau}} S(V)^W, \\ \Omega^1_{S_W} & = & \mathcal{F} \otimes_{S(V)^{W,\tau}} S(V)^W. \end{array}$$

The $\mathcal{G}$ is closed under the bracket product and acts naturally on $S(V)^W$ as derivations. In fact, $\mathcal{G}$ is an Abelian extension of $Der_{S(V)^{W,\tau}}$:

$$(3.2.5)\qquad 0 \to S(V)^{W,\tau}D \to \mathcal{G} \to Der_{S(V)^{W,\tau}} \to 0.$$

Combining (3.2.4) with (3.1.3), one gets the "descent expressions":

$$(3.2.6)\qquad \begin{array}{rcl} Der_{S_W}(-\log\Delta) & = & I(\mathcal{F}) \otimes_{S(V)^{W,\tau}} S(V)^W, \\ \Omega^1_{S_W}(\log\Delta) & = & I^*(\mathcal{G}) \otimes_{S(V)^{W,\tau}} S(V)^W. \end{array}$$

NOTE. The inclusion: $S(V)^{W,\tau} \subset S(V)^W$ induces the projection

$$(3.2.7)\qquad \pi_W : S_W \to \mathrm{Spec}(S(V)^{W,\tau})$$

forgetting the last coordinate P_l. By "descent", we mean that some geometric structure on S_W is a pull-back of that on $\mathrm{Spec}(S(V)^{W,\tau})$. We do not use explicitly the morphism π_W until §5 .

3.3. Metrics J and J^*.

We introduce non-degenerate symmetric bilinear forms J and J^* on the tangent and cotangent bundle of S_W, respectively. In fact, instead of introducing $S(V)^W$-bilinear forms on Der_{S_W} and $\Omega^1_{S_W}$, we introduce their descent $S(V)^{W,\tau}$-bilinear forms on the descent modules $\mathcal{G}$ and $\mathcal{F}$.

DEFINITION. The Lie derivative $L_D I$ defines a $S(V)^{W,\tau}$ -bilinear form:

$$(3.3.1)\qquad J^* : \mathcal{F} \times \mathcal{F} \to S(V)^{W,\tau}, \quad \omega_1 \times \omega_2 \to DI(\omega_1, \omega_2).$$

LEMMA. *The form J^* is nondegenerate everywhere on S_W. That is:*

$$\det(J^*(dP_i, dP_j)_{ij=1,\cdots,l}$$

is a non zero constant.

PROOF. One has an expression (recall 2.7 Lemma 2 and (2.9.5)):

$$\det\left((I(dP_i, dP_j))_{ij}\right) = \Delta = A_0 P_l^l + A_1 P_l^{l-1} + \cdots + A_l$$

where $A_i \in S(V)^{W,\tau}$. On the other hand, since

$$\deg(I^*(dP_i, dP_j)) = m_i + m_j < 2h \ (= 2\deg(P_l))$$

(recall (2.8.4), (2.8.1)), each entry $I^*(dP_i, dP_j)$ (as an element of $S(V)^W$) contains P_l at most linearly. Comparing these two facts, one obtains $\det\left((DI(dP_i, dP_j))_{ij}\right) = A_0$. But it was shown in 2.9 Lemma 1. that $A_0 \neq 0$. □

The degree of the ij-entries of the matrix expression of J^* is given by

(3.3.2) $$\deg(J^*(dP_i, dP_j)) = m_i + m_j - h.$$

So, if $m_i + m_j - h < 0$ then the entry vanishes. In view of the duality of the exponents (2.8.2), the matrix is a "skew lower triangular" matrix.

Since J^* is $S(V)^{W,\tau}$-nondegenerate, it induces $S(V)^{W,\tau}$-isomorphism

$$J^* : \mathcal{F} \simeq \mathcal{F}^* = \mathcal{G},\ J^*(dP_i) := J^*(dP_i, d\cdot) = \sum_{j=1}^{l} J^*(dP_i, dP_j)\frac{\partial}{\partial P_j}.$$

By this isomorphism, J^* induces a $S(V)^{W,\tau}$-symmetric-bilinear form on the dual module $\mathcal{G}$, which we shall denote by J. That is:

(3.3.3) $$J : \mathcal{G} \times \mathcal{G} \to S(V)^{W,\tau}, \quad J(\delta_1, \delta_2) := J^*((J^*)^{-1}\delta_1, (J^*)^{-1}\delta_2).$$

Again, J is a nondegenerate form on $\mathcal{G}$. Due to (3.1.6), one has

(3.3.4) $$J^*(dP_1) = D \quad \text{and} \quad J(D) = dP_1.$$

NOTE. The form J is identified with the residue pairing from a view point of the primitive form theory [S3] (c.f. (5.2.10)).

We shall show that *the metric J is flat, i.e. the curvature for J is zero.* This fact is a part of the *flat structure on S_W* given in 3.8 and 3.9. The following subsections 3.4 - 3.7 are devoted to the preparation.

3.4. Relationship between I and J.

The nondegeneracy of J^* implies the following quite important decomposition lemma, which leads to a reconstruction of I from J.

LEMMA. *One has a direct sum decomposition as $S(V)^{W,\tau}$-module:*

(3.4.1) $$Der_{S_W} = \mathcal{G} \oplus Der_{S_W}(-\log \Delta).$$

PROOF. We prove a more precise formula: for $k \in \mathbf{Z}_{\geq 0}$, one has

(3.4.2) $$S(V)^W_{\leq k}\, \mathcal{G} = \mathcal{G} \oplus S(V)^W_{\leq k-1} I(\mathcal{F}),$$

where $S(V)^W_{\leq k} := \{P \in S(V)^W \mid D^{k+1}P = 0\}$ $(k \in \mathbf{Z}_{\geq -1})$ is the module of polynomials in P_l of coefficients in $S(V)^{W,\tau}$ of degree $\leq k$.

Recall (3.1.3) that $I(dP_i) = \sum_{i=1}^{l} I(dP_i, dP_j)\frac{\partial}{\partial P_j}$ $(i = 1, \cdots, l)$ form $S(V)^W$-basis of $Der_{S_W}(-\log \Delta)$. Furthermore, the coefficient $I(dP_i, dP_j)$ is at most linear in P_l and $[D, I(dP_i)]$ $(i = 1, \cdots, l)$ are linearly independent over $S(V)^{W,\tau}$ (nondegeneracy of J^*).

We prove $\mathcal{G} \cap Der_{S_W}(-\log \Delta) = \{0\}$: if $\delta = \sum_{i=1}^{l} e_i I(dP_i) \in \mathcal{G}$ for $e_i \in S(V)^W_{\leq k}$ then we prove $e_i = 0$ by induction on k. The case $k = -1$ is true by definition $S(V)^W_{\leq -1} = 0$. Suppose $k \geq 0$. By assumption $\delta \in \mathcal{G}$, one has $0 = \mathrm{ad}(D)^{k+1}\delta = \sum_{i=1}^{l}\sum_{j=0}^{k+1} C(k+1, j)\ (D^j e_i)(\mathrm{ad}(D)^{k+1-j} I(dP_i))$ where all terms except for $j = k$ vanishes. So, we obtain $(k+1)\sum_{i=1}^{l}(D^k e_i)[D, I(dP_i)] = 0$. Then the linear independence of $[D, I(dP_i)]$ implies the vanishing of the coefficients $D^k e_i$. So, $e_i \in S(V)^W_{\leq k-1}$ and the induction applies.

The LHS of (3.4.2) includes RHS since $I(dP_i)$ is at most linear in P_l. We prove the opposite inclusion relation by an induction on k. Case $k = 0$ is clear (note $S(V)^W_{\leq -1} = \{0\}$ and $S(V)^W_{\leq 0} = S(V)^{W,\tau}$). Let $\delta = \sum_{i=1}^{l} f_i \frac{\partial}{\partial P_i} \in S(V)^W_{\leq k}\mathcal{G}$ for $k > 0$, where the coefficient of P_l^k in f_i is denoted by $f_i^{(k)}$. One can find $g_j \in S(V)^{W,\tau}$

$(j = 1, \cdots, l)$ such that $f_i^{(k)} = \sum_{j=1}^{l} J^*(dP_i, dP_j)g_j$. Then $\delta - P_l^{k-1}\sum_{j=1}^{l} I(dP_j)g_j$ belongs to $S(V)^W_{\leq k-1}\mathcal{G}$, so that one applies the induction hypothesis. □

The next corollaries shall be used in §4 to lift $\mathcal{G}$ to the normalization of the discriminant D_W. First, note that the ideal $(\partial\Delta) := Der_{S_W} \cdot \Delta$ in $S(V)^W$ contains the ideal (Δ) since $E\Delta = hl\Delta$.

COROLLARY. 1. *The expression* $\mathcal{G} \simeq Der_{S_W}/Der_{S_W}(-\log \Delta)$ *gives on* $\mathcal{G}$ *a* $S(V)^W$*-module structure of homological dimension* ≤ 1.
2. *The correspondence* $\delta \in \mathcal{G} \mapsto \delta\Delta$ *induces a* $S(V)^W$*-isomorphism:*

$$\mathcal{G} \simeq (\partial\Delta)/(\Delta). \tag{3.4.3}$$

PROOF. 1. Trivial. 2. Surjectivity: Due to the decomposition (3.4.1), one has $Der_{S_W} \cdot \Delta = \mathcal{G} \cdot \Delta + (\Delta)$, which implies the surjectivity.
Injectivity: Suppose $\delta \in \mathcal{G}$ is mapped to 0. This means $\delta\Delta \in (\Delta)$ and $\delta \in Der_{S_W}(-\log \Delta)$. The direct sum (3.4.1) implies $\delta = 0$. □

Denote by P_l* the multiplication of $P_l \in S(V)^W$ on $\mathcal{G}$. Define

$$w : \mathcal{G} \to Der_{S_W}(-\log \Delta), \quad w(\delta) := P_l\delta - P_l * \delta \in I(\mathcal{F}). \tag{3.4.4}$$

So, the decomposition (3.4.2) of the element $P_l\delta$ for $\delta \in \mathcal{G}$ is given by

$$P_l\delta = P_l * \delta + w(\delta). \tag{3.4.5}$$

Assertion. *i)* $w(\delta)$ *is the unique element in* $Der_{S_W}(-\log(\Delta))$ *with*

$$[D, w(\delta)] = \delta. \tag{3.4.6}$$

ii) The w *maps* $S(V)^{W,\tau}$*-free basis of* $\mathcal{G}$ *to* $S(V)^W$*-free basis of*

$$Der_{S_W}(-\log(\Delta))$$

(e.g. $w(D) = E$, *c.f.* (3.1.6) *and* (3.3.4)*).*

PROOF. i) The $w(\delta)$ obviously satisfies (3.4.6). If $w_1, w_2 \in Der_{S_W}(-\log \Delta)$ satisfies $[D, w_1] = [D, w_2]$. Then $w_1 - w_2 \in \mathcal{G}$ and is 0 by (3.4.1).
ii) Due to i), one has $w(J^*(dP_i)) = I(dP_i)$. □

LEMMA. *For* $\omega \in \mathcal{F}$ *and for* $\delta_1, \delta_2 \in \mathcal{G}$, *one has the formulae*

$$I(\omega) = w(J^*(\omega)) \quad \text{and} \quad J(\delta_1, \delta_2) = I^*(w(\delta_1), \delta_2) \tag{3.4.7}$$

For a $S(V)^{W,\tau}$*-basis* $\delta_1, \cdots, \delta_l$ *and its* J*-dual basis* $\delta_1^*, \cdots, \delta_l^*$, *one has*

$$I = \textstyle\sum_{i=1}^{l} \delta_i \otimes w(\delta_i^*). \tag{3.4.8}$$

PROOF. By definition of J^* (3.3.1), one has $J^*(\omega) = L_D I(\omega) = [D, I(\omega)]$. Then applying the characterization (3.4.6), one gets the first formula of (3.4.7). Using this, the second formula of (3.4.7) is calculated as: $I^*(w(\delta_1), \delta_2) = I^*(w(J^*J\delta_1), \delta_2) = I^*(IJ\delta_1, \delta_2) = J(\delta_1, \delta_2)$.
The coupling of the first tensor factor of RHS of (3.4.8) with $\omega \in \mathcal{F}$ is given by $\sum_{i=1}^{l} \delta_i(\omega) \cdot w(\delta_i^*) = w(\sum_{i=1}^{l} \delta_i(\omega) \cdot \delta_i^*) = w(J^*(\omega)) = I(\omega)$, which is equal to the coupling of $\omega \in \mathcal{F}$ with LHS of (3.4.8). □

3.5. The Levi-Civita connection ∇ on Der_{S_W}.

The Levi-Civita connection attached to the metric I^* [3] induces a connection ∇ on Der_{S_W} which is singular along the discriminant. We describe the singularity of ∇ in terms of logarithmic vector fields.

First we list up the properties, which should be satisfied by ∇.
i) $\nabla : Der_{S_W} \times Der_{S_W} \to Der_{S_W}(*\Delta)$ *(= the localization of* Der_{S_W} *by* Δ*) is a covariant differentiation. That is:* $\nabla_{\delta_1}\delta_2$ *for* $\delta_1,\ \delta_2 \in Der_{S_W}$ *is* $S(V)^W$*-linear in* δ_1 *and additive in* δ_2 *satisfying the Leibniz rule:*

$$\nabla_{\delta_1}(P\delta_2) = (\delta_1 P)\delta_2 + P(\nabla_{\delta_1}\delta_2) \qquad (P \in S(V)^W). \tag{3.5.1}$$

ii) ∇ *preserves* I^**:* $\nabla I^* = 0$. *That is: for* $\delta_1, \delta_2, \delta_3 \in Der_{S_W}$ *one has*

$$\delta_1 I^*(\delta_2, \delta_3) = I^*(\nabla_{\delta_1}\delta_2, \delta_3) + I^*(\delta_2, \nabla_{\delta_1}\delta_3). \tag{3.5.2}$$

iii) ∇ *is torsion free. That is: for* $\delta_1, \delta_2 \in Der_{S_W}$ *one has*

$$\nabla_{\delta_1}\delta_2 - \nabla_{\delta_2}\delta_1 = [\delta_1, \delta_2]. \tag{3.5.3}$$

The ∇ is determined uniquely by i),ii) and iii) by the formula:

$$\begin{aligned} 2I^*(\nabla_{\delta_1}\delta_2, \delta_3) = &+\delta_1 I^*(\delta_2, \delta_3) + \delta_2 I^*(\delta_3, \delta_1) - \delta_3 I^*(\delta_1, \delta_2) \\ &-I^*(\delta_1, [\delta_2, \delta_3]) + I^*(\delta_2, [\delta_3, \delta_1]) + I^*(\delta_3, [\delta_1, \delta_2]) \end{aligned} \tag{3.5.4}$$

for $\delta_1, \delta_2, \delta_3 \in Der_{S_W}$. Conversely, the (3.5.4) defines a connection ∇ satisfying i), ii) and iii): the *Levi-Civita connection* attached to I^*.

Using properties of logarithmic vector fields in 3.1, one checks that RHS of (3.5.4) belongs to $S(V)^W$ if two of δ_1, δ_2 and δ_3 are logarithmic. This means that the domain and the range of ∇ can be chosen as

$$\nabla : \begin{cases} Der_{S_W} \times Der_{S_W}(-\log(\Delta)) \to Der_{S_W}, \\ Der_{S_W}(-\log(\Delta)) \times Der_{S_W} \to Der_{S_W}, \\ Der_{S_W}(-\log(\Delta)) \times Der_{S_W}(-\log(\Delta)) \\ \qquad\qquad \to Der_{S_W}(-\log(\Delta)). \end{cases} \tag{3.5.5}$$

For short, we say that *the connection* ∇ *has logarithmic singularities.* In particular, the second line implies that the connection form for ∇ belongs to $End(Der_{S_W}) \otimes \Omega^1_{S_W}(\log(\Delta))$, i.e. the connection form for ∇ as the connection on the tangent bundle of S_W has the logarithmic pole along the discriminant. We shall not use this fact explicitly.
iv) ∇ *is integrable:* $\nabla^2 = 0$. *That is: for* $\delta_1, \delta_2, \delta_3 \in Der_{S_W}$

$$\nabla_{\delta_1}\nabla_{\delta_2}\delta_3 - \nabla_{\delta_2}\nabla_{\delta_1}\delta_3 = \nabla_{[\delta_1,\delta_2]}\delta_3. \tag{3.5.6}$$

(*Proof.* This can be seen by extending the domain and range of ∇ to $Der_V(*\Delta) = \sum_{i=1}^l S(V)_\Delta \frac{\partial}{\partial X_i}$ (such extension exists as the metric connection on Der_V). Then RHS of (3.5.4) vanishes by the substitution of $\delta_1, \delta_2, \delta_3$ by $\frac{\partial}{\partial X_i}, \frac{\partial}{\partial X_j}$ and by $\frac{\partial}{\partial X_k}$. This implies $\nabla \frac{\partial}{\partial X_i} = 0$ for $i = 1, \cdots, l$. Expressing $Der_{S_W} \subset \sum_{i=1}^l S(V)_\Delta \frac{\partial}{\partial X_i}$, we obtain the result.)

[3] There is an unfortunate disagreement on the notation I and I^* between [S3] and [S6]. We employed notation of [S6] since it agree with that of root systems.

3.6. Key Lemma.

We want to show that the action of ∇_D is invertible on Der_{S_W}. This is achieved by showing an isomorphism $\nabla_D : Der_{S_W}(-\log\Delta) \simeq Der_{S_W}$ and taking its inverse map in 3.8. For this end, we give a key lemma:

LEMMA. **1.** *The ∇_δ for $\delta \in \mathcal{G}$ maps $I(\mathcal{F})$ into $\mathcal{G} \oplus I(\mathcal{F})$. Furthermore, ∇_D induces an $S(V)^{W,\tau}$-isomorphism:*

$$\nabla_D : I(\mathcal{F}) \simeq \mathcal{G}. \tag{3.6.1}$$

2. *The image of the basis $I(dP_i) \in I(\mathcal{F})$ $(i = 1, \cdots, l)$ is given by*

$$\begin{aligned}&\nabla_D I(dP_i)\\ &= \textstyle\sum_{j=1}^{l} \left(\frac{h+m_i-m_j}{2h} J^*(dP_i, dP_j) + \sum_{k=1}^{l-1} \frac{m_k+1}{2h}(DA_{ij}^k)P_k\right) \frac{\partial}{\partial P_j}.\end{aligned} \tag{3.6.2}$$

Here DA_{ij}^k is an element of $S(V)^{W,\tau}$ given by

$$DA_{ij}^k = \textstyle\sum_{p,q=1}^{l} (J_{ip}J_{jq} - J_{jp}J_{iq}) \frac{\partial}{\partial P_q} J^{pk} \tag{3.6.3}$$

with $J_{ij} := J^*(dP_i, dP_j)$ and $J^{ij} := J(\frac{\partial}{\partial P_i}, \frac{\partial}{\partial P_j})$. One has

$$\nabla_D E = \tfrac{1}{h} D. \tag{3.6.4}$$

PROOF. 1. Let $\delta \in \mathcal{G}$. Because of (3.5.5), we know that $\nabla_\delta(I(\mathcal{F})) \subset Der_{S_W}$. In view of (3.2.6), it is sufficient to prove $I^*(\nabla_\delta(I(\mathcal{F})), I(\mathcal{F})) \subset S(V)^W_{\leq 1}$ in order to prove $\nabla_\delta(I(\mathcal{F})) \subset \mathcal{G} \oplus I(\mathcal{F}) = \mathcal{G} \otimes_{S(V)^{W,\tau}} S(V)^W_{\leq 1}$. So, for $U, V \in S(V)^W$, by using (3.5.4), we calculate

$$\begin{aligned}&2I(\nabla_\delta I(dU), I(dV))\\ =\ &+\delta I(dU, dV) + I(dU)\delta V - I(dV)\delta U\\ &-I^*(\delta, [I(dU), I(dV)]) + [I(dV), \delta]U + [\delta, I(dU)]V\end{aligned}$$

Developing the last two brackets, one cancels the second and the third terms in the RHS. By eliminating I and I^* in the LHS, one obtains:

$$2\langle \nabla_\delta I(dU), dV\rangle = \delta I(dU, dV) - I^*(\delta, [I(dU), I(dV)]). \tag{3.6.5}$$

In particular, for $U = P_i$ and $V = P_j$, one obtains

$$2\langle \nabla_\delta I(dP_i), dP_j\rangle = \delta I(dP_i, dP_j) - \textstyle\sum_{k=1}^{l} A_{ij}^k \cdot \delta P_k \tag{(3.6.5)*}$$

where A_{ij}^k shall be defined in the formula (3.6.6) and the above formula is calculated since one has $I(\delta, [I(dP_i), I(dP_j)]) = \sum_{k=1}^{l} A_{ij}^k \cdot \delta P_k$.

Since the module of logarithmic vector fields are closed under bracket product 3.1, there exists $A_{ij}^k \in S(V)^W$ for $i, j, k = 1, \cdots, l$ such that

$$[I(dP_i), I(dP_j)] = \textstyle\sum_{k=1}^{l} A_{ij}^k I(dP_k). \tag{3.6.6}$$

Since the degree in P_l of the coefficients in LHS is at most 2, the A_{ij}^k in the RHS has degrees in P_l at most 1 (recall (3.4.2)). Thus the RHS of (3.6.5)* has degree in P_l at most 1. This proves the first half of 1.

In order to prove $\nabla_D(I(\mathcal{F})) \subset \mathcal{G}$, we show $\langle \nabla_D(I(\mathcal{F})), \mathcal{F}\rangle \subset S(V)^{W,\tau}$. Substitute δ by D in (3.6.5)*, and one has

$$2\langle \nabla_D I(dP_i), dP_j\rangle = J^*(dP_i, dP_j) - A_{ij}^l. \tag{*}$$

Therefore, we have only to show $A_{ij}^l \in S(V)^{W,\tau}$ for $i, j = 1, \cdots, l$. Apply the both hand side of (3.6.6) to the invariant P_1 of lowest degree (2.8.5). Recalling (3.1.6), the LHS gives $h[I(dP_i), I(dP_j)]P_1 = I(dP_i)((m_j+1)P_j) - I(dP_j)((m_i+1)P_i) = (m_j - m_i)I(dP_i, dP_j)$ and the RHS gives $h\sum_{k=1}^{l} A_{ij}^k (m_k+1)P_k$. So, one gets an equality:

$$(**) \qquad (m_j - m_i)I(dP_i, dP_j) \;=\; \sum_{k=1}^{l} A_{ij}^k (m_k+1)P_k.$$

Apply D^2 to $(**)$. Since LHS vanishes and $D^2 A_{ij}^k = 0$, one obtains $0 = DA_{ij}^l(m_l + 1)$. That is: $A_{ij}^l \in S(V)^{W,\tau}$ and $\nabla_D(I(dP_i)) \in \mathcal{G}$.

2. Apply D once to $(**)$ and noting $m_l + 1 = h$, one has the equality:

$$(3.6.7) \qquad A_{ij}^l \;=\; \tfrac{m_j - m_i}{h} J^*(dP_i, dP_j) - \sum_{k=1}^{l-1} \tfrac{m_k+1}{h}(DA_{ij}^k)P_k.$$

Combining (3.6.7) with $(*)$, one obtains the formula (3.6.2).

Since ∇_D is a $S(V)^{W,\tau}$-homomorphism, in order to show the isomorphy (3.6.1), it is sufficient to show the non-degeneracy of the matrix $M = (M_{ij})_{ij=1}^l$, where M_{ij} is the coefficient of $\frac{\partial}{\partial P_j}$ of the RHS of (3.6.2). First, one notes that the weighted degree of M_{ij} is $m_i + m_j - h$ (3.3.2). So, $\det(M)$ should be of degree $\sum_{i=j=1}^{l}(m_i + m_j - h) = 0$, and hence, $\det(M)$ is a constant. Thus, the positive degree entries of M has no contributions to $\det(M)$ and, hence, one may calculate the determinant modulo the maximal ideal $S(V)_+^{W,\tau}$ in $S(V)^{W,\tau}$. So, the second term in (3.6.2) (containing the positive weighted factor P_k) can be ignored. Also, $J^*(dP_i, dP_j) \equiv 0$ modulo $S(V)_+^{W,\tau}$, if $m_i + m_j \neq h$. The remaining are only ij-entries of the matrix $(\frac{h+m_i-m_j}{2h} J^*(dP_i, dP_j))$ for $m_i + m_j = h$. Then the factor $\frac{h+m_i-m_j}{2h}$ in the ij-entry is $\frac{m_i}{h}$. Therefore, $\det(M) = (\prod_{i=1}^{l} \frac{m_i}{h}) \det((J^*(dP_i, dP_j))_{ij}) \neq 0$ (c.f. 3.3 Lemma).

Finally, let us prove (3.6.3). Take the pairing of the vector field (3.6.6) with a form dP_p for a $1 \le p \le l$. So, we obtain:

$$I(dP_i, dI(dP_j, dP_p)) - I(dP_j, dI(dP_i, dP_p)) = \sum_{k=1}^{l} A_{ij}^k I(dP_k, dP_p).$$

This is of degree at most 2 in P_l. Apply D^2 and one obtains

$$(3.6.8) \qquad \sum_{q=1}^{l} J_{iq} \tfrac{\partial}{\partial P_q} J_{jp} - \sum_{q=1}^{l} J_{jq} \tfrac{\partial}{\partial P_q} J_{ip} = \sum_{k=1}^{l} (DA_{ij}^k) J_{kp}$$

Multiply J^{pm} and sum for $p = 1, \cdots, l$. Since $\sum_{p=1}^{l} J_{jp}J^{pm} = \delta_j^m$ (Kronecker's delta), one replaces $\sum_{p=1}^{l}(\frac{\partial}{\partial P_q} J_{jp})J^{pm}$ by $-\sum_{p=1}^{l} J_{jp}(\frac{\partial}{\partial P_q} J^{pm})$. Finally, replacing the index m by k, one gets (3.6.3). The (3.6.4) is obtained by substituting $i = 1$ in (3.6.2) and applying (3.1.6). □

NOTE. These A_{ij}^k and DA_{ij}^k depend on a choice of the coordinate system $P_1, \cdots, P_l$. In particular, $DA_{ij}^k = 0$ if all J_{ij} are constant (c.f. (3.6.3)). This occurs if $P_1, \cdots, P_l$ is a *flat coordinate system* (c.f. 3.11).

3.7. Star product $*$, connection $\nabla\!/$ and exponent N.

As consequences of 3.6 Lemma, we introduce some structures on $\mathcal{G}$. They shall form building blocks of the flat structure in 3.9.

1. **Star product $*$ and Connection $\nabla\!\!\!/$.**

One has a bilinear map: $\mathcal{G} \times \mathcal{G} \to \mathcal{G} \oplus \nabla_D^{-1}\mathcal{G}$ by letting $\delta_1, \delta_2 \mapsto \nabla_{\delta_1}\nabla_D^{-1}\delta_2$ (use 3.6 Lemma 1). Then, by decomposing the image into the direct summands as:

$$(3.7.1) \qquad \nabla_{\delta_1}\nabla_D^{-1}\delta_2 = \delta_1 * \delta_2 + \nabla_D^{-1}\nabla\!\!\!/_{\delta_1}\delta_2,$$

we introduce two binary operations:

$$(3.7.2) \qquad * : \mathcal{G} \times \mathcal{G} \to \mathcal{G},$$

$$(3.7.3) \qquad \nabla : \mathcal{G} \times \mathcal{G} \to \mathcal{G}.$$

We give two direct consequences of the definition. A complete list of properties of $*$ and ∇ (and of N defined below) shall be given in 3.9.

i) *The* $*$ *is* $S(V)^{W,\tau}$*-bilinear and, so, is regarded as a distributive product structure on* $\mathcal{G}$. *The primitive vector field* D *is the (left) unit.*

$$(3.7.4) \qquad D * \delta = \delta.$$

Using (3.2.4), we extend the $*$ to a $S(V)^W$-algebra structure on Der_{S_W}.

ii) *The* $\nabla_{\delta_1}\delta_2$ *is a covariant differentiation of* δ_2 *by* δ_1. *One has*

$$(3.7.5) \qquad \nabla_D\delta \equiv 0 \quad \textit{for} \quad \delta \in \mathcal{G}.$$

By the extension (3.2.4), one may regard ∇ as a connection on Der_{S_W}.

2. **Exponent map** N.

Recall the principal part $w(\delta)$ of $P_l\delta$ for $\delta \in \mathcal{G}$ (see (3.4.4) and (3.4.5)). We introduce a $S(V)^{W,\tau}$-endomorphism N of $\mathcal{G}$ by the composition:

$$(3.7.6) \qquad N : \mathcal{G} \to \mathcal{G}\,, \quad N(\delta) := \nabla_D(w(\delta)).$$

and call it the *exponent map* (in analogy with Fuchsian type equations). By a use of N, the decomposition formula (3.4.5) is rewritten as

$$(3.7.7) \qquad P_l\,\delta \;=\; P_l * \delta \;+\; \nabla_D^{-1}N(\delta).$$

iii) *The exponent map* N *is independent of choices of* P_l *and* D.

Proof. One has to check that the changes $D \mapsto cD$ and $P_l \mapsto c^{-1}P_l + g$ for $c = const. \neq 0$ and $g \in S(V)^{W,\tau}$ may not cause a change of N. □

LEMMA. *The eigenvalues of the exponent map* N *are* $\frac{m_i}{h}$ $i = 1, \cdots, l$.

PROOF. Combining (3.7.6) with (3.4.7) and (3.6.2), one calculates:

$$N(J^*(dP_i)) = \nabla_D(w(J^*(dP_i))) = \nabla_D I(dP_i) = \tfrac{m_i}{h}J^*(dP_i)$$
$$+ \textstyle\sum_{j=1}^{l}\left(\frac{h-m_j-m_i}{2h}J^*(dP_i, dP_j) + \sum_{k=1}^{l-1}\frac{m_k+1}{2h}(DA_{ij}^k)P_k\right)\frac{\partial}{\partial P_j}.$$

Coefficients of $\frac{\partial}{\partial P_j}$ in the RHS consist only of positively graded terms (since the degree of $J^*(dP_i, dP_j)$ equals to 0 if and only if $m_j + m_i = h$ (3.3.2)). This implies that $N - \mathrm{diag}[\frac{m_i}{h}]_{i=1}^l$ is nilpotent: the eigenvalues of N coincide with that of the diagonal matrix $\mathrm{diag}[\frac{m_i}{h}]_{i=1}^l$. □

3.8. Fourier transformation (gravitational descendent).

We transform the $S(V)^{W,\tau}[P_l]$-free-module Der_{S_W} to a $S(V)^{W,\tau}[D^{-1}]$-free-module. This is a formal Fourier-Laplace transformation (called the gravitational descendent in the topological field theory).

THEOREM. **1.** *The covariant derivative* ∇_D *with respect to the primitive vector field* D *induces a* $S(V)^{W,\tau}$*-isomorphism*

$$(3.8.1) \qquad \nabla_D : Der_{S_W}(-\log\Delta) \xrightarrow{\sim} Der_{S_W}.$$

Therefore, ∇_D^{-1} *is a well-defined* $S(V)^{W,\tau}$*-endomorphism of* Der_{S_W}.

2. *The correspondence* $\sum_k (\nabla_D)^{-k}\delta_k \mapsto \sum_k D^{-k}\otimes\delta_k$ ($\delta_k \in \mathcal{G}$ *and is* $=0$ *except for finite number of* $k \in \mathbf{Z}_{\geq 0}$) *gives isomorphisms:*

$$(3.8.2)\qquad \begin{array}{rcll} Der_{S_W} & \simeq & \mathbf{Z}[D^{-1}]\otimes\mathcal{G} & \text{and} \\ Der_{S_W}(-\log\Delta) & \simeq & D^{-1}\mathbf{Z}[D^{-1}]\otimes\mathcal{G} & \end{array}$$

as $S(V)^{W,\tau}[D^{-1}]$*-modules, where* $\mathbf{Z}[D^{-1}]$ *is the polynomial ring with the indeterminate* D^{-1} *which commutes with the elements of* $S(V)^{W,\tau}$.

3. *The left-multiplication of* P_l *and the covariant differentiation* ∇_δ *with respect to* $\delta \in \mathcal{G}$ *on the RHS of* (3.8.2) *are given as follows.*

$$(3.8.3)\qquad P_l(D^{-k}\otimes\delta) = D^{-k}\otimes(P_l*\delta) + D^{-k-1}\otimes(N(\delta)+k\delta)$$

$$(3.8.4)\qquad \nabla_{\delta_1}(D^{-k-1}\otimes\delta_2) = D^{-k}\otimes(\delta_1*\delta_2) + D^{-k-1}\otimes\nabla_{\delta_1}\delta_2.$$

The $D^{-k}\mathbf{Z}[D^{-1}]\otimes\mathcal{G}$ *for* $k\in\mathbf{Z}_{\geq 0}$ *is a* $S(V)^W$*-free module of rank* l.

PROOF. **1.** and **2.** Let us prove by induction on $k \in \mathbf{Z}_{>0}$ the following:

A(k): *There exists a sequence* $\mathcal{G}_0,\cdots,\mathcal{G}_k$ *of* $S(V)^{W,\tau}$*-free submodules of* Der_{S_W} *of rank* l *such that* $S(V)^W_{\leq i}\mathcal{G} = \mathcal{G}_0\oplus\cdots\oplus\mathcal{G}_i$, $S(V)^W_{\leq i-1}I(\mathcal{F}) = \mathcal{G}_1\oplus\cdots\oplus\mathcal{G}_i$ *and* $\nabla_D : \mathcal{G}_i \simeq \mathcal{G}_{i-1}$ *for* $i=1,\cdots,k$.

A(1) is proven by (3.6.1). Assume A(k) for $k\geq 1$. Put $\mathcal{G}_{k+1} := \{\delta \in S(V)^W_{\leq k}I(\mathcal{F}) \mid \nabla_D(\delta)\in\mathcal{G}_k\}$. In the following i)-iv), we prove that $\mathcal{G}_0,\cdots,\mathcal{G}_k,\mathcal{G}_{k+1}$ satisfies the conditions for A(k+1).

i) $(\mathcal{G}_1\oplus\cdots\oplus\mathcal{G}_k)\cap\mathcal{G}_{k+1} = \{0\}$. Suppose δ were non-trivial element of LHS. Then, by induction hypothesis, $\nabla_D\delta \in \mathcal{G}_0\oplus\cdots\oplus\mathcal{G}_{k-1}$ and $\neq 0$. This contradicts to the facts $\nabla_D\delta\in\mathcal{G}_k\cap\mathcal{G}_0\oplus\cdots\oplus\mathcal{G}_{k-1}=\{0\}$.

ii) $\mathcal{G}_1\oplus\cdots\oplus\mathcal{G}_k$ and $\mathcal{G}_{k+1}$ generates $S(V)^W_{\leq k}I(\mathcal{F})$. Let δ be any element of $S(V)^W_{\leq k}I(\mathcal{F})$. Using (3.4.2) and 3.6 Lemma, one observes $\nabla_D\delta\in S(V)^W_{\leq k}\mathcal{G}$. By the induction hypothesis, one can find $\delta'\in\mathcal{G}_1\oplus\cdots\oplus\mathcal{G}_k$ so that $\nabla_D(\delta-\delta')\in\mathcal{G}_k$. This means $\delta-\delta'\in\mathcal{G}_{k+1}$.

iii) The above i) and ii) implies: $S(V)^W_{\leq k}I(\mathcal{F}) = \mathcal{G}_1\oplus\cdots\oplus\mathcal{G}_k\oplus\mathcal{G}_{k+1}$ and then, by a use of (3.4.2), $S(V)^W_{\leq k+1}\mathcal{G} = \mathcal{G}_0\oplus\cdots\oplus\mathcal{G}_k\oplus\mathcal{G}_{k+1}$.

iv) $\nabla_D : \mathcal{G}_{k+1}\to\mathcal{G}_k$ is bijective. We prove this in two steps:

a) to construct a $S(V)^{W,\tau}$-isomorphism $w_k : \mathcal{G}_k\to\mathcal{G}_{k+1}$,

b) to show that $\nabla_D\cdot w_k$ is an $S(V)^{W,\tau}$-isomorphism of $\mathcal{G}_k$.

a) For $\delta_i\in\mathcal{G}_i$ $0\leq i\leq k$, put $w_i(\delta_i) :=$ the projection of $P_l\ \delta_i\in S(V)^W_{\leq i+1}\mathcal{G}$ to the component $\mathcal{G}_{i+1}$ (w.r.t. the decomposition in iii)). In view of (3.4.2), w_i is a $S(V)^{W,\tau}$-isomorphism: $\mathcal{G}_i\simeq\mathcal{G}_{i+1}$.

b) By the induction hypothesis, one has the isomorphism: $\nabla^i_D : \mathcal{G}_i\simeq\mathcal{G}_0=\mathcal{G}$ for $0\leq i\leq k$. By a use of this isomorphism, let us show that $w_i(\nabla_D^{-i}(\delta)) = P_l\nabla_D^{-i}(\delta) - \nabla_D^{-i}(P_l*\delta)$ for $\delta\in\mathcal{G}$ by induction on $0\leq i\leq k$ (ie. one has to show that the RHS belongs to $\mathcal{G}_{i+1}$). For the purpose, it is sufficient to show a formula: $\nabla_D\left(P_l\nabla_D^{-i}(\delta)-\nabla_D^{-i}(P_l*\delta)\right) = \nabla_D^{-i}((N+i)(\delta))$ for $0\leq i\leq k$. But this is shown by induction on i, where the case $i=0$ is the formula (3.7.7). Then for $i\geq 1$, $\nabla_D\left(P_l\nabla_D^{-i}(\delta)-\nabla_D^{-i}(P_l*\delta)\right) = \nabla_D^{-i}(\delta)+P_l\nabla_D^{-i+1}(\delta)-\nabla_D^{-i+1}(P_l*\delta) = \nabla_D^{-i}((N+i)(\delta))$. On the other hand, N is an endomorphism of $\mathcal{G}$ whose eigenvalues are non-negative rational numbers (3.7 Lemma). So, $\nabla_D\cdot w_i = \mathrm{ad}(\nabla^i_D)(N+i)$ is an invertible endomorphism of G_i. In particular, the case for $i=k$ proves the isomorphism in iv). Thus A(k) for $k\in\mathbf{Z}_{>0}$ is proven. This, in particular, implies the isomorphism (3.8.1). Then, as a consequence, the subspaces $\mathcal{G}_k$ are unique since $\mathcal{G}_0 = \mathcal{G}$, $\mathcal{G}_1 = I(\mathcal{F})$ and $\mathcal{G}_{k-1} = \nabla_D(\mathcal{G}_k)$ for $k \in \mathbf{Z}_{>0}$.

3. For the case $k = 0$, use (3.7.7) and (3.7.1). Then, a successive applications of ∇_D^{-1} with the rule: $[\nabla_D, P_l] = 1$ and $[\nabla_D, \nabla_\delta] = 0$ for $\delta \in \mathcal{G}$ (use the definitions (3.2.1) and (3.2.3) of D and $\mathcal{G}$, respectively, and the integrality (3.5.6) of ∇) implies the formulae (3.8.3) and (3.8.4). □

NOTE. 1. The formula (3.6.4) implies $E = \frac{1}{h}D^{-1} \otimes D$. Since $E = w(D)$ (3.4 Assertion ii)), this means $N(D) = \frac{1}{h}D$. That is:
Fact. *The D is an eigenvector of N of the eigenvalue $\frac{1}{h}$.*
2. (3.8.3) and (3.8.4) decompose into two left-multiplicative formulae:

$$(3.8.3)* \quad \begin{aligned} P_l\, D^{-k} &= D^{-k}P_l + kD^{-k-1} && \text{for } k \in \mathbf{Z} \\ P_l\, \delta &= P_l * \delta + D^{-1} \otimes N(\delta) && \text{for } \delta \in \mathcal{G} \end{aligned}$$

$$(3.8.4)* \quad \begin{aligned} \nabla_\delta D^{-k} &= D^{-k}\nabla_\delta && \text{for } k \in \mathbf{Z} \text{ and } \delta \in \mathcal{G} \\ \nabla_{\delta_1}\delta_2 &= D \otimes (\delta_1 * \delta_2) + \boldsymbol{\nabla}_{\delta_1}\delta_2 && \text{for } \delta_1, \delta_2 \in \mathcal{G}. \end{aligned}$$

3. Filtration $D^{-k}\mathbf{Z}[D^{-1}] \otimes \mathcal{G}$ for $k \in \mathbf{Z}$ is identified with the Hodge filtration attached to the universal unfolding of simple singularities [S3] (c.f. §5). They can be also interpreted as the sequence of modules of vector fields having higher order contact with the discriminant [Te][Yo].

3.9. Flat structure.

Recall that $S_W := \mathrm{Spec}(S(V)^W)$ is the quotient variety of a vector space V by a finite reflection group W (with a W-invariant symmetric bilinear form I^* on V). On S_W, we have introduced several structures and objects such as $D, E, J, P_l*, w, *, \boldsymbol{\nabla}$ and N (3.1-3.7). In this paragraph, we clarify the relationships among them and reduce them to some basic ones. Regarded as a differential geometric structure on S_W, we call them altogether the *flat structure* on S_W (see [S3] and [S6]).
First, we recall once again the above mentioned structures and objects.

1. The *primitive vector field* D on S_W (3.2.1). Using D, we reduce the coefficient $S(V)^W$ to the subring $S(V)^{W,\tau}$ (3.2.2) and the tangent sheaf Der_{S_W} of S_W to a Lie algebra $\mathcal{G}$ (3.2.3) $S(V)^{W,\tau}$-free of rank l.

2. The *Euler vector field* E (3.1.6) attached to the graded ring structure on $S(V)^W$. It is normalized as $L_E(D) = -D$ and $L_E(I) = -\frac{2}{h}I$.

3. The *nondegenerate symmetric bilinear form* J (3.3.3) on $\mathcal{G}$, defined as the dual form of $J^* := L_D(I)$ (3.3.1).

4. The $S(V)^W = S(V)^{W,\tau}[P_l]$*-module structure on* $\mathcal{G}$ (3.4.3) and the attached section $w : \mathcal{G} \to Der_{S_W}(-\log \Delta)$ (3.4.4).

5. The graded $S(V)^{W,\tau}$*-algebra structure* $*$ *on* $\mathcal{G}$ (3.7.2).

6. The *covariant differentiation* $\boldsymbol{\nabla} : \mathcal{G} \times \mathcal{G} \to \mathcal{G}$ (3.7.3).

7. The *exponent map* $N := \nabla_D w \in End_{S(V)^{W,\tau}}(\mathcal{G})$ (3.7.6).

We shall show in the next theorem the following reductions:

a) The 4. and 5. are unified and strengthened to a $S(V)^W$*-algebra structure* on $\mathcal{G}$ (Theorem I.1 and (3.9.7)).

b) The section w is given by $E*$ (Theorem I.6).

c) The $\boldsymbol{\nabla}$ is the Levi-Civita connection of J (Theorem I.3).

d) The map N is given by $-\boldsymbol{\nabla} E + \frac{1+h}{h}\mathrm{id}_{\mathcal{G}}$ (Theorem I.9).

Thus, after these reductions, $w, \nabla\!\!\!/$ and N become superfluous, and *what remains as the basic ingredients of the flat structure are the vector fields D and E, a flat metric J and the $S(V)^W$-algebra structure $*$.*

THEOREM. *(Flat structure.) One has the relations I.1-I.9 among*

$$D, E, J, P_l*, w, *, \nabla\!\!\!/ \quad \text{and } N$$

(where, $\delta, \delta_1, \delta_2, \delta_3$ are arbitrary elements in $\mathcal{G}$).

I.1. *The product $*$ and the $S(V)^W$-module structure P_l* on $\mathcal{G}$ together define a commutative and associative algebra structure over $S(V)^W$.*

I.2. *The primitive vector field D is a) the unit element of the algebra, b) horizontal w.r.t. $\nabla\!\!\!/$, i.e. $\nabla\!\!\!/ D = 0$ and c) an eigenvector for N.*

I.3. *The $\nabla\!\!\!/$ is the Levi-Civita connection for the metric J, which is integrable. That is: the J is a flat metric and the $\nabla\!\!\!/$ has the properties*
i) the torsion freeness of $\nabla\!\!\!/$: $\nabla\!\!\!/_{\delta_1}\delta_2 - \nabla\!\!\!/_{\delta_2}\delta_1 = [\delta_1, \delta_2]$,
ii) the horizontality of J with respect to connection $\nabla\!\!\!/$: $\nabla\!\!\!/ J = 0$,
iii) the integrability of $\nabla\!\!\!/$: $\nabla\!\!\!/_{\delta_1}\nabla\!\!\!/_{\delta_2} - \nabla\!\!\!/_{\delta_2}\nabla\!\!\!/_{\delta_1} = \nabla\!\!\!/_{[\delta_1,\delta_2]}$.

I.4. *The $\nabla\!\!\!/ *$ is a symmetric (3,1)-tensor. That is:*

$$T(\delta_1, \delta_2, \delta_3) := \nabla\!\!\!/_{\delta_1}(\delta_2 * \delta_3) - (\nabla\!\!\!/_{\delta_1}\delta_2) * \delta_3 - \delta_2 * (\nabla\!\!\!/_{\delta_1}\delta_3) \tag{3.9.1}$$

is a symmetric $S(V)^{W,\tau}$-tri-linear form in $\delta_1, \delta_2, \delta_3$ (values in $\mathcal{G}$).
I.5. *The product $*$ is self-adjoint with respect to J. That is:*

$$J(\delta_1, \delta_2 * \delta_3) = J(\delta_1 * \delta_2, \delta_3). \tag{3.9.2}$$

I.6. *Let us extend the domain of the product $*$ from $\mathcal{G}$ to Der_{S^W} by a use of* (3.2.4) *$S(V)^W$-linearly. Then the section w is given by*

$$w(\delta) = E * \delta. \tag{3.9.3}$$

I.7. *Let us extend the domain of $\nabla\!\!\!/$ from $\mathcal{G}$ to Der_{S^W} by a use of* (3.2.4) *and Leibniz rule. Then the exponent map N is given by*

$$N(\delta) = -\nabla\!\!\!/_{\delta} E + \tfrac{1+h}{h}\delta. \tag{3.9.4}$$

I.8. *The N is horizontal with respect to the connection $\nabla\!\!\!/$. That is:*

$$\nabla\!\!\!/ N = 0. \tag{3.9.5}$$

I.9. *Let N^* be the adjoint action of N with respect to J, i.e.*

$$J(N(\delta_1), \delta_2) = J(\delta_1, N^*(\delta_2)).$$

Then, one has the duality formula:

$$N + N^* = 1. \tag{3.9.6}$$

PROOF. A guiding principle of the proof: We develop in powers of D^{-1} the properties of the connection ∇ using (3.8.3) and (3.8.4). The term-wise equality (due to the uniqueness (3.8.2)) yields the seeked results.

a) Consider the torsion-free condition of ∇: for $\delta_1,\ \delta_2 \in \mathcal{G}$

$$\nabla_{\delta_1}(D^{-1} \otimes \delta_2) - \nabla_{\delta_2}(D^{-1} \otimes \delta_1) = D^{-1} \otimes [\delta_1, \delta_2].$$

By a use of (3.8.4), the LHS is $\delta_1 * \delta_2 - \delta_2 * \delta_1 + (\nabla\!\!\!/_{\delta_1}\delta_2 - \nabla\!\!\!/_{\delta_2}\delta_1)D^{-1}$.

The constant term implies the commutativity of $*$ (I.1). This and the left-unitness (3.7.4) imply D is the unit of the $*$-algebra (I.2 a)).

The linear term implies the torsion-freeness of $\nabla\!\!\!/$ (I.3. i)). This and $\nabla\!\!\!/_D \mathcal{G} = 0$ (3.7.5) imply that the horizontality of D (I.2 b)).

That D is an eigenvector of N is shown in 3.8 Note 1. Fact. (I.2 c)).

b) Consider the integrability condition of ∇: for $\delta_1,\ \delta_2,\ \delta_3 \in \mathcal{G}$

$$(\nabla_{\delta_1}\nabla_{\delta_2} - \nabla_{\delta_2}\nabla_{\delta_1})\,\delta_3 D^{-2} = \nabla_{[\delta_1,\delta_2]}\delta_3 D^{-2}$$

By a use of (3.8.4), the LHS becomes

$$\begin{array}{l}\delta_1 * (\delta_2 * \delta_3) - \delta_2 * (\delta_1 * \delta_3) + \big(\nabla\!\!\!/_{\delta_1}(\delta_2 * \delta_3) - \nabla\!\!\!/_{\delta_2}(\delta_1 * \delta_3)\\ +\delta_1 * \nabla\!\!\!/_{\delta_2}\delta_3 - \delta_2 * \nabla\!\!\!/_{\delta_1}\delta_3\big)D^{-1} + \big(\nabla\!\!\!/_{\delta_1}\nabla\!\!\!/_{\delta_2}\delta_3 - \nabla\!\!\!/_{\delta_2}\nabla\!\!\!/_{\delta_1}\delta_3\big)D^{-2}.\end{array}$$

The RHS is $[\delta_1,\delta_2] * \delta_3 D^{-1} + \nabla\!\!\!/_{[\delta_1,\delta_2]}\delta_3 D^{-2}$.

The constant terms imply the associativity of $*$ (I.1).

The quadratic terms imply integrability of $\nabla\!\!\!/$ (I.3. iii)).

The linear terms imply a relation: for $\delta_1,\ \delta_2,\ \delta_3 \in \mathcal{G}$

1) $$\nabla\!\!\!/_{\delta_1}(\delta_2 * \delta_3) - \nabla\!\!\!/_{\delta_2}(\delta_1 * \delta_3) + \delta_1 * \nabla\!\!\!/_{\delta_2}\delta_3 - \delta_2 * \nabla\!\!\!/_{\delta_1}\delta_3 = [\delta_1,\delta_2] * \delta_3.$$

This means $T(\delta_1,\delta_2,\delta_3) = T(\delta_2,\delta_1,\delta_3)$, where $T(\delta_1,\delta_2,\delta_3)$ is the tensor (3.9.1). Since $T(\delta_1,\delta_2,\delta_3) = T(\delta_1,\delta_3,\delta_2)$ by definition, I.4 is shown.

c) Recall the Leibniz rule for ∇: for $\delta_1,\ \delta_2 \in \mathcal{G}$

$$\nabla_{\delta_1}(P_l\nabla_D^{-1}\delta_2) = (\delta_1 P_l)\nabla_D^{-1}\delta_2 + P_l\nabla_{\delta_1}\nabla_D^{-1}\delta_2.$$

By a use of (3.8.3), the LHS and RHS are developed in

$$\delta_1 * (P_l * \delta_2) + \big(\nabla\!\!\!/_{\delta_1}(P_l * \delta_2) + \delta_1 * (N+1)(\delta_2)\big)D^{-1} + \nabla\!\!\!/_{\delta_1}(N+1)(\delta_2)D^{-2},$$

$$P_l * (\delta_1 * \delta_2) + \big((\delta_1 P_l)\delta_2 + N(\delta_1 * \delta_2) + P_l * (\nabla\!\!\!/_{\delta_1}\delta_2)\big)D^{-1} + (N+1)(\nabla\!\!\!/_{\delta_1}\delta_2)D^{-2},$$

respectively. The constant terms give the equality:

(3.9.7) $$P_l * (\delta_1 * \delta_2) = \delta_1 * (P_l * \delta_2).$$

This means that *the $S(V)^W$-module structure on $\mathcal{G}$ was in fact a $S(V)^W$-algebra structure.* We note that the following are equivalent:

i) *The multiplication P_l* defines a $S(V)^W$-algebra structure on $\mathcal{G}$.*

ii) *There exists an element $[P_l] \in \mathcal{G}$ such that $P_l * \delta = [P_l] * \delta$.*

iii) *The section w is given by $w(\delta) = E * \delta$ for $\delta \in \mathcal{G}$.*

(*Proof.* i) $\Rightarrow$ ii): Substitute δ_2 by D and put $[P_l] := P_l * D$. ii) $\Rightarrow$ i): Due to the associativity of $*$. ii) $\Rightarrow$ iii): $E*\delta = w(D)*\delta = (P_l D - P_l * D)*\delta = P_l\delta - P_l*\delta = w(\delta)$. iii) $\Rightarrow$ ii): $P_l * \delta = P_l\delta - w(\delta) = P_l\delta - E*\delta = P_l\delta - (P_l D - P_l * D)*\delta) = (P_l * D)*\delta$. $\square$)

Therefore, the I.6 (3.9.3) is also shown.

The quadratic terms imply the horizontality of N (I.7. (3.9.5)).

The linear terms give an equality:

2) $$\begin{array}{rl} & N(\delta_1 * \delta_2) - \delta_1 * (N+1)(\delta_2)\\ = & \nabla\!\!\!/_{\delta_1}(P_l * \delta_2) - (\delta_1 P_l)\delta_2 - P_l * (\nabla\!\!\!/_{\delta_1}\delta_2).\end{array}$$

Specializing this for $\delta_2 = D$ and noting the fact $\nabla\!\!\!/ D = 0$, one obtains

$$N(\delta) = -(\delta P_l)D + \nabla\!\!\!/_\delta([P_l]) + \delta * (1+N)(D) = -\nabla\!\!\!/_\delta E + (1+\tfrac{1}{h})\delta.$$

This proves (3.9.4). Using (3.9.4), the equality 2) is rewritten as:

$$3)\qquad \begin{aligned} T(\delta_1,\delta_2,[P_l]) &= \tfrac{1}{h}\delta_1 * \delta_2 + N(\delta_1 * \delta_2) - \delta_1 * N(\delta_2) - \delta_2 * N(\delta_1) \\ &= -\delta_1 * \delta_2 - \nabla_{\delta_1 * \delta_2} E + \delta_1 * \nabla_{\delta_2} E + \delta_2 * \nabla_{\delta_1} E. \end{aligned}$$

Using the symmetry $T(\delta_1,\delta_2,[P_l]) = T([P_l],\delta_1,\delta_2)$ (I.4), this reduces to the homogeneity of $*$-produce: $[E,\delta_1 * \delta_2] - \delta_2 * [E,\delta_1] - \delta_1 * [E,\delta_2] = \delta_1 * \delta_2$.

d) For $\delta_1,\delta_2 \in \mathcal{G}$, using w (3.4.4) and applying (3.4.7), one has

$$4)\qquad I^*(w(\delta_1),w(\delta_2)) = I^*(P_l\delta_1 - P_l * \delta_1, w(\delta_2)) = P_l J(\delta_1,\delta_2) - J(P_l * \delta_1,\delta_2).$$

Apply D to BHS of 4). Using (3.7.6) and (3.4.7), the LHS is equal to

$$\begin{aligned} & I^*(\nabla_D w(\delta_1), w(\delta_2)) + I^*(w(\delta_1),\nabla_D w(\delta_2)) \\ = \;& I^*(N(\delta_1),w(\delta_2)) + I^*(w(\delta_1),N(\delta_2)) = J(N(\delta_1),\delta_2) + J(\delta_1,N(\delta_2)). \end{aligned}$$

The RHS is equal to $J(\delta_1,\delta_2)$. This implies the duality (3.9.6).

e) We show the horizontality of J (I.3.ii)). Let us apply $\delta \in \mathcal{G}$ to the both hand sides of 4) in d). Using (3.7.7), (3.7.1), (3.9.5) and apply 4) again for $\nabla_\delta \delta_1$ and δ_2,...etc., LHS is calculated as

$$\begin{aligned} &\delta I^*(w(\delta_1),w(\delta_2)) = I^*(\nabla_\delta \nabla_D^{-1} N(\delta_1), w(\delta_2)) + I^*(w(\delta_1), \nabla_\delta \nabla_D^{-1} N(\delta_2)) \\ &= I^*(\delta * N(\delta_1) + \nabla_D^{-1}(\nabla_\delta N(\delta_1)), w(\delta_2)) \\ &\quad + I^*(w(\delta_1), \delta * N(\delta_2) + \nabla_D^{-1}(\nabla_\delta N(\delta_2))) \\ &= J(\delta * N(\delta_1),\delta_2) + I^*(\nabla_D^{-1}(N(\nabla_\delta\delta_1), w(\delta_2)) \\ &\quad + J(\delta_1, \delta * N(\delta_2)) + I^*(w(\delta_1), \nabla_D^{-1}(N(\nabla_\delta\delta_2)) \\ &= J(\delta * N(\delta_1),\delta_2) + I^*(w(\nabla_\delta\delta_1), w(\delta_2)) \\ &\quad + J(\delta_1, \delta * N(\delta_2)) + I^*(w(\delta_1), w(\nabla_\delta\delta_2)) \\ &= J(\delta * N(\delta_1),\delta_2) + P_l J(\nabla_\delta\delta_1,\delta_2) - J(\nabla_\delta\delta_1, P_l * \delta_2) \\ &\quad + J(\delta_1, \delta * N(\delta_2)) + P_l J(\delta_1,\nabla_\delta\delta_2) - J(P_l * \delta_1, \nabla_\delta\delta_2). \end{aligned}$$

The RHS is given by $P_l \delta J(\delta_1,\delta_2) + (\delta P_l)J(\delta_1,\delta_2) - \delta J(P_l * \delta_1,\delta_2)$.
The leading terms in P_l in BHS give the horizontality of J (I.3. ii)). The remaining terms: $(\delta P_l)J(\delta_1,\delta_2) - \delta J(P_l * \delta_1,\delta_2) = J(\delta * N(\delta_1),\delta_2) - J(\nabla_\delta\delta_1, P_l * \delta_2) + J(\delta_1, \delta * N(\delta_2)) - J(P_l * \delta_1, \nabla_\delta\delta_2)$ is reduced to 2).

f) We show the self-adjointness of the $*$-product (I.5.(3.9.2)). We have only to prove the special case when $\delta_3 = D$ (and $\delta_1,\delta_2 \in \mathcal{G}$):

$$(3.9.8)\qquad J(\delta_1,\delta_2) = J(\delta_1 * \delta_2, D),$$

since, by a use of (3.9.8), the LHS and RHS of (3.9.2) are equal to $J(\delta_1 * (\delta_2 * \delta_3), D)$ and $J((\delta_1 * \delta_2) * \delta_3, D)$, respectively, and then, the associativity of the $*$-product implies the equality and hence (3.9.2).

First, we show a formula (see 3.10 1. for a meaning): for $\delta \in \mathcal{G}$:

$$(3.9.9)\qquad \delta = \nabla_\delta \nabla_D^{-1} D$$

Proof. This is the torsion-freeness of ∇. According to (3.7.1), the RHS is equal to $\delta * D + \nabla_D^{-1}\nabla_\delta D$. The first term is equal to δ, since D is the unit. The second term vanishes due to the horizontality of D. □

We return to a proof of (3.9.8). Using (3.9.9) and (3.4.7), one has

$$\begin{aligned} J(\delta_1,\delta_2) &= I^*(\nabla_{\delta_1}\nabla_D^{-1}D, w(\delta_2)) \\ &= \delta_1 I^*(\nabla_D^{-1}D, w(\delta_2)) - I^*(\nabla_D^{-1}D, \nabla_{\delta_1}w(\delta_2)) \end{aligned}$$

Since $\nabla_D^{-1}D = hE = hw(D)$ where $\frac{1}{h}$ is the eigenvalue of N on D (recall 3.8 Note 1.), one calculates further (again using (3.4.7))

$$\begin{aligned} \tfrac{1}{h}J(\delta_1,\delta_2) &= \delta_1 J(D, w(\delta_2)) - J(D, \nabla_{\delta_1}w(\delta_2)) \\ &= J(\boldsymbol{\nabla}_{\delta_1}D, w(\delta_2)) + J(D, \boldsymbol{\nabla}_{\delta_1}w(\delta_2)) - J(D, \nabla_{\delta_1}w(\delta_2)) \\ &= J(D, \boldsymbol{\nabla}_{\delta_1}w(\delta_2) - \nabla_{\delta_1}w(\delta_2)) \end{aligned}$$

where the first term in the second line vanishes due to the horizontality of D. Here the form J in the RHS is extended $S(V)^W$-linearly (3.2.4). Using the expression $w(\delta) = P_l\delta - P_l * \delta$ (3.4.4) and (3.7.1), one has

$$\boldsymbol{\nabla}_{\delta_1}w(\delta_2) = \delta_1 P_l \cdot \delta_2 + P_l\boldsymbol{\nabla}_{\delta_1}\delta_2 - \boldsymbol{\nabla}_{\delta_1}(P_l * \delta_2)$$

and

$$\begin{aligned} \nabla_{\delta_1}w(\delta_2) &= \delta_1 P_l \cdot \delta_2 + P_l(\nabla_D(\delta_1 * \delta_2) + \boldsymbol{\nabla}_{\delta_1}\delta_2) \\ &\quad -(\nabla_D(\delta_1 * (P_l * \delta_2)) + \boldsymbol{\nabla}_{\delta_1}(P_l * \delta_2)) \end{aligned}$$

(In fact, these calculations should be justified. See §4). So, we obtain

$$\begin{aligned} \boldsymbol{\nabla}_{\delta_1}w(\delta_2) - \nabla_{\delta_1}w(\delta_2) &= -P_l\nabla_D(\delta_1 * \delta_2) + \nabla_D(\delta_1 * (P_l * \delta_2)) \\ &= -\nabla_D(P_l(\delta_1 * \delta_2) - P_l * (\delta_1 * \delta_2)) + \delta_1 * \delta_2 \\ &= -\nabla_D(w(\delta_1 * \delta_2)) + \delta_1 * \delta_2 \\ &= -N(\delta_1 * \delta_2) + \delta_1 * \delta_2 \;=\; (1-N)(\delta_1 * \delta_2). \end{aligned}$$

Substituting this in the previous formula, one obtains

$$\tfrac{1}{h}J(\delta_1,\delta_2) = J(D, (1-N)(\delta_1 * \delta_2)) = J((1-N^*)(D), \delta_1 * \delta_2)$$

Recalling the duality (3.9.6) and the fact that $\frac{1}{h}$ is the eigenvalue of N for the vector D (3.8 Note 1.), one has $(1-N^*)(D) = N(D) = \frac{1}{h}D$. Multiplying h on BHS, we obtain (3.9.8). □

REMARK. 1. The $S(V)^W$-algebra on $\mathcal{G}$ defines the normalization: $\widetilde{D}_W = \mathrm{Spec}(\mathcal{G}) \to D_W = \mathrm{Spec}(S(V)^W/(\Delta)$ of the discriminant (c.f. §4).

2. The (3.9.8) together with the facts $w(D) = E = I(dP_1)$ implies

$$J(\delta_1,\delta_2) = \delta_1 * \delta_2\ (P_1). \tag{3.9.10}$$

3. The $*, P_l*, \boldsymbol{\nabla}, N$ and the metric J on the module $\mathcal{G}$ are sufficient to recover the $S(V)^W$-module Der_{S_W} with the connection ∇ and the metric I through (3.8.2), (3.8.3), (3.8.4) and (3.4.8).

3.10. Relations to primitive forms and Frobenius manifold.

We discuss some relations of the flat structure in 3.9 to primitive forms and to Frobenius manifold structure.

1. Primitive form. Since the ∇ is the Levi-Civita connection and torsion free, we have confused the module $\mathcal{H}^{(0)} := Der_{S_W}$ of operands (sections of a bundle) with the algebra Der_{S_W} of operators (covariant differentiation). In order to separate two roles, we introduce a sequence:

$$\mathcal{H}^{(-k)} := \nabla_D^{-k}Der_{S_W} \simeq D^{-k}\mathbf{Z}[D^{-1}] \otimes \mathcal{G}$$

of the modules of the operands (recall (3.8.2)) for $k \in \mathbf{Z}_{\geq 0}$, on which one has the action of the covariant differentiations:

$$\begin{aligned} \nabla : \qquad\qquad Der_{S_W} \times \mathcal{H}^{(-k-1)} &\to \mathcal{H}^{(-k)}, \\ \nabla : Der_{S_W}(-\log(\Delta^2)) \times \mathcal{H}^{(-k)} &\to \mathcal{H}^{(-k)}. \end{aligned}$$

Then the formula (3.9.9) (based on the torsion-freeness of ∇, used at several crucial steps of proof of Theorem) implies the following key fact:

Assertion. *The covariant differentiation of the element $\zeta^{(-1)} := \nabla_D^{-1} D$ (i.e. the correspondence: $\delta \mapsto \nabla_\delta \zeta^{(-1)}$) induces a bijection between the Lie algebra Der_{S_W} of operators and the module $\mathcal{H}^{(0)}$ of operands. More generally, the covariant differentiation of the element $\zeta^{(-k)} := \nabla_D^{-k} D$ ($k \in \mathbf{Z}_{\geq 0}$) induces an isomorphism of $S(V)^W$-modules:*

$$*) \qquad \nabla \zeta^{(-k-1)} \ : \ Der_{S_W} \simeq \mathcal{H}^{(-k)}.$$

Such $\zeta^{(-k)}$ (satisfying some additional conditions corresponding to (3.8.3), (3.8.4) and orthogonality, see [S6]) is called a *primitive form.*

The isomorphism $*$) has meanings in two directions:

1. Through the correspondence, cohomological structures on $\mathcal{H}^{(0)}$ (such as the polarization I, Hodge filtration) and the algebra structure on Der_{S_W} (such as the Lie algebra str., integrable system) are combined together in one object. This amalgamation creates some rich structures such as the flat structure in 3.9, the Frobenius structure and the flat coordinates (discussed in the following).

2. The module $\mathcal{H}^{(-k)}$ (the Hodge filter) is obtained by covariant differentiation of a single element $\zeta^{(-k-1)}$. That is: the data of the module $\mathcal{H}^{(-k)}$ are condensed in the single element $\zeta^{(-k-1)}$, the primitive form, and by differentiating the primitive form, they are resolved again. This view point is an origin of the word "primitive form". In fact, before the primitive form was introduced, one had studied the "primitive (i.e. potential) function" as a solution of second order uniformization equation such that its derivatives give the system of solutions of a torsion free connection [S2].

In the present paper, reversing the history, we start from the flat structure, and then we shall reconstruct the uniformization equations in §4, whose solutions (i.e. the primitive functions) are partially obtained by the period integral of the primitive form §5.

2. Frobenius manifold. A tuple $(M, \circ, e, E, g)$ of a manifold M with a metric g, two global vector fields e and E on M and a ring structure $\circ$ on the tangent bundle of M, subject to some axioms, is called a Frobenius manifold ([Do],[Ma, 0.4.1, I.1.3],[He, def 9.1],[Sab],[Ta]). As an immediate consequence of 3.9 Theorem, we show the next.

THEOREM. *The tuple $(S_W, *, D, E, J)$ form a Frobenius manifold.*

PROOF. In view that ∇ is the Levi-Civita connection of J I.3, we check the the axioms 1) - 5) for a Frobenius manifold [He def 9.1] as follows.

1) *The $*$ is self-adjoint with respect J.* This is shown in (3.9.2).
2) *The tensor $\nabla *$ is symmetric.* This is shown in I.4.
3) *The metric J is flat, i.e. $\nabla^2 = 0$.* This is shown in I.3. iii).
4) *The D is horizontal, i.e. $\nabla D = 0$.* This is shown in I.2.b).

5) *The $*$ is homogeneous of degree 1 and the J is homogeneous w.r.t. the Euler vector field E, respectively.* The first half follows from the defining formula (3.7.1) in view of the normalizations (2.8.5), (3.1.6) and (3.2.1) (see also the formula at the end of c) of the proof of 3.9 Theorem). Alternatively, one shows as follows: since $D* = 1$ one has $0 = \deg(D*) = \deg(D) + \deg(*) = -1 + \deg(*)$. The J is homogeneous since J^* is homogeneous by definition (3.3.1). One calculates: $\deg(J) = -\deg(J^*) = -\deg(L_D I) = -(\deg(D) + \deg(I)) = 1 + \frac{2}{h}$. □

3.11. Flat coordinates.

We introduce coordinates of S_W whose differentials are horizontal sections of the dual connection ∇^*. We call them *flat coordinates.*

The dual connection $\nabla^* : \mathcal{F} \to \mathcal{F}\otimes\mathcal{F}$ (3.2.3) is defined by $\delta\langle\delta',\omega\rangle = \langle\nabla_\delta\delta',\omega\rangle + \langle\delta,\nabla^*_\delta\omega\rangle$ for $\delta,\delta' \in \mathcal{G}$ and $\omega \in \mathcal{F}$. Let us denote by $\wedge\nabla^*$ the composition of ∇^* with the wedge product $\mathcal{F}\otimes\mathcal{F} \to \mathcal{F}\wedge\mathcal{F}$. The torsion freeness of ∇ I.3. i) implies $\wedge\nabla^*(\omega) = d\omega$ for $\omega \in \mathcal{F}$, since

$$\begin{aligned}
\wedge\nabla^*(\omega) &= \textstyle\sum_{i=1}^l dP_i \wedge \nabla^*_{\frac{\partial}{\partial P_i}}(\omega) = \sum_{ij=1}^l dP_i \wedge dP_j \langle \frac{\partial}{\partial P_j}, \nabla^*_{\frac{\partial}{\partial P_i}}(\omega)\rangle \\
&= \textstyle\sum_{ij=1}^l dP_i \wedge dP_j (\frac{\partial}{\partial P_i}\langle\frac{\partial}{\partial P_j},\omega\rangle - \langle\nabla_{\frac{\partial}{\partial P_i}}\frac{\partial}{\partial P_j},\omega\rangle) \\
&= d\omega - \textstyle\sum_{ij=1}^l dP_i \wedge dP_j \langle\nabla_{\frac{\partial}{\partial P_i}}\frac{\partial}{\partial P_j},\omega\rangle = d\omega.
\end{aligned}$$

The same formula holds for $\omega \in \mathcal{O}_{S_W} \otimes_{S(V)^W} \mathcal{F}$ where $\mathcal{O}_{S_W}$ is the sheaf of local (w.r.t. the classical topology on $S_{W,K}$) analytic functions on $S_{W,K}$ where K is either $\mathbf{R}$ or $\mathbf{C}$ (recall (2.9.2)). Thus, *a local analytic horizontal section $\omega \in \ker(\nabla^*)$ is a closed form: $d\omega = 0$ and there is a local analytic function u such that $\omega = du$,* where u satisfies the second order differential equation:

$$\nabla^* du = 0. \tag{3.11.1}$$

The space of global analytic solutions (including constants) on $S_{W,K}$ of the equation has rank $l+1$, since ∇^* is integrable and non-singular on $S_{W,K}$ and $S_{W,K}$ is simply connected. A solution, say u, is, in fact, a polynomial: develop u in a Taylor series $u = \sum_{d\in\mathbf{Z}_{\geq 0}} u_d$ for $u_d \in S(V)^W_d$ at the origin. Since ∇ is homogeneous and continuous w.r.t. the adic topology on the space of power serieses on S_W at the origin, each graded piece u_d is again a solution of (3.11.1). Since the solution space is finite dimensional, the expansion of u should be a finite sum (i.e. u is a polynomial). These imply the part i) of the next lemma.

LEMMA. *Consider the space of the polynomial solutions of* (3.11.1)*:*

$$\Omega := \{u \in S(V)^W \mid \nabla^* du = 0\}. \tag{3.11.2}$$

Then i) the Ω is a graded vector space of rank $l+1$ with the splitting: $\Omega = \mathbf{R}\cdot 1 \oplus \Omega_+$ for $\Omega_+ := \Omega \cap S(V)^W_+$, ii) Ω_+ generates freely the algebra $S(V)^W$ over $\mathbf{R}$, and iii) one has the bijections:

$$\begin{aligned}
S_{W,K} &\simeq \mathrm{Hom}^{alin}_{\mathbf{R}}(\Omega, K) \\
&\simeq \mathrm{Hom}_{\mathbf{R}}(\Omega^+, K).
\end{aligned} \tag{3.11.3}$$

Here, $\mathrm{Hom}^{alin}_{\mathbf{R}}$ *is the set of all $\mathbf{R}$-linear maps bringing $1 \in \Omega$ to $1 \in K$ ("alin" stands for "affine linear"), and* $\mathrm{Hom}_{\mathbf{R}}$ *is the set of all $\mathbf{R}$-linear maps. By this bijection, we regard S_W the linear flat space dual to Ω_+.*

PROOF. ii) Let $1, Q_1, \cdots, Q_l \in S(V)^W$ be homogeneous $\mathbf{R}$-basis of Ω with $\deg(Q_1) \leq \cdots \leq \deg(Q_l)$. The differentials $dQ_1, \cdots, dQ_l$ span $\ker(\nabla^*)$ over $\mathbf{R}$. As a general fact, one has $S(V)^{W,\tau} \otimes_{\mathbf{R}} \ker(\nabla^*) = \mathcal{F}$. Since dP_i (for coordinates $P_1, ..., P_l$ of S_W) are $S(V)^{W,\tau}$-basis of $\mathcal{F}$, the determinant$(\frac{\partial(Q_1,\cdots,Q_l)}{\partial(P_1,\cdots,P_l)})$ is a unit on S_W, and is a non-zero constant. Then, one sees that $\deg(Q_i) = \deg(P_i)$ $(i = 1, ..., l)$ and that each homogeneous block of the Jacobi matrix is constant invertible. Then the ring homomorphism: $\mathbf{R}[Q_1, ..., Q_l] \to S(V)^W$ is an isomorphism.

iii) The map $S_{W,K} \to \mathrm{Hom}^{alin}_{\mathbf{R}}(\Omega, K)$ is defined by the correspondence $\varphi \mapsto \varphi|\Omega$ for $\varphi \in \mathrm{Hom}^{alg}_{\mathbf{R}}(S(V)^W, K)$. Since Ω generates the ring of invariants $S(V)^W$, the map is injective. Conversely, since $Q_1, \cdots, Q_l$ are algebraically independent (2.5 Theorem), the map is surjective. □

The (3.11.1) is explicitly given by the system of second order equations

$$(\delta_1\delta_2 - \nabla_{\delta_1}\delta_2)u = 0 \qquad \text{for } \delta_1, \delta_2 \in \mathcal{G}. \tag{3.11.4}$$

Elements of Ω are called *flat invariants*. Homogeneous basis $Q_1, \cdots, Q_l$ of Ω_+ are called a system of *flat coordinates* of S_W. They are unique up to a weighted linear transformation. The flat coordinates for all finite reflection groups except for types E_7 and E_8 are calculated in [S-Y-S]. The flat coordinates for type E_7 is calculated in [Ya]. The flat coordinates for type E_8 is calculated in [Ka][No][No].

A translation $\xi \in \Omega_+^* := \mathrm{Hom}^{lin}_K(\Omega/K \cdot 1, K)$ of the affine space S_W extends naturally to a derivation of $S(\Omega_+) \simeq S(V)^W$. So, one has the embedding $\Omega_+^* \subset Der_{S_W}$ with the characterization: $\Omega_+^* = \{\delta \in \mathcal{G} \mid \nabla\delta = 0\}$ (*Proof.* For an element $\delta \in \mathcal{G}$, one has: $\nabla\delta = 0 \Leftrightarrow \langle\nabla_\xi\delta, \omega\rangle = 0$ for $\forall\xi \in \mathcal{G}$ and $\forall\omega \in \Omega \Leftrightarrow \xi\langle\delta,\omega\rangle - \langle\delta, \nabla^*_\xi\omega\rangle = 0$ for $\forall\xi \in \mathcal{G}$ and $\forall\omega \in \Omega \Leftrightarrow \langle\delta,\omega\rangle$ is a constant for $\forall\omega \in \Omega \Leftrightarrow \delta \in \Omega_+^*$. □) Therefore, we shall sometimes call an element of Ω_+^* (regarded as an element of Der_{S_W}) *flat tangent vector* of S_W (e.g. D is a flat tangent).

Here are some immediate uses of the flat coordinates.

1. *The quadratic form P_1 (2.8.5) is a flat invariant.*

This follows from (3.3.4) and 3.9 Theorem I.2.b) as below.

$$\begin{aligned}(\delta_1\delta_2 - \nabla_{\delta_1}\delta_2)P_1 &= \delta_1 J^*(J(\delta_2), dP_1) - J^*(J(\nabla_{\delta_1}\delta_2)), dP_1) \\ &= \delta_1 J(\delta_2, D) - J(\nabla_{\delta_1}\delta_2, D) = J(\delta_2, \nabla_{\delta_1}D) = 0.\end{aligned}$$

2. *Let $\Omega_+^* = \oplus\Omega_d^*$ be the graded decomposition of Ω_+^*. Then,*

a) The piece Ω_d^* is a vector space of rank $\#\{1 \leq i \leq l \mid d = -\frac{m_i+1}{h}\}$.

b) If $d_1 + d_2 + 1 + \frac{2}{h} = 0$ then $J|_{\Omega_{d_1}^* \times \Omega_{d_2}^*}$ is the duality between the two pieces. If $d_1 + d_2 + 1 + \frac{2}{h} \neq 0$ then $J|_{\Omega_{d_1}^* \times \Omega_{d_2}^*} = 0$.

c) The N is constant on each piece: $N|_{\Omega_d^*} = (d + \frac{1+h}{h})\mathrm{id}_{\Omega_d^*}$.

3. *The star product $*$ is calculated by an explicit formula:*

$$\frac{\partial}{\partial Q_i} * \left(\frac{h-m_j}{h} J^*(dQ_j)\right) = \sum_{k=1}^{l} \left(\frac{\partial}{\partial Q_i} I(dQ_j, dQ_k) - A^i_{jk}\right) \frac{\partial}{\partial Q_k} \tag{3.11.5}$$

Here, the coefficients A^i_{jk} is given by (3.6.6) by replacing $P_1, \cdots, P_l$ by the flat $Q_1, \cdots, Q_l$. Note $A^i_{jk} \in S(V)^{W,\tau}$ (i.e. $DA^i_{jk} = 0$ for (3.6.3)).

4. Flat potential. *Put $J_3(\delta_1, \delta_2, \delta_3) := J(\delta_1 * \delta_2, \delta_3)$ for δ_1, δ_2 and $\delta_3 \in \mathcal{G}$. Then a) J_3 is a symmetric tensor in the three variables, and b) there exists a polynomial $F \in \S(V)^W$ such that*

$$\delta_1\delta_2\delta_3 F = J(\delta_1, \delta_2, \delta_3) \tag{3.11.6}$$

for δ_1, δ_2 *and* $\delta_3 \in \Omega^*_+$.

PROOF. Due to the commutativity and associativity of $*$ and the self-adjointness of $*$ (3.9.2), J_3 is a symmetric tensor on $\mathcal{G}$ over $S(V)^{W,\tau}$. Therefore, we have only to show the following

Fact. *The 4-forms* $\delta_0 J_3(\delta_1, \delta_2, \delta_3)$ *in* $\delta_0, \cdots, \delta_3 \in \Omega^*_+$ *is symmetric.*

Put $J_4(\delta_0, \delta_1, \delta_2, \delta_3) := \delta_0 J_3(\delta_1, \delta_2, \delta_3)$. Let us specialize (3.9.1) to flat tangent vectors $\delta_1, \delta_2, \delta_3 \in \Omega^*_+$. Then one obtains $\nabla_{\delta_1}(\delta_2 * \delta_3)$. This implies that $\nabla_{\delta_1}(\delta_2 * \delta_3)$ is a symmetric cubic form on Ω^*_+. On the other hand, by using the metric property I.3. ii) of ∇ the definition of J_4 is rewritten as: $J_4(\delta_0, \delta_1, \delta_2, \delta_3) = J(\nabla_{\delta_0}(\delta_1 * \delta_2), \delta_3)$. This expression is now symmetric w.r.t. the letters $\{0, 1, 2\}$. Since it was already symmetric w.r.t. the letters $\{1, 2, 3\}$, altogether, J_4 is symmetric w.r.t. all letters $\{0, 1, 2, 3\}$. □

3.12. Dimensions.

Before ending this section, we clarify that some data, such as E, ∇, N and exponents, are absolutely defined independent of scaling factors.

Recall that in the study of the flat structure on the variety S_W (except for type A_1), there were two ambiguities of scaling factors for

1. a choice of an invariant form I in 2.2, and
2. a choice of a primitive derivation D in (3.2.1).

Let us call an object X has dimension (α, β) if the scale changes $I^* \mapsto \lambda I^*$ and $D \mapsto \mu D$ induce the scale change $X \mapsto \lambda^\alpha \mu^\beta X$ (in case of type A_1, one has $\lambda\mu = 1$ and, hence, only the difference $\alpha - \beta$ has a meaning).

The following table is easy to calculate.

object	α	β	deg	*object*	α	β	deg
I^*	1	0	$\frac{2}{h}$	E	0	0	0
D	0	1	-1	∇	0	0	0
I	-1	0	$-\frac{2}{h}$	$*$	0	-1	1
J	1	-1	$1+\frac{2}{h}$	N	0	0	0
J^*	-1	1	$-1-\frac{2}{h}$	Δ^2	0	$-l$	l
P_l	0	-1	1	P_1	1	0	$\frac{2}{h}$
P_l*	0	-1	1	J_3	1	-2	$2+\frac{2}{h}$
$[P_l]$	0	0	0	F	1	-1	$2+\frac{2}{h}$

The table shows that E, ∇, N and exponents are absolutely defined.

4. Uniformization equations

Following [S6,§5], we introduce the system of uniformizing equation $\mathcal{M}_{W,s}$ on S_W attached to the flat structure in 3.9. It is a system of second order holonomic system such that the first order derivatives of its solution gives the system of the solutions of the Levi-Civita connection ∇ in §3 on the lattice $\mathcal{H}^{(s)}$. The $\mathcal{M}_{W,s}$ has new features:

i) the equation contains a spectral parameter s shifting exponents,

ii) the equation is a $\mathcal{D}_{S_W}$-module instead of a $\mathcal{E}_{S_W}$-module,

iii) the homogeneity condition is omitted from the equation to allow non-constant solutions for the invariant cycles.

Compared with [S6,§5], we skipped a construction of the integral lattice on the local system of the solutions and the monodromy group action on it for the case of crystallographic reflection group (for details of them, see [ibd]). Explicit equations $\mathcal{M}_{W,s}$ and some solutions for types A_1, A_2 and A_3 were discussed in [S2] (see footnote 2.).

4.1. $S(V)^W$-modules $\mathcal{H}_W^{(s)}$ for $s \in \mathbf{C}$.
Let $\{D^{\mathbf{C}}\} := \{D^s \mid s \in \mathbf{C}\}$ be a group isomorphic to the complex number field $\mathbf{C}$ with the multiplicative rule: $D^s D^t = D^{s+t}$. Let $\mathbf{Z} \cdot \{D^{\mathbf{C}}\}$ be its group ring, which contains the Laurent algebra $\mathbf{Z}[D, D^{-1}]$. Put

$$\mathcal{H}_W := \mathbf{Z} \cdot \{D^{\mathbf{C}}\} \otimes_{\mathbf{Z}} \mathcal{G}, \tag{4.1.1}$$

where one recalls (3.2.3) for the $S(V)^{W,\tau}$-module $\mathcal{G}$. The $\mathcal{H}_W$ has natural $S(V)^{W,\tau}$-module structure by a commutation rule

$$f \cdot D^s - D^s \cdot f = 0 \qquad \text{for } f \in S(V)^{W,\tau} \text{ and } s \in \mathbf{C}. \tag{4.1.2}$$

Further, the $\mathcal{H}_W$ is equipped with a $S(V)^W = S(V)^{W,\tau}[P_l]$-module structure and a covariant differentiation action ∇_δ for $\delta \in \mathcal{G}$ as follows:
i) the multiplication of P_l is defined by the formula (3.8.3) and (3.8.3)*,
ii) the action ∇_δ for $\delta \in \mathcal{G}$ is defined by the formula (3.8.4) and (3.8.4)*,
where we replace $k \in \mathbf{Z}_{\geq 0}$ by a complex number $s \in \mathbf{C}$. It is a routine to check that the $S(V*)^W$-module structure does not depend on a choice of P_l, and the ∇-action satisfies the Leibniz rule and the integrability condition (i.e. $\nabla^2 = 0$).
For any $s \in \mathbf{C}$, let us introduce the submodule of $\mathcal{H}_W$:

$$\mathcal{H}_W^{(s)} := D^s \mathbf{Z}[D^{-1}] \otimes_{\mathbf{Z}} \mathcal{G}. \tag{4.1.3}$$

The $\mathcal{H}_W^{(s)}$ is a $S(V)^W$-submodule (i.e. closed under the left multiplication of P_l) easily seen from (3.8.3) and (3.8.3)*. Thus, by definition, one has an increasing sequence of the $S(V)^W$-modules:

$$\cdots \subset \mathcal{H}_W^{(s-2)} \subset \mathcal{H}_W^{(s-1)} \subset \mathcal{H}_W^{(s)} \subset \mathcal{H}_W^{(s+1)} \subset \mathcal{H}_W^{(s+2)} \subset \cdots, \tag{4.1.4}$$

and a short exact sequence of $S(V)^W$-modules:

$$0 \to \mathcal{H}_W^{(s-1)} \to \mathcal{H}_W^{(s)} \to \mathcal{G} \to 0, \tag{4.1.5}$$

where $\mathcal{G}$ is equipped with $S(V)^W$-module structure by the action of P_l* (3.9 Theorem I.1). One observes $\Delta^2 \mathcal{H}_W^{(s)} \subset \mathcal{H}_W^{(s-1)}$ since $\Delta^2 \mathcal{G} = 0$ due to the 3.4 Corollary. An explicit formula can be given as follows.

$$\begin{aligned} \Delta^2(D^s \otimes \partial_j) &= \textstyle\sum_{i=1}^l b_{ji}(P_l D^s \otimes \partial_i - D^s \otimes P_l * \partial_i) \\ &= \textstyle\sum_{i=1}^l b_{ji} D^{s-1} \otimes (N - s)\partial_i). \end{aligned} \tag{4.1.6}$$

Here $\partial_i = \frac{\partial}{\partial P_i}$ and (b_{ji}) is the adjoint matrix of $(P_l \delta_{ij} - a_{ij})$ with $w(\partial_i) = P_l \partial_i - P_l * \partial_i = \sum_{j=1}^l (P_l \delta_{ij} - a_{ij})\partial_j$ so that $\sum_{i=1}^l b_{ji} w(\partial_i) = \Delta^2 \partial_j$.

NOTE. The modules $\mathcal{H}_W^{(-i)}$ for $i \in \mathbf{Z}_{\geq 0}$ in 3.10 are identified with the above defined one through (3.8.2). But the identification changes according to a choice of a primitive form (c.f. (4.3.1)).

The covariant differentiation by $\delta \in \mathcal{G}$ defines a map $\nabla_\delta : \mathcal{H}_W^{(s)} \to \mathcal{H}_W^{(s+1)}$ (use (3.8.4) and (3.8.4)*). Since $Der_{S_W} = S(V)^W \otimes \mathcal{G}$, the action extends to a covariant differentiation $\nabla : Der_{S_W} \times \mathcal{H}_W^{(s)} \to \mathcal{H}_W^{(s+1)}$. Particularly, ∇_D induces a $S(V)^{W,\tau}$-bijection:

$$\nabla_D : \mathcal{H}_W^{(s)} \simeq \mathcal{H}_W^{(s+1)}, \quad D^s \otimes \delta \mapsto D^{s+1} \otimes \delta. \tag{4.1.7}$$

NOTE. The power action $\nabla_D^t : \mathcal{H}_W^{(s)} \to \mathcal{H}_W^{(s+t)}$ for $t \in \mathbf{C}$ can be defined by letting $\nabla_D^t(D^s \otimes \delta) := D^{s+t} \otimes \delta$ satisfying the relation $[\nabla_D^t, P_l] = t\nabla_D^{t-1}$.

4.2. Polarization element I_s.

We introduce the *polarization element* $I_s \in \mathcal{H}_W \otimes_{S(V)^W} \mathcal{H}_W$ (the tensor of left $S(V)^W$-modules) for $s \in \mathbf{C}$ by the next three formulae:

$$\begin{aligned} I_s &:= \textstyle\sum_{i=1}^l (D^s \otimes \delta_i) \otimes (P_l D^{-s} \otimes \delta_i^* - D^{-s} \otimes P_l * \delta_i^*) \\ &= \textstyle\sum_{i=1}^l (D^s \otimes \delta_i) \otimes (D^{-s} \otimes w(\delta_i^*) + sD^{-s-1} \otimes \delta_i^*) \\ &= \textstyle\sum_{i=1}^l (D^s \otimes \delta_i) \otimes (D^{-s-1} \otimes (N+s)(\delta_i^*)), \end{aligned} \tag{4.2.1}$$

where i) $\delta_1, \cdots, \delta_l$ are $S(V)^{W,\tau}$-free basis of $\mathcal{G}$ and $\delta_1^*, \cdots, \delta_l^*$ are their dual basis w.r.t. J so that the definition does not depend on the basis, and ii) the equivalence of the three definitions follows from (3.8.3)*.

NOTE. Through the embedding $Der_{S_W} \subset \mathcal{H}_W$ (3.8.2), the polarization I (3.4.8) is identified with the above defined I_0.

LEMMA. **1.** *For any $s \in \mathbf{C}$, one has the formula:*

$$I_s = {}^tI_{-s} \quad \text{and} \quad I_s = -{}^tI_{-s-1}, \tag{4.2.2}$$

where the transposition ${}^tI_s$ of the polarization I_s is defined by transposing the tensor factors in the definition (4.2.1).

2. *Let the covariant differentiation ∇_δ for $\delta \in \mathcal{G}$ act on the tensor $\mathcal{H}_W \otimes_{S(V)^W} \mathcal{H}_W$ naturally (by the Leibniz rule). Then one has*

$$\nabla I_s = 0. \tag{4.2.3}$$

PROOF. First, we note the following general formula:

$$*) \qquad \textstyle\sum_{i=1}^l f(A\delta_i) \otimes_{S(V)^{W,\tau}} g(\delta_i^*) = \sum_{i=1}^l f(\delta_i) \otimes_{S(V)^{W,\tau}} g(A^*\delta_i^*)$$

where f, g are $S(V)^{W,\tau}$-homomorphisms defined on $\mathcal{G}$, A and A^* are $S(V)^{W,\tau}$-endomorphism on $\mathcal{G}$ and its adjoint w.r.t. J, and $\delta_1, \cdots, \delta_l$ and $\delta_1^*, \cdots, \delta_l^*$ are $S(V)^{W,\tau}$-free basis of $\mathcal{G}$ and the dual basis w.r.t. J.

1. Take the first line of (4.2.1). The multiplication of P_l in the first term can be transposed to that in the left side of the tensor. The left action P_l* in the second term can be transposed to the left due to *) and the self adjointness of P_l* ((3.9.2) and II.6. i)). So, one obtains

$$\begin{aligned} I_s &= \textstyle\sum_{i=1}^l (D^s \otimes \delta_i) \otimes (P_l D^{-s} \otimes \delta_i^* - D^{-s} \otimes P_l * \delta_i^*) \\ &= \textstyle\sum_{i=1}^l (P_l D^s \otimes \delta_i - D^s \otimes (P_l * \delta_i)) \otimes (D^{-s} \otimes \delta_i^*) = {}^tI_{-s} \end{aligned}$$

This proves the first formula in (4.2.2).

Let us employ the third line of (4.2.1). Use the fact that $(N+s)^* = N^* + s = -N + s + 1$ ((3.9.6)), and apply *) so that one obtains:

$$\begin{aligned} I_s &= \textstyle\sum_{i=1}^l (D^s \otimes \delta_i) \otimes (D^{-s-1} \otimes (N+s)(\delta_i^*)) \\ &= \textstyle\sum_{i=1}^l (D^s \otimes (-N+s+1)\delta_i) \otimes (D^{-s-1} \otimes \delta_i^*) \\ &= -\textstyle\sum_{i=1}^l (D^{-u-1} \otimes (N+u)\delta_i) \otimes (D^u \otimes \delta_i^*) \quad h \text{ for } -u := s+1 \\ &= -{}^tI_u = -{}^tI_{-s-1}. \end{aligned}$$

This proves the second formula in (4.2.2).

2. Let δ_i $(i=1,\cdots,l)$ be flat basis (i.e. $\nabla\delta_i = 0$) of $\mathcal{G}$. Then the dual basis δ_i^* and their N-images $N(\delta_i)$, $N(\delta_i^*)$ are also flat. The covariant differentiation of I_s vanishes using the duality (3.9.6) again.

$$\begin{aligned}
&\nabla_\delta I_s\\
&= \textstyle\sum_{i=1}^l \nabla_\delta(D^s\otimes\delta_i)\otimes(P_l D^{-s}\otimes\delta_i^* - D^{-s}\otimes P_l*\delta_i^*)\\
&\quad + \textstyle\sum_{i=1}^l (D^s\otimes\delta_i)\otimes\nabla_\delta(D^{-s-1}\otimes(N+s)(\delta_i^*))\\
&= \textstyle\sum_{i=1}^l (D^{s+1}\otimes(\delta*\delta_i) + D^s\otimes\nabla_\delta\delta_i)\otimes(P_l D^{-s}\otimes\delta_i^* - D^{-s}\otimes P_l*\delta_i^*)\\
&\quad + \textstyle\sum_{i=1}^l (D^s\otimes\delta_i)\otimes(D^{-s}\otimes(\delta*(N+s)\delta_i^*) + D^{-s-1}\otimes\nabla_\delta(N+s)\delta_i^*))\\
&= \textstyle\sum_{i=1}^l (P_l D^{s+1}\otimes(\delta*\delta_i) - D^{s+1}\otimes(P_l*\delta*\delta_i))\otimes(D^{-s}\otimes\delta_i^*)\\
&\quad + \textstyle\sum_{i=1}^l (D^s\otimes(\delta*(N+s))^*\delta_i)\otimes(D^{-s}\otimes\delta_i^*)\\
&= \textstyle\sum_{i=1}^l (D^s\otimes(N-s-1)(\delta*\delta_i))\otimes(D^{-s}\otimes\delta_i^*)\\
&\quad + \textstyle\sum_{i=1}^l (D^s\otimes(N^*+s)(\delta*\delta_i))\otimes(D^{-s}\otimes\delta_i^*)\\
&= \textstyle\sum_{i=1}^l (D^s\otimes(N+N^*-1)(\delta*\delta_i))\otimes(D^{-s}\otimes\delta_i^*) = 0.
\end{aligned}$$

These complete a proof of Lemma. □

Combining the two formulae (4.2.2), one obtains the semi-periodicity $I_{s+1} = -I_s$. On the other hand, the formulae (4.2.2) imply, $I_0 = {}^tI_0$ and $I_{-\frac{1}{2}} = -{}^tI_{-\frac{1}{2}}$. Therefore, we obtain the next corollary.

COROLLARY. **1.** *One has $I_{s+n} = (-1)^n I_s$ for $s\in\mathbf{C}$ and $n\in\mathbf{Z}$.*

2. *The polarization I_s is a symmetric form for an integer $s\in\mathbf{Z}$ and is a skew symmetric form for a half integer $s\in\frac{1}{2}\mathbf{Z}\setminus\mathbf{Z}$.*

4.3. Connection $\nabla^{(s)}$ and uniformization equation.

Put $\zeta^{(s-1)} := D^{s-1}\otimes D\in\mathcal{H}_W$ for $s\in\mathbf{C}$ (c.f. 3.10 Assertion.). The covariant differentiation of $\zeta^{(s-1)}$ induces an $S(V)^W$-homomorphism:

$$(4.3.1)\qquad \varphi^{(s)} : Der_{S_W} \to \mathcal{H}_W^{(s)},\quad \delta\mapsto\nabla_\delta\zeta^{(s-1)}.$$

LEMMA. *If $s\notin\{exponents\}+\mathbf{Z}_{\geq 0}$, then $\varphi^{(s)}$ is an isomorphism.*

PROOF. By induction on $n\in\mathbf{Z}_{\geq 0}$, we prove that $\varphi^{(s)}$ induces isomorphy:

$$S(V)^W_{\leq n}\otimes_{S(V)^{W,\tau}}\mathcal{G} \;\simeq\; D^s\otimes\mathcal{G}\oplus D^{s-1}\otimes\mathcal{G}\oplus\cdots\oplus D^{s-n}\otimes\mathcal{G}.$$

For $n=0$, one has $\nabla_\delta(D^{s-1}\otimes D) = D^s\otimes\delta$ for $\delta\in\mathcal{G}$ and hence $\varphi^{(s)}:\mathcal{G}\simeq D^s\otimes\mathcal{G}$. Applying (3.8.3) n-times for $n>0$, one obtains

$$\begin{aligned}
P_l^n(D^s\otimes\delta) \;=\;& D^s\otimes(P_l*)^n\delta\\
+\;& D^{s-1}\otimes((P_l*)^{n-1}(N-s)+\cdots)\delta\\
+\;& \cdots\cdots\cdots\cdots\cdots\cdots\cdots\cdots\cdots\cdots \;+\\
+\;& D^{s-n}\otimes(N-s+n-1)\cdots(N-s+1)(N-s)\delta
\end{aligned}$$

where $(N-s+n-1)\cdots(N-s+1)(N-s)$ is an isomorphism of $\mathcal{G}$ to itself since each factor is invertible by the assumption on $s\in\mathbf{C}$. □

NOTE. The isomorphism for $s=0$ is the identification (3.8.2). In general, *for any $s\in\mathbf{C}$, $\mathcal{H}_W^{(s)}$ is a finitely generated $S(V)^W$-module of homological dimension ≤ 1.* (*Proof.* For a large $n\in\mathbf{Z}_{\geq 0}$, $\varphi^{(s-n)}$ is an isomorphism due to Lemma above. Then one obtains the statement on $\mathcal{H}_W^{(s-n+i)}$ by induction on $i\geq 0$, applying (4.1.5) and 3.4 Corollary 1.)

One has an embedding $\varphi^{(s)}\ :\ \mathcal{H}_W^{(s+n)}/\Delta^{2n}$-torsions $\subset \frac{1}{\Delta^{2n}} Der_{S_W}$ for $s \notin \{exp.\} + \mathbf{Z}_{\geq 0}$ and $n \in \mathbf{Z}_{\geq 0}$. In the sequel, we study this "generic part" as a logarithmic connection along the discriminant D_W.

Let $\nabla^{(s)}$ be the one parameter family of connection on Der_{S_W} satisfying

$$\varphi^{(s)}\nabla^{(s)}_{\delta}\delta' = \nabla_{\delta}(\varphi^{(s)}\delta').$$

It is *a logarithmic flat torsion free connection on the logarithmic tangent bundle of* S_W, i.e. one has (3.5.5), (3.5.6) and (3.5.3) replacing ∇ by $\nabla^{(s)}$. In fact $\nabla^{(s)}$ depends on s only linearly shifting the exponent. Let us consider the dual connection: $\nabla^{(s)*}\ :\ Der_{S_W} \times \Omega^1_{S_W} \to \Omega^1_{S_W}(\log(\Delta))$ defined on $\Omega^1_{S_W}$ by the rule $\delta\langle\delta',\omega\rangle = \langle\nabla^{(s)}_{\delta}\delta',\omega\rangle + \langle\delta',\nabla^{(s)*}_{\delta}\omega\rangle$ for the natural $S(V)^W$-pairing $\langle *\rangle$.

LEMMA. *A local analytic 1-form* ω *on* S_W *is horizontal w.r.t.* $\nabla^{(s)*}$ *(i.e.* $\nabla^{(s)*}\omega = 0$*) iff there exists a local analytic function* u *satisfying i)* $\omega = du$ *(the primitivity), and ii) the system of equations*

$$P(\delta_1,\delta_2)u = 0 \ \ \textit{and}\ \ Q_s(\delta_1,\delta_2)u = 0 \ \ \textit{for}\ \ \delta_1,\delta_2 \in \mathcal{G}, \tag{4.3.2}$$

where

$$P(\delta_1,\delta_2) := \delta_1\delta_2 - (\delta_1 * \delta_2)D - \nabla_{\delta_1}\delta_2 \tag{4.3.3}$$

$$Q_s(\delta_1,\delta_2) := \delta_1 w(\delta_2) - w(\nabla_{\delta_1}\delta_2) - \delta_1 * (N-s)\delta_2. \tag{4.3.4}$$

PROOF. We still denote by $\langle\cdot,\cdot\rangle$ the extension of the natural pairing between the module Der_{S_W} and the module $\Omega^{1an}_{S_W}$ of local analytic 1-forms (which takes values in the ring of local analytic functions). We denote by $\langle\langle\cdot,\cdot\rangle\rangle$ the pairing between $\mathcal{H}_W$ and $\Omega^{1an}_{S_W}$ via $\varphi^{(s)}$ (4.3.1) so that $\langle\delta,\omega\rangle = \langle\langle D^s\otimes\delta,\omega\rangle\rangle$ for $\delta \in \mathcal{G}$.

The horizontal section $\omega \in \Omega^{1an}_{S_W}$ of $\nabla^{(s)*}$ is a closed form, since

$$\begin{aligned} d\omega &= \textstyle\sum_{ij=1}^{l} dP_i \wedge dP_j\ \partial_i\langle\partial_j,\omega\rangle \\ &= \textstyle\sum_{ij=1}^{l} dP_i \wedge dP_j\ \langle\langle\nabla_{\partial_i}\nabla_{\partial_j}D^{s-1}\otimes D,\omega\rangle\rangle \\ &\quad +\textstyle\sum_{ij=1}^{l} dP_i \wedge dP_j\ \langle\partial_j,\nabla^{(s)*}_{\partial_i}\omega\rangle \ = \ 0 \end{aligned}$$

where the second line vanishes since $\nabla_{\partial_i}\nabla_{\partial_j}D^{s-1}\otimes D$ is symmetric w.r.t. the indices i and j, and the third line vanishes due to the horizontality: $\nabla^{(s)*}_{\partial_i}\omega = 0$ of ω.

That ω is horizontal implies $\nabla^{(s)*}\omega$ is perpendicular to Der_{S_W}. So,

$$\begin{aligned} 0 &= \langle\delta',\nabla^{(s)*}_{\delta}\omega\rangle - \langle\delta*\delta',\nabla^{(s)*}_{D}\omega\rangle \\ &= \delta\langle\delta',\omega\rangle - \langle\langle\nabla_{\delta}(D^s\otimes\delta'),\omega\rangle\rangle - \langle\delta*\delta',\nabla^{(s)*}_{D}\omega\rangle \\ &= \delta\langle\delta',\omega\rangle - \langle\langle D^{s+1}\otimes(\delta*\delta') + D^s\otimes\nabla_{\delta'}\delta,\omega\rangle\rangle \\ &\qquad -\langle\langle D^s\otimes(\delta*\delta'),\nabla^{(s)*}_{D}\omega\rangle\rangle \\ &= \delta\langle\delta',\omega\rangle - D\langle\langle D^s\otimes(\delta*\delta'),\omega\rangle\rangle - \langle\nabla_{\delta'}\delta,\omega\rangle \\ &= \delta\langle\delta',\omega\rangle - D\langle\delta*\delta',\omega\rangle - \langle\nabla_{\delta'}\delta,\omega\rangle \end{aligned}$$

This yields (4.3.3). One obtains (4.3.4) as follows.

$$\begin{aligned} 0 &= \langle w(\delta),\nabla^{(s)*}_{\delta'}\omega\rangle \\ &= \delta'\langle w(\delta),\omega\rangle - \langle\langle\nabla_{\delta'}(P_lD^s\otimes\delta - D^s\otimes(P_l*\delta)),\omega\rangle\rangle \\ &= \delta'\langle w(\delta),\omega\rangle - \langle\langle\nabla_{\delta'}(D^{s-1}\otimes(N(\delta)-s\delta)),\omega\rangle\rangle \\ &= \delta'\langle w(\delta),\omega\rangle - \langle\langle D^s\otimes(\delta'*(N-s)\delta)),\omega\rangle\rangle \\ &\qquad -\langle\langle D^{s-1}\otimes(\nabla_{\delta'}(N-s)\delta),\omega\rangle\rangle. \end{aligned}$$

The last term is modified as: $D^{s-1} \otimes (\nabla_{\delta'}(N-s)\delta) = D^{s-1} \otimes ((N-s)\nabla_{\delta'}\delta) = P_l(D^s \otimes \nabla_{\delta'}\delta) - D^s \otimes (P_l * \nabla_{\delta'}\delta) = \varphi^{(s)}(w(\nabla_{\delta'}\delta))$. □

REMARK. 1. One can prove that $\zeta^{(s-1)}$ satisfies the following equations by similar calculations as in the proof of the lemma (c.f. [S6,§5]).

$$(4.3.5)\qquad \begin{array}{llcl} i) & P(\delta_1,\delta_2)\,\zeta^{(s-1)} & = & 0 \quad \text{for } \delta_1,\delta_2 \in \mathcal{G} \\ ii) & Q_s(\delta_1,\delta_2)\,\zeta^{(s-1)} & = & 0 \quad \text{for } \delta_1,\delta_2 \in \mathcal{G} \\ iii) & (E-(1/h-s))\,\zeta^{(s-1)} & = & 0. \end{array}$$

2. We remark that the equations (4.3.4) includes in particular

$$(4.3.6)\qquad Q_s(\delta, D) = \delta(E-(1/h-s)) \quad \text{for } \delta \in \mathcal{G}.$$

3. For each fixed $s \in \mathbf{C}$, one may regard (4.3.2) as a family of ordinary differential equations in the indeterminate P_l parameterized by $\mathrm{Spec}(S(V)^{W,\tau})$ where the monodromy is preserved by the deformation on the complement of the bifurcation locus in $\mathrm{Spec}(S(V)^{W,\tau})$.

4.4. The characteristic variety of $\mathcal{M}_{W,s}$.

We call (4.3.2) *the one-parameter family of system of uniformization equations of type* W. The system is holonomic since the characteristic variety is the conormal bundle of the discriminant. The homogeneity condition iii) is deleted from (4.3.5), and the space of local solutions of the system has rank $l+1$. This gives a quite important freedom in a study of the system when 0 is an exponent and there is a nilpotent monodromy (see 4.5 Example). We refer to [S-K-K] for terminologies.

Let $\mathcal{D}_{S_W}$ be the algebra of polynomial coefficients differential operators on S_W (= the enveloping algebra of Der_{S_W} over $S(V)^W$= the algebra generated by P_i and ∂_j $(1 \le i,j \le l)$ with the commutation relations $[\partial_i, P_j] = \delta^i_j$ (Kronecker's delta)). Let $\mathcal{D}_{S_W,\le d}$ be the $S(V)^W$ submodule of differential operators of degree (i.e. the total degree w.r.t. $\partial_1,\cdots,\partial_l$) less or equal than $d \in \mathbf{Z}$. Then the $gr\mathcal{D}_{S_W} := \oplus_{d=0}^{\infty}\mathcal{D}_{S_W,\le d}/\mathcal{D}_{S_W,\le d-1}$ is a commutative graded algebra isomorphic to $S(V)^W[\xi_1,\cdots,\xi_l]$ (where ξ_i stands for the image of ∂_i), and is the coordinate ring for the cotangent space $T^*_{S_W}$ of S_W.

Let $\mathcal{I}_{W,s} := \Sigma_{\delta_1,\delta_2\in\mathcal{G}}P(\delta_1,\delta_2) + \Sigma_{\delta_1,\delta_2\in\mathcal{G}}Q_s(\delta_1,\delta_2)$ be the left ideal of $\mathcal{D}_{S_W}$ generated by $P(\delta_1,\delta_2)$ and $Q_s(\delta_1,\delta_2)$ for $\delta_1,\delta_2 \in \mathcal{G}$. Put

$$(4.4.1)\qquad \mathcal{M}_{W,s} := \mathcal{D}_{S_W}/\mathcal{I}_{W,s}$$

and call it also the one parameter family of the system of uniformization equations of type W, confusing with the system (4.3.2).

The *characteristic variety* of $\mathcal{M}_{W,s}$ is the subvariety of the cotangent space $T^*_{S_W}$ defined by the ideal $\sigma(\mathcal{I}_{W,s})$ in $S(V)^W[\xi_1,\cdots,\xi_l]$ generated by the principal symbols of all elements of $\mathcal{I}_{W,s}$, where the principal symbol $\sigma(X)$ for $X \in \mathcal{D}_{S_W}$ with $\deg(X) = d$ is, as usual, defined as the degree d-part of X, associating to ∂_i the commutative variable ξ_i. Recall that a left $\mathcal{D}_{S_W}$-module is called *holonomic* if its characteristic variety is a Lagrangian subvariety, and is *simple holonomic* if all, except for the zero section of the cotangent space, of irreducible components of its characteristic variety are simple.

THEOREM. *For each fixed $s \in \mathbf{C}$, the $\mathcal{M}_{W,s}$ is a simple regular holonomic system, whose characteristic variety is the union of the conormal bundle $N^*(D_W)$ of the discriminant D_W and the zero section $T^*_{S_W}(0)$ of the cotangent bundle $T^*_{S_W}$ of the multiplicity $l+1$.*

PROOF. The conormal bundle $N^*(D_W)$ of the D_W is defined as the Zariski closure in $T^*_{S_W}$ of the conormal bundle $N^*(D_W \setminus Sing(D_W))$ of the smooth part $D_W \setminus Sing(D_W)$, where the conormal bundle of a smooth subvariety $X \subset S_W$ is the subvariety of $T^*_{S_W}|X$ consisting of the covectors perpendicular to the subbundle T_X of $T_{S_W}|X$ at each point of X. Conormal bundle is always a middle dimensional subvariety of the cotangent bundle $T^*_{S_W}$, and is a Lagrangian subvariety.

For the proof of the theorem, we prepare two lemmas, whose proofs use the flat structure 3.9 on S_W. The first lemma states about the involutivity of the generator systems of the ideal $\mathcal{I}_{W,s}$.

LEMMA. *Let P be an element of $\mathcal{I}_{W,s}$ of degree $m \in \mathbf{Z}_{\geq 0}$. Then $m \geq 2$ and there exists elements R_{ij} $(ij = 1, \cdots, l-1)$ and S_i $(i = 1, \cdots, l)$ of $\mathcal{D}_{S_W}$ of degrees less or equal than $m - 2$ such that*

$$P = \textstyle\sum_{1\leq i\leq j\leq l-1} R_{ij} P(\partial_i, \partial_j) + \sum_{1\leq i\leq l} S_i Q_s(\partial_i). \tag{4.4.2}$$

PROOF. The $P(\delta_1, \delta_2)$ is a $S(V)^{W,\tau}$-symmetric bilinear form in $\delta_1, \delta_2 \in \mathcal{G}$ and $P(D, \delta) = 0$ due to 3.9 Theorem I.1, I.2 and I.3. Put

$$Q_s(\delta) := Q_s(D, \delta) = Dw(\delta) - (N - s)\delta. \tag{4.3.4)*$$

Then, by a use of 2) in the proof of 3.9 Theorem, the bilinear operator $Q_s(\delta_1, \delta_2)$ is reduced to a single variable operator $Q_s(\delta * \delta_2)$:

$$Q_s(\delta_1, \delta_2) = Q_s(\delta_1 * \delta_2) + P_l P(\delta_1, \delta_2) - P(\delta_1, P_l * \delta_2). \tag{4.3.4) **$$

Due to the symmetric $S(V)^{W,\tau}$-bilinearities of $P(\delta_1, \delta_2)$ and the reduction (4.3.4)**, the ideal $\mathcal{I}_{W,s}$ is generated by $P(\partial_i, \partial_j)$ $(1 \leq i \leq j \leq l-1)$ and $Q_s(\partial_i)$ $(1 \leq i \leq l)$. We show that the generator system is involutive: the symbol ideal $\sigma(\mathcal{I}_{W,s})$ is generated by the symbols of them.

For a $P \in \mathcal{I}_{W,s}$, let $P = \sum_{1\leq i\leq j\leq l-1} R_{ij} P(\partial_i, \partial_j) + \sum_{1\leq i\leq l} S_i Q_s(\partial_i)$ be an expression. Consider the maximal degree of the coefficients of the expression: $d := \max\{\deg(R_{ij})(1 \leq i \leq j \leq l-1), \deg(S_i)(1 \leq i \leq l)\}$. Of course $\deg(P) \leq d + 2$. We want to show that if $d + 2 > \deg(P)$, then there exists a new expression for P such that the maximal degree d' of the new coefficients is strictly less than d. For the purpose, we consider two relations among the generators of the ideal $\mathcal{I}_{W,s}$:

$$\begin{aligned}
*) \qquad & \delta P(\delta', \delta'') - \delta' P(\delta, \delta'') + D\left(P(\delta, \delta' * \delta'') - P(\delta', \delta * \delta'')\right) \\
& \quad + P(\delta, \nabla_{\delta'} \delta'') - P(\delta', \nabla_{\delta} \delta'') = 0, \\
**) \qquad & \delta Q_s(\delta') - D Q_s(\delta * \delta') - Q_s(\nabla_{\delta} \delta') \\
& \quad - E P(\delta, \delta') - \delta P(P_l * D, \delta') + D P(P_l * D, \delta * \delta') \\
& \quad - P([P_l * D, \delta], \delta') - P(\delta, \nabla_{P_l * D} \delta') \\
& \quad + P(\delta, (N + s - 1)\delta') + P(P_l * D, \nabla_{\delta} \delta') = 0
\end{aligned}$$

for $\delta, \delta', \delta'' \in \mathcal{G}$. The proof of them is easily reduced to 3.9 Theorem.

We return to an expression (4.4.2) of an element P of the ideal $\mathcal{I}_{W,s}$. If $d := \max\{\deg(R_{ij}), \deg(S_i)\} > \deg(P) - 2$, then one has

$$\textstyle\sum_{ij=1}^{l-1} \sigma_d(R_{ij}) \sigma(P(\partial_i, \partial_j)) + \sum_{i=1}^{l} \sigma_d(S_i) \sigma(Q_s(\partial_i)) = 0,$$

where $\sigma_d(X)$ means the homogeneous of degree d part of $X \in \mathcal{D}_{S_W, \leq d}$. Subtracting a suitable linear combination of the above elements *) and **), we obtain new coefficients R'_{ij} and S'_i such that $\sigma_d(R'_{ij}) = r_{ij} \xi_D^d$ and $\sigma_d(S'_i) = s_i \xi_D^d$ for $r_{ij}, s_i \in S(V)^W$. The new symbols satisfy the relation $\sum_{ij=1}^{l-1} r_{ij} \sigma(P(\partial_i, \partial_j)) + \sum_{i=1}^{l} s_i \sigma(Q_s(\partial_i)) = 0$.

Because of the algebraic independence of $\xi_1, \cdots, \xi_l$, this implies $r_{ij} = s_i = 0$. So, $\sigma_d(R'_{ij}) = \sigma_d(S'_i) = 0$, i.e. the degrees of R'_{ij} and S'_i are less than d. □

COROLLARY. *The symbol ideal $\sigma(\mathcal{I}_{W,s})$ is generated by principal symbols:*

$$\begin{array}{lrcll} (4.4.3) & \sigma(P(\partial_i,\partial_j)) & = & \xi_i\xi_j - \xi_D(\xi_i * \xi_j) & (1 \le i,j \le l-1), \\ & \sigma(Q_s(\partial_i)) & = & \xi_D(P_l\xi_i - P_l * \xi_i) & (1 \le i \le l), \end{array}$$

where the $*$-product on ξ_i is defined by the identification $\xi_i \leftrightarrow \partial_i \in \mathcal{G}$, and we put $\xi_D := \sigma(D)$. This induces a $S(V)^W$-algebra isomorphism:

$$(4.4.4) \qquad \mathcal{G} \simeq \mathcal{O}_{Ch(\mathcal{M}_{W,s}) \cap \{\xi_D = 1\}}$$

The next lemma determines the symbol ideal geometrically.

LEMMA. *Let $(\xi_1, \cdots, \xi_l)$ and $\mathcal{I}(N^*(D_W))$ be ideals in $S(V)^W[\xi_1, \cdots, \xi_l]$ defining the zero section of the cotangent bundle $T^*_{S_W}$ and the conormal bundle of the discriminant, respectively. Then one has:*

$$(4.4.5) \qquad \mathcal{I}(N^*(D_W)) \cap (\xi_1, \cdots, \xi_l)^2 \; = \; \sigma(\mathcal{I}_{W,s}) \; .$$

PROOF. We first show the inclusion $\supset$. Clearly, $\sigma(P(\delta_1,\delta_2))$ and $\sigma(Q_s(\delta))$ belong to $(\xi_1, \cdots, \xi_l)^2$ (see (4.4.3)). To show that the symbols vanish on the conormal bundle $N^*(D_W)$, it is sufficient to show

$$\delta_1(\Delta)\,\delta_2(\Delta) - (\delta_1 * \delta_2)(\Delta)\,D(\Delta), \;\; P_l\delta(\Delta) - P_l * \delta(\Delta) \;\in\; (\Delta)$$

for $\delta_1, \delta_2, \delta \in \mathcal{G}$, since the conormal vectors at the smooth points of the discriminant are given by constant multiple of the differential $d\Delta = \sum_i \partial_i \Delta dP_i$. The second formula is obvious since the difference $P_l\delta - P_l * \delta = w(\delta)$ is a logarithmic vector field (c.f. (3.4.4) and (3.1.1)).

The first formula: it is enough to show the formula for $\delta_1 = \delta \in \mathcal{G}$ and $\delta_2 = \delta_U := \nabla_D I(dU)$ for $U \in S(V)^W_{\le 1}$. Recalling (3.6.5), one has

$$2\delta_U(\Delta) = 2\langle \nabla_D I(dU), d\Delta\rangle = DI(dU, d\Delta) - I(D, [I(dU), I(d\Delta)]).$$

Recall (3.7.1) whose first term is the $*$-product and second term belongs to $Der_{S_W}(-\log(\Delta)$. Therefore, using the formula (3.6.5) again, one has

$$\begin{aligned} 2(\delta * \delta_U)(\Delta) \; mod \; (\Delta) &= 2\langle \nabla_\delta \nabla_D^{-1}\delta_U, d\Delta\rangle \\ &= 2\langle \nabla_\delta I(dU), d\Delta\rangle \\ &= \delta I(dU, d\Delta) - I(\delta, [I(dU), I(d\Delta)]). \end{aligned}$$

So,

$$\begin{aligned} & 2\delta(\Delta)\,\delta_U(\Delta) - 2(\delta * \delta_U)(\Delta)\,D(\Delta) \quad mod \;(\Delta) \\ = \;& \delta(\Delta)\,(DI(dU, d\Delta) - I(D, [I(dU), I(d\Delta)])) \\ + \;& D(\Delta)\,(\delta I(dU, d\Delta) - I(\delta, [I(dU), I(d\Delta)])) \\ = \;& XI(dU, d\Delta) - I(X, [I(dU), I(d\Delta)])) \end{aligned}$$

where $X := \delta(\Delta)D - D(\Delta)\delta$. By definition, $X\Delta = 0$ and hence X is a logarithmic vector field. Then, each of the last two terms is zero modulo the ideal (Δ) as follows. The first term: using I (3.1.3), one has $I(dU, d\Delta) \in (\Delta)$ and then $X(\Delta) = (\Delta)$. The second term: put $X = I(\omega)$ for some $\omega = \sum_{i=1}^l F_i dP_i \in \Omega^1_{S_W}$ (3.1.3). Then

$$\begin{aligned} & I^*(X, [I(dU), I(d\Delta)]) = \langle [I(dU), I(d\Delta)], \omega\rangle \\ = \;& \textstyle\sum_{i=1}^l F_i \langle [I(dU), I(d\Delta)], dP_i\rangle \\ = \;& \textstyle\sum_{i=1}^l F_i \,(I(dU)(I(d\Delta, dP_i)) - I(d\Delta, I(dU, dP_i)))\,. \end{aligned}$$

Using (3.1.3), one checks that each term belongs to the ideal (Δ).

Let us prove now the opposite inclusion: any homogeneous element X of degree $d \geq 2$ w.r.t.(ξ) in the LHS of (4.4.3) belongs to the RHS. First, by a successive application of the first relations of (4.4.3), we may reduce X to such element that any monomial in $\xi_1, \cdots, \xi_l$ contains (at least) $d-1$th power of ξ_D. So, put $X = \xi_D^{d-1} Y$, where Y is a linear form in (ξ). If the coefficients of the linear form Y contains the variable P_l, then by successive applications of the second relations of (4.4.3), we may reduce that the coefficients of Y belongs to $S(V)^{W,\tau}$. So, $X = \xi_D^{d-1} Y$ for $Y = \sigma(\delta)$ with $\delta \in \mathcal{G}$. The fact $X \in \mathcal{I}(N^*_{D_W})$ means $(D(\Delta))^{d-1}\delta(\Delta) \in (\Delta)$. Since $D(\Delta)$ is coprime to the discriminant Δ, this implies $\delta(\Delta) \in (\Delta)$. This is possible only when $\delta = 0$ due to the splitting (3.4.1) (c.f. (3.4.3)). These end the proof of Lemma. □

As a consequence, one has an exact sequence:

$$0 \to \mathcal{O}(T^*_{S_W}) \oplus \mathcal{O}_{S_W} \to gr(\mathcal{M}_{W,s}) \to \mathcal{O}_{N^*(D_W)} \to 0.$$

These complete a proof of Theorem. □

REMARK. The conormal bundle $N^*(D_W) \backslash \{\text{zero section of } T^*_{S_W}\}$ is smooth.

4.5. Duality between the solutions $Sol(\mathcal{M}_{W,s})$ and $Sol(\mathcal{M}_{W,-s})$.

We consider the solution sheaf $Sol(\mathcal{M}_{W,s}) := \mathcal{E}xt^0_{\mathcal{D}_{S_W}}(\mathcal{M}_{W,s}, \mathcal{O}_{S_W})$. Due to Theorem in the previous subsection, its restriction to $S_W \setminus D_W$ is a local system of rank $l+1$ containing the constant function sheaf. The exterior differentiation induces an exact sequence at each $x \in S_W \setminus D_W$

$$0 \to \mathbf{C}_{S_W,x} \to Sol(\mathcal{M}_{W,s})_x \xrightarrow{d} T^*_{S_W,x} \to 0. \tag{4.5.1}$$

Assertion 1. *One has the following splitting of the local system:*

$$Sol(\mathcal{M}_{W,s}) = \begin{cases} \mathbf{C}_{S_W} \oplus Sol(\widetilde{\mathcal{M}}_{W,s}) & \text{if} \quad 1/h - s \neq 0 \\ \mathbf{C}_{S_W}\lambda \oplus Sol(\widetilde{\mathcal{M}}_{W,S}) & \text{if} \quad 1/h - s = 0, \end{cases} \tag{4.5.2}$$

where $\widetilde{\mathcal{M}}_{W,s} := \mathcal{M}_{W,s}/\mathcal{D}_{S_W}(E-(1/h-s))$ is the equation for $\zeta^{(s-1)}$ (recall (4.3.5)) and λ is a solution of the equation $\mathcal{M}_{W,s}$ with $E\lambda = 1$.

PROOF. Let u be a (local) solution of $\mathcal{M}_{W,s}$. In view of (4.3.6) one has $(E - (1/h - s))u = c$ for $c \in \mathbf{C}$. If $1/h - s \neq 0$, then put $u' := u + c/(1/h - s)$ and one has $(E - (1/h - s))u' = 0$. That is: $u' \in Sol(\widetilde{\mathcal{M}}_{W,s})$. If $1/h - s = 0$, then one always has a solution with $c \neq 0$. □

REMARK. The same proof of (4.4.5) shows the formula:

$$\mathcal{I}(N^*(D_W)) = (\sigma(P(\partial_i, \partial_j)), \sigma(w(\partial_i)) \ (1 \leq i, j \leq l)) = \sigma(\mathcal{I}(\tilde{M}_{W,s})).$$

EXAMPLE. We illustrate the difference of $\mathcal{M}_{W,s}$ and $\widetilde{\mathcal{M}}_{W,s}$ by the example of type A_1. Since $l = 1$, $S(V)^W$ is generated by a single element $P_1 := \frac{1}{2}X^2$ with $I(dP_1, dP_1) = 2P_1$. Put $z := P_1$ and $D = \frac{d}{dz}$. Then,

$$\mathcal{M}_{A_1,s} : \tfrac{d}{dz}(z\tfrac{d}{dz} - (\tfrac{1}{2} - s))u = 0 \ \text{ and } \ \widetilde{\mathcal{M}}_{A_1,s} : (z\tfrac{d}{dz} - (\tfrac{1}{2} - s))u = 0.$$

Therefore, the solutions are given by

$$Sol(\mathcal{M}_{A_1,s}) : \quad u = \begin{cases} c\, z^{\frac{1}{2}-s} + d & (c, d \in \mathbf{C}) & \text{for } s \neq \frac{1}{2}, \\ c\, \log(z) + d & (c, d \in \mathbf{C}) & \text{for } s = \frac{1}{2}, \end{cases}$$

$$Sol(\widetilde{\mathcal{M}}_{A_1,s}) : \quad u = \begin{cases} c\, z^{\frac{1}{2}-s} & (c \in \mathbf{C}) & \text{for } s \neq \frac{1}{2}, \\ c\, 2\pi\sqrt{-1} & (c \in \mathbf{C}) & \text{for } s = \frac{1}{2}. \end{cases}$$

REMARK. We shall see (5.3 Example) that the non-constant solution λ of $\mathcal{M}_{W,s}$ in (4.5.2) for type A_1 is given by the indefinite integral of the primitive form $\frac{dz}{z}$ of type A_1 and the constant solution $2\pi\sqrt{-1}$ of $\widetilde{\mathcal{M}}_{W,s}$ is its period (a similar description by a use of $\mathfrak{p}$-function holds for type A_3, and presumably for type D_4). We introduced the system $\mathcal{M}_{W,s}$ in order to include such log-type function λ (=an indefinite integral of a primitive form) in the construction of the period map. The relationship between λ and its period goes much deeper. E.g. the transcendentality of $2\pi\sqrt{-1}$ is shown by a use exponential function = the inverse of log function for type A_1. How about the $\mathfrak{p}$-functions for types A_3 and D_4?

Assertion 2. *One has a left $\mathcal{D}_{S_W}$-homomorphism:*

$$\mathcal{M}_{W,s+1} \to \mathcal{M}_{W,s} \tag{4.5.3}$$

by the right multiplication of D. It induces a homomorphism:

$$D : Sol(\mathcal{M}_{W,s}) \to Sol(\mathcal{M}_{W,s+1}). \tag{4.5.4}$$

The kernel of this homomorphism is a vector subspace spanned by the flat coordinates of degree s and by the constant function 1_{S_W}.

PROOF. In order to show the well definedness of (4.5.3) and (4.5.4), it is sufficient to prove the relations:

$$\begin{aligned} P(\delta_1,\delta_2)D &= DP(\delta_1,\delta_2), \\ Q_{s+1}(\delta_1,\delta_2)D &= DQ_s(\delta_1,\delta_2). \end{aligned}$$

The verifications are left to the reader. The kernel of (4.5.4) is given by solutions of the system of the equations (4.3.3), (4.3.4)* and D. It is easy to see that the system is reduced to the system of equations:

$$(\delta_1\delta_2 - \nabla_{\delta_1}\delta_2)u = 0, \ \ ((N-s)\delta)u = 0 \text{ and } Du = 0.$$

This is the equation for a flat coordinate of degree s (if $s \neq 0$). □

Assertion 3. *The polarization $I \in Der_{S_W} \otimes Der_{S_W}$* (3.4.8) *induces a pairing between the two solution systems for s and $-s$:*

$$\begin{aligned} I : Sol(\mathcal{M}_{W,s})/\mathbf{C}_{S_W} \times Sol(\mathcal{M}_{W,-s})/\mathbf{C}_{S_W} &\to \mathbf{C}_{S_W} \\ du \times dv \quad \mapsto \quad I(du,dv) &= \textstyle\sum_{i=1}^{l} \delta_i u \cdot w(\delta_i^*)v \end{aligned} \tag{4.5.5}$$

which is nondegenerate at every point of $S_W \setminus D_W$. One has:

$$I(dDu, dv) = -I(du, dDv) \tag{4.5.6}$$

for $u \in Sol(\mathcal{M}_{W,s-1})$ and $v \in Sol(\mathcal{M}_{W,-s})$.

Proof. Combining the homomorphism $\varphi^{(s)}$ (4.3.1) , one obtains

$$I_s = \varphi^{(s)} \otimes \varphi^{(-s)}(I) = \varphi^{(-s)} \otimes \varphi^{(s)}(I).$$

Then the horizontality of I_s (4.2.3) implies further

$$\nabla^{(s)} \otimes \nabla^{(-s)}(I) = \nabla^{(-s)} \otimes \nabla^{(s)}(I) = 0.$$

This implies the value $I(du,dv)$ is a constant and (4.5.5) is defined. Recall that $\det(I(dP_i,dP_j)) = \Delta$ (3.1.5) so that I is nondegenerate on $S_W \setminus D_W$. So, the exact sequence (4.5.1) implies the nondegeneracy of the pairing (4.5.5). We show (4.5.6) by recalling (4.3.4)* and (3.9.6):

$$\begin{aligned} & I(du, dDv) = \sum_{i=1}^{l} \delta_i u \cdot w(\delta_i^*) Dv = \sum_{i=1}^{l} \delta_i u \cdot ((N+s-1)\delta_i^*)v \\ = \ & \sum_{i=1}^{l} ((N+s-1)^*\delta_i)u \cdot \delta_i^* v = -\sum_{i=1}^{l}((N-s)\delta_i)u \cdot \delta_i^* v \\ = \ & -\sum_{i=1}^{l} w(\delta_i)Du \cdot \delta_i^* v = -I(dDu, dv). \qquad \square \end{aligned}$$

REMARK. The nondeneracy does not hold for the pairing on $Sol(\widetilde{\mathcal{M}}_{W,s}) \times Sol(\widetilde{\mathcal{M}}_{W,-s})$ when $s \in \{exponents\} - \mathbf{Z}_{\geq 0}$ due to the splitting (4.5.2).

We are interested in the case when the application

$$D^n : Sol(\mathcal{M}_{W,s}) \to Sol(\mathcal{M}_{W,s+n})$$

for $n \in \mathbf{Z}_{\geq 0}$ and $s \in \mathbf{C}$ brings a solution space to its dual space, that is the case $-s = s + n$, since then we obtain a bilinear form:

$$\text{(4.5.7)} \qquad \begin{array}{ccc} I : Sol(\mathcal{M}_{W,s})/\mathbf{C}_{S_W} \times Sol(\mathcal{M}_{W,s})/\mathbf{C}_{S_W} & \to & \mathbf{C}_{S_W} \\ du \times dv & \mapsto & I(du, dD^n v). \end{array}$$

So, $s = -\frac{n}{2} \in \frac{1}{2}\mathbf{Z}_{\leq 0}$. According to the parity of n, there are two cases:

Case i) $s = 0, -1, -2, \cdots$. In this case (4.5.7) is a nondegenerate symmetric bilinear form as follows: applying the operator D, one has a sequence of isomorphism (for $\{\text{exponents}\} \cap \mathbf{Z} = \emptyset$) of the local systems:

$$\begin{aligned} \cdots \overset{\sim}{\to} Sol(\mathcal{M}_{W,-2})/\mathbf{C}_{S_W} \overset{\sim}{\to} Sol(\mathcal{M}_{W,-1})/\mathbf{C}_{S_W} \overset{\sim}{\to} Sol(\mathcal{M}_{W,0})/\mathbf{C}_{S_W} \\ \overset{\sim}{\to} Sol(\mathcal{M}_{W,1})/\mathbf{C}_{S_W} \overset{\sim}{\to} Sol(\mathcal{M}_{W,2})/\mathbf{C}_{S_W} \overset{\sim}{\to} \cdots \end{aligned}$$

which is equivariant with the inner product (4.5.7) (up to sign). One has the isomorphism: $Sol(\mathcal{M}_{W,0})/\mathbf{C}_{S_W} \simeq Sol(\widetilde{\mathcal{M}}_{W,0}) \simeq V_{\mathbf{C}}^*$ where (4.5.7) is identified with the Killing form I on V. For any covector $X \in V_{\mathbf{C}}$, let $X^{(n)} := D^n X \in Sol(\widetilde{\mathcal{M}}_{W,n})/\mathbf{C}_{S_W}$ for $n \in \mathbf{Z}$ so that $I(X^{(n)}, Y^{(n)}) = I_n(X^{(n)}, Y^{(-n)}) = (-1)^n I(X, Y)$ for $X, Y \in V_{\mathbf{C}}$. In fact, these spaces are identified with the middle homology group

Case ii) $s = -1/2, -3/2, -5/2, \cdots$. In this case (4.5.7) is a skew-symmetric bilinear form which may degenerate as follows: applying the operator D, one has a sequence of homomorphism of the local systems:

$$\begin{aligned} \cdots \overset{\sim}{\to} Sol(\mathcal{M}_{W,-\frac{5}{2}})/\mathbf{C}_{S_W} \overset{\sim}{\to} Sol(\mathcal{M}_{W,-\frac{3}{2}})/\mathbf{C}_{S_W} \overset{\sim}{\to} Sol(\mathcal{M}_{W,-\frac{1}{2}})/\mathbf{C}_{S_W} \\ \to Sol(\mathcal{M}_{W,\frac{1}{2}})/\mathbf{C}_{S_W} \overset{\sim}{\to} Sol(\mathcal{M}_{W,\frac{3}{2}})/\mathbf{C}_{S_W} \overset{\sim}{\to} \cdots \end{aligned}$$

which is equivariant with the inner product (4.5.7) (up to sign). The homomorphisms are isomorphic except at the middle step:

$$Sol(\mathcal{M}_{W,-\frac{1}{2}})/\mathbf{C}_{S_W} \to Sol(\mathcal{M}_{W,\frac{1}{2}})/\mathbf{C}_{S_W}.$$

The rank of the kernel of this step is equal to the multiplicity of exponent $\frac{h}{2}$ due to Assertion 2. So, it is non-degenerate iff $\frac{h}{2} \notin \{m_1, \cdots, m_l\}$. Recall 2.8 *Remark* i) a criterion for this condition in terms of the Coxeter graph. We shall investigate these cases more in details in §6.

Fix a base point, say $*$, of the universal covering space $(S_{W,\mathbf{C}} \setminus D_{W,\mathbf{C}})^{\sim}$ and branches at $*$ of the functions in $Sol(\mathcal{M}_{W,s})$. Then the analytic continuation of the evaluation homomorphism induces a map

$$\text{(4.5.8)} \qquad (S_{W,\mathbf{C}} \setminus D_{W,\mathbf{C}})^{\sim} \ \to \ \mathrm{Hom}_{\mathbf{C}}(Sol(\mathcal{M}_{W,s})_*, \mathbf{C})$$

which we shall call the period map of type W and weight s. By choosing a basis $u_1, \cdots, u_l$ and 1 of $Sol(\mathcal{M}_{W,s})$, the period map is given by the l-tuple $(u_1, \cdots, u_l)$ of functions and the jacobian is calculated by

$$\frac{\partial(u_1, \cdots, u_l)}{\partial(P_1, \cdots, P_l)} = c\Delta_W^{-s-\frac{1}{2}} \quad \text{for some } c \in \mathbf{C} \setminus \{0\}. \tag{4.5.9}$$

5. Period integrals for primitive forms

The connection $\nabla^{(s)}$ for $s = n/2$ $(n \in \mathbf{Z}_{\geq 0})$ is realized as the Gauss-Manin connection for a universal unfolding of a simple singularity of dimension n if W is a Weyl group of a Lie algebra. Then the solutions of the uniformization equation $\mathcal{M}_W^{(s)}$ are realized by *the period integrals of the primitive form* of the unfolding. Thus the period map is defined.

5.1. Universal unfoldings of simple singularities.

We recall briefly the description of universal unfoldings of simple singularities due to Brieskorn [Br1] (see [Sl] for non-simply-laced cases).

A simply-laced simple singularity of dimension n $(n \in \mathbf{Z}_{\geq 1})$ is the singular point of a hypersurface of dimension n defined by the equation (5.1.1) in coordinates $(x, y, z_3, \cdots, z_{n+1})$. The type, in LHS of Table, is given by the dual graph Γ of the exceptional divisors of the minimal resolution of the singularity for $n = 2$ (c.f. 2.4). We note that the equation is a weighted homogeneous polynomial of weight $(w_x, w_y, \frac{1}{2}, \cdots, \frac{1}{2} : 1)$ for suitable weights $w_x, w_y \in \mathbf{Q}_{>0}$ for x and y such that $w_x + w_y = \frac{1}{2} + \frac{1}{h}$.

$$(5.1.1) \qquad \begin{array}{llcccl} A_l: & x^{l+1} & - & y^2 & + & z_3^2 + \cdots + z_{n+1}^2 \\ D_l: & x^{l-1} & - & xy^2 & + & z_3^2 + \cdots + z_{n+1}^2 \\ E_6: & x^4 & + & y^3 & + & z_3^2 + \cdots + z_{n+1}^2 \\ E_7: & x^3y & + & y^3 & + & z_3^2 + \cdots + z_{n+1}^2 \\ E_8: & x^5 & + & z^3 & + & z_3^2 + \cdots + z_{n+1}^2 \end{array}$$

Let G be a group acting linearly on the two variables x, y such that the G-action leaves the equation for a simple singularity (5.1.1) invariant and the induced G-action on the dual graph Γ is faithful. The simple singularity with the G-action is called a simple singularity of non simply-laced type Γ/G, where Γ/G is given in the next table ([Ar2][Sl]).

(5.1.2)

Γ/G	Γ	G	The action of a generator of G
B_l	A_{2l-1}	$\mathbf{Z}/2\mathbf{Z}$	$:\ (x, y) \mapsto (-x, y)$
C_l	D_{l+1}	$\mathbf{Z}/2Z$	$:\ (x, y) \mapsto (x, -y)$
F_4	E_6	$\mathbf{Z}/2Z$	$:\ (x, y) \mapsto (-x, y)$
G_2	D_4	$\mathbf{Z}/3Z$	$:\ (x, y) \mapsto \frac{1}{2}(x + y, -3x + y)$

Let $\mathfrak{g}$, $\mathfrak{h}$ and W be a simple Lie algebra over $\mathbf{C}$, a Cartan subalgebra of $\mathfrak{g}$ and its Weyl group, respectively. The ring $S(\mathfrak{g}^*)^{\mathrm{ad}(\mathfrak{g})}$ of invariant polynomials on $\mathfrak{g}$ by the adjoint group action is isomorphic to $S(\mathfrak{h}^*)^W$ (Chevalley), and one obtains the flat adjoint quotient morphism: $\mathfrak{g} \to \mathfrak{h}//W \simeq S_W$, whose fiber over 0 is the nilpotent variety $N(\mathfrak{g})$ of $\mathfrak{g}$. For a subregular element $x \in N(\mathfrak{g})$, consider an affine subspace $\mathfrak{X}$ of $\mathfrak{g}$ which is transversal at x to the adjoint group orbit of x in $\mathfrak{g}$ (e.g. put $\mathfrak{X} := x + \mathfrak{z}_{\mathfrak{g}}(y)$ where $y \in N$ such that $\langle x, y, [x, y] \rangle$ form a sl_2 triplet).

THEOREM. *(Brieskorn [Br1][Sl]) Let $\varphi_2 : \mathfrak{X} \to S_W$ be the restriction of the adjoint quotient map to the slice $\mathfrak{X}$ as defined above. Then the fiber $X_0 := \varphi_2^{-1}(0)$ is the corresponding simple singularity of dimension $n = 2$ and φ_2 is its semi-universal unfolding. If $\mathfrak{g}$ is of a non-simply-laced type, then there is a G-action on $\mathfrak{X}$ such that φ_2 is G-invariant and the G-action on X_0 defines the singularity of non-simply-laced type.*

We collect some facts on the universal unfolding $\varphi_2 : \mathfrak{X} \to S_W$, which we shall use in 5.2 as a building (or supporting) data for a primitive form. Their proofs are either referred to or easy and omitted.

i) *The φ_2 is a weighted homogeneous polynomial map of the weights $(w_x, w_y, 1/2, d_1/h, \cdots, d_{l-1}/h : d_1/h, \cdots, d_l/h)$ (see 2.5 for d_i, [Br1]).*

ii) *The composition $\pi_W \circ \varphi_2$ with π_W (3.2.7) is a submersion. Hence the fiber product $Z := \mathfrak{X} \times_{\mathrm{Spec}(S(\mathfrak{h}^*)^{W,\tau})} S_W \simeq \mathbf{C}^3 \times S_W$ is a smooth affine variety with projections $p : Z \to S_W$ and $\hat{\pi}_W : Z \to \mathfrak{X}$. We realize $\mathfrak{X}$ as a hypersurface in Z: define $F_2(x, y, z_3, P_1, \cdots, P_l) := P_l \circ p - P_l \circ \varphi_2 \circ \hat{\pi}_W$.*

Then, a) $\hat{\pi}_W$ identifies the hypersurface $\{F_2 = 0\}$ in Z with $\mathfrak{X}$.

b) φ_2 is given by the restriction of the projection $p : Z \to S_W$ to $\mathfrak{X}$.

iii) *Let $\mathcal{O}_C := \mathcal{O}_Z/(F_2, \partial_x F_2, \partial_y F_2, \partial_{z_3} F_2)$ be the ring of polynomial functions on the critical set C of φ_2. Then, $\delta \mapsto \delta F_2 \mid_C$ induces a $S(V)^{W,\tau}$-isomorphism: $\mathcal{G} \simeq \mathcal{O}_C$ ($\simeq \mathcal{O}_C^G$ for non-simply-laced case).*

iv) *Put $r := w_x + w_y + (n-1)\frac{1}{2} = \frac{1}{h} + \frac{n}{2}$ and $s := \max\{d_i/h - d_j/h \mid i, j = 1, \cdots, l\} = \deg(P_l) - \deg(P_1)$. Then one has [Sl]: $2r + s = n + 1$ (duality) and $s < 1$ (a characterization of a simple singularity).*

5.2. The primitive form ζ_{F_n}.

We introduce the primitive form attached to the universal unfolding of a simple singularity. By a use of the covariant differentiation of the primitive form, the module of relative abelian differentials of the family is identified with the module of vector fields on S_W. This identifies the flat structure studied in §3 with that defined by the primitive form[S5-7].

Let $n \in \mathbf{Z}_{\geq 1}$. Consider a weighted homogeneous polynomial

(5.2.1) $$F_n(x, y, z_3, \cdots, z_{n+1}, P_1, \cdots, P_l)$$

of total degree 1 defined on the space $Z_n := \mathbf{C}^{n+1} \times S_W$ with the weight $(w_x, w_y, 1/2, \cdots, 1/2, d_1/h, \cdots, d_l/h)$. We call F_n a universal unfolding of a simple singularity of dimension n, if it satisfies the i), ii) and iii):

i) The restriction $F_n \mid_{P_1 = \cdots = P_l = 0}$ is a polynomial given in (5.1.1).

ii) Put $\mathcal{O}_{C_n} := \mathcal{O}_{Z_n}/(F_n, \partial_x F_n, \partial_y F_n, \partial_{z_1} F_n, \cdots, \partial_{z_{n+1}} F_n)$. Then, one has a $S(V)^{W,\tau}$-isomorphism: $\mathcal{G} \simeq \mathcal{O}_{C_n}$ by the correspondence $\delta \mapsto \delta F_n \mid_{C_n}$.

iii) The constant factor of F_n is normalized to: $\frac{\partial}{\partial P_l} F_n = 1$.

Under the data i), ii) and iii), the universal unfolding morphism

(5.2.2) $$\varphi_n : \mathfrak{X}_n \to S_W$$

is defined by the restriction $\varphi_n := p_n | \mathfrak{X}_n$ of the projection $p_n : Z_n \to S_W$ to the hypersurface $\mathfrak{X}_n := \{F_n = 0\} \subset Z_n$. The φ_n defines a flat family of n-dim. affine variety parameterized by S_W, where the 0-fiber $X_0 :=$ $\varphi_n^{-1}(0)$ is the simple singularity of dimension n. It is well-known (Milnor) that over the complement of the discriminant $S_W \setminus D_W$, φ_n defines a locally trivial fibration whose fiber, called a Milnor fiber, is homotopic to a bouquet of l-copies of n-dimensional spheres. In case of non- simply-laced type, the G-action (5.1.2) extends to Z_n leaving F_n invariant such that the map in ii) induces $\mathcal{G} \simeq \mathcal{O}^G_{C_n}$. So, φ_n is G-invariant.

Introduce the *module of Abelian differentials of* deg n *relative to* φ_n:

(5.2.3) $$\mathcal{H}^{(0)}_{F_n} := \Omega^{n+l+1}_{Z_n}/dF_n \wedge dP_1 \wedge \cdots \wedge dP_l \wedge d\Omega^{n-1}_{Z_n}$$

where $\Omega^p_{Z_n}$ is the module of polynomial coefficient p-forms on Z_n. In fact, an element $\omega \in \mathcal{H}^{(0)}_{F_n}$ defines a n-form (modulo closed forms)

(5.2.4) $$\mathrm{Res}_t[\omega] := \mathrm{Res}\left[\frac{\omega}{F_n, P_1-t_1, \cdots, P_l-t_l}\right]$$

on each fiber $X_t := \varphi_n^{-1}(t)$ for $t = (t_1, \cdots, t_l) \in S_{W,\mathbf{C}}$, and hence the de-Rham cohomology class $\mathrm{Res}_t[\omega] \in H^n(X_t, \Omega^*_{X_t})$ (see [Ha,§4] for a definition of the residue symbol Res). So, we denote ω also by $\mathrm{Res}[\omega]$.

REMARK. A justification to call $\mathcal{H}^{(0)}_{F_n}$ the module of abelian differentials is the following fact (which is an easy result of a study on the family φ_1, but, since we shall not use it, we do not give a proof of it) for the case $n = 1$ when the Milnor fiber X_t is a punctured curve.

Fact. *Let* $\zeta_1, \cdots, \zeta_l$ *be homogeneous* $S(V)^W$*-free basis of* $\mathcal{H}^{(0)}_{F_1}$ *such that* $\deg(\zeta_i) = -\frac{1}{2} + \frac{m_i}{h}$ $(i = 1, \cdots, l)$. *Then* $\mathrm{Res}_t[\zeta_i]$ *for* $0 < m_i < \frac{h}{2}$ *form basis of abelian differential of the first kind on* $\bar{X}_t$, $\mathrm{Res}_t[\zeta_i]$ *for* $\frac{h}{2} < m_i < h$ *form basis of abelian differential of the second kind on* $\bar{X}_t$, *and* $\mathrm{Res}_t[\zeta_i]$ *for* $m_i = \frac{h}{2}$ *form basis of abelian differential of the third kind having poles on the punctures on* $\bar{X}_t$.

We recall some structures a)-f) equipped on the module $\mathcal{H}^{(0)}_{F_n}$ as a lattice in $\mathbf{R}^n q_*(\Omega^{\cdot}_{\mathfrak{X}_n/T}[D, D^{-1}])$ (see [S7,§2(2.6.2)] for details and proofs).

a) *The* $S(V)^W$*-module structure on* $\mathcal{H}^{(0)}_{F_n}$ free of rank l.

b) *A decreasing filtration:* $\mathcal{H}^{(0)}_{F_n} \supset \mathcal{H}^{(-1)}_{F_n} \supset \mathcal{H}^{(-2)}_{F_n} \supset \cdots$ by free $S(V)^W$-modules of rank l (which we call the Hodge filtration).

c) *the Gauss-Manin connection:* $\nabla : Der_{S_W} \times \mathcal{H}^{(-k-1)}_{F_n} \to \mathcal{H}^{(-k)}_{F_n}$ such that $\nabla_D : \mathcal{H}^{(-k-1)}_{F_n} \to \mathcal{H}^{(-k)}_{F_n}$ is an $S(V)^{W,\tau}$-isomorphism.

d) *The* $S(V)^{W,\tau}[D^{-1}]$*-module structure* on $\mathcal{H}^{(0)}_{F_n}$ by putting $D^{-1} := \nabla_D^{-1}$, so that one has $\mathcal{H}^{(-k)}_{F_n} = D^{-k}\mathcal{H}^{(0)}_{F_n}$ $(k \in \mathbf{Z}_{\geq 0})$.

e) *The identification of graded pieces of the filtration with the module of Kähler differentials of degree* $n+1$ *on* $\mathfrak{X}_n$ *relative to* φ_n*:*

$$\mathcal{H}^{(-k)}_{F_n}/\mathcal{H}^{(-k-1)}_{F_n} \overset{D^k}{\simeq} \mathcal{H}^{(0)}_{F_n}/\mathcal{H}^{(-1)}_{F_n} \simeq \Omega^{n+1}_{\mathfrak{X}_n/S_W} := \Omega^{n+1}_{\mathfrak{X}_n}/\Omega^1_{S_W} \wedge \Omega^n_{\mathfrak{X}_n}.$$

f) *The higher residue pairing*:

$$K_{F_n} = \sum_{i=0}^{\infty} K^{(i)}_{F_n} D^{-i} : \mathcal{H}^{(0)}_{F_n} \times \mathcal{H}^{(0)}_{F_n} \to S(\mathfrak{h}^*)^{W,\tau}[D^{-1}] ,$$

which is a $S(V)^{W,\tau}[D^{-1}]$-sesqui-linear form compatible (in a suitable sense) with a)-e),whose leading term is given by the residue pairing:

$$K^{(0)}_{F_n}(\omega_1, \omega_2) = \mathrm{Res}\left[\frac{\omega_1 \cdot (\omega_2/dP_1 \wedge \cdots \wedge dP_l \wedge dx \wedge dy \wedge dz_3 \wedge \cdots \wedge dz_{n+1})}{\partial_x F_n,\ \partial_y F_n,\ \partial_{z_3} F_n, \cdots,\ \partial_{z_{n+1}} F_n}\right]$$

inducing a perfect selfdual-pairing on the piece $\Omega^{n+1}_{\mathfrak{X}_n/S_W} \simeq \mathcal{H}^{(0)}_{F_n}/\mathcal{H}^{(-1)}_{F_n}$.

We now consider the most basic element among all Abelian differentials:

(5.2.5) $$\zeta_{F_n} := \mathrm{Res}[dP_1 \wedge \cdots \wedge dP_l \wedge dx \wedge dy \wedge dz_3 \wedge \cdots \wedge dz_{n+1}],$$

which is the element in $\mathcal{H}_{F_n}^{(0)}$ of lowest degree. Using notation 5.1 iv),
$*)$ $\deg(\zeta_{F_n}) = \sum_{i=1}^{l} \frac{d_i}{h} + w_x + w_y + \frac{n-1}{2} - 1 - \sum_{i=1}^{l} \frac{d_i}{h} = r - 1 = \frac{1}{h} + \frac{n-2}{2}$.
Denote by $\zeta_{F_n}^{(-k)} := \nabla_D^{-k} \zeta_{F_n} \in \mathcal{H}_{F_n}^{(-k)}$ the element shifted k-times by D^{-1}. Under these setting and notation, we have theorems ([S5-7]).

THEOREM. *The element ζ_{F_n} is a primitive form for the family F_n.*

PROOF. This means that ζ_{F_n} satisfies the following properties 0)-iv).
0) One has the bijection: $\mathcal{G} \simeq \Omega^{n+1}_{\mathfrak{X}_n/S_W}$, $\delta \mapsto \nabla_\delta \zeta^{(-1)} \bmod \mathcal{H}_{F_n}^{(-1)}$.
i) ζ_{F_n} is homogeneous: $\nabla_E\, \zeta_{F_n}^{(0)} = (r-1)\, \zeta_{F_n}^{(0)}$.
ii) Orthogonality: $K_{F_n}^{(k)}(\nabla_\delta \zeta_{F_n}^{(-1)}, \nabla_{\bar\delta} \zeta_{F_n}^{(-1)}) = 0$ for $\delta, \bar\delta \in \mathcal{G}$ and $k \geq 1$.
iii) $K_{F_n}^{(k)}(\nabla_\delta \nabla_{\delta'} \zeta_{F_n}^{(-2)}, \nabla_{\delta''} \zeta_{F_n}^{(-1)}) = 0$ for $\delta, \delta', \delta'' \in \mathcal{G}$ and $k \geq 2$.
iv) $K_{F_n}^{(k)}(P_l \nabla_\delta \zeta_{F_n}^{(-1)}, \nabla_{\delta'} \zeta_{F_n}^{(-1)}) = 0$ for $\delta, \delta' \in \mathcal{G}$ and $k \geq 2$.
Let us briefly veryfy the conditions. The 0) follows directly from the next formula of covariant differentiation of ζ_{F_n}, and i) follows from $*)$.
$**)$ $\nabla_\delta \zeta^{(-1)} = \mathrm{Res}[\, \delta F_n\, dP_1 \wedge \cdots \wedge dP_l \wedge dx \wedge dy \wedge dz_3 \wedge \cdots dz_{n+1}]$.
The remaining ii), iii) and iv) are verified by degree check as follows.

One has $\deg(K_{F_n}^{(k)}) = -n-1-k$ and $\deg(\zeta_{F_n}^{(0)}) = r$. One may assume $\delta, \delta', \delta'' \in \mathcal{G}$ is $\frac{\partial}{\partial P_i}$ $(i = 1, \cdots, l)$. Then $\deg(\nabla_\delta \nabla_D^{-1}) = \deg P_l - \deg P_i \leq s$. Recalling the duality $2r + s = n+1$ and the characterization $s < 1$ (see 5.1 iv)), and taking the range of k in account, one calculate.

$$\begin{aligned}
&\deg(K_{F_n}^{(k)}(\nabla_\delta \zeta_{F_n}^{(-1)}, \nabla_{\delta'} \zeta_{F_n}^{(-1)})) \\
&\qquad \leq (-n-1-k) + (s+r) + (s+r) = s - k < 0, \\
&\deg(K_{F_n}^{(k)}(\nabla_\delta \nabla_{\delta'} \zeta_{F_n}^{(-2)}, \nabla_{\delta''} \zeta_{F_n}^{(-1)})) \\
&\qquad \leq (-n-1-k) + (2s+r) + (s+r) = 2s - k < 0, \\
&\deg(K_{F_n}^{(k)}(P_l \nabla_\delta \zeta_{F_n}^{(-1)}, \nabla_{\delta'} \zeta_{F_n}^{(-1)})) \\
&\qquad \leq (-n-1-k) + (1+s+r) + (s+r) = s + 1 - k < 0.
\end{aligned}$$

The negativity of degrees implies that these elements are zero. □

Let us state some immediate consequences of Theorem.

The 0) and ii) of the condition for a primitive form implies that one has a $S(V)^{W,\tau}[D^{-1}]$-module isomorphism:

$$\mathcal{G} \otimes \mathbf{Z}[D^{-1}] \simeq \mathcal{H}_{F_n}^{(0)}. \tag{5.2.6}$$

defined by the correspondence $\sum_{i \geq 0} \delta_i \otimes D^{-i} \mapsto \sum_{i \geq 0} \nabla_{\delta_i} D^{-i} \zeta_{F_n}^{(-1)}$.

NOTE. The formula (5.2.6) implies that the choice of a primitive form defines a splitting of the Hodge filtration (in the sense in b)) into an orthogonal sum of graded pieces $\mathfrak{g} \otimes D^{-k} \simeq \mathcal{H}_{F_n}^{(-k)}/\mathcal{H}_{F_n}^{(-k-1)}$. In analogy with the classical Hodge theory, where a choice of a Kähler metric defines an orthogonal splitting of the Hodge filtration, we may call (5.2.6) a Hodge decomposition. Note that we have not yet discussed on the $S(V)^W$- and Der_{S_W}- module structures on the modules in (5.2.6).

The iii) and iv) implies that the multiplication of P_l and the covariant differentiation by an element $\delta \in \mathcal{G}$ in the RHS of (5.2.6) is given by two terms in the

LHS. Let us write down explicitly the actions:

$$P_l(D^{-k}\otimes\delta)=D^{-k}\otimes(P_l *_n\delta)+D^{-k-1}\otimes(N_n(\delta)+k\delta), \tag{5.2.7}$$

$$\nabla_{\delta_1}(D^{-k-1}\otimes\delta_2)=D^{-k}\otimes(\delta_1 *_n\delta_2)+D^{-k-1}\otimes\nabla_{n\ \delta_1}\delta_2. \tag{5.2.8}$$

where $*_n:\mathcal{G}\times\mathcal{G}\to\mathcal{G}$ is a $S(V)^{W,\tau}$-bilinear map, $N_n:\mathcal{G}\to\mathcal{G}$ is a $S(V)^{W,\tau}$-endomorphism and $\nabla_n:\mathcal{G}\times\mathcal{G}\to\mathcal{G}$ is a covariant differentiation over $S(V)^{W,\tau}$.

1. *The $*_n$-product is induced from the ring structure in $\mathcal{O}_{C_n}$:*

$$(\delta_1 *_n\delta_2)F_n\equiv\delta_1F_n\cdot\delta_2F_n \ \ in \ \ \mathcal{O}_{C_n} \tag{5.2.9}$$

using the identification $\mathcal{G}\simeq\mathcal{O}_{C_n}$.

2. *Using the identification: $\mathcal{G}\simeq\Omega^{n+1}_{\mathfrak{X}_n/S_W}$, the flat metric J is given by the first residue pairing $K^{(0)}$.* That is: for $\delta_1,\delta_2\in\mathcal{G}$, one has

$$J(\delta_1,\delta_2)=c\cdot\mathrm{Res}\Big[\frac{\delta_1F_n\ \delta_2F_n\ \zeta_{F_n}}{\partial_xF_n,\partial_yF_n,\partial_{z_3}F_n,\cdots,\partial_{z_{n+1}}F_n}\Big] \tag{5.2.10}$$

3. *The intersection form on the homology group $H_n(X_t,\mathbf{Z})$ of a Milnor fiber X_t is now calculated by the flat metric (i.e. the residue pairing) and by the period of primitive forms:*

$$\langle\gamma,\gamma'\rangle=\frac{(-1)^{\frac{n}{2}-k}}{(2\pi)^n}\sum_{i=1}^n\delta_i\int_\gamma\zeta_{F_n}^{(k-1)}\cdot w(\delta^{i*})\int_{\gamma'}\zeta_{F_n}^{(\frac{n}{2}-k-1)} \tag{5.2.11}$$

where $\delta_1,\cdots,\delta_l$ and $\delta^{1*},\cdots,\delta^{l*}$ are $S(V)^{W,\tau}$-dual basis of $\mathcal{G}$ w.r.t. the inner product J_{F_n} [S6 §3].

4. *Let W be the Weyl group of the Lie algebra $\mathfrak{g}$ corresponding to the simple singularity F_n. The identification of the primitive form ζ_{F_n} with the $\zeta_W^{(-n/2)}=D^{-n/2}\otimes D$ in $\mathcal{H}_W$ 4.3 induces an isomorphism*

$$\mathcal{H}_{F_n}^{(k)}\simeq\mathcal{H}_W^{(k-n/2)} \tag{5.2.12}$$

compatible with the structures a)-f) up to the shift by $\frac{n}{2}$. In particular, one has $*_n=*$, $N_n=N+\frac{n}{2}$ and $\nabla_{F_n}=\nabla$.

We first show the result for $n=2$, where Brieskorn [Br1] has identified the intersection form on $H_2(X_t,\mathbf{Z})$ with the Killing form (up to a constant) of the corresponding Lie algebra $\mathfrak{g}$. For general n, the result is reduced to an unpublished result [S5,§5-6]: *Let $F_n+z_{n+2}^2+z_{n+3}^2=F_{n+2}$. The correspondence $\omega\in\mathcal{H}_{F_n}^{(k)}\mapsto\nabla_D(\omega\wedge dz_{n+2}\wedge dz_{n+3})\in\mathcal{H}_{F_{n+2}}^{(k+1)}$ for $k\le-1$ and $n\ge1$ defines a bijection: $\rho:\mathcal{H}_{F_n}^{(k)}\simeq\mathcal{H}_{F_{n+2}}^{(k+1)}$, which is compatible with the structures i)-iv) and $\rho(\zeta_{F_n}^{(k)})=\zeta_{F_{n+2}}^{(k+1)}$.*

REMARK. Another direct proof of the theorem for $n=1$ without using the result of [S5] may conjecturely be given by the following approach.

Let $\mathfrak{g}$ be a semi-simple algebra over $\mathbf{C}$. Let θ be an involution of $\mathfrak{g}$ inducing the eigenspace decomposition $\mathfrak{g}=\mathfrak{k}+\mathfrak{p}$. The restriction of the adjoint quotient morphism to the subspace $\mathfrak{p}$ induces a flat morphism whose fiber over 0 is the nilpotent variety $N(\mathfrak{p})$ (c.f. [K-R]). Then one wants to study again the restriction of the morphism on the subspace $\mathfrak{X}$ of $\mathfrak{p}$ which is transversal to singularity of $N(\mathfrak{p})$. This was studied in details by Sekiguchi [Se], whose result is not so simple as in Brieskorn's theorem. One reason is that $N(\mathfrak{p})$ may no longer be irreducible, and the other reason is that the singularity $N(\mathfrak{p})_{sing}$ may not be irreducible and each component may have different codimension in $N(\mathfrak{p})$. Nevertheless, we expect Brieskorn type theorem for the real split case as follows.

Conjecture: if θ is the involution associated to a real split form of $\mathfrak{g}$, then the restriction of the adjoint quotient morphism to a transversal slice at a point of a subregular nilpotent orbit in $N(\mathfrak{p})$ gives a real form of the universal unfolding of a simple singularity of dimension 1.

5.3. Period integral and period map.

As a consequence of 5.1 and 5.2, we can express the solutions of the equation $\widetilde{\mathcal{M}}_{W,-n/2}$ for $n \in \mathbf{Z}_{\geq 0}$ by the integrals of the primitive form for the family φ_n. Namely, let $\gamma(t)$ be a horizontal section $\mathbf{R}^n\varphi_{n*}(\mathbf{Z}_{\mathfrak{X}})$ where t runs in the universal covering $(S_{W,\mathbf{C}} \setminus D_W)^{\sim}$. Then the integral $\int_{\gamma(t)} \mathrm{Res}_t[\zeta_{F_n}^{(k)}]$ as a function on $(S_{W,\mathbf{C}} \setminus D_W)^{\sim}$ is a solution of $\widetilde{\mathcal{M}}_{W,k-n/2}$ for $\delta \int_{\gamma(t)} \mathrm{Res}_t[\zeta_{F_n}^{(k)}] = \int_{\gamma(t)} \nabla_\delta \mathrm{Res}_t[\zeta_{F_n}^{(k)}]$ $(\delta \in Der_{S_W})$. We regard $\int_{\gamma(t)} \mathrm{Res}_t[\zeta_{F_n}^{(k)}]$ defines further, at each point t, a linear functional on the flat tangent vector space Ω_+^*: $\delta \in \Omega_+^* \mapsto \delta \int_{\gamma(t)} \mathrm{Res}_t[\zeta_{F_n}^{(k)}]$. This defines the *period map* $t \in (S_{W,\mathbf{C}} \setminus D_W)^{\sim} \mapsto \int \mathrm{Res}_t[\zeta_{F_n}^{(k)}] \in \mathrm{Hom}(\mathbf{R}^n\varphi_{n*}(\mathbf{Z}_{\mathfrak{X}}), \Omega_{+\mathbf{C}})$. Applying D on Ω, we obtain *the projection to the first factor* $\mathrm{Hom}(\mathbf{R}^n\varphi_{n*}(\mathbf{Z}_{\mathfrak{X}}), \mathbf{C})$.

A description of the period map for $n = 2$ is achieved by [Br1][Lo] and [Y]. Namely, the simultaneous resolution $\tilde{\mathfrak{X}} \to \mathfrak{h}$ of the family $\mathfrak{X}_2 \to S_W$ is constructed in terms of Lie theory in [Br1], where the Kostant-Kirillov form on the (co-)adjoint orbit is identified with the primitive form [Y]. This implies that $\mathfrak{h}$ is the period domain and the inverse to the period map is nothing but the quotient morphism $\mathfrak{h} \to \mathfrak{h}/W \simeq S_W$. (For details, see the references. For an analytic treatment, see [Lo]).

We describe now the period map and its inverse for $n = 1$ of type A_1,A_2,A_3 or B_2 (where the projection to the first factor suffices). The types A_1 and A_2 are classical. The type A_3 is taken from [S2], and B_2 is a consequence of A_3 (which recover the classical well-known results).

Example. 1. A_1 : $F := xy - t$. The primitive form is given by the differential $\zeta = dx/x$. The base $\gamma \in H_1(X_t, \mathbf{Z})$ is invariant (under the monodromy action) and the period integral $\int_\gamma \zeta = 2\pi\sqrt{-1}$ is a constant independent of t. The indefinite integral $u = \int_1^t \zeta$ and 1 form basis of $Sol(\mathcal{M}_{A_1,-\frac{1}{2}})$. The period domain of u is the plane $\mathbf{C}$ with the translation action by $2\pi\sqrt{-1}\mathbf{Z}$. The inverse map is given by the exponential function $t = \exp(u)$.

2. A_2 : $F := y^2 - (4x^3 - g_2x - g_3)$ with $\Delta_{A_2} = g_3^2 - \frac{1}{27}g_2^3$. The primitive form is given by the elliptic integral of the first kind $\zeta = dx/\sqrt{4x^2 - g_2x - g_3}$. The Milnor fiber X_g is a punctured (at ∞) elliptic curve. The integrals $u_i = \int_{\gamma_i} \zeta$ $(i = 1, 2)$ over (oriented) basis $\gamma_1, \gamma_2 \in H_1(X_g, \mathbf{Z})$ and 1 give basis of $Sol(\mathcal{M}_{A_2,-\frac{1}{2}})$. The period domain of (u_1, u_2) is $\tilde{\mathcal{H}} := \{(u_1, u_2) \in C^2 \mid \Im(u_1/u_2) > 0\}$ with the modular group $\Gamma(A_2) = \mathrm{SL}(2, \mathbf{Z})$ action. The inverse map $\tilde{\mathcal{H}} \to S_{a_2} \setminus D_{A_2}$ to the period map is given by $g_2 := 60E_4(u_1, u_2)$ and $g_3 := 140E_6(u_1, u_2)$ where $E_{2i} := \sum_{(m,n)\in\mathbf{Z}^2\setminus\{(0,0)\}}(nu_1 + mu_2)^{-2i}$ is the elliptic Eisenstein series. The discriminant Δ_{A_2}, given by $\eta(\tau)^{24}u_2^{-12}$, generates the ideal of cusp forms, where $\eta(\tau) = \exp(2\pi\sqrt{-1}\tau/24)\Pi_{n=1}^{\infty}(1 - q^n)$ is the Dedekind η-function, $\tau = u_1/u_2$ and $q = \exp(2\pi\sqrt{-1}\tau))$.

3. A_3 : $F := s^2 - (t^4 + x_2t^2 + x_3t + x_4)$. The primitive form is given by the hyperelliptic integral of the lowest degree $\zeta = dt/\sqrt{t^4 + x_2t^2 + x_3t + x_4}$. The Milnor fiber $X_{\underline{x}}$ is a two distinct punctured elliptic curve. Let γ_1, γ_2 and γ be the basis of $H_1(X_x, \mathbf{Z})$ such that γ_1, γ_2 form (oriented) basis of the first homology group of the compactified elliptic curve and γ is the (monodromy) invariant cycle presented by a closed path in X_t turning once around one of the punctures. Then the integrals $\int_{\gamma_i} \zeta$ $(i = 1, 2)$ are the periods on $\bar{X}_{\underline{x}}$

but $\int_\gamma \zeta$ is realized by an integral over an interval connecting the two punctures on $X_{\underline{x}}$. In order to get a description of the third integral and the inverse map, we proceed the following reduction to A_2 [S2,Theorem]: regarding the polynomial $t^4+x_2t^2+x_3t+x_4$ as a binary quartic form, the invariants of weight 4 and 6 are given by $L(\underline{x}):=16x_4+\frac{4}{3}x_2^2$ and $-\frac{1}{9}M(\underline{x}):=-\frac{8}{27}x_2^3-4x_3^2+\frac{32}{3}x_2x_4$. The two punctures on the elliptic curve $X_{\underline{x}}$ correspond to the two points $p(\underline{x}):=(z=-\frac{2}{3}x_2, w=2x_3)$ and ∞ on the elliptic curve $w^2=4z^3-L(\underline{x})z+\frac{1}{9}M(\underline{x})$. The correspondence $\underline{x}\in S_{A_3}\mapsto(p(\underline{x}),g_2=L(\underline{x}),g_3=-\frac{1}{9}M(\underline{x}))$ gives a biregular morphism from S_{A_3} to the total space of Weierstrass family of (affine) elliptic curves: $w^2=4z^3-g_2z-g_3$ such that $\Delta_{A_3}=-\frac{27}{16}\Delta_{A_2}$. Consider the associated elliptic integrals:

$$u_i=\int_{\gamma_i}dz/\sqrt{4z^3-Lz+\tfrac{1}{9}M}\ (i=1,2)\ \&\ \ v=\int_\infty^{p(\underline{x})}dz/\sqrt{4z^3-Lz+\tfrac{1}{9}M}.$$

THEOREM. ([S2, Theorem])

1. The v,u_1,u_2 and 1 form basis of $Sol(\mathcal{M}_{A_3,-\frac{1}{2}})$.

2. The period domain of (v,u_1,u_2) and the modular group are given by

$$\begin{array}{lcl}\mathcal{B}(A_3) & = & \{(v,u_1,u_2)\in\mathbf{C}^3\mid\Im(u_1/u_2)>0,\ v\neq mu_1+nu_2\forall m,n\in\mathbf{Z}\},\\ \Gamma(A_3) & = & \mathrm{SL}(2,\mathbf{Z})\ |\!\times\ \mathbf{Z}^2.\end{array}$$

3. The inverse map $\mathcal{B}(A_3)\to S_{A_3}\setminus D_{A_3}$ is given by the following system:

$$\begin{array}{rcl}-\frac{2}{3}x_2 & = & \mathfrak{p}(v,u_1,u_2)=\ v^{-2}+\sum'\{(v-mu_1-nm_2)^{-2}-(nu_1+mu_2)^{-2}\},\\ -x_3 & = & -\frac{1}{2}\mathfrak{p}'(v,u_1,u_2)\ =\ v^{-3}+\sum'(v-mu_1-nm_2)^{-3},\\ L & = & 16x_4+\frac{4}{3}x_2^2\ =\ 60E_4(u_1,u_2).\end{array}$$

4. $B_2: F:=s^2-(t^4+x_2t^2+x_4)$. This is a subfamily of A_3 fixed by the $\mathbf{Z}/2$-action $(s,t)\mapsto(s,-t)$ with $\Delta_{B_2}=x_4(x_4-x_2^2/4)$. The cycles vanishing along $x_4=0$ defines the short root and the cycles vanishing along $x_4-x_2^2/4=0$ defines the long root of type B_2, respectively. The primitive form is given by the classical Legendre-Jacobi form of elliptic integral $dt/\sqrt{t^4+x_2t^2+x_4}$. Let us use the same reduction and notation as in the previous example A_3. Because of the $\mathbf{Z}_2$-symmetry, the difference, in the elliptic curve, of the two punctures on $X_{\underline{x}}$ is a two-torsion element, and, hence, the point $p(\underline{x})$ in the elliptic curve $w^2=4z^3-g_2z-g_3$ is a branching point of the double cover to the z-plane (i.e. $-\frac{2}{3}x_2$ is a solution of the cubic equation $4z^3-g_2z-g_3=0$). Then we can choose a base γ_1 of the homology group such that $2v=u_1$. As immediate consequences of this description, we obtain:

THEOREM. *1. The u_1,u_2 and 1 form basis of $Sol(\mathcal{M}_{B_2,-\frac{1}{2}})$.*

2. The period domain and the modular group acting on it are given by

$$\begin{array}{lcl}\mathcal{B}(B_2) & = & \{(v,u_1,u_2)\in\mathcal{B}(A_3)\mid 2v=u_1\}\simeq\{(u_1,u_2)\in\mathbf{C}^2\mid\Im(u_1/u_2)\geq 0\},\\ \Gamma(B_2) & = & \Gamma_0(2)\ :=\ \{\begin{bmatrix}a & b\\ c & d\end{bmatrix}\in\mathrm{SL}(2,\mathbf{Z})\mid b\equiv 0\bmod 2\}.\end{array}$$

3. The inverse map $\mathcal{B}(B_2)\to S_{B_2}\setminus D_{B_2}$ is given by the following system:

$$-\tfrac{2}{3}x_2=\mathfrak{p}(\tfrac{1}{2}u_1,u_1,u_2)\quad\text{and}\quad 16x_4+\tfrac{4}{3}x_2^2=60E_4(u_1,u_2).$$

4. There are two orbits of cusps on the boundary of $\mathcal{B}(B_2)$. Each factor $x_4=\frac{3}{16}(20E_4(u_1,u_2)-\mathfrak{p}^2(\frac{1}{2}u_1,u_1,u_2))$ and $4x_4-x_2^2=15E_4(u_1,u_2)-3\mathfrak{p}^2(\frac{1}{2}u_1,u_1,u_2)$ of the discriminant vanishes on each orbit of cusps, respectively, but vanishes nowhere on the period domain. Therefore, the factor is (up to a constant factor) given by $\eta(\tau)^8u_2^{-4}$ and $\eta(-\frac{1}{2\tau})^8u_1^{-4}=2^4\eta(2\tau)^8u_2^{-4}$, respectively, and the discriminant $\Delta_{B_2}=x_4(x_4-x_2^2/4)$ generates the ideal of cusp forms.

5. The flat coordinates are given by $x_2=-\frac{2}{3}\mathfrak{p}$ and $x_4-\frac{1}{8}x_2^2=\frac{15}{32}(4E_4-\mathfrak{p}^2)$.

6. Inverse maps from the Period domain

We try to understand the inverse map to the period map for n=1. There are some partial results for the types A_1, A_2, A_3 and B_2 (recall 5.3 Example). However, we are still unsure how to give a general conjectural description of the ring of inverse functions and of the flat structures in 3.9 in terms of the inverse functions on the period domain.

We first study in 6.1 an abstract group $\Gamma(M)$ attached to a Coxeter matrix M, which shall play the role of the modular group in 6.3 by letting it act on the symplectic vector space introduced in 6.2. The period domain for the period map and its inverse map are conjecturely described in 6.4 and 6.5. The goal of this section is 6.6 Conjecture 6, which states that certain power root (prescribed in 6.1) of the discrim-
inant is automorphic with the character ϑ_W on $\Gamma(W)$ given in 6.1.

6.1. Group $\Gamma(M)$.

Let $M = (m(\alpha,\beta))_{\alpha,\beta\in\Pi}$ be a Coxeter matrix (2.3). We introduce a group $\Gamma(M)$ attached to M by the relations (6.1.1), (6.1.2) and (6.1.3).

Generators: γ_α for $\alpha\in\Pi$

Relations: 1. For the pair $\alpha,\beta\in\Pi$ such that $m(\alpha,\beta)=2$

(6.1.1) $$\gamma_\alpha\gamma_\beta=\gamma_\beta\gamma_\alpha.$$

2. For the pair $\alpha,\beta\in\Pi$ such that $m(\alpha,\beta)\geq 3$

(6.1.2) $$\underbrace{\gamma_\alpha\gamma_\beta\gamma_\alpha\cdots}_{m(\alpha,\beta)}=\underbrace{\gamma_\beta\gamma_\alpha\gamma_\beta\cdots}_{m(\alpha,\beta)}$$

3. Let Γ' be any irreducible finite type subdiagram of the Coxeter graph Γ of M, and let h' be the Coxeter number of Γ'. Suppose there does not exist an exponent of Γ' which is equal to $h'/2$, then

(6.1.3) $$(\Pi_{\alpha\in\Gamma'}\gamma_\alpha)^{\langle h'\rangle}=1$$

for any order of the product (see Remark 1. below), where

(6.1.4) $$\langle h'\rangle := \begin{cases} 2h' & \text{if } h' \text{ is odd,}\\ h' & \text{if } h' \text{ is even and } h'/2 \text{ is even,}\\ \frac{h'}{2} & \text{if } h' \text{ is even and } h'/2 \text{ is odd.}\end{cases}$$

REMARK. 1. Different orders in the product in (6.1.3) define conjugate elements due to (6.1.1) [B,Ch.v,§6,1.Lemme 1.].

2. In the above Relations 3., the condition that there does not exist an exponent of Γ' equal to the half of its Coxeter number h' is equivalent to that the symplectic form I'_{odd} attached to Γ' (see 6.2 Assertion 1.) is nondegerate (c.f. 2.8 Remark and 6.3 Assertion 4.).

By definition, we have some obvious homomorphisms and characters.

i) The correspondence $\gamma_\alpha\mapsto -1$ ($\alpha\in\Pi$) induces a character

(6.1.5) $$\theta_M:\Gamma(M)\to\{\pm1\}.$$

Proof. We need to verify that $\langle h'\rangle\#(\Gamma')$ is an even number for any irreducible finite type Coxeter graph Γ' without an exponent equal to the half-Coxeter number. In view of (6.1.4), we have to check only the third case in the formula. According to the classification, one check that this occurs only for the types E_8, H_4 and $I_2(p)$ for $p = 2 \cdot odd$. In all cases, $\#(\Gamma')$ is an even number (c.f. also 6.2 Assertion 1.). □

ii) Put $k(M){:=}\gcd\{\langle h'\rangle\#(\Gamma')/2 \mid \Gamma'$: irreducible finite type subdiagrams of M s.t. $h'/2$ is not an exponent of Γ' (h':the Coxeter # of Γ')$\}$. The correspondence $\gamma_\alpha \mapsto \exp(\frac{\pi\sqrt{-1}}{k(M)})$ $(\alpha \in \Pi)$ induces a character

$$\vartheta_M : \Gamma(M) \to \mathbf{C}^\times \tag{6.1.6}$$

such that $\vartheta_M^{k(M)} = \theta_M$ and $\vartheta_M^{2k(M)} = 1$. If the graph Γ of M does not contain $I_2(p)$ for $p{=}2 \cdot odd$, then $k(M)$ is an even number. Except for $k(A_1) = \infty, k(A_2) = k(A_3) = 6, k(B_2) = k(C_2) = 4, k(D_4) = 6, k(G_2) = 3$ and $k(I_2(p)){=}\langle p\rangle$, one has $k(W){=}2$ for all finite Coxeter group W.

iii) The Artin group $A(M)$ is defined by the relations (6.1.1) and (6.1.2) on the generator system g_α $(\alpha \in \Pi)$ [B-S]. Then, the correspondence $g_\alpha \mapsto \gamma_\alpha$ for $\alpha \in \Pi$ induces a homomorphism from $A(M)$ onto $\Gamma(M)$:

$$\gamma : A(M) \to \Gamma(M). \tag{6.1.7}$$

iv) Suppose the Coxeter graph of M contains neither E_8, H_4 nor $I_2(p)$ for $p = 2 \cdot odd$ (crystallographic groups except E_8 and G_2 satisfy the assumption). Then the correspondence $\gamma_\alpha \mapsto \sigma_\alpha$ $(\alpha \in \Pi)$ induces a homomorphism σ from $\Gamma(M)$ onto the Coxeter group with the diagram:

$$\begin{array}{ccc} \Gamma(M) & \overset{\sigma}{\to} & W(M) \\ \downarrow \theta_M & & \downarrow \det \\ \{\pm 1\} & = & \{\pm 1\}. \end{array} \tag{6.1.8}$$

(*Proof.* Except for the third case of (6.1.4), all defining relations of $\Gamma(M)$ are satisfied in $W(M)$ (recall 2.8).)

Let us call the kernel of σ *the principal congruence subgroup of* $\Gamma(M)$.

REMARK. Our original intension, explained at the introduction, was to use the flat structure to understand the Artin group $A(M)$. By the use of the flat structure, we arrived at a group $\Gamma(M)$ which lies between $A(M)$ and $W(M)$. Except for the first few types A_2, B_2, G_2 or $I_2(p)$, the group $\Gamma(M)$ is close to the group $W(M)$ and we are only at the first stage (after $W(M)$) to understand the Artin group. It is interesting to construct Eilenberg-MacLane space for the group $\Gamma(M)$ (c.f. [D-S]).

6.2. Symplectic space $(\tilde{F}, \tilde{I}_{odd})$.

We construct a symplectic vector space on which $\Gamma(M)$ acts symplectic linearly. First, we recall the orthogonal representation of the Coxeter group $W(M)$ ([B, Ch.V,§4, theor.1]). Consider vector spaces

$$F := \oplus_{\alpha\in\Pi}\mathbf{R}e_\alpha$$

equipped with the symmetric bilinear form I defined by

$$I(e_\alpha, e_\beta) \quad = -2\cos(\pi/m(\alpha,\beta))/d_\alpha d_\beta,$$

where $d_\alpha \in \mathbf{R}\backslash\{0\}$ $(\alpha\in\Pi)$ are arbitrary scaling constants (we regard that the pair (F, I) is determined by M independent of the scaling constants and the basis e_α moves in F according to the scaling). A cobasis is introduced by $e_\alpha^\vee := 2e_\alpha/I(e_\alpha, e_\alpha)$ for $\alpha \in \Pi$. Then the Coxeter group $W(M)$ acts freely on F by letting a generator

$a_\alpha \in W(M)$ act on it by reflections $\sigma_\alpha : u \mapsto u - I(u, e_\alpha^\vee)e_\alpha$ so that $I(\sigma(u), \sigma(v)) = I(u,v)$.

¿From now on, we assume that the Coxeter graph (2.4) associated to M is a tree. Up to order, Π decomposes uniquely into a disjoint union:

$$\Pi = \Pi_1 \cup \Pi_2, \tag{6.2.1}$$

such that each Π_i is discrete, i.e. e_α for $\alpha \in \Pi_i$ are mutually orthogonal to each other. We introduce a skew symmetric form I_{odd} on V.

$$I_{odd}(e_\alpha, e_\beta) := \begin{cases} I(e_\alpha, e_\beta) & \text{if } \alpha \in \Pi_1 \text{ and } \beta \in \Pi_2, \\ -I(e_\alpha, e_\beta) & \text{if } \alpha \in \Pi_2 \text{ and } \beta \in \Pi_1, \\ 0 & \text{if } \alpha, \beta \in \Pi_1 \text{ or } \alpha, \beta \in \Pi_2. \end{cases} \tag{6.2.2}$$

Note that the I_{odd} does not depend on the scaling factors but only on (F, I) and (6.2.1). Interchanges of Π_i $(i = 1, 2)$ induces a sign change $I_{odd} \mapsto -I_{odd}$. In the sequel, we fix a decomposition (6.2.1) once for all.

EXAMPLE. Let W be a finite reflection group acting on (V, I) as in §2. By the choice of a chamber C, one obtains a Coxeter system $(W, \Pi(C))$ with the basis Π_W (2.3 Theo.5.), whose associated graph is a tree (2.4). Thus the above (6.2.2) defines the form $I_{odd,C}$ depending on C, where one has $I_{odd,C}(x,y) = I_{odd,wC}(wx, wy)$ for $w \in W$. Accordingly, many of our later constructions (such as the symplectic space $\tilde{F}$, the realization ρ of $\Gamma(M)$ in the symplectic group, the lattice $\tilde{Q}$, the period domain $\tilde{B}$, etc) depend on the choice of a chamber. However, they are conjugate to each other in suitable sense by the W-action. So, once for all, we choose and fix one chamber and denote the associated form by I_{odd}. We shall not mention explicitly the dependence on the chamber.

In general, I_{odd} may not be nonsingular. A more precise formula is:

Assertion 1. *Let the notation be as above. One has*

$$\operatorname{rank}(I_{odd}) = 2\min\{\#\Pi_1, \#\Pi_2\}, \tag{6.2.3}$$

$$\operatorname{corank}(I_{odd}) = |\#\Pi_1 - \#\Pi_2| = \operatorname{rank}(\ker(c+1)). \tag{6.2.4}$$

Here $c := \Pi_{\alpha\in\Pi}\sigma_\alpha$ is a Coxeter element in a generalized sense. In case $W(M)$ is a finite reflection group, the number (6.2.4) is equal to

$$\#\{1 \le i \le l \mid m_i = h/2\}. \tag{6.2.5}$$

PROOF. Put $\ker(I_{odd}) := \{\xi \in F \mid I_{odd}(\xi, x) = 0 \ \forall x \in F\}$ and $\mathbf{R}e_{\Pi_i} := \sum_{\alpha\in\Pi_i} \mathbf{R}e_\alpha$ $(i = 1,2)$. We may assume $\#\Pi_1 \ge \#\Pi_2$ without loss of generality. For a proof of (6.2.3) and (6.2.4), it is sufficient to prove

$$\ker(I_{odd}) = \mathbf{R}e_{\Pi_1} \cap (\mathbf{R}e_{\Pi_2})^\perp = \ker(c+1).$$

The inclusion $\ker(I_{odd}) \supset \mathbf{R}e_{\Pi_1} \cap (\mathbf{R}e_{\Pi_2})^\perp$ is obvious. We show the converse (i.e. $\xi = \sum_{\beta\in\Pi_2} c_\beta e_\beta$ and $I(\xi, e_\alpha) = 0$ for all $\alpha \in \Pi_1$ implies $\xi = 0$) by induction on $\#\Pi_2$. ¿From the inequality $\#\Pi_2 \le \#\Pi_1$, one sees that there is a vertex $\alpha \in \Pi_1$ which is connected with only one vertex, say β, of Π_2. Then $I(\xi, e_\alpha) = 0$ implies that the coefficient c_β vanishes. The diagram $\Pi \setminus \{\alpha, \beta\}$ may decompose into components but for each component Γ, one still has $\#(\Pi_2 \cap \Gamma) \le \#(\Pi_1 \cap \Gamma)$ so that one can proceed further by induction to show $\xi = 0$.

Let us show $\ker(c+1) \subset \ker(I_{odd})$. Choose the Coxeter element to be $c = c_1c_2$ for $c_i = \prod_{\alpha\in\Pi_i} \sigma_\alpha$. Let $\xi \in \ker(c+1)$. Then $c\xi = -\xi$ implies $c_1\xi + c_2\xi = 0$ (*). Let $\xi = \xi_1 + \xi_2$ with $\xi_i \in \mathbf{R}e_{\Pi_i}$ $(i = 1,2)$ so that $c_i\xi_i = -\xi_i$ $(i = 1,2)$ and $c_1\xi_2 - \xi_2 \in \mathbf{R}e_{\Pi_1}$, $c_2\xi_1 - \xi_1 \in \mathbf{R}e_{\Pi_2}$. Then the condition (*) implies $(c_1\xi_2 - \xi_2) + (c_2\xi_1 - \xi_1) = 0$ and hence $c_1\xi_2 - \xi_2 = -(c_2\xi_1 - \xi_1) = 0$. These imply that $I(\xi_2, e_\alpha) = 0$ for all $\alpha \in \Pi_1$ and $I(\xi_1, e_\alpha) = 0$ for all $\alpha \in \Pi_2$. Thus $I_{odd}(\xi_1, e_\alpha) = I_{odd}(\xi_2, e_\alpha) = 0$ for all $\alpha \in \Pi$, i.e. $\xi \in \ker(I_{odd})$. Conversely, suppose $\xi \in \mathbf{R}e_{\Pi_1} \cap (\mathbf{R}e_{\Pi_2})^{\perp}$. Then $c_1\xi = -\xi$ and $c_2\xi = \xi$ and hence $c\xi = -\xi$.
The (6.2.5) follows from the definition of the exponents 2.8. □

Assertion 1. implies that the form I_{odd} is non-degenerate $\Leftrightarrow \#\Pi_1 = \#\Pi_2 \Leftrightarrow$ there is no eigenvalue -1 of the Coxeter element. In fact, $\mathrm{corank}(I_{odd})$ is positive for types A_l, B_l, C_l with l =even, D_l with $l \geq 4$, E_7 and H_3 (recall the remark at the end of 2.8).

The *nonsingular-hull* $(\tilde{F}, \tilde{I}_{odd})$ of (F, I_{odd}) is the smallest nonsingular symplectic vector space containing (F, I_{odd}) (unique up to isomorphism). The explicit model of the nonsingular-hull is constructed as follows.

Assertion 2. *Consider the vector space*

$$\tilde{F} := F \oplus \ker(I_{odd}) \tag{6.2.6}$$

equipped with the skew symmetric form:

$$\tilde{I}_{odd}(x_1 \oplus y_1, x_2 \oplus y_2) := I_{odd}(x_1, x_2) + I(x_1, y_2) - I(x_2, y_1). \tag{6.2.7}$$

Then, $\tilde{I}_{odd}$ is nondegenerate, and $(\tilde{F}, \tilde{I}_{odd})$ is the nonsingular-hull.

PROOF. Without loss of generality, we assume $\#\Pi_1 \geq \#\Pi_2$. We saw in the proof of Assertion 1 that $\ker I_{odd} = \mathbf{R}e_{\Pi_1} \cap (\mathbf{R}e_{\Pi_2})^{\perp}$. Let us see that

$$\tilde{F} \;=\; \mathbf{R}e_{\Pi_1} \;\oplus\; (\mathbf{R}e_{\Pi_2} \oplus \ker I_{odd})$$

is a holonomic decomposition. That is: i) $\mathbf{R}e_{\Pi_1}$ and $\mathbf{R}e_{\Pi_2} \oplus \ker I_{odd}$ are totally isotropic w.r.t. $\tilde{I}_{odd}$ (*proof*: each of the spaces $\mathbf{R}e_{\Pi_1}$, $\mathbf{R}e_{\Pi_2}$ and $\ker I_{odd} \subset \mathbf{R}e_{\Pi_1}$ are totally isotropic. Then $I \mid \mathbf{R}e_{\Pi_2} \times \ker I_{odd} = 0$ implies the statement). ii) The coupling $I \mid \mathbf{R}e_{\Pi_1} \times (\mathbf{R}e_{\Pi_2} \oplus \ker I_{odd})$ is non-degenerate (*proof*: it suffices to show $I|\ker I_{odd}$ is non-degenerate, which follows since it is a restriction of the positive definite form $I|\mathbf{R}e_{\Pi_1}$). □

6.3. Symplectic linear representations of $\Gamma(M)$.

For each $\alpha \in \Pi$, we consider two transvections

$$s_\alpha^{\pm}: \quad u \in \tilde{F} \;\mapsto\; u \mp \tilde{I}_{odd}(e_\alpha^{\vee}, u)e_\alpha \in \tilde{F}. \tag{6.3.1}$$

One can directly show: i) $\tilde{I}_{odd}(s_\alpha^{\pm}(u), s_\alpha^{\pm}(v)) = \tilde{I}_{odd}(u,v)$, ii) $s_\alpha^{+}s_\alpha^{-} = 1$, and therefore the transvections belong to the symplectic group:

$$s_\alpha^{\pm} \in \mathrm{Sp}(\tilde{F}, \tilde{I}_{odd}) := \{g \in \mathrm{GL}(\tilde{F}) \mid \tilde{I}_{odd} \circ g = \tilde{I}_{odd}\}.$$

LEMMA. *The correspondences $\gamma_\alpha \mapsto s_\alpha^{\pm}$ $(\alpha \in \Pi)$ induce representations:*

$$\rho^{\pm} : \Gamma(M) \to \mathrm{Sp}(\tilde{F}, \tilde{I}_{odd}). \tag{6.3.2}$$

NOTE. ρ^+ and ρ^- are different and $\rho^- \neq (\rho^+)^{-1}$ (except for type A_1).

PROOF. We prove that the system $s_\alpha := s_\alpha^+$ ($\alpha \in \Pi$) satisfy the relations (6.1.1), (6.1.2) and (6.1.3). The other case is proven similarly.

We normalize the basis e_α for $\alpha \in \Pi$ as $I(e_\alpha, e_\alpha) = 2$ and $e_\alpha^\vee = e_\alpha$. Then one has $I(e_\alpha, e_\beta) = -2\cos(\pi/m(\alpha,\beta))$ for $\alpha, \beta \in \Pi$

and $s_\alpha(u) = u - I_{odd}(e_\alpha, u)e_\alpha$ for $u \in \tilde{F}$.

If $m(\alpha,\beta) = 2$, then $\tilde{I}_{odd}(e_\alpha, e_\beta) = 0$ and one has $s_\alpha s_\beta(u) = s_\beta s_\alpha(u) = u - \tilde{I}_{odd}(e_\alpha^\vee, u)e_\alpha - \tilde{I}_{odd}(e_\beta^\vee, u)e_\beta$. This proves (6.1.1).

For a proof of (6.1.2) and (6.1.3), we prepare two Assertions. The first one studied the case of two vertices in a slightly generalized form.

Let h and m be a pair of integers with $h \geq 3$ and $\gcd(h, m) = 1$. Consider a two dimensional vector space $\mathbf{R}e_1 + \mathbf{R}e_2$ together with a symplectic form I_{odd} with $I_{odd}(e_1, e_2) = -\lambda$ for $\lambda := 2\cos(\pi m/h)$. Let

$$A := \begin{bmatrix} 1 & \lambda \\ 0 & 1 \end{bmatrix} \quad \text{and} \quad B := \begin{bmatrix} 1 & 0 \\ -\lambda & 1 \end{bmatrix}$$

be the matrix expression of the transvections s_1 and s_2 on $\mathbf{R}e_1 + \mathbf{R}e_2$.

Assertion 3. *Put* $e := \exp(\pi\sqrt{-1}m/h)$. *Then, one has*

$$\text{(6.3.3)} \qquad \begin{bmatrix} 1 & 1 \\ e^{-1} & e \end{bmatrix}^{-1} AB \begin{bmatrix} 1 & 1 \\ e^{-1} & e \end{bmatrix} = \begin{bmatrix} -e^2 & 0 \\ 0 & -e^{-2} \end{bmatrix}$$

$$\text{(6.3.4)} \qquad \begin{bmatrix} 1 & 1 \\ -e & -e^{-1} \end{bmatrix}^{-1} BA \begin{bmatrix} 1 & 1 \\ -e & -e^{-1} \end{bmatrix} = \begin{bmatrix} -e^2 & 0 \\ 0 & -e^{-2} \end{bmatrix}$$

$$\text{(6.3.5)} \qquad \operatorname{order}(AB) = \operatorname{order}(BA) = \langle h \rangle.$$

$$\text{(6.3.6)} \qquad \underbrace{ABA\cdots}_{h} = \underbrace{BAB\cdots}_{h} = \begin{cases} 1 & \text{if } h \text{ is even and } h/2 \text{ is odd,} \\ -1 & \text{if } h \text{ is even and } h/2 \text{ is even,} \\ (-1)^{(h-1)/2}\begin{bmatrix} 0 & -1 \\ 1 & 0 \end{bmatrix} & \text{if } h \text{ is odd.} \end{cases}$$

PROOF. For (6.3.3) and (6.3.4), we calculate eigenvectors of AB and BA:

$$AB \begin{bmatrix} 1 \\ e^{\mp 1} \end{bmatrix} = -e^{\pm 2} \begin{bmatrix} 1 \\ e^{\mp 1} \end{bmatrix} \text{ and } BA \begin{bmatrix} 1 \\ -e^{\pm 1} \end{bmatrix} = -e^{\pm 2} \begin{bmatrix} 1 \\ -e^{\pm 1} \end{bmatrix}.$$

The (6.3.5) is an immediate consequence of (6.3.3) and (6.3.4). If h is even, use (6.3.3) and (6.3.4) to calculate $(AB)^{h/2} = (BA)^{h/2} = \operatorname{diag}[-(-1)^{h/2}, -(-1)^{h/2}]$ where we applied $e^h = -1$. This gives the first two formulae of (6.3.6). If h is odd, put $h = 2n+1$. Then,

$$\begin{aligned}
\underbrace{ABA\cdots}_{2n+1} &= \begin{bmatrix} 1 & 1 \\ e^{-1} & e \end{bmatrix} \begin{bmatrix} -e^2 & 0 \\ 0 & -e^{-2} \end{bmatrix}^n \begin{bmatrix} 1 & 1 \\ e^{-1} & e \end{bmatrix}^{-1} A \\
&= \tfrac{(-1)^n}{e-e^{-1}} \begin{bmatrix} 0 & e^{-1} - e \\ e - e^{-1} & e^{-2} - e^2 \end{bmatrix} \begin{bmatrix} 1 & e + e^{-1} \\ 0 & 1 \end{bmatrix} \\
&= (-1)^n \begin{bmatrix} 0 & -1 \\ 1 & 0 \end{bmatrix}. \\
\underbrace{BAB\cdots}_{2n+1} &= B \begin{bmatrix} 1 & 1 \\ e^{-1} & e \end{bmatrix} \begin{bmatrix} -e^2 & 0 \\ 0 & -e^{-2} \end{bmatrix}^n \begin{bmatrix} 1 & 1 \\ e^{-1} & e \end{bmatrix}^{-1} \\
&= \tfrac{(-1)^n}{e-e^{-1}} \begin{bmatrix} 1 & 0 \\ -e - e^{-1} & 1 \end{bmatrix} \begin{bmatrix} 0 & e^{-1} - e \\ e - e^{-1} & e^{-2} - e^2 \end{bmatrix} \\
&= (-1)^n \begin{bmatrix} 0 & -1 \\ 1 & 0 \end{bmatrix}.
\end{aligned}$$

Comparing two expressions, we obtains the last formula of (6.3.6). □

The (6.3.6) proves (6.1.2). For a proof of (6.1.3), we prepare second Assertion which reduces the general problem to the case of two vertices.

Assertion 4. *Let Γ' be an irreducible finite type subdiagram of the Coxeter graph of M with the Coxeter number h'. Put $V' := \Sigma_{\alpha\in\Gamma'}\mathbf{R}e_\alpha$. Then i)* $\mathrm{order}((\Pi_{\alpha\in\Gamma'}s_\alpha)|V') = \langle h'\rangle$ *except for* $2\langle h'\rangle$ *if Γ' is of type A_{4k+1}, and ii)* $\mathrm{order}(\Pi_{\alpha\in\Gamma'}s_\alpha) = \langle h'\rangle$ *if $h'/2$ is not an exponent of Γ' and $=\infty$ if $h'/2$ is an exponent of Γ' (recall* (6.1.4) *for $\langle h\rangle$).*

PROOF. Since Γ' is a tree, we decompose its vertices into a disjoint union $\Gamma'_1 \amalg \Gamma'_2$ such that each Γ'_i is totally disconnected. Put $c = c_1c_2$ for $c_i := \Pi_{\alpha\in\Gamma'_i}s_\alpha$ and $V' = V'_1 \oplus V'_2$ for $V'_i := \Sigma_{\alpha\in\Gamma'_i}\mathbf{R}e_\alpha$ for $i = 1,2$. Since $I_{odd}(e_\alpha, e_\beta) = 0$ for $\alpha, \beta \in \Gamma'_i$, one has $c_i(u) = u - \sum_{\alpha\in\Gamma'_i}\tilde{I}_{odd}(e_\alpha^\vee, u)e_\alpha$.

Since $(c_1 - c_2)(e_\beta) = -\sum_{\alpha\in\Gamma'_1}I_{odd}(e_\alpha^\vee, e_\beta)e_\alpha + \sum_{\alpha\in\Gamma'_2}I_{odd}(e_\alpha^\vee, e_\beta)e_\alpha = 2e_\beta - \sum_{\alpha\in\Gamma'}I(e_\alpha^\vee, e_\beta)e_\alpha$, the matrix expression of $2 - c_1 + c_2$ w.r.t. the basis e_α ($\alpha \in \Gamma'$) is the Cartan matrix of type Γ', which is up to a multiplication of a diagonal matrix $(I(e_\alpha, e_\alpha))_{\alpha\in\Gamma'}$ symmetric. Due the classical result on Cartan matrices (e.g. [B,chV,§6.2]) (or, it is easy to see directly), the eigenvalues are $2 - 2\cos(\pi m'_i/h')$ where m'_i are the exponents for Γ'. Let $e = e_1 + e_2$ with $e_i \in V'_i$ be an eigenvector ($\neq 0$) of $c_1 - c_2$ belonging to an eigenvalue $\lambda = 2\cos(\pi m'_i/h')$. Combining the equalities: $c_1(e_1) = e_1$, $c_2(e_2) = e_2$ and $(c_1 - c_2)(e_1 + e_2) = \lambda(e_1 + e_2)$, one has $c_1e_2 = e_2 + \lambda e_1$ and $c_2e_1 = e_1 - \lambda e_2$ and, hence, $e_1 - e_2$ is an eigenvector of $c_1 - c_2$ belonging to the eigenvalue $-\lambda$. If we assume $\lambda \neq 0$ (i.e. $2m'_i \neq h'$), then $e_1 \neq 0 \neq e_2$ and $V_\lambda := \mathbf{R}e_1 + \mathbf{R}e_2$ has rank 2. Then, c_1 and c_2 act on the space V_λ as transvections of the vectors e_1 and e_2 with respect to the skew symmetric form $\left[\begin{smallmatrix}0 & -\lambda\\ \lambda & 0\end{smallmatrix}\right]$. We saw in Assertion 3. that the order of $c|V_\lambda$ is equal to $\langle h'/\gcd(h', m'_i)\rangle$.

Put $\lambda_i := 2\cos(\pi m'_i/h')$ for $0 < m'_i < h'/2$. One has the decomposition:

$$V' = \ker(I'_{odd}) \oplus V_{\lambda_1} \oplus \cdots \oplus V_{\lambda_m},$$

where $I'_{odd} := I_{odd}|V'$ so that i) $\ker(I'_{odd})$ is the fixed point subspace of V' by the action of c, and ii) the action of c preserves the splitting.

i) Since the smallest exponent is $m'_1 = 1$ (recall (2.8.3)), one has $\mathrm{order}(c|V_{\lambda_1}) = \langle h'\rangle$. The order $\langle h'/\gcd(h', m'_i)\rangle$ of another factor $c|V_{\lambda_i}$ is not a divisor of $\langle h'\rangle$, only when h' is $2\times$odd and $\gcd(h', m'_i) = 2$. This occurs only when Γ' is of type A_{4k+1} ($k \geq 2$).

ii) If $\ker(I'_{odd}) = \{0\}$, then $\tilde{F} = V' \oplus V'^{\perp}$ so that $\mathrm{order}(c|V') = \mathrm{order}(c)$. Assume $\ker(I'_{odd}) = V'_1 \cap V_2'^{\perp} \neq \{0\}$. Then for $x \in \cap_i V_{\lambda_i}^\perp \setminus (\ker(I'_{odd}))^\perp$ one has $c^n x - x = n(\sum_{\alpha\in\Gamma'_1}I_{odd}(e_\alpha^\vee, x)e_\alpha) \neq 0$ for $n \neq 0$. □

Assertion 4. proves (6.1.3) and, hence, Lemma is proven. □

We state another consequence (on odd roots) of Assertion 3.

COROLLARY. *Let $m(\alpha, \beta) \in \mathbf{Z}_{\geq 3}$ be odd. Then $s_{\alpha\beta} := s_\alpha s_\beta s_\alpha \cdots$ (m-factors) is of order 4. It acts transitively on the set $\{\pm e_\alpha, \pm e_\beta\}$.*

Proof. Use the third line of the formula (6.3.6). □

Conjecture 1. *The homomorphisms* (6.3.2) *are injective.*

Assertion 5. *Conjecture 1. is true for Coxeter matrices with $l \leq 2$.*

Proof. This is trivial for types A_1 and $A_1 \times A_1$. Thus, we prove for type $I_2(m)$ ($m \geq 3$) (including A_2, B_2, C_2, H_2 and G_2). It is achieved by a use of the fundamental domain on the complex upper-half-plane.

Assertion 6. *Let $\bar{\rho} : \Gamma(I_2(m)) \to \mathrm{PSp}(F_{I_2(m)}) = \mathrm{Sp}(F_{I_2(m)})/\{\pm 1\}$ be the projectivization of $\rho = \rho^+$ for $m \geq 3$. Then the image of $\bar{\rho}$ is isomorphic to the group* $\langle \bar{s}_\alpha, \bar{s}_\beta \mid i)\ (\bar{s}_\alpha \bar{s}_\beta)^{m/2} = 1$ if m is even or $(\bar{s}_\alpha \bar{s}_\beta)^m = 1$ if m is odd, and *ii)* $\bar{s}_\alpha \bar{s}_\beta \bar{s}_\alpha \cdots = \bar{s}_\beta \bar{s}_\alpha \bar{s}_\beta \cdots$ (m-factors in both sides)$\rangle$.

Proof. It is obvious that the images $\bar{s}_\alpha$ and $\bar{s}_\beta$ in $\mathrm{PSp}(F_{I_2(m)})$ of s_α and s_β for $\Pi = \{\alpha, \beta\}$ satisfy the relations i) and ii) due to Assertion 3.

Let $\omega_1 = \langle e_\alpha, z \rangle$ and $\omega_2 = \langle e_\beta, z \rangle$ be homogeneous coordinates of $\mathbf{P}(\mathrm{Hom}_{\mathbf{R}}(F, \mathbf{C}))$. Consider the action of $\Gamma(I_2(m))$ on a connected component $\mathcal{H}$ of $\mathbf{P}(\mathrm{Hom}_{\mathbf{R}}(F, \mathbf{C})) \setminus \mathbf{P}(F^*)$, which is isomorphic to the complex upper half plane $\mathcal{H}$ with respect to the inhomogeneous coordinate $\tau := \omega_1/\omega_2$. Define a subset of $\mathcal{H}$:

$$\begin{aligned} \mathcal{F} \ &:= \ \{\tau \in \mathcal{H} \mid -\cos(\pi/m) \leq \Re(\tau) < \cos(\pi/m), |\tau| > 1\} \\ &\quad \cup \{e^{\sqrt{-1}\theta} \mid \pi/2 \leq \theta \leq \pi(1 - 1/m)\}. \end{aligned}$$

Then, the fundamental domain of the action is either $\mathcal{F}$ if m is odd or $\mathcal{F} \cup \mathcal{F}'$ if m is even where $\mathcal{F}'$ is the image of $\mathcal{F}$ by the transformation $\tau \mapsto -1/\tau$. (In fact, the two vertices e and $-e^{-1}$ of $\mathcal{F}$ are the fixed point of the action of $\bar{s}_\alpha \bar{s}_\beta$ and $\bar{s}_\beta \bar{s}_\alpha$, respectively, and the cusps $\sqrt{-1}\infty$ and 0 are fixed point by the actions of $\bar{s}_\alpha$ and $\bar{s}_\beta$, respectively.) $\square$

We return to a proof of Assertion 5. Let us "lift" the fundamental relations in $\mathrm{PSp}(F_{I_2(m)})$ to the elements in $\mathrm{Sp}(F_{I_2(m)})$. Then, due to Assertion 3, one has $s_\alpha s_\beta s_\alpha \cdots = s_\beta s_\alpha s_\beta \cdots$ (m-times) for all m, $(s_\alpha s_\beta)^{m/2} = 1$ if m is even and $m/2$ is odd, $(s_\alpha s_\beta)^{m/2} = -1$ if m is even and $m/2$ is even and $(s_\alpha s_\beta)^m = -1$ if m is odd. Then, by killing the sign factor, one obtains (6.1.2) and (6.1.3) as the fundamental relations for the image of $\Gamma(I_2(m))$ in $\mathrm{Sp}(\tilde{F}_{I_2(m)})$. So, Assertion 5. is proven. $\square$

EXAMPLE. There are three cases when $I_2(p)$ is crystallographic.

$\Gamma(I_2(3)) = \Gamma(A_2) = \Gamma_0(1) := \mathrm{SL}_2(\mathbf{Z})$,

$\Gamma(I_2(4)) = \Gamma(B_2) = \Gamma_0(2) := \{[\begin{smallmatrix} a & b \\ c & d \end{smallmatrix}] \in \mathrm{SL}_2(\mathbf{Z}) \mid b \equiv 0 \bmod 2\}$,

$\Gamma(I_2(6)) = \Gamma(G_2) = \Gamma_0(3) := \{[\begin{smallmatrix} a & b \\ c & d \end{smallmatrix}] \in \mathrm{SL}_2(\mathbf{Z}) \mid b \equiv 0 \bmod 3\}$.

REMARK. Let Z be a central element of the Artin group $A(W)$. Then $\rho(Z)|V$ is either 1 or -1. *Proof.* Put $z := \rho(Z)$. The relation $g_\alpha Z = Z g_\alpha$ for $\alpha \in \Pi$ implies $s_\alpha(z(u)) = z(u - \tilde{I}_{odd}(e_\alpha^\vee, u) e_\alpha)$ for $u \in \tilde{V}$. This implies $\tilde{I}_{odd}(e_\alpha^\vee, z(u)) e_\alpha = \tilde{I}_{odd}(e_\alpha^\vee, u) z(e_\alpha)$. Choosing u such that $\tilde{I}_{odd}(e_\alpha^\vee, u) \neq 0$, one observes that $z(e_\alpha) = c_\alpha e_\alpha$ for some constant c_α. Substituting u by e_β for $\beta \in \Pi$ and assuming $z(e_\beta) = c_\beta e_\beta$ for some constant c_β, one has $c_\beta \tilde{I}_{odd}(e_\alpha^\vee, e_\beta) = c_\alpha \tilde{I}_{odd}(e_\alpha^\vee, e_\beta)$. Since the graph for Π is connected, $c_\alpha = \varepsilon$ is independent of $\alpha \in \Pi$ and $\varepsilon \in \{\pm 1\}$.

6.4. Period domain $\mathcal{B}(W)$.

Suppose W is crystallographic. That is: there is a finite root system w.r.t. (F, I) whose Weyl group coincides with W (see [B,chap.VI,§2,n°5 and §4] for details). Then, for a suitable choice of the scaling constants d_α ($\alpha \in \Pi$) in 6.2, the matrix $(I(e_\alpha^\vee, e_\beta))_{\alpha,\beta \in \Pi}$, so called the Cartan matrix, is integral. In fact, up to an overall constant factor on the scaling constants, there are one or two choices according as M is simply laced or not. Choose one such scaling and consider the set $R := \cup_{\alpha\in\Pi} W\ e_\alpha$ and a lattice $Q := \Sigma_{\alpha\in\Pi}\mathbf{Z}e_\alpha$ in V. Then R forms a root system with simple basis $\{e_\alpha \mid \alpha \in \Pi\}$, $W = W(R)$ is the Weyl group and $Q = Q(R)$ is the root lattice of the root system. By a choice of an over all constant factor on the scaling constants, $I \mid Q \times Q$ is integral valued. We normalize $\min\{I(e_\alpha, e_\alpha) \mid \alpha \in \Pi\} = 2$. We shall sometimes denote $\Gamma(R)$ instead of $\Gamma(M)$ or $\Gamma(W)$ (here recall that the action of $\Gamma(R)$ on $Q(R)$ depends on a choice of a chamber (5.2 Example)). The group

$$\tilde{Q} := Q \oplus (Q \cap \ker(I_{odd})) \tag{6.4.1}$$

is a full sublattice in $\tilde{F}$ and $\tilde{I}_{odd} \mid \tilde{Q} \times \tilde{Q}$ is an integral symplectic form.

Problem. Let $\tilde{t}$ be a point in the chamber C and t be its image in S_W. Show that the first homology lattice $H_1(X_t, \mathbf{Z})$ of the Milnor fiber of φ_1 is canonically isomorphic to the lattice $(Q, I_{odd,C})$ (use [S8]).

Recall the flat vector space Ω_+ (3.11.2) attached to the reflection group W. Due to 1., the period map attached to the primitive form ζ_{F_1} (recall 5.3) is a holomorphic local submersion given by

$$P_W \quad : t \in (S_W \backslash D_W)^\sim \ \mapsto\ \textstyle\int \mathrm{Res}_t[\zeta_{F_1}^{(0)}] \in \mathrm{Hom}(Q, \Omega_{+\mathbf{C}}).$$

The projection to the first factor is given by local covering map

$$\langle D, P_W \rangle : \ t \in (S_W \backslash D_W)^\sim \ \mapsto\ \textstyle\int \mathrm{Res}_t[\zeta_{F_1}^{(0)}] \in \mathrm{Hom}(Q, \mathbf{C}).$$

Conjecture 2. *The period map P_W is a global immersion of $(S_W\backslash D_W)^\sim/\ker(\rho\circ\gamma)$ into a locally closed submanifold $\mathcal{B}(W)$ of an open cone $\tilde{\mathcal{B}}(W)$ in $\mathrm{Hom}(Q, \Omega_{+\mathbf{C}})$. Determine $\mathcal{B}(W)$ and $\tilde{\mathcal{B}}(W)$ explicitly without using period integral but only in terms of the reflection group.*

These problems are somehow asking for analogues of the Riemann's relations and Schottkey's relations for the classical periods of Abelian integrals on compact Riemann surfaces. Here the period domain is "homogenized" (i.e. admits $\mathbf{C}^\times$-action) due to the flat structure on Ω.

6.5. Inverse functions on $\mathcal{B}(W)$.

We observed in 5.3 Example that for types A_2, A_3 and B_2 the inverse map from the period domain $\mathcal{B}(W)$ (to be exact, from its projection to the first factor) to $S_W \backslash D_W$ is described by a system of elliptic Eisenstein series (including some specializations of the $\wp$-function).

There are four more types D_4, B_3, C_3 and G_2, where the similar considerations work to construct the inverse map. Namely, the Milnor fibers for types D_4, C_3 and G_2 are elliptic curves with 3 punctures either without symmetry, with $\mathbf{Z}/2$-symmetry or with $\mathbf{Z}/3$-symmetry, respectively. The Milnor fiber for type B_3 is a double cover of a once punctured elliptic curve.

Problem Work out explicitly the inversion problem for the period maps of types D_4, B_3, C_3 and G_2.

Beyond these examples, the genus of Milnor fiber increases (without a "reduction" to an elliptic curve). Then the indefinite integral $\int_{\infty_1}^{\infty_2} \zeta$ becomes multivalued by the periods $\int_\gamma \zeta$ for $\gamma \in H_1(\bar{X}_t, \mathbf{Z})$, where the periods are dense in the 1-dimensional complex plane. Then, it is already a problem to find a reasonable formulation for the period map and the inverse map. We need to introduce a concept of a regular function on the period domain $\mathcal{B}(W)$ (with a suitable polynomial growth condition on the boundary) which is automorphic for the group $\Gamma(W)$, and also a concept of pull-back to a function on $(V^*//W)_{\mathbf{C}}$. For instance, one may consider either a single indefinite integral $\int_\infty^p$ and its (multi-valued-)inverse function as in the classical $\mathfrak{p}$-function theory, or a system of derivatives $\delta_i \int_\infty^p \zeta$ of the periods to obtain the Jacobian variety as in the classical Abel-Jacobi theory? (See also a related work [Ko][Mu][To]).

Conjecture 3. *There exists a ring* $\mathbf{C}[E]$ *of regular functions on* $\mathcal{B}(W)$ *which are automorphic for the group* $\Gamma(W)$. *Any element of* $\mathbf{C}[E]$ *is a pull-back of a polynomial function in* $S(V)^W$ *to the period domain. This correspondence induces a natural isomorphism:*

$$\mathbf{C}[E] \simeq S(V)^W \otimes \mathbf{C}. \qquad (6.5.1)$$

The isomorphism implies that the Spec($\mathbf{C}[E]$) carries the flat structure studied in §3. Then we may naturally ask how to describe the flat structure in terms of $\mathbf{C}[E]$. In particular, $\mathbf{C}[E]$ is generated by algebraically independent homogeneous elements, say $E_4, \cdots, E_{2h}$ of degree $2d_1 = 4, \cdots, 2d_l = 2h$ such that $J(dE_{2d_i}, dE_{2d_j}) = const$, which we may call *flat automorphic forms* or *primitive automorphic forms.*

Conjecture 4. *Let W be crystallographic, which is not of type G_2. Then there exists an extension* $\mathbf{C}[E] \subset \mathbf{C}[\tilde{E}]$ *by a ring of regular functions on* $\mathcal{B}(W)$ *which are automorphic for the principal congruence subgroup of* $\Gamma(W)$ *(c.f. 6.1iv)) such that the commutative diagram holds:*

$$\begin{array}{ccc} \mathbf{C}[E] & \simeq & S(V)^W \otimes \mathbf{C} \\ \cap & & \cap \\ \mathbf{C}[\tilde{E}] & \simeq & S(V) \otimes \mathbf{C}. \end{array} \qquad (6.5.5)$$

We remark that, by definition, Spec($\mathbf{C}[\tilde{E}]$) carries a linear space structure with a flat metric, which is isomorphic to the pair (V^*, I^*).

6.6. Power root of the Discriminant Δ_W.

We turn our attention to the discriminant Δ_W and its power root. After a choice of the primitive vector field D in (3.2.1), the discriminant was normalized to be a monic polynomial of degree l in P_l (c.f. 2.9 Lemma 1.). However, in the sequel, we disregard tentatively the constant factors and proceeds the caluculations up to constant factor.

Recall the anti-invariants $\delta_W \in S(V)$ in 2.7 and the disriminant $\Delta_W = \delta_W^2 \in S(V)^W$ in 2.9. Let us denote by the same notation the corresponding elements in $\mathbf{C}[\tilde{E}]$ (resp. $\mathbf{C}[E]$) in RHS of (6.5.5). Owing to Conjectures 3 and 4, we have Jacobian expressions of Δ_W and δ_W as below. Namely, let $\tilde{E}_1, \cdots, \tilde{E}_l$ be a system of generator of $\mathbf{C}[\tilde{E}]$ corresponding to a linear coordinate system of V, and let $e_1, \cdots, e_l \in \Pi$ be the linear coordinate system on $\tilde{\mathcal{B}}(W)$. Then by a use of (4.5.9), up to a constant factor c, one has

$$dE_4 \wedge \cdots \wedge dE_{2h} = c\, \Delta_W \, de_1 \wedge \cdots \wedge de_l, \qquad (6.6.1)$$

$$d\tilde{E}_1 \wedge \cdots \wedge d\tilde{E}_l = c\, \delta_W \, de_1 \wedge \cdots \wedge de_l. \qquad (6.6.2)$$

One should define the meaning of "boundary components" of $\mathcal{B}(W)$ and "cusp forms" in $\mathbf{C}[E]$ or in $\mathbf{C}[\tilde{E}]$ such that the discriminant Δ_W (resp. δ_W) generates the ideal of cusp forms in $\mathbf{C}[E]$ (resp. $\mathbf{C}[\tilde{E}]$). The discriminant, as a function on $\mathcal{B}(W)$, vanishes nowhere. The δ_W is a square root of Δ_W, and is anti-invariant with respect to θ_W:

$$\gamma^* \delta_W = \theta_W(\gamma)\, \delta_W. \tag{6.6.3}$$

Here we recall (6.1.5) for the definition of the character θ, and recall the commutative diagram (6.1.8).

Conjecture 5. *Does δ_W have a suitable infinite product expression?*

REMARK. The conjectures and problems in 6.5 and 6.6 seem to have close relations with odd the root system: $R_{odd} := \cup_{\alpha \in \Pi_W} \Gamma(W) e_\alpha$. For instance, as a generalization of Eisenstein series, one may consider, for $d \in \mathbf{Z}_{\geq 2}$, the sum of partial fractions such as $\sum_{e \in R_{odd}} e^{-2d}$, $\sum_{e \in R_{odd,s}} e^{-2d}$, $\sum_{e \in R_{odd,l}} e^{-2d}$ or a sum of suitable combinations $e_s^{-2d} - e_l^{-2d}$ for $e_s \in R_{odd,s}$ and $e_l \in R_{odd,l}$ and also $e \in \Gamma(W) \cdot (\ker(I_{odd}) \cap Q)$ in case whenI_{odd} is degenerate on the period domain. There are supporting examples that the boundary component and the infinite product might reasonably be described in terms of the odd root system. In spite of the examples, it is still unclear to the author what are natural formulation of the partial fractional sums fitting to our setting and we leave them as an open problem.

EXAMPLE. **1.** Type A_2. We have the expression $\mathcal{B}(A_2) = \{(u_1, u_2) \in \mathbf{C}^2 \mid Im(u_1 \bar{u}_2 - \bar{u}_1 u_2) > 0\}$ where (u_1, u_2) are the homogeneous linear coordinates of the period domain. The ring $\mathbf{C}[E]$ is the ring of automorphic forms for the full modular group $\Gamma_0(1)$, and is generated by the classical Eisenstein series E_4 and E_6 (e.g. [Ko,p111]). The ring $\mathbf{C}[\tilde{E}]$ is the ring of automorphic forms for the principal congruence subgroup $\Gamma(2)$ of level 2. All conjectures are positively solved for type A_2.

The discriminant Δ_W (in homogeneous form) is given by

$$\Delta_{A_2}(u_1, u_2) = \eta(\tau)^{24} u_2^{-12},$$

where $\tau := u_1/u_2$ and $\eta(\tau)$ is the Dedekind eta-function (e.g. [Ko,p121]). It generates the ideal of cusp forms in the ring $\mathbf{C}[E]$. The generator δ_{A_2} of the anti-invariants with respect to the character θ_{A_2} is given by

$$\delta_{A_2}(u_1, u_2) = \eta(\tau)^{12} u_2^{-6},$$

(for this proof and for a finer statement see *Assertion 7.* below).

2. Type B_2. We have seen that the discriminant for the family φ_{B_2} decomposes into irreducible components corresponding to short and long roots (5.3 Example 4), and, hence, it is given by

$$\Delta_{B_2} = \eta(\tau)^8 \eta(2\tau)^8 u_2^{-8}.$$

Then one can show by a similar calculation as in type A_2 that

$$\delta_{B_2} = \eta(\tau)^4 \eta(2\tau)^4 u_2^{-4}.$$

is an anti-invariant with respect to the character θ_{B_2} on $\Gamma(B_2) = \Gamma_0(2)$.

3. Type G_2. Since the discriminant for the family φ_{G_2} decomposes into components corresponding to short and long roots, one may apriori describe the discriminant (up to a constant factor) to be of the form

$$\Delta_{G_2} = \eta(\tau)^6 \eta(3\tau)^6 u_2^{-6}.$$

Then one can show by a similar calculation as in type A_2 that

$$\delta_{G_2} = \eta(\tau)^3\eta(3\tau)^3u_2^{-3}$$

is an anti-invariant with respect to the character θ_{G_2} on $\Gamma(G_2) = \Gamma_0(3)$.

Throughout the examples, we observe further the following. Recall the integer $k(W)$ and the character ϑ_W of the group $\Gamma(W)$ (see 6.1 ii)).

Assertion 7. *For type A_2, B_2 and G_2, consider the $k(W)$th power root, say λ_W, of δ_W. Since $k(A_2) = 6$, $k(B_2) = 4$ and $k(G_2) = 3$, up to a constant factor, they are explicitly given by*

$$\lambda_{A_2} := \eta^2(\tau)u_2^{-1},\ \lambda_{B_2} := \eta(\tau)\eta(2\tau)u_2^{-1} \text{ and } \lambda_{G_2} := \eta(\tau)\eta(3\tau)u_2^{-1}.$$

Then, λ_W is an automorphic form for $\Gamma(W)$ with the character ϑ_W. That is:

$$g^*(\lambda_W) = \vartheta_W(g)\lambda_W \quad \text{for } g \in \Gamma(W).$$

PROOF. We have only to verify that λ_W is equivariant with the character ϑ_W. The verification of this fact is achieved by an elementary but slightly subtle use of the transformation formula for the eta-function, where we use the sign conventions in [Kob, p121].

Put $W_1 := W_{A_2}$, $W_2 := W_{B_2}$ and $W_3 := W_{G_2}$. Then, the modular group $\Gamma(W_p)$ is given by $\Gamma_0(p)$ for $p = 1, 2$ and 3, and is generated by $s_\alpha : (u_1, u_2) \mapsto (u_1, u_2)\begin{bmatrix} 1 & 0 \\ 1 & 1 \end{bmatrix}$ and $s_\beta^{-1} : (u_1, u_2)\begin{bmatrix} 1 & p \\ 0 & 1 \end{bmatrix}$ (recall 6.3 Example). On the other hand, the explicit formula (in the above examples) says that $\lambda_{W_p} = \eta(\tau)\eta(p\tau)u_2^{-1}$ $(p = 1, 2, 3)$.

Further more, one observes directly $k(W_p) = \frac{12}{1+p}$ $(p = 1, 2, 3)$. Therefore, by putting $\zeta := \exp(\pi\sqrt{-1}/12)$, the character ϑ_{W_p} is given by $\vartheta_{W_p}(s_\alpha) = \vartheta_{W_p}(s_\beta) = \zeta^{p+1}$ (the notation here is incoherent with that in 6.3). So,

$$\begin{aligned} s_\alpha^*(\lambda_{W_p}) &:= \eta(\tau+1)\ \eta(p\tau+p)\ u_2^{-1} \\ &= \zeta\eta(\tau)\ \zeta^p\eta(p\tau)\ u_2^{-1} = \zeta^{p+1}\lambda_{W_p}, \end{aligned}$$

$$\begin{aligned} s_\beta^{*-1}(\lambda_{W_p}) &:= \eta(\tau/(p\tau+1))\ \eta(p\tau/(p\tau+1))\ (pu_1+u_2)^{-1} \\ &= \sqrt{\tfrac{-(p\tau+1)/\tau}{\sqrt{-1}}}\ \eta(-p-1/\tau)\ \zeta\eta(-1/(p\tau+1))\ (pu_1+u_2)^{-1} \\ &= \sqrt{\tfrac{-(p\tau+1)/\tau}{\sqrt{-1}}\tfrac{\tau}{\sqrt{-1}}\tfrac{p\tau+1}{\sqrt{-1}}}\ \zeta^{-p}\eta(\tau)\ \zeta\eta(p\tau+1)\ (p\tau+1)^{-1}u_2^{-1} \\ &= \zeta^{-p+2}\sqrt{\tfrac{1}{\sqrt{-1}}}\ \eta(\tau)\ \eta(p\tau)\ u_2^{-1} \\ &= \zeta^{-p-1}\eta(\tau)\ \eta(p\tau)\ u_2^{-1} = \zeta^{-p-1}\lambda_{W_p}. \end{aligned}$$

□

We observe also the following nearly trivial remark, which is still interesting since the form I_{odd} is degenerate for the type A_3.

Fact. *A $k(A_3)$th power root $\lambda_{A_3} := \eta(\tau)^2u_2^{-1}$ of δ_{A_3} is automorphic for $\Gamma(A_3)$ with the character ϑ_{A_3} (here, recall $k(A_3)=k(A_2)=6$).*

Proof. This is shown by a reduction to A_2. Recall 5.3 Ex.3. for the setting. The morphism $S_{A_3}\backslash D_{A_3} \to S_{A_2}\backslash D_{A_2}$ induces an equality $\Delta_{A_3} = c\Delta_{A_2}$ and a homomorphism $A(A_3) \to A(A_2)$ bringing the generators a_1, a_2, a_3 of $A(A_3)$ to b_1, b_2, b_1 of $A(A_2) = \langle b_1, b_2 \mid b_1b_2b_1 = b_2b_1b_2\rangle$ ([S2, Theorem III]) so that the characters ϑ_{A_3} and

ϑ_{A_2} commutes with the homomorphism $\Gamma(A_3) \to \Gamma(A_2)$. This implies the result. $\square$

This fact encourages us to ask the next question:

Conjecture 6. *Let W be a crystallographic finite reflection group. Is the $k(W)$th power root of δ_W, say λ_W (up to a constant factor), an automorphic form for the group $\Gamma(W)$ with the character ϑ_W? Can one find an infinite product expression for λ_W compatible with Conjecture 4?*

In view of the fact that $k(D_4) = 6$ is the last largest power (among all finite crystallographic group) and that the period map for D_4 can still be described by elliptic integrals, it is interesting to have an exact and explicit expression of λ_{D_4} as a distinguished Jacobi form on $\tilde{\mathcal{B}}(D_4)$.

A concluding Remark.

Except for the discussed types $A_1, A_2, A_3, B_2, B_3, C_2, C_3, D_4$ and G_2, in all further cases, the character ϑ_W takes values in $\mathbf{Z}_4$=$\{\pm1,\pm\sqrt{-1}\}$. However, we have no information on the primitive automorphic forms for them at this stage, since they are beyond elliptic integrals. Actually, Conjecture 6 seems not be understandable only from the geometry of the family $\varphi_W : \mathfrak{X}_W \to S_W$. Instead, the conjecture seems reasonably understandable if one finds a suitable "mirror object" to the family φ_W such that it gives arise a suitable construction of the inversion maps and the flat structure on S_W. Perhaps, finding such mirror object (based on odd root systems?) may be the main question and goal of the present section and hence the main problem of the the present article.

References

[A] Aleksandrov, A.G.: Meromorphic connections with logarithmic poles along a free divisor, MPI Preprint Series 90 (2002).

[Ar1] Arnold,V.I.: Normal forms for functions near degenerate critical points, the Weyl groups of A_k, D_k, E_k and Lagrangian singularities; Funct. Anal. **27**(1972),254-272.

[Ar2] Arnold,V.I.: Critical points of functions on a manifold with boundary, the simple Lie groups B_k, C_k and F_4 and singularities of evolutes; Russian Math. Surveys **33** (1978), 99-116.

[B] Bourbaki, N.: Elements de mathematique, Fasc. XXXIV, Groupes et algebres de Lie, Chs. 4-6, Hermann, Paris 1968.

[Br1] Brieskorn, Egbert: Singular elements of semisimple algebraic groups, Actes Congres Int. Math. **2**, Nice (1970), 279-284.

[Br2] Brieskorn, Egbert: Die Fundamental Gruppe des Raumes der Regulären Orbits einer endlichen komplexen Spiegelungsgruppe, Inventiones Math. **12**, 57-61 (1971).

[Br3] Brieskorn, Egbert: Sur les groupes de tresses. Sém. Bourbaki (1971/71); Lect.Notes Math., **317**, 21-44 (1973).

[B-S] Brieskorn, Egbert & Saito, Kyoji: Artin Gruppen und Coxeter Gruppen, Inventiones Math. **17**, 245-271 (1972).

[Ch1] Chevalley, Claude: The Betti numbers of the exceptional simple Lie groups, Proc. Intern. Congress of Math. (Cambrdge, Mass., 1950), vol.2, Amer. Math.Soc., Province RI, 1952,pp.21-24.

[Ch2] Chevalley, Claude, Invariants of finite groups generated by reflections, Amer. J. Math. **77** (1955), 778-782.

[C] Coleman, A.J.: The Betti numbers of the simple Lie groups, Canad. J. Math. **10**, 349-356 (1958).

[Co1] Coxeter, H.S.M.: Discrete groups generated by reflections. Ann. of Math. **35** (1934), 588-621.

[Co2] Coxeter, H.S.M.: The product of the generators of a finite group generated by reflections, Duke Math. J. **18** (1951), 765-782.

[D-S] De Contini,C. & Salvetti, Mario: Cohomology of Coxeter groups and Artingroups, Math. Res. Lett. 7 (2000).

[De] Deligne, Pierre: Les immeubles des tresses généralizé, Inventiones Math. **17** 273-302 (1972).

[De2] Deligne, Pierre: Action du groupe des tresses sur categorie, Invent. math. **128**, 159-175 (1997).

[Du] Dubrovin, Boris: Geometry of 2D-Topological Field Theory, LN in Math. **1620**, 120-348 (1996).

[Gi] Givental, Alexander: A_{n-1} singularites and nKdV hierarchies, math.AG/0209205.

[Ha] Hartshorne, Robin: Residue and Duality, Lecture Notes in Math., , Springer, Berlin-Heidelberg-New York,196.

[He] Hertling, Claus: Frobenius Manifolds and Moduli Spaces for Singularities, Cambridge University Press, 2002.

[Ko] Komatsu, Makoto: Period map of hyperelliptic curves, Preprint 1998

[Kob] Koblitz, Niels: Introduction to elliptic curves and modular forms, 2nd ed., Grad. Texts in Math., vol.97, Springer-Verlag, New York, 1993.

[K1] Kostant, Beltram: The principal three-dimensional subgroup and the Betti numbers of a complex simple group, A. J. M., **81**, 973-1032 (1959).

[K2] Kostant, Beltram: Lie group representations on polynomial rings, Amer. J. Math., **85**, 327-404 (1963).

[KR] Kostant, Beltram & Rallis, S.: Orbits and representaions associated with , Amer. J. Math., **93**, (1971) 753-809.

[L] Looijenga, Eduard: A period mapping for certain semi-universal deformations, Com. Math., **30** (1975), 299-316.

[Ma1] Manin, Yuri: Three constructions of Frobenius manifolds: a comparative study, Asian J. Math.

[Ma2] Manin, Yuri: Frobenius manifolds, Quantum Cohomology, and Moduli Spaces, AMS Colloquim Publications, Vol. **47** (1999).

[Mu] Mumford, David: Tata Lectures on Theta II, Jacobian thera functions and differential equations, Birkh"auser (1984).

[Sab] Sabbah, Claude: Déformations isomnodromiques et variétés de Frobenius, Svoirs Actuels, EDP Sciences/CNRS Édition (2002).

[S1] Saito, Kyoji: Einfach Elliptische Singularitáten, Inv. math., (1974).

[S2] Saito, Kyoji: On the uniformization of complements of discriminant loci, AMS Summer Institute, Williams college, 1975, RIMS Kokyuroku **287**, 117-137 (1977).

[S3] Saito, Kyoji: On a Linear Structureof the Quotient Variety by a Finite Reflexion Group, RIMS Preprint 288 (1979), Publ. RIMS, Kyoto Univ. **29**, 535-579 (1993).

[S4] Saito, Kyoji: Theory of logarithmic differential forms and logarithmic vector fields, J. Fac. Sci. Univ. Tokyo Sect. IA Math.**27**, 265-291 (1981).

[S5] Saito, Kyoji: On the period of primitive integrals I, RIMS-412 (1982).

[S6] Saito, Kyoji: Period Mapping Associated to a Primitive form, Publ. RIMS, Kyoto Univ., **19**, 1231-1264 (1983).

[S7] Saito, Kyoji: The higher residue pairings $K_F^{(k)}$ for a family of hypersurface singularities, Proc. Symp. Pure Math., Vol **40** (1983), 441-463.

[S8] Saito, Kyoji: The polyhedron dual to the chamber decomposition for a finite Coxeter group, preprint (1997).

[Sa] Salvetti, Mario: Topology of the complement of real hyperplanes in $\mathbf{C}^n$, Invent. math. **83**, (1987) 603-618.

[S-Y-S] Saito, Kyoji &Yano, Tamaki & Sekiguchi, Jiro: On a ceertain generator system of the ring of invariants of a finite reflexion group, Comm. Algebra,**8**, 373-408 (1980).

[S-S] Sekiguchi, Jiro & Shimizu, Yasuhiro: Simple Singularities and Infinitesimally Symmetric Spaces, Proc. Japan Acad. **57**, Ser. A (1981), pp42-46.

[Sl] Slodowy, Peter: Simple singularities and Simple Algebraic Groups, Lecture Notes in Math., **815**, Springer, Berlin-Heidelberg-New York,1980.

[Sp] Springer, Tony A.: Reductive Groups, Proc. of Symposia in Pure Math. Vol. 33 (1979), Part 1, pp.3-27.

[Ta] Takahashi, Atsushi: Lecture at Kinosaki School on Mirror Symmetry, September 8-12 (2002).

[Te] Terao, Hiroaki: The Hodge filtration and contact-order filtration of derivations of Coxeter arrangements, (math.CO/0205058).

[Th] Thomae, J.: Beitrag zur Bestimmung von $\theta(0, 0, \ldots, 0)$ durch die Klassenmoduln algebraischer Funktionen, J. für die reine und angew. Math., Bd. **71** (1870) 201–222.

[Y] Yamada, Hiroshi: Lie group theoretic construction of period mappings, Math.Z. **220**, 231-255 (1995).

[Ya] Yano, Tamaki: On the invariant ring of the Weyl group of type E_7, preprint 1981.

[Yo] Yoshinaga, Masahiko: The primitive derivation and freeness of multi-Coxeter arrangements, Proc. Japan Acad., **78**, Ser. A (2002).

RIMS, Kyoto university

The Laplacian for a Frobenius manifold

Ikuo Satake

1. Introduction

In this paper, we define the holomorphic Laplacian for a Frobenius manifold. We give the description of the Laplacian in terms of the prepotential. This will be used to characterize the flat coordinates for the universal unfolding of the function with a simple elliptic singularity. This characterization enables us to solve the so-called "Jacobi's inversion problem," i.e. the description of the flat coordinates as the automorphic functions on the period domain w.r.t. the period mapping for the primitive forms. More explicitly the Laplacian relates the flat coordinates with the theta functions on the period domain (see 10.3). About the description of flat coordinates as the automorphic functions, we shall write in a forthcoming paper. The author would like to thank Prof. Claus Hertling and Prof. Atsushi Takahashi for valuable discussions. The auther also thanks the referee for the valuable advices.

2. Frobenius manifold

For the contents of this section, we refer to Hertling [**3**].

2.1. Definition of Frobenius manifold. Here all manifolds will be complex manifolds. $\mathcal{O}_S$, Θ_S are the structure sheaf and the holomorphic tangent sheaf of a manifold S respectively. A (k,l)-tensor is an $\mathcal{O}_S$-linear map $T : \Theta_S^{\otimes k} \to \Theta_S^{\otimes l}$. A metric g is a symmmetric non-degenerate $(2,0)$-tensor, a multiplication on Θ_S is a commutative and associative $(2,1)$-tensor.

DEFINITION 2.1. [**3**, p146 Definition 9.1] *A Frobenius manifold is a tuple $(S, \circ, e, E, g)$ where S is a complex manifold of dimension $m \geq 1$ with metric g and multiplication $\circ$ on Θ_S, e is a global unit field and E is another global vector field, subject to the following condition:*

(1) *the metric is multiplication invariant, $g(X \circ Y, Z) = g(X, Y \circ Z)$,*
(2) *(potentiality) the $(3,1)$-tensor $\nabla \circ (X,Y,Z) := \nabla_X(Y \circ Z) - \nabla_X(Y) \circ Z - Y \circ \nabla_X Z$ is symmetric (here ∇ is the Levi-Civita connection of the metric),*
(3) *the metric g is flat*
(4) *the unit field e is flat, $\nabla e = 0$,*
(5) *the Euler field E satisfies $Lie_E(\circ) = 1 \cdot \circ$ and $Lie_E(g) = \mathrm{D} \cdot g$ for some $\mathrm{D} \in \mathbf{C}$.*

In this paper we also assume that S is connected and the multiplication $\circ$ is generically semisimple.

This means that outside the set $\mathcal{K}$ (which is introduced in [**3**, p13], called " caustic" and is a hypersurface or an empty set) the multiplication $\circ$ is semisimple.

We remark that in a neighborhood of a point with semisimple multiplication, we can take canonical coordinates $u_1, \cdots, u_m$ which satisfy $e_i \circ e_j = \delta_{ij} e_i$, $e = \sum_{i=1}^m e_i$ for $e_i = \frac{\partial}{\partial u_i}$.

The Euler field E is described as

$$E = \sum_{i=1}^{m} (u_i + r_i) e_i \tag{2.1}$$

for some $r_i \in \mathbf{C}$.

There exists locally a holomorphic function $\widetilde{F}$ which satisfies

$$g(\delta, \delta' \circ \delta'') = \delta\delta'\delta''\widetilde{F} \tag{2.2}$$

for any local sections δ, δ', δ'' of Θ_S with $\nabla\delta = \nabla\delta' = \nabla\delta'' = 0$. We call $\widetilde{F}$ a "prepotential".

2.2. The $\mathcal{O}_S$-homomorphism $\mathcal{U}$ and the discriminant Δ.

Definition 2.2. *The $\mathcal{O}_S$-homomorphism $\mathcal{U}$ is defined by the multiplication of Euler field E:*

$$\mathcal{U} : \Theta_S \to \Theta_S, \qquad \delta \mapsto E \circ \delta. \tag{2.3}$$

Definition 2.3. *We define*

$$\Delta := \det(\Theta_S \overset{\mathcal{U}}{\to} \Theta_S), \tag{2.4}$$

$$\mathrm{trace}\,\mathcal{U} := \mathrm{trace}(\Theta_S \overset{\mathcal{U}}{\to} \Theta_S), \tag{2.5}$$

where " det, trace" mean the determinant and the trace of the $\mathcal{O}_S$-homomorphism respectively. The analytic subspace $(D, \mathcal{O}_D)$ is defined by $\mathcal{O}_D := \mathcal{O}_S/(\Delta)$, $D := Supp(\mathcal{O}_D)$.

D is an empty set or hypersurface. For the case $D = \emptyset$, the discussions below are trivial or parallel to the case when D is a hypersurface. Thus we assume that D is a hypersurface.

2.3. The sheaf of logarithmic vector fields $\Theta_S(-\log D)$. We define the sheaf of logarithmic vector fields $\Theta_S(-\log D)$ by

$$\Theta_S(-\log D) := \{\delta \in \Theta_S | \delta\Delta \in (\Delta)\}.$$

Proposition 2.4. [**3**, p47 Theorem 4.9] *$\mathcal{U}$ gives the $\mathcal{O}_S$-isomorphism:*

$$\mathcal{U} : \Theta_S \overset{\sim}{\to} \Theta_S(-\log D). \tag{2.6}$$

Thus $\Theta_S(-\log D)$ is an $\mathcal{O}_S$-locally free module.

By this proposition, we see that $\frac{\mathcal{U}(\delta)\Delta}{\Delta}$ is holomorphic for a local section δ of Θ_S.

Proposition 2.5. *For a local section δ of Θ_S, we have*

$$\frac{\mathcal{U}(\delta)\Delta}{\Delta} = \delta(\mathrm{trace}\,\mathcal{U}). \tag{2.7}$$

PROOF. We need to prove only at a point with semisimple multiplication. Take the canonical coordinate. Then using a basis $e_1, \cdots, e_m$, the $\mathcal{O}_S$-homomorphism $\mathcal{U}$ is represented by a diagonal matrix with entries $u_1 + r_1, \cdots, u_m + r_m$ by (2.1). Then $\Delta = \prod_{i=1}^m (u_i + r_i)$ and trace$\mathcal{U} = \sum_{i=1}^m (u_i + r_i)$ and we obtain the proposition by an easy calculation. □

2.4. $(D, \mathcal{O}_D)$ is reduced.

PROPOSITION 2.6. *The analytic space $(D, \mathcal{O}_D)$ is reduced.*

PROOF. First we assume that $m \geq 2$. We prove that for any $x \in \mathcal{K} \cap D$,

$$\dim_x(\mathcal{K} \cap D) \leq m - 2. \tag{2.8}$$

If there exists a point $p \in \mathcal{K} \cap D$ s.t. $\dim_x(\mathcal{K} \cap D) = m - 1$, then $(\mathcal{K} \cap D)_x$ and $\mathcal{K}_x$ have a common irreducible component as a germ of analytic set [**2**, p103 Theorem]. Since $\mathcal{K}$ is e-invariant ([**3**, p16]), this common irreducible component is also e-invariant. This contradicts to the transversality of D and e [**3**, p41 Theorem 4.1 and Remark 4.2]. Thus we obtain (2.8).

Since D is a hypersurface, D is pure dimensional of dimension $m - 1$. Thereby $D \setminus \mathcal{K} \cap D$ is dense in D.

At the points of $q \in D \setminus \mathcal{K} \cap D$, the multiplication is semisimple. Then the discriminant Δ is $\Delta = \prod_{i=1}^m (u_i + r_i)$ as in the proof of Proposition 2.5, thereby $\mathcal{O}_{S,q}/(\Delta_q)$ is reduced.

Since $\mathcal{O}_S/(\Delta)$ is reduced at each point of an open dense subset $D \setminus \mathcal{K} \cap D$ of D, $\mathcal{O}_S/(\Delta)$ is reduced at every point of D by the discussion of [**6**, p65].

For the case $m = 1$, then $e \circ e = e$ gives that $\mathcal{K} = \emptyset$. Then the assertion is obtained. □

2.5. The sheaf of logarithmic 1-forms $\Omega_S^1(\log D)$.

PROPOSITION 2.7.

$$\Omega_S^1(\log D) := \{\eta \in \Omega_S^1(*D) | \eta\Delta \in \Omega_S^1, \ \Delta d\eta \in \Omega_S^2\}$$

has the natural $\mathcal{O}_S$-complete pairing:

$$\Omega_S^1(\log D) \otimes_{\mathcal{O}_S} \Theta_S(-\log D) \to \mathcal{O}_S. \tag{2.9}$$

Thus $\Omega_S^1(\log D)$is an $\mathcal{O}_S$-locally free module.

For the proof, we refer to Saito [**10**, p.268].

3. Definition of I on S

3.1. Definition of I. In this subsection, we review the definition of I introduced by Dubrovin [**1**, p191 (3.13)]. The metric g induces the $\mathcal{O}_S$-isomorphism and the non-degenerate $\mathcal{O}_S$-symmetric bilinear form:

$$g : \Omega_S^1 \overset{\sim}{\to} \Theta_S, \qquad g : \Omega_S^1 \times \Omega_S^1 \to \mathcal{O}_S. \tag{3.1}$$

We remind that $\mathcal{U} : \Theta_S \to \Theta_S(-\log D)$ is an $\mathcal{O}_S$-isomorphism.

The $\mathcal{O}_S$-bilinear form:

$$I : \Omega_S^1(\log D) \otimes_{\mathcal{O}_S} \Omega_S^1 \to \mathcal{O}_S \tag{3.2}$$

is defined by the composite of

(3.3)

$$\Omega_S^1(\log D) \otimes_{\mathcal{O}_S} \Omega_S^1 \overset{1\otimes g}{\to} \Omega_S^1(\log D) \otimes_{\mathcal{O}_S} \Theta_S \overset{1\otimes \mathcal{U}}{\to} \Omega_S^1(\log D) \otimes_{\mathcal{O}_S} \Theta_S(-\log D) \overset{<,>}{\to} \mathcal{O}_S.$$

This gives the $\mathcal{O}_S$-complete pairing. We remark that I is flat which is proved by Saito and Dubrovin.

For local sections ω, ω' of Ω_S^1,

$$I(\omega,\omega') = <\omega, E\circ g(\omega')> = g(g(\omega), E\circ g(\omega')) = g(E, g(\omega)\circ g(\omega')) \tag{3.4}$$

which is the same as [**1**, p191 (3.13)]. Therefore I on $\Omega_S^1 \otimes_{\mathcal{O}_S} \Omega_S^1$ is symmetric. The $\mathcal{O}_S$-bilinear form:

$$I : \Omega_S^1(\log D) \otimes_{\mathcal{O}_S} \Omega_S^1(\log D) \to \mathcal{O}_S(*D) \tag{3.5}$$

which extends (3.2) is unique.

Thus the $\mathcal{O}_S$-bilinear form (3.5) defines the $\mathcal{O}_S$-complete pairing:

$$I : \Omega_S^1 \otimes_{\mathcal{O}_S} \Omega_S^1(\log D) \to \mathcal{O}_S. \tag{3.6}$$

3.2. Description of I in terms of the prepotential $\widetilde{F}$. Let $t^1, \cdots, t^m$ be local coordinates of S s.t.

$$\partial_i := \frac{\partial}{\partial t^i} \qquad (i = 1, \cdots, m) \tag{3.7}$$

gives flat vector fields. We call these coordinates "flat coordinates". We prepare the notation:

$$\partial^i := g(dt^i). \tag{3.8}$$

PROPOSITION 3.1.

$$I(dt^i, dt^j) = E\partial^i\partial^j\widetilde{F} \tag{3.9}$$

where E is the Euler field and $\widetilde{F}$ is a prepotential.

PROOF. This is a direct consequence (3.4). □

REMARK 3.2. *Saito introduced this metric I in* [**11**, p.1257, (5.4.1)] *for a universal unfolding of a function with an isolated singularity. He also showed that multiplied by known constant it gave the intersection number for vanishing cycles of the Milnor fiber through the primitive form.*

4. The construction of the ramified double cover of $(S \setminus D^{sing}, D \setminus D^{sing})$

4.1. D^{sing}.

DEFINITION 4.1. *We define the analytic subset*

$$D^{sing} := \{x \in D | \mathcal{O}_{D,x} \text{ is not a regular local ring}\} \tag{4.1}$$

of D.

Since the analytic space $(D, \mathcal{O}_D)$ is reduced, we have $\dim_x D^{sing} \leq \dim_x D - 1$ for any $x \in D^{sing}$. We remind that $\dim_x D = \dim_x S - 1$ because D is a hypersurface. Thus we have

$$\dim D^{sing} \leq \dim S - 2. \tag{4.2}$$

4.2. The space of ramified double cover of $S \setminus D^{sing}$.

LEMMA 4.2. *The closed analytic subspace:*

$$D \setminus D^{sing} \subset S \setminus D^{sing}$$

is a smooth analytic submanifold of codimension 1.

DEFINITION 4.3. *We define the analytic space $(S', \mathcal{O}_{S'})$ as a closed subspace of $(S \setminus D^{sing}) \times \mathbf{C}$ defined by the ideal generated by $\Delta - y^2$, where y is the coordinate of $\mathbf{C}$. As a set, we have*

$$S' = \{(z, y) \in (S \setminus D^{sing}) \times \mathbf{C} | \Delta = y^2\}.$$

Also the closed subspace D' of S' is defined by the ideal generated by y. Let

$$\alpha : S' \to S \setminus D^{sing}$$

be the composite of the closed immersion $S' \to (S \setminus D^{sing}) \times \mathbf{C}$ and the natural projection.

The squares in the following diagram are all Cartesian:

$$\begin{array}{ccccc} D' & \xrightarrow{\subset} & S' & \xrightarrow{y} & \mathbf{C} \\ \downarrow & & \downarrow \alpha & & \downarrow \\ D \setminus D^{sing} & \xrightarrow{\subset} & S \setminus D^{sing} & \xrightarrow{\Delta} & \mathbf{C} \end{array} .$$

PROPOSITION 4.4. *S' is a smooth analytic manifold and its closed subspace:*

$$D' \subset S'$$

is a smooth analytic submanifold of codimension 1.

PROPOSITION 4.5. *$\alpha : S' \to S \setminus D^{sing}$ is a flat ramified double covering, i.e. flat, surjective, proper with finite fiber and the degree of the covering is 2. D' is the set of ramification locus and $D \setminus D^{sing}$ is the set of branch locus. The restriction $\alpha|_{S' \setminus D'} : S' \setminus D' \to S \setminus D$ is the étale covering of degree 2.*

For the proof of Lemma 4.2, Propositions 4.4, 4.5, we only need the following Lemma:

LEMMA 4.6. *For a ramification point $p \in D'$ and $\alpha(p) \in D \setminus D^{sing}$, we can take a coordinate neighborhood $U_p \subset S'$ and $V_{\alpha(p)} \subset S \setminus D^{sing}$ whose coordinates of U_p and $V_{\alpha(p)}$ are $z'_1, \cdots, z'_{m-1}, y$ and $z_1, \cdots, z_{m-1}, \Delta$, respectively s.t.*

$$\alpha|_{U_p} : U_p \to V_{\alpha(p)}$$

is the ramified double cover and

$$z_1 \circ \alpha = z'_1, \cdots, z_{m-1} \circ \alpha = z'_{m-1}, \ \Delta \circ \alpha = y^2.$$

PROOF. Since $\mathcal{O}_{D,\alpha(p)} = \mathcal{O}_{S,\alpha(p)}/(\Delta)_{\alpha(p)}$ is a regular local ring, we could take a neighborhood $V_{\alpha(p)} \subset S \setminus D^{sing}$ of $\alpha(p)$ s.t. Δ could be taken as a part of the coordinates in $V_{\alpha(p)}$. Thus we take coordinates: $z_1, \cdots, z_{m-1}, \Delta$. Take $U_p := \alpha^{-1}(V_{\alpha(p)})$. Then we see that $z'_1 := z_1 \circ \alpha, \cdots, z'_{m-1} := z_{m-1} \circ \alpha$, y give the coordinates on U_p. □

The morphism:

$$(S \setminus D^{sing}) \times \mathbf{C} \to (S \setminus D^{sing}) \times \mathbf{C}, \qquad (z,y) \to (z,-y)$$

induces the involution:

$$\sigma : S' \to S'.$$

This gives the covering transformation of $\alpha : S' \to S \setminus D^{sing}$. By the coordinates introduced in Lemma 4.6, σ is expressed as

$$(4.3) \qquad \sigma|_{U_p} : U_p \to U_p, \qquad (z'_1, \cdots, z'_{m-1}, y) \mapsto (z'_1, \cdots, z'_{m-1}, -y).$$

5. $\mathcal{O}_S$-modules

The involution $\sigma : S' \to S'$ defines the $\mathcal{O}_{S\setminus D^{sing}}$-isomorphism:

$$\sigma^* : \alpha_*\Omega^p_{S'} \to \alpha_*\Omega^p_{S'}.$$

DEFINITION 5.1. *We define the $\mathcal{O}_{S\setminus D^{sing}}$-modules:*

$$(5.1) \qquad (\alpha_*\Omega^p_{S'})^{even} := \{\eta \in \alpha_*\Omega^p_{S'} | \sigma^*(\eta) = \eta\},$$

$$(5.2) \qquad (\alpha_*\Omega^p_{S'})^{odd} := \{\eta \in \alpha_*\Omega^p_{S'} | \sigma^*(\eta) = -\eta\}.$$

We have a decomposition:

$$(5.3) \qquad \alpha_*\Omega^p_{S'} = (\alpha_*\Omega^p_{S'})^{even} \oplus (\alpha_*\Omega^p_{S'})^{odd}.$$

PROPOSITION 5.2. *The natural $\mathcal{O}_{S\setminus D^{sing}}$-homomorphism:*

$$\Omega^p_{S\setminus D^{sing}} \to (\alpha_*\Omega^p_{S'})^{even}$$

gives an $\mathcal{O}_{S\setminus D^{sing}}$-isomorphism. Thus $(\alpha_*\Omega^p_{S'})^{even}$ is an $\mathcal{O}_{S\setminus D^{sing}}$- locally free module of rank $\binom{m}{p}$.

PROOF. We show it on each stalk on $q \in S \setminus D^{sing}$. On $q \in S \setminus D$, this is clear. On $q \in D \setminus D^{sing}$, put $p = \alpha^{-1}(q) \in D'$ and take the neighborhood U_p, $V_{\alpha(p)}$ and coordinates as in Lemma 4.6. Then any local section of $\alpha_*\Omega^p_{S'}$ and the action of σ^* on it can be written explicitly by the coordinates $z'_1, \cdots, z'_{m-1}, y$, using (4.3). By the uniqueness of the Taylor expansion, we see that this local section of $(\alpha_*\Omega^p_{S'})^{even}$ can be written by the coordinates $z_1, \cdots, z_{m-1}, \Delta$. □

PROPOSITION 5.3. *$(\alpha_*\Omega^p_{S'})^{odd}$ is an $\mathcal{O}_{S\setminus D^{sing}}$- locally free module of rank $\binom{m}{p}$.*

PROOF. On $S \setminus D$, this is clear. On a neighborhood of the point of $D \setminus D^{sing}$, we could construct an $\mathcal{O}_{S\setminus D^{sing}}$-free basis of $(\alpha_*\Omega^p_{S'})^{odd}$ by the parallel discussion of the proof of Proposition 5.2. □

DEFINITION 5.4. *We define the $\mathcal{O}_S$-modules:*

$$(5.4) \qquad \Omega^{p,even}_S := \iota_*(\alpha_*\Omega^p_{S'})^{even},$$

$$(5.5) \qquad \Omega^{p,odd}_S := \iota_*(\alpha_*\Omega^p_{S'})^{odd},$$

where we put

$$\iota : S \setminus D^{sing} \to S.$$

By the Proposition 5.2, we have $\iota^{-1}\Omega_S^p = \Omega^p_{S\setminus D^{sing}} \simeq (\alpha_*\Omega^p_{S'})^{even}$. Taking the direct image by ι, we have $\iota_*\iota^{-1}\Omega_S^p \simeq \Omega_S^{p,even}$. Since Ω_S^p is an $\mathcal{O}_S$-locally free module and $\dim D^{sing} \leq \dim S - 2$, we obtain an $\mathcal{O}_S$-isomorphism $\Omega_S^p \simeq \iota_*\iota^{-1}\Omega_S^p$. Thus we have

$$\Omega_S^p \simeq \Omega_S^{p,even}. \tag{5.6}$$

Especially $\Omega_S^{p,even}$ is an $\mathcal{O}_S$-locally free module of rank $\binom{m}{p}$. For an $\mathcal{O}_S$-module $\Omega_S^{p,odd}$, we will see the $\mathcal{O}_S$-isomorphism (8.7). Thus $\Omega_S^{p,odd}$ is also an $\mathcal{O}_S$-locally free module of rank $\binom{m}{p}$.

6. $\mathcal{O}_{S'}$-modules

6.1. $\mathcal{O}_{S'}$-free basis. On $S' \setminus D'$, the $\mathcal{O}_{S'}$-modules:

$$(\iota\circ\alpha)^*\Omega_S^1,\ (\iota\circ\alpha)^*\Omega_S^1(\log D),\ \Omega_{S'}^1$$

are all isomorphic to $\Omega^1_{S'}(\log D')$. We construct locally an $\mathcal{O}_{S'}$-free basis of the following $\mathcal{O}_{S'}$-modules in the neighborhood of $p \in D'$ discussed in Lemma 4.6.

(1) The pull back of the $\mathcal{O}_S$-free basis $dz_1, \cdots, dz_{m-1}, \frac{1}{2}d\Delta$ of Ω_S^1 on $V_{\alpha(p)}$:

$$(\iota\circ\alpha)^*dz_1 = dz_1', \ \cdots, (\iota\circ\alpha)^*dz_{m-1} = dz'_{m-1},\ (\iota\circ\alpha)^*\frac{1}{2}d\Delta = ydy$$

give an $\mathcal{O}_{S'}$-free basis of $(\iota\circ\alpha)^*\Omega_S^1$ on U_p.

(2) The pull back of the $\mathcal{O}_S$-free basis $dz_1, \cdots, dz_{m-1}, \frac{1}{2}\frac{d\Delta}{\Delta}$ of $\Omega^1_{S\setminus D^{sing}}(\log(D\setminus D^{sing}))$ on $V_{\alpha(p)}$:

$$(\iota\circ\alpha)^*dz_1 = dz_1', \ \cdots, (\iota\circ\alpha)^*dz_{m-1} = dz'_{m-1},\ (\iota\circ\alpha)^*\frac{1}{2}\frac{d\Delta}{\Delta} = \frac{dy}{y}$$

give an $\mathcal{O}_{S'}$-free basis of $\alpha^*\Omega^1_{S\setminus D^{sing}}(\log(D\setminus D^{sing}))$ on U_p.

(3) The forms:

$$dz_1',\ \cdots, dz'_{m-1},\ dy$$

give an $\mathcal{O}_{S'}$-free basis of $\Omega^1_{S'}$.

(4) The forms:

$$dz_1',\ \cdots, dz'_{m-1},\ \frac{1}{y}dy$$

give an $\mathcal{O}_{S'}$-free basis of $\Omega^1_{S'}(\log D')$.

6.2. Relation of the $\mathcal{O}_{S'}$-modules. By the facts of the previous subsection, we have

PROPOSITION 6.1. *$\alpha : S' \to S\setminus D^{sing}$ is a log étale morphism, i.e.*

$$\alpha^*\Omega^1_{S\setminus D^{sing}}(\log(D\setminus D^{sing})) \to \Omega^1_{S'}(\log D')$$

is an isomorphism. Composing this with the isomorphism:

$$\iota^*\Omega^1_S(\log D) \xrightarrow{\sim} \Omega^1_{S\setminus D^{sing}}(\log(D\setminus D^{sing})),$$

we have:

$$(\iota\circ\alpha)^*\Omega^1_S(\log D) \xrightarrow{\sim} \Omega^1_{S'}(\log D').$$

By this proposition and the injections:

$$(\iota\circ\alpha)^*\Omega^1_S \to \Omega^1_{S'} \to \Omega^1_{S'}(\log D'),$$

we can regard the modules

$$(\iota\circ\alpha)^*\Omega^1_S,\ \Omega^1_{S'},\ (\iota\circ\alpha)^*\Omega^1_S(\log D)$$

as the $\mathcal{O}_{S'}$-submodules of $\Omega^1_{S'}(\log D')$.

LEMMA 6.2. *For any $p\in S'$, take an open neighborhood $U\subset S'$, $\mathcal{O}_{S'}$-free basis $\{e_i\}_{1\le i\le m}$ of $(\iota\circ\alpha)^*\Omega^1_S$, $\mathcal{O}_{S'}$-free basis $\{e'_i\}_{1\le i\le m}$ of $\Omega^1_{S'}$, $\mathcal{O}_{S'}$-free basis $\{e''_i\}_{1\le i\le m}$ of $(\iota\circ\alpha)^*\Omega^1_S(\log D)$.*

The inclusions $(\iota\circ\alpha)^*\Omega^1_S\subset\Omega^1_{S'}\subset(\iota\circ\alpha)^*\Omega^1_S(\log D)$ give the relations

$$e_j=\sum_{i=1}^m a_{ij}e'_i,\quad e'_j=\sum_{i=1}^m b_{ij}e''_i$$

for $a_{ij},\ b_{ij}\in\Gamma(U,\mathcal{O}_{S'})$.

: *(1) The change of choice of $\{e_i\}$, $\{e'_i\}$, $\{e''_i\}$ shifts $\det(a_{ij})$, $\det(b_{ij})$ to those multiplied by elements of $\Gamma(U,\mathcal{O}^\times_{S'})$.*

: *(2) $\det(a_{ij})$, $\det(b_{ij})$ are equal to y up to multiplication of an element of $\Gamma(U,\mathcal{O}^\times_{S'})$.*

LEMMA 6.3. *As an $\mathcal{O}_{S'}$-submodule of $\Omega^1_{S'}(\log D')\otimes_{\mathcal{O}_{S'}}\Omega^1_{S'}(\log D')$,*

$$\Omega^1_{S'}\otimes_{\mathcal{O}_{S'}}\Omega^1_{S'}$$

is an $\mathcal{O}_{S'}$-submodule of

$$(\iota\circ\alpha)^*\Omega^1_S\otimes_{\mathcal{O}_{S'}}(\iota\circ\alpha)^*\Omega^1_S(\log D)+(\iota\circ\alpha)^*\Omega^1_S(\log D)\otimes_{\mathcal{O}_{S'}}(\iota\circ\alpha)^*\Omega^1_S.$$

Proof of Lemma 6.2 and Lemma 6.3. Lemma 6.2(1) is easy. For proof of Lemma 6.2(1) and Lemma 6.3, we need only to prove them at the points of D'. At the points of D', we obtain the proof by the direct calculation using these $\mathcal{O}_{S'}$-free bases in Subsection 6.1. □

7. The lift of I on the ramified double cover S'

Taking the pull-back $(\iota\circ\alpha)^*$ of the $\mathcal{O}_S$-bilinear form (3.5), we have

$$I:\Omega^1_{S'}(\log D')\otimes_{\mathcal{O}_{S'}}\Omega^1_{S'}(\log D')\to\mathcal{O}_{S'}(*D'). \tag{7.1}$$

Also taking the pull-back $(\iota\circ\alpha)^*$ of the $\mathcal{O}_S$-complete pairings (3.2), (3.6), we have the following lemma.

LEMMA 7.1. *The restrictions of (7.1) to $(\iota\circ\alpha)^*\Omega^1_S\otimes_{\mathcal{O}_{S'}}(\iota\circ\alpha)^*\Omega^1_S(\log D)$ and $(\iota\circ\alpha)^*\Omega^1_S(\log D)\otimes_{\mathcal{O}_{S'}}(\iota\circ\alpha)^*\Omega^1_S$ give the $\mathcal{O}_{S'}$-complete pairings:*

$$I:(\iota\circ\alpha)^*\Omega^1_S\otimes_{\mathcal{O}_{S'}}(\iota\circ\alpha)^*\Omega^1_S(\log D)\ \to\ \mathcal{O}_{S'}, \tag{7.2}$$

$$I:(\iota\circ\alpha)^*\Omega^1_S(\log D)\otimes_{\mathcal{O}_{S'}}(\iota\circ\alpha)^*\Omega^1_S\ \to\ \mathcal{O}_{S'}. \tag{7.3}$$

PROPOSITION 7.2. *The restriction of (7.1) to $\Omega^1_{S'}\otimes_{\mathcal{O}_{S'}}\Omega^1_{S'}$ induces the $\mathcal{O}_{S'}$-symmetric bilinear form:*

$$I:\Omega^1_{S'}\otimes_{\mathcal{O}_{S'}}\Omega^1_{S'}\to\mathcal{O}_{S'}. \tag{7.4}$$

PROOF. By Lemma 7.1, the restriction of I to

$$(\iota\circ\alpha)^*\Omega^1_S\otimes_{\mathcal{O}_{S'}}(\iota\circ\alpha)^*\Omega^1_S(\log D)+(\iota\circ\alpha)^*\Omega^1_S(\log D)\otimes_{\mathcal{O}_{S'}}(\iota\circ\alpha)^*\Omega^1_S$$

has values in $\mathcal{O}_{S'}$. Combining this with Lemma 6.3, we have the result. □

PROPOSITION 7.3. *The $\mathcal{O}_{S'}$-bilinear form (7.4) gives the non-degenerate $\mathcal{O}_{S'}$-symmetric bilinear form on $\Omega^1_{S'}$.*

PROOF. We use the notations as in Lemma 6.2. The statement (7.2) gives $\det I(e_i,e''_j)\in\Gamma(U,\mathcal{O}^\times_{S'})$. By Lemma 6.2, we have $\det I(e'_i,e'_j)\in\Gamma(U,\mathcal{O}^\times_{S'})$. This proves the Proposition. □

DEFINITION 7.4. *The $\mathcal{O}_{S'}$-symmetric bilinear form (7.4) induces the non-degenerate $\mathcal{O}_{S'}$-symmetric bilinear form:*

$$I:\Theta_{S'}\otimes_{\mathcal{O}_{S'}}\Theta_{S'}\to\mathcal{O}_{S'}. \tag{7.5}$$

8. The definition of the Laplacian

8.1. General theory. We start with the general framework of holomorphic metric. Let M be a complex manifold with $\dim M=m$. Let $g_M:\Theta_M\times\Theta_M\to\mathcal{O}_M$ be a non-degenerate $\mathcal{O}_M$-symmetric bilinear form. We call g_M a "holomorphic metric." For the holomorphic metric, we will define a volume form, Hodge's $*$ operator and a Laplacian which are parallel to the case of Riemannian metric by changing $\mathbf{R}$ to $\mathbf{C}$. We will give the exact definitions. A holomorphic m-form ω is called a "volume form" if ω is expressed locally as

$$\omega=\sqrt{\det g_{ij}}\,dx_1\wedge\cdots\wedge dx_m$$

by coordinates $x_1,\cdots,x_m$ on the open set $U\subset M$, where

$$g_{ij}=g_M\left(\frac{\partial}{\partial x_i},\frac{\partial}{\partial x_j}\right),$$

and $\sqrt{\det g_{ij}}$ means a holomorphic function on U s.t. $(\sqrt{\det g_{ij}})^2=\det g_{ij}$.

If ω is a volume form, then so is $-\omega$. As in the real case, the volume form always exists locally.

In this subsection, we assume the existence of a volume form on M and fix it and denote by $vol.g_M$.

We prepare some $\mathcal{O}_M$-isomorphisms. For the integer $p\,(0\le p\le m)$, the wedge product defines the $\mathcal{O}_M$-isomorphism:

$$f_1:\Omega^p_M\to\underline{\mathrm{Hom}}_{\mathcal{O}_M}(\Omega^{m-p}_M,\Omega^m_M),\qquad \eta\mapsto(\omega\mapsto\eta\wedge\omega).$$

The volume form $vol.g_M$ defines the $\mathcal{O}_M$-isomorphism:

$$f_2:\mathcal{O}_M\to\Omega^m_M,\qquad f\mapsto f\cdot vol.g_M.$$

This isomorphism f_2 defines the $\mathcal{O}_M$-isomorphism:

$$f_3:\underline{\mathrm{Hom}}_{\mathcal{O}_M}(\Omega^{m-p}_M,\Omega^m_M)\to\underline{\mathrm{Hom}}_{\mathcal{O}_M}(\Omega^{m-p}_M,\mathcal{O}_M),\qquad \varphi\mapsto(f_2)^{-1}\circ\varphi.$$

The holomorphic metric g_M induces the $\mathcal{O}_M$-complete pairing: $g_M:\Omega^{m-p}_M\times\Omega^{m-p}_M\to\mathcal{O}_M$ (which we denote by the same notation as a holomorphic metric), and gives the $\mathcal{O}_M$-isomorphism:

$$f_4:\Omega^{m-p}_M\to\underline{\mathrm{Hom}}_{\mathcal{O}_M}(\Omega^{m-p}_M,\mathcal{O}_M),\qquad \eta\mapsto(\omega\mapsto g_M(\eta,\omega)).$$

DEFINITION 8.1. *We define the $\mathcal{O}_M$-isomorphism*

$$*: \Omega_M^p \to \Omega_M^{m-p}$$

by the composite of

$$\Omega_M^p \xrightarrow{f_1} \underline{\mathrm{Hom}}_{\mathcal{O}_M}(\Omega_M^{m-p}, \Omega_M^m) \xrightarrow{f_3} \underline{\mathrm{Hom}}_{\mathcal{O}_M}(\Omega_M^{m-p}, \mathcal{O}_M) \xrightarrow{(f_4)^{-1}} \Omega_M^{m-p}.$$

We call this $\mathcal{O}_M$-isomorphism $*$, the "Hodge's $*$ operator". We also define

$$\delta := *d* : \Omega_M^p \to \Omega_M^{p-1}, \tag{8.1}$$

$$\mathcal{D} := d\delta + \delta d : \Omega_M^p \to \Omega_M^p. \tag{8.2}$$

We call the operator $\mathcal{D}$ the Laplacian.

As in the case of Riemannian geometry (see for example, Rosenberg [**9**, p18 and p21]), $\mathcal{D} : \mathcal{O}_M \to \mathcal{O}_M$ in the coordinate neighborhood U with coordinates $x_1, \cdots, x_m$ is described as

$$\begin{aligned} \mathcal{D}(f) &= (*d*d + d*d*)f \\ &= (*d*d)f \\ &= \sum_{i,j=1}^{m} Gram(g_M)^{-1/2}\frac{\partial}{\partial x_i}[Gram(g_M)^{1/2} g_M(dx_i, dx_j)\frac{\partial}{\partial x_j} f] \end{aligned} \tag{8.3}$$

for $Gram(g_M) := \det(g_M(\frac{\partial}{\partial x_i}, \frac{\partial}{\partial x_j}))$. By this description (8.3), we have

$$\mathcal{D}(h_1 h_2) = \mathcal{D}(h_1)h_2 + h_1\mathcal{D}(h_2) + 2g_M(dh_1, dh_2) \tag{8.4}$$

for any local sections h_1, h_2 of $\mathcal{O}_M$.

We also remark that

$$* \circ * = (-1)^{p(m-p)} \text{ on } \Omega_M^p. \tag{8.5}$$

8.2. Volume form on S'. We come back to the situation of the previous section. We assert the existence of the volume form for the holomorphic metric I on S'.

We assume the existence of the volume form on S for the holomorphic metric g. We fix the volume form for g and denote it by $vol.g$.

PROPOSITION 8.2.

$$\frac{1}{y}(\iota \circ \alpha)^*(vol.g) \in \Gamma(S', \Omega_{S'}^m)$$

gives the volume form on S' for the holomorphic metric I. We denote it by $vol.I$.

PROOF. We need only to check that $\frac{1}{y}(\iota\circ\alpha)^*(vol.g)$ gives the volume form $vol.I$ on $S' \setminus D'$. Since $\alpha|_{S'\setminus D'}$ is unramified, we can take on each point $p \in S' \setminus D'$ the open neighborhood $U \subset S' \setminus D'$ s.t. $\alpha|_U$ is the isomorphism. By this isomorphism, $\frac{1}{y}$ is a single valued holomorphic function on $\alpha(U)$. We need to check that $\frac{1}{y}vol.g$ gives the volume form $vol.I$ on $\alpha(U)$. This can be easily checked by the relation of I, g, $\mathcal{U}$ and the fact $\det\mathcal{U} = y^2$. □

COROLLARY 8.3. *$vol.I \in \Gamma(S \setminus D^{sing}, \alpha_*\Omega_{S'}^m)$ has the property:*

$$\sigma^* vol.I = -vol.I$$

for $\sigma^* : \alpha_*\Omega_{S'}^m \to \alpha_*\Omega_{S'}^m$.

8.3. Laplacian on the Frobenius manifold. By applying the general theory of Laplacian to the holomorphic metric I on S', we have $* : \Omega^p_{S'} \to \Omega^{m-p}_{S'}$ and $\mathcal{D} : \Omega^p_{S'} \to \Omega^p_{S'}$.

PROPOSITION 8.4. *The induced morphisms:*

$$(8.6) \qquad * : \alpha_*\Omega^p_{S'} \to \alpha_*\Omega^{m-p}_{S'}, \qquad \mathcal{D} : \alpha_*\Omega^p_{S'} \to \alpha_*\Omega^p_{S'}$$

and the morphism $\sigma^* : \alpha_*\Omega^{p'}_{S'} \to \alpha_*\Omega^{p'}_{S'}$ *have the relations* $* \circ \sigma^* + \sigma^* \circ * = 0$, $\mathcal{D} \circ \sigma^* = \sigma^* \circ \mathcal{D}$.

PROOF. As in the proof of Proposition 8.2, we need to prove it only on each neighborhood $U \subset S \setminus D$ of $p \in S \setminus D$ because $\alpha_*\Omega^p_{S'}$ is a locally free module. □

By this proposition and the property (8.5), we have the $\mathcal{O}_{S\setminus D^{sing}}$-isomorphisms:

$$* : (\alpha_*\Omega^p_{S'})^{even} \to (\alpha_*\Omega^{m-p}_{S'})^{odd}, \qquad * : (\alpha_*\Omega^p_{S'})^{odd} \to (\alpha_*\Omega^{m-p}_{S'})^{even}$$

and the morphisms:

$$\mathcal{D} : (\alpha_*\Omega^p_{S'})^{even} \to (\alpha_*\Omega^p_{S'})^{even}, \qquad \mathcal{D} : (\alpha_*\Omega^p_{S'})^{odd} \to (\alpha_*\Omega^p_{S'})^{odd}.$$

Taking the direct image by ι, we have the $\mathcal{O}_S$-isomorphisms:

$$(8.7) \qquad * : \Omega^{p,even}_S \to \Omega^{m-p,odd}_S, \qquad * : \Omega^{p,odd}_S \to \Omega^{m-p,even}_S$$

and the morphisms:

$$\mathcal{D} : \Omega^{p,even}_S \to \Omega^{p,even}_S, \qquad \mathcal{D} : \Omega^{p,odd}_S \to \Omega^{p,odd}_S.$$

By Proposition 8.2, the following composition of $\mathcal{O}_S$-isomorphisms:

$$(8.8) \qquad \mathcal{O}_S \overset{\times vol.g}{\to} \Omega^n_S \simeq \Omega^{n,even}_S \overset{*}{\to} \Omega^{0,odd}_S$$

equals to the $\mathcal{O}_S$-homomorphism:

$$\iota_*\mathcal{O}_{S\setminus D^{sing}} \to \iota_*(\alpha_*\Omega^p_{S'})^{odd}$$

which is the direct image of the multiplication map $\mathcal{O}_{S\setminus D^{sing}} \to (\alpha_*\Omega^p_{S'})^{odd}$ multiplied by $y \in \Gamma(S \setminus D^{sing}, (\alpha_*\Omega^0_{S'})^{odd})$. Thus we merely denote the $\mathcal{O}_S$-isomorphism (8.8) by

$$(8.9) \qquad \mathcal{O}_S \overset{\times y}{\to} \Omega^{0,odd}_S.$$

DEFINITION 8.5.

$$\mathcal{D} : \quad \mathcal{O}_S \simeq \Omega^{0,even}_S \overset{\mathcal{D}}{\to} \Omega^{0,even}_S \simeq \mathcal{O}_S,$$

$$\mathcal{D}^{Jac} : \quad \mathcal{O}_S \overset{\times y}{\simeq} \Omega^{0,odd}_S \overset{\mathcal{D}}{\to} \Omega^{0,odd}_S \overset{(\times y)^{-1}}{\simeq} \mathcal{O}_S.$$

They give 2nd order differential operators.

9. Description of the Laplacian in terms of flat coordinates

9.1. The $\mathcal{O}_S$-endomorphism $\mathcal{V}$.

DEFINITION 9.1. [3, p147] *The $\mathcal{O}_S$-endomorphism:$\mathcal{V} : \Theta_S \to \Theta_S$ is defined by*

$$(9.1) \qquad \mathcal{V}(X) = \nabla_X E - \frac{\mathrm{D}}{2} X$$

for a local vector field X. We remark that $\mathrm{D} \in \mathbf{C}$ is defined in Def.2.1.

PROPOSITION 9.2. *The following equation holds:*

$$\nabla_\delta(E\circ\delta')-E\circ(\nabla_\delta\delta')=\mathcal{V}(\delta\circ\delta')-\delta\circ\mathcal{V}(\delta')+\delta\circ\delta' \tag{9.2}$$

for local vector fields δ, δ'.

PROOF. First we remind the following relation (see Hertling [**3**, p147 (9.2)]):

$$\nabla_X(Y\circ Z)-Y\circ\nabla_X Z-\nabla_Y(X\circ Z)+X\circ\nabla_Y Z-[X,Y]\circ Z=0.$$

We substitute $X=\delta, Y=E, Z=\delta'$. Then we have

$$\nabla_\delta(E\circ\delta')-E\circ(\nabla_\delta\delta')=\nabla_E(\delta\circ\delta')-\delta\circ\nabla_E\delta'+[\delta,E]\circ\delta'. \tag{9.3}$$

The torsion freeness of ∇ and the definition of $\mathcal{V}$ imply

$$\nabla_E(\delta\circ\delta') = \mathcal{V}(\delta\circ\delta')+\frac{\mathrm{D}}{2}\delta\circ\delta'+Lie_E(\delta\circ\delta'), \tag{9.4}$$

$$\delta\circ\nabla_E\delta' = \delta\circ\mathcal{V}(\delta')+\delta\circ(\frac{\mathrm{D}}{2}\delta')+\delta\circ Lie_E\delta'. \tag{9.5}$$

Therefore the equations (9.3), (9.4), (9.5) with $Lie_E\circ=\circ$ give the equation (9.2). □

COROLLARY 9.3. *For* δ,δ' *are local vector fields with* $\nabla\delta=\nabla\delta'=0$, *we have*

$$[\delta,\mathcal{U}(\delta')]=\mathcal{V}(\delta\circ\delta')-\delta\circ\mathcal{V}(\delta')+\delta\circ\delta'. \tag{9.6}$$

PROOF. By the torsion freeness of ∇ and the assumption of flatness of δ, δ', we have $\nabla_\delta(E\circ\delta')-E\circ(\nabla_\delta\delta')=[\delta,\mathcal{U}(\delta')]$. □

For any $\mathcal{O}_S$-endomorphism: $\varphi:\Theta_S\to\Theta_S$, we could define φ^* by the property:

$$g(\varphi^*(\delta),\delta')=g(\delta,\varphi(\delta')), \tag{9.7}$$

because g is non-degenerate. For the flat coordinates introduced in (3.7), (3.8), we have

$$\sum_{i=1}^m\varphi(\partial_i)\otimes\partial^i=\sum_{i=1}^m\partial_i\otimes\varphi^*(\partial^i)=\sum_{i=1}^m\partial^i\otimes\varphi^*(\partial_i). \tag{9.8}$$

PROPOSITION 9.4. *We have*

$$(1)\quad \partial_k(\mathrm{trace}\,\mathcal{U})=g(\partial_k,\sum_{i=1}^m\partial_i\circ\partial^i), \tag{9.9}$$

$$(2)\quad \sum_{k=1}^m\{\partial^k(\mathrm{trace}\,\mathcal{U})\}\partial_k=\sum_{i=1}^m\partial_i\circ\partial^i, \tag{9.10}$$

$$(3)\quad \partial_k(\mathrm{trace}\,\mathcal{U})=\sum_{i=1}^m\partial_k\partial_i\partial^i\tilde{F}. \tag{9.11}$$

PROOF. The torsion freeness of ∇ and the flatness of ∂_i give

$$\begin{aligned}\partial_k(\mathrm{trace}\,\mathcal{U}) &= \sum_{i=1}^m\partial_k g(\mathcal{U}(\partial_i),\partial^i)\\ &= \sum_{i=1}^m g(\nabla_{\partial_k}(\mathcal{U}(\partial_i)),\partial^i)\\ &= \sum_{i=1}^m g([\partial_k,\mathcal{U}(\partial_i)],\partial^i).\end{aligned}$$

Using the equation (9.6), we have

$$\begin{aligned}\sum_{i=1}^{m} g([\partial_k, \mathcal{U}(\partial_i)], \partial^i) &= \sum_{i=1}^{m} g(\mathcal{V}(\partial_k \circ \partial_i) - \partial_k \circ \mathcal{V}\partial_i + \partial_k \circ \partial_i, \partial^i) \\ &= \sum_{i=1}^{m} \left[g(\partial_k, \partial_i \circ \mathcal{V}^*(\partial^i)) - g(\partial_k, \mathcal{V}(\partial_i) \circ \partial^i) + g(\partial_k, \partial_i \circ \partial^i)\right] \\ &= \sum_{i=1}^{m} g(\partial_k, \partial_i \circ \partial^i)\end{aligned}$$

because of the equation (9.8) for $\varphi = \mathcal{V}$. Thus we have the equation (9.9).

The equations (9.10), (9.11) are direct consequences of (9.9). □

PROPOSITION 9.5. *The following equation holds:*

$$\sum_{i=1}^{m}[\partial^i, \mathcal{U}(\partial_i)] = \sum_{i=1}^{m}[(E + \frac{\mathrm{D}}{2})\partial^i(\mathrm{trace}\,\mathcal{U})]\partial_i. \tag{9.12}$$

PROOF. First we prove

$$\sum_{i=1}^{m} \partial^i \circ \mathcal{V}(\partial_i) = 0. \tag{9.13}$$

The commutativity of the product $\circ$ and the equation (9.8) for $\varphi = \mathcal{V}$ give $\sum_{i=1}^{m} \partial^i \circ \mathcal{V}(\partial_i) = \sum_{i=1}^{m} \partial^i \circ \mathcal{V}^*(\partial_i)$. The sum of both hand sides $\sum_{i=1}^{m} \partial^i \circ \mathcal{V}(\partial_i) + \sum_{i=1}^{m} \partial^i \circ \mathcal{V}^*(\partial_i)$ equals 0 by the equation $\mathcal{V} + \mathcal{V}^* = 0$ (which is obtained by $Lie_E g = \mathrm{D} \cdot g$, $\nabla g = 0$ and the torsion freeness of ∇, see Hertling [**3**, p147 (9.4)]). Thus we have the equation (9.13).

The following equation

$$\sum_{i=1}^{m}[\partial^i, \mathcal{U}(\partial_i)] = (\mathcal{V} + 1)(\sum_{i=1}^{m} \partial^i \circ \partial_i) \tag{9.14}$$

is a consequence of (9.6) and (9.13). Next we have

$$Lie_E(\sum_{i=1}^{m} \partial^i \circ \partial_i) = (1 - \mathrm{D}) \sum_{i=1}^{m} \partial^i \circ \partial_i \tag{9.15}$$

by the facts: $Lie_E(\sum_{i=1}^{m} \partial^i \otimes \partial_i) = -\mathrm{D} \sum_{i=1}^{m} \partial^i \otimes \partial_i$ (which is the consequence of the equation $Lie_E g = \mathrm{D}g$) and $Lie_E \circ = \circ$.

The equation (9.4) gives

$$\nabla_E(\sum_{i=1}^{m} \partial^i \circ \partial_i) = \mathcal{V}(\sum_{i=1}^{m} \partial^i \circ \partial_i) + \frac{\mathrm{D}}{2} \sum_{i=1}^{m} \partial^i \circ \partial_i + Lie_E(\sum_{i=1}^{m} \partial^i \circ \partial_i). \tag{9.16}$$

The equations (9.15) and (9.16) give the equation

$$(\mathcal{V} + 1)(\sum_{i=1}^{m} \partial^i \circ \partial_i) = (\nabla_E + \frac{\mathrm{D}}{2})(\sum_{i=1}^{m} \partial^i \circ \partial_i). \tag{9.17}$$

By the equation (9.10), we have

$$
\begin{aligned}
(\nabla_E+\frac{\mathrm{D}}{2})(\sum_{i=1}^{m}\partial^i\circ\partial_i) &= (\nabla_E+\frac{\mathrm{D}}{2})[\{\sum_{i=1}^{m}\partial^i(\mathrm{trace}\,\mathcal{U})\}\partial_i] \\
&= \sum_{i=1}^{m}[(E+\frac{\mathrm{D}}{2})\{\partial^i(\mathrm{trace}\,\mathcal{U})\}]\partial_i. \qquad (9.18)
\end{aligned}
$$

The equations (9.14), (9.17), (9.18) give the equation (9.12). □

9.2. Description of the Laplacian in terms of the operator $\mathcal{U}$. We will describe the Laplacian and the twisted Laplacian:

$$
\begin{aligned}
\mathcal{D}:\mathcal{O}_S &\to \mathcal{O}_S,\\
\mathcal{D}^{Jac}:\mathcal{O}_S &\to \mathcal{O}_S,
\end{aligned}
$$

as 2nd order differential operators in terms of flat coordinates. For the description of $\mathcal{D}^{Jac}$, we have (for a local section f of $\mathcal{O}_S$)

$$
\begin{aligned}
\mathcal{D}^{Jac}(f) &:= y^{-1}\mathcal{D}(yf)\\
&= y^{-1}[y\mathcal{D}(f)+\mathcal{D}(y)f+2I(dy,df)]\\
&= \mathcal{D}(f)+\frac{\mathcal{D}(y)}{y}f+I(\frac{2dy}{y},df)\\
&= \mathcal{D}(f)+\frac{\mathcal{D}(y)}{y}f+I(\frac{d\Delta}{\Delta},df)\\
&= \mathcal{D}(f)+\frac{\mathcal{D}(y)}{y}f+I(\frac{d\Delta}{\Delta})f,
\end{aligned}
$$

where $I:\Omega^1_S(\log D)\to\Theta_S$ is the $\mathcal{O}_S$-isomorphism induced by the $\mathcal{O}_S$-complete pairing (3.2). Thus we need only to describe $\mathcal{D}$, $\frac{\mathcal{D}(y)}{y}$ and $I(\frac{d\Delta}{\Delta})$.

THEOREM 9.6. *The Laplacian $\mathcal{D}$ for a local section f of $\mathcal{O}_S$, $\frac{\mathcal{D}(y)}{y}$ and $I(\frac{d\Delta}{\Delta})$ are represented by the operator $\mathcal{U}$ and flat coordinates $t^1,\cdots,t^m$:*

$$
\mathcal{D}(f) = \sum_{i=1}^{m}\left[\mathcal{U}(\partial_i)\partial^i+\{(E+\frac{\mathrm{D}-1}{2})\partial^i(\mathrm{trace}\,\mathcal{U})\}\partial_i\right]f, \qquad (9.19)
$$

$$
\frac{\mathcal{D}(y)}{y} = \frac{1}{2}\mathcal{D}_g(\mathrm{trace}\,\mathcal{U}), \qquad (9.20)
$$

$$
I(\frac{d\Delta}{\Delta}) = g(d(\mathrm{trace}\,\mathcal{U})), \qquad (9.21)
$$

where $\mathcal{D}_g$ denote the Laplacian w.r.t. the holomorphic metric g, the notations $\partial_i=\frac{\partial}{\partial t^i}$, ∂^j are as in Section 3.2.

PROOF. For the proof of (9.19), we need only calculate on each simply connected neighborhood U of any $p\in S\setminus D$. By the equation (8.3), we have

$$
\mathcal{D}(f) = \sum_{i,j=1}^{m} Gram(I)^{-1/2}\partial_i[Gram(I)^{1/2}I(dt^i,dt^j)\partial_j f] \qquad (9.22)
$$

for $Gram(I):=\det(I(\partial_i,\partial_j))$. We easily have $\sum_{j=1}^{m}I(dt^i,dt^j)\partial_j=\mathcal{U}(\partial^i)$ and $Gram(I)=const.\Delta^{-1}=const.y^{-2}$. By using the fact that the metric g is constant for ∂_i, ∂^j, we have

$$\begin{aligned} & \text{(R.H.S.) of (9.22)} \\ = & \sum_{i=1}^{m} y\partial_i[y^{-1}\mathcal{U}(\partial^i)f] \\ = & \sum_{i=1}^{m} y\partial^i[y^{-1}\mathcal{U}(\partial_i)f] \\ = & \sum_{i=1}^{m} \partial^i\mathcal{U}(\partial_i)f + \sum_{i=1}^{m} y(\partial^i y^{-1})\mathcal{U}(\partial_i)f \\ = & \sum_{i=1}^{m} \mathcal{U}(\partial_i)\partial^i f + \sum_{i=1}^{m}[\partial^i,\mathcal{U}(\partial_i)]f + \sum_{i=1}^{m} y(\partial^i y^{-1})\mathcal{U}(\partial_i)f. \end{aligned} \tag{9.23}$$

We already obtained the following equation in (9.12):

$$\sum_{i=1}^{m}[\partial^i,\mathcal{U}(\partial_i)] = \sum_{i=1}^{m}\{(E+\frac{\mathrm{D}}{2})\partial^i(\mathrm{trace}\,\mathcal{U})\}\partial_i. \tag{9.24}$$

We show the following equation:

$$\sum_{i=1}^{m} y(\partial^i y^{-1})\mathcal{U}(\partial_i) = -\sum_{i=1}^{m}\frac{1}{2}(\partial^i(\mathrm{trace}\,\mathcal{U}))\partial_i. \tag{9.25}$$

By the multiplication invariance of g, the operator $\mathcal{U}$ is self-adjoint w.r.t.g. Thus we have

$$\sum_{i=1}^{m}\partial^i\otimes\mathcal{U}(\partial_i) = \sum_{i=1}^{m}\mathcal{U}(\partial^i)\otimes\partial_i \text{ in } \Theta_S\otimes_{\mathcal{O}_S}\Theta_S.$$

Thereby we have

$$\sum_{i=1}^{m}(\partial^i y^{-1})\mathcal{U}(\partial_i) = \sum_{i=1}^{m}(\mathcal{U}(\partial^i)y^{-1})\partial_i.$$

The equation (2.7) gives

$$\sum_{i=1}^{m} y(\mathcal{U}(\partial^i)y^{-1})\partial_i = -\sum_{i=1}^{m}\frac{1}{2}(\partial^i(\mathrm{trace}\,\mathcal{U}))\partial_i.$$

Therefore we obtain (9.25). The equations (9.22), (9.23), (9.24), (9.25) give the equation (9.19).

For the proof of (9.20), using (2.7), we have

$$\begin{aligned} y^{-1}\mathcal{D}(y) &= y^{-1}\sum_{i=1}^{m} y\partial^i[y^{-1}\mathcal{U}(\partial_i)y] \\ &= \sum_{i=1}^{m}\partial^i[\frac{1}{2}\partial_i(\mathrm{trace}\,\mathcal{U})] \\ &= \frac{1}{2}\mathcal{D}_g(\mathrm{trace}\,\mathcal{U}). \end{aligned}$$

For the proof of (9.21), we have

$$<\mathcal{U}(\delta),\frac{d\Delta}{\Delta}> = \frac{\mathcal{U}(\delta)\Delta}{\Delta} = \delta(\mathrm{trace}\,\mathcal{U}) = <\delta, d(\mathrm{trace}\,\mathcal{U})>$$

for any local vector field δ. Put $\eta = g^{-1}(\delta)$, we have

$$< \mathcal{U}(g(\eta)), \frac{d\Delta}{\Delta} >=< g(\eta), d(\mathrm{trace}\mathcal{U}) > .$$

This could be rewritten as

$$I(\eta, \frac{d\Delta}{\Delta}) = g(\eta, d(\mathrm{trace}\mathcal{U})).$$

This gives the equation (9.21). □

9.3. Description of the Laplacian in terms of flat coordinates.

THEOREM 9.7. *The Laplacian $\mathcal{D}$ for a local section f of $\mathcal{O}_S$, $\frac{\mathcal{D}(y)}{y}$ and $I(\frac{d\Delta}{\Delta})$ are represented by flat coordinates:*

$$(9.26) \quad \mathcal{D}(f) = \sum_{i,j=1}^{m} \left[(E\partial^i\partial^j\widetilde{F})\partial_i\partial_j + \{(E+\frac{\mathrm{D}-1}{2})\partial^i\partial_j\partial^j\widetilde{F}\}\partial_i \right] f,$$

$$(9.27) \quad \frac{\mathcal{D}(y)}{y} = \frac{1}{2}\sum_{i,j=1}^{m} \partial_i\partial^i\partial_j\partial^j\widetilde{F},$$

$$(9.28) \quad I(\frac{d\Delta}{\Delta}) = \sum_{i,j=1}^{m} (\partial^j\partial_i\partial^i\widetilde{F})\partial_j.$$

PROOF. First we show

$$(9.29) \qquad \sum_{i=1}^{m} \mathcal{U}(\partial_i)\partial^i f = \sum_{i,j=1}^{m} (E\partial^i\partial^j\widetilde{F})\partial_i\partial_j f.$$

This is shown by the equations

$$\begin{aligned} \mathcal{U}(\partial_i) &= E \circ \partial_i \\ &= \sum_{k=1}^{m} < E \circ \partial_i, dt^k > \partial_k \\ &= \sum_{k=1}^{m} g(E \circ \partial_i, \partial^k)\partial_k \\ &= \sum_{k=1}^{m} (E\partial_i\partial^k\widetilde{F})\partial_k \end{aligned}$$

and the assumption that ∂_i are flat. The equation (9.11) gives the equation

$$(9.30) \qquad (E+\frac{\mathrm{D}-1}{2})\partial^i(\mathrm{trace}\mathcal{U}) = \sum_{j=1}^{m}(E+\frac{\mathrm{D}-1}{2})\partial^i\partial_j\partial^j\widetilde{F}.$$

The equations (9.29) and (9.30) give the equation (9.26).

We have the equation (9.27) by the elementary fact that the Laplacian $\mathcal{D}_g : \mathcal{O}_S \to \mathcal{O}_S$ for the holomorphic metric g is represented by $\sum_{i=1}^{m} \partial_i\partial^i$.

The equation (9.28) is obtained by the equation $g(dt_i) = \partial^i$ and the assumption that ∂^i are flat. □

10. The cases of a universal unfolding of a function with a simple elliptic singularity

In this section, we consider the Frobenius manifold which is obtained as a parameter space of a universal unfolding of a function with a simple elliptic singularity.

The Frobenius manifold structure on a parameter space S of a universal unfolding of a function with an isolated singularity is a consequence of the existence of the primitive form defined by K. Saito and studied by K. Saito [**11**] and M. Saito [**13**].

For a simple elliptic singularity case, the global construction of primitive form by K. Saito gives the Frobenius manifold structure on a parameter space S globally.

We give explicitly the relation of flat coordinates and the Laplacian for the case of simple elliptic singularity. For proofs of the following propositions, we will publish elsewhere. For flat coordinates and the prepotential of the case of simple elliptic singularity, the explicit calculations are done by Noumi-Yamada [**8**], Verlinde-Warner [**16**], Maassarani [**7**], Klemm-Theisen-Schmidt [**5**].

For each case of the simple elliptic singularities, we denote by J_E, the J-invariant and by τ, the flat function defined by Saito [**11**, p.1245 (3.3.16)]. We put

$$\eta := \left[\frac{(\frac{dJ_E}{d\tau})^6}{J_E^4(J_E-1)^3} \right]^{1/24}. \tag{10.1}$$

Up to constant factor, this will be identified with Dedekind's η function through the period mapping. We remark that the following results should be compared with the results of Satake [**15**].

10.1. $\tilde{E}_6$ case.

PROPOSITION 10.1. *Let t be one of flat coordinates of the universal unfolding of a function with a simple elliptic singularity of type $\tilde{E}_6$, with degree $1/3$ ($Et = (1/3)t$). Then*

$$\mathcal{D}(\eta^8 t) = 0, \tag{10.2}$$

$$\mathcal{D}^{Jac}(\eta^6) = 0, \tag{10.3}$$

$$\mathcal{D}^{Jac}(\eta^6\eta^2 t) = 0. \tag{10.4}$$

This result is already obtained by the result of Satake [**14**], using the identification of $\mathcal{D}$ with the Laplacian defined on the period domain through the period mapping. For flat coordinates with other degrees, we can describe them by degree $1/3$ flat coordinates, the degree 0 (known) functions and the Laplacian.

10.2. $\tilde{E}_7$ case.

PROPOSITION 10.2. *Let t be one of flat coordinates of the universal unfolding of a function with a simple elliptic singularity of type $\tilde{E}_7$, with degree $1/4$ ($Et = (1/4)t$). Then*

$$\mathcal{D}(\eta^9 t) = 0, \quad \mathcal{D}^{Jac}(\eta^7) = 0, \quad \mathcal{D}^{Jac}(\eta^7\eta^2 t) = 0. \tag{10.5}$$

For flat coordinates with other degrees, we can describe them by degree $1/4$ flat coordinates, the degree 0 (known) functions and the Laplacian.

10.3. Further remarks. The theory of period mapping for a primitive form for a universal unfolding of a function with a simple elliptic singularity is constructed by K. Saito [**11**], [**12**].

We denote by S the parameter space of a universal unfolding of a function with a simple elliptic singularity. We also denote the period domain by Ω and the monodromy group by Γ.

K. Saito's theory of period mapping gives the isomorphism between the space S (except the points which correspond to simple elliptic singularities) and the quotient space Ω/Γ. Thus a holomorphic function on S (homogeneous for the $\mathbf{C}^*$-action) can be regarded as a Γ-invariant holomorphic function on Ω (with suitable growth condition).

Moreover the Laplacian on S which we construct in this paper can be regarded as the Laplacian on Ω defined in [**15**]. The latter is essentially the same as the Laplacian which was studied in [**4**, §13.3].

The space of Γ-invariant holomorphic functions (with suitable growth condition) on Ω annihilated by the Laplacian is exactly the linear span of theta functions which is Weyl group invariant.

Thereby for $\tilde{E}_6$ case, $\eta^8 t$ can be regarded as a linear combination of Weyl group invariant theta functions on Ω by the equation (10.2). Also $\eta^6 y$ can be regarded as A_ρ, the Weyl denominator for affine Lie algebra of type $E_6^{(1)}$, up to a constant factor by the equation (10.3). Therefore $\eta^2 t$ can be regarded as a linear combination of characters of irreducible highest weight modules with dominant integral highest weight of an affine Lie algebra of type $E_6^{(1)}$ because of the equation (10.4) and the Weyl-Kac character formula.

For $\tilde{E}_7$ case, we could also discuss the results (10.5) parallel to the above.

References

[1] B. Dubrovin, *Geometry of 2D topological field theories*, Integrable Systems and Quantum Groups (ed. by R. Donagi, et al.), Lecture Notes in Math. **1620**, Springer-Verlag, 1996 pp. 120–348.

[2] H. Grauert, R. Remmert, *Coherent Analytic Sheaves*, Grundlehren der mathematischen Wissenschaften **265**, Springer-Verlag, 1984.

[3] C. Hertling, *Frobenius manifolds and moduli spaces for singularities*, Cambridge Tracts in Math. **151**, Cambridge Univ. Press, 2002.

[4] V. G. Kac, *Infinite dimensional Lie algebra*, Cambridge University Press, third edition, 1995.

[5] A. Klemm, S. Theisen, M.G. Schmidt, *Correlation functions for topological Landau-Ginzburg models with $c \leq 3$*, International Journal of Modern Physics A. **7**, No.25 (1992) 6215–6244.

[6] E. Looijenga, *Isolated Singular Points on Complete Intersections*, London Math. Soc. Lecture Note Series **77**, Cambridge University Press, 1984.

[7] Z. Maassarani, *On the solution of topological Landau-Ginzburg models with $c = 3$*, Physics Letters B **273** (1991) 457–462.

[8] M. Noumi, Y. Yamada, *Notes on the flat structures associated with simple and simply elliptic singularities*, Integrable Systems and Algebraic Geometry, Proceedings of the Taniguchi Symposium 1997, World Scientific, 1998.

[9] S. Rosenberg, *The Laplacian on a Riemannian Manifold*, London Mathematical Society Student Texts **31**, 1997.

[10] K. Saito, *Theory of logarithmic differential forms and logarithmic vector fields* , J. Fac. Sci. Univ. Tokyo, Sect. IA Math, **27** (1981) 265–291.

[11] K. Saito, *Period Mapping Associated to a Primitive Form*, Publ. RIMS. Kyoto Univ. **19** (1983) 1231–1264.
[12] K. Saito, *Extended Affine Root System II*, Publ. RIMS. Kyoto Univ. **26** (1990) 15-78.
[13] M. Saito, *On the structure of Brieskorn lattice*, Ann. Inst. Fourier Grenoble **39** (1989) 27-72.
[14] I. Satake, *Flat Structure for the Simple Elliptic Singularity of Thpe $\tilde{E}_6$ and Jacobi Form*, Proceedings of the Japan Academy, vol.**69**, Ser. A, No. 7 (1993) 247–251.
[15] I. Satake, *Flat Structure and the Prepotential for the Elliptic Root System of Type $D_4^{(1,1)}$*, Topological field theory, primitive forms and related topics (ed. by M. Kashiwara, et al.), Progress in Math. **160**, Birkhäuser, 1998, 427–452.
[16] E. Verlinde, N.P. Warner, *Topological Landau-Ginzburg matter at $c = 3$*, Physics Letters B **269** (1991) 96–102.

DEPARTMENT OF MATHEMATICS, OSAKA UNIVERSITY, TOYONAKA, OSAKA, 560-0043, JAPAN

Virtual fundamental classes, global normal cones and Fulton's canonical classes

Bernd Siebert

Introduction

This note, written in January 1997[1], grew out of an attempt to understand references [**Be**], [**BeFa**] and [**LiTi**]. In these papers two related but different methods are presented for the construction of a certain Chow class on moduli spaces of stable (parametrized) curves in a projective manifold V, called virtual fundamental class. This class replaces the usual fundamental class of these spaces in the definition of basic enumerative invariants of V involving curves, called Gromov-Witten (GW-) invariants. They are invariant under smooth deformations of V.

Both approaches are based on a globalization of the concept of normal cones of germs of the space under study inside some modelling space, that is $C_{U|M}$ for $U \subset X$ open with $\iota : U \hookrightarrow M$ and M smooth over k. The essential idea of using bundles of cones inside a vector bundle for globalizing virtual fundamental classes is due to Li and Tian. The data needed to glue differs however somewhat in the two constructions.

A proper understanding of the relationship between the two approaches seemed necessary for finding the natural framework for comparison of algebraic virtual fundamental classes with the author's definition in [**Si1**] of virtual fundamental classes in the symplectic context [**Si3**].

In a first step Behrend and Fantechi use a generalization of the concept of scheme, called Artin stacks, to make sense of the quotient $C_{U|M}/T_M|_U$. These quotients being unique up to canonical isomorphism they glue to an Artin (cone) stack $\mathcal{C}_X$ intrinsically associated to any X. In a second step they need a morphism $\varphi^\bullet : [\mathcal{F}^{-1} \to \mathcal{F}^0] \to \mathcal{L}^\bullet_X$ (in the derived category) from a two-term complex of locally free sheaves to the cotangent complex, inducing an isomorphism in H^0 and an epimorphism in H^{-1}, to cook up an ordinary cone $C(\varphi^\bullet) \subset F_1$, F_1 the vector bundle associated to $\mathcal{F}^{-1}$. Intersection with the zero section finally produces the virtual fundamental class. So a priori the latter depends on the choice of $\varphi^\bullet$.

This is a very natural and mature approach, that clearly separates the globalization process of the normal cone from the construction of the virtual fundamental class. A possible disadvantage is that in dealing with Artin stacks some of the necessary verifications become rather technical, non-geometric in nature.

[1]February 5, 1997; revised May 18, 1997, November 29, 2002, February 19, 2003

Li and Tian circumvent the morphism to the cotangent complex by introducing the notion of "perfect tangent obstruction complex". This is a morphism $\mathcal{F}^{-1} \to \mathcal{F}^0$ of locally free sheaves on X with kernel and cokernel being tangent and obstruction spaces for morphisms to X, compatible with base change. Using relative, formal, "Kuranishi families" as an intermediate object they construct a well-defined cone $C \subset F_1$. Another, less important difference to [**BeFa**] is the use of an absolute obstruction theory instead of one relative to the space of pre-stable curves.

In a previous version of [**LiTi**] the slightly stronger claim was made that already a presentation $\mathcal{F}^{-1} \to \mathcal{F}^0$ of Ω_X should suffice to construct the cone. In trying to understand this statement I was lead to the problem of reformulating [**BeFa**] from the point of view of gluing local cones. Since in the latter reference Artin stacks are used only as book-keeping device rather than as actual spaces it should not come as surprise that one can get along without them (this has already been indicated in op.cit.). Contrary to what I expected, things can be formulated in a rather elegant but direct way via some *yoga of cones bundles.* This part of the paper (Sections 2 and 3) is just a down-to-earth reformulation of (parts of) Sections 2–4 of [**BeFa**]. Section 1 presents the necessary notations concerning cones and linear spaces, the latter being a convenient way of looking at coherent sheaves for our purposes.

In Section 4 we establish a closed formula for virtual fundamental classes involving only the scheme-theoretic structure of X via Fulton's canonical class (Definition 4.3) and the Chern class of the virtual bundle $F_0 - F_1$, F_i the vector bundle associated to $\mathcal{F}^{-i}$ (Theorem 4.6). This formula was actually found by the author in summer 1995 while searching for a purely algebraic definition of GW-invariants. It should be useful for computations, see the author's recent little survey [**Si2**].

A few remarks on GW-theory are in order. First, today I consider the yoga of cone bundles in Sections 2 and 3 as one ingredient for the most economic path to algebraic Gromov-Witten invariants. The other ingredients are going over to Deligne-Mumford stacks, and replacing the morphism to the cotangent complex by an obstruction theory. If the latter is defined similar to [**Ar**], 2.6, rather than in [**LiTi**], one can show [**Si4**] that it is locally nothing but a morphism to the cotangent complex as in [**BeFa**]. Hence the yoga of cone bundles applies to produce the virtual fundamental class.

Second, I would like to illustrate the perspective of the content of Section 4 by the following formula for virtual fundamental classes in Gromov-Witten theory.

THEOREM 0.1. *Let V be a projective variety, smooth over a field K of characteristic 0, and $R \in A_1(V)$, the first Chow group. If $g = 0$ or $\mathcal{C} := \mathcal{C}_{R,g,k}(V)$, the moduli space of stable curves $(C, \mathbf{x}, \varphi : C \to V)$ in V of genus g with k marked points $\mathbf{x} = (x_1, \ldots, x_k)$ and $\varphi_*[C] = R$, is embeddable into a space smooth over $\mathfrak{M}_{g,k}$, then the virtual fundamental class relevant for GW-invariants is*

$$[\![\mathcal{C}]\!] = \left\{c(\operatorname{ind}^V_{R,g,k})^{-1} \cap c_F(\mathcal{C}/\mathfrak{M}_{g,k})\right\}_{d(V,R,g,k)}.$$

Here $\{\,.\,\}_d$ denotes the d-dimensional part of a cycle, $\mathfrak{M}_{g,k}$ is the Artin stack of k-pointed pre-stable curves of genus g, $d(V,R,g,k) = c_1(V)\cdot R + (1-g)\dim V + 3g - 3$ is the expected dimension, and $c_F(\mathcal{C}/\mathfrak{M}_{g,k})$ is Fulton's canonical class for $\mathcal{C}$ relative $\mathfrak{M}_{g,k}$. ◇

Here $\operatorname{ind}^V_{R,g,k} = F_0 - F_1$ is the virtual vector bundle associated to the partial resolution $\varphi^\bullet : [\mathcal{F}^{-1} \to \mathcal{F}^0] \to \mathcal{L}^\bullet_{\mathcal{C}/\mathfrak{M}_{g,k}}$ mentioned in the introduction. It represents the (domain of the) perfect relative obstruction theory $(R\pi_*(f_V^T))^\vee$ of Behrend [**Be**] in $K^0(\mathcal{C})$.

A note on categories: To keep things simple we work here in the category of schemes of finite type over a field k, not necessarily algebraically closed or of characteristic 0. The extension to other base schemes is straightforward. For the purpose of GW-theory one also has to replace schemes by (generalizations of) orbifolds, that is Deligne-Mumford stacks in the algebraic category or analytic orbispaces in an analytic context. Again, our results can be easily adapted to these categories.

For GW-theory this is still not sufficient, because $\mathfrak{M}_{g,k}$ is an Artin stack rather than Deligne-Mumford. One can nevertheless give a construction of the relevant cone without ever really using Artin stacks. For instance, Fulton's canonical class relative $\mathfrak{M}_{g,k}$ has the following simple definition: Embed $\mathcal{C}$ into a smooth Deligne-Mumford k-stack N. In the important case $g = 0$ one could take $\mathcal{C}_{\iota_* R,g,k}(\mathbb{P}^N)$, if $\iota : V \hookrightarrow \mathbb{P}^N$ is a closed embedding. Let $q : \mathcal{U} \to \mathcal{C}$ be the universal curve. Then the pull-back of the (virtual) tangent bundle of $\mathfrak{M}_{g,k}$ is

$$T_{\mathfrak{M}} := \mathcal{E}xt^1_q(\omega_{\mathcal{U}/\mathcal{C}}, \mathcal{O}_{\mathcal{U}}) - \mathcal{E}xt^0_q(\omega_{\mathcal{U}/\mathcal{C}}, \mathcal{O}_{\mathcal{U}}),$$

as an element of $K^0(\mathcal{C})$ (the $\mathcal{E}xt^i_q$ are the derived functors of $q_* \circ \mathcal{H}om$) and

$$\begin{aligned} c_F(\mathcal{C}/\mathfrak{M}_{g,k}) &= \Big(c(T_{\mathfrak{M}})^{-1} \cup c(T_N)\Big) \cap s(C_{\mathcal{C}/N}) \\ &= c(T_{\mathfrak{M}})^{-1} \cap c_F(\mathcal{C}). \end{aligned}$$

Here all sheaves and classes have to be understood in the sense of Deligne-Mumford stacks. With these remarks understood the theorem is a special case of Theorem 4.6.

In preparing this paper discussions with H. Flenner and S. Schröer have been helpful. I am grateful to the referee for a very attentive reading of the manuscript and several competent suggestions.

1. Cones

1.1. Linear spaces. For any algebraic k-scheme X, $\mathbb{A}^1_X = X \times \mathbb{A}^1_k$ has the structure of a *ring over* X: There are morphisms

$$\alpha : \mathbb{A}^1_X \times_X \mathbb{A}^1_X \longrightarrow \mathbb{A}^1_X, \quad \iota : \mathbb{A}^1_X \longrightarrow \mathbb{A}^1_X, \quad \mu : \mathbb{A}^1_X \times_X \mathbb{A}^1_X \longrightarrow \mathbb{A}^1_X,$$

and sections $n, e : X \to \mathbb{A}^1_X$ fulfilling the usual commutative ring axioms with α as addition, ι as additive inverse, μ as multiplication and n, e as neutral elements for α, μ.

A *linear space over* X is an $\mathbb{A}^1_X$-module of finite type over X, that is, an affine morphism $\pi : L \to X$ of finite type together with morphisms

$$a : L \times_X L \longrightarrow L, \quad m : \mathbb{A}^1_X \times_X L \longrightarrow L$$

and a zero section $z : X \to L$, fulfilling the usual module axioms relative X, that is $m \circ (\mu \times \mathrm{Id}_L) = m \circ (\mathrm{Id}_{\mathbb{A}^1_X} \times m)$ as maps from $\mathbb{A}^1_X \times_X \mathbb{A}^1_X \times_X L$ to L etc.[2] By abuse of notation we just write L for the tuple (π, a, m). In the sequel we will restrict to linear spaces that are *representable*, that is, which locally are closed subspaces of vector bundles with induced linear structure. With the obvious notion of homomorphism

[2] "Linear space (over X)" or "linear fiber space" ("Linearer Faserraum") seem to be the classical notation for the "abelian cones" of [**BeFa**]

of linear spaces over X we get the category $\mathrm{Lin}(X)$ of representable linear spaces over X.

There is an anti-equivalence of categories

$$\mathrm{Lin}(X) \longrightarrow \mathrm{Coh}(X)$$

to the category of coherent $\mathcal{O}_X$-modules [**EGA-II**],§1.7: On objects this associates to $L \in \mathrm{Lin}(X)$ the sheaf $\mathrm{Hom}_{\mathrm{Lin}(X)}(L, \mathbb{A}^1_X)$; in the other direction, $\mathcal{F} \in \mathrm{Coh}(X)$ corresponds to

$$L(\mathcal{F}) := \mathrm{Spec}_{\mathcal{O}_X} S^\bullet \mathcal{F},$$

where $S^\bullet\mathcal{F}$ is the symmetric algebra over the $\mathcal{O}_X$-module $\mathcal{F}$. For example, the addition operation $a : L(\mathcal{F}) \times_X L(\mathcal{F}) \to L(\mathcal{F})$ comes from the diagonal morphism $\mathcal{F} \to \mathcal{F} \oplus \mathcal{F}$, $f \mapsto (f, f)$ by application of the functor $L = \mathrm{Spec}_{\mathcal{O}_X} \circ S^\bullet$. Note that a vector bundle E corresponds to the locally free sheaf $\mathcal{O}(E^\vee)$, $E^\vee$ the *dual* bundle.

Representable linear spaces are thus just another way to look at coherent sheaves. We will jump freely between both descriptions and use whichever seems more appropriate in a particular context. Note also that $\mathrm{Lin}(X)$ is an abelian category, so it makes sense to talk about monomorphisms, epimorphisms and exact sequences. A monomorphism $\Phi : E \to F$ of linear spaces corresponds to an epimorphism $\varphi : \mathcal{F} \to \mathcal{E}$ of sheaves and is thus a closed embedding of schemes. An epimorphism $\psi : F \to G$ of linear spaces, however, need not be a surjection of schemes (consider the inclusion $\psi : \mathcal{I} \hookrightarrow \mathcal{O}_X$ for any nontrivial ideal sheaf $\mathcal{I}$).

1.2. Cones. A *cone* C over X is a scheme of the form $\mathrm{Spec}_{\mathcal{O}_X} S^\bullet$ where $S^\bullet = \oplus_{d\ge 0} S^d$ is a graded $\mathcal{O}_X$-module with $S^0 = \mathcal{O}_X$ and $S^\bullet$ generated by $S^1 \in \mathrm{Coh}(X)$. $S^\bullet$ as graded algebra is not in general determined up to isomorphism by the scheme C over X. For the grading one needs to distinguish the generating submodule $S^1 = \mathcal{F}$, or, equivalently, a closed embedding $C \hookrightarrow L(\mathcal{F})$ into a linear space. Such datum could be called *polarization* of C. We will only deal with polarized cones in the sequel.

EXAMPLE 1.1. If X is a closed subscheme of an algebraic k-scheme M with ideal sheaf $\mathcal{I}$ then the cone

$$C_{X|M} = \mathrm{Spec}_{\mathcal{O}_X}\big(\oplus_{d\ge 0} \mathcal{I}^d/\mathcal{I}^{d+1}\big)$$

over X is called *normal cone* to X in M. $C_{X|M}$ is naturally embedded into the *normal space* $N_{X|M} = L(\mathcal{I}/\mathcal{I}^2)$ of X in M (to avoid confusion with $(\mathcal{I}/\mathcal{I}^2)^\vee$, I would rather not call $N_{X|M}$ normal sheaf as in [**BeFa**]). ◇

If C, C' are cones over X, then so is $C \oplus C' := C \times_X C'$.

To a polarized cone $C = \mathrm{Spec}\, S^\bullet$ is associated a Chow class on X, its Segre class

$$s(C) := \sum_{r\ge 0} p_*(\xi^r \cap [\mathbb{P}(C)]),$$

where $p : \mathbb{P}(C) := \mathrm{Proj}\, S^\bullet \to X$ is the projection and $\xi = c_1(\mathcal{O}_{\mathbb{P}(C)}(1))$.

We propose the following formulation of the concept of exact sequence of cones [**Fu**, Expl.4.1.6].

DEFINITION 1.2. Let

$$0 \longrightarrow E \xrightarrow{\ \Phi\ } F \xrightarrow{\ \Psi\ } Q \longrightarrow 0 \qquad (*)$$

be an exact sequence of linear spaces. Let $C \subset Q$ be a cone and set $\tilde{C} := \Psi^{-1}(C)$. Then $(*)$ restricts to

$$0 \longrightarrow E \longrightarrow \tilde{C} \longrightarrow C \longrightarrow 0 \quad .$$

Sequences of cones of this form will be called *exact.* ◇

REMARK 1.3. Exact sequences of cones might not be very useful unless $(*)$ splits locally. In this case $\tilde{C}$ is locally of the form $C \oplus E$, and as in [**Fu**, Expl.4.1.6] one can show $s(\tilde{C}) = s(C \oplus E)$. In the non-split case a convenient way to relate the Segre classes of $\tilde{C}$ and C seem to be unknown.

But note that if E is a vector bundle $(*)$ always splits locally, and so we retrieve the definition of exact sequences of cones as in [**Fu**]. ◇

For an exact sequence of cones as in the definition $\tilde{C}$ is preserved by the additive action of E on F. In other words, $\tilde{C}$ wears the structure of an E-module. More generally, if $\Phi : E \to F$ is a homomorphism of linear spaces and $C \subset F$ is a cone then C is called *E-cone* if C is an E-module via Φ, that is if C is preserved by the additive action of E on F induced by Φ.

EXAMPLE 1.4. In the situation of Example 1.1 $C_{X|M}$ is a $T_M|_X$-cone via the natural homomorphism $\Phi : T_M|_X \to N_{X|M}$. On the sheaf level this action of $T_M|_X$ is

$$\bigoplus_d \mathcal{I}^d/\mathcal{I}^{d+1} \quad \longrightarrow \quad S^\bullet\Omega_M|_X \otimes \bigoplus_d \mathcal{I}^d/\mathcal{I}^{d+1},$$

where for $f_i \in \mathcal{I}/\mathcal{I}^2$ the image of $f_1 \cdot \ldots \cdot f_d$ in the direct summand $S^e\Omega_M|_X \otimes \mathcal{I}^{d-e}/\mathcal{I}^{d-e+1}$ of the target is the sum over all partitions $\{i_1, \ldots, i_e\}$, $\{j_1, \ldots, j_{d-e}\}$ of $\{1, \ldots, n\}$ of terms

$$\mathrm{d}f_{i_1} \cdot \ldots \cdot \mathrm{d}f_{i_e} \otimes f_{j_1} \cdot \ldots \cdot f_{j_{d-e}}.$$

◇

There are many examples of morphisms of linear spaces $E \to F$ and E-cones $C \subset F$ that do not descend to the quotient $Q = E/F$, for instance the examples in Remark 2.18,3 and in Remark 3.5. However, there is one important class of morphisms where it is always possible, namely for locally split monomorphisms. We first treat the split case:

LEMMA 1.5. *Let $\mathcal{E}$, $\mathcal{F} \in \mathrm{Coh}(X)$ and $E = L(\mathcal{E})$, $F = L(\mathcal{F})$ the corresponding linear spaces over X and $C \subset E \oplus F$ an E-invariant closed subscheme with respect to the action of E on the first summand.*

Then C is of the form $E \oplus \bar{C}$ for some uniquely determined closed subscheme $\bar{C} \subset F$.

Proof. The statement is local in X, so we may assume $X = \operatorname{Spec} A$, $E = \operatorname{Spec} A[\mathbf{X}]/\langle\mathbf{e}\rangle$, $F = \operatorname{Spec} A[\mathbf{Y}]/\langle\mathbf{f}\rangle$ with $\mathbf{X} = (X_1, \ldots X_r)$, $\mathbf{Y} = (Y_1, \ldots Y_s)$ and $\mathbf{e} = (e_1, \ldots, e_k)$, $\mathbf{f} = (f_1, \ldots, f_l)$ tuples of linear forms with coefficients in A, $\langle\mathbf{e}\rangle$, $\langle\mathbf{f}\rangle$ the ideals generated by their entries. Then $C = \operatorname{Spec} A[\mathbf{X}, \mathbf{Y}]/I$ with I an ideal containing $\langle\mathbf{e}\rangle + \langle\mathbf{f}\rangle$.

The only possible candidate for $\bar{C}$ is the intersection of C with $0 \oplus F$, that is $\bar{C} = \operatorname{Spec} A[\mathbf{X}, \mathbf{Y}]/(I + \langle\mathbf{X}\rangle) = \operatorname{Spec} A[\mathbf{Y}]/\bar{I}$, with $\bar{I} = \{f(0, \mathbf{Y}) \mid f(\mathbf{X}, \mathbf{Y}) \in I\}$. We have to show that $I = \langle \bar{I} \rangle + \langle\mathbf{e}\rangle$.

C to be E-invariant means that for any $f(\mathbf{X},\mathbf{Y}) = \sum_{M,N} a_{MN}\mathbf{X}^M\mathbf{Y}^N \in I$

$$f(\mathbf{X}+\mathbf{X}',\mathbf{Y}) \;=\; \sum_{M,N} a_{MN}(\mathbf{X}+\mathbf{X}')^M\mathbf{Y}^N \;\in\; \langle I\rangle + \langle\varphi(\mathbf{e})\rangle \qquad (*)$$

holds in $A[\mathbf{X},\mathbf{X}',\mathbf{Y}]$, where $\varphi : A[\mathbf{X}] \to A[\mathbf{X}']$, $X_\mu \mapsto X'_\mu$. Modulo $\langle\mathbf{X}\rangle$ this says

$$f(\mathbf{X}',\mathbf{Y}) \in \langle \bar{I}\rangle + \langle\varphi(\mathbf{e})\rangle$$

in $A[\mathbf{X}',\mathbf{Y}]$. Replacing $\mathbf{X}'$ by $\mathbf{X}$ we thus get $I \subset \langle\bar{I}\rangle + \langle\mathbf{e}\rangle$.

For the other direction we look at $(*)$ modulo $\mathbf{X}+\mathbf{X}'$ to conclude

$$f(0,\mathbf{Y}) \;=\; \sum_N a_{0N}\mathbf{Y}^N \;\in\; I+\mathbf{e} \;=\; I$$

for any $f \in I$, that is $\bar{I} \subset I$. ◇

Proposition 1.6. *Let*

$$0 \longrightarrow F \longrightarrow E \overset{q}{\longrightarrow} Q \qquad (*)$$

be an exact sequence of linear spaces with F a vector bundle, and let $C \subset E$ be an F-cone.

Then there exists a unique cone $\bar{C} \subset Q$ such that $(*)$ induces an exact sequence of cones

$$0 \longrightarrow F \longrightarrow C \longrightarrow \bar{C} \longrightarrow 0\,.$$

In particular, C descends to Q: $C = q^{-1}(\bar{C})$.

Proof. By replacing Q by the closed subspace $E/F \subset Q$ we may assume q to be an epimorphism. Then, since F is a vector bundle, locally $(*)$ splits and we may apply the previous lemma to construct $\bar{C} \subset Q$. ◇

In other words the proposition says that $\bar{C}$ is the scheme theoretic quotient of C by the free action of F. This is a convenient way to think about $\bar{C}$.

2. Going up and down for E-cones

In this section we investigate the behavior of E-cones under morphisms of two-term complexes, that is commutative squares, in $\mathrm{Lin}(X)$. If $\Phi_\bullet = (\Phi_0,\Phi_1) : F_\bullet = (F_0 \to F_1) \to (E_0 \to E_1)$ is such a morphism the corresponding morphism of coherent sheaves will be written $\varphi^\bullet = (\varphi^{-1},\varphi^0) : \mathcal{E}^\bullet = (\mathcal{E}^{-1} \to \mathcal{E}^0) \to \mathcal{F}^\bullet = (\mathcal{F}^{-1} \to \mathcal{F}^0)$. Then $\Phi_i = L(\varphi^{-i})$, $E_i = L(\mathcal{E}^{-i})$, $F_i = L(\mathcal{F}^{-i})$ for $i = 0,1$.

2.1. Going up.

Lemma 2.1. *Let $\Phi_\bullet : F_\bullet \to E_\bullet$ be a commutative square in $\mathrm{Lin}(X)$, and $C \hookrightarrow E_1$ an E_0-cone. Then $\Phi_1^{-1}(C) \hookrightarrow F_1$ is an F_0-cone.*

Proof. Consider the diagram

$$\begin{array}{ccc} F_0 \oplus F_1 & \xrightarrow{\ \alpha\ } & F_1 \\ {\scriptstyle \Phi_0\oplus\Phi_1}\big\downarrow & & \big\downarrow{\scriptstyle \Phi_1} \\ E_0 \oplus E_1 & \xrightarrow{\ \alpha'\ } & E_1 \end{array}$$

with horizontal arrows the morphisms defining the F_0- and E_0-module structures on F_1 and E_1 respectively. By hypothesis $E_0 \oplus C$ is a closed subscheme of $(\alpha')^{-1}(C)$. Thus $F_0 \oplus \Phi_1^{-1}(C) = (\Phi_0 \oplus \Phi_1)^{-1}(E_0 \oplus C)$ is a closed subscheme of $\alpha^{-1}(\Phi_1^{-1}(C))$. ◇

By this lemma we are able to make the following definition.

DEFINITION 2.2. (going up) Let $\Phi_\bullet : F_\bullet \to E_\bullet$ be a commutative square in $\mathrm{Lin}(X)$ and $C \subset E_1$ an E_0-cone. Then the F_0-cone

$$\Phi_\bullet^!(C) := \Phi_1^{-1}(C)$$

in F_1 is called *pull-back* of C under $\Phi_\bullet$. ◇

The pull-back depends only on the homotopy class of $\varphi^\bullet$ (or $\Phi_\bullet$).

PROPOSITION 2.3. *Let $\varphi^\bullet, \psi^\bullet : \mathcal{E}^\bullet = [\mathcal{E}^{-1} \xrightarrow{d} \mathcal{E}^0] \to \mathcal{F}^\bullet$ be homotopic commutative squares in* $\mathrm{Coh}(X)$. *Then for any E_0-cone $C \subset E_1$*

$$\Phi_\bullet^!(C) \;=\; \Psi_\bullet^!(C)\,.$$

Proof. Let $k : \mathcal{E}^0 \to \mathcal{F}^{-1}$ be the homotopy: $\psi^{-1} = \varphi^{-1} + k \circ d$, $\psi^0 = \varphi^0 + d \circ k$. Writing $K = L(k)$ and $\alpha : E_0 \oplus E_1 \to E_1$ for the structure map, Ψ_1 may be decomposed into

$$F_1 \xrightarrow{(K,\Phi_1)} E_0 \oplus E_1 \xrightarrow{\alpha} E_1\,.$$

Since $E_0 \oplus C \subset \alpha^{-1}(C)$, $(K,\Phi_1)^{-1}(E_0 \oplus C) = \Phi_1^{-1}(C)$ is a closed subscheme of $\Psi_1^{-1}(C)$. But the claim is symmetric in $\Phi_\bullet$, $\Psi_\bullet$, hence $\Phi_1^{-1}(C) = \Psi_1^{-1}(C)$. ◇

It is well-known that for general morphisms $\varphi^\bullet$, $\psi^\bullet$ of complexes in $\mathrm{Coh}(X)$ the implications "$\varphi^\bullet$ homotopic $\psi^\bullet$" implies "$\varphi^\bullet = \psi^\bullet$ as morphisms in the derived category", implies "$H^\bullet(\varphi^\bullet) = H^\bullet(\psi^\bullet)$" are strict [**Ha**, §1.4]. For split two-term complexes, however, all these notions coincide.

LEMMA 2.4. *Let $\mathcal{E}^\bullet = [\mathcal{E}^{-1} \to \mathcal{E}^0]$, $\mathcal{F}^\bullet = [\mathcal{F}^{-1} \to \mathcal{F}^0]$ be two-term complexes of coherent $\mathcal{O}_X$-modules, and $\varphi^\bullet = (\varphi^{-1}, \varphi^0)$, $\psi^\bullet = (\psi^{-1}, \psi^0) : \mathcal{E}^\bullet \to \mathcal{F}^\bullet$ be two commutative squares inducing the same map in cohomology: $H^i(\varphi^\bullet) = H^i(\psi^\bullet)$, $i = -1, 0$.*

If $\mathcal{E}^\bullet$ is split then $\varphi^\bullet$ and $\psi^\bullet$ are homotopic, that is there exists a homomorphism $k : \mathcal{E}^0 \to \mathcal{F}^{-1}$ with

$$d' \circ k \;=\; \varphi^0 - \psi^0, \quad k \circ d \;=\; \varphi^{-1} - \psi^{-1}\,.$$

The *proof* is by chasing the diagram with exact rows, which defines $\mathcal{K}$, $\mathcal{K}'$, $\mathcal{Q}$, $\mathcal{Q}'$:

$$\begin{array}{ccccccccc}
0 & \longrightarrow & \mathcal{K} & \xrightarrow{i} & \mathcal{E}^{-1} & \xrightarrow{d} & \mathcal{E}^0 & \xrightarrow{q} & \mathcal{Q} & \longrightarrow & 0 \\
 & & \downarrow{\scriptstyle H^{-1}(\varphi^\bullet) = H^{-1}(\psi^\bullet)} & & \downarrow{\scriptstyle \varphi^{-1},\, \psi^{-1}} & \swarrow{\scriptstyle k} & \downarrow{\scriptstyle \varphi^0,\, \psi^0} & & \downarrow{\scriptstyle H^0(\varphi^\bullet) = H^0(\psi^\bullet)} & & \\
0 & \longrightarrow & \mathcal{K}' & \xrightarrow{i'} & \mathcal{F}^{-1} & \xrightarrow{d'} & \mathcal{F}^0 & \xrightarrow{q'} & \mathcal{Q}' & \longrightarrow & 0\,.
\end{array}$$

Namely, since $q' \circ (\varphi^0 - \psi^0) = 0$, there is a map $\alpha : \mathcal{E}^0 \to \mathcal{F}^{-1}$ with $d' \circ \alpha = \varphi^0 - \psi^0$. Then $d' \circ (\varphi^{-1} - \psi^{-1} - \alpha \circ d) = (\varphi^0 - \psi^0) \circ d - d' \circ \alpha \circ d = 0$ and hence there exists $\beta : \mathcal{E}^{-1} \to \mathcal{K}'$ with $\varphi^{-1} - \psi^{-1} - \alpha \circ d = i' \circ \beta$. Let $\bar\beta : \mathcal{E}^0 \to \mathcal{K}'$ be the composition of β with a split of d. Then $k := \alpha + i' \circ \bar\beta$ is the desired homotopy. ◇

From this lemma and the proposition we conclude

COROLLARY 2.5. *Let $\Phi_\bullet$, $\Psi_\bullet : F_\bullet \to E_\bullet$ be commutative squares of linear spaces over X with $H^\bullet(\Phi_\bullet) = H^\bullet(\Psi_\bullet)$ and $E_\bullet$ locally (!) split. Then for any E_0-cone $C \subset E_1$*

$$\Phi_\bullet^!(C) = \Psi_\bullet^!(C)\,.$$

$\diamond$

Due to the local nature of the claim local splittings suffice.

The next result about functoriality of going up follows directly from the definition.

PROPOSITION 2.6. *Let $\Phi_\bullet : E_\bullet \to F_\bullet$, $\Psi_\bullet : F_\bullet \to G_\bullet$ be commutative squares of linear spaces and $C \subset G_1$ a G_0-cone. Then*

$$(\Psi_\bullet \circ \Phi_\bullet)^!(C) \;=\; \Phi_\bullet^! \circ \Psi_\bullet^!(C)\,. \qquad \diamond$$

2.2. Going down. Going down, or push-forward, of F_0-cones in F_1 to E_1 is a little more subtle. The central tool will be Proposition 1.6. To make this proposition applicable we need a little lemma.

LEMMA 2.7. *Let $\varphi^\bullet : (\mathcal{E}^{-1} \xrightarrow{d} \mathcal{E}^0) \to (\mathcal{F}^{-1} \xrightarrow{d'} \mathcal{F}^0)$ be a commutative square in* $\mathrm{Coh}(X)$. *Then the complex*

$$0 \longrightarrow \mathcal{E}^{-1} \stackrel{(d,\varphi^{-1})}{\longrightarrow} \mathcal{E}^0 \oplus \mathcal{F}^{-1} \stackrel{\varphi^0\circ\mathrm{pr}_1 - d'\circ\mathrm{pr}_2}{\longrightarrow} \mathcal{F}^0 \longrightarrow 0$$

is exact at

i) $\mathcal{F}^0$ *iff $H^0(\varphi^\bullet)$ is surjective*
ii) $\mathcal{E}^0 \oplus \mathcal{F}^{-1}$ *iff $H^0(\varphi^\bullet)$ is injective and $H^{-1}(\varphi^\bullet)$ is surjective*
iii) $\mathcal{E}^{-1}$ *iff $H^{-1}(\varphi^\bullet)$ is injective.*

Proof. Chase the diagram

$$\begin{array}{ccccccccccc}
0 & \longrightarrow & \mathcal{K} & \longrightarrow & \mathcal{E}^{-1} & \stackrel{d}{\longrightarrow} & \mathcal{E}^0 & \longrightarrow & \mathcal{Q} & \longrightarrow & 0 \\
 & & \downarrow{\scriptstyle H^{-1}(\varphi^\bullet)} & & \downarrow{\scriptstyle \varphi^{-1}} & & \downarrow{\scriptstyle \varphi^0} & & \downarrow{\scriptstyle H^0(\varphi^\bullet)} & & \\
0 & \longrightarrow & \mathcal{K}' & \longrightarrow & \mathcal{F}^{-1} & \stackrel{d'}{\longrightarrow} & \mathcal{F}^0 & \longrightarrow & \mathcal{Q}' & \longrightarrow & 0\,.
\end{array} \qquad \diamond$$

If $\varphi^\bullet$ is a quasi-isomorphism we thus get exactness of the stated complex. And $\varphi^\bullet$, viewed as a commutative square, is cartesian ($\mathcal{E}^{-1} = \mathcal{E}^0 \oplus_{\mathcal{F}^0} \mathcal{F}^{-1}$) iff $H^0(\varphi^\bullet)$ is injective and $H^{-1}(\varphi^\bullet)$ is an isomorphism, and it is cocartesian ($\mathcal{F}^0 = (\mathcal{E}^0 \oplus \mathcal{F}^{-1})/\mathcal{E}^{-1}$) iff $H^0(\varphi^\bullet)$ is an isomorphism and $H^{-1}(\varphi^\bullet)$ is surjective. Assume now that F_0 is a vector bundle and that $\Phi_\bullet : [F_0 \xrightarrow{D'} F_1] \to [E_0 \xrightarrow{D} E_1]$ induces an isomorphism on H^0 and a closed embedding of linear spaces on H^1. If these conditions are satisfied we say that *going down is applicable to $\Phi_\bullet$*. Then

$$0 \longrightarrow F_0 \stackrel{(\Phi_0,-D')}{\longrightarrow} E_0 \oplus F_1 \stackrel{q}{\longrightarrow} E_1$$

is exact (Lemma 2.7, $q = D \circ \mathrm{pr}_1 + \Phi_1 \circ \mathrm{pr}_2$) and we may apply Proposition 1.6.

DEFINITION 2.8. (going down) Let $\Phi_\bullet : F_\bullet \to E_\bullet$ be a commutative square in $\mathrm{Lin}(X)$, to which going down is applicable (see above), and let $C \subset F_1$ be an F_0-cone. The unique cone $\bar{C} \subset \mathrm{im}\, q \subset E_1$ with $q^{-1}(\bar{C}) = E_0 \oplus C$, which exists by Proposition 1.6, is called *push-forward* of C by $\Phi_\bullet$, denoted $(\Phi_\bullet)_!(C)$. ◇

Note that $(\Phi_\bullet)_!(C)$ is actually an E_0-cone because $E_0 \oplus C$ is one. And by Proposition 1.6:

PROPOSITION 2.9. *If going down is applicable to $\Phi_\bullet : F_\bullet \to E_\bullet$, and $C \subset F_1$ is an F_0-cone, there is an exact sequence of cones*

$$0 \longrightarrow F_0 \longrightarrow E_0 \oplus C \longrightarrow (\Phi_\bullet)_!(C) \longrightarrow 0\,.$$ ◇

REMARK 2.10. Local freeness of F_0 (or local splittability of the relevant exact sequence of linear spaces) seems to be indispensable, since otherwise $E_0 \oplus C$ need not descend to E_1. See Remark 2.18,3 for a related example. ◇

As with going up, going down depends only on the homotopy class of $\Phi_\bullet$.

PROPOSITION 2.11. *Let $\Phi_\bullet$, $\Psi_\bullet : F_\bullet \to E_\bullet$ be homotopic morphisms of commutative squares in* $\mathrm{Lin}(X)$ *and $C \subset F_1$ an F_0-cone. If going down is applicable to $\Phi_\bullet$ (or, equivalently, to $\Psi_\bullet$) then*

$$(\Phi_\bullet)_!(C) \;=\; (\Psi_\bullet)_!(C)\,.$$

Proof. Let $K : F_1 \to E_0$ be a homotopy between $\Phi_\bullet$ and $\Psi_\bullet$, that is $\Psi_0 = \Phi_0 + K \circ D'$, $\Psi_1 = \Phi_1 + D \circ K$ ($D : E_0 \to E_1$, $D' : F_0 \to F_1$ the differentials). Then the following diagram

$$\begin{array}{ccccccc}
0 & \longrightarrow & F_0 & \xrightarrow{(\Phi_0,-D')} & E_0 \oplus F_1 & \xrightarrow{q_\Phi} & E_1 \\
 & & \Big\downarrow{\scriptstyle \mathrm{Id}} & & \Big\downarrow{\scriptstyle \chi} & & \Big\downarrow{\scriptstyle \mathrm{Id}} \\
0 & \longrightarrow & F_0 & \xrightarrow{(\Psi_0,-D')} & E_0 \oplus F_1 & \xrightarrow{q_\Psi} & E_1
\end{array}$$

with $\chi = (\mathrm{pr}_1 - K \circ \mathrm{pr}_2, \mathrm{pr}_2)$, $q_\Phi = D \circ \mathrm{pr}_1 + \Phi_1 \circ \mathrm{pr}_2$ and $q_\Psi = D \circ \mathrm{pr}_1 + \Psi_1 \circ \mathrm{pr}_2$, is commutative. Now $\chi^{-1}(E_0 \oplus C) = E_0 \oplus C$ and the conclusion follows from the definition of going down. ◇

$F_\bullet$ always splits locally by local freeness of $\mathcal{F}^0$. Taking into account Lemma 2.4 and in view of the local nature of going down, we conclude:

COROLLARY 2.12. *Let $\Phi_\bullet$, $\Psi_\bullet : F_\bullet \to E_\bullet$ be commutative squares in* $\mathrm{Lin}(X)$, *to which going down is applicable, with $H^\bullet(\Phi_\bullet) = H^\bullet(\Psi_\bullet)$, and let $C \subset F_1$ be an F_0-cone. Then $(\Phi_\bullet)_!(C) = (\Psi_\bullet)_!(C)$.* ◇

We observe also that since $q^{-1}(\bar{C}) = E_0 \oplus C \subset E_0 \oplus F_1$ and $q|_{0 \oplus F_1} = \Phi_1$, $\Phi_1^{-1}(\bar{C}) = C$. In other words:

PROPOSITION 2.13. *Whenever going down is applicable to $\Phi_\bullet : F_\bullet \to E_\bullet$ then $\Phi_\bullet^!$ is a left inverse to $(\Phi_\bullet)_!$, that is*

$$\Phi_\bullet^!(\Phi_\bullet)_!(C) \;=\; C$$

for any F_0-cone $C \subset F_1$. ◇

Note that $\Phi_\bullet^!$ is generally not right-inverse to $(\Phi_\bullet)_!$. For example consider $\Phi_\bullet = (\mathrm{Id}, \iota) : (F_0 \to F_1) \to (F_0 \to F_1 \oplus N)$ for any linear space N over X, $\iota : F_1 \to F_1 \oplus N$ the inclusion of the first factor and F_0 acting trivially on N. Then for an F_0-cone of the form $C \oplus N$ it holds $(\Phi_\bullet)_! \Phi_\bullet^!(C \oplus N) = C \oplus 0$. Compare however Proposition 2.15.

As a formal consequence of the last proposition and Proposition 2.6 we obtain functoriality of going down.

PROPOSITION 2.14. *Let $\Psi_\bullet : G_\bullet \to F_\bullet$, $\Phi_\bullet : F_\bullet \to E_\bullet$ be commutative squares of linear spaces to which going down is applicable, and let $C \subset G_1$ be a G_0-cone. Then*

$$(\Phi_\bullet \circ \Psi_\bullet)_!(C) \;=\; (\Phi_\bullet)_! \circ (\Psi_\bullet)_!(C)\,.$$

Proof. Insert $C = (\Phi_\bullet \circ \Psi_\bullet)^!(\Phi_\bullet \circ \Psi_\bullet)_!(C) = \Psi_\bullet^! \Phi_\bullet^! (\Phi_\bullet \circ \Psi_\bullet)_!(C)$ into the right-hand side and apply Proposition 2.13 twice to arrive at the left-hand side! $\diamond$

2.3. The case of quasi-isomorphisms. By definition a morphism $\Phi_\bullet$ of two-term complexes is a quasi-isomorphism if $H^i(\Phi_\bullet)$ is an isomorphism for $i = 0, 1$. This is equivalent to requiring that $\Phi_\bullet$ viewed as a commutative square is cartesian and cocartesian, see Lemma 2.7. Going up and down behaves well with respect to quasi-isomorphisms:

PROPOSITION 2.15. *Let $\Phi_\bullet : F_\bullet \to E_\bullet$ be a quasi-isomorphism of two-term complexes of linear spaces with F_0 locally free. Then going up and down induces a functorial one-to-one correspondence between F_0-cones $C \subset F_1$ and E_0-cones $\bar{C} \subset E_1$.*

Proof. In view of Proposition 2.13 it remains to show that if $\bar{C} \subset E_1$ is an E_0-cone then $\bar{C} = (\Phi_\bullet)_! \Phi_\bullet^!(\bar{C})$. This is a local problem. We may thus assume that there exists a local splitting $\sigma : E_0 \oplus F_1 \to F_0$ of the exact sequence

$$0 \longrightarrow F_0 \longrightarrow E_0 \oplus F_1 \xrightarrow{\;q\;} E_1 \longrightarrow 0, \quad q = D \circ \mathrm{pr}_1 + \Phi_1 \circ \mathrm{pr}_2$$

from Lemma 2.7. Then $\chi = (\sigma, q) : E_0 \oplus F_1 \to F_0 \oplus E_1$ is an isomorphism mapping the diagonal F_0-action on $E_0 \oplus F_1$ to the action on the first factor of $F_0 \oplus E_1$. Since σ is a splitting, χ induces an isomorphism $\ker(\sigma) \to E_1$. Therefore

$$\chi\big(\ker(\sigma) \cap q^{-1}(\bar{C})\big) = \chi(q^{-1}(\bar{C})) \cap (0 \oplus E_1) = 0 \oplus \bar{C}.$$

But $\chi(q^{-1}(\bar{C}))$ is an F_0-cone, and hence Proposition 1.6 implies $\chi(q^{-1}(\bar{C})) = F_0 \oplus \bar{C}$. By definition this says $\bar{C} = (\Phi_\bullet)_! \Phi_\bullet^!(\bar{C})$. $\diamond$

Using the nice behavior under quasi-isomorphisms we may now define *going down for morphisms in the derived category* $D(\mathrm{Coh}(X))$ of the category of coherent sheaves. In the language of linear spaces a morphism of two-term complexes in the derived category $\Phi_\bullet : F_\bullet \to E_\bullet$ consists of

1. another two-term complex $G_\bullet$
2. a quasi-isomorphism $\Theta_\bullet : E_\bullet \to G_\bullet$
3. rm and a morphism $\Psi_\bullet : F_\bullet \to G_\bullet$.

Two morphisms defined by tuples $(G_\bullet, \Theta_\bullet, \Psi_\bullet)$ and $(G'_\bullet, \Theta'_\bullet, \Psi'_\bullet)$ are considered equivalent if there exists a two-term complex $\tilde{G}_\bullet$ and quasi-isomorphisms $\Lambda_\bullet : G_\bullet \to$

$\tilde{G}_\bullet$, $\Lambda'_\bullet : G'_\bullet \to \tilde{G}_\bullet$ making the following diagram commutative up to homotopy:

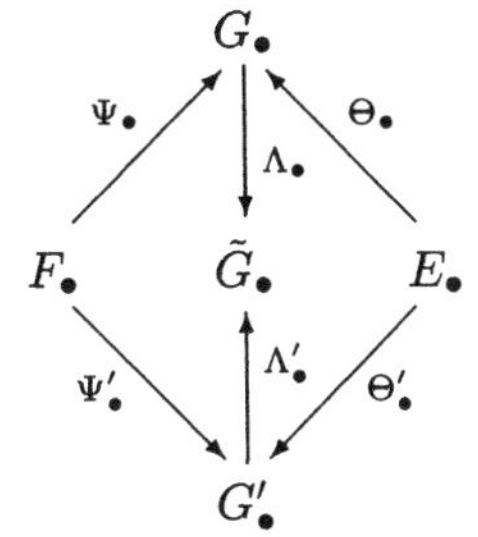

(∗)

DEFINITION 2.16. (going down in the derived category) Let $\Phi_\bullet : F_\bullet \to E_\bullet$ be a morphism of two-term complexes of linear spaces in the derived category, inducing an isomorphism on H^0 and a closed embedding on H^1. Moreover, we require E_0 and F_0 to be locally free. When these assumptions are satisfied we say that *going down is applicable to* $\Phi_\bullet$. In this case the *push-forward* of an F_0-cone $C \subset F_1$ is defined to be the E_0-cone

$$(\Phi_\bullet)_!(C) := (\Theta_\bullet)^!(\Psi_\bullet)_!(C) \subset E_1\,,$$

whenever $(G_\bullet, \Psi_\bullet, \Theta_\bullet)$ is a representative of $\Phi_\bullet$. ◇

Using the previous results it is easy to check that this is well-defined.

Again, for two-term complexes, the morphism class in the derived category can be checked on the level of cohomology:

PROPOSITION 2.17. *Let* $\Phi_\bullet$, $\Phi'_\bullet : F_\bullet \to E_\bullet$ *be as in the definition with* $H^i(\Phi_\bullet) = H^i(\Phi'_\bullet)$, $i = 0, 1$.

Then, locally, $\Phi_\bullet = \Phi'_\bullet$ *as morphisms in the derived category. In particular, for any* F_0*-cone* $C \subset F_1$ *it holds*

$$(\Phi_\bullet)_!(C) \;=\; (\Phi'_\bullet)_!(C)\,.$$

Proof. Locally, we have to produce a two-term complex $\tilde{G}_\bullet$ and quasi-isomorphisms $\Lambda_\bullet : G_\bullet \to \tilde{G}_\bullet$, $\Lambda'_\bullet : G'_\bullet \to \tilde{G}_\bullet$ making (∗) commutative up to homotopy. If $E_i = L(\mathcal{E}^i)$, $G_i = L(\mathcal{G}^i)$, $G'_i = L(\mathcal{G}'^i)$, $i = 0, 1$, we set $\tilde{G}_i := L(\mathcal{G}^i \times_{\mathcal{E}^i} \mathcal{G}'^i)$ with differential $L(D \times D')$, and $\Lambda_\bullet$, $\Lambda'_\bullet$ induced by the projections $\mathcal{G}^i \times_{\mathcal{E}^i} \mathcal{G}'^i \to \mathcal{G}^i$ and $\mathcal{G}^i \times_{\mathcal{E}^i} \mathcal{G}'^i \to \mathcal{G}'^i$ respectively. It is straightforward to check that $\Theta_\bullet$, $\Theta'_\bullet$ induce cocartesian squares in cohomology. Hence $\Lambda_\bullet$, $\Lambda'_\bullet$ inherit to be quasi-isomorphisms from $\Theta_\bullet$, $\Theta'_\bullet$. With this choice (∗) commutes in cohomology, hence locally up to homotopy by Lemma 2.4. ◇

REMARK 2.18. In view of the previous results one might wonder if there exists "going down" when being only given maps on the level of cohomology. There are three remarks I want to make on this.

(1) A map in cohomology is considerably weaker than a map of complexes, even for two-term complexes. For instance, let $E_\bullet = [E_0 \xrightarrow{D} E_1]$ be a non-split epimorphism of linear spaces and $K = \ker D$. Then $H^\bullet(E_\bullet) = H^\bullet([K_\bullet \to 0])$, but the identity map in cohomology is not induced by a morphism of complexes.

(2) There is going down for "cones coming from cohomology": By such cones we mean cones of the form $C = p^{-1}(\bar{C})$ for some $\bar{C} \subset H^1(E_\bullet)$, $p : E_1 \to H^1(E_\bullet)$ the cokernel of $E_\bullet$. Namely, if $\rho : H^1(E_\bullet) \to H^1(F_\bullet)$ is a closed embedding and $p' : F_1 \to H^1(F_\bullet)$ is the cokernel of $F_\bullet$, one may set $\rho_!(C) := p'^{-1}(\rho(\bar{C}))$. In case $\rho = H^1(\Phi_\bullet)$ with $\Phi_\bullet : E_\bullet \to F_\bullet$ a morphism to which going down is applicable, then $\rho_!(C)$ obviously coincides with $(\Phi_\bullet)_!(C)$.

(3) Not every E_0-cone in E_1 comes from cohomology. As a simple example take $X = \mathbb{A}^1_k = \operatorname{Spec} k[T]$, $E_0 = E_1 = X \times \mathbb{A}^2_k = \operatorname{Spec} k[T, U, V]$, $D : E_0 \to E_1$ corresponding to the homomorphism of $k[T]$-algebras sending U, V to TU, TV respectively, and C the line $V(T, U) \subset E_1$.

See also Remark 3.5 for another, less artificial example. ◇

3. Global normal cones

If $\iota : X \hookrightarrow M$ is a closed embedding of algebraic k-schemes the normal cone $C_{X|M} \subset N_{X|M}$ is a $TM|_X$-cone (Example 1.4). With nonsingular M theses normal cones are essentially unique, namely up to vector bundle factors. In fact, if $\iota' : X \hookrightarrow M'$ is another such embedding we may consider the diagonal $(\iota, \iota') : X \hookrightarrow M \times M'$ to reduce to the case where $\iota = \pi \circ \iota'$, $\pi : M' \to M$ a smooth morphism. But then there is an exact sequence of cones

$$0 \longrightarrow (\iota')^* T_{M'|M} \longrightarrow C_{X|M'} \longrightarrow C_{X|M} \longrightarrow 0. \tag{3.1}$$

Based on this observation Behrend and Fantechi show that to any X there is associated a cone stack (a certain Artin stack) over X of pure relative dimension zero, the *intrinsic normal cone* $\mathcal{C}_X$. Locally, $\mathcal{C}_X$ is nothing but the stack-theoretic quotient $C_{X|M}/T_M|_X$, and the above exact sequence of cones is responsible for the fact that these quotients glue.

One essential insight of Behrend and Fantechi is that one can retrieve an actual cone over X by giving a morphism $\varphi^\bullet : \mathcal{F}^\bullet \to \mathcal{L}^\bullet_X$ in the derived category inducing an isomorphism in H^0 and an epimorphism in H^{-1}, and such that $\mathcal{F}^\bullet = [\mathcal{F}^{-1} \to \mathcal{F}^0]$ is a two-term complex of locally free sheaves. Here $\mathcal{L}^\bullet_X$ is the cotangent complex of X. (In the language of [**BeFa**], $\varphi^\bullet$ is a "global resolution" of a "perfect obstruction theory" for X.)

The cotangent complex is a complicated and largely mysterious object canonically associated to any scheme, or even ringed topos [**Il**]. However, here we will work exclusively with the truncated complex $\tau_{\ge -1}\mathcal{L}^\bullet_X$. This is simply an object of the derived category that has the following explicit local description: If $U \subset X$ is an open subscheme and $U \hookrightarrow M$ is a closed embedding into a smooth scheme M then

$$\tau_{\ge -1}\mathcal{L}^\bullet_X = [\mathcal{I}/\mathcal{I}^2 \to \Omega_M|_U],$$

where the complex on the right hand side has entries at -1 and 0 (this follows from the exact triangle for the cotangent complex, see below). In particular, if X is globally embedded into a smooth scheme we can avoid the cotangent complex at all.

Using our study of going up and down for E-cones we will see that the object needed is the following.

DEFINITION 3.1. A *global normal space* for X is a morphism $\varphi^\bullet : \mathcal{F}^\bullet = [\mathcal{F}^{-1} \to \mathcal{F}^0] \to \tau_{\geq -1}\mathcal{L}_X^\bullet$ in the derived category with $\mathcal{F}^0$ locally free and inducing an isomorphism in H^0 and an epimorphism in H^{-1}. ◇

Given a global normal space $\Phi_\bullet : (\tau_{le1}L_X)_\bullet \to F_\bullet$, now written in terms of linear spaces $\tau_{\leq 1}(L_X)_\bullet = L(\tau_{\geq -1}\mathcal{L}_X^\bullet)$ etc., we may construct a cone $C = C(\Phi_\bullet) \subset F_0$ as follows: Let $U \subset X$ be an open set embedded into a nonsingular M, $\iota : U \hookrightarrow M$. The exact triangle of relative cotangent complexes associated to $U \to M \to \operatorname{Spec} k$ yields a morphism in the derived category

$$\Lambda_\bullet : [T_M|_U \to N_{U|M}] \longrightarrow \tau_{\leq 1}(L_X)_\bullet$$

that induces isomorphisms in H^i, $i = 0, 1$. As F_0 and $T_M|_U$ are vector bundles the composition $\Phi_\bullet \circ \Lambda_\bullet$ thus fulfills the assumptions of Definition 2.16. We may define

$$C|_U \; := \; (\Phi_\bullet \circ \Lambda_\bullet)_!(C_{U|M}) \,. \tag{3.2}$$

It remains to show

LEMMA 3.2. $(\Phi_\bullet \circ \Lambda_\bullet)_!(C_{U|M}) \subset F_1|_U$ *is independent of choices.*

Proof. It suffices to treat the case of another embedding $\iota' : U \to M'$ s.th. $\iota = \pi \circ \iota'$ for some smooth morphism $\pi : M' \to M$, see above. We have a commutative diagram with exact rows and columns

$$\begin{array}{ccccccccc}
 & & 0 & & 0 & & & & \\
 & & \downarrow & & \downarrow & & & & \\
 & & \iota'^*T_{M'|M} & = & \iota'^*T_{M'|M} & & & & \\
 & & \downarrow & & \downarrow & & & & \\
0 & \longrightarrow T_U \longrightarrow & \iota'^*T_{M'} & \longrightarrow & N_{U|M'} & \longrightarrow & T_1(U) & \longrightarrow & 0 \\
 & \downarrow & \downarrow & & \downarrow & & \downarrow & & \\
0 & \longrightarrow T_U \longrightarrow & \iota^*T_M & \longrightarrow & N_{U|M} & \longrightarrow & T_1(U) & \longrightarrow & 0\,. \\
 & & \downarrow & & \downarrow & & & & \\
 & & 0 & & 0 & & & &
\end{array}$$

Here $T_1(U)$ is the linear space associated to the first higher cotangent sheaf of U. This shows that $D\pi$ induces a quasi-isomorphism $\Theta_\bullet : [T_{M'}|_U \to N_{U|M'}] \to [T_M|_U \to N_{U|M}]$ with $\Lambda'_\bullet = \Lambda_\bullet \circ \Theta_\bullet$. Moreover, from the exact sequence of cones (3.1)

$$C_{U|M'} \; = \; \Theta_1^{-1}(C_{U|M}) \; = \; \Theta_\bullet^!(C_{U|M}) \,.$$

Thus by Proposition 2.15 we conclude

$$(\Lambda'_\bullet)_!(C_{U|M'}) \; = \; (\Lambda_\bullet)_!(\Theta_\bullet)_!\Theta_\bullet^!(C_{U|M}) \; = \; (\Lambda_\bullet)_!(C_{U|M}) \,.$$

◇

In view of Proposition 2.17 we obtain the first of the two main results of this paper.

THEOREM 3.3. *Let X be an algebraic k-scheme. To any global normal space $\Phi_\bullet : \tau_{\le 1}(L_X)_\bullet \to F_\bullet$ for X is associated an F_0-cone $C(\Phi_\bullet) \subset F_1$, locally of the form (3.2), and of pure dimension equal to* $\mathrm{rk} F_0$. *Moreover, $C(\Phi_\bullet)$ depends only on the map induced in cohomology.*

Proof. It remains to check the statement on the dimension. Locally, we may choose a representation of $\Phi_\bullet \circ \Lambda_\bullet : [T_M|_U \to N_{U|M}] \to F_\bullet$, where $\iota : U \hookrightarrow M$ is a closed embedding of an open $U \subset X$, by $(G_\bullet, \Theta_\bullet, \Psi_\bullet)$ with

- $\Psi_\bullet : [T_M|_U \to N_{U|M}] \to G_\bullet$
- $\Theta_\bullet : F_\bullet \to G_\bullet$ a quasi-isomorphism
- $G_\bullet = [G_0 \to G_1]$ with G_0 a vector bundle (!).

We get two exact sequences of cones (Proposition 2.9)

$$\begin{array}{ccccccccc} 0 & \longrightarrow & T_M|_U & \longrightarrow & G_0 \oplus C_{U|M} & \longrightarrow & (\Psi_\bullet)_! C_{U|M} & \longrightarrow & 0 \\ 0 & \longrightarrow & F_0 & \longrightarrow & G_0 \oplus C(\Phi_\bullet)|_U & \longrightarrow & (\Psi_\bullet)_! C_{U|M} & \longrightarrow & 0 . \end{array}$$

The first one shows that $(\Psi_\bullet)_! C_{U|M}$ is pure dimensional of dimension equal to $\mathrm{rk} G_0$, and then by the second one $C(\Phi_\bullet)$ is pure $(\mathrm{rk} F_0)$-dimensional. ◇

DEFINITION 3.4. $C(\Phi_\bullet)$ is called the *global normal cone* associated to the global normal space $\Phi_\bullet$. ◇

REMARK 3.5. 1) The picture would be especially simple if for any closed embedding $\iota : X \hookrightarrow M$ into a nonsingular M, $C_{X|M}$ came from a cone in the intrinsically defined linear space $T_1(X)$. This is, however, generally wrong:

Consider the fat point $X = \mathrm{Spec}\, R$, $R = k[X,Y]/(X^2, XY, Y^2)$, with its embedding into $M = \mathbb{A}^2_k = \mathrm{Spec}\, k[X,Y]$. Letting A, B, C correspond to the generators X^2, XY, Y^2 of the ideal, $C_{X|M} = \mathrm{Spec}\, R[A,B,C]/(B^2 - AC, XC - YB, XB - YA)$. $T_1(X)$ is the linear space corresponding to the kernel I_T of

$$R[A,B,C] \longrightarrow \Omega_{M|k} \otimes_{k[X,Y]} R, \quad A \mapsto 2X dX,\ B \mapsto Y dX + X dY,\ C \mapsto 2Y dY ,$$

which is $(XA, YA, XB, YB, XC, YC) = (X,Y) \cdot (A,B,C)$. A cone comes from $T_1(X)$ iff its ideal is contained in I_T. But obviously $B^2 - AC \notin I_T$, and so $C_{X|M}$ does not come from a cone in $T_1(X)$.

2) By uniqueness of minimal free resolutions of modules over a local ring (see e.g. [**Ei**, Thm.20.2]) it is not hard to show that for any, not necessarily closed point $x \in X$ there is a *minimal* germ of global normal spaces at x. This is constructed by embedding a neighborhood of x in X into a smooth k-scheme M of dimension $\mathrm{embdim}_x X = \dim_{k(x)} \Omega_x \otimes k(x)$. We assume k perfect here to assure that regular k-schemes are smooth over k. A germ of global normal space at x can then be defined by selecting a minimal set of generators for the ideal defining $X \hookrightarrow M$. This germ of global normal space is minimal in the sense that any other germ of global normal space at x can be obtained by adding trivial factors.

As a consequence, the germ at x of any global normal cone is isomorphic to $C_{X|M}$ plus a vector bundle factor. In particular, the "nonlinear parts" of global normal cones are locally unique. ◇

4. Virtual fundamental class and Fulton's canonical class

4.1. Virtual fundamental classes. If X is an algebraic k-scheme and $\Phi_\bullet : \tau_{\leq 1}(L_X)_\bullet \to F_\bullet$ is a global normal space for X with also F_1 *a vector bundle* we speak of a *free* global normal space of *rank* $\mathrm{rk}(\Phi_\bullet) = \mathrm{rk}\, F_0 - \mathrm{rk}\, F_1$. We may then intersect the zero section $s : X \to F_1$ of F_1 with the global normal cone $C(\Phi_\bullet) \subset F_1$ to produce a class on X.

DEFINITION 4.1. Let $\Phi_\bullet : \tau_{\leq 1}(L_X)_\bullet \to F_\bullet$ be a free global normal space. The Chow class

$$[X, \Phi_\bullet] \;:=\; s^![C(\Phi_\bullet)] \in A_{\mathrm{rk}(\Phi_\bullet)}(X)$$

is called *virtual fundamental class* of X with respect to $\Phi_\bullet$. ◇

Note that $[X, \Phi_\bullet]$ contains as much information as $[C(\Phi_\bullet)] \in A_{\mathrm{rk}(F_0)}(F_1)$, for

$$[C(\Phi_\bullet)] \;=\; p^! s^! [C(\Phi_\bullet)] \;=\; p^! [X, \Phi_\bullet]\,,$$

where $p : F_1 \to X$ is the projection.

One of the most important property of such classes is their compatibility with specializations. In the application to the construction of invariants from moduli spaces associated to a projective manifold V, say (as in Gromov-Witten or Donaldson-theory), this property implies invariance under smooth deformations of V. There are two versions of the specialization theorem, one involving global normal spaces of the total space of a family, and the other working with *relative* global normal spaces (that is, a morphism $\Phi_\bullet : \tau_{\leq 1}(L_{\mathcal{X}|S})_\bullet \to F_\bullet$, where $\mathcal{X} \to S$ is the family of algebraic k-schemes under consideration). We do not have anything to add to the presentation in [**BeFa**, Proposition 5.10 and Proposition 7.2], which translates almost literally into our language.

Before turning to an explicit formula for the computation of $[X, \Phi_\bullet]$ in hopefully more accessible terms, we want to add the following point of view: For any pure-dimensional cone C in a vector bundle F there is a formula for the intersection with the zero locus in terms of the Segre class of C and the total Chern class of F [**Fu**, Expl. 4.1.8]. Applied to $C(\Phi_\bullet) \subset F_1$ it says

$$[X, \Phi_\bullet] \;=\; \{c(F_1) \cap s(C(\Phi_\bullet))\}_{\mathrm{rk}(\Phi_\bullet)}\,,$$

where $\{\,.\,\}_d : A_*(X) \to A_d(X)$ denotes the projection to the d-dimensional part. Now for any $r > 0$, the image of $[C(\Phi_\bullet)]$ under the monomorphism $\iota_r : F_1 \hookrightarrow F_1 \oplus \mathbb{A}^r_X$ becomes rationally trivial, while $c(F_1 \oplus \mathbb{A}^r_X) = c(F_1)$. Thus letting $\tilde{s}^r : X \hookrightarrow F_1 \oplus \mathbb{A}^r_X$ be the zero section we see

$$\{c(F_1) \cap s(C(\Phi_\bullet))\}_{\mathrm{rk}(\Phi_\bullet)-r} \;=\; (\tilde{s}^r)^! (\iota_r)_* [C(\Phi_\bullet)] \;=\; 0\,.$$

This teaches us two things: First, if F_1 splits off a trivial factor $F_1 = \bar{F}_1 \oplus \mathbb{A}^1_X$ with $\mathrm{im}\,\Phi_1 \subset \bar{F}_1$ then $[X, \Phi_\bullet] = 0$. So the result is trivial if F_1 is not chosen small enough. And second, if $[X, \Phi_\bullet] \neq 0$ then $\mathrm{rk}(\Phi_\bullet)$ is the smallest number d such that

$$\{c(F_1) \cap s(C(\Phi_\bullet))\}_d \;\neq\; 0\,.$$

EXAMPLE 4.2. Let X be smooth of dimension n. Then the cotangent complex of X is exact at $\mathcal{L}_X^{-1}$. So the natural morphism $\Phi_\bullet : \tau_{\leq 1}(L_X)_\bullet \to [T_X \to O]$, $O = X$ the trivial linear space over X, is an isomorphism in H^0 and H^1. Then $C(\Phi_\bullet) = X = O$ and $c(F_1) \cap s(C(\Phi_\bullet)) = s(C(\Phi_\bullet)) = [X]$ has vanishing components in dimensions smaller than n. ◇

4.2. Fulton's canonical class. If an algebraic k-scheme X is globally embeddable into a smooth k-scheme M (e.g. X quasi-projective) then

$$c_F(X) \;:=\; c(T_M|_X)\cap s(C_{X|M}) \in A_*(X)$$

is a Chow-class on X that is independent of the choice of embedding [**Fu**, Expl. 4.2.6].

DEFINITION 4.3. $c_F(X)$ is called *Fulton's canonical class.* ◇

Note that if X is smooth one may choose $X = M$ and so $c_F(X) = c(T_X)\cap[X]$. For comparison of $c_F(X)$ with Mather's and MacPherson's Chern classes see [**Al**].

Given a (not necessarily free) global normal space $\Phi_\bullet : \tau_{\le 1}(L_X)_\bullet \to F_\bullet$ for X, $c_F(X)$ can also be expressed as follows:

PROPOSITION 4.4. *Let $\Phi_\bullet : \tau_{\le 1}(L_X)_\bullet \to F_\bullet$ be a global normal space for a quasi-projective X. Then*

$$c_F(X) \;=\; c(F_0)\cap s(C(\Phi_\bullet))\,.$$

Proof. By quasi-projectivity there exists a global closed embedding $\iota : X \hookrightarrow M$ of X into a smooth M. This yields a globally defined morphism in the derived category $\Lambda_\bullet : [T_M|_X \to N_{X|M}] \to \tau_{\le 1}(L_X)_\bullet$. Also by quasi-projectivity any sheaf is the quotient of a locally free sheaf. Hence there is a global representative $(G_\bullet, \Theta_\bullet, \Psi_\bullet)$ of $\Phi_\bullet \circ \Lambda_\bullet$ in the construction of Theorem 3.3, that is, with G_0 a vector bundle, $\Theta_\bullet : F_\bullet \to G_\bullet$ a quasi-isomorphism, $\Psi_\bullet : [T_M|_X \to N_{X|M}] \to G_\bullet$. We get two exact sequences of cones with vector bundle kernels (see Proposition 2.9)

$$\begin{array}{ccccccccc} 0 & \longrightarrow & T_M|_X & \longrightarrow & G_0 \oplus C_{X|M} & \longrightarrow & (\Psi_\bullet)_!(C_{X|M}) & \longrightarrow & 0 \\ 0 & \longrightarrow & F_0 & \longrightarrow & G_0 \oplus C(\Phi_\bullet) & \longrightarrow & (\Psi_\bullet)_!(C_{X|M}) & \longrightarrow & 0\,, \end{array}$$

which by the multiplicativity of Segre classes in exact sequences of cones with vector bundle kernels imply

$$c(T_M|_X)\cap s(C_{X|M}) \;=\; c(G_0)\cap s\Big((\Psi_\bullet)_!(C_{X|M})\Big) \;=\; c(F_0)\cap s(C(\Phi_\bullet))\,. \qquad \diamond$$

REMARK 4.5. If X is any algebraic k-scheme with global normal spaces one could take the right-hand side of the formula in the proposition as definition for a generalization of Fulton's canonical class on projective schemes. However, I was not able to prove independence of this class from the choice of $\Phi_\bullet$. And in case X is not quasi-projective but embeddable into a smooth scheme, in the construction of Theorem 3.3 we might not not be able to choose G_0 locally free. Then the coincidence of this class with $c_F(X)$ is not clear either. The problem is that on one hand the globally defined complex linking two global normal spaces $\Phi_\bullet : \tau_{\le 1}(L_X)_\bullet \to F_\bullet$, $\Phi'_\bullet : \tau_{\le 1}(L_X)_\bullet \to F'_\bullet$ is the cotangent complex, which need not be globally representable by a complex $L_\bullet$ with L_0 a vector bundle, while on the other hand Segre classes do not behave well in exact sequences unless the kernels are vector bundles. ◇

We are now ready to deduce the announced formula for the virtual fundamental class.

THEOREM 4.6. *Let X be a projective k-scheme and $\Phi_\bullet : \tau_{\leq 1}(L_X)_\bullet \to F_\bullet$ a free global normal space for X of constant rank d. Then*

$$[X, \Phi_\bullet] = \left\{ c(\operatorname{ind} F_\bullet)^{-1} \cap c_F(X) \right\}_d,$$

where $\operatorname{ind} F_\bullet$ *is the virtual bundle* $F_0 - F_1 \in K^0(X)$.

Proof. As remarked at the end of the last subsection the virtual fundamental class can be computed by the formula

$$s^![C(\Phi_\bullet)] = \left\{ c(F_1) \cap s(C(\Phi_\bullet)) \right\}_d.$$

Now just insert $c(F_0)^{-1} \cup c(F_0)$ and use Proposition 4.4 ◇

REMARK 4.7. 1) This formula enlightens the dependence of virtual fundamental classes on the choice of global normal spaces: Interestingly, $[X, \Phi_\bullet]$ depends only on the index bundle of $F_\bullet$ rather than on any of the finer data used to construct $C(\Phi_\bullet)$. But note also that for another choice $\Phi'_\bullet : \tau_{\leq 1}(L_X)_\bullet \to F'_\bullet$ of global normal space, $[X, \Phi'_\bullet]$ can not in general be computed from $[X, \Phi_\bullet]$ and $\operatorname{ind} F_\bullet$, $\operatorname{ind} F'_\bullet$ alone.

2) One can take this formula as *definition* of the virtual fundamental class of X without knowing anything about the more sophisticated theory of global normal cones in the non-projective case. This was the point of view of the author in summer 1995 in an attempt to define GW-invariants in algebraic geometry, when I observed it from formal considerations. Unfortunately, I was not aware of Vistoli's rational equivalence [**Vi**], from which the crucial independence of the invariants under smooth deformations can be derived. I learnt also that the same formula has independently be discovered by Brussee for complex spaces constructed as zero locus of holomorphic Fredholm sections of holomorphic Banach bundles over complex Banach manifolds, as occurring for example in Seiberg-Witten theory [**Bs**] (the interpretation of Brussee's $c_*(X)$ as Fulton's canonical class is not quite clear, though).

3) At the beginning of Section 3 we mentioned Behrend and Fantechi's intrinsic normal cone $\mathcal{C}_X$, which locally was the stack-theoretic quotient of $C_{X|M}$ by the action of $T_M|_X$ for some embedding $X \hookrightarrow M$ into a smooth space.

Now if X is globally embedded into a smooth scheme M, $\mathcal{C}_X$ is globally the quotient of $C_{X|M}$ by $T_M|_X$. Hence in view of the multiplicative behavior of Segre classes in exact sequences of cones with vector bundle kernels, $c_F(X)$ could with some right considered as *Segre class of* $\mathcal{C}_X$.

Conversely, if there was a theory of Segre classes for cone stacks, the Segre class of $\mathcal{C}_X$ would generalize Fulton's canonical class to arbitrary algebraic k-schemes. ◇

References

[Al] P. Aluffi: *Chern classes for singular hypersurfaces*, Trans. Amer. Math. Soc. **351** (1999), 3989–4026

[Ar] M. Artin: *Versal deformations and algebraic stacks*, Inv. Math. **27** (1974) 165–189

[Be] K. Behrend: *GW-invariants in algebraic geometry*, Inv. Math. **127** (1997), 601–617

[BeFa] K. Behrend, B. Fantechi: *The intrinsic normal cone*, Inv. Math. **128** (1997), 45–88

[Bs] R. Brussee: *The canonical class and the C^∞-properties of Kähler surfaces*, New York Journ. Math. **2** (1996), 103–146

[EGA-II] A. Grothendieck: *Éléments de géométrie algébrique II: Étude globale élémentaire de quelques classes de morphismes.* Publ. Math. Inst. Hautes Étud. Sci. 8 (1961).

[Ei] D. Eisenbud: *Commutative algebra. With a view toward algebraic geometry*, Springer 1995

[Fu] W. Fulton: *Intersection theory*, Springer 1984

[Il] L. Illusie: *Complexe cotangent et déformations*, 2 vols., Lecture Notes in Mathematics 239/283, Springer 1971/1972

[Ha] R. Hartshorne: *Residues and duality*, Lecture Notes in Math. 20, Springer 1966

[LiTi] J. Li, G. Tian: *Virtual moduli cycles and GW-invariants of algebraic varieties*, in: Topics in symplectic 4-manifolds, 1st International Press lectures presented in Irvine, CA, USA, March 28-30, 1996 (R. Stern ed.), 47–83

[Si1] B. Siebert: *Gromov-Witten invariants for general symplectic manifolds*, preprint dg-ga 9608005, accepted as Habilitationsschrift, Bochum 1998.

[Si2] B. Siebert: *An update on (small) quantum cohomology*, in: Mirror symmetry III, Proceedings of the conference on Geometry and Physics, Montreal 1995 (D.H. Phong, L. Vinet, S.T. Yau eds.), 279–312, AMS/IP Stud. Adv. Math. **10**, Amer. Math. Soc. 1999

[Si3] B. Siebert: *Algebraic and symplectic Gromov-Witten invariants coincide*, Ann. Inst. Fourier **49** (1999) 1743–1795.

[Si4] B. Siebert: *Logarithmic Gromov-Witten invariants*, unfinished manuscript 2001.

[Vi] A. Vistoli: *Intersection theory on algebraic stacks and on their moduli spaces*, Inv. Math. **97** (1989) 613–670

Institut für Mathematik, Universität Freiburg, D-79102 Freiburg
E-mail address: bernd.siebert@math.uni-freiburg.de

A Note on BPS Invariants on Calabi–Yau 3-folds

Atsushi Takahashi

1. Introduction

Let X be a Calabi–Yau 3-fold. Since the virtual dimension of the moduli space of 0-pointed stable maps is given by

$$\text{vdim}_{\mathbb{C}}\overline{\mathcal{M}}_{g,0}(X,\beta) := \dim_{\mathbb{C}} X \cdot (1-g) - \int_{\beta} K_X + 3g - 3 = 0, \tag{1.1}$$

one can define rational numbers called Gromov–Witten invariants of 0-pointed stable maps for all $g \in \mathbb{Z}_{\geq 0}$ and $\beta \in H_2(X,\mathbb{Z})$ as follows:

$$N_g(\beta) := \deg\left[\overline{\mathcal{M}}_{g,0}(X,\beta)\right]^{virt} \in \mathbb{Q}, \quad \text{where } A_0(\overline{\mathcal{M}}_{g,0}(X,\beta),\mathbb{Q}) \overset{deg}{\simeq} \mathbb{Q}. \tag{1.2}$$

It is very useful to introduce the following generating functions of Gromov–Witten invariants:

$$F_g^X := \sum_{\beta \in H_2(X,\mathbb{Z})} N_g(\beta) q^{\beta}. \tag{1.3}$$

F_g^X are well-defined as formal power series of q since $N_g(\beta) = 0$ for $\beta \notin \overline{NE(X)}$ where $\overline{NE(X)}$ is the Mori cone of X (the cone of curves in X).

Over the last 10 years a large numbers of studies have been made on F_0^X, however, little is known about higher genus potentials F_g^X $(g \geq 1)$. The following are our motives for studying BPS invariants:

- **Structure theorem**:
  We expect that Gromov–Witten invariants are written in terms of more fundamental *integral* invariants. It is known in all examples that there exists $n_0^X(\beta) \in \mathbb{Z}$ such that

$$F_0^X = \sum_{\beta \in H_2(X,\mathbb{Z})} n_0^X(\beta) \frac{1}{k^3} q^{k\beta}, \tag{1.4}$$

  (Aspinwall–Morrison formula). This fact suggests the existence of fundamental integral invariants.

  We also expect the analogy of the gauge theory on 4-manifolds such as the structure theorem by Kronheimer-Mrowka and the equivalence between Donaldson and Seiberg–Witten invariants.

- **Definition of Gromov–Witten invariants by $D^b(Coh(X))$:**
  Dubrovin observed in [**D**] that the monodromy data for $\mathbb{P}^2$ (which reconstruct $F_0^{\mathbb{P}^2}$) are essentially given by the bounded derived category $D^b(Coh(\mathbb{P}^2))$ of coherent sheaves. For example, the Stokes matrix comes from the Euler number of an exceptional collection of $D^b(Coh(\mathbb{P}^2))$. By the Givental's reconstruction theorem for semisimple quantum cohomologies [**G**], $F_g^{\mathbb{P}^2}$ $(g \geq 1)$ are determined by $F_0^{\mathbb{P}^2}$.
  Although the quantum cohomology of a Calabi–Yau 3-fold is not semisimple, we expect the analogs of Dubrovin's and Givental's theorems.
- **Gopakumar–Vafa formula**
  Based on the string duality between Type IIA and M-theory, physicists Gopakumar and Vafa [**GV**] introduced a remarkable formula for F_g^X's which generalize the Aspinwall–Morrison formula:

$$\sum_{g\geq 0} F_g^X \lambda^{2g-2} = \sum_{k>0, h\geq 0, \beta\in H_2(X,\mathbb{Z})} n_h^X(\beta)\frac{1}{k}\left(2\sin(\frac{k\lambda}{2})\right)^{2h-2} q^{k\beta}. \tag{1.5}$$

  They conjectured the existence of integral invariants $n_h^X(\beta)$ called BPS invariants defined by M2-branes wrapped around the curves in X, and the equivalence between Gromov–Witten invariants and BPS invariants is given by (1.5).

REMARK. One can always define rational numbers called **conjectural BPS invariants** $n_h^{conj}(\beta)$ in terms of Gromov–Witten invariants $N_g(\beta/k), k \in \mathbb{Z}_{>0}, g \leq h$, by the GV formula (1.5) and the Möbius inversion formula (Proposition 2.1 in [**BP**]). In this approach, the problems are to prove that $n_h^{conj}(\beta) \in \mathbb{Z}$ (**integrality conjecture**) and to clarify the geometric meaning of $n_h^{conj}(\beta)$. The integrality conjecture was checked by Bryan–Katz–Leung for contractible rational curves in [**BKL**] and by Bryan and Pandharipande for some super-rigid curves in [**BP**]. Also, we are informed that Fukaya–Ono [**FO**] proved the genus 0 part of the integrality conjecture in the symplectic category.

Our aims are to define BPS invariants geometrically and to show the Gopakumar-Vafa formula later as the equivalence between Gromov–Witten and BPS invariants. So far, we have been studying a mathematical definition of BPS invariants in [**HST2**] by the moduli space of sheaves and formulated the Gopakumar–Vafa conjecture. In particular, there we have proved the existence of natural $(sl_2)_L \times (sl_2)_R$-action on the intersection cohomology of the moduli spaces.

The purpose of this paper is to clarify the "naturality" of our moduli spaces of stable sheaves used to define BPS invariants based on the results in [**HST2**] and new evidences for the Gopakumar–Vafa formula (1.5). From these results, we see that our moduli spaces $M_\beta(X)$ are necessary and sufficient to define the BPS invariants as far as we use as D-brane moduli spaces the moduli spaces of semi-stable sheaves of dimension 1. We have not yet succeeded to define our D-brane moduli spaces from $D^b(Coh(X))$ since at this moment we have no mathematical notion of stability on triangulated categories. However, once one could define some notion of stability and moduli space of stable objects on $D^b(Coh(X))$, then we expect our moduli spaces will be naturally derived from them since there are several phenomena that only our moduli spaces could explain.

The main results in this paper are Proposition 4.1, Theorems 5.2, 5.5, 5.6 and 5.8. We shall recall our definition of BPS invariants given in [**HST2**] in section 2, and give our mathematical conjectures in section 3. Then we shall explain above "naturality" (section 4) with some new evidences for Gopakumar–Vafa conjecture (found in section 5). We refer to [**FP**], [**BP**], [**P**], [**KZ**] the details of calculations of Gromov–Witten invariants.

Acknowledgement

I am deeply grateful to Professor Kyoji Saito for his encouragement. I also would like to thank S. Hosono and M. -H. Saito for valuable discussions. Part of this work are done during my stay at the Max Plank Institute für Mathematik. I would like to thank for hospitality. This work was partly supported by Grant-in Aid for Scientific Research grant numbers 14740042 of the Ministry of Education, Science and Culture in Japan.

2. A Mathematical Definition of BPS Invariants

2.1. D-brane moduli spaces. It is not well-understood how to define mathematically the BPS D-branes and moduli spaces of them. One may regard the category of (B-type) D-branes as $D^b(Coh(X))$, the bounded derived category of coherent sheaves on X. It is necessary to find D-branes satisfying the BPS condition to define good (e.g., Hausdorff) moduli spaces for them. Mathematically speaking, BPS condition is nothing but the stability condition. However we have not succeeded to define the suitable notion of stability on objects of $D^b(Coh(X))$. Due to absence of stability on $D^b(Coh(X))$, we shall use here instead the abelian category of coherent sheaves and the usual (μ-)stability on sheaves.

Let us first recall the notion of stability.

Definition 2.1. (**Stability**)
Let $\mathcal{E}$ be a coherent sheaf which is pure of dimension l on X, i.e., $\dim_{\mathbb{C}} \operatorname{Supp}(\mathcal{F}) = l$ for any nontrivial coherent subsheaf $\mathcal{F} \subset \mathcal{E}$. Then $\mathcal{E}$ is called stable (resp. semistable) if for any proper subsheaf $\mathcal{F}$,

$$p(\mathcal{F}, m) < p(\mathcal{E}, m), \text{ for } m >> 0.$$

$$(\text{resp. } \; p(\mathcal{F}, m) \leq p(\mathcal{E}, m), \text{ for } m >> 0),$$

where

$$P(\mathcal{E}, m) := \chi(X, \mathcal{E}(m)) = \sum_{i=0}^{l} \alpha_i(\mathcal{E}) \frac{m^i}{i!}$$

is the Hilbert polynomial of $\mathcal{E}$ and $p(\mathcal{E}, m) := P(\mathcal{E}, m)/\alpha_l(\mathcal{E})$.

One can construct the following moduli spaces $M_{d,\chi}(X)$ of semistable sheaves by the Simpson's construction (see, for example [**HL**]):

Definition 2.2. Let us fix an ample line bundle L on X. We denote by $M_{d,\chi}(X)$ the moduli space of semistable sheaves $\mathcal{E}$ on X with Hilbert polynomial

$$P(\mathcal{E}, m) = dm + \chi.$$

Here we have restricted to sheaves with 1-dimensional supports, since we are only interested in D2-branes (or M2-branes). Note that for 1-dimensional sheaves, the stability defined above coincides with the $\mu-$stability defined by $\alpha_{l-1}(\mathcal{E})/\alpha_l(\mathcal{E})$ instead of $p(\mathcal{E}, m)$ in the above definition. It is known that $M_{d,\chi}(X)$ is a projective scheme (Theorem 4.3.4 [**HL**]).

REMARK. Recently, Aspinwall–Douglas [**AD**] have proposed the notion of Π-stability on $D^b(Coh(X))$. They proposed to use instead of P in the definition 2.1 the periods Π of the mirror manifold of X. Note that the corresponding objects on X to Π in general receive quantum corrections, (i.e., contain information of Gromov–Witten invariants) since we have the following expression of Π by genus 0 Gromov–Witten potential F_0^X in the symplectic basis: $\Pi = (1, t, \partial_t F_0^X, t\partial_t F_0^X - 2F_0^X)$ where for simplicity we deal with the case $H^2(X) = \mathbb{C}$ and set $t := \log q$. Thus one might think that there is a problem of self-reference, more precisely, one might think that we are using Gromov–Witten invariants to define Gromov–Witten invariants!

However, since we are working near the the large radius limit in the complexified Kähler moduli space, as far as we deal with D2-branes $\mathcal{E}$ corresponding to vector $(0, 0, \alpha_1(\mathcal{E}), \alpha_0(\mathcal{E}))$, the stability condition receives no quantum corrections ($\Pi(\mathcal{E}) = \alpha_1(\mathcal{E}) \cdot t + \alpha_0(\mathcal{E})$). It is also conjectured that Π-stability coincides with the usual (μ-)stability near the large radius limit. Hence we have no contradictions here.

2.2. Support morphism. We next define the support morphism from the moduli space of stable sheaves to a suitable moduli space parameterizing the deformation of curves in X. A problem happens since we can not define the canonical scheme structure on $\mathrm{Supp}(\mathcal{E})$ for $\mathcal{E}$. Hence it might be impossible to use the Hilbert scheme which is the most natural object as the deformation space of subschemes in X. Our solution in [**HST2**] to this problem is to use the Chow variety $Chow(X)$, which is a projective variety parameterizing algebraic cycles on X. As a result, our support morphism becomes a generalization of the Hilbert-Chow morphism, and coincides with it when we consider stable sheaves supported on families of rational curves in X.

Consider a coherent sheaf $\mathcal{E}$ on X pure of dimension 1. Let $\mathrm{Supp}(\mathcal{E})$ be the support of $\mathcal{E}$, $Y_1, \cdots, Y_l$ be the irreducible components of $\mathrm{Supp}(\mathcal{E})$ and v_i be the generic point of Y_i. Then the stalk $\mathcal{E}_{v_i} = \mathcal{E} \otimes_{\mathcal{O}_X} \mathcal{O}_{X,v_i}$ is an Artinian module of finite length $length(\mathcal{E}_{v_i})$. One can define an algebraic cycle $s(\mathcal{E})$ by

$$s(\mathcal{E}) := \sum_{i=1}^{l} length(\mathcal{E}_{v_i}) \cdot Y_i. \tag{2.1}$$

DEFINITION 2.3. (**D-brane moduli space**)

$$M_{\beta,\chi}(X) := \{\mathcal{E} \in M_{d,\chi}(X) \mid [s(\mathcal{E})] = \beta \in H_2(X, \mathbb{Z})\}. \tag{2.2}$$

In particular, we set

$$M_\beta(X) := M_{\beta,1}(X). \tag{2.3}$$

Note that if $\mathcal{E} \in M_\beta(X)$ then $\chi(\mathcal{E}) = 1$ and $\int_{s(\mathcal{E})} c_1(L) = d$.

REMARK. We shall restrict our attentions to $M_\beta(X)$. This corresponds to the statement in [**GV**] that the degeneracy of BPS states should be independent on χ. One may think this restriction is not natural but we shall see in section 4 these moduli spaces are sufficient to our mathematical formulation of the Gopakumar–Vafa formula.

Let us assume that $M_\beta(X)$ is normal (or take the normalization of $M_\beta(X)$ and denote it by $M_\beta(X)$ again for simplicity).

PROPOSITION 2.1. ([**HST2**])
The natural map

$$\begin{array}{rccc} \pi_\beta : & M_\beta(X) & \to & Chow(X) \\ & \mathcal{E} & \mapsto & s(\mathcal{E}) \end{array} \tag{2.4}$$

becomes a morphism of projective schemes. □

More precisely, we can construct π_β as follows. Since the constant term of $P(\mathcal{E}, m)$ is 1, any semi-stable sheaf $\mathcal{E} \in M_\beta(X)$ is stable and there exists a universal sheaf $\mathcal{F}_\beta$ over $M_\beta(X) \times X$ (Corollary 4.6.6 [**HL**]).

Let us consider the algebraic cycle $s(\mathcal{F}_\beta) \subset M_\beta(X) \times X$ and the natural morphism $p_\beta : s(\mathcal{F}_\beta) \to M_\beta(X)$. Since p_β is a proper morphism all of whose fibers have dimension 1, $s(\mathcal{F}_\beta) \to M_\beta(X)$ is a "well defined family of algebraic cycles" (see for example [**Ko**]). We can obtain the unique morphism $M_\beta \to Chow(X)$ by the universality of the Chow variety, i.e., we have the following commutative diagram:

$$\begin{array}{ccccc} s(\mathcal{F}_\beta) & \longrightarrow & \mathcal{U} & \subset Chow(X) \times X \\ {\scriptstyle p_\beta}\downarrow & & \downarrow{\scriptstyle p} & \\ M_\beta(X) & \longrightarrow & Chow(X), & \end{array} \tag{2.5}$$

where $p : \mathcal{U} \to Chow(X)$ is the universal well defined family of algebraic cycles.

Let us denote by $S_\beta(X)$ the normalization of the image of $M_\beta(X)$ in $Chow(X)$. Since $M_\beta(X)$ is normal, the morphism factors through $S_\beta(X)$ from the universal property of the normalization, and we obtain the natural morphism

$$\pi_\beta : M_\beta(X) \to S_\beta(X). \tag{2.6}$$

Note that π_β is projective since $M_\beta(X)$ and $S_\beta(X)$ are projective.

REMARK. If we consider sheaves in a smooth projective surface in X, then π_β coincides with the support morphism given by Le Portier [**LeP**]. We shall use this property when we calculate local BPS invariants, BPS invariants for sheaves restricted in a surface embedded in a Calabi–Yau 3-fold.

2.3. BPS Invariants. We have the following theorem and definition of BPS invariants.

THEOREM 2.2. ([**HST2**]) *Let $\pi_\beta : M_\beta(X) \to S_\beta(X)$ be the projective morphism defined in* (2.6). *Let us fix a relative ample line bundle L_1 on $M_\beta(X)$ and an ample line bundle L_2 on $S_\beta(X)$ respectively.*

Then $IH^*(M_\beta(X))$ is a representation of the $(sl_2)_L \times (sl_2)_R$ defined by $\omega_L := c_1(L_1)$ and $\omega_R := c_1(L_2)$:

$$IH^*(M_\beta(X)) = \sum_{j_1, j_2} N_{j_1, j_2} (j_1)_L \otimes (j_2)_R, \tag{2.7}$$

where by (j) we denote the irreducible $(2j+1)$-dimensional representation of sl_2. As a $\mathbb{C}$-vector space, one can decompose $IH^*(M_\beta(X))$ as

$$IH^*(M_\beta(X)) = \sum_{j_1, j_2} N_{j_1, j_2} (j_1)_L \otimes (j_2)_R = \bigoplus_{h \geq 0} \left[(\frac{1}{2})_L \oplus 2(0)_L \right]^{\otimes h} \otimes R_h(\beta). \tag{2.8}$$

where we denote by $R_h(\beta)$ a virtual representation (a linear combination of representations over $\mathbb{Z}$) of $(sl_2)_R$-action. □

Since neither $M_\beta(X)$ nor the morphism π_β may be smooth in general, we cannot prove the existence of such an action on $H^*(M_\beta(X), \mathbb{C})$ by the usual Leray's spectral sequence. However, the "perverse" Leray's spectral sequence tells us the origin of the $(sl_2)_L \times (sl_2)_R$-action on intersection cohomology $IH^*(M_\beta(X))$. Not only physics but also mathematics can explain $(sl_2)_L \times (sl_2)_R$ quite naturally!

DEFINITION 2.4. **(BPS invariants)**
One can define *integers* $n_h(\beta)$ by the following formula:

$$n_h(\beta) := \begin{cases} Tr_{R_h(\beta)}(-1)^{2H_R} & \text{for } \beta \neq 0 \\ \frac{1}{2}Tr_{R_h(\beta)}(-1)^{2H_R} & \text{for } \beta = 0, \end{cases} \tag{2.9}$$

where we mean by $Tr(-1)^{2H_R}$ the graded trace defined by

$$Tr_{(j)_R}(-1)^{2H_R} := (-1)^{2j}(2j+1), \quad j \in \frac{1}{2} \cdot \mathbb{Z}. \tag{2.10}$$

$n_h(\beta)$ will be called *BPS invariants*.

REMARK. If $M_\beta(X)$ is not normal, one should consider also the contributions from non-normal subvarieties of $M_\beta(X)$, i.e., repeat the same procedure and define $n_h(\beta)$ for each subvariety. $n_h(\beta)$ should be defined by the total sum of these.

3. Conjectures

Now we can formulate the conjecture proposed by Gopakumar–Vafa as follows:

CONJECTURE 3.1. *$n_h(\beta) \in \mathbb{Z}$ defined in* (2.9) *should be deformation invariants satisfying the Gopakumar–Vafa formula* (1.5)*. More precisely,*

$$n_h(\beta) = n_h^{conj}(\beta), \quad \text{for all } \beta \in H_2(X, \mathbb{Z}), \tag{3.1}$$

where $n_h^{conj}(\beta)$ are the conjectural BPS invariants

$$\sum_{g \geq 0} F_g^X \lambda^{2g-2} = \sum_{k>0, h\geq 0, \beta \in H_2(X,\mathbb{Z})} n_h^{conj}(\beta) \frac{1}{k} \left(2\sin(\frac{k\lambda}{2})\right)^{2h-2} q^{k\beta}. \tag{3.2}$$

defined by Gromov–Witten invariants via Gopakumar–Vafa formula.

On the other hand, we can consider the another integral invariants associated to $M_\beta(X)$ called holomorphic Casson invariants defined by Thomas [**Th**]. The holomorphic Casson invariant for $M_\beta(X)$ is defined by the degree of the virtual fundamental class of $M_\beta(X)$.

CONJECTURE 3.2.

$$n_0(\beta) = \deg\left[M_\beta(X)\right]^{virt}, \quad {}^\forall \beta \in H_2(X, \mathbb{Z}). \tag{3.3}$$

Since for all $\mathcal{E} \in M_\beta(X)$ is stable, $[M_\beta(X)]^{virt}$ is well-defined. We see that the virtual dimension of $M_\beta(X)$ is 0, hence we have

$$\deg[M_\beta(X)]^{virt} \in \mathbb{Z}, \quad \text{where } A_0(M_\beta(X), \mathbb{Z}) \overset{\deg}{\simeq} \mathbb{Z}.$$

Thus we obtain integers for every $M_\beta(X)$. Note that $n_0(\beta) = \deg\left[M_\beta(X)\right]^{virt}$ whenever $M_\beta(X)$ is smooth since both are given by $(-1)^{\dim_{\mathbb{C}} M_\beta(X)}\chi(M_\beta(X))$.

4. Discussions on moduli spaces $M_\beta(X)$

We shall explain that our moduli spaces $\{M_\beta(X)\}_{\beta\in H_2(X,\mathbb{Z})}$ are natural ones to define BPS invariants. We shall discuss

(i) reduction from semistable sheaves to stable sheaves,
(ii) $\{M_{\beta,\chi}(X)\}_{\beta\in H_2(X,\mathbb{Z}),\chi\in\mathbb{Z}}$ to $\{M_\beta(X)\}_{\beta\in H_2(X,\mathbb{Z})}$,
(iii) nice properties of $M_\beta(X)$.

4.1. From semistable sheaves to stable sheaves. Gopakumar–Vafa formula (1.5) gives
(4.1)

$$\exp\left(\sum_{g\geq 0} F_g^X \lambda^{2g-2}\right) = \exp\left(\sum_{k>0,h\geq 0,\beta\in H_2(X,\mathbb{Z})} n_h(\beta)\frac{1}{k}\left(2\sin(\frac{k\lambda}{2})\right)^{2h-2} q^{k\beta}\right).$$

Note that since we have taken exponential, the LHS of (4.1) represents the number of all holomorphic maps from curves that are non-connected in general. Motivated by the physical statement in [**GV**], we expect the correspondence

$$\{\text{all holomorphic maps from curves}\} \overset{GV(4.1)}{\Longleftrightarrow} \{\text{all semi-stable sheaves of dimension } 1\}.$$

Note that any semi-stable sheaf is S-equivalent to a direct sum of stable sheaves and that there is a similarity between the RHS of (4.1) and the Macdonald formula on the Poincaré polynomial χ_t of symmetric products of a variety V:

$$\sum_{l\geq 0}\chi_t(\mathrm{Sym}^l V)q^l = \exp\left(\sum_{k\geq 1}\frac{1}{k}\chi_{t^k}(V)q^k\right). \tag{4.2}$$

Recall that BPS invariants are defined by $n_h(\beta) := Tr_{R_h(\beta)}(-1)^{2H_R}$ and

$$\left(2\sin(\frac{\lambda}{2})\right)^{2h} = Tr_{I_h}(-1)^{2H_L}(e^{i\lambda}), \quad I_h := \left[(\frac{1}{2})_L \oplus 2(0)_L\right]^{\otimes h}.$$

The relation between Poincaré polynomial and our $(sl_2)_L\times(sl_2)_R$-decomposition are already discussed in [**HST1**]: Poincaré polynomial is a character for the diagonal $(sl_2)_{diag}$, $((sl_2)_{diag}\subset(sl_2)_L\times(sl_2)_R)$ and our generating function of BPS invariants is the character for $(sl_2)_R$. Thus the RHS of (4.1) can be regarded as an analog of the Macdonald formula.

When we count the number of holomorphic maps, we can restrict to maps from connected curves (i.e., we replace $\exp(\sum_{g\geq 0}F_g^X\lambda^{2g-2})$ to $\sum_{g\geq 0}F_g^X\lambda^{2g-2}$.). Hence by the equation (4.1) we expect $n_h(\beta)$ should count only stable sheaves.

4.2. From $\{M_{\beta,\chi}(X)\}_{\beta,\chi}$ to $\{M_\beta(X)\}_\beta$. The discussion above tells us that for the definition of BPS invariants not all moduli spaces $\{M_{\beta,\chi}(X)\}_{\beta,\chi}$ may be necessary. From the facts given below, we see that moduli spaces $\{M_\beta(X)\}_\beta$ are necessary and sufficient (see section 5 for the statements of Theorems).

- When $\beta=0$, stable sheaves are only skyscraper sheaves with $\chi=1$, and Theorem 5.2 shows that $M_0(X)=X$ is the correct moduli space.
- Theorem 5.5 implies that $M_\beta(X)$ (or $M_{\beta,\chi}(X)$ with fixed χ) is sufficient to define BPS invariants. If one considers $M_{\beta,\chi}(X)$ for all χ, then one gets too many stable sheaves for the Gopakumar–Vafa formula (1.5) to hold.

- Theorem 5.7 implies that one has to consider the moduli of sheaves with $\chi = 1$, i.e., $M_\beta(X)$, since the number of stable sheaves depends on χ. In this case, one also sees that if one considers $M_{\beta,\chi}(X)$ for all χ, then one gets too much stable sheaves for the Gopakumar–Vafa formula (1.5) to hold.

Now we are sure that $\{M_\beta(X)\}_{\beta\in H_2(X,\mathbb{Z})}$ are natural moduli spaces to define BPS invariants, as far as we consider the moduli spaces of semi-stable sheaves of dimension 1.

Thus we see that the Gopakumar–Vafa conjecture should relate the following moduli spaces:

$$\{\overline{\mathcal{M}}_{g,0}(X,\beta)\}_{g\geq 0,\beta\in H_2(X,\mathbb{Z})} \overset{\text{GV conjecture}}{\Longleftrightarrow} \{M_\beta(X)\}_{\beta\in H_2(X,\mathbb{Z})}.$$

We expect the Gopakumar–Vafa formula (1.5) will be proved by constructing a correspondence on some product spaces of $\overline{\mathcal{M}}_{g,0}(X,\beta)$ and $M_\beta(X)$, and the correspondence would be given by some moduli spaces of stable maps from curves with sheaves to X [1].

4.3. Nice properties of $M_\beta(X)$. Moduli space $M_\beta(X)$ has nice properties since $\chi(\mathcal{E}) = 1$, for all $\mathcal{E} \in M_\beta(X)$. First we note that as discussed in section 2, any semi-stable sheaf $\mathcal{E} \in M_\beta(X)$ is stable. This is because for an ample line bundle L on X, $s(\mathcal{E})\cdot L$ and 1 are always relatively coprime. As a result, $[M_\beta(X)]^{virt}$ can be defined for all $\beta \in H_2(X,\mathbb{Z})$ and we could formulate Conjecture 3.2. We do not know whether $[M_{\beta,\chi}(X)]^{virt}$ exists or not for $\chi \neq 1$. We should also remark that our moduli space $M_\beta(X)$ naturally contains stable sheaves on non-reduced schemes by the construction.

The most important thing is the existence of a non-trivial homomorphism $\mathcal{O}_X \to \mathcal{E}$. This fact is often used to calculate BPS invariants in [**HST2**] and in section 5. In particular, by combining with the stability condition we get the following statement:

PROPOSITION 4.1. *Let $\mathcal{C} \to S \subset Hilb(X)$ be a flat family of curves with $H^0(\mathcal{C}_s) = \mathbb{C}$ and $H^1(\mathcal{C}_s) = 0$ at $^\forall s \in S$. Then $S \subset M_\beta$ for some β and $\pi_\beta|_S$ coincides with the Hilbert–Chow morphism.* □

This follows from the fact that the morphism $\mathcal{O}_X \to \mathcal{E}$ is unique and induces an isomorphism $\mathcal{O}_{s(\mathcal{E})} \simeq \mathcal{E}$ by the stability condition. It means that BPS invariants of the family of rational curves should be counted by the multiplicity in the Hilbert scheme or holomorphic Casson invariant $\deg[S]^{virt}$ as expected. In this case, deformation invariance of the relevant BPS invariants follows from that of $\deg[S]^{virt}$. Evidences for this are given as Theorems 5.5 and 5.6.

Another nice property is the compatibility with the proposal by Katz–Klemm–Vafa [**KKV**]. They gave under some assumptions the way to calculate BPS invariants based on the Abel–Jacobi map and applied this method to obtain local BPS invariants for del Pezzo surfaces. Let C_g be a smooth curve of genus g. The Abel–Jacobi map is a surjective morphism defined by

$$\text{(4.3)} \qquad \mathrm{Hilb}^g(C_g) \to \mathrm{Jac}^g(C_g), \quad I \mapsto [\mathcal{H}om_{\mathcal{O}_{C_g}}(I, \mathcal{O}_{C_g})].$$

Then one can relate $H^*(\mathrm{Jac}^g(C_g))$ with $H^*(\mathrm{Hilb}^g(C_g))$ by using the theta divisor. Katz–Klemm–Vafa considered the family $\mathcal{C}_g \to S$ of curves of genus g and relative

[1]The author would like to thank M.-H.Saito for valuable discussions on this.

version of (4.3), and then they expressed BPS invariants by $\chi(\mathrm{Hilb}^k(\mathcal{C}_g))$ for $0 \leq k \leq g$ and by some mysterious "correction terms"(see [**KKV**] for details).

Recall the role of $\chi = 1$ condition in the Abel–Jacobi map (4.3). The surjectivity holds since $\chi(\mathcal{E}) = 1$ for all $\mathcal{E} \in \mathrm{Jac}^g(\mathcal{C}_g)$ and the birationality holds since $H^0(\mathcal{E}) = \mathbb{C}$ is an open condition. Hence we may expect the analog for $M_\beta(X)$ since $\chi(\mathcal{E}) = 1$ for all $\mathcal{E} \in M_\beta(X)$. If Abel–Jacobi type morphism from the relative Hilbert scheme of points on the support curves to $M_\beta(X)$ exists, then it will describe a good approximation of $M_\beta(X)$ since the relative Hilbert scheme of points on curves are often easier to deal with.

However, we see that when the support curve is reducible we have at most an Abel–Jacobi type birational map in general since the morphism $\mathcal{O}_X \to \mathcal{E}$ may factor through some irreducible component. For example, let us consider a reducible curve $C = C_1 \cup C_2$ where C_1 is a smooth rational curve and C_2 is a smooth elliptic curve and C_1 and C_2 intersect at one point [2]. In particular, there is a stable sheaf $\mathcal{E}$ defined by the nontrivial extension $0 \to \mathcal{O}_{C_2} \to \mathcal{E} \to \mathcal{O}_{C_1} \to 0$. It is discussed in [**KKV**] that the 0-th BPS invariant n_0 is the sum of the Euler number of the relative Hilbert scheme of points $\chi(C) = 1$ and a correction term -1 (and hence $n_0 = 0$).

The geometric meaning of the correction term was not clear in [**KKV**], but now we have a nice observation based on $M_\beta(X)$. One sees that the correction terms in [**KKV**] are due to the non-existence of the Abel–Jacobi type morphism. Note that each point on the component C_1 in the relative Hilbert scheme of points C gives a unstable sheaf. In order to recover $M_\beta(X)$ ($\simeq C_2$ in this example) from the relative Hilbert scheme of points C, one must first subtract the component C_1, and then instead of the unstable sheaf $\mathcal{F}$ defined by the non-trivial extension $0 \to \mathcal{O}_{C_1} \to \mathcal{F} \to \mathcal{O}_{C_2} \to 0$ which corresponds to the intersection point $C_1 \cap C_2$, we have to add the stable sheaf $\mathcal{E}$ defined by the nontrivial extension $0 \to \mathcal{O}_{C_2} \to \mathcal{E} \to \mathcal{O}_{C_1} \to 0$ which does not correspond to any point in the relative Hilbert scheme of points. In this way, we get the correction term $-\chi(C_1) + \chi(C_1 \cap C_2) = -1$.

Since in general we have at most a birational map from the relative Hilbert scheme of points on curves to our moduli spaces, we have to consider the contributions from unstable sheaves to be subtracted and stable sheaves to add in order to get $M_\beta(X)$ from the relative Hilbert scheme of points. This is the geometric meaning of the correction terms by [**KKV**]. We shall discuss the local $\mathbb{P}^2$ case in the next section (Theorem 5.8).

5. Evidences

In this section we shall give some evidences for Conjecture 3.1 which we used in the previous section.

5.1. Constant Maps. Let us first consider the BPS invariants for $\beta = 0 \in H_2(X, \mathbb{Z})$ which should correspond to Gromov–Witten invariants for constant maps via the Gopakumar–Vafa formula (1.5).

It is necessary to introduce the notion of positivity of D-branes to discuss the cases $\beta = 0$ and $\beta \neq 0$ at the same level. Note that in [**GV**] the methods used to obtain BPS invariants $\beta = 0$ and $\beta \neq 0$ are different. As positive roots in the Lie

[2]This example can be found when one consider the family of elliptic curves in del Pezzo surfaces dP_5. See [**KKV**]

algebra theory, the notion of positivity of D-branes are naturally defined as follows (see also [**HM**]):

DEFINITION 5.1. D2-D0 bound state with a D-brane charge $(\beta, m) \in H_2(X,\mathbb{Z})\oplus H_0(X,\mathbb{Z})$ is positive (we denote this by $(\beta,m)>0$) if

$$(\beta,m)>0 \quad \overset{\text{def}}{\Longleftrightarrow} \quad \beta>0, m\in\mathbb{Z}$$
$$\text{or } \beta=0, m>0,$$
$$\text{where} \quad \beta>0 \quad \overset{\text{def}}{\Longleftrightarrow} \quad \beta\in\overline{NE(X)}\cap H_2(X,\mathbb{Z})\backslash\{0\}.$$

In the physical derivation of the Gopakumar–Vafa formula (1.5) (see [**GV**]), one has to count BPS states over positive D-branes:

(5.1)
$$\sum_{g\geq 0} F_g^X\lambda^{2g-2} = \sum_{(\beta,m)>0,h\geq 0} Tr_{R_h(\beta)}(-1)^{2H_R}\int_\epsilon^\infty \frac{ds}{s}(2i\sinh\frac{s}{2})^{2r-2}q^{\frac{s}{\lambda}\cdot\beta}\exp(-2\pi i m\frac{s}{\lambda}).$$

Some careful calculations with identity

$$\sum_{m\in\mathbb{Z}}\exp(-2\pi i m\frac{s}{\lambda}) = \sum_{k\in\mathbb{Z}}\delta(\frac{s}{\lambda}-k).$$

(where we denote by δ the Dirac's delta function) give 1/2-factor in Definition 2.4:

$$\sum_{g\geq 0}F_g^X\lambda^{2g-2} = \sum_{k>0,h\geq 0,\beta\in H_2(X,\mathbb{Z})} n_h(\beta)\frac{1}{k}\left(2\sin(\frac{k\lambda}{2})\right)^{2h-2}q^{k\beta},$$

where $n_h(\beta) := Tr_{R_h(\beta)}(-1)^{2H_R}$ for $\beta\neq 0$ and $n_h(0) := Tr_{R_h(0)}(-1)^{2H_R}/2$.

In the BPS invariants side, we should consider the contribution from skyscraper sheaves with Euler number 1 on X. $M_0(X)$ and $S_0(X)$ is nothing but X itself and hence the support morphism $\pi_0 : M_0(X)\to S_0(X)$ is identity. It is almost clear that the only non-vanishing BPS invariant is $n_0(0)$ given by

$$n_0(0) = \frac{1}{2}Tr_{H^*(X)}(-1)^{2H_R} = (-1)^{\dim_{\mathbb{C}}X}\frac{\chi(X)}{2} = -\frac{\chi(X)}{2}. \tag{5.2}$$

Gromov–Witten invariants for constant maps ($N_g(0)$, $g\geq 2$) have the following expression, which was proved by Faber and Pandharipande:

THEOREM 5.1. ([**FP**])

(5.3)
$$-\frac{\chi(X)}{2}\zeta(3)+\sum_{g\geq 2}N_g(0)\lambda^{2g-2} = -\frac{1}{2}\chi(X)\sum_{k\geq 1}\frac{1}{k}\left(2\sin(\frac{k\lambda}{2})\right)^{-2}+\frac{\chi(X)}{2}\frac{1}{12}\zeta(1).$$

In particular, $n_0^{conj}(0) = -\frac{\chi(X)}{2}$. □

REMARK. We usually add the term $+\frac{\chi(X)}{2}\frac{1}{12}\zeta(1)$ into right hand side to cancel the obvious divergence. The expression will be more beautiful after setting formally,

$$N_0(0) := -\frac{\chi(X)}{2}\zeta(3),\ N_1(0) := -\frac{\chi(X)}{2}\zeta(1). \tag{5.4}$$

Thus we have

THEOREM 5.2.

$$n_h(0) = \begin{cases} -\frac{1}{2}\chi(X) & \textit{for } h=0 \\ 0 & \textit{otherwise,}\end{cases} \tag{5.5}$$

and we have $n_0(0) = n_0^{conj}(0)$. □

5.2. Contractible rational curve in X. Let $C \subset X$ be a smooth irreducible curve in a Calabi-Yau 3-fold. Let $N = N_{C/X}$ be the normal bundle of C in X. The curve $C \subset X$ is called *rigid* if $H^0(C, N) = 0$, and the curve $C \subset X$ is called *super-rigid* if $H^0(C', \mu^*(N)) = 0$ for all non-constant maps of nonsingular curves $\mu : C' \to C$. For a super-rigid curve C, one can define the local contributions of C to the Gromov-Witten invariants $N_g(n[C])$. Denote corresponding Gromov–Witten invariants by $N_g(n \cdot C)$ and call them the **local Gromov–Witten invariants.**

If C is a curve of genus 0, then the super-rigidity and the rigidity coincides and these imply that $N_{C/X} = \mathcal{O}_C(-1) \oplus \mathcal{O}_C(-1)$. The relevant local Gromov–Witten invariants are given by

$$N_g(n \cdot C) := \int_{[\overline{\mathcal{M}}_{g,0}(\mathbb{P}^1, n[\mathbb{P}^1])]^{virt}} c_{\text{top}}(R^1\pi_*\mu^*N), \ n \geq 0,$$

where

$$\pi : \overline{\mathcal{M}}_{g,1}(\mathbb{P}^1, n[\mathbb{P}^1]) \to \overline{\mathcal{M}}_{g,0}(\mathbb{P}^1, n[\mathbb{P}^1])$$

is the universal family,

$$\mu : \overline{\mathcal{M}}_{g,1}(\mathbb{P}^1, n[\mathbb{P}^1]) \to \mathbb{P}^1, \ (f : \Sigma_g \to \mathbb{P}^1, x \in \Sigma_g) \mapsto f(x) \in \mathbb{P}^1$$

is the evaluation map and $N := \mathcal{O}_{\mathbb{P}^1}(-1) \oplus \mathcal{O}_{\mathbb{P}^1}(-1)$.

Faber–Pandharipande proved the following theorem for the generating function of local Gromov–Witten invariants $N_g(n \cdot C)$:

THEOREM 5.3. ([**FP**])

$$\sum_{g \geq 0, n \geq 1} N_g(n \cdot C) q^n \lambda^{2g-2} = \sum_{k \geq 1} \frac{1}{k} \left(2 \sin(\frac{k\lambda}{2}) \right)^{-2} q^k. \tag{5.6}$$

One can define the **conjectural local BPS invariants** $n_g^{conj}(d \cdot C) \in \mathbb{Q}$ by the Gopakumar–Vafa formula (1.5)

$$\sum_{g \geq 0, n \geq 0} N_g(n \cdot C) q^n \lambda^{2g-2} = \sum_{k > 0, h \geq 0, n \geq 0} n_h^{conj}(n \cdot C) \frac{1}{k} \left(2 \sin(\frac{k\lambda}{2}) \right)^{2h-2} q^{kn}. \tag{5.7}$$

From this formula, the conjectural local BPS invariants $n_g^{conj}(n \cdot \mathbb{P}^1)$ are given by

$$n_h^{conj}(n \cdot \mathbb{P}^1) = \begin{cases} 1 & \text{for } h = 0 \text{ and } n = 1 \\ 0 & \text{otherwise.} \end{cases} \tag{5.8}$$

The generalization of Theorem 5.3 to a contractible smooth rational curve C is given in [**BKL**]. First let us recall some facts on contractible smooth rational curves.

DEFINITION 5.2. Let C be a smooth rational curve in Calabi–Yau 3-fold X and suppose that there exists a contraction morphism $f : X \to Y$ with $f(C) = p \in Y$ which induces an isomorphism $f : X \backslash C \simeq Y \backslash p$. ($f : X \to Y$ is a small resolution of an isolated singularity $p \in Y$ with the exceptional set $f^{-1}(p) = C$). Kollár's length l is defined to be the length at the generic point of C of the sheaf $\mathcal{O}_X / f^{-1} m_{Y,p}$ where $m_{Y,p}$ is the maximal ideal sheaf of $p \in Y$.

It is known that $p \in Y$ is a compound DuVal singularity and $N_{C/X}$ is $\mathcal{O}_C(-1)\oplus \mathcal{O}_C(-1)$, $\mathcal{O}_C \oplus \mathcal{O}_C(-2)$ or $\mathcal{O}_C(1) \oplus \mathcal{O}_C(-3)$.

Let Y_0 be a generic hyperplane section and let X_0 be the proper transform of Y_0. It is known by the Reid's Lemma [**R**] that the minimal resolution Z_0 of Y_0 factors through X_0 and hence the length l can be computed by the length of $\mathcal{O}_{X_0}/f^{-1}|_{X_0}(m_{Y_0,p})$ and coincides with the multiplicity of C in the fundamental cycle of the corresponding ADE singularity. We refer to [**KM**] for details.

Let $C_n \subset X_0$ be subschemes defined by the symbolic power $\mathcal{I}_C^{(n)}$ of the ideal $\mathcal{I}_C$ defining $C \subset X_0$, and let k_n be the multiplicities of C_n in Hilbert scheme. The theorem by [**BKL**] gives the following conjectural local BPS invariants.

THEOREM 5.4. ([**BKL**])
Let C be a contractible smooth rational curve in a Calabi–Yau 3-fold X. C_n deforms to k_n smooth rational curves with normal bundle $\mathcal{O}(-1)\oplus\mathcal{O}(-1)$ and homology class $n[C]$ under a generic deformation of X. Since Gromov–Witten invariants are deformation invariants, conjectural local BPS invariants are given by

$$n_h^{conj}(n\cdot C) = \begin{cases} k_i & \text{for } h=0, n=1,2,\dots,l \\ 0 & \text{otherwise.}\end{cases} \tag{5.9}$$

□

Let us consider the subspace $M_{n\cdot C}(X)$ of $M_{n[C]}(X)$ defined by

$$M_{n\cdot C}(X) := \{\mathcal{E}\in M_{n[C]}(X) \mid s(\mathcal{E}) = n\cdot C\} \subset M_{n[C]}(X).$$

In order to define **local BPS invariants**, we only have to replace $M_\beta(X)$ by $M_{n\cdot C}(X)$ in Theorem 2.2. By (2.9) one has local BPS invariants $n_h(n\cdot C)$. What we would like to do is to compare $n_h(n\cdot C)$ with $n_h^{conj}(n\cdot C)$.

THEOREM 5.5. *Let $C \subset X$ be a contractible smooth rational curve on a Calabi–Yau 3-fold X and let l be the Kollár's length for C. Then our local BPS invariants are given by*

$$n_h(n\cdot C) = \begin{cases} k_i & \text{for } h=0, n=1,2,\dots,l \\ 0 & \text{otherwise.}\end{cases} \tag{5.10}$$

In particular, $n_h(n\cdot C) = n_h^{conj}(n\cdot C)$ for all h and n.

PROOF. Let us first notice that if $\mathcal{E}\in M_{n\cdot C}(X)$, then there exists an ideal sheaf J of $\mathcal{O}_X$ such that $\mathcal{O}_X/J \subset \mathcal{E}$ since $\chi(\mathcal{E}) = 1$. The contraction morphism $f: X\to Y$ being rational, we have $R^1f_*\mathcal{O}_X = 0$ and hence $R^1f_*(\mathcal{O}_X/J) = 0$ and $H^0(\mathcal{O}_X/J) = \chi(\mathcal{O}_X/J) \geq 1$. Thus by the stability condition, one easily sees that $\mathcal{O}_X/J \simeq \mathcal{E}$.

Conversely, if J is an ideal sheaf of $\mathcal{O}_X$ such that $I_C^k \subset J \subset I_C$ for some k and $\chi(\mathcal{O}_X/J) = 1$, then $\mathcal{O}_X/J$ is stable. Let us show this fact. Assume that there exists a destabilizing (proper) subsheaf $\mathcal{E}$ of $\mathcal{O}_X/J$. One sees that the destabilizing property of $\mathcal{E}$ and $\chi(\mathcal{O}_X/J) = 1$ imply that $\chi(\mathcal{E}) = 1$ and the existence of an ideal sheaf K of $\mathcal{O}_X$ such that $\mathcal{O}_X/K \subset \mathcal{E} \subset \mathcal{O}_X/J$. However this gives $K = J$, which contradicts the assumption that $\mathcal{E}$ is a proper subsheaf of $\mathcal{O}_X/J$.

Now we have the identification

$$M_{n\cdot C}(X) \simeq \{J\subset \mathcal{O}_X \mid J\subset I_C,\ \chi(\mathcal{O}_X/J) = 1\} \subset Hilb(X)$$

as a set, however by the way of Simpson's construction of moduli space of semi-stable sheaves (see for example, [**HL**]), we can conclude that $M_{n\cdot C}(X)$ is exactly the component of the Hilbert scheme, in other words, the isomorphism given above is also an isomorphism between schemes.

We see that if $J \subset \mathcal{O}_X$ satisfies $I_C^k \subset J \subset I_C$ and $\chi(\mathcal{O}_X/J) = 1$, then $f^{-1}m_{Y,p} \subset J$. On the other hand, in the proof of Lemma 2.8. [**BKL**], one can find that $\chi(\mathcal{O}_{C_n}) = 1$ for $C_n \subset X_0$, $n = 1, \ldots, l$, where C_n is the subscheme defined by the symbolic power $\mathcal{I}_C^{(n)}$ of the ideal $\mathcal{I}_C$ defining $C \subset X_0$ and X_0 is the proper transform of a generic hyperplane section through the singular point p. Thus the moduli space $M_{n\cdot C}(X)$ is the component of the Hilbert scheme containing ideal sheaves of C_n for $n = 1, \ldots, l$, and it is empty set otherwise.

Now let us calculate $n_h(n \cdot C)$ by our method. Since $C \subset X$ is isolated, one sees that $S_{n[C]} \subset Chow(X)$ consists of just one point $\{n \cdot C\}$ and the projective morphism

$$\pi_{n\cdot C} : (M_{n\cdot C})_{red} \to S_{nC}$$

is the usual Hilbert–Chow morphism. Since $\pi_{n\cdot C}$ is identity and the multiplicity of $M_{n\cdot C}$ is k_n by definition, we have the desired result. □

5.3. Conic bundle in X. We shall calculate the local BPS invariants for homology classes of curves contracted under a Type III primitive contraction. Let $E \subset X$ be a conic bundle over a smooth curve C of genus g. We denote the class of the fiber of $E \to C$ by $2f$. Let consider the moduli space given by

$$M_d := \{\mathcal{E} \in M_{[df]}(X) \mid s(\mathcal{E}) \in |df|\} \subset M_{[df]}(X).$$

The natural support map is given by

$$\begin{array}{rccc} (5.11) & \pi_d : M_d & \to & |df| \\ & \mathcal{E} & \mapsto & s(\mathcal{E}), \end{array}$$

and the local BPS invariant $n_h(d)$ is defined by

$$IH^*(M_d) = \bigoplus_{h\geq 0} I_h \otimes R_h(d)$$

$$n_h(d) := Tr_{R_h(d)}(-1)^{2H_R}.$$

THEOREM 5.6. *We have the following local BPS invariants:*

$$(5.12) \qquad n_h(d) = \begin{cases} 2\sum_{r\geq 0} a_r(r+1) & \text{for } h = 0 \text{ and } d = 1 \\ 2g-2 & \text{for } h = 0 \text{ and } d = 2 \\ 0 & \text{otherwise,} \end{cases}$$

where a_r is the number of fibers $E \to C$ which are line pairs with an A_r singularity.

In particular $n_h(d) = n_h(d)^{conj}$, where $n_h(d)^{conj}$ are the conjectural BPS invariants calculated by Wilson [**W**].

PROOF. Since all fibers are $(-2, 0)$-curves, we have $n_h(d) = 0$ for $d \geq 3, h \geq 0$ by the similar argument discussed in [**HST2**] (section 4). We can easily see that $\mathcal{E} \in M_d$ is isomorphic to $\mathcal{O}_{s(\mathcal{E})}$ by the same argument as in the case of contractible curves. Hence we have $M_2 \simeq |2f| \simeq C$ and $n_0(2) = -\chi(M_2) = -\chi(C) = 2g-2$, $n_h(2) = 0$ if $h \neq 0$. Moreover, we can count the multiplicity in the Hilbert scheme of $\mathcal{O}_{s(\mathcal{E})}$ for $\mathcal{E} \in M_1$ (see [**Tz**]), and we have that the multiplicity of a smooth rational curve passing an A_r-singularity is $r+1$. Since each fiber which is line pairs has two smooth rational curves passing an A_r-singularity for some r, the total contribution to $n_0(1)$ is $2 \cdot \sum_{r\geq 0} a_r(r+1)$. □

5.4. Super-rigid Elliptic curve in X. Let $E \subset X$ be a super-rigid elliptic curve. It is known that the normal bundle N of E in X is isomorphic to $N \simeq L \oplus L^{-1}$ where L is a non-torsion element of the Picard group of E. In this case, $M_{n \cdot E}$ and local BPS invariants $n_h(n \cdot E)$ are given as follows:

THEOREM 5.7. ([**HST2**])
$M_{n \cdot E} \simeq E$ for $n \geq 1$ and

$$n_h(n \cdot E) = \begin{cases} 1 & \text{for } h = 1, n \geq 1 \\ 0 & \text{otherwise.} \end{cases} \tag{5.13}$$

In particular we have $n_h(n \cdot E) = n_h(n \cdot E)^{conj}$ where $n_h(n \cdot E)^{conj}$ is the conjectural BPS invariants calculated by Pandharipande [**P**]. □

5.5. $\mathbb{P}^2$ in X. We shall calculate local BPS invariants for some homology classes in $\mathbb{P}^2$ in a smooth Calabi–Yau 3-fold X based on our mathematical Definition 2.4. Calculations of BPS invariants are easily applied to other del Pezzo surfaces but here we shall restrict to $\mathbb{P}^2$ case to ease the notations.

Let us denote by H the hyperplane class and consider the moduli space given by

$$M_d := \{\mathcal{E} \in M_{[dH]}(X) \mid s(\mathcal{E}) \in |dH|\} \subset M_{[dH]}(X).$$

The natural support map is given by

$$\begin{array}{rcl} \pi_d : M_d & \to & |dH| \\ \mathcal{E} & \mapsto & s(\mathcal{E}), \end{array} \tag{5.14}$$

and the local BPS invariant $n_h(d)$ is defined by

$$IH^*(M_d) = \bigoplus_{h \geq 0} I_h \otimes R_h(d)$$

$$n_h(d) := Tr_{R_h(d)}(-1)^{2H_R}.$$

We shall calculate our local BPS invariants for $d \leq 4$, which coincides with the conjectural BPS invariants determined by the localization technique [**KZ**].

THEOREM 5.8. *$n_h(d) = n_h^{conj}(d)$ for $d \leq 3, h \geq 0$ and $d = 4, h = 0, h \geq 3$, where $n_h^{conj}(d)$ is the conjectural local BPS invariants obtained in* [**KZ**]. □

It is known that there are also two ways to obtain BPS invariants physically, B-model calculation and Katz–Klemm–Vafa proposal (see for example, [**KKV**][**KZ**]), which all give the same BPS invariants. In particular, we shall find the origin of contributions from reducible curves proposed in [**KKV**] section 5. This fact supports our definition of moduli spaces and local BPS invariants.

First of all, one can easily see that $|dH| \simeq \mathbb{P}^{d(d+3)/2}$ and the arithmetic genus of C is $g_a(C) = (d-1)(d-2)/2$ for $C \in |dH|$. Note also that for $\mathcal{E} \in M_d$ there exists $C \in |kH|, k \leq d$ such that

$$0 \to \mathcal{O}_C \hookrightarrow \mathcal{E} \tag{5.15}$$

since $\chi(\mathcal{E}) = 1$.

(i) $d = 1$: $|H| \simeq \mathbb{P}^2$ and all fibers $C \in |H|$ are smooth rational curves with normal bundles $N_{C/X} \simeq \mathcal{O}_{\mathbb{P}^1}(-3) \oplus \mathcal{O}_{\mathbb{P}^1}(1)$. One can easily show that $\pi_1 : M_1 \simeq |H|$ and

$$n_h(1) = \begin{cases} 3 & \text{for } h = 0 \\ 0 & \text{otherwise.} \end{cases} \tag{5.16}$$

(ii) $d = 2$: The stability condition for $\mathcal{E} \in M_2$ and the fact that $\chi(\mathcal{O}_C) = 1$ for all $C \in |L|, L = H, 2H$ imply that the inclusion (5.15) gives an isomorphism $\mathcal{O}_{s(\mathcal{E})} \simeq \mathcal{E}$. Conversely, one can show that all sheaves $\mathcal{O}_C$, $C \in |2H|$ are stable. If $\mathcal{O}_C$ were unstable, since $\chi(\mathcal{O}_C) = 1$ there exists a subscheme D and an inclusion which satisfy $\mathcal{O}_D \subset \mathcal{F} \subset \mathcal{O}_C$ for a destabilizing proper subsheaf $\mathcal{F}$ of $\mathcal{O}_C$, but this is impossible. Moreover, one can conclude that $M_2 \simeq |2H|$ as a scheme by recalling the Simpson's construction of the moduli space of stable sheaves. Thus we have $\pi_2 : M_2 \simeq |2H|$ and

$$n_h(2) = \begin{cases} -6 & \text{for } h = 0 \\ 0 & \text{otherwise.} \end{cases} \tag{5.17}$$

(iii) $d = 3$: The inclusion (5.15) induces the following exact sequence by the stability of $\mathcal{E} \in M_3$:

$$0 \to \mathcal{O}_{s(\mathcal{E})} \to \mathcal{E} \to \mathbb{C}_p \to 0, \ p \in \mathbb{P}^2. \tag{5.18}$$

The point p in the above exact sequence is uniquely defined by $\mathcal{E}$ since by the Serre duality and the stability of $\mathcal{E}$ and $\mathcal{O}_{s(\mathcal{E})}$ we have

$$H^1(\mathcal{E}) \simeq Hom(\mathcal{E}, \mathcal{O}_{s(\mathcal{E})})^* = 0,$$

and hence $Hom(\mathcal{O}_{s(\mathcal{E})}, \mathcal{E}) = \mathbb{C}$. Let $\mathcal{C}_3$ be the Hilbert scheme of one point on fibers of $|3H|$, i.e., let $\mathcal{C}_3 \to |3H|$ be the universal family of the complete linear system $|3H|$. Then by applying $\mathcal{H}om_{\mathcal{O}_{s(\mathcal{E})}}(\cdot, \mathcal{O}_{s(\mathcal{E})})$ to (5.18), we have an isomorphism $M_3 \simeq \mathcal{C}_3$. Indeed, one can show that $\mathcal{E}xt^1_{\mathcal{O}_C}(\mathbb{C}_p, \mathcal{O}_C) = \mathbb{C}_p$ for $C \in |3H|$, $p \in C$ and that the natural morphism $\mathcal{E} \to \mathcal{H}om_{\mathcal{O}_{s(\mathcal{E})}}(\mathcal{H}om_{\mathcal{O}_{s(\mathcal{E})}}(\mathcal{E}, \mathcal{O}_{s(\mathcal{E})}), \mathcal{O}_{s(\mathcal{E})})$ is an isomorphism.

Note that $\mathcal{C}_3$ is isomorphic to a $\mathbb{P}^8$-bundle over $\mathbb{P}^2$. Thus we have the following $sl_2 \times sl_2$-decomposition of $H^*(M_3)$:

$$H^*(M_3) = (\frac{1}{2})_L \otimes (\frac{9}{2})_R \oplus (0)_L \otimes (3)_R \tag{5.19}$$

$$= [(\frac{1}{2})_L \oplus 2(0)_L] \otimes (\frac{9}{2})_R \oplus (0)_L \otimes [(3)_R - 2(\frac{9}{2})_R], \tag{5.20}$$

since $H^*(M_3) = (5) \oplus (4) \oplus (3)$ as a (usual) Lefschetz sl_2-representation. Now we have obtained the following local BPS invariants:

$$n_h(3) = \begin{cases} 10 & \text{for } h = 1 \\ -27 & \text{for } h = 0 \\ 0 & \text{otherwise.} \end{cases} \tag{5.21}$$

(iv) $d = 4$: As in the case $d = 3$, let us approximate M_4 by a relative Hilbert scheme. Since $g_a(C) = 3$ for $C \in |4H|$, it is natural to expect that there exist a morphism

$$\begin{array}{ccc} Hilb^3(\mathcal{C}_4) & \to & M_4 \\ \{0 \to I \to \mathcal{O}_C \to \mathcal{F} \to 0\} & \mapsto & [\mathcal{H}om_{\mathcal{O}_C}(I, \mathcal{O}_C)], \end{array} \tag{5.22}$$

where by $[\mathcal{H}om_{\mathcal{O}_C}(I, \mathcal{O}_C)]$ we mean the isomorphism class of $\mathcal{H}om_{\mathcal{O}_C}(I, \mathcal{O}_C)$ and $Hilb^3(\mathcal{C}_4)$ is the relative Hilbert scheme of 3-points on fibers of the universal family of the complete linear system $\mathcal{C}_4 \to |4H|$, which is a $\mathbb{P}^{11}$-bundle over $Hilb^3(\mathbb{P}^2)$. There is not such a morphism but we have a birational map. Indeed, if the support

$s(\mathcal{E})$ of $\mathcal{E} \in M_4$ is irreducible, the equation (5.15) implies that there is an exact sequence

$$0 \to \mathcal{O}_{s(\mathcal{E})} \to \mathcal{E} \to \mathcal{F} \to 0, \ \dim_{\mathbb{C}} \mathcal{F} = 0, \ \chi(\mathcal{F}) = 3,$$

and we can apply the argument similar to that of $d = 3$.

Let us consider the locus $|H| \times |3H| \subset |4H|$ and the point $C \cup C' \in |H| \times |3H|$. We have the following exact sequence for $I \in Hilb^3(C) \subset Hilb^3(C \cup C')$

$$(5.23) \qquad 0 \to \mathcal{O}_C \to \mathcal{H}om_{\mathcal{O}_{C\cup C'}}(I, \mathcal{O}_{C\cup C'}) \to \mathcal{O}_{C'} \to 0, \quad C \in |H|, C' \in |3H|.$$

Since $\chi(\mathcal{O}_{C'}) = 0$, $\mathcal{H}om_{\mathcal{O}_{C\cup C'}}(I, \mathcal{O}_{C\cup C'})$ is unstable and this is the reason that we have no morphism from $Hilb^3(\mathcal{C}_4)$ to M_4. On the other hand, we have stable sheaves by "flipping" the above exact sequence with nontrivial extension class which do not give points in $Hilb^3(\mathcal{C}_4)$:

$$(5.24) \qquad 0 \to \mathcal{O}_{C'} \to \mathcal{E} \to \mathcal{O}_C \to 0, \quad C \in |H|, C' \in |3H|.$$

Since we only have a birational map from $Hilb^3(\mathcal{C}_4)$ to M_4, we have to subtract unstable sheaves given by (5.23) and add stable sheaves given by (5.24) in order to have our moduli space M_4 from $Hilb^3(\mathcal{C}_4)$. More precisely, for each point in $|H| \times |3H|$, we should replace $\mathbb{P}^3 \simeq Hilb^3(C)$ by $\mathbb{P}^2 \simeq \mathbb{P}\mathrm{Ext}^1_{\mathcal{O}_{\mathbb{P}^2}}(\mathcal{O}_C, \mathcal{O}_{C'})$. Note that Euler number changes by $-\chi(\mathbb{P}^3) + \chi(\mathbb{P}^2) = -1$, which is exactly the same contribution as the "correction term" discussed in [**KKV**].

Moreover, as in the theory of the Abel–Jacobi map, in order to relate $Hilb^3(\mathcal{C}_4)$ with M_4, we have to consider the locus in M_4 where $\dim_{\mathbb{C}} H^0(\mathcal{E}) \geq 2$ and blow down $Hilb^3(\mathcal{C}_4)$ along the corresponding locus (in this case, isomorphic to $\mathcal{C}_4$) in $Hilb^3(\mathcal{C}_4)$.

Thus we can construct M_4 by blowing down $Hilb^3(\mathcal{C}_4)$ along the $\mathbb{P}^1$-bundle over $\mathcal{C}_4$ and by replacing the $\mathbb{P}^3$-bundle over $|H| \times |3H|$ by the $\mathbb{P}^2$-bundle over $|H| \times |3H|$. Hence

$$\begin{aligned}
\chi(M_4) =& \chi(Hilb^3(\mathcal{C}_4)) - \chi(\mathcal{C}_4) \\
& - \chi(\mathbb{P}^3) \cdot \chi(|H| \times |3H|) + \chi(\mathbb{P}^2) \cdot \chi(|H| \times |3H|) \\
=& \chi(\mathbb{P}^{11}) \cdot \chi(Hilb^3(\mathbb{P}^2)) - \chi(\mathbb{P}^{13}) \cdot (\mathbb{P}^2) - 3 \cdot 10 \\
=& 12 \cdot 22 - 14 \cdot 3 - 30 = 192
\end{aligned}$$

Note that $30 = n_0(1) \cdot n_1(3)$ is in agreement with the "correction term" in [**KKV**].

We have the following local BPS invariants since M_4 and $|4H| \simeq \mathbb{P}^{14}$ are smooth.

$$(5.25) \qquad n_h(4) = \begin{cases} 15 & \text{for } h = 3 \\ -192 & \text{for } h = 0 \\ 0 & h \geq 4 \end{cases}$$

REMARK. Unfortunately, we do not know the Lefschetz $(sl_2)_L \times (sl_2)_R$-decomposition and we can not determine $n_1(4), n_2(4)$ by our Definition 2.4.

References

[AD] P. Aspinwall, M. Douglas, *D-brane Stability and Monodromy*, JHEP 0205 (2002) 031.

[BKL] J. Bryan, S. Katz, N. Leung *Multiple covers and the integrality conjecture for rational curves in Calabi-Yau threefolds*, J. Alg. Geom. **10** (2001), no. 3, 549–568.

[BP] J. Bryan, R. Pandharipande, *BPS states of curves in Calabi-Yau 3-folds*, Geometry and Topology, **5** (2001) Paper no. 9, 287-318.

[D] B. Dubrovin, *Geometry and analytic theory of Frobenius manifolds*, Proceedings of the International Congress of Mathematicians, Vol. II (Berlin, 1998). Doc. Math. 1998, Extra Vol. II, 315–326.

[FP] C. Faber, R. Pandharipande, *Hodge integrals and Gromov-Witten theory*. Invent. Math. **139** (2000), no. 1, 173–199.

[FO] K. Fukaya, K. Ono, *Gromov–Witten invariants over $\mathbb{Z}$ of general symplectic manifolds, (a summary)*, in Proceedings of Taniguchi Symposium, 1999, Nara.

[G] A. Givental, *Gromov–Witten Invariants and Quantization of Quadratic Hamiltonians*, Mosc. Math. J. **1** (2001), no. 4, 551–568, 645.

[GV] R. Gopakumar, C. Vafa, *M-Theory and Topological Strings-I,II*, hep-th/9809187, hep-th/9812127.

[HM] A. Harvey, G. Moore, *On the algebras of BPS states*, Commun. Math. Phys. **197** (1998) 489-519.

[HL] D. Huybrechts, M. Lehn, *The Geometry of Moduli Spaces of Sheaves*, Aspects of Mathematics, Vol E31, Vieweg.

[HST1] S. Hosono, M.-H. Saito, A. Takahashi, *Holomorphic Anomaly Equation and BPS State Counting of Rational Elliptic Surface*, Adv. Theoret. Math. Phys. **3** (1999) 177-208.

[HST2] S. Hosono, M.-H. Saito, A. Takahashi, *Relative Lefschetz action and BPS State Counting*, Internat. Math. Res. Notices **15** (2001) 783-816.

[KKV] S. Katz, A. Klemm, C. Vafa, *M-theory, Topological String and Spinning Black Holes*, Adv. Theor. Math. Phys. **3** (1999), no. 5, 1445–1537.

[KM] S. Katz and D. Morrison, *Gorenstein Threefold Singularities with Small Resolutions via Invariant Theory for Weyl Groups*, J. Alg. Geom. **1** (1992) 449-530.

[Ko] J. Kollar, *Rational Curves in Algebraic Variety*, A Series of Modern Surveys in Mathematics, Springer-Verlag, Berlin, 1996.

[KY] T. Kawai, K. Yoshioka, *String Partition Functions and Infinite Products*, Adv. Theoret. Math. Physics. **4** (2000), 397–485.

[KZ] A. Klemm and E. Zaslow, *Local Mirror Symmetry at Higher Genus*, Winter School on Mirror Symmetry, Vector Bundles and Lagrangian Submanifolds (Cambridge, MA, 1999), 183–207, AMS/IP Stud. Adv. Math., 23, Amer. Math. Soc., Providence, RI, 2001.

[LeP] J. Le Potier, *Systèmes Cohérents et Structures de Niveau*, Astérisque **214** (1993).

[P] R. Pandharipande, *Hodge integrals and degenerate contributions*, Commun. Math. Phys. **208** (1999), no. 2, 489–506.

[R] M. Reid, *Minimal Models of Canonical 3-folds*, Adv. Stud. Pure Math. 1 (1983), Algebraic Varieties and Analytic Varieties, 131–180.

[Th] R. Thomas, *A holomorphic Casson invariant for Calabi-Yau 3-folds, and bundles on K3 fibrations*, J. Differential Geom. **54** (2000), no. 2, 367–438.

[Tz] N. Tziolas, *Multiplicities of Smooth Rational Curves on Singular Local Complete intersection Calabi–Yau Threefolds*, J. Alg. Geom. **10** (2001) 497-513

[W] P. M. H. Wilson, *Flops, Type III contractions and Gromov–Witten invariants of Calabi–Yau 3-folds*, London Math. Soc. Lecture Note Ser., **264**, Cambridge Univ. Press, Cambridge, 1999.

Research Institute for Mathematical Sciences, Kyoto University, Kyoto 606-8502, Japan

E-mail address: `atsushi@kurims.kyoto-u.ac.jp`

List of Participants

P. Albers
Leipzig

S. Barannikov
Paris

K. Behrend
Vancouver

M. Bertola
Montreal

B. Dubrovin
Trieste

C. Faber
Stockholm

B. Fantechi
Udine

J. Fernandez
Salt Lake City

A. Gathmann
Princeton

A. Gerasimov
Moskau

E. Getzler
Evanston

A.B. Givental
Berkeley

C. Hertling
Bonn

E.-N. Ionel
Madison

T.J. Jarvis
Provo

D. Joe
Postech

R. Kaufmann
Los Angeles

B. Kim
Pohang

T. Kimura
Boston

J. Kock
Nice

Y.P. Lee
Los Angeles

X. Liu
Notre Dame

Yu. Manin
Bonn

E. Mann
Strassbourg

M. Marcolli
Bonn

A. Matsuo
Tokyo

S. Merkulov
Glasgow

M. Noumi
Kobe

Y. Ohyama
Osaka

H. Ohta
Nagoya

K. Ono
Sapporo

A. Okounkov
Berkeley

R. Pandharipande
Princeton

G. Pearlstein
Irvine

A. Polishchuk
Boston

P. Pribik
Berkeley

M. Rosellen
Bonn

C. Sabbah
Paris

M.-H. Saito
Kobe

K. Saito
Kyoto

I. Satake
Osaka

B. Siebert
Freiburg

I.A.B. Strachan
Hull

A. Takahashi
Kyoto

H. Terao
Tokyo

A. Tsuchiya
Nagoya

T. Valdesalici
Florence

Y. Zhang
Beijing

Graph theory, geometry and combinatorics brought together to generate a wealth of beauty in ideas

Stefan Felsner

Geometric Graphs and Arrangements

Some Chapters from Combinatorial Geometry

2004. x, 170 pp. (Advanced Lectures in Mathematics, ed. by Aigner, Martin/Mehrmann, Volker/Wüstholz, Gisbert/Gritzmann, Peter) Softc.

Approx. € 29,90 ISBN 3-528-06972-4

Contents: Geometric Graphs: Turán Problems - Schnyder Woods or How to Draw a Planar Graph? - Topological Graphs: Crossing Lemma and Applications- k-Sets and k-Facets - Combinatorial Problems for Sets of Points and Lines - Combinatorial Representations of Arrangements of Pseudolines - Triangulations and Flips - Rigidity and Pseudo-Triangulations

Among the intuitively appealing aspects of graph theory is its close connection to drawings and geometry. The development of computer technology has become a source of motivation to reconsider these connections, in particular geometric graphs are emerging as a new subfield of graph theory. Arrangements of points and lines are the objects for many challenging problems and surprising solutions in combinatorial geometry. The book is a collection of beautiful and partly very recent results from the intersection of geometry, graph theory and combinatorics.

vieweg

Abraham-Lincoln-Straße 46
65189 Wiesbaden
Fax 0611.7878-400
www.vieweg.de

Stand 1.7.2004. Änderungen vorbehalten.
Erhältlich im Buchhandel oder im Verlag.

MIX
Papier aus verantwortungsvollen Quellen
Paper from responsible sources
FSC® C105338

If you have any concerns about our products,
you can contact us on
ProductSafety@springernature.com

In case Publisher is established outside the EU,
the EU authorized representative is:
Springer Nature Customer Service Center GmbH
Europaplatz 3, 69115 Heidelberg, Germany

Printed by Libri Plureos GmbH
in Hamburg, Germany